HYDRAULICS/HYDROLOGY of ARID LANDS

(H²AL)

Proceedings of the International Symposium

Sponsored by the

Hydraulics Division of the American Society of Civil Engineers
Irrigation and Drainage Division of the American Society of Civil Engineers
Association of State Flood Plain Managers

Hosted by the San Diego Section, ASCE

Catamaran Resort Hotel
San Diego, California
July 30—August 2, 1990

Edited By:
Richard H. French
Water Resources Center
Desert Research Institute
Las Vegas, Nevada

Published by the
American Society of Civil Engineers
345 East 47th Street
New York, New York 10017-2398

ABSTRACT

This proceedings consists of papers presented at the International Symposium on the Hydraulics and Hydrology of Arid Lands which was held in conjunction with the 1990 National Conference on Hydraulics Engineering in San Diego, California from July 30-August 3, 1990. The goal of the International Symposium was to provide a forum for researchers and practitioners to discuss the state-of-the art of hydraulics and hydrology as they relate to arid regions. To this end, papers on such topics as water needs, alluvial fan flooding, water marketing, remote sensing, debris flows, precipitation frequency, soil erosion, and nitrate transport were discussed. Thus, this proceedings provides the engineer with a review of engineering problems in the arid environment and a guide for future investigations.

Library of Congress Cataloging-in-Publication Data

Hydraulics/hydrology of arid lands (H^2AL): proceedings of the international symposium/sponsored by the Hydraulics Division of the American Society of Civil Engineers, Irrigation and Drainage Division of the American Society of Civil Engineers, Association of State Flood Plain Managers: hosted by the San Diego Section, ASCE, Catamaran Resort Hotel, San Diego, California, July 30-August 2, 1990: edited by Richard H. French.

p. cm.

"Proceedings of the International Symposium on the Hydraulics and Hydrology of Arid Lands (H^2AL)"—Pref.

Includes indexes.

ISBN 0-87262-771-3

1. Hydraulic engineering—Congresses. 2. Hydrology—Arid lands—Congresses. I. French, Richard H. II. American Society of Civil Engineers. Hydraulics Division. III. American Society of Civil Engineers. Irrigation and Drainage Division. IV. Association of State Flood Plain Managers. V. International Symposium on the Hydraulics and Hydrology of Arid Lands (H^2AL) (1990: San Diego, Calif.)

TC5.H845 1990

627—dc20 90-40322

CIP

Library of Congress Catalog Card No: 90-40322

ISBN 0-87262-771-3

Manufactured in the United States of America

PREFACE

This is the Proceedings of the International Symposium on the Hydraulics and Hydrology of Arid Lands (H^2AL) held in conjunction with the 1990 National Conference on Hydraulic Engineering (NCHE). Both the Symposium and the National Conference were held at the Catamaran Hotel in San Diego, California from July 30 to August 3, 1990. H^2AL was organized by the Task Committee on Flood Hazard Analysis on Alluvial Fans whose parent committee is the Surface Water Hydrology Committee

The goal of H^2AL was to provide a forum for researchers and practitioners to discuss the state-of-the-art of hydraulics and hydrology as they relate to arid regions. Some of the problems in arid regions such as alluvial fan flooding are unique while other problems such as water supply and quality are universal. However, in arid regions, water resources is the paramount concern. As is the case with all civil engineering challenges, the solution of water resources challenges in the arid environment requires the interaction of engineers and scientists, economists and all levels of government. This symposium provided a forum to discuss and debate these challenges and the opportunity for discussion among those doing research, those applying research results, and those who must finally recommend and enunciate local, regional and national policy.

Although many of the papers in this volume were solicited by the members of the Task Committee, there were also many fine papers volunteered. It was difficult to select among the many papers contributed and produce a program of limited size. All papers presented here were peer reviewed by one or more members of the Task Committee and are eligible for discussion in the ASCE Journal of Hydraulic Engineering and for ASCE awards.

Special appreciation is due to Ibrahim M. Elassiouti who endowed the Arid Lands Hydraulic Engineering Award and the past awardees David A. Woolhiser and Mamdouh M.A. Shahin and the 1989 awardee, Marvin Jensen, for delivering keynote lectures.

Special thanks is also due to Cheng-Lung Chen who personally solicited the sessions debris flows primarily involving our colleagues from the Peoples Republic of China; and Lonnie C. Roy who provided significant assistance in organizing the papers into sessions. Many others also deserve thanks for their efforts but space precludes listing them individually here. The substantial financial support provided me by the Water Resources Center of the Desert Research Institute, University of Nevada System to organize this Symposium is gratefully acknowledged.

Finally, the hours devoted by my wife, Darlene, who acted as Symposium Secretary, are very gratefully acknowledged. In any effort, there are those conceive what should be done and there are those who accomplish the job. Far more than ASCE or me, it is my wife who made this Symposium a success. Many more than myself owe Darlene thanks for her dedication to this probject.

CONTENTS

Plenary Session of the International Symposium on the Hydraulics/Hydrology of Arid Lands

SESSION 1

Keynote Lectures on the Hydraulics/Hydrology of Arid Lands
Presiding: Ibrahim M. Elassiouti

SESSION 2

Alluvial Fans and the National Flood Insurance
Program's Floodplain Management Policies
Presiding: David J. Greenwood

SESSION 3

Water in the Desert, Part 1
Presiding: Terry Katzer

---

*Manuscript not available at the time of printing.

SESSION 4

Debris Flows: Field Observations
Presiding: Jeffrey Keaton

SESSION 5

Development and Implementation of Drainage Criteria
For Urban Areas in Arid Regions
Presiding: Kenneth L. Edwards

SESSION 6

Water in the Desert, Part 2
Presiding: Les K. Lampe

SESSION 7

Debris Flows I: Fluvial Processes
Presiding: Toshiyuki Shigemura and Cheng-lung Chen

SESSION 8

Flood Control Facilities: Design Considerations
Presiding: Lonnie C. Roy

SESSION 9

Technical Aspects of Transfer of Water From Northern to Southern California: Part 1
Presiding: Edward F. Sing

SESSION 10

Debris Flows II: Rheological Modeling
Presiding: Jeffrey Bradley

SESSION 11

Modeling of Flood Flows on Alluvial Fans
Presiding: Virginia Bax-Valentine

SESSION 12

Applications of Remote Sensing in Forecasting
Water Supply and Flooding
Presiding: Thomas R. Carroll

SESSION 13

Technical Aspects of Transfer of Water From
Northern to Southern California: Part 2
Presiding: Michael Mulvihill

SESSION 14

Hydrologic/Hydraulic Processes on Alluvial Fans
Presiding: David R. Dawdy

SESSION 15

Water Banking and Conjunctive Use
Presiding: Catalino Cecilio

SESSION 16

Debris Flows III: Unsteady Flows
Presiding: Jeffrey Holland

SESSION 17

Alluvial Channel Behavior
Presiding: Joseph Hill

SESSION 18

Water Supply in the Desert: Part 3
Presiding: Philip H. Burgi

SESSION 19

Streamflow: Frequency Analysis in the Arid Environment
Presiding: Muna Faltas

SESSION 20

Precipitation in Arid Lands: Part 1
Presiding: Herb Osborn

SESSION 21

Debris Flows IV: Mechanisms and Characteristics, Part I
Presiding: Ray B. Krone

SESSION 22

Frequency and Risk Analysis in the Arid Environment
Presiding: George V. Sabol

SESSION 23

Precipitation in Arid Lands: Part 2
Presiding: David A. Woolhiser

SESSION 24

Debris Flows IV: Mechanisms and Characteristics, Part 2
Presiding: Robert C. MacArthur

SESSION 25

Applications of Geomorphology to Civil Engineering Problems
Presiding: James Slosson

SESSION 26

Debris Flows V: Hyperconcentrated Flows
Presiding: Harvey Jobson

SESSION 31

Hydraulic/Hydrologic Arid Region Research Needs
Presiding: L. Douglas James

SESSION 32

Debris Flows VII: Hazard Mitigation Measures
Presiding: David T. Williams

Environmental Aspects of Water Resources Development

Ibrahim M. Elassiouti[1], Member ASCE

Abstract

Long-term management of resources requires a comprehensive analysis of the effects that development and management of one resource may be expected to exert upon other resources. However, when the High Aswan Dam in Egypt was planned in the early sixties, some environmental considerations were not foreseen. This paper discusses the negative and positive consequences of the High Aswan Dam on the Nile River.

Introduction

The past 25 to 30 years has been a period of unprecedented water resources development in many countries of the world. Water projects have been built to provide societies with water and hydroelectric power needed for development and economic growth. Although water resources developments are still progressing at a rapid rate and many beneficial effects are being recorded, the environmental impacts of some water projects have not been what water resources planners and political leaders had expected. The High Aswan Dam (HAD) of Egypt, one of the major dams of the world, has been the environmentalists' favorite example of development gone wrong. The scheme, built primarily for providing large-scale storage of the Nile water and for generating hydropower, was criticized for contributing to environmental disruptions. It was promptly condemned for many reasons, among which were loss of the Mediterranean fishery, increase in schistosomiasis, rising salinity, erosion of the Nile bed and banks, reduction in the

----

1 Professor of Hydraulic Engineering, Cairo University, Giza, Egypt.

fertility of the Nile valley through the absence of silt deposition, and coastal erosion of the river delta. Facts were unavailable at that time. Now after almost twenty years of monitoring and analysis, the positive and negative effects of HAD may be ascertained.

The Aswan Controversy

Since the completion of the HAD in 1970 and the subsequent controversy on its environmental effects, many questions were posed concerning the project. Is it an economic blessing, an engineering wonder or an environmental disaster? How do the benefits of economic development measure against the cost of ecological change? Should the dam be dismantled, as extremists in Egypt have urged, or should it be preserved? If preserved, how should it be managed with respect to the conflicting demands of multi-purpose water use, and what measures must be taken to minimize the economic benefits and minimize the adverse environmental impacts? What, in short, are the facts?

The HAD has received a lot of media exposure but facts do not speak for themselves. Facts are usually either not available or they are used from this or that perspective to speak for this or that other party or interest. The HAD is important in many ways. It is a landmark in Egyptian history. It is also a landmark in the history of river and lake control. Internationally, it caused a war-the Suez war in 1956, precipitated by Egyptian President Nasser's seizure of the Suez Canal. Nasser's act prompted Britain, France, and Israel to declare war on Egypt. Nasser had nationalized the canal because the Western nations had suddenly withdrawn their promise to help Egypt build a dam at Aswan on the Nile, and he hoped to use proceeds from the canal to build the dam. Nationally, the completed dam has stood as a symbol of Egypt's struggle for survival and prosperity.

Yet, for all its importance and controversial interests, facts about the benefits and risks were not well known. The reasons had been largely political. HAD was constructed without aid from the West and with massive financial and technical assistance from the Soviet Union. Because of national pride and the association of the dam with Nasser's prestige, scientific evaluation of the dam and its effects were discouraged in Egypt during Nasser's tenure. Only after President Sadat succeeded Nasser in 1972 was discussion of the dam gradually tolerated in Egypt. This liberalization of discussion was part of Sadat's overall program of de-Nasserization, de-Sovietization, and re-alignment with the West.

Environmental Impact of the High Aswan Dam

The construction of HAD caused distinct modifications in hydraulic regime of the Nile River. River flow control eliminated the Nile flood which used to flush and clean the river once a year. Many kilometers of canals and drains were added as a sequence of reclaiming more agricultural land on the water stored in the HAD reservoir, (Elassiouti et al. 1989). Millions of people are brought into contact with the added water areas in the new irrigation and drainage system and around the reservoir area which covers more than 6000 square kilometers at high water level. After some two decades of monitoring and evaluation, the effects of the HAD may be summarized as follows.

Positive effects of High Aswan Dam

In recent years, studies provided hard facts in support of the HAD and proved that it has done more good than ill, (Abu-Zeid 1989). River flow controls and generation of hydropower have significantly contributed towards Egypt's economic and social development. The following are some of the benefits:

- Increase of Egypt's annual share of the Nile water from 48 to 55 milliard (10^9) cubic meters;
- Transfer of about one million acres from seasonal to perennial irrigation;
- Agricultural expansion in 1.2 million acres of new land due to increased water availability;
- Expansion in rice cultivation;
- Generation of hydropower;
- Improvement of navigation;
- Protection from high floods as in 1964 and 1975 , and from low floods and droughts as in 1972, 1979, 1982, 1985, 1986 and 1987.

Adverse side effects of High Aswan Dam

- Siltation
  Prior to the construction of the HAD, the Nile carried an annual load of suspended silt in the range of 80 to 110 million tons. Since 1964, this silt has been deposited in the reservoir, and the water released from the dam is practically silt-free. The change in siltation pattern has led to several concerns: first, the effect of siltation on the capacity of the reservoir; and second, various problems associated with loss of silt deposits, namely, the diminishment of soil fertility, river bed erosion, and coastal erosion of the Nile Delta.

- Water quality changes
  Studies indicate that the main changes in river water quality following the construction of the dam have been primarily in suspended solids, dissolved solids, and algal content, (Mancy and Hafez, 1979). These changes can be directly related to physico-chemical and biological transformations taking place in the reservoir. At the present time, water released from the dam is particularly silt-free and its suspended solid content consists largely of phytoplanktons. Although the dam has significantly changed the downstream river condition, research indicates that downstream deterioration in water quality is attributable to the impoundment of water by local barrages and to increased pollution rather than directly to the HAD.

- Ecosystem alteration
  The impact of river-controls on the Nile ecosystem can be classified according to three regions : 1) the HAD reservoir, 2) the river from Aswan to Cairo, 3) the Delta. The assessment of ecosystem characteristics was based on data from a comprehensive surveillance and monitoring programs which include measurements of such indicators of water quality as benthos (bottom aquatic organisms), phytoplankton and zooplankton (minute floating plants and animals which form the lowest level of the aquatic food chain), hydrophytes (plants which thrive in stagnant water or water logged soil), and fishes. As the Nile water enters the HAD reservoir, it undergoes a gradual change from typical riverain conditions at the southern end to typical lacustrine (lake-like) conditions in the northern end of the reservoir. The fish population of HAD reservoir has significantly increased. In contrast to other African lakes and swamps, the reservoir is far devoid of the water-hyacinth (a hydrophyte). Other types of hydrophytes occur sporadically in khors (embayments). A few miles downstream from HAD, the Nile water starts to revert to riverain conditions. The river section from Aswan to Cairo exhibits fairly good water quality conditions . Population density and intensified river-use north Cairo are causing water pollution problems in certain locations on both the Rosetta and Demietta branches of the Nile . In this region there is a significant reduction in the fish population as well as excessive production of hydrophytes. The levels of nutrients, organic residues, and industrial waste products are alarming in the Rosetta branch.

- Health implications
Following the construction of the HAD, environmentalists predicted that schistomiasis (bilharzia), already endemic in Egypt, would spread with the expansion of perennial irrigation. Contrary to these predictions, the prevalence of the disease among the rural population has been generally on the decline, (Hussein 1981). The researchers attributes their findings to the following factors: 1) while HAD has ensured a more dependably water supply for perennial irrigation, there has been no significant expansion of agricultural land, hence no net expansion of infection sources, 2) a large shift of population from rural high transmission areas to urban low transmission areas, and 3) the effect of environmental health programs, including protected water supply, chemotherapy, delivery of health services, and health education.

- Archaeological implications
The banks of the Nile is the repository of some of the most magnificent relics of ancient civilization. These monuments are an irreplaceable part of the human heritage and as such their preservation is of deep concern to the international community. Egyptian planners of HAD realized from the beginning that the waters of the reservoir would drown some of the monuments upstream. In 1963, the temple of Abu Simbel was cut into gigantic blocks and later meticulously reassembled at a higher elevation. Similarly, the Ptolemaic temple of Kalabsha was moved 5 km from its original site and reassembled on a small island facing the HAD. Despite these effects, a number of lesser monuments are permanently submerged under the water of the reservoir. What the Egyptian planners did not foresee was that the rise of the underground water table would damage the foundations of some of the archaeological sites downstream. In Luxor, 230 km downstream from the dam, one wall of the Karnak temple has fallen and several other structures are in danger. The solution to this problem seems to lie in artificially lowering the water table in areas of archaeological significance. This can be done by the continual pumping out of underground water. The cost of sustaining such protective measures will be high.

- Socio-economic implications
  The socio-economic changes brought about by the HAD are: 1) the displacement and resettlement of the Nubians, 2) urbanization and industrial development (e.g. fisheries and tourism in the lake), and 3) impact of the nomadic populations newly settled along the shores of the reservoir.

Conclusions

Water development projects invariably have adverse effects, and thus the real question is not whether such development will affect the environment, but how much change is acceptable to the society as a whole, and what counter measures should be taken to keep the adverse changes to a minimum and within that acceptable range. The High Aswan Dam in Egypt has received its share of criticisms for contributing to environmental disruption. Twenty years of monitoring and evaluation indicate that the project is essential for social and economic development of Egypt even though it has contributed to some environmental problems. The case study of High Aswan Dam and its effect on the Nile provides information helpful to scientists and engineers who are faced with the controversial problems attendant on the building of large dams.

Appendix I. - References

Abu-Zeid, M. (1989). "Environmental Impact of the Aswan High Dam.' Water Resources Development, Vol. 5 No. 3

Elassiouti, I., et al. (1989). "Environmental and health Impact of The Irrigation and Drainage System in Egypt." Report, Advisory Panel for Land Drainage in Egypt.

Hussein, M. (1981). "Public Health Implication of Water Resources Management in Egypt." Proc. Inter. Conf. on Water Resources Management in Egypt, Cairo.

Mancy, K. H., and Hafez, M. (1979). "Water Quality and Ecosystem Considerations in Integrated Nile Resource Management." Proc. Inter. Conf. on Water Resources Planning in Egypt, Cairo.

Annual Flow Variations in the Nile River System

Mamdouh Shahin[1]

Abstract

The annual flow series of the Nile River System at some key gauging stations for the period 1912-1988 are presented and discussed. The relatively stable period, pre-1961, has helped establish relationships between the flows at the different sites for flood forecasting. The rise and the decline in the annual rainfall, both after 1961, in the Equatorial Lakes Plateau and the Ethiopian Plateau, respectively have resulted in non-homogeneity in the series. The long-term trend in the water level series since 621 AD near Cairo, Egypt is briefly discussed.

Brief Description of the Nile River System

The Nile River spans some 35 degrees of latitude between 5°S and 30°N. The river basin covers an area of about 2.9×10^6 km^2, shared by the Sudan, Ethiopia, Egypt, Uganda, Tanzania, Kenya, Zaire, Rwanda and Burundi.

Almost 70% of the annual natural flow that used to reach Aswan, Egypt (see Figure 1) orignate in the Ethiopian Highlands and flow in the Nile by the way of the Sobat, the Blue Nile and the Atbara. The remaining 30% arrive by the way of the While Nile.

The flow below Aswan is carried by the Nile a distance of 900 km before reaching Cairo, north of which the Nile bifurcates into two branches. These used to discharge the excess water into the Mediterranean Sea after having the agricultural land in the Delta area irrigated.

Rainfall and Evaporation

The Nile River, its tributaries and lakes cover a wide range of topographic and climatic conditions.

[1] Shahin, Senior Lecturer, Int'l. Inst. for Hydr. and Environ. Engg., Oude Delft 95, Delft, The Netherlands.

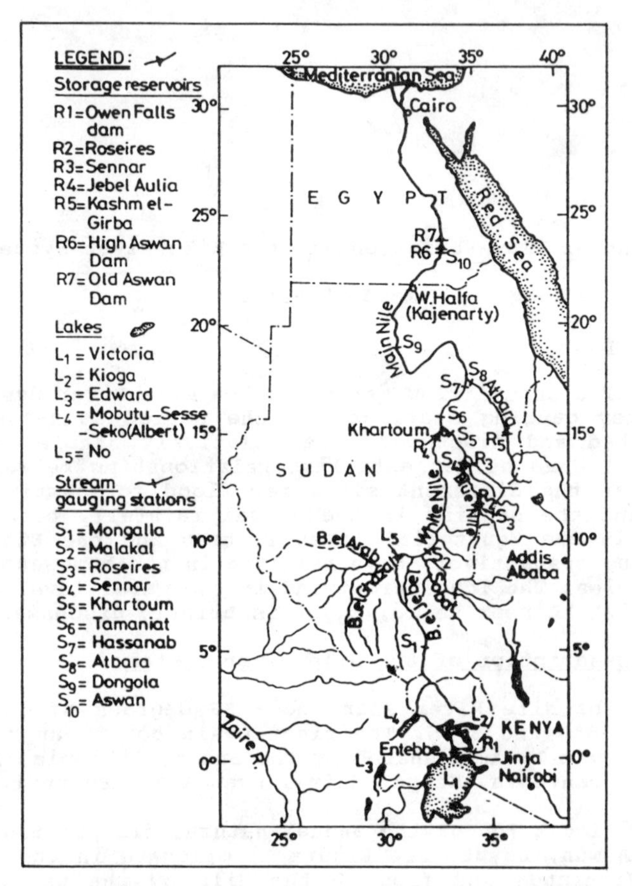

Figure 1 - The Nile River System and the Key Stream Gauging Stations

A large surface, mainly in Egypt and northern Sudan is practically dry. The remaining surface of the basin receives an annual rainfall varying from say 150 mm at Khartoum to more than 1400 mm in the Ehiopian Highlands. The rain depth in the area extending from say 20°N up to the equator is influenced by the latitude and the geographic location of the point where rainfall is measured. The basic statistics of the rainfall at all stations in the Nile Basin for the period 1931-1960 have been worked out and published (Shahin, 1985).

The rainy season 1961-1962 was reported to have

produced more rainfall than any other season during this century over Eastern Africa (Flohn and Burkhardt, 1985). The same reference adds that large areas in Kenya, Tanzania and Uganda received during five months 300 to 500% of the average precipitation.

Measurements and estimates of free water evaporation and potential evapotranspiration have been elaborated and discussed by several investigators ((Haude, 1959), (Shahin, 1970 and 1985), (Bakry, 1977), and (Flohn and Wittenberg, 1980)). From these references as well as other sources one can conclude that the annual evaporation along the southern coast of the Mediterranean Sea is about 1100 mm, at the apex of the Nile Delta is at 1500 mm and at Aswan about 2600 mm. The evaporation reaches a maximum of about 3000 mm/yr in the reach from Halfa to Atbara. The annual rate is reduced to 2800 mm at Khartoum and further to about 1600 mm at Malakal.

The surface water evaporation from Lake Tana, Roseires and Sennar reservoirs is 1100, 2200 and 2500 mm/yr, respectively. The annual evaporation from the swamps is nearly 150% times the evaporation from the open water at the same location, bringing it to about 2700 mm/yr in the Sudd area and in the Bahr el-Ghazal basin and to 2000 mm/yr in the Machar swamps in the Sobat basin.

River Discharge

1a - Short-term Variation: 1912-1960

The period 1912-1960 was a fairly stable period as far as the outcome of the hydrologic processes is concerned. The annual maxima and so the annual minima levels of Lake Victoria, for example, used to vary within limited ranges.

The basic statistics of the Nile flows for the period 1912-1960 at the ten key stations are listed in Table 1.

Table 1 - Basic Statistical Decriptors of the Nile Annual Flow Series in the Period 1912-1960

Key station	x	s	Key station	x	s
Mongalla	26.02	6.84	Tamaniat	75.96	10.89
Malakal	27.49	3.62	Hassanab	74.32	9.52
Roseires	49.94	8.83	Atbara	12.27	3.91
Sennar	48.98	9.46	Dongola	86.38	12.46
Khartoum	52.32	9.39	Aswan	84.37	12.26

x = mean, 10^9 m^3/yr and s = standard deviation, 10^9 m^3/yr.

The limited variability of the river flow in that rather stable period has helped develop simple regression models for flood forecasting (Szalay, 1973). As an example the annual flow $(V_{MN})_k$ at Kajenarty on the Main Nile for the period 1918-1957 could be expressed as

$$(V_{MN})_k = 34.0 + 6.07\ (F_{BN})_R \qquad \ldots (1)$$

where $(F_{BN})_R$ is the flood flow of the Blue Nile at Roseires in June and July, $10^9 m^3$.

The extent of the stability and the underlying homogeneity of the river discharge data were questioned for the first time while designing the High Aswan Dam. The mean annual flow at Aswan for the period 1870-1955 was $92.3 \times 10^9\ m^3$. The mean for the period 1870-1898 was $110 \times 10^9\ m^3$/yr and for the period 1899-1955 was $83.3 \times 10^9\ m^3$/yr, i.e with a negative jump of about 29%. By removing the jump the characteristics of the Nile River may be reduced to a long stationary stochastic process (Yevjevich, 1972).

1b. Short-term Variation: after 1960

The water level of Victoria Lake rose between October 1959 and May 1964 by about 2.5 m. Following a slight fall, it began to rise again in 1978 and by mid-1979 had again reached almost the level of 1964 (Kite, 1981). As a result, the average outflow from the lake at Jinja increased from $20.8 \times 10^9\ m^3$/yr for 1900-1960, to $41.2 \times 10^9\ m^3$/yr for 1962-1978. This extra supply caused the average for 1912-1973 to be 21% and 7% larger than that for 1912-1960, at Mongalla and Malakal respectively.

The increase in the flow of the Equatorial Plateau rivers was not reinforced by a similar rise in the flow of the Ethiopian Plateau rivers. Accordingly, the flow in the Main Nile showed a slight decline in the mean annual flow, furthered by the storage on the Blue Nile and the Atbara.

The average water level, $(V)_{9\text{-}12}$, of Lake Victoria in September-December was found to be in linear relationship with the volume of water, $(A)_{2\text{-}6}$, in $10^9\ m^3$/yr, reaching Aswan in February-June. This relationship (Flohn and Burkhardt, 1985) can be written as

$$(V)_{9\text{-}12} = 10.0 + 0.047\ (A)_{2\text{-}6} \qquad \ldots (2)$$

Since $(V)_{9\text{-}12}$ is highly correlated with $(V)_{annual}$ it has been possible to reconstruct the missing lake levels for the period 1870-1900. The high water levels around 1878 and 1892 were also reported earlier.

The flow series of the Nile above Aswan for the

period 1870-1975 and the annual high river levels at Cairo since the year 1700 have been used for examining the possibility of short-term climatic changes in the source regions of the Nile water. The relationship obtained is

$$(A)_{annual} = -308 + 20.8\,H \qquad \ldots (3)$$

where $(A)_{annual}$ is the annual flow above Aswan in $10^9\,m^3$ and H is the annual maximum water level, in m. The conclusion is that, except for a few short periods, of 10-20 years duration, the river flow was never "at rest" with respect to the average flow in the 275 years covered by the investigation. (Riehl et al, 1976).

Periodic fluctuation of the level of Lake Victoria and its relation to the number of the sunspots has long been a subject of study. More recently, analysis of the Nile flow at Aswan by means of the correlation function has emphasized the existence of periodicities of 21, 7, 6, 4.2 and 2.7 years (Andel and Balek, 1971).

After the second peak which followed that of 1962, has been reached around the year 1979, the level of Lake Victoria kept fluctuating. Though the general trend is a falling one, the pre-1960 levels have not been recovered. This has brought the mean for 1961-1988 to about 47.3 x 10^9 m^3/yr or 182% times the mean for 1912-1960. Due to the increase in the losses incurred in the Sudd region, the ratio of the two means has fallen to 121% at Malakal.

The drought that swept over Ethiopia in the nineteen sixties kept recurring in the seventies and the eighties. Subsequently, the mean annual natural flow for 1961-1988 became 90% at Roseires and 80% at Khartoum, both on the Blue Nile, of the annual mean for 1912-1960. As such, the mean flow at Aswan fell from about 84.4 x 10^9 m^3/year for 1912-1960 to 76.7 x 10^9 m^3/yr for 1961-1988.

A summary of the basic statistics for the period 1961-1988 is given in Table 2, while the flow series at a few selected stations for 1912-1988 are given in Figure 2.

Table 2 - Basic Statistical Decriptors of the Nile Annual Flow Series in the Period 1912-1960

Key station	x	s	Key station	x	s
Mongalla	47.29	7.09	Tamaniat	68.69	13.75
Malakal	33.38	5.42	Hassanab	68.35	14.42
Roseires	44.09	9.56	Atbara	9.17	3.83
Sennar	36.81	9.51	Dongola	79.97	12.74
Khartoum	39.69	10.08	Aswan	76.72	14.14

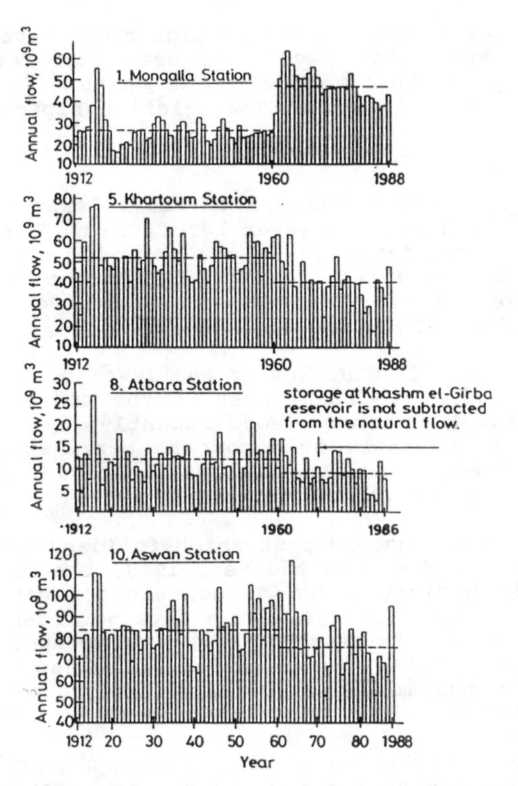

Figure 2 - The Nile Flow Series in the Period 1912-1988

2 - Long-term Variation

The Nile levels at Roda opposite Cairo have been recorded since the year 622 AD. In spite of the intriguing nature of these series they have been used with various degrees of cautiousness for detecting the possible climatic changes in the past thirteen centuries or more.

The annual minima and maxima level sequences were analyzed and auto-regressive models fitted to the data. The model fitted to the minima sequence (622-1285 AD) is

$$y_t = 0.42y_{t-1} + 0.10y_{t-2} + 0.06y_{t-3} + 0.10y_{t-5} + 0.02y_{t-6} + 0.04y_{t-11} + 0.11y_{t-13} + \epsilon_t \quad \ldots (4)$$

where y = x - 10.8 is the deviation from the mean value, and x is the water level in m (Balek, 1977).

The minima sequence of 829 years (621-1449 AD) has been recently re-analyzed (Aguado, 1987) using the autocorrelation function. The ARIMA model (2,1,1) that has been fitted to that particular sequence can be written as

$$x_t = 1.36x_{t-1} - 0.26x_{t-2} - 0.10x_{t-3} - 0.93\epsilon_{t-1} + \epsilon_t \quad \ldots (5)$$

Recent investigations including both annual minima and annual maxima sequences are currently underway.

References

Aguado, E. (1987). "A time series analysis of the Nile River low-flows." Ann. Ass. Amer. Geog.: 72(1), 109-119.
Andel, J. and J. Balek, (1971). "Analysis of periodicity in hydrological sequences." Jour. of Hydro.: 14,66-82, North Holland Publishing Company, Amsterdam.
Balek, J. (1977). "Hydrology and water resources in tropical Africa." Developments in Water Science 8. Elsevier, Sci. Pub. Co. Amsterdam, Oxford, New York.
El-Bakry, M. (1975). "Evaporation from Lake Victoria." Met. Res. Bull.: 7 (1), 21-47, General Organization for Government Printing Offices, Cairo.
Flohn, H. and T. Burkhardt, (1985). "Nile runoff at Aswan and Lake Victoria." Zeitschrift für Gletscherkunde und Glazialgeologie: 21,125-130, Universitätsverslag Wagner, Innsbruck.
Flohn, H. and H. Wittenberg, H. (1980). "Die Verdunstung als wasserwirtschaftliche Schüsselgröße zum Qattara-Projekt." Wasser und Boden: 8,352-258.
Haude, W. (1959). "Die Verteilung der potentiellen Verdungstung in Ägypten." Erdkunde: 13 (3), 214-224.
Kite, G.W. (1981). "Recent changes in level of Lake Victoria." Bull. Hydro. Sc. : 26 (3), 233-243.
Riehl, H., M. El-Bakry and J. Meitin (1979). "Nile River discharge." Monthly Weather Review: 107,1546-1553, Amer. Met. Soc.
Shahin, M. (1970). "Analysis of evaporation pan data in U.A.R. (Egypt)." Ann. Bull. ICID, 53-69, New Delhi.
Shahin, M. (1985). "Hydrology of the Nile Basin." Developments in water science: 21, Elsevier Sci. Pub. Co., Amsterdam, Oxford, New York and Tokyo.
Szalay, M. (1973). "Forecasting of flood characteristics of the Nile." Proc. of Int. Symp. on Riv. Mechs., IAHR, B38 1-11, Bangkok, Thailand.
Yevjevich, V. (1972). "Stochastic processes in hydrology." Wat. Res. Publications, Fort Collins, Colorado.

Arid Lands - Impending Water-Population Crises

Marvin E. Jensen [1] Honorary Member, ASCE

ABSTRACT

Many developing countries in arid lands will face major food production problems because of population growth rates. Irrigation water requirements will soon exceed available renewable water resources. The performance of many irrigation projects must be improved and production on rainfed lands must be increased to avoid major future conflicts in many arid areas.

INTRODUCTION

The United Nations Population Crisis Committee (PCC) recently updated its projection of world population (Science, 1990). PCC indicates that instead of leveling off at 10.2 billion, the ultimate population number will be 14 billion <u>if</u> business continues as usual. Unconstrained population growth in some arid areas with limited fresh water supplies is on a collision course with disaster. Improved hydrological data will be needed for planning and managing scarce water resources under increasing competition for existing water supplies. The purpose of this brief paper is to review recent irrigation development trends and call to your attention serious emerging problems facing developing countries in arid regions.

Long-term trends in irrigation development in the western USA indicate what can be expected in other arid areas. Recent projections of increased food production in developing countries include continuous expansion of irrigated lands to meet increasing food demands. Population growth in some Middle East developing countries is creating increasing water demands that soon will exceed the renewable supplies. Desalinized sea water now provides for basic human needs in several areas. However, it is much too expensive to use in irrigated agriculture for food production. Long-term resolution of water supply-food production problems will require sustainable agricultural development, but future water management must involve managing demand as well as water supplies. Population management is not a role for hydrologists. Hydrologists will need to develop new and better techniques for providing reliable hydrological data where only limited historical data are available.

[1] Director, Colorado Institute for Irrigation Management, Colorado State University, Fort Collins, CO 80523, USA.

EXPANDING WATER NEEDS

Population Growth and Increasing Potential for Conflict

In arid areas, new sources of water are being sought and developed to meet increasing needs for food and fiber production, domestic, municipal and industrial uses. Internationally shared rivers are being developed independently by countries without water treaties or water-sharing agreements. Middle East rivers that currently are nearly fully developed are the Jordan, the Nile and the Tigris-Euphrates. Population growth rates in most of the countries sharing these waters are near 3 percent per year with an average doubling time of 27 years (Table 1). These arid areas may become the future sites of major conflicts as countries independently develop their water supplies for irrigated agriculture and power production.

TABLE 1. Populations, Natural Increase Rates and Doubling Times for Selected Middle East Countries. (Source: Pop. Ref. Bur., 1989)

Country	Estimated 1989 population (millions)	Natural annual rate (percent)	Doubling time (years)
Egypt	54.8	2.8	24
Ethiopia	49.8	2.1	33
Iraq	18.1	3.8	18
Israel	4.5	1.6	43
Jordan	4.0	3.5	20
Saudi Arabia	14.7	3.4	20
Syria	12.1	3.8	18
Sudan	24.5	2.8	24
Turkey	55.4	2.2	32
Total	237.9		
Weighted average		2.7	27

Turkey is developing plans to use Euphrates River water for its large southeastern $21 billion Anatolia development project. Its nearly completed Ataturk Dam will produce power and enable irrigating 18,600 ha (Moffett III, 1990). Turkey closed the Euphrates for a month in January 1990 to begin filling the reservoir behind the dam. The terms of international conventions indicate that water rights are to be shared according to population and needs taking into account historical allocations. However, international law acknowledges absolute sovereignty of nations over resources they control. Turkey maintains it has no international obligations and indicates that it will do its best not to harm its neighbors. Syria plans to irrigate 40,400 ha of new land along the Euphrates and needs 15 million cubic meters of water to avoid a food crisis by the year 2000 (Moffett III, 1990). Iraq, at the end of the Euphrates, is facing even more severe water problems.

Israel and Syria are the two main users of the water from the Jordan River. Jordan is mining ground water to meets its current needs. Syria and Jordan are planning to construct a dam on the Yarmuk, a tributary of the Jordan River, but this may provide only a short period of relief for Jordan because of its rapidly expanding population (Moffett III, 1990).

The Nile River is the world's longest river, extending 6,800 km from high rainfall areas to the very arid regions of Sudan and Egypt. Egypt and Sudan have an agreement to share the waters of the Nile, but Ethiopia, the source of the Blue Nile and 80 percent of the Nile water, is not receptive to a river-sharing agreement. Flow records on the Nile have been maintained since 1871. Average flow rates for 15 to 40-year periods may vary by as much as ± 10 percent. Recent analyses indicate that the average flow rate may be declining. The runoff was above normal in 1989 and if this continues several years, the current apparent declining trend may not be real. Egypt now imports about 60 percent of its food and its population will double in about 24 years.

Environmental Concerns

Environmental pressures have increased as impacts of developing water supplies for food and fiber production have become more visible. The most dramatic example impact of a change in water use is the decline of the Aral Sea in the USSR. Waters from two rivers provided 55 km^3 per year to the sea from 1920 to 1960. Since then, most of this water has been diverted to agriculture for producing mainly cotton causing the sea to shrink drastically. An estimated 30-35 km^3 per year of water would be required just to stabilize the lake (Sun, 1988; Ellis, 1990). On a smaller scale in the USA, the diversion of a major portion of the Truckee River flow for irrigated agriculture started in 1902. This caused the level of Pyramid Lake in Nevada to recede about 24 m over an 80-year period. An estimated annual inflow of 495 million m^3 would be required to maintain the present level of the lake. The lake's salinity content will still increase about 0.1 percent per year due to evaporation and concentration of dissolved solids in the inflow water.

In arid areas, all changes in consumptive water uses represent trade-offs -- enhancing one consumptive use will directly affect other consumptive uses. There is no free lunch. Development of renewable water supplies for food and fiber production on irrigated lands in most developing countries in arid regions has been essential to avoid massive human starvation, malnutrition and declining quality of life. Agriculture cannot be sustained on mined ground water. Ground water reserves should be used to enhance production during drought cycles. Unfortunately, once developed, fossil water usually continues to be depleted. The land subsidence that occurs with ground water mining may not cause major damage except in certain areas.

Short-term Solutions

In the near-term, expansion or development of irrigated agriculture must be integrated with increased production on rainfed lands. The performance of many relatively new irrigation projects must be improved. Technology exists for increasing retention of precipitation on rainfed cultivated lands. This technology, involving the use of crop residues, can greatly reduce evaporation losses. Unfortunately, the competition for crop residues in many poor developing countries is very great because it is often a primary source of livestock feed and fuel for cooking. Modern rainfed agricultural technology must be adapted to cultivated rainfed lands in arid areas, and the best rainfed technology must be combined with the best irrigation technology to avoid major food crises in the next two decades.

TRENDS IN IRRIGATION DEVELOPMENT

Western USA

Irrigation development experience in the western USA may indicate changes to be expected in other arid areas. Development and expansion of irrigated land during the past century in the western USA continued until about 1980. The most rapid rate of expansion occurred during 1970s as thousands of center pivot sprinkler systems were installed in the Great Plains (mainly in Nebraska) where ground water was readily available (Fig. 1). The area irrigated in the Texas High Plains increased rapidly during the drought years of the 1950s and reached a peak in 1978. The area of land irrigated in the Texas High Plains has decreased greatly since 1978 as water table levels dropped, pumping costs increased and farm commodity prices declined. The area irrigated in the Pacific Northwest increased steadily for many decades, but has decreased slightly since 1978. The irrigated land in the Mountain states was some of the first to be developed and has remained relatively constant.

The annual rate of expansion from 1944 to 1949 was about 4.5 percent. It declined to about 1 percent in the 1960s. The large rate of expansion in the 1970s has been offset by a decrease from 1978 to 1987. These changes represent the effects of many complex factors, but basically they show that the rate of development of irrigated land will decrease as available renewable water supplies are developed. They also indicate that once developed, full irrigated agriculture cannot be sustained on mined ground water.

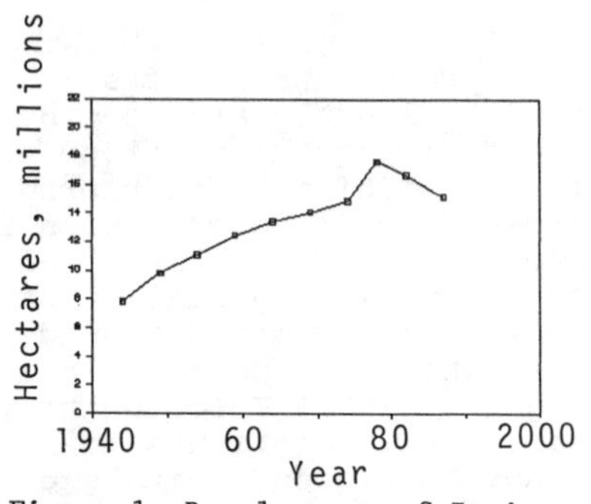

Figure 1. Development of Irrigated Land in Western USA. (Sources: USDC, 1989; and USDA, 1983)

World and Middle East Developing Countries

FAO statistics are not complete or consistent prior to 1972. The world-wide area of irrigated land increased from 1972 to 1987 at an annual declining rate. The total in 1987 was 227 million hectares. The global rate of expansion of irrigated land was over 2 percent per year during the 1960s and 1970s. Since 1980, the rate of expansion has decreased to about 1 percent per year even though populations in many developing countries are continuing to increase at about 3 percent per year. There are many reasons for the decline in the rate of expansion. Declines are expected as the better lands and readily available water supplies are developed. However, poor performance of many irrigation projects, lack of adequate maintenance, waterlogging, and high costs have been major contributing factors.

The area of irrigated land in Middle East developing countries has also been increasing at a declining rate. The total in 1987 was 116.5 million hectares. The rate of development decreased from 2.2 percent per year from 1972 to 1977 to about 1.5 percent per year from 1982 to 1987. Clearly, continued expansion of irrigated land will not provide the increases in food production needed to keep up with population growth.

ACTIONS NEEDED

During the green revolution, expansion of irrigated land played a major role in increasing food production to meet population needs. FAO estimates that 36 percent of the total food production is on 15 percent of the arable land that is irrigated. Recent projections of actions needed to increase food production include contributions from expanded irrigated lands to meet population growth needs assuming an expansion rate of 2.25 percent per year (Alexandratos, 1989). The data that I have presented indicate that these projections may be at least two times greater than the rates that can be expected in the 1990s.

The performance of existing irrigated projects in many developing countries must be improved. This will require major investments in rehabilitation, improved operations and maintenance, expanded adaptive research and technology transfer, and training at all levels. Production on rainfed lands must also be increased by adapting proven water conservation practices, such as crop residue management, and increased use of fertilizers, especially phosphorus. In the long-term, demand for water must be managed within the resources available for producing or purchasing needed food and fiber.

REFERENCES

Alexandratos, N. (ed.). 1989. World Agriculture: Toward 2000, an FAO Study. Belhaven Press, London.

Ellis, W. C. 1990. The Aral: A Soviet sea lies dying. Nat'l. Geographic 177(2):73-92.

FAO. 1989. The State of Food and Agriculture 1989. FAO Agricultural Series No. 22, Rome.

Moffett III, G. D. 1990. Middle East Water - Time is running out. Parts 1-4, The Christian Science Monitor, March 8, 13, 14 & 16.

Population Reference Bureau, Inc. 1989. World population data sheet. Washington, DC.

Science. 1990. No more "babies as usual." Briefings, Science 247:1183.

Sun, M. 1988. Environmental awakening in the Soviet Union. Science 241:1033-1035.

U.S. Department of Agriculture. 1983. Agricultural Statistics. 568 pp.

U.S. Department of Commerce, Bureau of Census. 1989. 1987 Census of Agriculture. Vol. 1, Part 51, 424 pp.

FLOODPLAIN MANAGEMENT FOR ALLUVIAL FAN AREAS

Cecelia Rosenberg[1]

INTRODUCTION

The National Flood Insurance Program (NFIP) floodplain management regulations (found at Chapter 44 of the Code of Federal Regulations (CFR) Section 60.3) are the minimum requirements that must be adopted and enforced by all communities wishing to participate in the NFIP. Because the regulations are applied nationwide, they were developed to cover a broad range of flooding and site conditions. Therefore, it is the responsibility of the community to enact regulations (which may be more restrictive than those of the NFIP) to address a flood condition or management objective specific to a particular location. Because current NFIP regulations do not address the special floodplain management requirements of Special Flood Hazard Areas (SFHAs) on alluvial fans, many fast-growing jurisdictions within the arid and semiarid sections of the United States need guidance to regulate new construction within these high hazard floodplains. This paper describes the approach that the Federal Insurance Administration (FIA)/Office of Loss Reduction (OLR) will take to the development of Federal floodplain management regulations that fit more closely the needs of arid west communities.

THE NEED FOR NEW REGULATIONS

NFIP floodplain management regulations require that all new construction and substantial improvements taking place within AO zones (shallow sheetflow areas) have the lowest floor (including basement) elevated above the highest adjacent grade at least to the depth indicated on the Flood Insurance Rate Map (FIRM). In addition, adequate drainage paths are required to be maintained in order to direct floodflow around and away from structures. While these requirements (found at 44 CFR, Section 60.3, Paragraphs (c) (7) and (11)) are sufficient to mitigate damage to structures from the low-velocity sheetflows found in both arid and nonarid regions, they do not address the severe hazards of erosion, scour, sediment and debris deposition, high velocities, shifting channels, mudflow, and floodwater, sediment, and debris impact forces that characterize alluvial fan flooding.

---

[1]Physical Scientist, Federal Insurance Administration, 500 C Street, SW, Washington, D.C. 20472.

Awareness of the need for regulations for alluvial fan SFHAs originates from the growth in new construction that continues to take place in the Western United States. Evidence of accelerated floodplain development is found in the noticeable increase in Letter of Map Revision requests to FIA's Office of Risk Assessment from communities in California, Nevada, and Arizona. As more SFHAs are identified in arid west communities, local floodplain managers will need guidelines to instruct developers in building safely within alluvial fan floodplains. Outside of California, Nevada, and Arizona, the states most affected by this issue are New Mexico, Colorado, Washington, Oregon, Utah, Idaho, Montana, and Wyoming.

The following issues, which arise during the course of building in areas of alluvial fan flood hazards, should be addressed in future Federal regulations:

1. The need for the appropriate application of AO zone elevation requirements (in particular, the need to define "highest adjacent grade" when the regrading of natural ground on a hillside or slope must be done to situate construction pads)

2. The need to identify the most appropriate method of elevating structures to minimize damage from sediment and debris deposition and high-velocity water and debris impacts; similarly, the need to indicate the depth to which foundations should be sunk to mitigate scour and erosion

3. The need to address protection of adjacent properties from floodwaters that are being diverted away from other structures

THE NEED FOR TECHNICAL GUIDANCE

In addition to the need for floodplain management regulations for construction in areas of alluvial fan flood hazards, technical guidance is needed to aid communities and builders in carrying out provisions of the regulations. The severity of the hazards and complexity of hydraulics are such that traditional riverine approaches to elevating and floodproofing individual structures do not work. Furthermore, the conventional principles used to site, plan, and grade subdivisions must be examined so that these activities use methods designed to minimize the effects of alluvial fan flood hazards.

Elevating Individual Structures

Consideration should be given to prohibiting the elevation of buildings on fill in alluvial fan SFHAs (as it is in coastal high hazard areas, called V zones) to address erosion and scour. At a minimum, the method of elevating single-family structures on fill must be evaluated to determine its effectiveness in providing protection from high-velocity flow, scour, erosion, and debris impact. Methods of armoring the fill pad and orienting the pad should be investigated to minimize erosion and to facilitate the flow of water and sediment without affecting structural foundations.

The use of piles or columns (required in V zones) should be evaluated for suitability in various types of alluvial fan hazard settings. Guidance is needed to inform builders of installation methods, use of appropriate materials (such as wood versus concrete), diameter of piles, anchoring/connection techniques, and embedment depth.

Research is needed to identify building materials that are effective in providing added strength to structures so they may withstand impact forces of debris and hydrodynamic loads; building shapes and floor plans that minimize exposure of the building's uphill face to water and debris loads and sediment deposition; and lot layouts that allow maximum water and debris flow around or underneath the building.

Protecting Subdivisions

The most cost-effective methods of protecting subdivisions from alluvial fan flood hazards often involve whole-fan and smaller-scale structural measures such as diversion dams, detention basins, weirs, drop structures, levees, and flood-control channels. Many nonstructural alternatives are available, however; such alternatives include orienting streets in a downhill direction to serve as conduits to flow; reserving open space within the subdivision plan to accommodate flow paths; aligning all fill pads in a downslope direction to minimize water and debris impacts and sediment deposition; and elevating all homes on piles or columns. Research is needed before FIA can establish criteria to ensure that these alternative measures are utilized in proper combinations and in the appropriate hazard settings. A large part of this research should focus on cost comparisons among the various approaches.

Implementing Local-Level Regulatory Tools

Local-level regulatory controls must be set in place for a community to ensure the effective application and operation of the aforementioned flood-mitigation tools. Such controls include zoning ordinances, subdivision regulations, open space/land acquisition plans (to reserve "floodways" for floodflows), building codes, and maintenance and inspection requirements for any structural flood-control measures. Communities need guidance to implement these tools effectively and legally.

FIA'S PROPOSAL

Many structural and nonstructural measures have been proposed to protect buildings from alluvial fan flood hazards. In most cases, these solutions have been developed within the context of a specific community's flood regime. Because there are as many different alluvial fan flood hazard scenarios as there are fans and pediments, FIA must research the full range of potential hazards as well as the geopolitical settings in which they may occur. FIA can then issue technical guidance and regulations relevant to all communities with such hazards.

Furthermore, because mandatory flood insurance purchase requirements are enforced within all SFHAs and sanctions are imposed on all participating communities that do not carry out the provisions of the NFIP, it is imperative that FIA carefully analyze the magnitude of alluvial fan flood hazards and accompanying risks. Therefore, FIA is proposing a rigorous program of research to examine the floodplain management issues affecting the arid west communities participating in the NFIP. If warranted, FIA will then implement regulations and standards for construction. The research program will be completed in three phases, discussed below.

Phase I: Development of Interim Consensus Standards

The Arid West Committee of the Association of State Floodplain Managers is participating with OLR in the formulation of interim standards for the management of development in areas of alluvial fan flood hazards. A study (consisting of a detailed questionnaire) is underway to secure a consensus among State and local floodplain managers in the arid regions, private engineering and planning professionals, developers, citizens groups, and other Federal agencies (U.S. Army Corps of Engineers, U.S. Geological Survey) regarding the issues that should be addressed in interim regulations or policies. This survey will also ascertain their experiences with the successes and failures of regulating development in areas of alluvial fan flood hazards. An equally important goal of this project is to identify the location and type of alluvial fan flood hazards in communities participating in the NFIP, and to determine the number of insurable buildings at risk. This information will help to define the scope of the larger-scale study described herein.

Interim Federal policy will enable communities to manage construction within alluvial fan SFHAs while the larger-scale study is underway. FIA anticipates that any resulting interim standards (or regulations) will expand on the AO zone regulations currently applied in alluvial fan SFHAs. A final report with recommendations for interim regulations and/or standards is expected in January 1991.

Phase II: Comprehensive Study of Alluvial Fan Flooding in the Western United States

During Phase II, OLR will undertake a multiyear comprehensive study of alluvial fan flooding in the arid west states. This study will identify the range of natural and manmade characteristics of areas subject to alluvial fan flood hazards. The six major research elements of this study are listed below. The results of the first five elements will yield data on the occurrence of alluvial fan flood hazards; the sixth element involves analysis of the results of elements one through five, and recommendations for guidance or regulations.

1. Identification and inventory of fans in arid west states
2. Documentation of human influences and characteristics of identified fans
3. Documentation of natural fan characteristics and related flood hazards
4. Chronicle of damage assessments of previous fan floods
5. Report on state-of-the-art of alluvial fan floodplain management (nonstructural and structural management approaches)
6. Recommendations for technical guidance and/or regulations

Phase III: Evaluation of Regulatory Need and Formulation of Guidance, Policy and Regulations

Phase III of the FIA research program will involve evaluating the impact of potential regulations. The evaluation will focus on the following:

1. A benefit/cost analysis of technical guidance recommendations
2. Communities likely to be affected by any new regulations
3. Standards needed for specific land uses
4. Successful and unsuccessful examples of community floodplain management
5. Appropriate locations on fans to apply any regulatory tools
6. Appropriate implementation of prescriptive or performance standards

CONCLUSION

The OLR strategy to implement a floodplain management program for construction in areas subject to alluvial fan flood hazards is being developed as part of the overall FIA commitment to reducing losses to life and property. OLR is working with other FIA offices to resolve issues of alluvial fan flood hazard identification, floodplain management standards, and insurance rating in a coordinated fashion. These efforts comprise a major element of the NFIP's goals for the U. S. Decade for Natural Disaster Reduction for the 1990s. By instituting a floodplain management program that addresses the requirements of arid communities faced with complex and unpredictable flood hazards, FIA will be closer to its goal of minimizing flood risks to flood-prone structures throughout the United States.

IMPROVEMENTS TO MAPPING OF ALLUVIAL FAN FLOODING

Alan A. Johnson[1]
Associate Member ASCE

Introduction

The Federal Emergency Management Agency (FEMA), which is responsible for identifying floodplains for the National Flood Insurance Program, is working to refine its assistance to engineers and floodplain managers by issuing revised and simplified guidance and modeling for mapping the regulatory floodplains of alluvial fan flooding. The appendix to the Flood Insurance Study Guidelines and Specifications for Study Contractors (SC Guidelines) that provides information regarding alluvial fan flooding analyses is being abridged to emphasize the production of high-quality data for use in the FEMA methodology. FEMA has also developed and released a model to be used on a personal computer (PC) that is intended to simplify the application of the FEMA alluvial fan flooding methodology. In an effort to improve the assessment of whether structural flood-control measures can effectively protect areas from the flooding associated with alluvial fan sources, FEMA has contracted with the U.S. Army Corps of Engineers (COE) to determine the parameters by which a structural measure can be evaluated.

Revising the Alluvial Fan Appendix in the SC Guidelines

The 1985 version of the SC Guidelines, Appendix 5, "Alluvial Fan Studies," focused primarily on the computations to be performed for the FEMA methodology. Therefore, users of the SC Guidelines also concentrated their efforts on the computations. Sometimes, they provided questionable input parameters because they had focused on calculating the depths and velocities without ensuring that the fundamental input parameters for the methodology, the discharge-frequency (Q-f) curve and the outside boundaries, were reasonable and defensible. The accuracy of such results has been challenged, either by FEMA or by appellants, causing a delay in the issuance of effective mapping of these flood risks.

----

[1]Hydraulic Engineer, Federal Insurance Administration, 500 C St., SW., Washington, D.C. 20472.

When assessing the alluvial fan flooding analysis guidance as part of the SC Guidelines update, FEMA determined that some of the published equations in the methodology were in error and that the procedure for developing accurate input data was not thoroughly explained. To simplify the guidance, FEMA has removed the computations and examples of how to apply the methodology. This deletion was possible because of the introduction of the PC model and accompanying user's manual. To speed the production of accurate mapping of the flood risks from alluvial fan flooding, FEMA is formalizing the requirement that descriptions of and documentation for the choices made to determine the Q-f curves and outside fan boundaries be submitted for review prior to modeling. This modification of the SC Guidelines should lead to the timely production of reasonable and defensible SFHAs representing flooding from alluvial fan sources. A final version of the revised SC Guidelines will be available in fiscal year 1991.

FEMA Alluvial Fan Model

FEMA has made available a PC model for developing the depths and velocity zones that are shown within alluvial fan flooding areas on Flood Insurance Rate Maps (FIRMs). This model, which is a computer version of the methodology published in the 1985 SC Guidelines including corrections of previously noted errors, is the best tool we have identified for developing the flood insurance zones related to flooding on alluvial fans. The model can provide the analyses for single-channel or multiple-channel portions of the alluvial fan.

The program for the model was written in BASIC and is designed to produce, after a series of input prompts, a set of tables that lists avulsion factors, discharges, velocities, depths, and widths. The program will also produce a graphic "pie wedge," which visually presents the approximate distance between the flood insurance zone boundaries. FEMA will provide the model and accompanying user's manual to requesters who submit 5-1/4-inch or 3-1/2-inch floppy disks and information on the BASIC package that will be used. By using the program, the user will obtain the data necessary to delineate the flood insurance zone boundaries, based on the previously identified outside boundaries.

The FEMA alluvial fan flooding model contains all of the key elements of the methodology; however, site specific concerns will need to be addressed by the engineer when applying the model. For example, the model and the user's manual do not specifically address modifications from an "ideal" alluvial fan, such as an entrenched channel, or structural flood-control measures on alluvial fans. The engineer would need to understand these features and address them as part of the input parameters before using the model. Also, the model does not provide analyses of coalescent flooding. The coalescence analyses must be done outside this model using the single-fan analyses that are completed with the model.

The model is only one of a minimum of six tools that are necessary for defining the SFHAs on an alluvial fan. The other five tools--a topographic map, a soils map, an aerial photogrammetric map, U.S. Water Resources Council Bulletin 17B, and a source of hydrologic data--provide the documentation to delineate the areas below the apex of an alluvial fan that are at risk from flooding. The model then provides an expedient means to use the FEMA methodology to define the zones that will be within the SFHA between the outside boundaries already developed.

FEMA is developing a training guide for analyzing alluvial fan flooding using actual field information and engineering assessment techniques in conjunction with the FEMA alluvial fan flooding model. All interested parties are encouraged to submit examples for inclusion in the training manual. This training manual is tentatively scheduled to be available in fiscal year 1991.

Assessment of Structural Approaches to Flood Control on Alluvial Fans

FEMA has contracted with the COE, Hydrological Engineering Center (HEC), to assess the effectiveness of structural flood-control measures used to protect areas from alluvial fan flooding. The tasks detailed in this contract require the COE to request information from other Federal, State, and local agencies on the types of flood-control measures (floodwalls, levees, dikes, channels, debris basins) that have been used, the designed levels of protection, whether the structures have worked or failed, and how the measures have been tested. Those who have experience in the design, construction, and historical success or failure of these types of projects are encouraged to contact the HEC. FEMA's goal in conducting this study is to develop a more definitive set of evaluation criteria and procedures for analyzing structural measures on alluvial fans to augment the currently effective requirements, presented in Section 65.13 of the NFIP regulations.

Conclusion

The timeliness and accuracy of mapping or remapping the flood risk for alluvial fan flooding sources shown as SFHAs on FIRMs has improved and will continue to improve. FEMA has taken steps in fiscal year 1990 to improve the quality of the draft FISs submitted by revising the SC Guidelines and by releasing a PC version of the alluvial fan model. In fiscal year 1991, FEMA will work to revise the guidance and regulations to more clearly define the documentation that would demonstrate that the protection provided by natural or structural flood-control measures warrants removing areas from alluvial fan SFHAs.

Entrenched Channels and Alluvial Fan Flooding

Edward R. Mifflin, A.M. ASCE[1]

Introduction

Because the individual flood paths on alluvial fans are unpredictable, determining flood hazards on such landforms is hindered by particular complications. The proper framework in which to make such determinations was presented by Dawdy in 1979 [Dawdy, 1979]. Subsequent to that publication, the Federal Emergency Management Agency [FEMA, 1985] adopted the methodology proposed by Dawdy to advance their efforts in showing flood hazard areas on Flood Insurance Rate Maps (FIRMs). As more alluvial fans were modeled, it became more apparent that in some cases, as Dawdy had cautioned, the site-specific conditions did not adhere to the simple boundary conditions set forth in Dawdy's paper. This paper was written to demonstrate that considering one such condition, the presence of an entrenched channel, does not preclude the use of Dawdy's methodology, but rather attests to its versatility.

Review of the Methodology

Let H be a random variable denoting the occurrence of flooding at a given point on an alluvial fan. That is,

$$H = \begin{cases} 1 \text{ if the point is flooded} \\ 0 \text{ otherwise} \end{cases} \tag{1}$$

Let Q be a random variable denoting the peak discharge resulting from a storm over the watershed. If f_Q is the probability density function (pdf) of Q, then the probability of the point being inundated by a flood with a peak discharge of at least q_0 cubic feet per second (cfs) is

$$P(H=1) = \int_{q_0}^{\infty} P_{H|Q}(1,q)\, f_Q(q)\, dq \tag{2}$$

where $P_{H|Q}(1,q)$ is the probability of the point being flooded, given that the peak discharge is q cfs.

---

[1]Hydrologic Engineer, Michael Baker, Jr., Inc., 1420 King Street, Alexandria, Virginia 22314.

Dawdy [1979] defines the conditional probability $P_{H|Q}(1,q)$ as the ratio of the channel width formed by q cfs, w(q), and the width of the area of the alluvial fan that is subject to flooding, W, at the elevation of the point of concern. When the latter width is much greater than the channel width [$W>>w(q)$], that definition is equivalent to saying that each point on the contour has the same probability of being flooded.

Dawdy defines the channel width by considering a constant discharge, q, that creates a rectangular channel, flows at critical depth, and erodes the sides of the channel until the change in width per change in depth equals -200.

It follows that, given a storm will produce a peak discharge of q cfs, the probability of a point being inundated by that flood is

$$P_{H|Q}(1,q) = \frac{w(q)}{W} = 9.408\,\frac{q^{2/5}}{W} \tag{3}$$

In addition, define the pdf of Q, f_Q, to be log-Pearson Type III. Thus, the probability that a storm producing a peak discharge of at least q_0 in any given year is

$$P(Q>q_0) = \int_{y_0}^{\infty} \frac{\lambda^k (y-m)^{k-1}}{\Gamma(k)}\, e^{-\lambda(y-m)}\, dy \tag{4}$$

where λ, k, and m are the three parameters of the Pearson Type III distribution, Γ(·) is the gamma function, and y is the base 10 logarithm of the discharge, q.

The 100-year discharge at any point on the alluvial fan is that discharge, q_{100}, that has a base 10 logarithm such that $\log_{10} q_{100} = y_{100}$ and

$$0.01 = P(H=1) = \int_{y_{100}}^{\infty} 9.408\,\frac{e^{0.92y}}{W}\,\frac{\lambda^k (y-m)^{k-1}}{\Gamma(k)}\, e^{-\lambda(y-m)}\, dy \tag{5}$$

Rearranging equation (5) yields

$$0.01 = \frac{9.408C}{W} \int_{y_{100}}^{\infty} \frac{(\lambda')^k (y-m)^{k-1}}{\Gamma(k)}\, e^{-\lambda'(y-m)}\, dy \tag{6}$$

where

$$C = \frac{e^{0.92m}\,\lambda^k}{(\lambda-0.92)^k} \tag{7}$$

and

$$\lambda' = \lambda - 0.92 \tag{8}$$

Note that the integrand above is the Pearson Type III distribution assumed in equation (4) for $\log_{10}Q$ with a change in the scaling parameter from λ to $\lambda' = \lambda - 0.92$.

If f_Q is log-normal with mean μ and standard deviation σ, we get

$$0.01 = \frac{9.408\,C}{W} \int_{y_{100}}^{\infty} \frac{e^{-\frac{1}{2}\left(\frac{y-\mu'}{\sigma}\right)^2}}{\sigma\sqrt{2\pi}}\,dy \qquad (9)$$

where

$$C = e^{0.92\mu + 0.42\sigma^2} \qquad (10)$$

and

$$\mu' = \mu + 0.92\sigma^2 \qquad (11)$$

Note that the integrand above is the normal distribution that describes the pdf of $\log_{10} Q$ with a change in the mean from μ to $\mu' = \mu + 0.92\sigma^2$.

The channel characteristics defined by Dawdy lead to expressions of depth and velocity as functions of discharge. Using those expressions, one can define 100-year flood depths and velocities for points on the fan. That is, the 100-year depth and velocity at some point on the fan are the depth and velocity related to the discharge, q_0, that yields a value of 0.01 for $P(H=1)$ in equation (2).

Considering Entrenched Channels

The definition of the conditional probability, $P_{H|Q}$, is fundamental to applying the methodology. For example, that all possible paths from all possible flows are equally probable may not be true. This paper discusses a simple situation where it is not.

Consider the following. Figure 1 shows an alluvial fan with an entrenched channel. The channel is stable and safely conveys flows (and their associated sediment loads) up to 1,000 cfs. Figure 2 is the flood frequency curve at the apex of the fan. The curve is log-normal with a mean and standard deviation of 1.0. Note that 1,000 cfs is about a 45-year flood.

Using the methodology and ignoring the entrenched channel results in identifying the flood hazards shown on Figure 3. In that figure, Zone AO is used to denote flood hazard areas subject to 100-year flooding with depths and velocities as shown. Zone B is used to denote flood hazard areas subject to 100-year flooding with a depth of less than 0.5 foot or subject to between 100- and 500-year flooding.

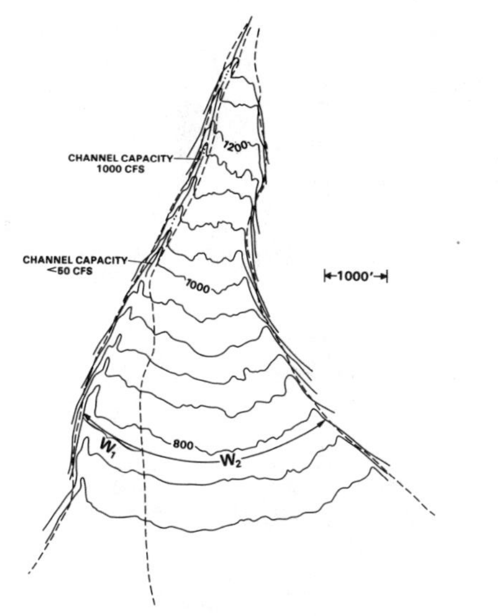

Figure 1. Example Alluvial Fan

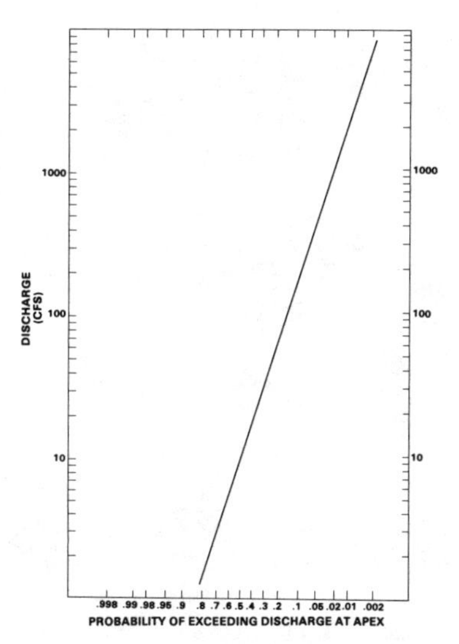

Figure 2. Flood Frequency Curve

If, as proposed above, the channel will safely convey floods up to 1,000 cfs and if at flows greater than 1,000 cfs the channel will fail at the apex, then our analysis must be broken down to incorporate the boundary conditions for flows less than 1,000 cfs and for flows greater than 1,000 cfs. Such an analysis is performed by rewriting equation (2) as

$$P(H = 1) = \int_{q_0}^{1000} P_{H|Q}^{W_1}(1, q) f_Q(q)\, dq + \int_{q_0}^{1000} P_{H|Q}^{W_2}(1, q) f_Q(q)\, dq \qquad (12)$$

where it is understood that the first integral vanishes for $q > 1000$. The superscripts W_1, and W_2 are the widths of the areas subject to flooding from flows less than and greater than 1,000 cfs, respectively (see Figure 1). They are used to show that the conditional probability defined by equation (3) is not the same in the two integrals. Specifically, W in equation (3) is replaced by W_1 or W_2.

Performing our analysis this way results in identifying the flood hazards shown on Figure 4. The zone designations are defined the same as above. Zone C is used to denote areas subject to flooding less than once in 500 years on the long-term average.

The example discussed above restricts channel failure to a specific point (the apex). If the point at which the channel will fail is unknown, then the analysis must be broken down further. For example,

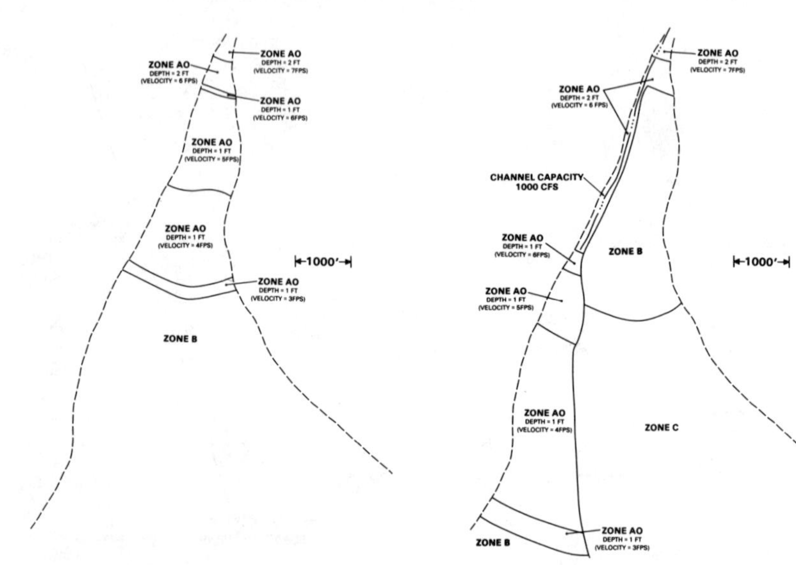

Figure 3. Flood Insurance Zones (No Entrenched Channel)

Figure 4. Flood Insurance Zones (Failure at Apex)

consider the situation where, given the flow exceeds 1,000 cfs, the probability of the channel failing in any reach is proportional to the length of the reach. In that case, the probability of a point being hit by a flood of magnitude greater than $q_0 > 1{,}000$ cfs that leaves the channel in reach n is

$$\frac{\ell_n}{L} \int_{q_0}^{\infty} P_{H|Q}^{W_n}(1,q)\, f_Q(q)\, dq \tag{13}$$

where L is the length of the entrenched channel, ℓ_n is the length of the nth reach and W_n is the width of the area subject to flooding from failures occurring in the nth reach.

Therefore, the probability of a point on the fan, with an entrenched channel as described above, being hit by a flood of magnitude greater than q_0 cfs is

$$P(H=1) = \int_{q_0}^{1000} P_{H|Q}^{W_{N+1}}(1,q)\, f_Q(q)\, dq + \sum_{n=1}^{N} \frac{\ell_n}{L} \int_{1000}^{\infty} P_{H|Q}^{W_n}(1,q)\, f_Q(q)\, dq \tag{14}$$

where W_{N+1} is the width of the area subject to flooding from less than 1,000 cfs and W_n is the width of the area subject to flooding from a failure in reach n.

If there are N reaches we must delineate N+1 areas that are conditionally subject to flooding--one for each reach and one below the entrenched channel. Figure 5 shows the boundaries of eight such areas. The length of the entrenched channel is 4,200 feet; the length of each reach is shown on Figure 5. Using equation (14) in our analysis results in identifying the flood hazards shown on Figure 6.

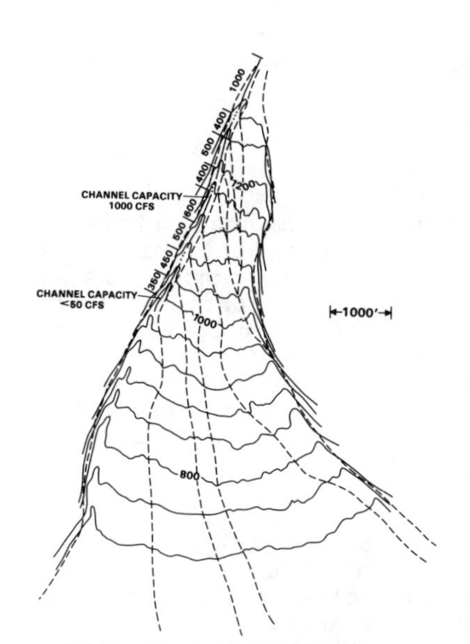

Figure 5. Areas Subject to Flooding From Failure at One of Eight Reaches

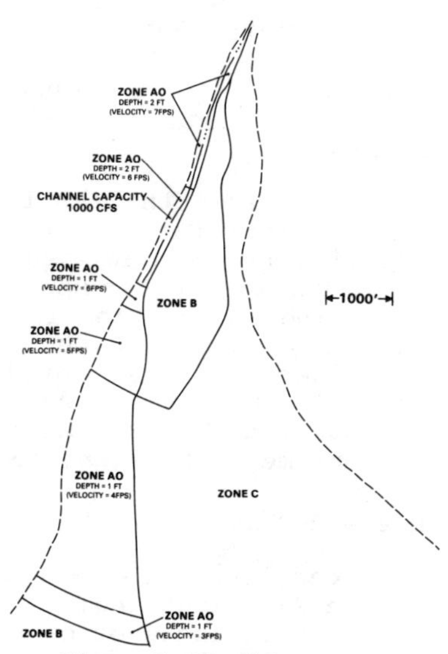

Figure 6. Flood Insurance Zones (Failure Probability Proportional to Reach Length)

References

1. Dawdy, D.R., "Flood Frequency Estimates on Alluvial Fans," Journal of the Hydraulics Division ASCE, Proceedings, Vol. 105, No. HYII, pp. 1407-1413, November 1979.

2. Federal Emergency Management Agency, "Flood Insurance Study Guidelines and Specifications for Study Contracts," September 1985.

Water for the San Diego Region: Past, Present and Future

Byron M. Buck[1]

Abstract

Ever since the earliest days of permanent settlement in the arid San Diego region, securing reliable water supplies has been an issue of critical importance. The San Diego region's water needs are expected to continue to grow requiring an additional 200,000 AF of water by 2010. The San Diego County Water Authority is in the midst of a comprehensive water resources planning effort which will determine the directions the agency will go to meet future demands and develop policies and implementation programs to implement the agency's plans.

Background and History

Ever since the earliest days of permanent settlement in the San Diego region, securing reliable water supplies has been an issue of critical importance. This issue continues today and promises to become more acute as population continues to grow and development of additional water supplies proves more difficult.

San Diego is an arid region which receives only about 10 inches of rain annually. This precipitation is highly variable from year to year and groundwater is scarce, complicating water development efforts. The Spanish missionaries were the first to develop water supply systems after nearly being forced from the region when the San Diego River ran dry during a drought. With the settlement of the region by anglos, additional local supplies were developed and substantial dams were emplaced on the region's short and steep drainages. Following a period of municipalization of many local water companies, city fathers began thinking about the need for supplemental

[1] Director of Water Resources Planning, San Diego County Water Authority, 3211 Fifth Avenue, San Diego, CA 92103

water to supply the region's future needs. Ridiculed at the time, a 112,000 AF filing was made by the City of for water from the Colorado River, although it was not at all clear how such water would be delivered.

With the advent of WWII, and a rapid influx in population, it was clear that the meager local water supplies were becoming exhausted. The San Diego County Water Authority was formed by the State in 1944 as a regional agency with the purpose of bringing in supplemental supplies to its retail member agencies (figure one). After evaluating the options it was decided that the U.S. Bureau of Reclamation would build a pipeline from the Colorado River Aqueduct and the Water Authority would join the Metropolitan Water District, owner of the Aqueduct.

Since the war, the Authority built additional pipelines connecting to the Metropolitan system and water from both the Colorado River and the State Water Project is being brought to San Diego. These waters now constitute over 90% of all the water consumed in the region.

The Water Resources Planning Environment

The San Diego region continues to grow and its current population of just over two million is expected to grow to 2.8 million by 2010. This will require a minimum addition of 200,000 AF water and the conveyance system to deliver it. Since the Authority's sole source of supply is the Metropolitan Water District of Southern California, future supplies for the region have thus far been dependent upon that agency's success in developing additional water supplies. The primary focus for these additional supplies has been the State Water Project authorized in 1960 but thus far only half complete and able to deliver only one-half its intended supply. Completion of this project was expected to ensure Southern California of adequate water supplies following the loss of reliable supplies from the Colorado River due to the Arizona vs. California decision. In 1990 Southern California may experience its first cut in supply due to this 1963 court decision. Expansion of the state project is bound up in an ongoing controversy over the future of the San Francisco Bay - San Joaquin Delta estuary, from which the State project receives its water. This controversy has forced all water users to rethink their demands for water and ways to reduce reliance on any one source.

The Planning Response

In recognition of the difficulty of acquisition of new sources of supply, the vulnerability the San Diego region has in relying so heavily on one source and the ultimate responsibility to supply water to the San Diego region, the

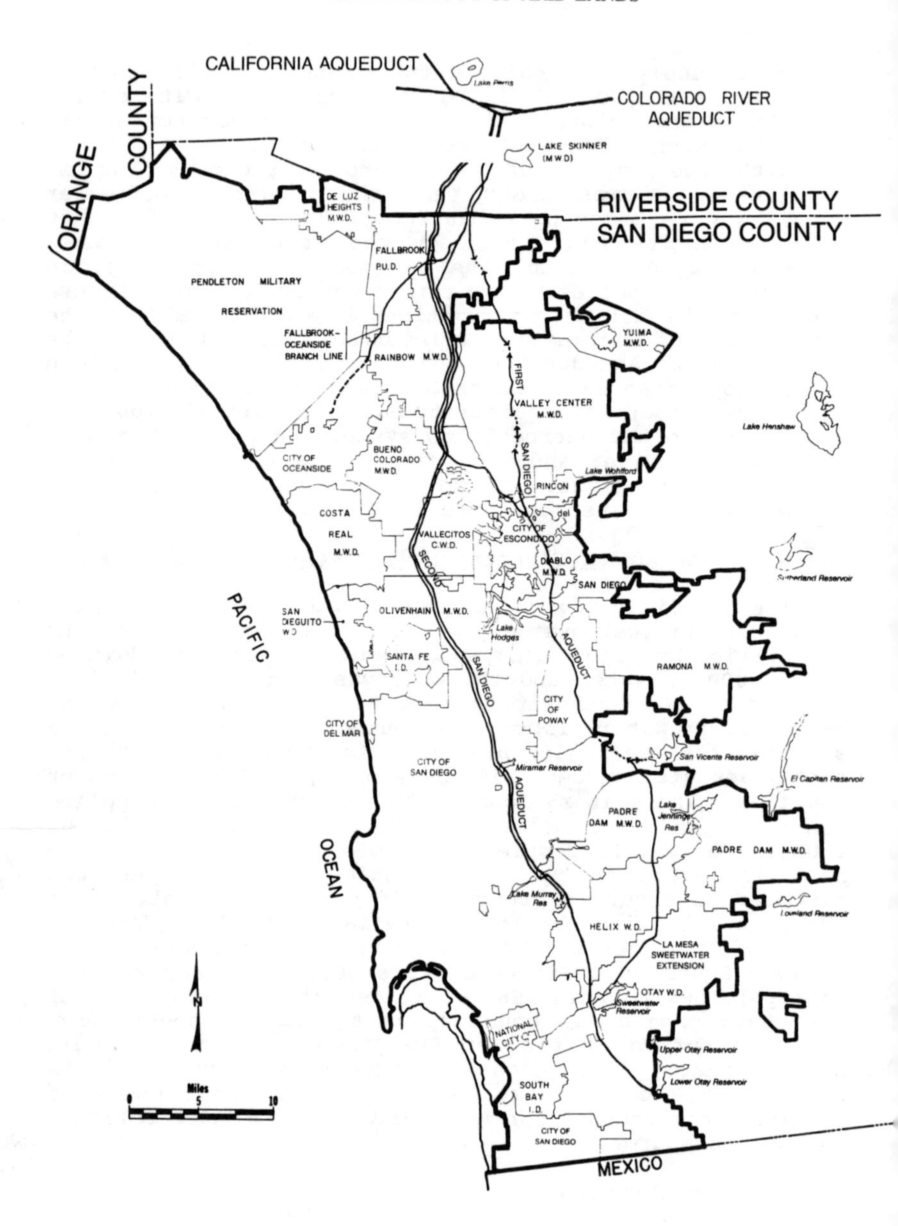

SAN DIEGO COUNTY WATER AUTHORITY

Figure 1

Authority created a Water Resources Planning department in late 1988 to plan for the region's long term water future. This department is in the midst of creating a water resources plan which will examine the water situation in the region and chart a course for the future. Specifically, this plan will analyze the resource situation, project water needs, analyze resource options recommend a resource strategy and develop implementation measures for the adopted strategy and policies of the Authority's Board of Directors. The plan will forecast water demands to the year 2030 and analyze the various alternatives for meeting these demands. The plan will identify long term policies for supply related to national statewide and regional water supply issues. Five year objectives for implementation will be developed for specific supply development activities. Given the rapidly changing arena of water demand and supply, it is expected that the plan will require updating in five years.

One of the critical features in any water resources planning effort is projection of water demands. Traditionally, most supply entities have used per capita demand forecasting methods due to the availability of population data and simplicity of the forecast method. This method has proved reasonably accurate in the past for most areas, including the Authority's service area. This method, however, is limited in that it cannot address the rapid economic and social changes which in turn affect water consumption. These flaws, coupled with demands by regulatory agencies requiring more accurate and sophisticated projections, have caused the Authority and the Metropolitan Water District to evaluate more sophisticated means of forecasting water demands. Metropolitan has developed a version of the IWR-MAIN forecasting model into a version known as MWD-MAIN. While Metropolitan has been refining this model for over three years, it has historically been inaccurate in the Authority's service area due to problems in assessing land use and water use patterns in the Authority's rural and semi-rural areas. The Authority is working with Metropolitan to refine the model's representation of these areas and it is expected that the MWD-MAIN model will then be used for the Authority's final forecasts.

Interim per capita forecasts indicate that the Authority will need between about 800,000 to 900,000 acre feet of water by the year 2010 (figure two). The range of demand is based upon assumptions made for per capita municipal and industrial use and growth or decline in the region's agricultural water consumption.

One of the serious problems the Authority and other urban water suppliers are encountering in California is a continuing increase in per capita consumption (figure three). In spite of conservation education efforts and

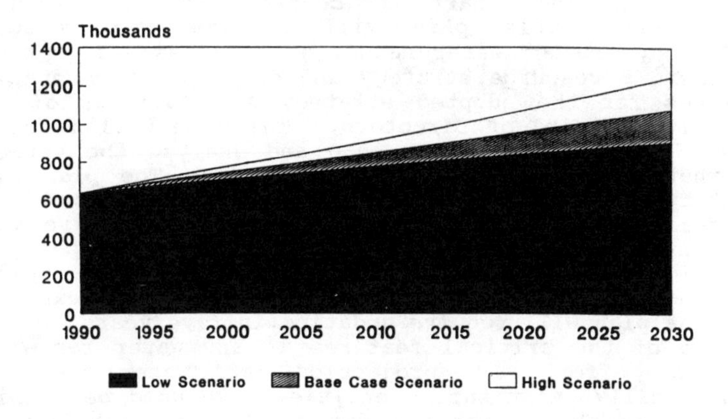

Figure 2

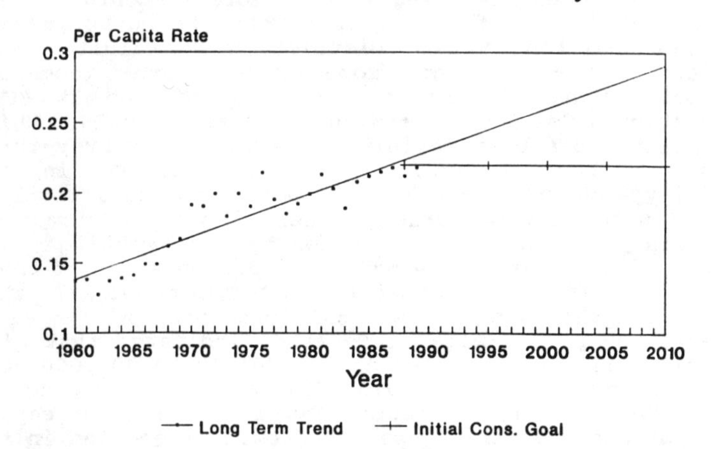

Figure 3

adoption of lower flow plumbing standards in 1981, per capita rates are not declining in the residential sector and are increasing overall. This is thought to be caused by more development in warmer inland regions, increased affluence, proliferation of water wasting devices such as home reverse osmosis systems and the expansion of use and poor management of automatic sprinkler devices. As part of the Authority's resource efforts, increased attention will be paid to improving long-term water use efficiency through a five-year $10 million program of plumbing retrofit programs, water use audits and technical assistance. It is recognized by the Authority that additional imported supplies will only be authorized when it can be demonstrated that existing supplies are being used efficiently.

In addition to developing a detailed conservation program the Authority will be evaluating a variety of alternative water supplies which may help in meeting the region's future needs. These alternative range from the expansion of existing sources via Metropolitan's activities to local supply development in the form of brackish water desalination, seawater desalination and water reclamation. Non-traditional supply sources are also being evaluated, such as innovative ways of shipping water as a commodity. The Authority is also examining the costs associated with an additional Colorado River connection which would allow water from agricultural areas in California to be marketed in San Diego. The Authority is already well underway with implementing its goal of 100,000 AF of reclaimed water being produced locally by the year 2010. All of these sources are being subjected to analysis under six criteria: Reliability, Cost, Environmental Impacts, Institutional Feasibility, Water Quality, and Independence.

While it is unlikely that the Water Authority will venture out to develop imported water supplies which involve the sources already developed and being pursued by the Metropolitan Water District, the comprehensive review of alternatives provided by the Water Resources Plan will enable the Authority to develop policies which support the efforts of Metropolitan. The plan will also enable the maximization of San Diego's local resources and provide alternatives for the region's long-term water future.

IMPLEMENTED SOLUTIONS TO WATER SUPPLY PROBLEMS THE TUCSON EXPERIENCE

R. Bruce Johnson[1]

ABSTRACT

During the past 30 years Tucson Arizona has been one of the fastest growing urban centers in the United States. During this period, the population served by the municipal water utility, Tucson Water, has increased from 180,000 people in 1960 to about 600,000 in 1989. For our community, the provision of an adequate water supply has been both complex and costly.

Throughout these years of accelerated growth, Tucson Water has responded to major problems associated with water supply development, peak demand management, water resource management and providing long-term assured water supplies for our community.

Beginning over 20 years ago, Tucson Water initiated an agricultural retirement program to secure and preserve water supply development options. In the mid-70's a precedent setting peak demand management program was implemented which has received national attention. More recently Tucson Water has performed a wastewater reuse assessment and built a reclaimed water treatment and delivery system which leads the state in this type of resource utilization. Parallel to these efforts is a major feasibility assessment program to evaluate the role which groundwater recharge will play in our future. To provide an overview of these water resource programs, Tucson Water has recently completed a 110 year Water Resources Master Plan which evaluates and prioritizes future water resources development options for our community.

----

[1] R. Bruce Johnson, Chief Hydrologist, Tucson Water, P.O. Box 27210, Tucson, AZ 85726-7219

INTRODUCTION

During the past 30 years Tucson, Arizona has faced a number of major problems associated with the development of an adequate long-term water supply. To a large extent these problems have been related to our tremendous growth in population and the natural occurrence of water in the Tucson area.

The temperate climate and lifestyle found in Tucson have resulted in one of the highest sustained growth rates in the United States. Providing for the coordinated expansion of our water supply system has been a major challenge.

Groundwater is currently the sole source of water supply for the Tucson metropolitan area. This source of supply also serves local agriculture, industry and mining operations. Historically, pumpage from local aquifers has been estimated to exceed natural replenishment by a factor of about five to one. As a result of the large imbalance between supply and demand, significant long-term declines in local groundwater levels have been observed. It is this imbalance which has been the principal cause of many of our local water supply problems.

Obviously, a community of our size has many problems and issues associated with its water supplies. A brief discussion of some of the more fundamental problems faced by Tucson Water together with the implemented solutions are highlighted on the following pages.

AGRICULTURAL LAND RETIREMENT PROGRAM

During the period 1945-1958 it became apparent that Tucson was growing quite rapidly. To meet expected needs for water, several studies were performed to evaluate alternative plans and to develop long-term water supply strategies. Several alternatives were implemented to meet short-term needs, however, one alternative had major long-term benefits to the water future of Tucson.

An adjacent groundwater basin, Avra Valley, was evaluated for its water supply potential. Properly developed, Avra Valley groundwater could be imported to the Tucson basin well into the future thus providing the first major long-tem element in the future water supply for the community.

The City began implementing the plan by purchasing well sites and right-of-ways beginning in 1951. By the early 1960's water was being produced for local service in Avra Valley. In 1968 and 1969 the City went ahead with its importation plan to bring Avra Valley water to Tucson. Six water supply wells and a large 42 inch diameter transmission line were constructed at a cost of $3.1 million. The transmission line had a design capacity of 30 million gallons per day (mgd) while the six wells totalled 8 mgd in capacity.

The City's efforts to implement the new import system, however, resulted in a series of three lawsuits, commonly referred to as Jarvis I, II and III. While not preventing the importation of Avra Valley water, the court decisions set up the framework and mechanism by which future supplies would be obtained and established the amount of water that could be transferred.

Elements of the Jarvis decision dictated that the City purchase lands actively used for agriculture and retire these lands from active production in order to establish the right to transport groundwater from Avra Valley to the Tucson basin. Following this decision the City today owns 22,522 acres of retired agricultural lands which have a combined associated right to groundwater of 64,288 acre feet per year.

The development and implementation of the ag-land retirement concept secured a long term future water right for the City and, additionally, preserved available groundwater that would have been consumed for agricultural purposes.

Currently, the issues surrounding this concept are being debated state wide. Tucson's importation program, however, remains intact and secure as a long term water supply element.

PEAK DEMAND MANAGEMENT

Prior to 1974 the prevailing water service philosophy in Tucson was to meet the unmanaged peak demand requirements of our customers by increased capital expenditures for expansion of the water system. The peak demand period of the summer of 1974, however, caused the City to re-evaluate that philosophy. Recoiling from one of the hottest, driest periods on record, the City well system proved to be incapable of meeting peak flow requirements. Local disruptions in service and chronic low pressures occurred throughout the distribution system. The experience convinced staff that a new philosophy was needed.

After extensive evaluation of water demand characteristics, capital costs and water rates studies, it was determined that the capital improvement program could be reduced from $150 million to under $100 million if the peak demands on the system could be reduced by 25 percent. Alternative plans to accomplish this goal were detailed and carefully evaluated by the water utility.

In the summer of 1977 the "Beat-the-Peak" program was implemented. This voluntary program of outdoor watering required that watering be limited to alternate days and never between 4:00 p.m. and 8:00 p.m. Its success was immediate and dramatic.

Peak demands on the distribution system dropped by over 25 percent. A secondary benefit of reduced overall water use for the single family residential class was also observed. This program has proven to be one of the most cost efficient and cost beneficial programs we have ever implemented. It has paved the way for today's water conservation programs and served as a model for other water utilities across the country.

RECLAIMED WATER

During the past 10 years, more and more interest in water resources management has been generated at the State and local level. In recognition of the potential role which effluent reuse would play in future efforts to manage basin-wide groundwater overdraft, the City of Tucson implemented a wastewater reuse system.

A major wastewater reuse assessment was done in 1982. The primary purpose of the assessment was to identify and evaluate the potential use of reclaimed water for turf irrigation needs throughout the community. In so doing, Tucson Water could eliminate major peak users on the potable distribution system and lay the foundation for the full utilization of our effluent water resources. Based upon this work, the City implemented its reclaimed water system. The system consists of an 8.2 mgd pressure filtration plant, a 3 million gallon reservoir, a 12 mgd booster facility and a major network of distribution lines to supply turf irrigation uses throughout the community. A major element of the system is a reclaimed water recharge facility which provides seasonal sub-surface storage of reclaimed water for use during peak demand periods.

By the summer of 1990 deliveries of reclaimed water will be about 6,000 to 7,000 acre feet per year. The reuse assessment identified a long term requirement for 35,000 acre feet of reclaimed water per year. Ours is the largest such system in the state. The reclaimed water system is a direct response to the problems associated with growing peak demands for potable water and the need to develop full utilization of effluent in the future.

RECHARGE FEASIBILITY ASSESSMENT

To resolve the problems associated with the long term need to fully utilize effluent and utilize the short term excess of water available from the Central Arizona Project, Tucson is conducting the Recharge Feasibility Assessment Program.

This study is designed to identify and test the recharge methodologies most likely to be successful in Tucson's hydrologic setting.

A comprehensive review and assessment of existing hydrogeologic data, water quality information, legal and institutional framework and recharge methods has been completed. Pilot studies testing both surface infiltration and well injection methods are now on-going. Six pilot injection well studies are being conducted to evaluate the operational efficiencies and maintenance requirements associated with these facilities. Three surface infiltration pilot projects are planned to evaluate the efficiencies of this recharge method. In all cases, water quality interactions are being closely studied. The results of this study will allow Tucson to implement the most cost-effective program to recharge, store and recover water available in the near term for ultimate use in the future.

WATER RESOURCES MASTER PLAN

The programs which have been discussed in this paper are in direct response to specific problems faced by our water utility. Most have arisen out of an operational need or concern and have been put into place with nearly immediate results. But how do these programs serve us in the future? What additional needs or policies should we be considering?

To respond to these future concerns, Tucson Water recently completed a one year study effort to define a 110 year Water Resources Master Plan. This effort involved Federal and State agencies, Tucson Water staff, City management, community leaders, water interest groups and the public at large. This massive analysis and identification of future needs pulls together all those factors which will dictate and shape our water supply future.

The plan is set up for annual updates with major reevaluation every five years. It has become our "blueprint" for the future.

SUMMARY

Tucson Water has faced a wide array of problems in providing long term assured water supply for the community. Many of the problems Tucson has responded to are the same as other communities will have to face in the future. We have done so, perhaps, at an earlier time because of our rapid growth rate and singular source of supply. For this reason our experiences may be of value to those City's facing similar conditions.

ACKNOWLEDGEMENTS

The assistance rendered by co-workers of Tucson Water during the preparation of this paper is gratefully acknowledged.

The Use of Wholesale Water Rates to Encourage the Groundwater Conjunctive Use

Richard W. Atwater *

The Southern California groundwater basins provide roughly one-third of the annual supply of the 15 million people within the Metropolitan Water District's service area. The conjunctive operation of the basins began in the late 1940s with the initial deliveries of Colorado River water. During the 1950s and 1960s, over 3 million acre-feet ($3.7 \times 10^9 m^3$) of Colorado River water was recharged into the basins to correct overdraft conditions.

Today, the basins are increasingly a vital part of the long-term strategy to imported water for either seasonal storage or for long-term drought and emergency reserves. The imported supplies have become increasingly less reliable as the region's water demands have increased dramatically. For example, Metropolitan's water sales have doubled during the past decade. Concurrently, the dependable supplies for Metropolitan from the State Water Project and Colorado River have decreased. In addition, the City of Los Angeles' imported supplies from the Owens Valley and Mono Basin have also decreased significantly. Obviously, expanding the groundwater conjunctive use with imported supplies will be an increasingly major factor in ensuring reliable water supplies to the 15 million population in Southern California.

Metropolitan's Board of Directors first adopted a policy of storing imported Colorado River water within groundwater basins in 1931. Initial deliveries of Colorado River suppplies started in 1940. In the late 1940s, Metropolitan sold water to the Orange County Water District for spreading in the Santa Ana River. In the early 1950s, a comprehensive program to replenish the overdrafted groundwater basins in Los Angeles and Orange Counties was initiated. This program was developed cooperatively by Metropolitan with several of its member agencies and local groundwater replenishment districts. During the 1950s, approximately 25 percemt of all Metropolitan's water sales were for groundwater replenishment purposes.

The availability and price of imported Colorado River water were critical components to the development of groundwater management programs in Southern California. Economical imported water was an incentive to develop conjunctive management schemes that allowed for groundwater extractions in excess of the natural safe yield of the basins. For example, basin overdrafting became a viable operating strategy as replenishment with excess imported water was economically available. The Orange County, Main San Gabriel, Central and West Coast, Chino, Raymond and San Fernando Basin management plans, which together make up 80 percent of the local groundwater production in the Metropolitan service area, have all integrated to some degree basin overdrafting rather than direct service from Metropolitan. All these basins production in in excess of the natural safe yield as a result of replenishment service from Metropolitan.

*Director of Resources, M.W.D., Los Angeles, CA

During the drought of 1977, the success of the replenishment program provided critical flexibility in meeting Southern California demands. Metropolitan was able to reduce imported supplies by temporarily suspending the groundwater replenishment service, and by shifting some demand to groundwater basins actively involved in the replenishment program. During this partial service interruption, many basins were able to be overdrafted beyond normal levels, further easing the effects of the drought.

In 1981, Metropolitan's Board of Directors adopted a comprehensive new water rate structure, the interruptible program, in recognition of the tremendous benefit to the region of the ability to tap groundwater basin storage during droughts, and to thereby free supplies for areas without such capability.

The success of the interruptible program is evidenced by significant increases in purchases of interruptible service as a result of local capital investment in groundwater production facilities. For example, from 1987-1988 to the 1989-1990 fiscal year, the City of Los Angeles has increased its interruptible in-lieu deliveries from approximately 45,000 acre-feet to 88,000 acre-feet. Likewise, the Chino Basin Municipal Water District increased its interruptible deliveries during the same period from 11,000 acre-feet to 45,000 acre-feet. The Los Angeles increase was the result of new well capacity being installed in the San Fernando Basin to overdraft that basin during droughts and emergencies. In Chino Basin, the increase reflects agency development of redundant capability to meet demands both with imported supplies when water is available, and with groundwater during shortages of imported supplies.

Metropolitan's Board of Directors, in July 1989, adopted an innovative wholesale water pricing program (called Seasonal Storage Service) to encourage greater utilization of the groundwater basins and a more efficient operation of the import delivery system. The objectives of this wholesale water pricing program are:

* Maximize the capability for Metropolitan to import additional water when available, thereby increasing the supply from the SWP and Colorado River;

* Increase member agency production of local water during peak-demand periods to enhance the ability to regulate flows on Metropolitan's aqueduct and distribution facilities, and lessen the seasonal drawdown of Metropolitan's storage reserves; and

* Increase member agency production of local water during droughts, making additional imported supplies available for member agencies that do not participate directly in the program.

It is expected that it will encourage additional capital investment in groundwater production facilities and the installation of wellhead treatment plants. Further, it is intended to encourage a reduction in the drought year <u>noninterruptible</u> demands of member agencies but also would encourage increased groundwater production during the summer and thereby reduce member agency peaking on Metropolitan's system. Through the seasonal storage program, Metropolitan would provide an incentive to operate the groundwater basins similar to Metropolitan's terminal reservoir, Lake Mathews. Just as Metropolitan uses a portion of Lake Mathews for seasonal regulation in addition to holding carryover storage, Metropolitan would encourage seasonal storage as well as drought storage of the groundwater basins.

The seasonal storage program specifically encourages increased production of local water in the summer and the development of a capability to take compensating deliveries from Metropolitan when Metropolitan has a greater capability to meet the demand. It is a price-based incentive to utilize local storage reserves when imported water is unavailable.

The Local Projects Program (LPP), created in 1981, provides financial rebates to local agencies in developing new local water supplies to reduce demands on our imported supplies. Recovering or treating nonpotable groundwater has been financially supported by Metropolitan to encourage the expanded conjunctive use of the groundwater basins. For example, Metropolitan is participating through LPP in the Arlington Groundwater Basin Desalter Project. Approximately 6,100 acre-feet per year of high salinity groundwater from the Arlington Basin will be treated for local use and groundwater recharge. With the implementation of the Arlington Desalter, it is expected that the restoration of the Arlington Basin will be completed in 20 years.

In another LPP agreement, Metropolitan is participating with Foothill Municipal Water District and Crescenta Vally County Water District in the implementation of the $1.7 million Glenwood Nitrate Water Reclamation Project. Under this project, which was completed at the beginning of the year and is now operational, excess nitrates are removed from Verdugo Basin groundwater before the treated water is blended for potable use. Complementing this project, the major source of nitrates (septic tanks) was reduced with local construction of an integrated sewer system allowing the complete restoration of the Verdugo Basin from nitrate contamination.

In conclusion, the use of innovative wholesale water rates and rebates by the Metropolitan Water District are providing the key financial incentives to implement new local groundwater management projects in Souther California. The goal of these programs are to better integrate the operation of the import system with local water resources to maximize the availability of the supplies to everyone in the region.

COMPETITION FOR WATER IN SOUTHERN NEVADA - LOCAL NEEDS VERSUS FEDERAL CONTROL

by
Richard H. French, M. ASCE[1]
Christine Chairsell[2]
Gilbert F. Cochran[3]

Abstract

The impacts of lands in southern Nevada, withdrawn by the federal government for defense and defense-related purposes on the water supply, are discussed. Nevada is an illustration of an area in which the most controversial issues of the decade, water availability and nuclear and defense strategies, compete.

Introduction

The development of water supply infrastructure in areas growing at an exponential rate is always difficult; but it is particularly difficult in the most arid state in the United States and in a state where most of the land is controlled by the federal government. The Las Vegas Metro area and Laughlin, both in Clark County, Nevada, are among the fastest growing urban areas in the United States and are within the driest portion of the most arid state in the United States. The availability of water resources, at reasonable cost, could be a brake on the rate of development and may ultimately define the limits of development in Southern Nevada. The largest land owner in Nevada is the federal government with the use of the land being controlled primarily by the Departments of Interior, Defense (DOD), and Energy (DOE). In this paper, the implications of federal withdrawal of land for defense and defense-related purposes in Southern Nevada are discussed from the perspective of water resources management and planning. The authors wish to note that the control of the land in Southern Nevada by Federal agencies for reasons other than defense activities also have impacts on water resources, but space limitations preclude a full discussion of these impacts in this paper.

Extent of the Federal Lands in Nevada

The State of Nevada is comprised of approximately 110,540 mi^2. Of this total area, the federal government controls approximately 94,340 mi^2 or 85.4% of the total area. In Table 1 the agencies and the lands for which they are responsible are summarized.

---

[1]Research Professor, Desert Research Institute, Las Vegas, NV
[2]Lecturer, Dept. of Political Science, University of Nevada, Las Vegas, NV
[3]Research Professor, Desert Research Institute, Reno, Nevada

Note, because of the size of the numbers involved and the memorandums of understanding between the various agencies, the data in Table 1 are approximate. Of these agencies, the ones that have the greatest impact on water resources in Nevada are those that are responsible for land withdrawn from the public domain for a specific purpose such as DOD and DOE that have major withdrawals in Southern Nevada (Air Force, 4,800 mi^2 and DOE, 1,200 mi^2) and in close proximity to the rapidly growing Las Vegas Metro area. While access to these withdrawn lands is possible and applications for easements across these lands can be made, such activities cannot interfere with the primary purpose for which the withdrawals were made. Given the nature of the activities on defense-related land withdrawals, access and easements are currently to all intents and purposes not available.

TABLE 1: Summary of Federal Agencies Responsible for the Lands of Nevada.

Federal Agency	% of NV Land	Area Controlled (mi^2)
Bureau of Land Management	67.9	75,000
Forest Service	7.2	8,000
Department of Defense	4.7	5,200
Department of Fish & Wildlife	3.1	3,400
Department of Energy	1.1	1,200
Bureau of Reclamation	0.9	990
Park Service	0.5	550
Total	85.4	94,340

Associated with federal land control is the question of federal water rights in Nevada. Although they are under not obligation to do so, the Air Force has followed Nevada water law. In contrast, the DOE on the Nevada Test Site (NTS) has asserted the Doctrine of Federal Reserve Water Rights. Given its importance, the legal foundation of this doctrine warrants some discussion.

The Reserve Doctrine

The birth of the Reserve Doctrine occurred in 1908 with the U.S. Supreme Court asserting the rights of the Indians to water in *Winters* v. *U.S.* (207 U.S. 564, 1908). The Supreme Court held that the waters flowing through or bordering on the Fort Berthold Indian Reservation, Montana, were reserved by the U.S. Government and the Indians by the treaty that established the reservation. In setting aside the land, the U.S. Government, as trustee, also set aside water for that land. The priority date for Indian rights was held to be the date that the reservation was actually established. The Winters Doctrine established itself as a precedent for the protection of Indian rights to water. It was originally perceived to be a special doctrine applicable to Indians only, but the Supreme Court continued to expand this doctrine to include all federal reserves.

In the *Federal Power Commission* v. *Oregon* (349 U.S. 435, 1955), also known as the Pelton Dam Decision, the Supreme Court declared that federal reserved water rights pertained to all types of reservations when it held that Oregon could not prevent the Pelton Project 2030 from being constructed, partially upon an Indian Reservation and partially upon land reserved by the federal government in 1909.

Six weeks after the Pelton Dam Decision, in July 1955, the commander of the Naval Munitions Depot at Hawthorne, Nevada informed the state engineer that the Navy, in accordance with the Pelton Dam Decision, would cease to apply for water rights under Nevada State Law. In *Nevada ex. rel. Shamburger* v. *United States*, 1960, the federal circuit court refused to reverse the Navy's decision. Thus, the Navy was free to withdraw groundwater without advising the State of Nevada of the quantity withdrawn. This facility was subsequently transferred to the control of the U.S. Army and the status of surface water rights is still clouded.

In *Arizona* v. *California*, (373 U.S. 546, 1963), the Supreme Court relied heavily upon *Winters* and expanded the Reserve Doctrine. The Court specifically addressed Indian rights by declaring Indian rights constituted a right to unspecified amounts of water and quantified these rights in accordance with the measure of the total practicable irrigable acres upon the reservations (Nelson, 1977). Also included in the Reserve Doctrine were various other reserves such as national recreation areas and national forests, with specific references to Lake Mead National Wildlife Refuge, Havasu Lake National Wildlife Refuge, Imperial National Wildlife Refuge, and Gila National Forest.

In 1971, the Supreme Court identified lands on which the Reserve Doctrine may be applied in *U.S.* v. *District Court for Eagle County* (401 U.S. 520, 1971). The Court defined any federal enclave as federally reserved lands on which federally reserved water may be used to fulfill federal purposes. Federal purpose would eventually be limited to the purpose for which the federal reserve was created (U.S. v Mexico, 436 U.S. 645, 1978). The Court ruled that the McCarran Act subjected these federal reserved rights to state court adjudication, but it granted a federal right to an indeterminate amount of water should the federal government decide to extract oil from shale Ranquist, 1975).

The Reserve Doctrine continues to pose some very difficult problems for western water rights. States enjoy the determination of future allocations, but federal use of western lands and water reserved by the federal government as in the case of the DOD and DOE, have an impact upon state planning.

Impacts/Conclusions

The impacts of federal withdrawals in Southern Nevada are fourfold and primarily associated with those lands withdrawn for defense-related purposes. First, there is the assertion of the Federal Reserve Right by DOE that guarantees DOE unquantified water rights sufficient to meet the purposes of the NTS withdrawal. The adverse impact is that the amount of water reserved is not quantified but only referred to as the "amount necessary to fulfill federal objectives." When the water is put to use, the U.S. right will have priority over any other water rights established after the date of the reservation. Federal use of water is not subject to state laws which regulate appropriations or water use (Trealease, 1971). In addition, while the Reserve Doctrine is applicable to unappropriated water, most western water is already fully appropriated and in many instances over-appropriated. Most unappropriated waters are in the form of spring run off and floods. Finally, Nevada state water management and planning is hampered by the uncertainty of DOE's need for water in the future as the mission of the NTS changes and evolves.

Second, an unavoidable result of the underground nuclear weapons testing program at NTS has been the introduction of radioactive materials into the groundwater system beneath portions of the site. Sufficient data are not available to precisely estimate the amount of water that has been contaminated by radionuclides; but the volume could be substantial depending on the assumptions made regarding the distribution of radionuclides surrounding the actual test

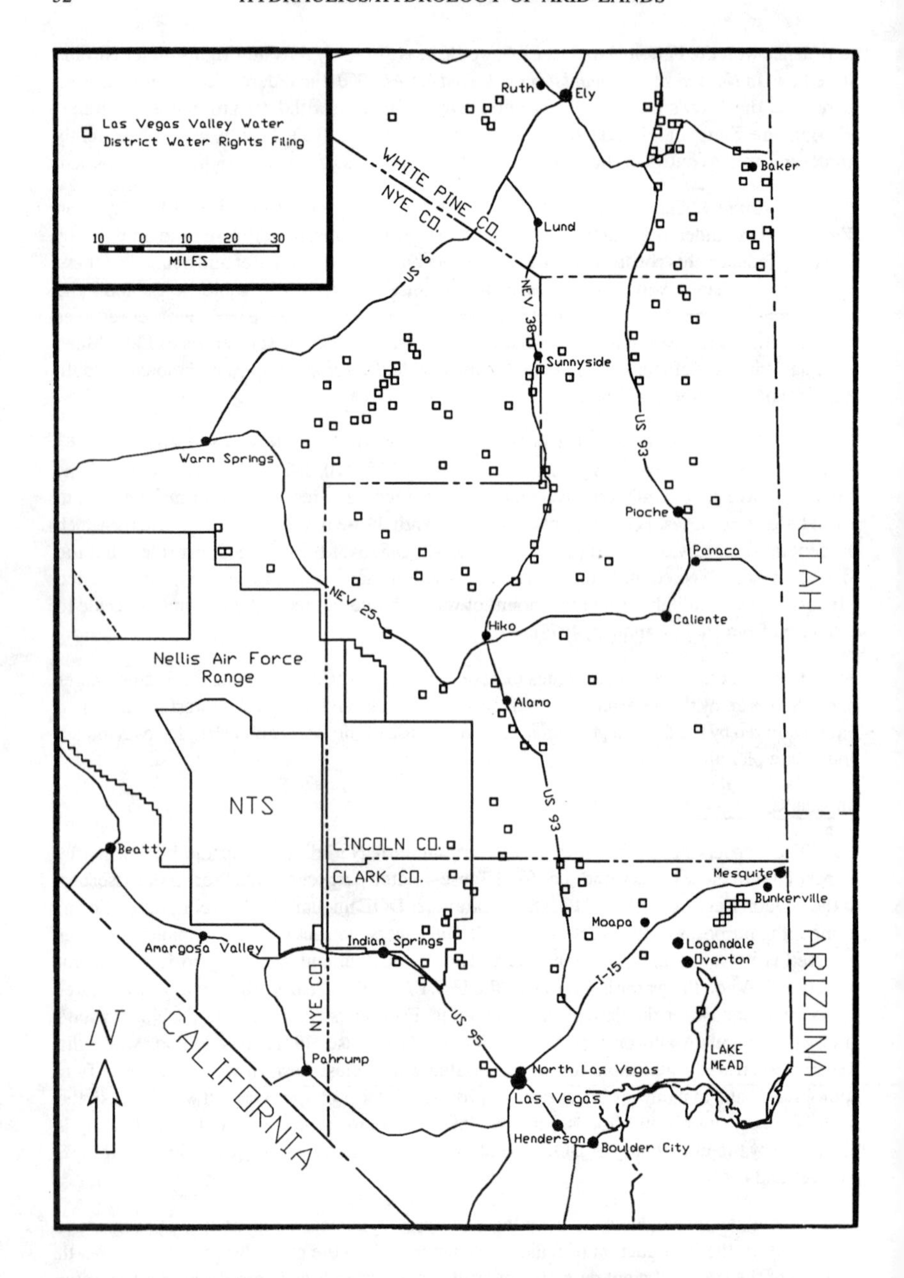

Figure 1. Relationship of Las Vegas Valley Water District Water Rights Filings to Withdrawn Lands in Southern Nevada.

points. While the contaminated groundwater poses no current public health or safety problems either on or off the NTS, a portion of a limited resource will not be available for development in the foreseeable future. Water resources development near the boundaries of the NTS could result in the movement of on-site contamination and the contamination of additional water.

Third, the NTS and Nellis Air Force Range Complex (NAFRC) withdrawals result in 13 hydrographic basins that are over 50 percent withdrawn. Of these 13 basins, 9 are over 80% withdrawn and 6 are over 90% withdrawn. According to reconnaissance level hydrologic studies performed by the U.S. Geologic Survey for the Nevada Department of Conservation and Natural Resources, these 13 basins collectively represent over 50,000 ac–ft per year of perennial groundwater yield and over 12,000,000 ac–ft of groundwater in storage in the upper 100 feet of saturated sediments. The quantification and development of the water underlying these areas requires access which is currently restricted. It is appropriate to note that the Air Force has in the past provided access to portions of NAFRC for studies of the regional carbonate aquifer. In addition, the hydrologic research and investigations support by DOE at the NTS have had a very positive impact on the development of a better understanding of the regional hydrology.

Fourth, the withdrawn lands lie between the Las Vegas Metropolitan area and areas where water is available to be developed, Figure 1. Figure 1 schematically shows where the Las Vegas Valley Water District has filed for available water rights and also the extent of defense–related lands in Southern Nevada. This figure demonstrates that water rights filings have generally avoided withdrawn lands. All water rights filings lie to the east and north of the withdrawn lands. This figure also suggests that the infrastructure required to convey water to the metro area will avoid the withdrawn areas. From a parochial civilian viewpoint, the lands of Southern Nevada that have been withdrawn for defense–related purposes may represent in the future a real and continuing cost as the infrastructure required to transport water to the metro area is built and put into operation. Further, the withdrawn lands are not part of the property tax base even though they provide a unique resources from the viewpoint of national security.

Nevada is an illustration of an area in which the most controversial issues of the decade; water availability and nuclear and defense strategies compete. The lands withdrawn for defense purposes with federal reserve rights implied in one instance and the escalating competition for water in Southern Nevada may put the state–federal relationship to the ultimate test.

References

Nelson, M.C., 1977. The Winters Doctrine: Seventy Years of Application of Reserved RIghts on Indian Reservations. University of Arizona, Office of Arid Lands Studies, Tucson, Az., p. 26.

Ranquist, H.A., 1975. The Winters Doctrine and How it Grew: Federal Reservation of Rights to Use Water. Brigham Young University Law Review, p.651.

Trealease, F., 1971. Federal–State Relations in Water Law. PB203–600, Legal Study No. 5. National Water Commission, Virginia, pp. 39–40.

DEBRIS FLOW POTENTIAL AND SEDIMENT YIELD ANALYSIS FOLLOWING WILD FIRE EVENTS IN MOUNTAINOUS TERRAIN

Craig V Nelson[1] and
Robert C. Rasely[2]

Abstract

The Pacific Southwest Inter-Agency Committee (PSIAC) 1968 Sediment Yield Rating Model provides a method to rapidly assess any increase in potential sediment yields following a wild fire event. Increased rates of erosion after a fire in mountainous areas significantly elevate the potential for debris flow/flood events. The PSIAC method provides decision makers with information needed to determine if erosion control mitigation measures are warranted. PSIAC input requires assigning values to nine parameters affecting the sediment yield: geology, soils, climate, runoff, topography, ground cover, land use, upland erosion, and channel erosion and transport. These values are then converted into the potential sediment yield in acre-feet per square mile of drainage area per year ($m^3/km^2/yr$).

Introduction

The September 1988 Affleck Park Fire consumed over 5600 acres (22.7 km^2) of mountainous watershed, mainly in urbanized Emigration Canyon about 18 miles (29 km) East of Salt Lake City. The more than 400 fire fighters brought in to fight the fire succeeded in preventing any canyon homes from being destroyed. After the fire was extinguished questions arose about the risk from debris flow/floods from the denuded hillsides. While U.S. Forest Service seeding efforts and natural regrowth were expected to revegetate and stabilize slopes in three to five years, thunderstorms and snowmelt runoff posed immediate concerns to residents below the burned slopes.

---

[1]County Geologist, Salt Lake County Public Works, 2001 South State Street #N3700, Salt Lake City, UT 84190-4200
[2]Geologist, USDA Soil Conservation Service, P.O. Box 11350, Salt Lake City, UT 84147

The situation facing local government officials as a direct result of the fire is summarized below:

* A steep watershed burned in an urbanized canyon with over 300 homes built in the drainages and along creeks.
* About 70% of the burn rated as moderate to high in intensity.
* Approximately 15% of the soil surface altered to a water repellant condition.
* September and October average about 5 inches (12.7 cm) of precipitation in canyon areas, with 1 inch (2.5 cm) thunderstorms not uncommon.

A method was needed to quickly predict the amount of sediment that could come off the burned slopes given a worst-case storm event, possibly in a debris flow or flood.

Sediment Yield Models

Several sediment yield prediction techniques have been developed for use in western U.S. rangelands. Renard (1980) applied four methods to the Walnut Gulch watershed in southeastern Arizona to get an estimated annual sediment yield. He tested the Dendy/Bolton method (Dendy and Bolton, 1976), the Flaxman regression equation (Flaxman, 1972), the Renard method (Renard, 1972), and the PSIAC method (PSIAC, 1968). By measuring the actual amount of sediment accumulating in 9 sub-watershed catch ponds, Renard was able to evaluate the accuracy of each model. Of the four models tested, Renard concluded that the PSIAC method provided the best results. This was the method chosen to evaluate the sediment yields following the Affleck Park Fire.

The PSIAC Methodology

The PSIAC procedure uses an interdisciplinary rating model to evaluate nine parameters that affect sediment yield. Ideally each PSIAC assessment is performed as a group effort, drawing participants from geology, soils, forestry, hydrology, geotechnical engineering, and other specialties that each case may require. Each factor is given a numerical rating based on a range of values within the PSIAC model. Vigorous discussion among the experts usually results, and tends to lead to more accurate PSIAC results.

The PSIAC factor ratings are performed on a work sheet carried to the field where the actual factor ratings are discussed. By using PSIAC, sediment yields can be estimated for both pre- and post-fire conditions to provide a comparison of increased erosion potential.

The following example of the PSIAC procedure used for Secret Canyon, a 450 acre (1.8 km^2) sub-watershed of Emigration Canyon, will help illustrate the actual application of the PSIAC method.

Geology

Geology is rated from 0 to 10. A marine shale would warrant a rating of 10; rocks of medium hardness would rate a 5; and massive, hard formations would be rated as 0. Secret Canyon is within the Cretaceous Kelvin formation, which in this area of the canyon is a massive conglomerate. The pre-fire geology factor was rated as 0, and since fire has no effect on the geology, the post fire rating was also rated as 0.

Soils

A fine textured, easily eroded soil would receive a soils factor rating of 10; medium textured soils with some rock fragments a 5; and soils high in rock fragments, aggregated clays, or organics would receive a 0. As mentioned earlier, discussion among the participants may compel interpolation between the PSIAC guidelines. Both pre- and post-fire soil factors were rated as 6 in Secret Canyon.

Climate

Areas subject to intense, long-duration storms would receive a climate factor rating of 10; storms of moderate length or infrequent thunder storms would rate a 5; and arid climates or areas where the precipitation is mainly snow would be rated 0. In Secret Canyon, both pre- and post-fire climate factors were placed at 7 based on the 30-40 inches (76-102 cm) of precipitation received annually, and fairly common intense 1-inch (2.5 cm) storms.

Runoff

Areas subject to high-peak flows would receive a runoff factor rating of 10; moderate flows a 5; and low flows a 0. Evaluating runoff factors is an area where the PSIAC team can benefit from the participation of a hydrologist. The Secret Canyon runoff pre-fire factor was rated at 3. The post-fire rating was increased to 8 because of the expected increased flows due to water repellant soils and the decrease in vegetation.

Topography

PSIAC rates topography on a scale between 0 and 20. Steep slopes or drainages with no floodplain development rate a 20; moderate slopes (less than 20% grade) rate a 10; and areas of gentle slope would rate a 0. Because the slopes of the hillsides in Secret Canyon are generally above 60%, both pre- and post-fire topography ratings were placed at 20.

Ground Cover

The PSIAC values for ground cover vary from -10 to 10. Areas where ground cover does not exceed 20% rate a 10; litter and understory not exceeding about 40% rate a 0; and if slopes are completely protected a rating of -10 would be given. The pre-fire ground cover in Secret Canyon was estimated at about 65%, or a -4 rating. The post-fire rating of 8 was based on about 25% remaining ground cover with a substantially intact grass root mat.

Land Use

Watersheds that have been intensively cultivated, grazed, or burned rate a land use value of 10; less intensive land use would warrant a rating of 0; and areas not cultivated or only slightly grazed would be rated at -10. The pre-fire rating for Secret Canyon was placed at -5 because although there were a few hiking trails and dirt roads, it was largely ungrazed. The post-fire rating of 10 was given due to the recent fire.

Upland Erosion

Upland Erosion is rated from 0 to 25. If more than 50% of the area has rill and gully topography or landslides a rating of 25 would be given; less evidence of active erosion would rate a 10; and no evidence of erosion in the area would be rated 0. Secret Canyon's pre-fire rating of 6 was based on a 15% estimated rill formation. The post-fire value was rated at the maximum 25 because of the potential for concentrated sheet and rill flow during a storm event.

Channel Erosion

The final PSIAC factor is channel erosion, rated on a scale of 0 to 25. A drainage with active headcutting and eroding banks would rate 25; occasional bank or bed erosion would rate 10; and flat gradient channels, channels in bedrock, or drainages with erosion control devices would be rated at 0. Based on the moderate past amount of erosion observed in Secret Canyon the pre-fire channel erosion factor was rated at 10. The post-fire rating was increased to 15 due to the potential for bank erosion and scouring in the main ephemeral channel during a storm.

Calculating Sediment Yield

The values from each factor are summed to give a PSIAC factor total. Secret Canyon totaled 43 in the pre-fire assessment and 99 for the post-fire conditions. These values are unit-less and are converted into sediment volume using relationships developed by Renard's experimental data (Figure 1).

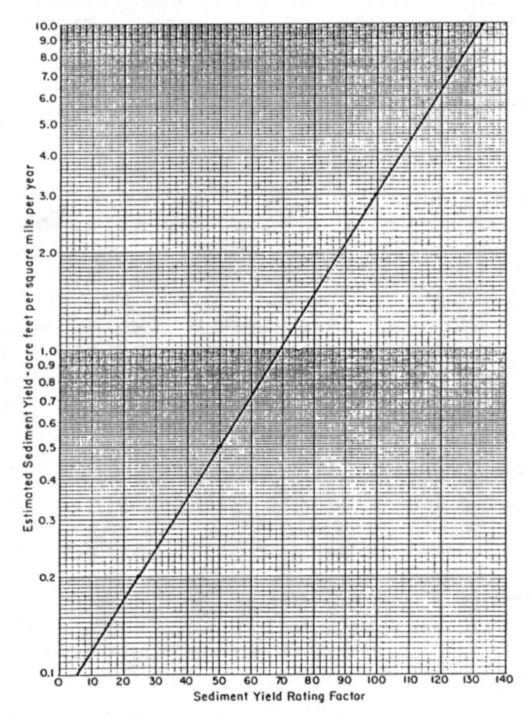

Figure 1. Conversion Chart

The pre-fire PSIAC total of 43 corresponds to an estimated moderate sediment yield of .37 acre-feet of sediment per square mile of drainage per year (456 m^3). The post-fire PSIAC total of 99 converts to a high to very high erosion rate of 2.9 acre-feet of sediment per square mile of drainage per year (3,576 m^3). The fire increased the potential sediment yield in Secret Canyon by 700 percent.

The total potential sediment yield volume of an area can be calculated by multiplying the size of the drainage area by the PSIAC sediment yield rate. Although the product is an annual sediment volume, it may be possible to move all this sediment during one intense storm-runoff event. Thus, the resulting sediment volume may be used in considering mitigation measure design, such as the size of debris catch basins.

Conclusion

The PSIAC method provides a valuable interdisciplinary, rapid response tool for assessing the increase in erosion related hazards following a wild fire. It produces both pre- and post-fire sediment yield comparisons, and provides decision makers with timely quantitative values for mitigation designs.

Appendix A - References Cited

Dendy, F.E., and Bolton, G.C., 1976, Sediment yield-runoff drainage area relationships in the United States: Journal of Soil and Water Conservation, Vol. 31, No. 6, p. 264-266.

Flaxman, E.M., 1972, Predicting sediment yield in Western United States: Journal of the Hydraulics Division, American Society of Civil Engineers, Vol. 98, No. 12, p. 2073-2085.

Pacific Southwest Inter-Agency Committee, 1968, Factors affecting sediment yield and measures for the reduction of erosion and sediment yield: Unpublished report of the Water Management Subcommittee, 13 p.

Renard, K.G., 1972, Sediment problems in the arid and semiarid Southwest: Soil Conservation Society of America 27th Annual Meeting Proceedings, p. 225-232.

Renard, K.G., 1980, Estimating erosion and sediment yield from rangeland: American Society of Civil Engineers Proceedings of the Symposium on Watershed Management, Boise, ID, p. 164-175.

POTENTIAL SEDIMENT YIELD FROM A BURNED DRAINAGE - AN EXAMPLE FROM THE WASATCH FRONT, UTAH

Robert M. Robison[1 2]

Abstract

Two debris flows hit Mapleton City, Utah, as a result of rainstorms on a steep drainage basin, which had recently been burned. Debris flow potential for the area was evaluated to determine if hazardous conditions existed. Also, the increase in sediment loss due to fire denudation of slopes could be observed in relation to similar contiguous unburned drainages. Numerical calculations for the burned drainage were based on accepted methods developed by the Pacific Southwest Inter-Agency committee (PSIAC, 1968). Yield estimates indicated about 4,700 yd^3 (3,600 m^3) of potentially available debris flow sediment. Reports addressing the potential for debris flows and possible mitigation measures were given to several government agencies. No action was taken before two thunderstorms hit several days later, which initiated debris flows. No specific path for the debris flows could be predicted, and the entire surface of the alluvial fan was considered at risk. The debris flows hit a residence, damaged fences, partially filled a canal, and covered several roads. Preliminary measurements by the U.S. Forest Service indicated that about 15,000 yd^3 (11,500 m^3) of sediments were deposited on the alluvial fan at the mouth of the drainage. However, an evaluation of "retainable-sediments", that which would remain in a debris basin, was somewhat less, about 5,000 to 7,000 yd^3 (3,800 to 5,400 m^3), which is close to the volume predicted by the PSIAC calculations. Special considerations for (PSIAC) debris flow analyses should include: a) drainages with excessively steep slopes; b) thin soils; c) the possibility of long duration/extremely intense storms; d) flow paths for "non-retainable" sediments in mud floods; and e) the necessity for quick action by "mitigating" agencies.

----

[1]Utah County Geologist, 100 E. Center, Rm 3800, Provo, UT, 84601
[2]Current Address: Sergent Hauskins, & Beckwith, 4030 South 500 West, Suite 90, Salt Lake City, UT 84123

Introduction

Debris flow potential is greatly increased when vegetation and other types of ground cover (litter) is removed (by fire) and soil is left exposed which is susceptible to erosion and mass movement. This condition existed in Middle Slide Canyon, east of Mapleton City, Utah, following the fire on September 2-4, 1989, and debris flows from this drainage were possible. The purpose of the investigation was two-fold: a) to determine the debris flow potential; and b) to observe a similar contiguous drainage for sediment loss.

Site Conditions

The majority of the 630 acres burned were located in the Middle Slide Canyon drainage (526 acres burned of a total 619 in the drainage) and will be referred to as the Middle Slide fire in this report (also in U.S. Forest Service report FSH 2509.13, FS-2500-A). The burned area in Middle Slide Canyon was practically devoid of litter after the fire. However, the areas in contiguous drainages, which were burned only along the upper ridges, were not included in sediment loss evaluations due to remaining ground cover which would buffer the effects of debris flows and should prevent damaging debris flows from these adjacent canyons. In addition, two other processes may contribute to debris flows following a fire (Wells, 1987). Dry ravel, the downslope movement of material without moisture, and the formation of water-repellent soil a few millimeters below the ground surface which induces rills and rapidly concentrates runoff. Both of these phenomenons add material to stream channels, and contribute to the sediment volume.

Middle Slide Canyon is formed in the Oquirrh Formation (Davis, 1983) which consists of limestone, sandstone, and orthoquartzite. This formation is relatively stable and no landslides involving bedrock were noted. Slopes in the drainage are generally very steep, ranging from about 30 percent to over 100 percent. In addition, thin soils cover the majority of the drainage. Deposits of alluvium/colluvium (glacial?) occur in the bottom of the canyon. These deposits consist of poorly cemented alluvium and colluvium which has previously been channeled by debris flows and flooding.

Middle Slide Canyon has had debris flows in the recent past. Old debris flow scars are present in the drainage and trench (fault) studies conducted by the U.S. Geological Survey and the Utah Geological and Mineral Survey at the top of the fan at the mouth of the canyon revealed evidence for historic debris flows, probably during the 1930's. The area was not burned at that time and the debris flows may have been the result of denudation from over-grazing in the higher elevations.

Procedures

Numerical calculations, using the PSIAC (1968) method, have been used with success along the Wasatch Front in similar environments (Nelson and Rasely, 1989, Robison, 1988), and were used to evaluate potential sediment yield. Potential debris flow volumes were calculated for Middle Slide Canyon using the PSIAC (1968) method. Nine parameters are used, and four evaluations were made for differing intensities of burned conditions. A total of 526 acres burned in Middle Slide Canyon of which 70 percent burned hot, 15 percent moderate, and 15 percent light (Paul Scabelund, USFS, personal communication, 1989). Potential debris flow contributions from the various areas are:

hot 70%	=	368 acres	=	3,990 yd^3
	(=	149 ha	=	3,050 m^3)
moderate 15%	=	79 acres	=	390 yd^3
	(=	32 ha	=	300 m^3)
light 15%	=	79 acres	=	185 yd^3
	(=	32 ha	=	140 m^3)
burn 100%	=	526 acres	=	4,565 yd^3
	(=	213 ha	=	3,500 m^3)
* natural		93 acres	=	105 yd^3
		(38 ha	=	90 m^3)
TOTAL (present condition)		619 acres	=	4,670 yd^3
		(250 ha	=	3,570 m^3)
		for Planning	=	4,700 CUBIC YARDS
			(=	3,600 CUBIC METERS)

*In addition to the material contributed from the burned areas, approximately 105 cubic yards could be added to the total yield from the unburned 93 acres within the drainage. This would make the total potential debris flow volume 4,670 yd^3 (3,570 m^3), which for planning purposes, rounds up to 4,700 yd^3 (3,600 m^3) (see Robison, 1989, for calculation work sheets).

The potential volume of sediment-per-unit-area gives a PSIAC classification of 1 (very-high). This is based on a potential average annual yield scale of 1 to 5, where 1 is most-severe, and 5 least-severe (PSIAC, 1968).

Debris Flows

The potential for damage from debris flows is based not only on the amount of material in the flow, but on its extent, direction, and a determination of what is at risk. A review of 1984 aerial photographs suggests that the most recent flows (either clear-water or debris) from Middle Slide Canyon occurred on the top center of its alluvial fan and flowed west. This is no guarantee, however, that future flows will do the same. Debris flows may block their own channel which can cause a change in direction of flow. In addition, a fault graben at the top of the fan could channelize a flow and direct it north or south, perpendicular to the canyon. The entire alluvial fan at the mouth of Middle Slide Canyon was considered equally at risk from debris flows.

The area below the mouth of Middle Slide Canyon has not yet been subjected to extensive residential development, in addition, no water collection systems were found. However, at least one home northwest of the canyon mouth was at risk (and was damaged) as well as roads, fences, and the Mapleton Lateral canal. The U.S. Forest Service report (FSH 2509.13; FS-2500-A) for the Middle Slide fire contains information as to the dollar value of these structures.

Several common debris flow mitigation methods were suggested in a report by Robison (1989). These included sediment basins, check dams, and deflection berms. Although other mitigation methods may have been possible, they were not immediately considered. In addition, cost/benefit calculations for structures were not conducted at that time. No mitigation measures were taken until after debris flows damaged a house, fences, a canal, and buried roads.

The first debris flow was initiated by a severe thunderstorm on September 18-19, 1989. A second debris flow occurred following another storm on September 19-20, 1989. The second event consisted primarily of reworked sediments form the first debris flow, although some new material was transported from the burned area. A preliminary estimate of the volume of sediments by U.S. Forest Service personnel (Paul Scabelund, personal communication, 1989) indicated that about 15,000 yd^3 (11,500 m^3) had been deposited by the flows. However, another estimate was made by Robison (1989, unpublished field data) to evaluate that material which would be retained in a debris basin. About 5,000 to 7,000 yd^3 (3,800 to 5,350 m^3) was considered "retainable". Both estimates were complicated by deviations in the surface of the alluvial fan and the varying thickness of the debris flow deposits.

The contiguous drainage on the south side of Middle Slide Canyon contains similar vegetation, area, bedrock, and slopes. Only a small portion of the upper rim of this canyon was damaged, and sufficient litter was in place to prevent debris flows from the same storms which produced the debris flows immediately to the north.

Conclusions

In conclusion, approximately 4,700 yd^3 (3,600 m^3) of material was predicted to mobilize as a debris flow at the mouth of Middle Slide Canyon. The after-the-fact evaluations of retainable sediments, as compared to the PSIAC estimates, indicate that the PSIAC method is valid for sediment yield. However, several considerations for extreme conditions should be included in debris flow analyses: a) drainages with excessively steep slopes, much greater than 30 percent; b) thin soils; c) the possibility of long duration/extremely intense storms; d) flow paths for "non-retainable" sediments in mud flows beyond mitigation structures; and e) the necessity for quick action by "mitigating" agencies.

It will be approximately 3 to 5 years before vegetation is re-established and reduces the threat of debris flow to pre-fire conditions. Therefore, it is important that reseeding of the burned area be started as soon as possible to re-establish vegetation and develop ground cover. However, the rip-rap protected deflection-berm now in place at the mouth of the drainage should prevent damage to currently developed areas.

The lack of debris flows from the contiguous drainage during the same storms which produced sediment from the burned drainage is significant. Although the concept may seem obvious, the evidence found here shows that fire damage greatly increases debris flow potential by about 700 to 800 percent (Robert Rasely, personal communications, 1990).

REFERENCES

Davis, F.D., 1983, Geologic map of the Southern Wasatch Front, Utah: Utah Geological and Mineral Survey map 55-A.

Pacific Southwest Inter-Agency Committee (PSIAC), 1968, Report of the Water Management Subcommittee on Factors affecting sediment yield in the Pacific Southwest area and selection and evaluation of measures for reduction of erosion and sediment yield, Unpublished technical report, October, 14 p.

Robison, R.M., 1989, Middle Slide Canyon fire: Unpublished report to the Utah County Emergency Services Department, 16 p.

Robison, R.M., 1988, Erosion and sedimentation geologic reconnaissance report: Unpublished report to Alpine City, Utah, 15 p.

Nelson, C.V., and Rasely, R.C., 1989, Evaluating the debris flow potential after a wildfire; rapid response using the PSIAC method, Salt Lake County, Utah: Geological Society of America Abstracts with Programs, Vol. 21, No. 5, p. 121.

Wells, W.G. II, 1987, The effects of fire on the generation of debris flows in southern California, in Costa, J.E., and Wieczorek, G.F., eds., Debris Flows/Avalanches: Process, Recognition, and Mitigation: Geological Society of America, Review in Engineering Geology, Vol. VII, p. 105-114.

PROCESS-BASED DEBRIS-FLOW PREDICTION METHOD

Scott R. Williams[1]
Mike Lowe[2]

Abstract

This paper briefly identifies problems with traditional approaches to debris-flow prediction and presents a simplified process-based model (Williams and others, 1988) to accomplish the task. It also explains the basis and necessity for the model's creation, and shares insights gained through its use in Wasatch Front canyons in Davis County, Utah.

Introduction

Debris flows and debris floods that have occurred in Davis County during the past 75 years have caused loss of life and significant damage to property. Local government has spent large sums of tax money to provide mitigation below those canyons which have produced significant debris events in the past. Study of debris production from all canyons during recent extreme climatic cycles has raised questions relating to the common assumptions of repeatability of these events. Canyons where historical events were recorded generally produced negligible debris during the 1983-84 wet cycle, but pristine drainages produced major debris flows (Wieczorek and others 1983).

Previous attempts at prediction have centered around clear-water hydrologic model methodology. Assumptions have been made about similarities on a region-wide basis and have related future debris-flow potential to recent measurable events on the basis of watershed size (Liou, 1988). This assumption may be relevant to clear-water runoff, but doesn't consider the process necessary to

[1]Assistant Director/Flood Mitigation Planner, Davis County Public Works, P.O. Box 618 Farmington, Utah 84025.

[2]Geologist, Utah Geological and Mineral Survey, 606 Blackhawk Way, Salt Lake City, Utah 84148.

generate an initial debris flow or what is required to put debris in the position to be mobilized by some triggering event (climatic or seismic). Historical observations of past debris flows and studies of vegetative effects on watershed runoff have generally not been considered in these approaches.

Triggering mechanisms for debris-flow/flood events are generally storm-induced erosion on denuded areas, or landslides. Soil characteristics, topography, and vegetative conditions are important in determining the effects of these probabilistic events. The magnitude of the debris flow beyond the canyon mouth is determined by the volume of debris produced from the flood source area and channel conditions. Measurements and calculations of sediment production from the flood source areas show that their debris contribution is usually less than 20% of the total and in many cases is negligible. Most debris is derived from the channel. Debris production and accumulation in channels are slow intermittent geologic processes; reaccumulation after an event may take a long time. A recently scoured stream channel cannot contribute the same volume of debris during subsequent events as it did during an initial event. Given this logic, any debris flow which scours a perennial stream channel represents a Probable Maximum Flood (for debris volume) rather than something rated under traditional climatic return period criteria.

Methods of Investigation

When evaluating debris-flow hazards, factors affecting the initiation, magnitude (volume of debris), and recurrence should be considered along with probable areas of deposition. The proposed method of study consists of the following main elements.

- Historical research of scientific reports on past events
- Aerial photo and/or helicopter reconnaissance of canyons to evaluate the magnitude of channel erosion.
- Aerial photo reconnaissance of the watershed to identify possible flood-source areas and landslides.
- Measurement of channel cross-sections for channel-condition assessment.
- Mapping of bedrock exposed in the stream channel to assess the relative accumulation (%) of debris.
- Assessment of watershed vegetation cover conditions and probability of deterioration through fire, drought, grazing, or human abuse.
- Study of length, slope, and other pertinent stream channel characteristics.

This model (Fig. 1) can provide a rational estimate of the probabilities associated with the triggering events and the deterministic debris flow that will be produced.

Determine Watershed Conditions	Assess Probabilities of Triggering Events	Analyze Channel Conditions	Predict Debris Volume (PMF)
-mapping -mgt.policy -use	-fire/climate -earthquake	-history -erosion -bedrock %	

Figure 1. Model

Investigation Results

Historical Data: Our research into the relationship of the triggering event and debris-flow magnitude has produced the following results. Debris volumes produced from events are largely a function of the length of the perennial stream channel involved. Calculations from historical events from over 50 years ago and the recent past (1983) have produced the following relationship:

$$(TDF - TE) / TCCL = CDF$$

where TDF is the total volume of debris measured at the canyon mouth, TE is the estimated volume from the flood source area (sheet erosion, landslide, etc.), TCCL is the total contributing channel length and CDF equals the channel debris production factor.

This not only applies to the thunderstorm-erosion-generated events of the 1920s and 1930s, but is also true of the 1983 Rudd Canyon landslide-generated debris-flow event (Table 1). It would appear that about 30 cubic meters of debris per lineal meter (12 yd^3/ft) of channel may represent a debris volume (PMF) for Davis County Canyons. It seems that it also could represent a physical limit of debris possible per unit of channel length for perennial stream canyons of this type. This criteria was applied to an event reported on a perennial stream in California and produced a resultant CDF in the same range (Grahm and others, 1978).

Table 1. Selected debris-flow data

Canyon/Creek	Farmington	Rudd	Steed	Davis	Parrish
Avg.stream slope	0.127	0.314	0.341	0.305	0.177
TCCL (m)	18,036	1,652	5,354	3,902	6006
TDF - TE (m^3)	524,208	48,708	155,247	111,980	167,676
Year	1923	1983	1923	1923	1930
CDF (m^3/m length)	29.06	29.48	28.99	28.69	27.92
1983 TE (m^3)	17,000	64,000	10,000	min.	1,000

<u>Stream-Channel Conditions:</u> Figure 2 shows a pristine canyon bottom (Centerville Canyon, no historical events) and two stream-channel profiles (Parrish Canyon, 1930 event, Rudd Canyon, 1983 event). Rudd Canyon and Parrish Canyon have similar shapes and side-channel slopes, in spite of the fact that the Parrish Canyon debris flows occurred more than 50 years ago. The Parrish Canyon channel is much deeper than Rudd Canyon because Parrish Canyon has a much larger drainage basin and flatter average channel slope. There was therefore more potential accumulation of colluvium and less chance for annual runoff cycles to clean the channel. The Centerville Canyon profile, however, has much gentler side slopes. Figure 2 illustrates the large volume of material which could be eroded from the Centerville Canyon stream channel during future debris-flow events.

The amount of bedrock exposed in the channel bottoms of the three stream channels varies dramatically. It is estimated that Rudd Canyon's stream channel is over 50, and perhaps as much as 70 percent bedrock, sometimes forming long rock chutes where water rushes down canyon at high velocities. Although the Parrish Canyon debris-flow event had occurred more than 50 years earlier, it is estimated that 40-50 percent of the stream channel had bedrock control with most other areas appearing to has only a thin deposit of debris overlying bedrock. The amount of bedrock exposed in the Centerville Canyon stream channel is

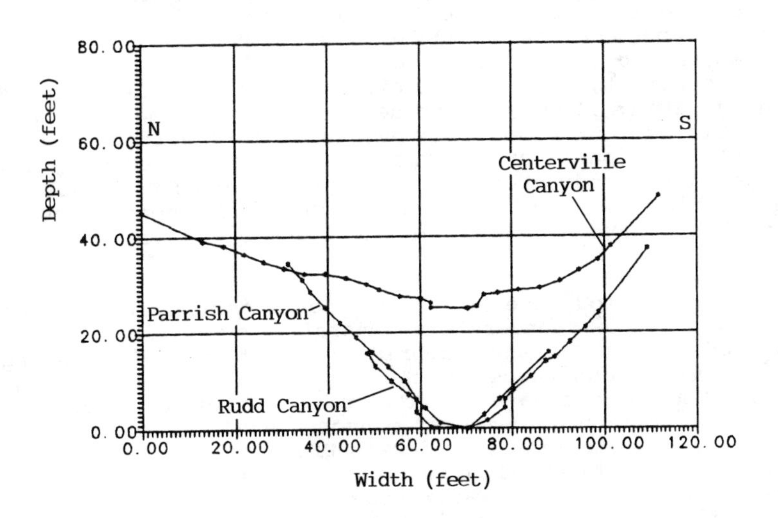

Figure 2. Stream channel profiles.

estimated at 10 percent or less. Mapping is currently underway that will allow better quantification of these percentages. These observations, when combined with the stream channel profiles, serve to illustrate the slow rate of debris accumulation in the central Davis County stream channels.

Watershed Conditions: Central Davis County drainage basins are currently heavily vegetated. Excluding bedrock outcrops, the cover of vegetation generally exceeds 80 percent. Runoff during cloudburst rainstorms is significantly reduced compared to that produced from the denuded flood-source areas which triggered the debris flows of 1923 and 1930 (Croft, 1967). In a 1945 report, it was stated that runoff was less than 0.5% of rainfall in the revegetated (65% cover) watersheds of Davis County, while devastating floods were generated from the same storm on burned areas above Salt Lake City (Craddock, 1945). In all cases, triggering storms were estimated to be 25-year or less return period climatic events. The U.S. Forest Service now manages the watershed above Davis County and prohibits commercial grazing by domestic animals. Access to the watershed is primarily from a road at the top of the drainage basin. Most of the canyons are steep with limited or difficult access, even by foot traffic which tends to limit much of the fire sensitive area to use by only the hardiest of hiker. These access problems help decrease the watershed fire risk. Terraces constructed in the 1930s to prevent erosion from overland flow are still reasonably intact, and will help reduce erosion even if cloudburst storms should occur following fires.

In contrast, the most significant deterioration of the central Davis County watershed in recent time has been landsliding. The landslide events of 1983 were caused by a heavy snow pack, an abnormally late rapid snowmelt, and an undrained bedrock aquifer (Mathewson and Santi, 1987). The combination of climatic and ground-water conditions is thought to have a return period of 500 years or more.

Conclusion

In our opinion, the most significant debris-flow threat exists from a thunderstorm-generated event in a fire-damaged canyon. Major debris-flows of the devastating first-event magnitudes in the past cannot occur in most Davis County drainages. The real danger lies in the pristine canyons. These canyons have little or no engineered protection from debris flows and all have highly developed residential communities near canyon mouths. There is also little watershed mitigation because there haven't been any significant problems in the past. Probabilistic prediction of future debris flows in these canyons must relate to the probabilities of triggering events like watershed deterioration (fire then rain), or climatic/ground-water conditions or earthquakes producing landslides that reach the main channel. We believe that this approach presents a more accurate appraisal of debris-flow potential in Davis County canyons than others previously used. The model, as well as the findings concerning the physical limit of debris production in an initial major event, may have applicability in other areas with perennial streams and somewhat similar conditions.

References

Craddock, G.W., 1945. Salt Lake City flood, 1945. Utah Academy of Sciences, Arts, and Letters, Vol. 23 pp.51-61.

Croft, A.R., 1967. Rainstorm debris floods; a problem in public welfare. University of Arizona Agricultural Experiment Station Report 248, Tucson, Arizona, 36 pages.

Grahm, A., Shuriman, G., Slosson, J.E., Yokum, D., 1978. Hidden Springs Flood, 1978, Southern California. 16 pages.

Liou, J., 1988. Mud flood and mud flow mapping in Davis County, Utah. Proceedings of the Conference on Arid West Floodplain Management Issues, Association of State Floodplain Managers, Las Vegas, 1988, pp. 111-146.

Mathewson, C.C., and Santi, P.M., 1987. Bedrock ground water; source of sustained post debris flow stream discharge. in McCalpin, James, ed., Proceedings of the 23rd Annual Symposium on Engineering Geology and Soils Engineering: Utah State University Press, Logan Utah, pp.253-265.

Williams, S.R., Lowe, Mike, and Smith, S.W., 1988. The discrete debris-mud Flow risk analysis method. Proceedings of the Conference on Arid West Floodplain Management Issues, Association of State Floodplain Managers, Las Vegas, 1988, pp. 157-168.

Wieczorek, G.F., Ellen, S., Lipps, E.W., Cannon, S.H., and Short, D.N., 1983. Potential for debris flow and debris flood along the Wasatch Front between Salt Lake City and Willard, Utah and measures for their mitigation. U.S. Geological Survey Open-File Report 83-635.

Olancha Debris Flow: An Example of an Isolated Damaging Event

J. E. Slosson,[1] Member, ASCE and T. L. Slosson[2]

Abstract

This paper presents the case history of an isolated flood/debris flow that damaged the City of Los Angeles Department of Water and Power Aqueduct and threatened serious damage to U.S. Highway 395, including any vehicles which might have been traversing the highway at the time. The case history represents another example of an isolated meteorologic event typical to arid/semi-arid regions of the American Southwest. Had this event occurred in a densely populated geographic location rather than the remote and very sparsely populated area of Olancha, California, both loss of life and property damage could have been very significant. This event again alerts us to the fact that we need to know more about alluvial fan environments and the meteorologic quirks of the arid/semi-arid environs of the Southwestern portion of the United States as well as other similar geographic settings. The authors have studied many such isolated events and have found each to be unique but, in many ways, similar.

Background

During the second week of August, 1989, a summer sub-tropical storm or hurricane-type storm, spawned in the east-central Pacific Ocean, swept ashore over northern Mexico and moved into Arizona, Nevada, and the southern and east-central portions of California. The storm prompted flash flood warnings along a 200-mile front that moved northeasterly from the Mexican border to the Rocky Mountains. Water up to two feet deep flooded homes at Winterhaven in the Imperial Valley and rose to three feet along portions of Highway I-8 in San Diego County. Winds gusted at velocities up to 100 mph

----

[1]Chief Engineering Geologist, Slosson and Associates, 15500 Erwin Street, #1123, Van Nuys, CA 91411

[2]Supervising Engineering Geologist, Slosson and Associates, 15500 Erwin Street, #1123, Van Nuys, CA 91411

in Yuma, 90 mph in Las Vegas, and 60 mph in Olancha as the storm moved along its north-easterly path toward the Rocky Mountains. Flash flooding caused damage in Yuma, Arizona; in Las Vegas, Nevada; and in Baker, Twenty-nine Palms, and Benton in the desert regions of California.

A particularly serious consequence of the torrential rainfall occurred along approximately 3 miles of the eastern front of the Sierra Nevada near Olancha. There, a relatively small storm cell stalled over the mountains on August 8 (repeating on the 9th), causing local, extremely high intensity rainfall over a short period of time. This localized downpour generated massive debris flows from four nearby creeks in close proximity. Debris was transported approximately two miles by the flood water, burying about two miles of the ≈31-foot wide and 15-foot deep Department of Water and Power Owens Valley Aqueduct. An estimated 250,000 cubic yards of debris, ranging in size from sand to boulders approximately 5 feet in diameter, were swept downslope toward the aqueduct. It was estimated that approximately 100,000 cy^3, composed chiefly of boulder, cobble, gravel and sand-size debris were deposited between the fan apexes and the aqueduct. Another 50,000 cy^3 of similar sized materials were deposited in the aqueduct, burying a 2-mile segment. Approximately 100,000 $cy^3\pm$ of predominantly gravel, sand, and silt were deposited down gradient between the aqueduct and U.S. 395.

If it had not been for the existence of the aqueduct with its containment and decelerating capabilities, structures in the town of Olancha would have been seriously impacted and vehicles (and motorists) traveling along U.S. 395 swept off of the roadway and/or buried.

This storm, with its isolated and seriously damaging effects, provides a good text book example of the fickle nature of sub-tropical storms that extend over the desert areas of the Southwest. Many historic flooding events of alluvial fans have been the product of such storms and many will occur in the future. The existence of an isolated storm cell was caused by topographic (orographic) controls, air mass directional flow patterns, and local temperature variation related to convectional heating (thermals). There is a valuable lesson to be learned from these isolated storm events, mainly that they should be considered in the planning and design of critical lifelines. A satellite picture of the storm vividly demonstrates the presence of a single, isolated storm cell in the vicinity of

Olancha at approximately the same time that the Olancha flood/debris flow was impacting the Los Angeles Water and Power aqueduct. Another storm cell was causing flooding in the Chalfant/Hammil Valley area ≈100 miles to the north where an estimated 50 homes were inundated. U.S. Forest Service records show precipitation at the Benton Station (northern end of Hammil Valley) was 4" in 5 hours on the 8th and 6" in 3 hours on the 10th of August. An interesting side note is that the compliance factor to the order to evacuate issued by the Fire Chief (also local Emergency Services Office for the region) was only 10%. This is surely an indication of the need to educate residents in alluvial fan areas of the potential life-threatening hazard posed by high intensity rainfall.

Site Conditions

The Olancha flood/debris flow was a unique phenomenon in the following respects:

1) It occurred as an isolated extreme event in conjunction with a rather wide (≈200 miles) subtropical storm.

2) Within the storm there were a large number of individual cells as well as cell masses.

3) Isolated cells were the source of site specific torrential rainfall at Olancha and over the Chalfant/Hammil Valley area, whereas Las Vegas, Phoenix, Winterhaven, 29 Palms, etc., were inundated by rainfall from cell masses or very large cells.

4) The storm cell or cells which approached the Sierra Nevada near Olancha were affected by stalling and orographic lifting. The orographic lifting caused a rapid cooling of the air mass as it was elevated to a zone of lower colder, atmospheric air pressure causing a reduction in moisture carrying capacity and, thus, precipitation. The White Mountains to the east of the Chalfant/Hammil Valley caused a similar stalling and orographic lifting causing flooding in that area.

5) The Sierra Nevada Mountains acted as a barrier or high ridge to the northeasterly-moving air mass or storm and forced the storm cell upward as it passed over the mountain barrier. The elevation at Olancha is approximately 3,800 feet, the Great Western Divide averages just

over 10,000 feet, and the high peak of the Sierra Nevada in the immediate vicinity rises to approximately 14,000 feet. Thus, the air must rise 6,000 feet to 10,000 feet just to rise over the barrier. Other things being equal, this air flow pattern should cause a drop in temperature of ≈40°F (the drop in temperature will range from 5.5°F for dry air to 3°F for wet air for each 1,000 feet of rise). Recognizing that as an air mass cools its moisture retention capability decreases, it becomes apparent that rainfall begins and increases as the air mass rises and cools.

6) During the peak of the storm, wind velocities of up to 60 mph were recorded at Olancha. Dependent on the direction of wind flow and continuity of the high velocity winds, the rate of rise could have caused the average rate of cooling of the rising air mass to remain at approximately 5.5°F for each 1,000 feet thus increasing the cooling rate and intensifying the precipitation.

7) Exacerbating the sudden drop in temperature from orographic control was the rise of the heated air mass as it passed over warmer terrain. Air temperatures in the mid- to high 90's prevailed in the Owens Valley area during the second week of August (1989) causing updrafts of warm air which assisted in the elevation of the storm cell(s). This phenomenon set the stage for thunderstorms with cumulonimbus clouds and intense local rainfall as the strong updrafts of warm air pushed the storm cell upward to cooler air. This rapid uplift of air, often referred to as convectional air circulation, may cause a drop in temperature of ≈5.5°F for each 1000 feet of rise. Dependent on the temperature differential, this uplifting or updraft effect may cause air within a cumulonimbus cloud to rise 20,000 feet to 40,000 feet. The more rapid and the higher the rise, the greater the loss of moisture or the more intense the rainfall. Rainfall rates of as much as 6" in 3 hours at Benton (or the Chalfant/Hammil Valley) and 5" in 5 hours at Independence, California were recorded. No rainfall records were available from Olancha, but it is estimated that for the very localized thunderstorm in the Sierra Nevada near Olancha, the rainfall rate was 8"-10" in 3 to 4 hours.

8) The gradient of the Cartago, Olancha, Falls, and Walker Creeks in the higher elevation within the zone of high rainfall intensity ranges from ≈25° to 50°. At lower elevations in the foothills it ranges from 10° to 25°; from the fan apex to the aqueduct it varies from 4° to 10°. This topographic control encourages high velocity flow even with an extreme roughness condition and a high load factor.

9) Falls Creek and Walker Creek have confluence at the lower elevations of the mountains (elevation 5,000'). The confluence of Walker and Olancha Creeks occurs on the alluvial fan apex (about elevation 4,500'). These three creeks have a combined watershed area of approximately 12 square miles. The major damage to the aqueduct came from the combined debris flow from these three creeks, the quantity of which was the aforementioned 250,000 cy^3. An undetermined quantity of materials was also produced from Cartago Creek, which had a water shed area of 9½ square miles.

10) The steepness of the eastern side of the Sierra Nevada is directly related to tectonic activity. The Sierra Nevada geomorphologically is a tilted fault block, with the Sierra Nevada fault system located at the eastern edge of the tilted block -- or between U.S. 395 and mountain front. The Owens Valley fault is active, having its latest episode of movement and earthquake activity in 1872.

11) The fault/tectonic origin of the steep eastern side of the Sierra Nevada also caused the granitic bedrock to be shattered and thus rather easily eroded. This inherent nature of the surficial bedrock allowed rapid erosion to occur with an estimated 250,000 cy^3 of rock debris ranging in size from 2- to 5- foot boulders to be transported from the apex of the fan 1 to 2 miles to the aqueduct and eventually to U.S. 395, a distance of 2 to 3 miles.

12) Debris flow material was apparently, from erosion and transport evidence, traveling at a high velocity as it exited the steep gradient channels. Upon leaving the confines of the canyons, the flow began to spread as it flowed across the fan apex and spread further as it approached the aqueduct some 2 miles from the

apex. At the aqueduct, the debris flow had spread and coalesced over a 2-mile wide zone as it filled in and crossed the aqueduct. Few boulders were carried beyond the aqueduct. The aqueduct was filled with a mixture of grain sizes ranging from fine sand to boulders up to 4 feet in maximum diameter. It was estimated that the volume of debris (rock) in the aqueduct (canal) was 50,000 cy^3. Rock debris scattered (deposited) from the apex to the aqueduct was estimated to be 100,000 cy^3 with material ranging from boulders to coarse-grained sand. Between the aqueduct and U.S. 395 an estimated volume of ≈100,000 cy^3 with a few cobbles and small boulders to fine-grained sand was deposited.

It can been seen that a series of conditions came together in the vicinity of Olancha during the second week of August of 1989, producing the debris flow. Greater damage and associated loss of life could have occurred if these same conditions were to come together in an area of higher population density in the region of an alluvial fan. Further study of the factors involved in flows and floods on alluvial fans is needed so that proper planning, design, and mitigation can be programmed to limit the damages and potential for loss of life. While each past event of this type has been found to be unique and different, there are in all instances some similarities. If, in future studies, a method of delineating and predicting what can happen on an alluvial fan can be compensated by design, it will still need data and mapping on a site-by-site basis to fully analyze the potential inasmuch as all fans differ in structure. Hopefully, these studies will be conducted, and we can all learn from the disasters.

The authors would like to thank Dennis Williams and Jim Snead, engineers with the City of Los Angeles Department of Water and Power; Donna Perez, Interagency Dispatch Center, U.S. Forest Service; Jerry Newton, editor of the Inyo Register; Gerard Shuirman, CEF, ASCE; and Will Gould of the National Weather Service for their invaluable source of data and photos and their courteous assistance and encouragement.

Deadly Debris Flows on I-5 near Grapevine, CA

Vincent S. Cronin[1], James E. Slosson, Thomas L. Slosson[2], and Gerald Shuirman[3]

Abstract

A series of debris flows inundated the southbound lanes of Interstate Highway 5 (I-5) ~110 km (70 mi) NW of Los Angeles, California, between Fort Tejon and Grapevine, early on February 5, 1978. These flows resulted in the death of a motorist and damage to several trucks and autos. The design of the highway and its drainage systems is insufficient to guarantee the public safety, because it fails to address the debris flow hazards that are present. An adequate geological site investigation undertaken during the design phase would have shown that landslides are ubiquitous along this segment of I-5, and that a series of Holocene debris fans extend into Grapevine Canyon. The steep topography, fractured bedrock and loose soil conditions are quite favorable for the continued development of debris flows. Although there were no recorded debris flows along this segment of I-5 or its predecessor (US 99) between 1933 and 1978, it was unreasonable to assume that slope failures would not occur in the future as they had in the past. Lack of adequate mitigation since 1978 makes it highly probable that debris flows will inundate I-5 in the future, with similar results.

Chronology

An old auto road, nicknamed *the Grapevine*, was improved in 1933 and designated US Highway 99. The construction of an improved highway was authorized by the California Highway

----

[1]Dept. of Geosciences, University of Wisconsin–Milwaukee, P.O. Box 413, Milwaukee, WI 53201.

[2]Slosson and Associates, 15500 Erwin Street, Suite 1123, Van Nuys, CA 91411

[3]15858 Sutton St., Encino, CA 91436

Commission in 1947. Preliminary designs for the Grapevine segment of Interstate Highway 5 (I-5) were approved in 1957, and construction was completed in 1960. I-5 was planned, engineered and constructed by the California Department of Transportation (CalTrans). During the design phase, highway alignment was planned first, and the drainage system was subsequently designed. The potential for debris flows was not recognized or considered in the designs of the highway or its drainage system. The drainage system was designed to handle only clear-water flow, without debris. No debris flows are reported to have occurred along the Grapevine segment of US 99 or I-5 from 1933 till 1978. The maximum flow of Grapevine Creek was assumed to be 85 m^3/s (3,000 ft^3/s) during highway design (deposition of Charles N. McKee [RCE] of CalTrans).

Drought conditions prevailed from 1975-77 in this area. A major wind storm occurred December 20-21, 1977, which stripped dry vegetation and topsoil from the summits and north-facing slopes of the local mountains. Grains ≤0.64 cm (0.25 in) in diameter were transported from some affected areas, while sheltered and lowland areas were sites of deposition. The lower slopes in Grapevine Canyon were essentially unchanged by the windstorm, although some eolian sediment was deposited in the area. Four utility towers, one of which was 4.8 km (3 mi) from the site of the February 5 debris flows, were blown down by winds measured at 240 km/hr (150 mi/hr) with peak gusts estimated at 309 km/hr (192 mi/hr; deposition of Steven Gouze). Although it has been asserted that the magnitude of the windstorm could not have been anticipated, meteorologist John W. James testified that wind storms of this magnitude are not unusual for this area. No engineers or geologists were asked to inspect the hills after the wind storm (deposition of Richard W. Johnson of CalTrans).

Rainfall records from gauges at two ranger stations located at higher elevations near the Grapevine segment of I-5 show that ~25 cm (10 in) of rain fell in the area between November and January 20. No rainfall was recorded between January 20 and February 4. The average annual rainfall at Fort Tejon is ~28 cm (11 in), so it is reasonable to assume that the shallow subsurface was saturated in early February. Approximately 2.3 cm (0.9 in) of rain fell at the Chuchupate station during the evening of February 4. A cell of heavy rain was recorded at Chuchupate from 1-3 am February 5, and probably traveled the ~13 km (8 mi) northeast to the Grapevine segment of I-5 by 4-5 am. The California Highway Patrol (CHP) estimated that ~13 cm (5 in) of rain fell in half an hour on the night of February 4-5 (deposition of Loren Kistrow of the CHP).

Just before dawn on February 5, debris flows with a total volume of ~35-55,000 m^3 (40-60,000 yd^3) moved simulta-

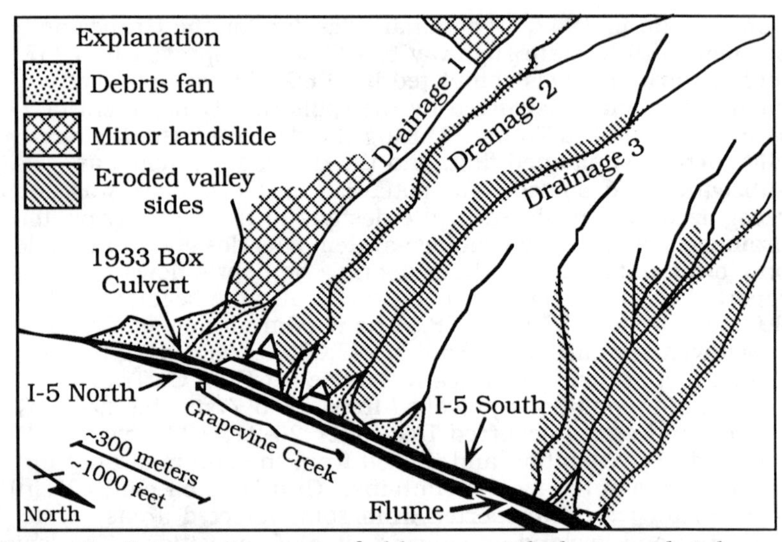

Figure 1. Interpretive map of oblique aerial photograph taken one month after the debris flow. The debris flows of February 5, 1978, originated in Drainages 1-3.

neously down three adjacent canyons west of I-5 between Fort Tejon and Grapevine (Figure 1). The drains at the base of the canyons were too small to accommodate the volume of material being transported, because they had been designed for clear-water runoff only. The southbound (uphill-directed) lanes of I-5 in this area were designed to be lower than the adjacent northbound lanes, so the southbound lanes became a flood channel after the drains were blocked by debris. In the absence of debris basins or adequate drains, the debris flowed onto the southbound lanes and coalesced into a single flow that was up to a meter in depth. Witnesses reported seeing boulders and small trees in the debris. The debris on I-5 extended from Drainage 1 northward ~1.25 km (0.78 mi). Debris flowed along I-5 with sufficient energy to move fully loaded tractor-trailer rigs, and to lift and carry automobiles. A car driven by a young woman was swept down the highway and into an open 3.0 x 2.4 m (10 x 8 ft) flume located between the north and south lanes of I-5. There was no guard rail between the highway and the open flume. The woman's body and the wreckage of her vehicle were eventually found >2 km (1.25 mi) from the site of initial impact.

Nine additional debris flows occurred in this area between February 5 and April 4, 1978. These flows were smaller than the February 5 flow, but were large enough to require clean-up crews from CalTrans to remove debris from the highway. Debris flow events ceased with the end of the rainy season, during which Fort Tejon received roughly twice as much rain as normal.

Observations and Interpretations

Subsequent site investigation and aerial photo interpretation showed the following major points.

[1] Youthful debris fans existed below each of the canyons before the highway was built, and their significance was either not recognized or ignored by the designers of I-5.

[2] Perhaps half of the debris that was mobilized on February 5, 1978, was eroded from the walls of the channel down which it flowed.

[3] Slopes above the Grapevine segment of I-5 are very steep, typically >1.5:1 (>34°), and tend to be gravitationally unstable.

[4] Local slopes are mantled in loose colluvium that is easily saturated and mobilized during moderate rainfall.

[5] Local bedrock is pervasively fractured, and loose boulders are common on the slopes.

[6] Utility poles on the slopes are tilted due to active down-slope creep.

[7] Clear geomorphic indicators of landsliding are present along a large percentage of the slope area adjacent to the Grapevine segment of I-5, both on the east and the west sides of the highway.

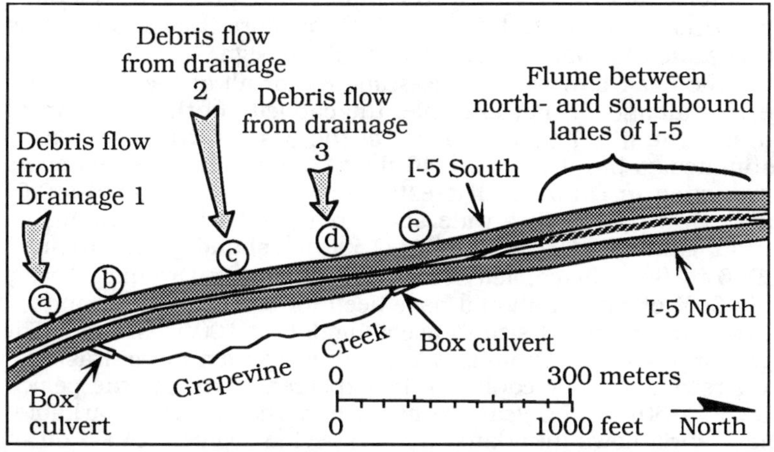

Figure 2. Map of the site of the fatality, adapted from the as-built plans for I-5. Critical drainage devices are located at points a-e: [a] 1.8 x 2.1 m box culvert from 1933; [b] 46 cm diameter corrugated metal pipe (CMP); [c] 107 cm CMP; [d] 91 cm CMP; [e] 61 cm CMP. Estimated watershed areas are: 2.4 km^2 for Drainage 1; 1.2 km^2 for Drainage 2; and 0.5 km^2 for Drainage 3.

[8] The areas that had been denuded by the wind storm of December 20-21 were not in the source regions for the material incorporated in the debris flows of February 5. Some of the material deposited during the wind storm may have been remobilized in the debris flows; however, most of the debris was locally eroded colluvium.

[9] Inlets to the I-5 drainage devices are typically located at the base of canyons, at the bottom of channels cut in older debris fans. Hence, there is loose, easily mobilized material immediately adjacent to the drains that can clog the inlet. Highway debris and dead plant material was also noted in the grates covering the drainage inlets.

The drainage facilities along Grapevine Creek were designed for a maximum flow (Q) of 85 m^3/s (3000 ft^3/s). Re-calculation using the 1939 design standards shows that a conservative estimate for a 50 year storm should have been Q_{50} = 163 m^3/s (5760 ft^3/s), or Q_{100} = 219 m^3/s (7750 ft^3/s) for a 100 year storm, based upon a watershed of 51.8 km^2 (20 mi^2). By 1958 when I-5 was being designed and built, a bulking factor should have been taken into account that would have increased the clear-water flow by at least 65%. A debris production assessment for this area indicates that drainage devices should be designed to accommodate 14,470 m^3/km^2 (45,000 yd^3/mi^2) of sediment from the corresponding watershed. The bulked Q_{100} for Grapevine Creek at the time I-5 was designed should have been ~362 m^3/s (12,800 ft^3/s), or more than 4 times the Q_{100} value that was actually used in the design.

The peak flow was underestimated for all of the drainage devices along I-5. For example, the capacity of the 1933-vintage 1.8 x 2.1 m (6 x 7 ft) box culvert at the base of Drainage 1 is estimated to be ~12 m^3/s (440 ft^3/s; Figure 2). The sediment production in Drainage 1 is estimated to be ~33,445 m^3 (40,000 yd^3). Drainage 1 has a watershed area of ~2.4 km^2 (0.92 mi^2), for an estimated Q_{50} = 42 m^3/s (1500 ft^3/s) and Q_{100} = 57 m^3/s (2000 ft^3/s). The capacity of the box culvert in Drainage 1 was only 30% of what it should have been for a 50 year flow, and 22% of the required size to accommodate a 100 year flow. The CMP drains below Drainages 2 and 3 were even more drastically undersized, so they could not be expected to handle the peak flow in a 50- or 100-year storm. It is perhaps less remarkable that debris flows inundated the Grapevine segment of I-5 on February 5, 1978, than that the highway had not been inundated earlier, given the undersized design of the drainage system.

Conclusions

The occurrence of the debris flows that caused a fatality on I-5 in 1978 was predictable and preventable. The fatality and damage associated with the debris flows could have been avoided through the proper design and maintenance of I-5 and its associated drainage systems. The design phase of the highway development should have included the collection of data by qualified specialists to accurately characterize the site's geology and soils, so that these data could be used in the design of an adequate drainage system. The fact that the slopes adjacent to the Grapevine segment of I-5 are on private property is insufficient justification for failing to conduct an adequate geological assessment.

The existing drainage system should be replaced or augmented so that it will have sufficient capacity to accommodate both water and debris. Debris basins should be established where there is sufficient space, as at the base of Drainage 1. Hillside areas adjacent to the highway should be examined on a regular schedule by qualified geologists and soils experts in order to describe slope conditions and identify potential hazards. Warning systems should be established to notify the California Highway Patrol when hazardous conditions exist that would justify temporary highway closure.

Adequate corrective measures have not yet been planned to mitigate the hazardous conditions that exist along the Grapevine segment of I-5. It is likely that debris flows will inundate I-5 with similar deadly results in the future.

References

Depositions referenced in this paper were taken during pre-trial proceedings for the case of William Watkins, et al., versus the State of California, et al., and related cross-actions.

Evolution of Drainage Criteria, Albuquerque, New Mexico

Clifford E. Anderson[1], M.ASCE
Richard J. Heggen[2], M.ASCE

Abstract

Storm drainage in Albuquerque, NM is a central issue in land development. To expedite municipal development while assuring public safety and preserving environmental attributes in a public arena where new information is constantly introduced requires dynamic regulation. Drainage criteria in Albuquerque have evolved from textbook computations to standards specifically developed for the region. The evolutionary process has not at all times been straightforward. To date, the process has unified a Rational Method approach for small watersheds with a modified HYMO model for large watersheds. Particular attention has been given to matching both peak rate and volume of runoff.

Background

The Albuquerque Metropolitan Arroyo Flood Control Authority (AMAFCA) formalized drainage policy in 1972 for the Albuquerque, NM area. A 100-year, 6-hour storm was established as the normal design event. For Albuquerque, such a storm ranges from 2.2 to 2.9 inches. Discharge from developed areas could exceed neither the undeveloped rate nor volume. The policy did not specify detailed hydrograph analysis procedures. Engineers generally used the Rational Method and simplified Soil Conservation Service (SCS) procedures for analysis of small projects. For large projects, hydrology was developed using proprietary computer programs or public-domain programs such as HEC-1 and the SCS TR-20. While the analyses tended to be textbook correct, they were widely recognized to be

----

[1]Drainage Engr., AMAFCA, 2600 Prospect NE, Albuquerque, NM 87107
[2]Assoc. Prof., Dept. of Civil Engrg., Univ. of New Mexico, Albuquerque, NM 87131.

inadequate for a rapidly-urbanizing alluvial fan system subject to ephemeral events.

In 1981, the Albuquerque Master Drainage Study was completed for the major urbanized areas. The study utilized the USDA Agricultural Research Service HYMO computer program (Williams & Hann, 1973), modified to include routing through storm sewers and hydrograph division at road intersections and inlet structures. HYMO uses the SCS curve number (CN) procedure modified to utilize an adjustable unit hydrograph. To obtain reasonable peak runoff rates, the unit hydrograph was calibrated locally, runoff from pervious and impervious portions of a watershed were computed separately and combined, and a front-weighted rainfall distribution was utilized.

The City of Albuquerque codified its first drainage ordinance in 1982. The ordinance recognized the 100-year 6-hour storm as the design standard. The City adopted a Development Process Manual (DPM) (City of Albuquerque, 1982), a specification of planning and engineering requirements. A major section of the DPM was "Drainage, Flood Control and Erosion Control". The DPM made official the use of the Rational Method and the SCS dimensionless unit hydrograph procedure. A unique feature of the Rational Method was the determination of a C coefficient based upon SCS hydrologic soil group and percentage of impervious area.

During the next several years, inaccuracies and inconsistencies with design criteria became apparent. Problems were noted at the methodological "boundaries", the interface between basins evaluated with different models. For a given basin, alternative analyses yielded conflicting conclusions. Between master plans and site-specific submittals, the whole would rarely correspond to the sum of its parts. Engineering judgement regarding rainfall distribution, CN's, and impervious percentages had major impact on the results. Methods that seemed to predict runoff volume failed to adequately predict rates. Methods that showed promise at predicting rates misestimated the volumes. Certain methods required adjusted data sets to achieve reasonable results. In retrospect, the sophistication of numerical computations had outpaced the basic data acquisition needed to substantiate the models. The City issued a "Notice of Emergency Rule" in 1986, deleting the use of Rational Method C's based on hydrologic soil group. Analysis reverted to conventional handbook computations, standards legally defensible, but hydrologically over-simplified.

Data Acquisition

In the late 1970's, the U.S. Geological Survey established 10 streamflow gages and 11 recording rain gages for undeveloped and urbanizing areas. One additional streamflow gage and six recording rain gages were added in the early 1980's. The program now provides rainfall-runoff relationships for major storms. The U.S. Weather Bureau, in connection with AMAFCA, established a volunteer rainfall reporter network, currently with 120 observers. Together, such data provides much better spatial and temporal resolution of major storms that the data of a decade past. In addition, 40 years of Albuquerque International Airport rainfall records were inspected. Reappraisal indicated that the 1-hour storm, not the 6-hour storm, was the event associated with critical runoff conditions for most urban areas.

In 1982, rainfall losses were studied using a 10 ft^2. rainfall simulator (Sabol *et al*, 1982). Tests were conducted at 10 sites representing both natural and developing conditions. The study concluded that rainfall-runoff data did "not follow a constant CN line", the "hydrologic soil group and SCS aids for the selection of CN does not in general indicate the most appropriate CN for tested plots", and the SCS rainfall-runoff equation "application, especially in Albuquerque, is based on assumptions which may be too generalized and may be in error".

In 1987, 102 split ring infiltration tests were carried out at 32 sites (Heggen, 1987). Data showed that infiltration depends upon land surface treatment and that infiltration rate over the first 30 minutes can be treated as a constant. Hydrologic soil group was not a strong indicator of rate. The infiltration rate for lawns varied excessively, depending upon the timing of irrigation.

Towards New Criteria

In 1988, the City, AMAFCA and local consulting engineers formed a committee to develop a new drainage manual for Albuquerque. New procedures must satisfy three criteria: 1) be justified in light of the improved rainfall and rainfall-loss data; 2) provide a unified approach for basins of all sizes; and 3) facilitate consistent analysis among the many engineers involved in drainage issues.

The metropolitan area was divided into four hydrologic zones. For each zone, a 100-year hyetograph was developed with depths consistent with NOAA data, the timing consistent with storm records. Initial abstraction and uniform infiltration parameters were developed for four land treatment classifications, as shown in Table 1.

TABLE 1.-Initial Abstraction and Infiltration Rate

Land treatment (1)	Initial abstraction (in) (2)	Infiltration (in/hr) (3)
Impervious	0.10	0.04
Compacted earth	0.25	1.15
Irrigated lawns	0.32	1.63
Natural	0.40	2.10

For sub-basins of 30 acres or less, most local drainage projects, Rational Method C's were developed based on the 10-min. runoff intensity equal to 45 percent of the 1-hr. depth, P_{60}. C's were found to depend upon the storm recurrence frequency and rainfall. For the central business district, C's are shown in Table 2.

TABLE 2.-Rational Method C's

Land treatment	Recurrence Interval		
(1)	100-yr (2)	10-yr (3)	2-yr (4)
Impervious	0.95	0.95	0.94
Compacted earth	0.64	0.53	0.27
Irrigated lawns	0.47	0.34	0.11
Natural	0.35	0.20	0.02

Based on the hyetograph and abstractions, excess precipitation was computed, as shown in Table 3.

TABLE 3.-Excess Precipitation (in)

Land treatment	Recurrence Interval		
(1)	100-yr (2)	10-yr (3)	2-yr (4)
Impervious	2.27	1.41	0.82
Compacted earth	1.05	0.50	0.16
Irrigated lawns	0.78	0.34	0.08
Natural	0.60	0.24	0.03

For sub-basins larger than 30 acres and major drainage projects, a unit hydrograph procedure was established, modifying HYMO to (a) compute runoff by initial abstraction and uniform rate infiltration, (b) compute hydrographs separately for impervious and pervious portions of a watershed and combine, and (c) shape the unit hydrograph separately for impervious and pervious portions to reflect a variable K/t_p ratio (HYMO parameters) where,

$$K/t_p = a + bP_{60} \quad (1)$$

for watersheds 0 to 30 acres, and

$$K/t_p = cP_{60}{}^d \qquad (2)$$

for watershed greater than 130 acres. The coefficients a and b are empirical functions of initial abstraction, infiltration rate and P_{60}. Coefficients c and d are functions of infiltration rate and P_{60}. From 30 to 130 acres, K/t_p transitions between the equations. Figure 1 illustrates the dependency of the unit hydrograph peak on the K/t_p ratio. For pervious areas, the K/t_p ratio varies from 0.42 to 1.25; for impervious areas, the ratio ranges from 0.40 to 0.75. The resultant unit hydrograph procedure agrees in volume and peak rate with the small-basin results shown in Tables 1-3.

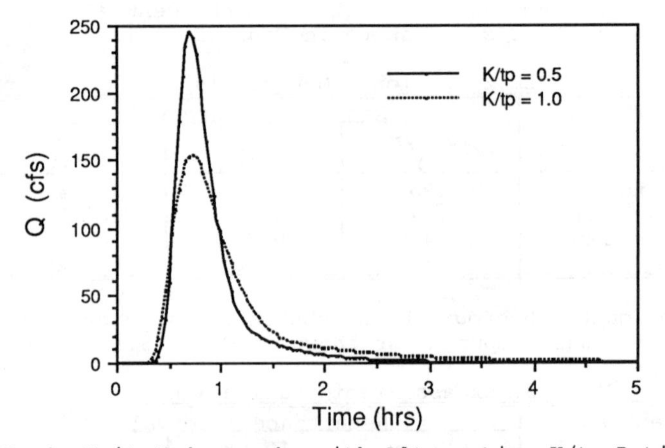

FIG. 1.-Unit Hydrographs with Alternative K/t_p Ratios

HYMO was chosen because of its extensive use in the area, its modularized subroutine structure allowing easy source code modification, its simplified input, its capacity to simulate street and storm sewer flows, and its suitability for a variety of computers. An enhanced version of HYMO, suitable for use on PC microcomputers, is available from AMAFCA.

The new procedure is being used for major public works projects in Albuquerque. The engineering community appears to welcome a step toward standardization. Some caution has been voiced that a change in criteria may expose in-place structures to requirements for redesign, an aspect of infrastructure management not unique to flood channels. The computational coefficients established for Albuquerque are being generalized as functions of abstraction, area and rainfall depth to allow the procedure to be utilized throughout New Mexico.

References

City of Albuquerque (1982). *Development Process Manual; Vol. 1. Procedures; Vol. 2. Design Criteria; Vol. 3. Policies and Plans,* Department of Public Works

Heggen, R.J. (1987). *Split Ring Infiltration Basic Data Collection and Interpretation,* Report No. PDS 110/210, Bureau of Engineering Research, Univ. of New Mexico

Sabol, G.V., Ward, T.J. and Seiger, A.D. (1982). *Phase II, Rainfall Infiltration of Selected Soils in the Albuquerque Drainage Area,* Civil Engineering Department, New Mexico State Univ.

Williams, J.R. and Hann, R.W. (1973). *HYMO: Problem-Oriented Computer Language for Hydrologic Modeling,* Users Manual, ARS-S-9, Agricultural Research Service, USDA.

DRAINAGE MANUAL FOR CLARK COUNTY, NEVADA

A. S. "Andy" Andrews, Member[1], P.E., Gale W. Fraser, II[2], P.E., and Alan J. Leak[3], P.E.

Abstract

The development of a flood control district and its associated capital improvement and regulatory programs does not by itself address drainage standards for a community. While the capital improvement and regulatory programs go a long way in solving existing flooding problems and minimizing future flooding problems, a common denominator is missing. This common denominator is drainage standards that are laid out in a drainage manual. Such a drainage manual covers all aspects of drainage planning including policy, design criteria, applicable laws, and guidelines for development of private and public stormwater management facilities. The preparation of a drainage manual in an arid region that is experiencing rapid growth warrants the inclusion of special features that attempt to address the concerns of all affected individuals.

Introduction

The preparation and acceptance of a drainage manual that spells out submittal requirements, hydrologic data, and hydraulic requirements for all future private and public flood control facilities is a formidable task. This is especially true when dealing in an area that is experiencing tremendous growth but averages only four inches of rainfall per year. Not only are growth and average annual rainfall important factors in preparation and implementation of a manual, the manual must also address the lack of historic

---

[1]President, WRC Engineering, Inc., 1660 South Albion Street, Suite 500, Denver, Colorado 80222.
[2]Assistant General Manager, Clark County Regional Flood Control District, 301 East Clark Avenue, Suite 300, Las Vegas, Nevada 89101.
[3]Water Resource Engineer, WRC Engineering, Inc., 1660 South Albion Street, Suite 500, Denver, Colorado 80222.

hydrologic information as well as the lack of historic construction and operation of existing flood control facilities. While existing criteria are available from various books and manuals for the planning and design of flood control facilities, these criteria have always been questioned for use in arid regions that experience infrequent flash flooding.

The flash flooding is due not only to the severe local summer thunderstorms, but also to the existing natural topography in Clark County. The natural topography typically causes runoff to flow at or near critical depth. The superimposing of development and flood control facilities on the natural topography, likewise, requires runoff to be analyzed in the supercritical flow regime. The design of facilities that limit flow to a subcritical flow regime mandates more expensive structures that occupy more right of way. The drainage design manual addresses criteria that takes into account hydrologic and hydraulic information that is site specific to Clark County.

Brief History

For the past 80 years, flooding in Clark County, Nevada, has resulted in the loss of life and many millions of dollars in property damage. The most recent flooding event occurred in 1984 when the flooding of Ceasars Palace made the front page headlines on a number of newspapers across the United States. The Nevada Legislature, in 1985, approved a bill that made substantial technical and policy changes to Nevada's law concerning flood control districts. The Clark County Regional Flood Control District (District) has been the first district to form under the new law. This required considerable marketing to the constituents of Clark County in order to become a District and to obtain a revenue source. The District encompasses Clark County and five cities contained therein. These cities include Las Vegas, North Las Vegas, Henderson, Boulder City, and Mesquite.

The District has enabled various programs that deal with capital improvements to regulations. One of the more encompassing programs has been the development of a drainage manual for use on both private and public flood control facilities. The increased urbanization and lack of adequate storm drainage facilities prompted the District to seek a uniform and standardized approach to stormwater management.

Manual Development

The development of a drainage manual in an area that is experiencing tremendous growth represents a major step towards bringing a uniform approach in dealing with drainage problems as well as in establishing uniform standards for designing and implementing drainage improvement facilities.

An important first step in the process was to identify the manual goals and the affected parties that would play an important role in the development and adoption of the manual. Early in the development process, the District identified the manual goals as 1) To develop a uniform and consistent criteria for development of rainfall/runoff models and design of flood control facilities, 2) To develop simplified procedures that would minimize the time required to prepare drainage studies and their review, 3) To prepare a single stand alone document that would provide all the necessary information to prepare and review drainage submittals, and 4) To provide a criteria that would assist engineers and designers in addressing flooding problems in a manner consistent with the goals and objectives of the local communities and the District.

The affected parties were included in the development of the manual by formation of a Technical Review Committee (TRC) comprising of representatives of four communities and a representative of the local Consulting Engineers Council. The TRC provided a forum for discussion of the diverse criteria used by different entities and the mechanism to forge a uniform criteria acceptable to all interested parties in Clark County.

The next step was to employ an engineering firm that has developed manuals for other communities and possessed the expertise required in arid regions. It was important to realize that the formation of a TRC and the employment of a consultant did not relieve the District of presenting various drafts of the manual to other interested parties. In fact, a Consulting Engineers Council subcommittee met independently of the TRC to review drafts of the manual.

It was also important that the consultant identify key items that are considered critical to the successful completion of the manual. For example, a thorough understanding of the existing practices in the area for hydrologic and hydraulic analysis is considered extremely important. It was mutually understood that people would resist drastic changes in such practices. Therefore, it was important for the consultant to understand that while new practices were introduced in the manual, some of the existing practices with minor modifications would still be allowed. Additional examples of such key items included adaptability of the manual to the existing policies, laws, and regulations; its compatibility with other Community's requirements; its ease of updating; its ease of understanding and use by both the engineer who is preparing drainage submittals and the reviewer who will approve the same submittals; the need to include in the manual numerous worked examples for different types of calculations; and others.

Nontechnical Issues

One of the most difficult sections of the manual to develop was the section on Drainage Planning and Submittal. This section, while the least technical, proved to be the most difficult due to the present written and unwritten requirements that each entity utilizes. The process of determining when a submittal is required and the content of the submittal was discussed throughout the development of the manual. The final manual contains submittal requirements that are uniform and accepted by all impacted entities.

Other important nontechnical issues included in the manual are sections that deal with drainage policy and drainage law. The drainage policy section was also discussed throughout the development of the manual due to the technical issues that were included in the manual which had a direct bearing on policy. The drainage law section was written by an attorney who was familiar with drainage issues in Nevada. The drainage law sections goes through a brief history of drainage law in Nevada, along with presenting recent court findings. The drainage law section, however, is not a blueprint for future law suits due to the general lack of cases heard and interpretations made by the Nevada Courts.

The issue of maintaining a balanced approach in development of different sections of the manual was of paramount importance. The need for strict criteria in certain areas was balanced by allowing for design flexibility in other areas, within acceptable limitations.

Technical Issues

The contents of the technical sections of the manual were each examined carefully to insure their applicability to the area's arid environment. The application of humid criteria would have accommodated the design flows but would have resulted in higher costs of the drainage improvements. These sections were thus prepared to deal with the many design aspects of arid area drainage including rainfall/runoff modeling; open channel, storm sewer, and storm inlet design; allowable street flow; culvert, bridge, and additional hydraulic structures design; detention pond sizing; erosion control criteria and facility design; and analysis criteria for development on alluvial fans. Each of these sections included criteria which is unique to arid area drainage analysis.

One of the more important sections of the manual dealt with hydrologic criteria. Extensive debate has occurred in the Clark County area over the proper design rainfall values to be used in drainage analysis. A review of these studies showed that much of the differences between different studies were due to the lack of adequate historic rainfall data on which design rainfall rates had to be established and that a single significant storm event occurring today could again change the design rainfall analysis. A more important feature of storm drainage design is to implement consistent data which will result in facilities which provide an equal level of protection. To avoid constant switching of rainfall data, the data presented in the criteria is recommended to be used for a minimum ten year period. The district has implemented a fast track schedule to install several rain gages in the Clark County area in hopes of obtaining detailed rainfall related data in the coming years. The ten year period should allow the District time to obtain and analyze this detailed data and prepare a statistically more precise set of rainfall data for use at that time.

Runoff analysis in the Clark County area has typically been prepared using several runoff methods. As a result, widely varying peak flow rates and volumes have been computed in many instances for the same drainage basin. The engineering community was reluctant, however, to limit all analysis to a single model realizing the need to not limit drainage design in the area to a limited number of firms, especially for the small, less risk developments (i.e. 1/4 acre parcels (0.10 hectares)). Therefore, an analysis was conducted to determine the range of drainage areas over which different models would typically produce comparatively similar results. The results showed that for small basins, all models could be used if consistent parameters associated with each model were adopted (i.e. time of concentration). Therefore the manual presents unified values and equations for the similar parameters and sets limits on the drainage basin size for the different models.

Other examples of specific criteria related to arid areas was to allow the occurrence of supercritical flow in open channels within given maximum permissible velocities. In this manner, economically feasible facilities could be built in the topographically steeper areas of Clark County.

Runoff in streets was also limited by using a velocity times depth or "vd" parameter. This limit more equally balanced the potential damage to streets and to the public safety with the economic need to use streets to convey a large portion of runoff in the urban areas. Storm sewers were also allowed to carry more flow than typically found in humid areas by allowing a maximum permissible velocity of 25 feet per second (7.6 meters per second) in reinforced concrete pipe. On-site detention was recommended for use in Clark County when existing downstream facilities are found to be inadequate. Regional detention would also be used to reduce peak flows from large drainage basins to maximize the conveyance capacity of existing and proposed conveyance facilities. Criteria was also provided to assist in the selection of subcritical and supercritical flow energy dissipators and debris structures for use in the Clark County area.

Finally, one of the least understood drainage problems facing developers and engineers in the Clark County area is the analysis and solutions for development on alluvial fans. Therefore, information was included in the manual to assist the engineer in identifying alluvial fans as well as to briefly demonstrate the effects of development on the properties of alluvial fan drainage.

Conclusion

A great deal of effort was placed on preparation of a hydrologic criteria and drainage design manual which would account for the complexities and differences in design in an arid environment without sacrificing the safety of the general public. The final manual presents a unified set of standards and hydrologic criteria for drainage design in an arid environment and addresses both the technical and nontechnical issues of drainage design and analysis. The use of the uniform standards in the manual will allow development in Clark County to proceed while obtaining consistent and unified drainage facilities for the economic benefit of Clark County residents. The completion and implementation of this manual provides the District with the necessary tool to tie together the regulatory and capital improvement programs in order to solve existing and future flooding problems in the Clark County area.

Urban Hydrology in the Desert, Antelope Valley, California

James C. Blodgett (Member, ASCE),[1] Iraj Nasseri (Member, ASCE),[2]
and Ann L. Elliott[1]

Abstract

A study of urban hydrology in Antelope Valley includes data collection, analysis of rainfall and runoff frequencies, and comparison of results from various rainfall-runoff models. This paper discusses only parts of the project that include data collection and frequency analyses.

Introduction

Storm runoff in drainage basins undergoing urbanization is a major environmental concern. Runoff volumes from newly urbanized drainage basins are significantly altered due to an increase in impervious areas. A reliable, regional method to predict storm runoff in drainage basins that will be urbanized is needed by land-use planning agencies in southern California in order to establish criteria for design of drainage facilities. Currently, these agencies use a variety of drainage-design methods that provide inconsistent estimates of runoff from basin to basin.

An analysis of the hydrology in Antelope Valley, California (Fig. 1), in the Mojave Desert, is the subject of a cooperative study being done by the U.S. Geological Survey and Los Angeles County Department of Public Works. The initial phase of the study, begun in October 1988, includes selecting and instrumenting nine representative drainage basins, measuring soil infiltration, and assembling historical rainfall and runoff data. Later phases of the project will include using synthesized flow records to calculate precipitation and runoff frequencies and determining which hydrologic factors are significant on the basis of model results. The final phase, planned to be completed by September 1992, will compare results from application of rainfall-runoff models by the U.S. Geological Survey and the Environmental Protection Agency, which simulate the actual physical processes occurring in the basin, with methods such as the Modified Rational and Soil Conversation Service models, using a common set of hydrologic data. This paper will address only the data collection and frequency analyses, because results from application of the models are not yet available.

[1]Hydrologists, U.S. Geological Survey, 2800 Cottage Way, Room W-2234, Sacramento, CA 95825.
[2]Head of Planning, Hydraulic/Water Conservation Division, Los Angeles County Department of Public Works, 900 South Fremont Ave., Alhambra, CA 91803.

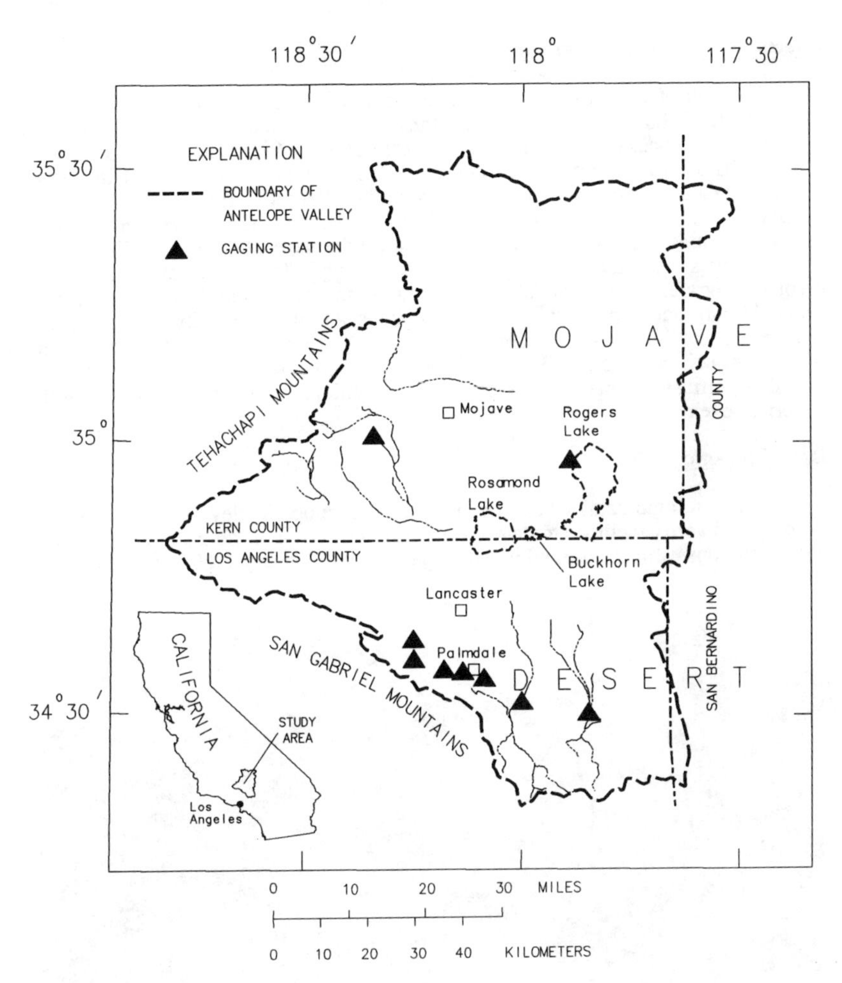

Figure 1. Location of Antelope Valley, California, and Gaging Stations

Geography

Antelope Valley is about 50 mi north of Los Angeles in the southwestern Mojave Desert (Fig. 1). The valley is a closed inland drainage basin covering about 2,400 mi^2. The Tehachapi Mountains, with elevations as high as 8,000 ft, are the northern and western borders and the San Gabriel Mountains, with elevations as high as 10,000 ft, are the southern border. Average elevation of the valley floor is about 2,500 ft. The present flood-control system is inadequate; for example, during March 1983 extensive overbank flooding occurred. Much of the valley is subject to floodflows, which follow unpredictable paths across the valley floor toward Rosamond, Buckhorn, and Rogers Lakes (Fig. 1).

Effects of Urbanization on Runoff

Much of the western and southern part of Antelope Valley, particularly along the foothills and on the alluvial fans, is being urbanized. The combined population of Lancaster and Palmdale has increased from 63,200 in 1970 to 118,000 in 1984 (U.S. Army Corps of Engineers, 1986) and to about 128,000 in 1989 (Lancaster Chamber of Commerce, oral commun., February 1990). Urbanization involves more than construction of buildings. It entails subdividing land parcels based on zoning requirements, altering existing drainage channels, grading land, constructing streets and buildings, and landscaping. Typical channel alteration involves regrading the channel bed, changing the alignment, and constructing channel linings, culverts, gutters, bridges, siphons, and flow-detention basins. Land preparation for building purposes involves leveling or benching the parcel and subsequent soil compacting (Fig. 2). The inevitable consequences of urbanization include significant changes in runoff magnitude, timing, and duration compared with historical events.

Data Collection

Hydrologic data, which are limited in Antelope Valley, are necessary to develop and verify methods to determine basin rainfall-runoff characteristics. Nine gaging stations were established in Antelope Valley in January 1989 to continuously

Figure 2. Land Preparation Prior to Construction of Homes as a Part of Urbanization in Antelope Valley, California

record streamflow and precipitation data (Fig. 1). The gaging station sites were selected to provide areal diversity in basin size, slope, exposure, soil types, and urbanization. Streamflow data from these stations are needed to calculate peak discharge, daily mean discharge, and flood hydrograph volumes. They are also needed for rainfall-runoff modeling.

Ten partial-record streamflow stations also were established to calculate peak discharges (not shown on Fig. 1). In addition, flows are measured at selected sites during floods to obtain downstream flow attenuation data.

The nine precipitation stations established for this study, others installed by Los Angeles County Department of Public Works, and the few hourly National Weather Service stations will provide the detailed records needed for modeling rainfall-runoff relations. Analysis of historical data from daily precipitation stations in the study area indicates wide areal variability in precipitation throughout the valley. Average annual precipitation ranges from 4 in. on the valley floor to 9 in. near the foothills and 20 in. in the mountains.

Soil-infiltration measurements, needed as input for rainfall-runoff models, were made in nine basins in autumn 1989. Soil samples from each measurement site were analyzed for particle-size distribution. Infiltration was measured with a portable double-cap infiltrometer, which was developed by Constantz (1983) as an alternative to a double-ring infiltrometer, for areas with limited access or with a scarcity of water. Several measurements were made in each basin to describe areal variation in soil infiltration rates. Infiltration rates in unpaved urbanized areas were found to be lower than in nonurbanized areas, mainly as a result of soil compaction.

Beginning in 1988, basin and stream channel characteristics of Antelope Valley were documented annually using aerial photographs. These photographs were used to identify land use, quantify areas of impermeable surfaces, and document significant landform and drainage channel changes with time.

Frequency Analyses of Precipitation and Streamflow

Mean precipitation at Palmdale, based on records from 1933 to date, is 7.5 in. with a standard deviation of the annual means of 4.23 in. The standard deviation was about 50 percent of the mean for most stations in the area. This indicates the need for records as long as 25 years in order to obtain a dependable value of the mean. The large degree of variability at Palmdale, shown in Fig. 3, indicates that during most decades since 1940, 7 of 10 years had annual precipitation below the mean. This suggests that most floods occur during years with above normal precipitation.

Maximum 24-hour precipitation at Palmdale shows a large variation about its mean of 1.4 in. (Fig. 4). Some years that had high 24-hour precipitation, such as 1971, did not have high annual precipitation; however, other years, such as 1941, 1944, 1952, and 1983 that had maximum 24-hour precipitation in excess of 2 in., were also years of high annual precipitation.

A frequency relation for annual 24-hour precipitation was developed by Miller and others (1973), on the basis of a regional analysis of climatic stations in the Western United States with records of 10 years or more. Another relation was

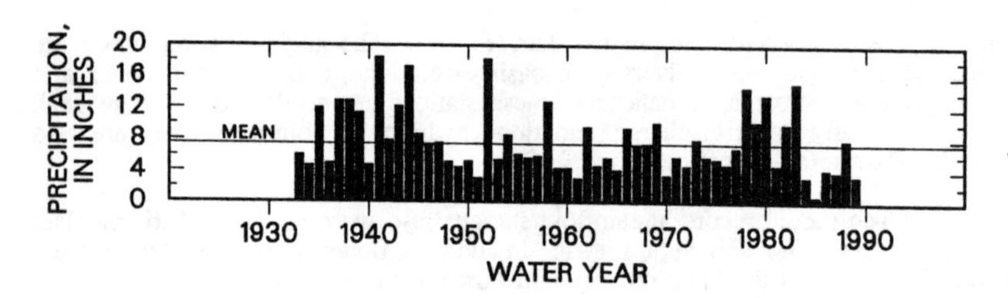

Figure 3. Annual Precipitation at Palmdale, California

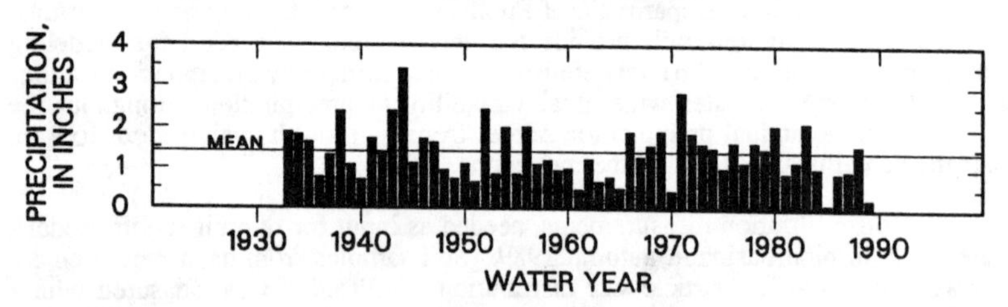

Figure 4. Maximum 24-Hour Precipitation at Palmdale, California

developed from the observed record at Palmdale, on the basis of the log-Pearson distribution (U.S. Water Resources Council, 1981) which is suitable for use in drier areas (Fig. 5). The relation based on the log-Pearson procedure gives values lower than that of Miller and others (1973) by 1.6 in. for a 100-year recurrence interval (Fig. 5). The recurrence interval of the 1944 storm could be estimated at 11 or 68 years, depending on which relation in Fig. 5 is used. This difference is attributed to the fact that the relation developed by Miller and others (1973) is regional, based on average data for several stations.

Large floods occurred in Antelope Valley in 1938, 1943, 1969, and 1978 (Fig. 6). The recurrence interval of the 1938 flood, discharge 8,300 ft^3/s on Big Rock Creek near Valyermo, is estimated to be 67 years, based on the log-Pearson frequency distribution (U.S. Water Resources Council, 1981) for the observed record of 66 years. The regional frequency relation developed by Waananen and Crippen (1977) indicates the recurrence interval of the 1938 flood to be over 100 years.

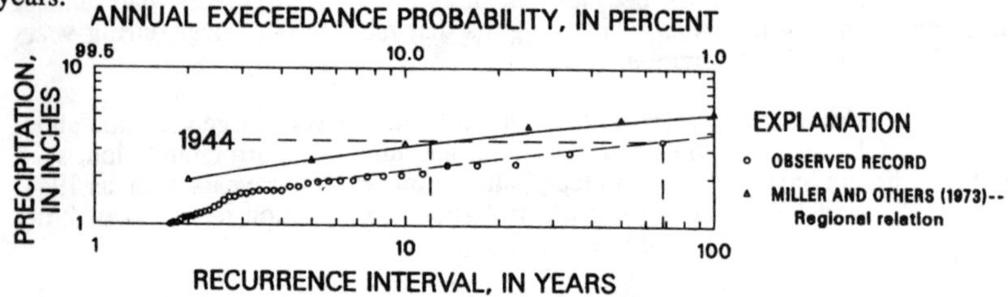

Figure 5. Frequency Curves of Annual Maximum 24-Hour Precipitation at Palmdale, California

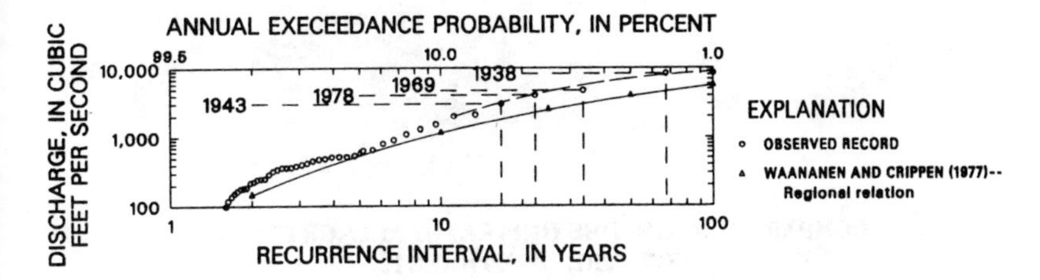

Figure 6. Frequency Curves of Peak Discharge at Big Rock Creek Near Valyermo, California

The differences between the regional and at-site frequency relations of Figs. 5 and 6 indicate the need for collection of more data in the valley. Development of relations between precipitation and runoff and their corresponding frequency as a part of the effort to evaluate various rainfall-runoff models will be an important part of this study.

Appendix I. References

Constantz, J. (1983). "Adequacy of a compact double-cap infiltrometer compared to the ASTM double-ring infiltrometer." National Conference on Advances in Infiltration, Chicago, IL, 1983, Proceedings, 5 p.

Miller J.F., Fredrick, R.H., and Tracey, R.J. (1973). "Precipitation-frequency atlas of the Western United States." National Oceanic and Atmospheric Administration, vol. XI - California, Atlas 2, 71 p.

U.S. Army Corps of Engineers. (1986). "Antelope Valley California, Hydrologic investigation for feasibility studies of the Los Angeles County Department of Public Works Master Drainage Plan." U.S. Army Corps of Engineers, Los Angeles District, 30 p.

U.S. Water Resources Council. (1981). "Guidelines for determining flood flow frequency." U.S. Water Resources Council, Bulletin 17B of the Hydrology Committee, Washington, D.C., 28 p.

Waananen, A.O., and Crippen, J.R. (1977). "Magnitude and frequency of floods in California." U.S. Geological Survey Water-Resources Investigations Report 77-21, 96 p.

Appendix II. Conversion Factors

foot (ft)	0.3048	meter
cubic foot per second (ft^3/s)	0.02832	cubic meter per second
inch (in.)	25.4	millimeter
mile (mi)	1.609	kilometer
square mile (mi^2)	2.590	square kilometer

COMPARISON OF DESIGN RAINFALL CRITERIA FOR THE SOUTHWEST

George V. Sabol[1]
Kenneth A. Stevens[1]

Abstract

The design of drainage and flood control facilities or the management of floodplains for alluvial fans is extremely sensitive to the design rainfall criteria that is used as input to the hydrologic model. The results of a study using several combinations of design rainfall criteria in deterministic rainfall-runoff models of watersheds is presented. The results indicate that some of the more commonly used design rainfall criteria may not adequately represent the rainfall characteristics of the southwest. It is concluded that design rainfall criteria for the southwest must represent both the spatial and temporal characteristics of regional severe storms if valid models for use on alluvial fans are to be developed and used.

Introduction

Rainfall induced floods are the result of a severe storm over the contributing watershed. Often, in flood hydrology, these storms are classified as either local storms or general storms. Local storms are typically short duration, high intensity rainfalls of limited areal distribution. They often are of 1-hour duration or less and are virtually always less than 6-hours unless associated with a larger storm system. In the southwest, they often are less than 25 square miles with 100 square miles as a large local storm. The size limit for an independent local storm is usually considered to be less than 500 square miles. General storms are large systems that are often associated with frontal activity. General storms are lower intensity, longer duration

1 George V. Sabol consulting Engineers, Inc., 1351 East 141 Ave., Brighton, CO 80601

storms that cover very large areas. In the southwest, the local storm is usually the critical design event except for large watersheds and major watercourses.

Since the majority of drainage and flood control facilities are for smaller drainage areas, there is the need to adequately define the spatial and temporal distribution of local storms. Alluvial fans and alluvial plains are common landforms in the southwest that are undergoing development. These watersheds are usually small and therefore the local storm would constitute the critical flood producing event.

Design rainfall criteria are often contained in regional drainage design criteria, but often such criteria are not available and the hydrologist or engineer must develop or adopt prudent design rainfall criteria. Often the criteria contained in regional drainage design criteria or the criteria that is adopted is from generalized relations that have been published by various federal agencies. Such generalized criteria may not have been developed for severe local storms in the southwest and the use of such criteria could result in overdesign or underdesign. No studies are known to have been performed or published that compare various design rainfall criteria for local storms in the southwest. The four selected design rainfall criteria are summarized in Table 1.

These criteria have been compared only at the 100-year return period using the rainfall depth-duration statistics for Phoenix, Arizona. These rainfall depths for durations from 1-hour to 24-hours were obtained from NOAA Atlas 2 (Miller and others, 1973), and the rainfall depths for durations less than 1-hour were derived by revised short-duration rainfall ratios by NOAA (Arkell and Richards, 1986).

The comparisons have been made by modeling eight synthetic watersheds using the HEC-1 Flood Hydrology Program (U.S. Army Corps of Engineers, 1988). The eight synthetic watersheds vary in size from 0.1 square mile to 500 square miles, and the watershed characteristics have been selected to be representative of natural (undeveloped) watersheds that typically occur in Arizona and much of the southwest. The SCS Dimensionless unit hydrograph was used for all watersheds, and the Green and Ampt infiltration equation with a surface retention loss was used based on information in the Maricopa County Hydrologic Design Manual.

The equivalent uniform depth of rainfall for each of the synthetic watersheds using the four design rainfall criteria are shown in Table 2. The difference in rainfall depths are due to two reasons. First, two of the distributions (HYP and SCS) are for 24-hour durations and the other two distributions (HRM and MC) are for 6-hour durations. Second, different depth-area reduction curves have been used as indicated in Table 1. From Table 2 it is noted that there is very little reduction in rainfall depth using the depth-area reduction curve from NOAA Atlas 2 (HYP and SCS criteria). The areal distribution for local storms in the southwest is much more limited than the NOAA Atlas 2 depth-area reduction curve represents. Both the HMR and MC have fairly comparable rainfall depths although the depth using HMR diminishes more quickly with increasing area than the MC criteria.

The rainfall excess from the HEC-1 models is shown in Table 3. The rainfall excess is a function of both the method to calculate rainfall losses and the temporal distribution of the rainfall itself. Several facts are observed from Table 3: First, using the SCS criteria, there is little difference in rainfall excess with size of drainage area. This is not reasonable for local storms in the southwest. Second, both the HYP and SCS criteria result in similar estimates of rainfall excess for watersheds larger than 100 square miles while the HYP results in greater rainfall excess for smaller watersheds. This is because of the greater rainfall intensities for short durations in the hypothetical distribution. Third, both the HMR and MC criteria result in similar rainfall excess as the HYP criteria for watersheds smaller than 1 square mile. Fourth, the rainfall excess using HMR criteria diminishes a little quicker than the MC criteria for larger watersheds. Both the HMR and MC criteria result in reduction of rainfall excess with increasing watershed area as would be anticipated for local storms on watersheds in the southwest.

Table 4 shows the maximum rainfall intensity for the computation interval that was used. These are areally averaged intensities and obviously are much greater for small areas with small computation intervals than large areas where larger computation intervals are used. Several facts are observed from Table 4: First, the HYP criteria has the highest, short-duration rainfall intensity. This is because depth-duration statistics are input for 5 minutes and 10 minutes, whereas the shortest interval of rainfall input that was digitized from the distributions for the other three criteria is

15 minutes. Second, the areally averaged maximum rainfall intensities for the SCS criteria are virtually uniform for all watersheds from 0.1 to 500 square miles. This is not reasonable for local storms. Third, the HMR criteria results in somewhat higher rainfall intensities than the MC criteria for small watersheds (less than 10 square miles), and the intensities are about the same for areas larger than about 50 square miles. Fourth, all four criteria result in similar rainfall intensities in the range of 25 to 100 square miles.

Table 5 shows the peak discharge for each synthetic watershed from the HEC-1 models. Notice that for both the HYP and SCS criteria that the peak discharges continually increase for increasingly larger watersheds. For both the HMR and MC criteria, the peak discharges reach a maximum for watersheds between 25 and 100 square miles. That size is a practical limit of the rainfall excess producing portion of local storms, and reduced peak discharges past 100 square miles is the result of areally averaging the storm rainfall over the entire watershed.

CONCLUSIONS

1. The depth-area reduction curve in NOAA Atlas 2 is inappropriate for local storms in the southwest.
2. The hypothetic distribution with the NOAA Atlas 2 depth-area reduction curve will probably result in overestimation of design discharges for watersheds larger than about 10 square miles.
3. The SCS Type II distribution with the NOAA Atlas 2 depth-area reduction curve will probably result in underestimation of design discharges for watersheds smaller than 25 square miles and overestimation of design discharges for watersheds larger than 100 square miles.
4. The procedure in Hydrometeorologic Report No. 49 can probably be used to develop reasonable design rainfall criteria for watersheds smaller than 25 square miles.
5. The procedure for developing local storm design rainfall criteria as contained in the Maricopa County Hydrologic Design Manual results in flood discharges that increase with increasing area up to about 100 square miles and then decreasing discharges for areas larger than about 100 square miles.
6. Design rainfall criteria that are based on the analysis of regional data and historic storms are superior to generalized criteria. Both the HMR and the MC criteria were developed in this manner.

7. Design rainfall criteria that are based on the analysi of an appropriate regional, severe storm will probably yield more reliable flood estimates than either genera ized criteria or regionalized criteria. The MC criter fits this conclusion. Specific design rainfall criter should be developed based on historic storms when data are available.

TABLE 1
Comparison of rainfall **depths**

Rainfall Criteria	Equivalent Uniform Depth of Rain, in inches — Area, in square miles							
	0.1	1	10	25	50	100	250	500
(1)	(2)	(3)	(4)	(5)	(6)	(7)	(8)	(9)
HYP	3.93	3.92	3.88	3.82	3.74	3.66	3.58	3.58
SCS	3.93	3.92	3.88	3.82	3.74	3.66	3.58	3.58
HMR	3.25	3.25	2.85	2.59	2.33	2.01	1.49	1.13
MC	3.22	3.22	3.03	2.87	2.77	2.58	2.22	1.84

TABLE 2
Comparison of rainfall **excesses**

Rainfall Criteria	Rainfall Excess, in inches — Area, in square miles							
	0.1	1	10	25	50	100	250	500
(1)	(2)	(3)	(4)	(5)	(6)	(7)	(8)	(9)
HYP	1.81	1.80	1.70	1.56	1.39	1.19	1.04	1.02
SCS	1.22	1.22	1.19	1.16	1.12	1.09	1.05	1.05
HMR	1.62	1.62	1.15	.86	.60	.36	.03	0.0
MC	1.70	1.58	1.19	.94	.80	.62	.34	0.1

TABLE 3
Comparison of rainfall **intensities**

Rainfall Criteria	Maximum Rainfall Intensity, in inches/hour — Area, in square miles							
	0.1	1	10	25	50	100	250	500
(1)	(2)	(3)	(4)	(5)	(6)	(7)	(8)	(9)
HYP	9.0	8.4	5.5	3.3	2.9	2.5	2.1	2.0
SCS	3.0	3.0	3.0	2.9	2.8	2.8	2.7	2.7
HMR	7.4	7.4	5.4	2.9	2.1	1.8	1.1	0.8
MC	5.8	4.8	2.6	2.0	1.8	1.5	1.1	0.8

TABLE 4
Comparison of peak **discharges**

Rainfall Criteria	Peak Discharge, in cfs							
	Area, in square miles							
	0.1	1	10	25	50	100	250	500
(1)	(2)	(3)	(4)	(5)	(6)	(7)	(8)	(9)
HYP	260	1,640	6,060	8,220	10,600	14,200	18,500	26,500
SCS	142	1,075	4,260	6,140	8,560	12,900	18,700	27,300
HMR	267	1,560	4,080	4,530	4,520	4,250	600	0
MC	250	1,370	4,050	4,960	6,050	7,270	6,140	2,300

REFERENCES

1. Arkell, R.E., and Richards, F., 1986, Short duration rainfall relations for the western United States: conference on Climate and Water Management - A Critical Era, Ashville, N.C., published by Amer. Meteorological Society, Boston, Mass.
2. Flood Control District of Maricopa County, 1989, Maricopa County Hydrologic Design Manual (draft), Phoenix, Arizona.
3. Hansen, E.M., Schwarz, F.K., and Riedel, J.T., 1984, Probable maximum precipitation estimates, Colorado River and Great Basin Drainages: Hydrometeorological Report No. 49, National Oceanic and Atmospheric Administration and U.S. Army Corps of Engineers.
4. Miller, J.F., Frederick, R.H., and Tracey, R.J., 1973, NOAA Atlas 2, Precipitation-frequency atlas of the western United States, Volume VIII, Arizona, National Weather Service, U.S. Dept. of Commerce, Silver Springs, MD.
5. Soil Conservation Service, 1973, A method for estimating volume and rate of runoff in small watersheds: SCS-TP-149.
6. U.S. Army Corps of Engineers, 1974, New River and Phoenix City Streams, Design Memorandum No. 2, Hydrology, Part 1: Los Angeles District, California.
7. U.S. Army Corps of Engineers, 1982, Hydrologic analysis of ungaged watersheds using HEC-1: Hydrologic Engineering Center, Davis, California.
8. U.S. Army Corps of Engineers, 1988, HEC-1 Flood Hydrograph Package (1988 Version): Hydrologic Engineering Center, Davis, California.

INCLUSION OF WETLANDS IN DESIGNATED FLOODWAY STUDIES

Lester S. Dixon[1], M. ASCE and H. Brooks Carter[2]

Abstract

The purpose of this paper is to propose a change in the traditional methods and procedures used by hydrologist and hydraulic engineers to define floodways and floodway fringe areas. We propose that designated floodways be adjusted to encompass wetlands, resulting in better protection of our nation's diminishing wetlands and a reduction in development problems encountered on a regular basis across the nation due to permit requirements pursuant to Section 404 of the Clean Water Act.

Introduction

On a daily basis, developers who have spent large sums of money for flood plain land, engineering, and authorizations find themselves in a position of having to scale down their project considerably to the chagrin of investment partners or absorb the costs to date due to abandonment of the project. What is the cause of this frustrating situation? Often it's the discovery at a late date that the property is wetland regulated by the U.S. Army Corps of Engineers (Corps).

The purpose of this paper is to propose a change in the traditional methods and procedures used by hydrologists and hydraulic engineers to define floodways and floodway fringe areas. We propose that designated floodways be adjusted to encompass wetlands, resulting in better protection of our Nation's wetlands and a reduction in development problems.

Background

Well planned construction within the 100-year

---

[1]Chief, Planning Division, U.S. Army Corps of Engineers, Pittsburgh District; and adjunct Assistant Professor University of Utah.
[2]Chief, Utah Regulatory Office, U.S. Army Corps of Engineers, Sacramento District.

flood plains is opposed by those who say the natural values are too valuable to lose, and promoted by those who argue that the land is too valuable not to develop.

Two national programs greatly influence development within the 100-year flood plain. Under the National Flood Insurance Program, the 100 year flood plain is divided into a floodway and a floodway fringe. The floodway must be kept free of encroachment in order to carry the 100 year flood. Development may occur within the floodway fringe provided that increases in flood height for the 100 year event are limited to 1.0 foot.

The Corps regulates the discharge of dredged or fill material into waters of the United States, which includes wetlands, under Section 404. In order to protect the values of this diminishing resource, strict regulations have been promulgated which prohibit the placement of fill material into wetlands if a practicable alternative exists, and for activities which are not water dependent (i.e. they don't have to be constructed in wetlands to fulfill their basic purpose) guidelines written by the EPA instruct the Corps to assume that other practicable alternatives exist unless clearly demonstrated otherwise.

Recommendation

Designation of a floodway fringe by local communities implies that an area is available for development and construction, but the existence of wetlands may render development unfeasible or greatly reduce its scope.

Therefore, we propose that hydrologists and hydraulic engineers adjust floodway boundaries to include wetlands within the 100-year flood plain. Hogan (1989) described a typical procedure for determining the floodway boundaries when using the Corps of Engineers HEC-2 water surface profile computer program. One of the steps in the procedure says that boundaries should be adjusted to be consistent with local needs. Therefore, with appropriate local official approval, this procedure could be amended to include, "Identify important wetlands within the 100-year flood plain and adjust the floodway boundary to encompass these areas." This proposal is simply illustrated in the Figure.

Conclusion

Hydrologic studies to define floodways and floodway fringe areas are an important tool in the management of flood plains. A minimum standard for defining floodways has been developed under the National Flood Insurance Program. Implementation of more stringent standards is a local community decision. By proposing the inclusion of wetlands in the determination of designated floodways, hydrologists and hydraulic engineers have an opportunity to contribute to the protection of our Nation's wetlands and to assist in better defining floodway fringe areas available for development.

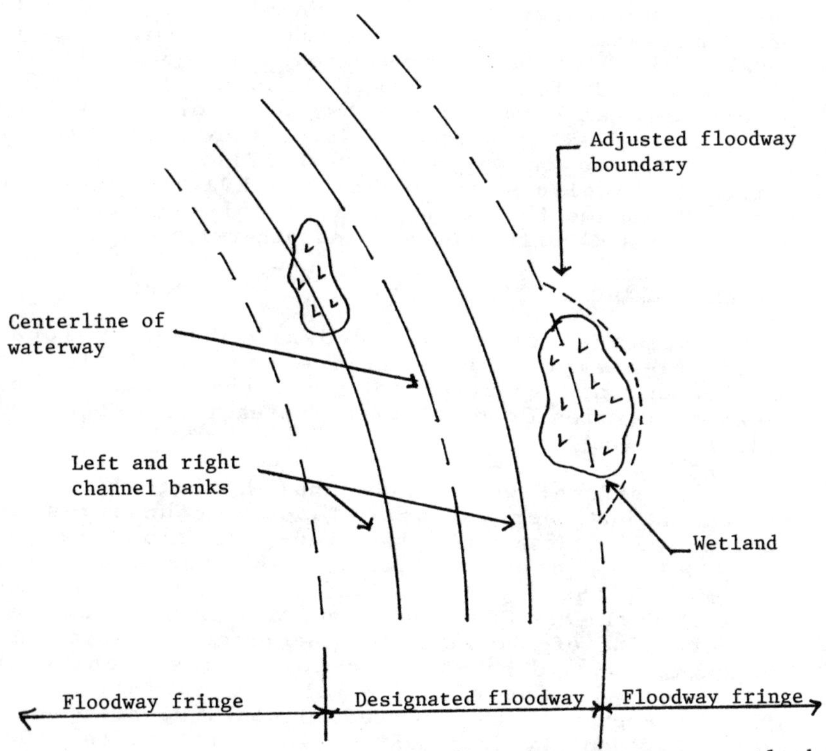

Example of a floodway boundary adjusted to encompass a wetland.

APPENDIX. - REFERENCES

Hogan, D.G., "Computer Assisted Floodplain Hydrology and Hydraulics Featuring the U.S. Army Corps of Engineers' HEC-1 and HEC-2 Software Systems," 1st ed., McGraw-Hill Publishing Company, New York, 1989

National Flood Insurance Act, 1 August 1968, Public Law 90-448

Clean Water Act, 27 October 1977, Public Law 95-217

THE COLORADO RIVER WATER RIGHTS: BACK TO THE FUTURE

Susan M. Trager, Senior Attorney*
Victor R. Sofelkanik, Associate*

A discussion about tomorrow's projected water shortages in California and efforts to meet the anticipated demands through the Colorado River supply requires at least a brief review of the law of the River and California's Seven-Party Agreement. The point of origin for an understanding of the principles, policies, and politics controlling the allocation of the water of the Colorado is the collection of treaties, compacts, and acts of Congress and state legislatures governing use of the River.

THE COLORADO RIVER COMPACT

The Colorado River Compact of 1922[1] ("Compact") enacted August 19, 1921, divided waters of the Colorado River system between the Upper Basin and Lower Basin. The Compact is the cornerstone of the law of the River. The Compact's purpose was to establish beneficial uses, promote interstate comity, and establish and improve the agricultural and industrial development of the River.

The Compact defined the "Upper Basin" to include those parts of Arizona, Colorado, New Mexico, Utah, and Wyoming within and from which waters naturally drain into the Colorado River System above Lee Ferry. "Lower Basin" states include those parts of Arizona, California, Nevada, New Mexico, and Utah within and from which waters naturally drain into the Colorado River System below Lee Ferry. The Basins are divided at a point one mile below the mouth of the Paria River, at Lee Ferry.[2]

The Upper and Lower Basins are each apportioned 7.5 million acre feet per year in perpetuity, for beneficial consumptive use. Additionally, the Lower Basin was apportioned the right to increase its consumption by one

* Law Offices of Susan M. Trager
Irvine, California

million acre-feet per year, if water is available. The Upper Basin is also obligated to release 75 million acre-feet of water every continuing 10-year period at Lee Ferry. Neither Basin may hoard or waste water. Water to satisfy Mexican treaty obligations would come from surplus waters, and should a deficiency occur, the allotment is to be made up equally by both Basins.

The Compact was not ratified by all parties until The Boulder Canyon Project Act of 1928 provided a ratification procedure approved by six states. The Compact went into effect in 1929, but Arizona did not ratify the Compact until 1944.[3]

THE BOULDER CANYON PROJECT ACT OF 1928

The Boulder Canyon Project Act of 1928[4] ("Act") authorized the construction and operation of a massive storage and hydroelectric project now known as Hoover Dam and the All-American Canal in Imperial County, California. The Act's purpose was to control flooding, regulate flows and improve navigation on the river.

The Act reserved Lower Basin water for Arizona and Nevada, and provided protection to the Upper Basin from unbridled water development benefitting the Lower Basin. The Act also provided that "no person shall have or be entitled to have the use for any purpose of water stored. . . [behind Hoover Dam] except by contract. . . ", as authorized by the Act.[5]

SEVEN-PARTY AGREEMENT

The Secretary of the Interior requested that California establish priority among its major users before execution of the water delivery contracts.[6]

The "California Seven-Party Agreement", effective 1931, accomplished this allocation. The first three priorities went to the major agricultural users: Palo Verde Irrigation District; Yuma Project of the United States Bureau of Reclamation; and Imperial Irrigation District. The next two priorities went to the Metropolitan Water District of Southern California.

The total entitlement to the first four priority holders is 4.4 million acre-feet, the total allocated to California by the California Limitation Act.[7]

THE COLORADO RIVER STORAGE PROJECT ACT

The Colorado River Storage Project Act[8], enacted in 1956, authorized development of water resources in the Upper Basin. This multi-purpose, comprehensive plan included irrigation projects, hydroelectric plants, accompanying facilities, and created an Upper Basin fund. The fund, established by Section 5 of the Act, was intended to repay costs of operations and maintenance of the facilities. Monies for the fund are collected from operation at the storage projects.

ARIZONA V. CALIFORNIA[9]
373 U.S. 546 (1963)

The inability of the three Lower Basin States to agree on the sharing of the Colorado River Compact water prompted Arizona to file suit in the Supreme Court.

The Supreme Court decision determined whether state or federal law applies in apportionment and delivery of Lower Basin waters. The Court held that the moment water passes below Lee Ferry it becomes water controlled by the United States and subject to the Boulder Canyon Project Act.

PRIVATE WATER TRANSFERS

Lower States Basin water is over allocated. This has prompted many proposals to transfer Upper Basin water, to thirsty Lower Basin purchasers. Of these, the "Galloway Proposal" is probably the most notorious.

Galloway was a scheme by an entrepreneurial group which proposed to deliver waters impounded in the Upper Basin to the San Diego County Water Authority.[10] The source of the proposed water was the White River and waters stored behind a dam on the Yampa River.[11] The proposal also required a "wheeling"[12] agreement with the Metropolitan Water District of Southern California for the use of its delivery facilities to transport the water to San Diego County.

The Galloway Proposal was doomed to failure because the "available" water in the Upper Basin, although subject to loss due to nonuse, was also subject to the priorities in effect among the Lower Basin States and among the parties to the Seven-Party Agreement. The Galloway Proposal was the first in a series of schemes highlighting some of the unanswered questions of how the resources of the Colorado might be reallocated. The legal obstacles to this type of proposal are numerous.

A major obstacle to "wheeling" water by private transferrors arises in connection with gaining access to public transfer facilities. Some claim that "wheeling" is already possible.[13] The California legislature has begun to address the issue but uncertainty, arising from unanswered questions, effectively blocks transfers for more parochial reasons.

PUBLIC WATER TRANSFERS

In California, agricultural users are generally regarded by municipal and industrial users as the water source of the future. Two parties to the Seven-Party Agreement, the primarily agricultural Imperial Irrigation District ("IID") and the primarily municipal and industrial Metropolitan Water District of Southern California ("MWD"), have attempted to fashion an intrastate water sales agreement.

This agreement involves the financing of conservation methods by MWD. Water is to be conserved by lining the All-American canal and by improving gate techniques. The water conserved would then be available to MWD to sell to its coastal customers.

It has long been recognized, as far back as the Special Master Report in Arizona v. California, that a substantial quantity of Colorado River water was lost due to ineffective or non-existent conservation methods. According to a recent report by the Environmental Defense Fund, the California Department of Water Resources and the U.S. Bureau of Reclamation estimate that as much as 438,000 acre-feet of water per year is lost to the Salton Sea and dissipates through evaporation. The remainder seeps into the groundwater through the All-American Canal.[14]

The concept of the selling of water by one agency to another is referred to generally as "water marketing." Although this concept is not new, it is currently an attractive topic in California at water agency conventions, and evidence on the topic has been presented to the State Water Resources Control Board in the ongoing Bay-Delta hearings. Water transfers (and "marketing," which relies on the ability to transfer) have been facilitated by the California legislature. Recent amendments to the Water Code[15] include conservation within the definition of beneficial use, and remove disincentives to the use of reclaimed water.[16]

The incentives behind water marketing are obvious. It is much less expensive in some instances to obtain water from existing sources than to create new ones.

Conserving and marketing water could eliminate the need to construct additional facilities, resulting in reduced capital facilities costs and, some assume, reduced environmental damage.[17]

With respect to the MWD/IID agreement, some commentators hold the view that the agreement conflicts with the provisions of the Seven-Party Agreement.

Under the Seven-Party Agreement, any water not used by one agency flows to the agency next in priority. Many commentators are of the view that the water conserved by the MWD/IID Agreement should flow to the next priority user in the Seven-Party Agreement, Palo Verde. Any water which the next priority user could not put to reasonable and beneficial use would then flow to MWD, the holder of the Fourth Priority and Fifth Priority.

An additional problem with the MWD/IID Agreement, apart from bypassing the priority use of the water, is its apparent conflict with Arizona v. California. The MWD/IID Agreement usurps the Watermaster authority provided by the Boulder Canyon Project Act. California law appears to allow the parties to act without the approval of higher priority parties.

CONCLUSION

However flawed, the MWD/IID Agreement is likely to be implemented, and will serve as a model for future transfers. It is but one chapter in the history of this great river which provides water and power to nearly 14 million people. As the resources of the River become more precious, the need for resolution of the issues involving water conservation, salinity control, transfers, and marketing will become increasingly important.

FOOTNOTES

[1] Colorado River Compact, 43 U.S.C. §617-618 (1982 and Supp. III 1985).

[2] M. Nathanson, Updating The Hoover Dam Documents. U.S. Gov't Printing Office, Denver, at 4 (1980).

[3] W. Abbott, California Colorado River Issues, 19 Pac. L.J. 1391, at 1395, fn. 12 (1988).

[4] Boulder Canyon Project Act, 43 U.S.C. §617-618 p (1982 and Supp. III 1985).

[5] M. Nathanson, supra at 5.

[6] M. Nathanson, supra at 8.

[7] California Limitation Act, 1929 Cal.Stat. Ch. 16, 48th Sess; Statutes and Amendments to Codes (1929).

[8] Colorado River Storage Project Act, 43 U.S.C. §1501-1556 (1982 and Supp. III 1985).

[9] Arizona v. California, 373 U.S. 546 (1963), 375 U.S. 892 (1964); 376 U.S. 340 (1964).

[10] C. Boronkay, Water Marketing, Cal.St.B.J. Real Prop., Fall 1986, Vol. 4, Number 4, at 27.

[11] W. Abbott, supra at 1411.

[12] The transportation of water, by man, through natural and artificial means.

[13] C. Boronkay, supra at 27.

[14] Environmental Defense Fund, Trading Conservation Investments for Water, at v, (1983).

[15] California Water Code §§1010 and 1011.

[16] S. Hori, Water Marketing, Cal. St. B.J. Real Prop., Fall 1986, Vol. 4, Number 4, at 27.

[17] T. Graff, Water Marketing, Cal.St.B.J. Real Prop., Fall 1986, Vol. 4, Number 4, at 27.

California-Nevada Water Marketing Issues

Jeanine Jones, AM, ASCE*

Introduction and Setting

Water marketing and water transfers are being promoted as one solution to the problem of continued growth in water demand without concomitant political and economic support for the construction of new water development projects. Water marketing becomes more complex when interstate issues are considered. Northern California and Nevada share several interstate ground water and surface water basins. Increasing competition among Nevada water users for the limited supply available in this arid region has led to the proposal of varied water transfer projects. These projects entail transfers of water from rural, low population density areas in California to rapidly urbanizing areas in Nevada.

In general the California side of the stateline is water rich but sparsely populated, while the reverse is true on the Nevada side. Figure 1 shows several areas where water marketing projects have been proposed. Locations considered include interstate rivers -- the Truckee and Carson Rivers -- and ground water basins bordering the stateline -- Surprise Valley, Sierra Valley/Long Valley, and Honey Lake Valley. Although the westernmost portion of this area lies within California's Sierra Nevada mountains, the majority of the region falls within the arid basin and range geomorphic province, and the entire area is a region of interior drainage. Thus the Truckee River flows from Lake Tahoe to its terminus at Pyramid Lake, the Carson River flows from California to a terminal sink in

* Senior Water Resources Engineer, California Department of Water Resources, 3251 S Street, Sacramento, CA 95816

Nevada, and Honey Lake is the terminus of the Susan River. Average annual precipitation is up to 40 inches per year near Lake Tahoe, but drops off rapidly eastward with Reno's average annual precipitation being only 7 inches per year.

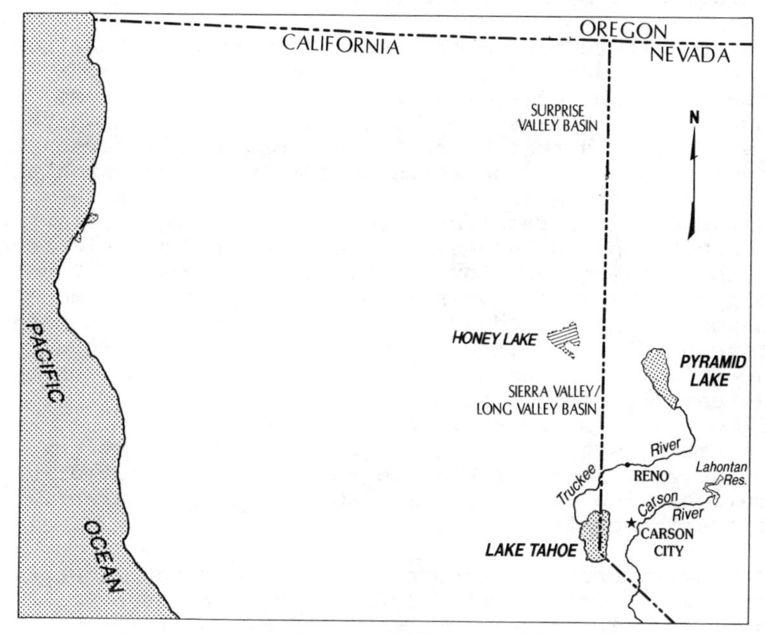

Figure 1. Location Map

Legal and Institutional Framework

Management of water resources in an interstate setting is complicated by the disparity of water rights laws in California and in Nevada. California's water rights laws deal with the administration of surface water rights based on riparian and appropriative (for rights initiated after 1914) doctrines. The California Water Resources Control Board does not regulate transfers of pre-1914 rights. Certain surface water rights transfers in California must take into account state statutes covering area of origin and county of origin protection provisions. California lacks a statewide regulatory process for ground water rights, although state law allows the legislative creation of ground water management districts which can be empowered to control ground water extraction within their boundaries. Nevada, in contrast, administers both ground and surface water rights under the prior appropriation doctrine. The Nevada State Engineer can regulate extraction from ground water basins, and normally limits extractions to the perennial yield of a basin. Both surface and ground water transfers are administered by the State Engineer.

The interstate nature of these water resources brings with it a federal role. A key element in interstate water transfer considerations is the U. S. Supreme Court's 1982 Sporhase v. Nebraska decision, in which the court held that water is an article of interstate commerce and that one state generally may not prohibit the export of water to another state. Interstate compacts or congressional legislation, however, may prohibit exports because Congress has given its consent to the prohibition (for example the Yellowstone River Compact). Surface water rights on the Truckee and Carson Rivers are adjudicated under three federal court decrees. Sale or transfer of a decreed right requires the approval of the court having jurisdiction over the decree. Settlement of both rivers' interstate allocations is presently being attempted in negotiations embodied in congressional legislation. Senate Bill 1554, if enacted as presently drafted, could provide some criteria regulating water transfers across the stateline.

The environmental review process for a water rights transfer across the stateline is another issue to consider. The requirements of the National Environmental Policy Act (NEPA) are clearly applicable if a federal agency is involved in the sale or transfer of a water right. Likewise in California the involvement of a state or local agency can trigger the California Environmental Quality Act (CEQA) review process. The State of Nevada, however, has no direct counterpart in its statutes to CEQA or NEPA.

Surface Water Transfers

The Carson and Truckee Rivers have been involved in litigation since the turn of the century, ultimately resulting in three federal court adjudications of water rights -- the Alpine Decree on the Carson River, the Orr Ditch Decree on the Truckee River, and the Sierra Valley Decree on a Truckee River tributary. The Alpine Decree adjudicates water rights among upstream California users and downstream Nevada users; the majority of the water involved is dedicated to the

Newlands Project, a U. S. Bureau of Reclamation agricultural project located at the river's terminus in Nevada. The Alpine Decree permits sale or transfer of adjudicated rights, and also provides for the conversion of the consumptive use portion of an agricultural right to municipal use. While some growth in the conversion of agricultural rights to municipal use is expected in the Carson City area, most transfer activity now is focused on acquiring agricultural rights to sustain wetlands in the Lahontan Valley adjacent to the Newlands Project. The Nevada Legislature has already provided funding for some transfers to the wetlands; more funding for transfers would be provided by the passage of S 1554. The Orr Ditch Decree on the Truckee River adjudicates only surface water uses in Nevada; California water users were not parties to the decree. The largest water users on the river are the rapidly growing Reno metropolitan area and Pyramid Lake, both located in Nevada. The Pyramid Lake Paiute Tribe has sought to obtain more water for the lake in a series of legal actions intended to reverse a decline in lake levels and to support endangered and threatened lake-dwelling fish.

One of the more difficult aspects in today's negotiations on the interstate portions of S 1554 is an attempt to incorporate provisions on water transfers, whether interstate or intrastate, into the bill. Some Nevada interests, for example, wish to preserve their ability to acquire an out-of-basin diversion of a pre-1914 appropriative right confirmed in the Sierra Valley Decree. This transfer could physically be accomplished by terminating the existing diversion and allowing the water to remain in the Truckee River and to flow to Nevada. While this right is owned by private individuals who are shareholders in a mutual water company, local government interests in California tend to oppose its transfer because of the transfer's possible impacts on Sierra Valley's agricultural lifestyle. On the Carson River, in contrast, the Alpine Decree already provides a mechanism for the voluntary interstate transfer of decreed rights, and some water users there appear to be receptive to the bill's continuing this practice for voluntary transfers. The federal government will become a significant water marketer on this river if the bill is enacted with its authorization to spend $16 million on acquisition of rights for Lahontan Valley wetlands.

Ground Water Transfers

Several ground water marketing projects have been proposed which would extract ground water from northern Nevada basins and transport it by pipeline to supply the metropolitan Reno area. Nevada interests proposed the Silver State project in the 1980s, a staged project to import about 63,000 acre-feet per year of water from six extraction areas extending north from Reno to the Surprise Valley Basin near the Oregon border; four of these extraction areas involved interstate basins or basins adjacent to the stateline. The Silver State project was envisioned as a major water development scheme for the Reno area, and included over 300 miles of pipeline, associated pumping stations, 40-50 high capacity production wells, and terminal storage facilities in Reno. Although the Silver State project did not go forward in its entirety, a portion of the project is being pursued today in the form of the Truckee Meadows project. This effort is backed by a Nevada public-private partnership and proposes to supply the Reno area with ground water exported from Honey Lake Valley, an interstate basin located just north of Reno. The amount of water to be extracted has not been finalized; various interest groups are considering amounts on the order of 10,000 to 20,000 acre-feet per year.

The importation projects studied to date have served to crystallize opposition to Reno-area water transfers among the rural California counties along the stateline. Although Nevada water developers are required to apply to the State Engineer to purchase or obtain ground water rights on the Nevada portion of the basins, there is no corresponding statewide requirement in California. Local officials in California often feel that this lack of control over ground water extraction coupled with the absence of a CEQA process in Nevada could lead to exploitation of the water resources available to their counties. Commonly voiced issues include third-party economic impacts, desiccation of existing wetlands, and impacts on future development ability in California. Technical concerns associated with the proposed ground water extractions have not yet been addressed for most of the interstate basins, except Honey Lake, where a study on the basin's response to pumping is nearing completion. Additional studies have been requested for this basin to examine the possible impacts of extraction on water quality.

These Nevada water development efforts have stimulated California entities to seek methods to manage water resources so that their transferability is limited or restrained. In the absence of a statewide mechanism for controlling ground water use and transfer, attention has been focused on legislative formation of ground water management districts or joint powers agencies. The California Legislature passed the Sierra Valley and Long Valley Ground Water Basin Act in 1980, enabling the Boards of Supervisors of Plumas and Sierra Counties to form the Sierra Valley Ground Water Management District and allowing the Boards of Supervisors of Lassen and Sierra Counties to enter into a joint powers agreement with the State of Nevada or with Washoe County, Nevada for the purpose of ground water management in the Long Valley basin. Within its boundaries the Sierra Valley District may: require ground water users to register their wells and to file annual extraction reports, issue permits for exports of ground water from the District, regulate extractions if overdraft occurs, and levy a pump tax. Last year the California Legislature enacted a similar statute for the Honey Lake basin, allowing the creation of a ground water management district modeled after the Sierra Valley example. Discussions have also been underway for some time regarding the formation of a joint powers agency among five northeastern California counties with Nevada government agencies to provide a forum for working on water policy and other regional issues.

Summary and Conclusions

Many California local agency interests tend to oppose, at least initially, transfers of water across the stateline to Nevada, and have been to some extent successful in creating provisions to limit or restrain transfers. Methods available include addressing transfers in federal legislation or court decrees, and formation of local ground water management districts to provide some degree of regulation absent other statewide controls. While a federal court decision has held that states cannot arbitrarily restrict the transfer of water across their boundaries, the conditions under which a transfer may occur can still be constrained by, for example, an environmental review process. Consent between a willing buyer and willing seller is not the only condition necessary for a successful water transfer, but regulatory and political concerns must be satisfied as well.

The views expressed above are those of the author, and do not reflect opinions or policy of the Department of Water Resources.

Moving Toward The Millennium

Terry Katzer[1] and Kay Brothers[2]

Abstract

The last decade of the 1900's is just beginning and paradoxically the water resources readily available to Clark County, Nevada are just ending. In a few short years Nevada's allocation from the Colorado River will be fully utilized and additional ground water, over and above existing permitted rights, from the Las Vegas Valley basin is not available under current Nevada water law.

A major water resources project, involving importing ground water from basins located in eastern and southern Nevada, has begun but awaits the dawn of the twenty first century to become operational. These basins have an estimated perennial yield of about 660 million cubic meters with about 40 percent already allocated, leaving 400 million cubic meters of potentially available water. Thus the purchasing of some portion of these existing water rights becomes a very important part of the project. A massive amount of legal, technical, and environmental work must be done prior to developing the first cubic meter of this vast water supply. Constraints on development are focused by the emotional upwelling of residents in the targeted ground-water basins.

Technically it is possible to build five to six hundred kilometers of pipeline/canals with pumping stations, pumping wells, treatment facilities, and perhaps electrical generating facilities. However, minimizing impacts on the environment, providing water as needed and available to communities and development along the pipeline route, and addressing emotional and political

----

[1]Director of Research

[2]Sr. Hydrologist, Las Vegas Valley Water District, 3700 W. Charleston Blvd., Las Vegas, NV 89153

issues are a few of the uncountable challenges awaiting this development project.

Introduction

Southern Nevada, in particular the Las Vegas area, has enjoyed a steadily rising economy for over fifty years. Growth has seen the once small-town stop on the Union Pacific Railroad grow into the largest urban area in Nevada and a major southwestern city. Concomitant with growth has been an ever increasing demand for water. The purpose of this paper is to look at southern Nevada's water supply and to present a brief overview of a water resource development project that will take the most arid part of the most arid state in the union well into the first century of the next millennium.

Location

Las Vegas Valley, shown in Figure 1, nestled in the Basin and Range Province and the Mojave Desert, just south of the Great Basin, is bounded on all sides by mountain ranges. The Spring Mountains to the west, with a crestline well over 2700 m in altitude, and precipitation well over 50 cm, provide the vast majority of ground water to Las Vegas Valley. The valley itself ranges from 550 m on the east to about 900 m at its western margin against the Spring Mountains. Precipitation on the valley floor is about 10 cm a year.

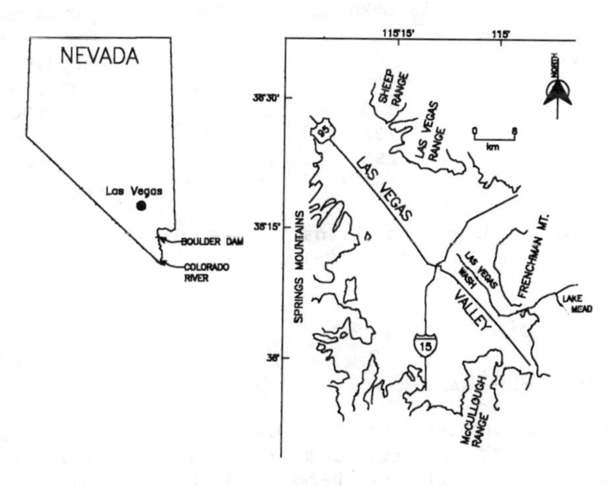

Figure 1.--Location and general features of the Las Vegas Valley area

Background

At the start of the twentieth century the pioneering population utilized the waters discharging from several springs in the valley. By the early 1940's the artesian pressure was declining and corresponding ground-water levels were quite literally dropping out of sight due to ground-water pumpage. To date (1990) the maximum decline in the water table has been about 90 m in the western part of the valley.

In 1955, state law was changed to allow ground-water mining to continue until the Southern Nevada Water System could be built. This system, which treats and brings Colorado River water into the valley from Lake Mead, was authorized in 1965 and made a reality by the construction of Hoover Dam in 1935. Nevada has a consumptive use allocation from the river of 369 million m^3. The current popular model indicates the actual diversion may be about half again as much, or nearly 554 million m^3. Table 1 shows historic ground-water pumping and water imports from the Colorado River and the spectacular rise in population.

Table 1.--Las Vegas Valley ground-water pumpage and imported Colorado River water, rounded to nearest 10^6 cubic meters[1] and population for select years

Calendar Year	Ground Water	Colorado River water via Lake Mead	Total	Population[2]
1970	106	42	148	255,000
1975	90	101	191	320,000[e)]
1980	87	146	233	442,000
1985	84	196	280	567,000
1989	82[e)]	299	381	736,000
2000	86[e)]	480[e)]	566[e)]	1,000,000[e)]

[1] Data from the files of the Nevada State Engineer
[2] Population figures from Clark County Comprehensive Planning
[e)] estimate

Hydrologic Setting

Steady State Conditions

Primary ground-water recharge occurs when water on the mountain block infiltrates through the soil and percolates downward to the soil-bedrock interface. The degree of fractures encountered in the bedrock then determine the course and fate of the infiltrating water. Water that originates in the alluvial areas from either snowmelt, rainfall, or runoff percolates directly into the basin fill.

Discharge from ground-water basins occurs by a variety of processes: as underflow, either through basin-fill sediments or bedrock or both, to the next down-gradient valley; as evaporation from bare soil surfaces; as transpiration from phreatophytes; as spring flow; and as surface flow.

Perennial Yield

Perennial yield is the average amount of water that can be removed annually from a ground-water system without causing an adverse affect on that system. Under steady-state conditions, ground-water recharge equals ground-water discharge (with no change in storage) and theoretically this volume of water is the perennial yield. However, basin geometry, the usual non-homogeneity of basin-fill sediments, the anisotropy of aquifer characteristics coupled with the uneven distribution of recharge and discharge, all combine to significantly reduce the potential for capturing all the discharge in any given ground-water system. The perennial yields of the targeted ground-water basins which are shown in Figure 2, reflect minor refinements by the U.S. Geological Survey based on the recent work of Harrill and others (1988).

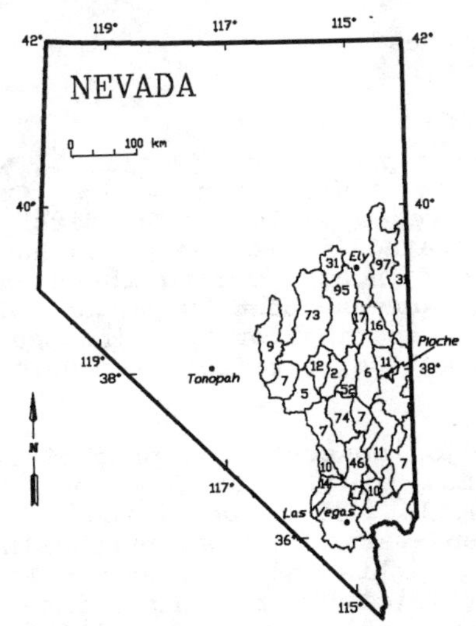

Figure 2.-Perennial yields in basins where ground water is potentially available, in millions m^3

It is with the perennial values in mind that the Las Vegas Valley Water District (LVVWD) in October, 1989, filed with the State Engineer 145 applications to appropriate ground water and one application to appropriate surface water. The filings, made in 27 basins in eastern and southern Nevada, total about one billion m^3 of water. However, the available perennial yield, over and above the existing permitted water rights as compiled by the Nevada Division of Water Resources (1988), is nearly 400 million m^3.

Storage

In addition to the amount of water available perennially there is a vast amount of water in storage. The U.S. Geological Survey estimates that in the top 30 m of the saturated basin-fill sediments there is about 55 billion m^3 of water. Much of the area is bounded and underlain by carbonate rocks which also contain large amounts of ground water. Dettinger (1989), estimated that the top 30 m of rock aquifers in the southern part of Nevada contains about 7 billion m^3 of water. This value, extrapolated to the targeted basins (Figure 2), may conservatively equal 12-18 billion m^3. Considerably less than the basin-fill aquifers, but an important source of water nonetheless. In general, over much of the area, the basin-fill aquifers and the carbonate aquifers are in hydraulic continuity with each other.

Meeting Future Demands

Short Term

The population projection (Table 1), with ongoing refinement, is the planning base for future water resources for Las Vegas Valley. The first major water shortage is anticipated in 3-5 years and has been identified as a peaking problem during the high summer use period. The LVVWD serves about 80 percent of the valley population and is preparing for upcoming peak demands in three ways; construction, artificial recharge, and conservation.

First, a major construction program is under way to add reservoirs, pipelines, pumping stations and wells. The new wells will be permitted by redistributing existing ground-water rights, purchasing, whenever possible, existing rights and using Colorado River water which has been artificially recharged into the ground-water system. Some existing wells will also withdraw recharged water.

Secondly, based on the work of Katzer and Brothers (1989) and Brothers and Katzer (in press), artificial recharge of treated Colorado River water is a viable technique for banking water. During winter months there is excess capacity in the distribution system and treated Colorado River water is imported into the valley and injected into the ground-water system. Water is currently (1990) being banked against the future peak demands as shown in Table 2. Thus by 1994, when the peak demand water shortage is first expected, there will be about 123 million m^3 available to be recovered through production wells.

Table 2.--Artificial recharge of treated Colorado River water

Year	Recharge Volume (10^3 m^3)
1987	2
1988	1,292
1989	4,551
1990	18,450[1)]
1991	24,600[2)]
1992	30,750[2)]
1993	30,750[2)]
1994	30,750[2)]

1) projected--actual recharge to March 1990 is approx. 4674x10^3 m^3
2) projected

The third part of the plan for meeting short term demands is to implement a conservation program. To help solve peak demands the program will target outside water use. An overall reduction might postpone the need for additional new water resources, but the fact remains additional water will be needed to sustain continued development.

Long Term

Future long-term development simply requires additional water. The conceptualization of developing ground water in other basins for import to Las Vegas Valley was first proposed by the Nevada State Engineer's office (1971) and by the Nevada Division of Water Planning (1982). Thus nearly 20 years ago water importation was realized as an option to allow continued development in southern Nevada.

Nevada water resources belong to the public. Those water users that can put this resource to beneficial use first, without impacting existing rights and continue to use the water beneficially, keep the right.

As this massive water resource project develops, the LVVWD will face numerous challenges to utilize this important resource to the benefit of the most with minimizing the impacts on the environment, wildlife, existing water rights holders, and rural residents. The challenges are truly of a magnitude that will require all the innovation that the twenty first century can bring for successful completion of this water resource development project.

References

Brothers, K., and Katzer, T., (1990), "Water banking in Las Vegas Valley, Clark County, Nevada": Journal of Hydrology, (in press)

Dettinger, M.D., (1989), "Distribution of carbonate-rock aquifers in southern Nevada and the potential for their development, summary of findings, 1985-88": U.S. Geological Survey, Summary Report No. 1, 37 p.

Harrill, J.R., Gates, J.S., and Thomas, J.M., (1988), "Major ground-water flow systems in the Great Basin Region of Nevada, Utah and adjacent states": U.S. Geological Survey, H.A. 694-C, 2 sheets

Katzer, T., and Brothers, K., (1989), "Artificial recharge to the Las Vegas Valley ground-water system, a demonstration project, Clark County, Nevada": Journal of Ground Water, January-February, 1989, p. 51-56

Nevada Division of Water Planning, (1982), "Water for southern Nevada": State of Nevada, Department of Conservation and Natural Resources, Water Supply Report 2, 402 p.

Nevada Division of Water Resources and Water Planning, (1988), "Hydrographic basin statistical summary, ground-water basins 001-232": State of Nevada, Department of Conservation and Natural Resources

Nevada State Engineer's office, (1971), "Water supply for the future in southern Nevada": State of Nevada, Department of Conservation and Natural Resources, 95 p.

WATER CONSERVATION IN LOS ANGELES

Richard F. Harasick [1]

arid /'ar-ed/adj 1: excessively dry; spec: having insufficient rainfall to support a city. syn see LOS ANGELES. Los angeles is definately a member of the arid lands community. With its relatively low rainfall compunded by lawsuits threatening future imported supplies of water, and the current drought, it is very apparent that water is not limitless. The Los Angeles Department of Water and Power shares the responsibility with its customers of managing water resources by implementing water conservation measures that can be applied at the home or workplace. By working together and carefully managing the use of water, Los Angeles will have adequate supplies to fill future needs.

WATER CONSERVATION PROGRAMS

Los Angeles has long encouraged water conservation. Efforts to achieve efficient use of the City's valuable resources began upon the founding of the municipal water system in 1902 with the installation of water meters for all customers. During the past year, the Department of Water and Power (DWP) has continued this tradition by implementing many additional conservation programs and policies. This document will outline conservation programs and strategies enacted during this past year.

LANDSCAPING

The Department of Water and Power has implemented an aggressive outdoor water conservation program. Emphasis had been placed on indoor water use in the past. Efforts will continue in that area; however, outdoor water use represents, on the average, 25 percent of the City's total

---

1Civil Engineering Associate, Los Angeles Department of Water and Power, Box 111, Room 1348, Los Angeles, CA 90051. Associate Member, American Society of Civil Engineers.

water use and nearly 50 percent of the residential water use. Since low water using plants require 30 to 60 percent less water than other prevalent types of landscape in the City, the DWP considers this aspect of its program extremely important. In the future, DWP will try to educate consumers in the use of efficient landscaping and irrigation systems through symposiums, brochures, exhibits, and technical instruction. The following is a description of DWP programs that were implemented during 1989:

o <u>Annual Spring Garden Expo:</u> In 1990, the Department presented its third Garden Expo. The purpose of this Expo was to provide participants with information on low water using plants and efficient irrigation techniques. Nearly 1000 participants have attended the Department's Garden Expo. The expo is a half day symposium in which three speakers are featured and at which participants can view exhibits and take garden tours. The Expo is offered free to all City residents.

o <u>Demonstration Gardens:</u> Another program designed to change residents' attitudes toward low water using plants are demonstration gardens sponsored by the Department. The Department has contributed $20,000 towards the completion of the Lummis Home Garden and $2,500 to the establishment of a water conservation garden at the Los Angeles Arboretum. In addition, Department facilities have also utilized these types of plant material. The purpose is to show residents that low water using gardens can be much more than merely rock gardens and cactus.

o <u>CIMIS Weather Station:</u> The California Irrigation Management Information System (CIMIS) is a California Department of Water Resources (DWR) program consisting of a statewide network of remote weather stations. Information from the stations assists farmers in efficient irrigation practices. The Metropolitan Water District of Southern California (MWD) and DWP have worked cooperatively to install a CIMIS weather station in Forest Lawn Memorial Park. The information from this station can be utilized to develop efficient irrigation schedules for large urban turf sites. Methodology is being developed to make this information available to greater numbers of large turf customers.

o <u>Large Turf Water Curtailment Program:</u> The Water Conservation/Sewer Flow Reduction Ordinance (#163532 of the municipal code of the City of Los Angeles) contains a provision mandating a 10% reduction in water used for the irrigation of large turf areas (i.e., in excess of 3 acres). These customers have been identified by the Water Engineering Design Division (WEDD) and apprised of their allowable usage (allotments are based on

90% of comparable billing periods in 1986). Approximately 40 customers representing parks, schools, cemeteries, and golf courses have been contacted. Failure to comply with these provisions can result in the assessment of a 10% surcharge on water costs. For the period 9/5/89 through 2/6/90, the overall frequency of large turf customer surcharge was 16.9%. That and additional information has been summarized in the attached table entitled "Large Turf Program". Large turf customers using reclaimed water for irrigation or who have installed water-saving devices specifically designed or manufactured (as determined by DWP) to reduce large turf irrigation water by 10 percent may qualify for exemption from this provision of the ordinance.

o Xeriscape Requirements for New Construction: The aforementioned ordinance further states that no building permit shall be issued to construct any industrial, commercial or multi-family residential structure unless the Department of City Planning first determines that Xeriscape will be installed on the lot on which the structure is located. A point system has been established to evaluate degree of compliance. The number of points necessary for compliance is based on the square footage of the lot in question. This review system is administered on an ongoing basis.

BUSINESS AND INDUSTRY

Approximately 30% of the water used in Los Angeles is used by commercial (excluding multiple family residential units), industrial, and governmental customers. The DWP emphasis on providing service and information to these customers is apparent in the distribution of brochures, presentation of symposiums, on-site services such as meter loans and facility audits, and the recognition of exemplary water conservation programs through award presentations.

o Business and Industry Bulletins and Brochures: Specific industrial and commercial information bulletins have been composed and are updated periodically by DWP. This literature is made available, upon request, to interested customers. These bulletins provide helpful water conservation tips for specific industries. Industries for which bulletins have been composed include: Commercial Buildings, Schools and Colleges, Restaurants, Laundries and Linen Suppliers, Hotels, Health Care Facilities, Golf Courses, Food Processing Industries, and Beverage Industries.
DWP also distributes "Water Conservation: A 10-Step Approach for the Business User".

- Business and Industry Symposium: On May 3, 1989, the DWP held its first water conservation symposium for its commercial and industrial customers. The half-day program included three presentations: Xeriscape Practices in the Commercial/Industrial Sector, Water Conservation through Cooling Tower Ozonation, and A Water Conservation Case Study of the Los Angeles Hilton Hotel. In addition, eight awards for excellence in water management were presented as part of the day's luncheon program.

RESIDENTIAL PROGRAMS

The Department has implemented programs aimed specifically at reducing indoor water use in both single family and multiple family residential units. Even before the implementation of a mandatory retrofit program in the City of Los Angeles, DWP provided water conservation devices to its residential customers through numerous programs over a 12 year period.

In January of 1988, the DWP in cooperation with the MWD of Southern California, initiated a pilot study to test the effectiveness of various methods of distributing water-saving devices as well as their water saving potential. A total of 1,500 water conservation kits were distributed in the Westchester area of the City. The effectiveness of the pilot study is currently being analyzed.

In April of 1988, the City of Los Angeles enacted the Water Conservation Ordinance to Reduce Sewer Flows which re-quired all water users in the City of Los Angeles to install low-flow showerheads (less than 3 gpm), to equip tank-type toilets with displacement devices, and to reduce flush volumes on valve-type toilets to 3.5 gallons. DWP was charged with the responsibility of providing low-flow showerheads and toilet tank displacement bags to all residential customers requesting them. As a result, over 1.2 million low-flow showerheads have been distributed since July 1988. The Department is continuing to distribute these devices to enable citizens to meet the requirements of this ordinance.

- Ultra-Low-Flush (ULF) Toilet Rebate Program: In an effort to promote the use of ULF toilets (i.e., toilets that flush at the rate of 1.6 gallons/flush or less), the Department of Water and Power has implemented a toilet rebate program. Customers replacing existing toilets with City-approved ULF toilets are eligible to receive a $100 rebate for each replacement.

PUBLIC INFORMATION PROGRAMS

DWP maintains an extensive public information program

to encourage its customers to conserve water. This program includes the distribution of brochures and bill inserts, the display of exhibits, and the implementation of educational programs for schools. The overall purpose is to inform customers about the sources of their water and the need to use it wisely.

- <u>Advertising:</u> To increase awareness of the need to conserve water an extensive advertising campaign is implemented during droughts. These efforts utilize television, radio, and print media to disseminate the message. Past campaigns have used Los Angeles area weathercasters, billboard ads, and billboards on DWP service vehicles as a means of expanding customer awareness.

- <u>Water Awareness Month:</u> May, 1990 - Water Awareness Month was a statewide effort to inform and educate the public about water issues. To achieve those goals, DWP organized numereous activities aimed at residential, school, commercial and industrial customers.

- <u>Water Bills and Bill Inserts:</u> Water bills from DWP include a water consumption comparison between the current and the immediately prior year to encourage water use efficiency. Substantially higher usage draws attention to the possibility of a water leak or other water waste problem. In addition, DWP prints messages on bills to notify customers of higher summer rates, the need to conserve, and of excessively high water use.

  Bill inserts are also used to pass on conservation tips and to announce special conservation programs.

- <u>Exhibits:</u> DWP has developed portable water conservation exhibits for use at schools, shopping malls, conferences, and other special gatherings.

- <u>Speakers' Bureau:</u> The DWP Speakers' Bureau provides speakers and films for schools, civic organizations, and business groups. A toll- free number is publicized for groups interested in obtaining speakers knowledgeable in the field of water conservation as well as other areas of interest to customers.

- <u>School Education Programs:</u> The DWP educational program concentrates on educating students in the area of water supply and quality as well as the need to conserve. DWP provides school districts with publications, reference materials, course outlines, and films. Additionally, tours of Department facilities are conducted for students and instructors.

Teacher workshops are held to assist instructional personnel in developing a water supply and conservation curriculum. Such workshops have been conducted in cooperation with the Los Angeles Unified School District. The most recent workshop was entitled "Project Water Science" and presented topics on water supply, water conservation, and water quality to secondary school teachers.

DWP also sponsors an annual Poster Contest entitled "Conserving Water at School and Home" for junior and senior high school students. Cash awards and certificates of merit are presented to the winners. The winning poster designs are incorporated into a conservation calendar which is distributed throughout the Los Angeles Unified School District (LAUSD).

Additionally, a Student Home Water Audit has been developed to envolve all of the LAUSD. Every fourth-grader will receive a water conservation checklist to take home to his family to complete. The questions allow families to assess their water use characteristics while providing them hints as to conserve water.

GENERAL

o Pricing: In November, 1985, the DWP adopted a seasonal pricing structure under which water is priced at a higher rate during the higher demand summer periods than during the winter. The current differential between summer and winter rates is nearly 25%. The higher water costs during the summer, when water use is greatest, will encourage conservation among DWP customers.

A RESERVOIR YIELD EVALUATION IN AN ARID CLIMATE

by Les K. Lampe, M. ASCE,[1]

INTRODUCTION

A recent paper (Smith and Lampe, 1988) summarizing the state of the art in practical reservoir yield evaluations indicated that "...hydrologic uncertainty is a way of life," and yield estimates of water supply reservoirs are by their nature imprecise. A review of approximately 20 yield estimates for the Middle and Eastern United States over the last fifteen years leads to the appraisal that yield calculations rarely can be more accurate than roughly plus or minus twenty percent. The same uncertainty is dramatically greater in arid climates where the average annual runoff is only a minor fraction of the average annual rainfall, and subtle changes in precipitation patterns or agricultural land use practices can lead to substantial decreases in streamflows. The purpose of this paper is to present a case study of a yield evaluation in an arid climate thereby explaining major factors in such a yield estimation.

The case cited is that of Cedar Bluff Reservoir in west-central Kansas. The reservoir location is shown on Figure 1. Average annual inflows to the reservoir have fallen from about 60,000 acre-fee (74 million cubic meters) historically to about 8,000 acre-feet (9.9 million cubic meters) per year from 1979 to 1989. During 1989 the inflow was only 1,900 acre-feet (2.3 million cubic meters). The dramatic flow decreases are primarily attributable to human activities.

INFLOWS

Determination of critical inflows are most subject to both the vagaries of nature and the practices of man. The example of Cedar Bluff Reservoir in west-central Kansas is illustrative of several problems. The project was formulated as a water supply reservoir for an irrigation project along the Smoky Hill River in Trego and Ellis Counties. Project planning during the 1940's was based on recorded flows of the Smoky Hill river which had averaged about 60,000 acre-feet (74 million cubic meters) per year, which is an average of one-quarter inch of runoff from the 4,980 square- mile (12,900 square kilometer) drainage area above the reservoir.

---

(1) Director of Water Resources Engineering, Black & Veatch Engineers-Architects, P.O. Box 8405, Kansas City, MO 64114

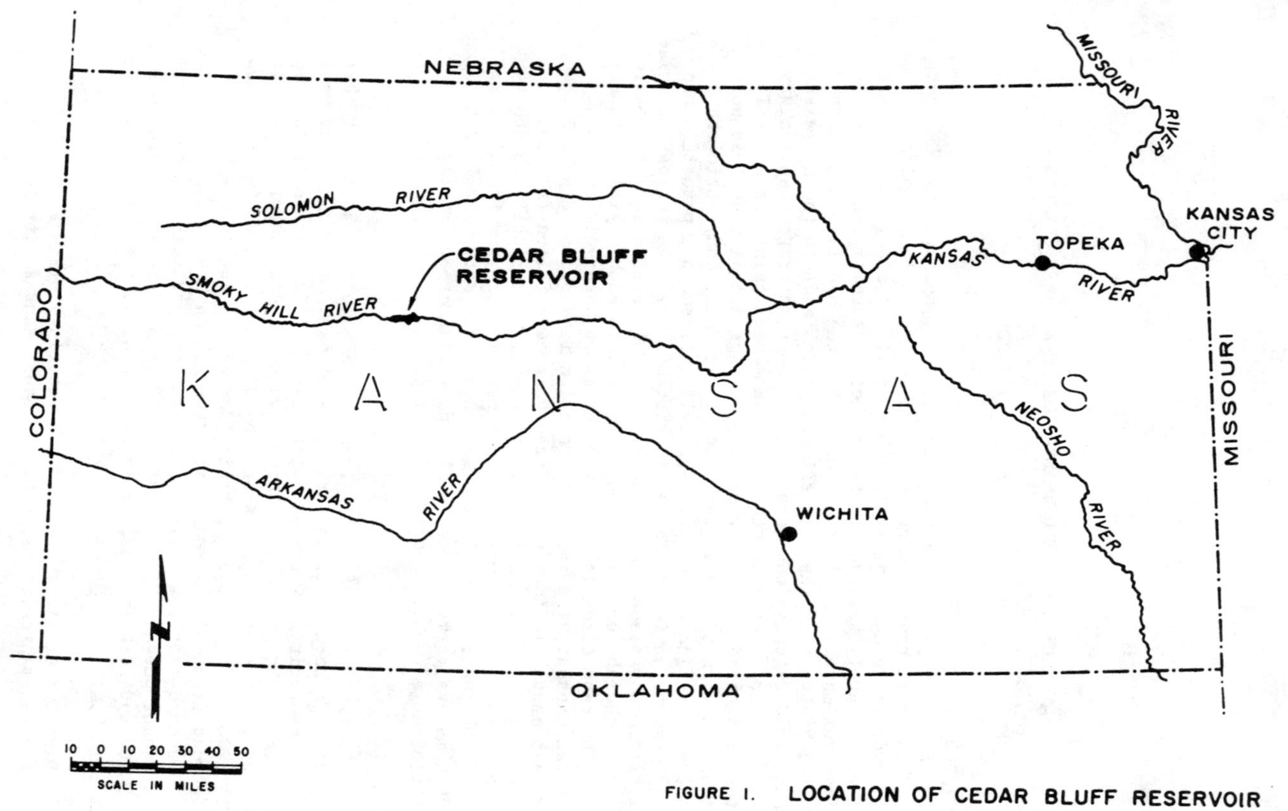

FIGURE I. LOCATION OF CEDAR BLUFF RESERVOIR

The reservoir was completed in September 1952 and filled in the early 1950's, which was just in time for the most severe drought in recorded history for much of the High Plains. Even during the extremely dry period from 1952 through 1956, average inflow to the reservoir averaged 21,000 acre-feet (26 million cubic meters) per year. This was more than enough to satisfy water needs for the newly-formed 6,000-acre (2,800 hectare) irrigation district and to maintain the volume of water in storage at about 70,000 acre-feet (86 million cubic meters).

By the 1960's, a noticeable downward shift in the mass curve of reservoir inflows was apparent. The causes of this shift are unknown, but leading contributors are suspected to be (1) increased use of conservation tillage practices for both irrigated and dry-land farming on the drainage area above the reservoir, (2) depletion of inflows through irrigation withdrawals from alluvial wells upstream of the reservoir, and (3) possibly a subtle shift in the climate. Of these three factors, the only one that is remotely quantifiable is the magnitude of upstream irrigation withdrawals. The Kansas State Board of Agriculture, Division of Water Resources, has tabulated irrigation water rights that allow withdrawal of as much as 30,000 acre-feet (37 million cubic meters) from the Smoky Hill River, its tributaries and associated alluvia above Cedar Bluff Lake. The magnitude of these withdrawals is insufficient to result in the reduction in mean annual inflows from 60,000 acre-feet (74 million cubic meters) before 1960 to about 8,000 acre-feet (9.9 million cubic meters) in recent years. The latter value is about 0.03 inch of runoff from the watershed. These inflows are barely enough to satisfy the net losses from lake evaporation. The inflow during 1989 was 1,900 acre-feet (2.3 million cubic meters). Irrigation releases were discontinued from the reservoir in 1978.

STORAGE

Cedar Bluff Reservoir was constructed with 191,000 acre-feet (236 million cubic meters) of flood control storage; 150,000 acre-feet (185 million cubic meters) of water supply storage to meet the needs of the Cedar Bluff Irrigation District; a fish hatchery, and minor municipal needs; 27,000 acre-feet (33 million cubic meters) of inactive storage; and 8,300 acre-feet (10 million cubic meters) of dead storage. These storage volumes include allowances for 100 years of sediment accumulation. Most of the water supply storage is not practicably usable because inflows are insufficient to overcome the effects of evaporation, and, as the reservoir stage increases with additional inflows, the area increases, evaporation increases, and a limit is reached when the total reservoir storage reaches about 20,000 acre-feet (25 million cubic meters).

LOSSES

Releases from the reservoir to the Smoky Hill River have averaged only 1,100 acre-feet (1.4 million cubic meters) year in the period from 1978 through 1989 and have fallen to zero for the last several years. The net lake evaporation averages 40 inches (102 centimeters) per year and results in losses of about 8,000 acre-feet (10 million cubic meters) per year when computed over the average lake surface area for the same period. No withdrawals have been made during this same period for irrigation or municipal purposes even though the City of Hays, Kansas, has an alluvial wellfield 22 miles (35 kilometers) downstream and has been in a precarious water supply situation through the 1980's. The City would have made use of the water, but the project was formulated for only a small amount of municipal water supply, and the storage committed to that purpose was obligated to use by the City of Russell, Kansas. Substantial regulatory requirements had to be met for releases to benefit the City of Hays.

YIELD

Studies performed by the Kansas State Board of Agriculture, Division of Water Resources (Watson, 1984), indicated the incongruous result that the greater the reservoir storage volume that was maintained, the lower the reservoir yield. This is attributable to the tenuous balance between inflow volumes and the amounts of net lake evaporation. Their studies showed that an annual yield of 1,200 acre-feet (1.5 million cubic meters) might be possible based on inflow records from the late 1970's and early 1980's.

Subsequently, the Kansas Water Office recognized that the project would no longer fill its intended purpose as an irrigation supply reservoir and initiated negotiations with the Bureau of Reclamation to reformulate the project. The primary reasons for the reformulation were to allow a minor amount of storage to be used for releases to recharge downstream alluvial aquifers, to relieve the Cedar Bluff Irrigation District of its financial obligations to the federal government for an irrigation project that was no longer feasible, and to allow stability in lake operation to protect the aquatic habitat and recreational resource. These negotiations resulted in a Memorandum of Understanding (Kansas Water Office, 1988) between the Bureau of Reclamation and the State of Kansas that accomplishes the foregoing objectives. It awaits only Congressional approval to go into effect.

During the negotiations process, another yield estimate was made by Black & Veatch Engineers to determine the annual amount of water that could be safely withdrawn for recharge of downstream alluvial wellfields. This resulted in a yield estimate of 700 acre-feet (860,000 cubic meters) per year. The value is considered highly uncertain because the lowest river flows have been induced by human activities and have occurred in recent years. A recurrence of precipitation patterns similar to those in the legendary droughts of the 1930's and the 1950's may well cause the actual reservoir yield to fall to zero.

SUMMARY

The foregoing brief description of the yield evaluation for a reservoir in arid western Kansas illustrates several significant points. First is that in environments where the mean annual runoff is a small percentage of the mean annual precipitation, changes in the hydrologic regime from human activities can have a dramatic effect on reservoir inflows and yields. The second conclusion is that this tenuous balance between precipitation, land uses, upstream withdrawals, and reservoir inflows makes accurate assessment of yields extremely difficult, and the practitioner must exercise great caution. Finally, project planning in arid regions should not be based on reservoir yield estimates completed 10, 20, or 30 years previously. New yield estimates should be completed to reflect the latest available data.

REFERENCES

1. Black & Veatch, Engineers-Architects. Unpublished reports on water supply for the City of Hays, Kansas. Kansas City, Missouri. 1986-1990.
2. Kansas Water Office. Memorandum of Understanding Between the U.S. Department of Interior, Bureau of Reclamation; Fish and Wildlife Service; the State of Kansas; and Cedar Bluff Irrigation District No. 6 Concerning Reformulation and Operation of the Cedar Bluff Unit. Topeka, Kansas. 1988
3. Koelliker, James K., Impact of Improved Agricultural Water Use Efficiency on Reservoir Storage in Sub-Humid Areas. Kansas Water Resources Research Institute, Manhattan, Kansas. September, 1984.
4. Smith, Robert L., "Surface Waters of Kansas." Presented at the Annual Meeting of the Kansas Section, American Water Works Association, Manhattan, Kansas. March 28, 1984.
5. Smith, Robert L., Unpublished memoranda on yield evaluations for Cedar Bluff Reservoir. Black & Veatch, Engineers-Architects, Kansas City, Missouri. 1987-1988.
6. Smith, Robert L. and Les K. Lampe, "Dependable Yield Evaluations - How Much Water is Really There?" Journal American Water Works Association. September 1988.
7. U.S. Department of the Interior, Bureau of Reclamation, Definite Plan Report - Cedar Bluff Unit. Denver, Colorado 1958.
8. Watson, Everett R., Preliminary Engineering Report - Proposed Smoky Hill River Intensive Groundwater Use Control Area. Kansas State Board of Agriculture, Division of Water Resources. Topeka, Kansas, February 1984.

EROSION PROCESSES IN UPLAND AREAS

Vida G. Wright[1] and Ray B. Krone[2]

I. INTRODUCTION

Erosion is caused by detachment of soil particles from the soil mass by the erosive forces of wind and water. Only erosion due to water is investigated herein. Erosive forces considered include the kinematic energy of the rainfall that results in soil splash and shear stress of the overland flow near the bed that will result in scour. After a soil particle is detached and entrained, it will be transported by the surface runoff. However, depending on the eroding and transport capacities of the surface runoff, some of the suspended particles may be deposited on the flow path. Eroding capacity is dependent on the flow depth, bed geometry, and critical shear strength of the soil. Transport capacity of the surface runoff is dependent on the flow depth and the bed geometry. Erosion analyses presented herein are for unprotected soils.

Due to the highly non-linear nature of the flow and sediment transport equations, an analytical solution for erosion at upland areas does not exist. Therefore, numerical analyses techniques were used to simulate erosion processes from the upland areas. This model has not yet been verified by measurements of erosion rates from actual slopes. The coefficients used are based on published data, however, and the representations appear to be within observed sediment production rates. The approach provides insight into erosion processes and promises to be useful for predicting soil erosion under various storm scenarios.

---

[1]Associate, Dames & Moore, Sacramento, California

[2]Professor of Civil Engineering Emeritus,
University of California, Davis, California

II. GOVERNING EQUATIONS

To describe the physics of eroding surfaces and transport of suspended soil particles by surface runoff, a surface runoff model based on the overland flow theory (conservation of mass and the linear momentum principle) has been developed *(Bennett, J.P., 1974)*.

$$\frac{y}{t} + \frac{q}{x} = I(x,t) - F(x,t) \quad (1)$$

$$S_o = S_f \quad (2)$$

Where:

y(x,t) = water depth (ft)
x = the spacial coordinate measured along the slope (ft)
t = time index (sec)
q(x,t) = discharge per unit width (ft^2/sec)
I(x,t) = rainfall intensity (ft/sec)
F(x,t) = infiltration rate (ft/sec)
S_o = bed slope
S_f = friction slope

To simulate worst case erosion scenario, it was assumed that F(x,t) = 0. The flow is described by friction law for turbulent flows and is given by Manning equation:

$$q = \frac{1.49\ y^{5/3}\ S^{1/2}}{n} \quad (3)$$

Where: n = Manning's roughness coefficient

Soil particles are detached by soil splash and flow scour. The conservation of mass for sediment transport process is:

$$\frac{\delta(cy)}{\delta t} + \frac{\delta(cq)}{\delta x} = E_r + \frac{S_p}{\rho_s} \quad (4)$$

Where: E_r = the rate of soil detachment due to overland flow and is described as $(1-P)\frac{\delta y_s}{\delta t}$; y_s is the soil depth from a fixed datum; and P is the soil porosity

Sp = is the rate of soil splash cause by kinetic energy of the rainfall droplet (lb/ft^2)

ρ_s = soil density (slugs/ft^3)

Soil splash is caused by the impact action of the raindrops. The rate of soil splash can be significant especially prior to commencement of surface water runoff. The quantity of the soil detached due to soil splash can be determined by the empirical equation proposed by Bubenzer and Jones (1971):

$$Sp = m(2.78 \times 10^{-7} I)^{\alpha} K_e^{\beta} P_c^{-\gamma} \quad (5)$$

Where:
- Sp = soil splash (kg/m^2)
- I = rainfall intensity (mm/hr)
- K_e = total kinetic energy of raindrops (J/m^2)
- P_c = percentage of clay in soil
- m = empirical coefficient (1.5 - 3.0)
- α = empirical coefficient (0.25 - 0.55)
- β = empirical coefficient (0.83 - 1.49)
- γ = empirical coefficient (0.40 - 0.60)

As suggested by Wischmeier & Smith (1958) the kinetic energy, K_e, can be estimated as:

$$K_e = 24.16D + 8.73D \log (I/25.4) \quad (6)$$

Where D is defined as rainfall depth in millimeters and is computed as rainfall intensity (I) multiplied by time (t). However, it is assumed that the kinetic energy of the rainfall droplet will be dissipated by the increasing depth of overland flow. Therefore, for the analysis presented in this paper, D is assumed to be an exponential function of rainfall depth and the overland flow depth and described as: $D = (I \times t) \times EXP(-y)$.

The sediment continuity equation (4) has two unknowns c and y_s and therefore, another equation is needed for closure. A first order reaction equation proposed by Foster and Meyer (*1975*) provides a relationship for erosion and transport capacities of the flow, erodibility of the soil particles; and the rate of sediment transport.

$$\frac{E_r}{E_{rc}} + \frac{cq}{T_c} = 1 \quad (7)$$

Where:
- E_r = the rate of soil detachment by overland flow
- E_{rc} = eroding capacity of the overland flow and is described as $C_d(\tau - \tau_{cr})^{1.5}$; τ is the flow shear stress described as $\gamma y S_o$ (N/m^2); τ_{cr} is the shear strength of the soil described as $(0.0493)10^{0.0183P_c}$ (N/m^2); and C_d is an empirical coefficient ranging from 0.001 to 0.8

T_c = transport capacity of the flow and is described as $C_t \tau^{1.5}$; C_t is an empirical coefficient ranging from 0.0001 to 0.5

Equations 1 through 7 describe the interrelated processes dominating erosion, overland flow, and sediment transport. Transported sediment and eroded profile resulting from a defined storm event with known intensity and duration is obtained from simultaneous solution of the above equations.

III. SELECTION OF EMPIRICAL COEFFICIENTS

Sediment transport is a highly empirical science and consequently the governing equations presented in Section II are associated with a host of empirical coefficients. The values selected for these coefficients are based on suggested values in the literature and they are generally based on the site and material properties. The empirical coefficients are described below.

- Manning's Roughness Coefficient (n): the value of n is dependent on a number of factors including the surface roughness, surface cover, flow depth, etc. For overland flow where the flow depth is only on the order of a few millimeters, the value of n may range from 0.03 to 0.12 (*Chow V.T., 1959*).

- Soil Splash Parameters: four empirical coefficients of m, α, β, and γ are associated with soil splash equations. The suggested values for these parameters are m = 3.0, β = 1.14, γ = 0.52, and α = 0.41 (*Akan and Ezen, 1982*).

- Erosion Capacity Coefficient C_d: this coefficient describes the erodibility potential of the soil and its suggested values range from 0.001 to 0.50.

- Transport Capacity Coefficient C_t: this coefficient describes the transport potential of the soil particles and its suggested value ranges from 0.001 to 0.80.

- Shear Strength of the Soil τ_{cr}: several empirical coefficient of τ_{cr} for a consolidated soil have been suggested in the literature. In the analyses presented here we have assumed that τ_{cr} is a function of clay content and described as follows (*Smerdon & Beasley, 1961*). $\tau_{cr} = (0.0493)10^{0.0183P}C$.

IV. CASE STUDY

The rate of soil erosion from a levee in its unprotected state during two storm events with durations of 24 hours and return periods of two and one hundred years was analyzed. The rainfall intensities for these storm events were assumed to be 0.080 and 0.180 inches per hour. Erosion computations were made based on the following assumptions:

- Surface roughness coefficient defined as Manning's n was estimated to be 0.05.
- The levee slope was 1:V to 5:H. Assuming a height of 25 feet, the levee length was estimated to be 127.5 ft.
- The average clay content of the soil was estimated to be 32 percent.
- The average porosity of the soil was estimated to be 42 percent.
- The average density of the soil was estimated to be 1,530 kg/m^3.

The sensitivity of the erosion processes to variations of Manning's n and levee slope were determined. The sensitivity analyses were only performed for the two year storm. However, the conclusions of the sensitivity analyses may be extended to the one-hundred year or any other storm with a different duration and intensity.

Levee Roughness Sensitivity Analysis: The surface roughness can vary considerably depending on the method of levee construction. The results of the sensitivity analysis indicated that the erosion rate on the levee is impacted considerably by the value of Manning's n selected for the analysis. Three n values of 0.04, 0.05, and 0.06 were selected for the sensitivity analyses. As the Manning's n was increased from 0.04 to 0.06, the corresponding erosion rates were increased from 0.794 to 1.150 cubic yards per unit width (ft) of levee face.

Levee Slope Sensitivity Analysis: A change in the bed slope from 10 percent to 30 percent resulted in an increase in the erosion rate from 0.471 to 1.488 cubic yards per unit width (ft) of levee face.

Erosion Depth and Transported Sediments on An Unprotected Slope: At the end of the two-year storm, approximately 0.973 cubic yards per unit width (ft) of levee face would be eroded. The erosion depth at the end of this storm ranged from 2.3 to 3 inches with the maximum erosion taking place at the top of the levee where soil splash erosive forces are largest.

In the event of the one-hundred year storm, the model estimated that approximately 4.7 to 6 inches of the levees would be eroded if the slope is not protected. The transported levee sediment was estimated to be 2 cubic yards per unit width (ft) of levee face.

Erosion Protection Measures: The results of the erosion analyses indicated that the major cause of erosion on the levees is soil splash resulting from impact of the rainfall droplets. Therefore, application of mulch on the levees to adsorb the kinetic energy of the rainfall and should eliminate erosion of the levees.

V. ACKNOWLEDGMENTS

Other individuals who contributed to the development of the computer model include Professor L. Kavvas and Dr. G. S. Govindareju of the University of California at Davis. The case study was performed during Dr. Wright's employment with Harding Lawson Associates in Sacramento, California.

VI. REFERENCES

Akan, A.S., and Ezen, S.C., 1982, *Mathematical Simulation of Erosion on Graded Terraces*. Proceedings of Ereter Symposium, IAHS, Publication No. 137.

Bennet, J.P., 1974, *Concepts of Mathematical Modeling of Sediment Yield*. Journal of Water Resources Research.

Bubenzer, G.D., and Jones, B.A. Jr. 1971, *Drop Size and Impact Velocity Effects on the Detachment of Soils Under Simulated Rainfall*. Transaction of American Society of Agricultural Engineers.

Chow, V.T., 1959, *Open Channel Hydraulics*, McGraw Hill, New York, New York. Foster, G.R., and Meyer, L.D. ______, *A Closed-Form Soil Erosion Equation for Upland Area*, Soil and Water Conservation Research Division, USDA, Lafayette, Indiana, Purdue Journal Series No. 4607.

Foster, G.R., and Meyer, L.D., 1975, *Mathematical Simulation of Upland Erosion by Fundamental Erosion Mechanics*, USDA Agricultural Research Division.

Pall, R., and et.al, 1982, *Impacts of Soil Characteristics on Soil Erodibility*. Proceeding of Ereter Symposium, IAHS Publication No. 137.

The United States Department of Commerce Publication, 1973.

Complex Geomorphic Response to Minor Climate Changes, San Diego County, CA.

R. Craig Kochel[1] and Dale F. Ritter[2]

Introduction

It is generally accepted that significant changes in the climate during the Quaternary Period have altered hydrologic regimes of geomorphic systems to the extent that physical changes and instability resulted. Systems crossing these geomorphic thresholds are typically characterized by changes in process rates and/or landform morphology that are directed toward establishment of a different systemic equilibrium. The new equilibrium state is adjusted to the modified climatic and hydrologic conditions. Many times, threshold responses are rapid and associated with significant geologic hazards. What is not known, however, is the magnitude of climatic change required to produce threshold-crossing events and their associated responses. The often-overlooked role of minor climatic fluctuations on the dynamics of physical geomorphic systems may become increasingly important given the growing concern over possible global climatic changes. Knowledge concerning the type and degree of climatic variation required to induce significant hydrologic and geomorphic response is essential for judicious planning in regions subject to cyclic climatic change.

Precipitation in San Diego County, California over the past several hundred years has been characterized by excursions between periods of relatively low precipitation having durations of several decades punctuated by short-term wet intervals of increased precipitation lasting less than a decade (Ganus, 1976; Schulman, 1947). The latest phase of above average precipitation began in 1978 and continued through 1983 when stations across the county experienced precipitation between 61% and 112% above normal. The effects of this recent wet episode have been well documented along coastal areas near San Diego (Emery and Kuhn, 1982), however, it is not clear how the increased precipitation may have altered the hydrologic regime and geomorphic processes farther inland. The focus of our study was a reconnaissance effort to investigate the geomorphic response of hillslopes and channels to the increased precipitation between 1978 and 1983 in a variety of physiographic settings throughout San Diego County where rainfall records indicate that the last wet episode penetrated inland to the Sonoran Desert.

Physiography

San Diego County hosts a wide range of topographic, geologic, and climatic environments from low desert basins in the east near the Salton Sea Trough to the

----

[1]Department of Geology, Bucknell University, Lewisburg, PA 17837
[2]Quaternary Sciences Center, Desert Research Institute, U.Nevada, Reno, NV 89506

high, relatively moist ridges of the Peninsular Ranges in the west-central area. The area was divided into physiographic zones distinguished by relief, lithology, and river channel charcteristics (Fig 1). Upland mountains (UM) are characterized by precipitous slopes, narrow divides, and dense forest cover composed of coniferous hardwoods on wet mountain crests and chapparal tree and shrub vegetation along lower slopes. Lesser mountains and upland rolling terrain comprise a large segment of the uplands between the western marine terraces and the desert. The rolling uplands (UR) are vegetated by chapparal, grasses, and orchards. The upland piedmont (UP) is dominated by extensive alluvial fans which extend into the high desert fringes. Desert mountains (DM) are similar to upland mountains except they are less extensive and receive less precipitation due to the rain shadow produced by the upland mountains of the Peninsular Ranges to their west. Desert piedmont (DP) areas throughout the Sonoran Desert area of eastern San Diego County are dominated by a variety of alluvial fans and associated bajadas. Many of the desert piedmont regions are transitional into playa lakes of the low desert basins (DB).

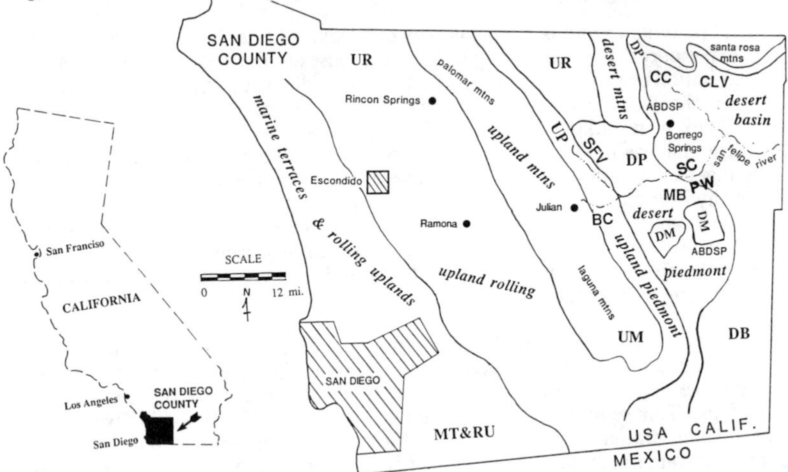

Figure 1. Index map of study area showing physiographic zonation of San Diego County. Average annual precipitation corresponds closely with topography. Precipitation increases eastward from 12 inches/year along the coast to 49 inches/year along the upland mountains and then decreases sharply to the east, to 12 inches along the piedmont and 3 inches in the desert basins. BC is Banner Canyon, MB is Mescal Bajada, PW is Pinyon Wash, SFV is San Felipe Valley, CLV is Clark Lake Valley, CC is Coyote Canyon, and ABDSP is Anza-Borrego Desert State Park.

Regional Geomorphic Response to 1978-1983 Wet Episode

Destabilizing hydrologic responses could occur in the form of mass movements along slopes as well as debris flows, floods, and channel erosion or deposition on alluvial fans and downstream reaches of the fluvial system. Our most detailed observational efforts focused on alluvial fans in the region because previous geomorphic research has shown that fans are particularly sensitive to minor changes in the external variables of sediment yield and discharge. Alluvial fans are likely to contain the best records of upstream hydrologic response to repeated minor climate changes.

The recent wet period (1978-1983) resulted in a wide variety of hydrologic responses which depended upon their ambient climatic conditions prior to 1978 and topographic characteristics. Markedly different geomorphic response to climate change could be anticipated if the theoretical relationship between effective annual precipitation and sediment yield predicted by Langbein and Schumm (1958) was accurate. Their study indicated that maximum erosion rates would coincide with semi-arid areas (having effective annual precipitation near 12 inches). Their model predicted that sediment yield would decline rapidly toward either more arid or more humid conditions. Upland regions typically had more than 15 inches of average annual precipitation prior to 1978, therefore, their ambient position would have been to the right of the peak in the sediment yield curve of Langbein and Schumm (1958). Upland regions showed insignificant response to the increased rainfall during the wet period. Inspection of aerial photographs and observations by local ranchers between 1976 and 1983 indicate that upland regions generally experienced increased vegetative growth throughout the wet period, which likely promoted increased slope stability.

Upland piedmont regions, dominated by extensive alluvial fans in the San Felipe Valley, experienced a mixture of hydrologic responses to the wet interval. The average precipitation regime for these regions preceding the wet period was close to the peak (12 inches) of the Langbein-Schumm curve. For at least one major stream, Banner Creek near Julian (Fig. 1), significant flooding and erosion occurred only following an intense rainfall event in February 1980. These preliminary observations indicate that a critical threshold of cumulative precipitation had to be exceeded before extensive hydrologic responses were triggered. After saturated conditions existed, a single heavy rain produced major flooding. Similar single-event storm precipitation totals in excess of the 1983 event at other times during recent decades failed to produce significant flooding because they were not preceeded by substantial antecedant rainfall amounts.

Inspection of aerial photographs taken prior to and near the end of the wet period showed significant flooding and erosion along several high-gradient channel reaches and extensive deposition of fine-grained sediments in proximal regions of large alluvial fans. This sediment, dominated by grus weathered from granitic bedrock in the headwater drainages of the Volcan Mountains, was deposited by assumed hyperconcentrated sediment flows. Field inspection in 1988 of the sites of fine-grained deposition on alluvial fans observed on aerial photos revealed little remaining evidence of the flood effects which occurred only 5 years previously. This is in sharp contrast with the expected recovery rates for fluvial systems in semi-arid regions suggested by Wolman and Gerson (1978). The relatively fine grain-size of the flood sediment appears to have facilitated rapidl revegetation by grasses within several years in spite of the semi-arid climate. This indicates the importance of the texture of flood sediments as substrate materials when considering recovery rates.

The San Felipe Valley

Extensive accumulation occurred on alluvial fans and downstream floodplains along the San Felipe River Valley between Borrego Springs and Julian (Fig. 1). Evidence from repairs made to ranch fences showed that extensive floodplain sedimentation commenced during the large flood of February 1980, but continued uninterrupted for several years following the termination of the wet period. Aggradation by the San Felipe River resulted in the burial of a major highway bridge just upstream from Sentenac Canyon during this same interval.

Figures 2a,b show four distinct episodes of high precipitation between January 1980 and October 1983, all coincident with the January-April winter months. Significant runoff (measured at the gaging station at Sentenac Canyon) occurred only during February-March 1980 and February-June 1983. Cumulative plots (Figs. 2c,d) reveal similar trends, but show the lack of runoff during wet seasons in 1981 and 1982 in spite of precipitation intensites similar to those in 1980 and 1983. Apparently, significant runoff occurs in the San Felipe River only following extended periods of substantial precipitation capable of exceeding the storage capacity of the surficial deposits (Fig. 2e). The most likely storage sites are the extensive alluvial fans along the western margin of San Felipe Valley, composed of very permeable gravels and grus deposits. Figure 2e indicates that in spite of the excessive precipitation during January-April 1980, much of the water entered storage, and was subsequently released during May-December when runoff was substantial during a period of very limited precipitation. Between January 1981 and May 1983 almost all of the accumulated precipitation entered storage. Apparently, the storage capacity was exceeded during the wet months in early 1983, triggering a steady release of water into the river throughout the remainder of 1983. The extensive aggradation observed in the San Felipe Valley bottom continued unabated for several years while stored water was being released during the dry period. This sequence of events illustrates how important antecedent events can be in the triggering of complex response phenomena in river and fan systems.

Interestingly, downstream from Sentenac Canyon, the San Felipe River experienced excessive channel widening and incision during this same time interval. The response of this segment of the river channel is in sharp contrast to the extensive aggradation upstream of the constricted reach at Sentenac Canyon in the San Felipe Valley and may reflect complex controls on sediment transport and storage within the San Felipe River system which need to be understood when generating models predicting hydrologic response to short-term climatic fluctuations.

Desert Regions

Channels in desert piedmont regions experienced considerable, but complex changes in response to the 1978-1983 wet interval. Numerous alluvial fans in the desert areas dominated by fluvial sedimentation showed evidence of significant episodes of recent aggradation and cutting along their active washes based on dendrogeomorphic observations of woody vegetation in channels and low terraces. These washes are characterized by complex terraces recording recent cutting and filling episodes which likely correspond to significant changes in upland piedmont portions of their watersheds. Virtually all of the desert channels showed signs of significant erosional activity from the period encompassing the recent wet interval. Figure 3 shows the presence of three well-developed terraces studied in detail on Pinyon Wash which we believe to be representative of climatically driven episodes of valley aggradation and channel entrenchment in the desert alluvial fan areas.

Our observations suggest that two episodes of aggradation occurred in Pinyon Wash since 1938. The first depositional phase, designated Qa2, is temporally bracketed with stratigraphic and dendrogeomorphic field relationships between 1938 and 1947. The second recent aggradational phase occurred within the last ten years. Both aggradtional episodes correlate with anomalously wet intervals in the precipitation records from the region.

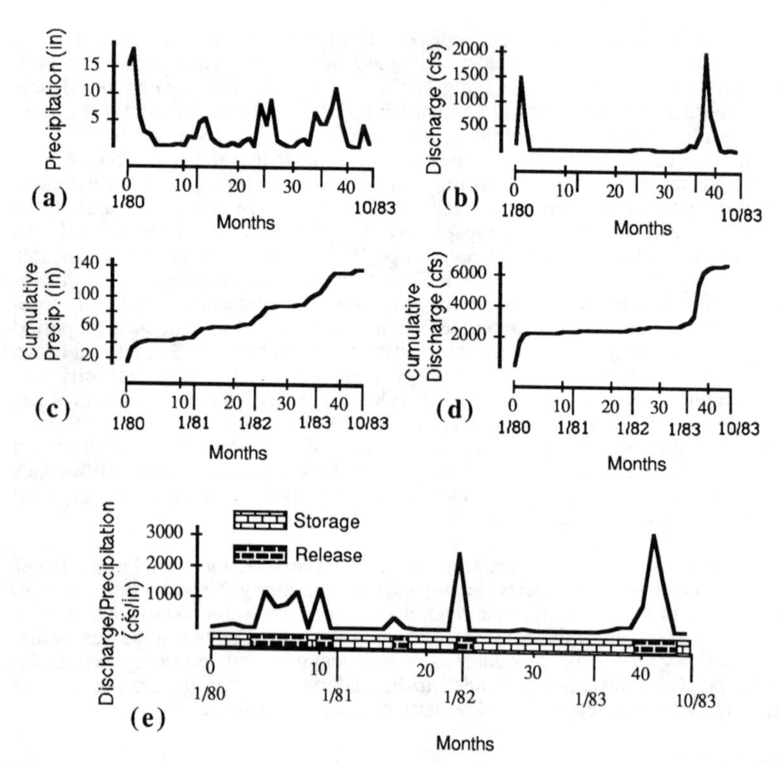

Figure 2. Precipitation and runoff in the San Felipe River Valley, 1980-1983. Precipitation was measured at Julian; mean daily discharge was measured at Sentenac Canyon. a) Monthly precipitation, b) Monthly mean discharge, c) Cumulative monthly precipitation, d) Cumulative monthly mean discharge.

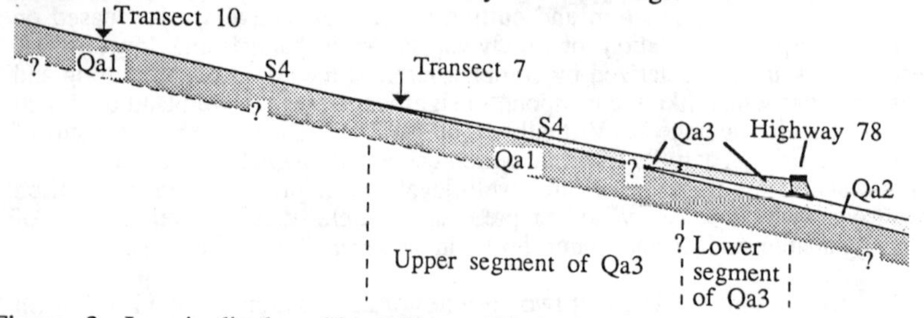

Figure 3. Longitudinal profile of Pinyon Wash showing the various Quaternary alluvial units underlying the modern wash (S4).

Summary

Reconnaissance studies in eastern San Diego County indicate that the recent wet interval (1978-1983) prompted different geomorphic and hydrologic responses in different areas. All physiographic subareas experienced similar increases in precipitation. The character of response to this climatic perturbation apparently depended on local, physiographically controlled variations of ambient precipitation.

Geomorphic response, manifested by erosion or deposition, can be linked in a general way to the increase or decrease in sediment yield predicted by the Langbein-Schumm model. Upland mountain regions having high ambient precipitation were unaffected by the increased precipitation. Detailed studies in a desert piedmont basin suggests that deposition occurred during the two previous wet periods (1938-1945; 1978-1983). Presumably, sediment stored on slopes in arid desert uplands was flushed into the lower subareas during conditions of greater precipitation. Surface flow needed to transport the debris probably became influent as it crossed the porous valley floor alluvium. Runoff was most likely stored in the underground system, resulting in a downstream decrease in discharge that ultimately caused deposition.

Semi-arid valley bottoms and piedmont zones underlain by alluvial fans experienced complex geomorphic and hydrologic responses. Runoff in the piedmont zone appears to be extremely sensitive to the volume of storage available in the alluvial fan deposits. As a result, intense precipitation did not always generate high discharge in the San Felipe River. Heavy precipitation occurring in the early part of the wet interval produced less runoff than expected because the system possessed large storage space after decades of dry conditions. In contrast, less intense precipitation produced significant San Felipe discharge during the latter phase of the wet interval because less underground storage space was then available.

Our observations reinforce the well-known fact that analysis of antecedent precipitation is critical in any interpretation of hydrologic and geomorphic responses to rainfall. Of significance here, however, is that antecedance might be important over longer time spans (years) rather than the days or weeks considered as the priming period for major floods. In addition, it may be particularly important to consider prior precipitation over longer terms in regions prone to alternating wet and dry episodes in order to explain hydrologic events set within either of those intervals. Finally, we stress that our conclusions are based on limited observations gathered in a preliminary study. However, the results do indicate that short-term, minor climate fluctuations are capable of generating significant geomorphic responses. Although these observations indicate great complexity can be expected in predicting the response of alluvial fan and river systems to climate changes, we feel that these responses may be used as a basis for predicting the types of environmental adjustments expected during climate shifts and underscore the need for research aimed at understanding the linkages between climate, hillslope, and channel systems.

Research was supported by The National Geographic Society (Grant NSG3763-88).

References Cited

Emery, K.O., and Kuhn, G.G., 1982, Sea cliffs: Their processes, profiles, and classifications: Geol. Soc. Amer. Bull., 93, 644-655.

Ganus, W.J., 1976, Is southern California ready for a wet period?: Environ. Southwest, 492, 9-13.

Langbein, W.B., and Schumm, S.A., 1958, Yield of sediment in relation to mean annual precipitation: Trans. Amer. Geophys. Union, 39, 1076-1084.

Schulman, E., 1947, Tree-ring hydrology in southern California: Univ. Arizona, Tree-Ring Res. Lab. Bull. 4.

Wolman, M.G., and Gerson, R., 1978, Relative scales of time and effectiveness of climate in watershed geomorphology: Earth Surf. Proc. Landf., 3, 180-208.

Design Storms and Sizing of Flood Control Facilities

Timothy E. Sutko[1] and Syndi Flippin[2]

Abstract

Design storms which embody the rainfall characteristics of an area are used for the sizing of drainage facilities. The selection of a design storm may significantly influence the design and therefore the cost of those facilities and the ability of public agencies to implement a flood control program in a timely manner. In Clark County, Nevada, the SCS Type II storm was utilized for most rainfall-runoff studies performed prior to 1979. Subsequent to that time, 3-hour design storms have been used for Flood Hazard Studies, Master Plan development and other drainage studies. The Corps of Engineers developed five 6-hour storm distributions for their 1988 feasibility study of the Las Vegas Wash and tributaries. A study was conducted to determine what impact the use of 3-, 6- and 24-hour design storms have on the sizing of flood control facilities in the Las Vegas Valley.

Purpose

The purpose of this study was to investigate how the sizing of flood control structures is affected by the duration of the design storm. Three design storms were investigated: 1) 3-hour, 2) 6-hour, and 3) 24-hour storms. A secondary purpose was to investigate what impact the use of curve numbers (CN) rather than a uniform loss rate to determine excess rainfall (runoff) has on the sizing of flood control facilities. The impact on facility sizing was determined by the magnitude of the peak discharge (cfs) and the volume of runoff (acre-feet) from selected subareas.

[1] Senior Hydrologist, Clark County Regional Flood Control District, Las Vegas, NV
[2] Staff Engineer, ESI, Consulting Engineers, Inc.

Background

In 1985, James M. Montgomery Engineers (JMM) prepared the Flood Control Master Plan for the Regional Flood Control District (RFCD). Included in this work was the estimation of 100-year peak flows in Master Plan study areas throughout Clark County. These discharges were determined using the HEC-1 Flood Hydrograph Package. For this effort, JMM used a 3-hour design storm with an areal extent of up to 200 square miles.

Subsequent to the adoption of the Master Plan, the U.S. Army Corps of Engineers (Corps) developed estimates of the 100-year expected probability peak discharges for drainages within the Las Vegas Valley as part of their Feasibility Study of Las Vegas Wash and tributaries. While the Corps utilized the basic framework of the model developed by JMM, significant changes in some of the hydrologic parameters were incorporated. While JMM used one overland flow plane to describe the runoff response of areas modeled using the kinematic wave option, the Corps determined that it would be more appropriate to use two overland flow planes for each subarea. Using this scheme, one overland flow plane represented impervious areas and the other represented pervious areas. The Corps used a 6-hour design storm because of their belief that "a storm of this duration will account for almost all of the volume produced by summer thunderstorms..."

At the request of the Regional Flood Control District, the Corps also determined estimates of the 100-year computed probability discharge values. These values were determined using a uniform loss rate of 0.50 in/hr. In their previous study the Corps used a uniform loss rate of 0.35 in/hr. All other hydrologic parameters used in this modeling effort were the same as those used in the determination of the expected probability discharge values. In nearly all other respects, the model used by the Corps in their work is identical to the model created by JMM for the Master Plan.

Because there are, in many cases, significant differences in the peak 100-year flood flow values estimated for the Master Plan and for each of the Corps' two studies, questions have been raised regarding the selection of design storms and rainfall loss functions for the design of drainage facilities in Clark County. In a region with little available data for the calibration of rainfall-runoff models, these are legitimate questions.

Methodology

In an attempt to determine what effect storm duration and loss function have on the estimation of peak flows and volumes used for the design of drainage facilities, a comparative study was undertaken. For this work, the Corps' HEC-1 model was used to described selected drainage areas within the Las Vegas Valley. This model provided information on basin sizes, times of concentration, kinematic wave parameters and routing parameters. The Corps' model was used because it was more readily available than the JMM model.

Inasmuch as it was impractical to investigate the effect of different loss functions and storms for each subarea included in the Corps' model, only certain areas regarded as being representative of conditions in the Las Vegas Valley were selected. The drainage basins used in this study are listed in Table 1.

The computer models for each of these drainage areas were run using the Corps' computed probability uniform loss rate (0.50 inches per hour) and the RFCD 3-hour design storm. The 100-year point depth rainfall at McCarran Airport adjusted by a factor of 1.43 and the appropriate depth-area reduction factor (per Hydro 40) was used for each basin. Upon completion of the computer run, the peak and volume of runoff from each area were recorded. The 3-hour storm data was replaced with data describing a 6-hour design storm and the process repeated. Similarly, the SCS Type II storm having a 24-hour duration was used. Rainfall data for each subarea are presented in Table 1.

Following these computer simulations, the Corps' loss function parameters were replaced with the parameters utilized by JMM. Simulations utilizing the same storm data as described above were then run and the results recorded.

In the course of substituting the Corps' rainfall loss parameters with the JMM parameters, it was noted that JMM used a mixture of CNs and initial/uniform loss rates in many areas. In order to determine the effect of using curve numbers versus a uniform loss rate, a model was created which used only curve numbers to describe the losses. Curve numbers were determined using SCS soil survey maps. Simulations utilizing the 3-, 6-, and 24-hour storm data were then run and the results recorded.

Table 1. Rainfall data used for computer simulations.

Drainage Subbasin	Area (sq. mi)	3-hour Storm	6-hour Storm	24-hour Storm
B5	3.11	2.19	2.55	2.96
B3	6.24	2.08	2.49	2.93
AP	14.57	1.84	2.22	2.81
R6	18.16	1.81	2.19	2.75
F7	36.66	1.66	2.05	2.66
A1C	87.68	1.49	1.83	2.55
D4	130.21	1.41	1.77	2.49

3-hour point depth = 2.48"
6-hour point depth = 2.77"
24-hour point depth = 2.96"

Results

The following conclusions were reached from the evaluation of the computer simulation results:

1) Regardless of the rainfall loss function used, the 24-hour storm resulted in a higher peak discharge and a larger volume of runoff than the 3- or 6-hour storm.

2) When the Corps' uniform loss rate parameters were employed, the 3-hour storm produced a higher peak discharge and a larger volume of runoff than the 6-hour storm.

3) When either JMM's loss rate parameters or the CNs were employed, the 6-hour storm produced higher peak discharge and a larger volume of runoff than the 3-hour storm.

4) When CNs were used to describe the loss function, the use of a 6-hour storm resulted in a 11-30 percent larger peak discharge and a 27-52 percent larger volume of runoff than a 3-hour storm. Generally, the larger the area, the larger the difference when using the two different durations.

Discussion

For the areas modeled in this study, the use of a longer duration design storm resulted in higher peak discharges and a larger volume of runoff than when a shorter duration design storm was used.

A noteable exception to this trend was the 6-hour storm using a uniform loss rate. This combination of parameters produced a lesser volume and lower peak discharge of runoff than the 3-hour storm used with the uniform loss rate. The reason for this exception is attributed to the fact that all of the rainfall at the beginning and the end of the storm are considered as losses. In fact, only when the intensity exceeds 0.50 in/hr is any runoff generated. Inasmuch as the total rainfall depth for the 6-hour storm is only 20-25% greater than the 3-hour storm depth and is distributed over an additional 3-hour period, it is not surprising that less runoff volume is generated by the 6-hour event.

The peak discharges, however, are a function of the rainfall intensity rather than the total depth. In nearly all cases the short duration (5, 10, 15 minute) rainfall intensities for the 3-hour storm are greater than those of the 6-hour storm.

On the other hand, when CNs are used to model rainfall losses, those losses become a function of the rainfall intensity and do not remain constant. The higher the rainfall intensity, the greater the losses. While the 3-hour design storm is more intense than the 6-hour event, the increased losses result in lower peak discharges. Because the losses are not constant throughout the storm, runoff is generated for nearly the entire duration of the event. Therefore, the 6-hour storm, which has a greater total depth, generates a larger volume of runoff.

Given that the longer duration design rainfalls tend to generate higher peak discharges and greater volumes of runoff, it should be expected that increased costs will be incurred for the construction of drainage facilities if these storms are used for the design of those facilities.

Conclusions

For the areas modeled in this study, the use of a uniform loss rate in combination with a 3-hour design storm results in higher peak discharges and a larger volume of runoff than resulted from a 6-hour design storm. When the SCS Curve Numbers were used to model the rainfall losses, the 6-hour design storm generated higher volumes and peak flows than did the 3-hour design storm. Regardless of the rainfall loss function used, the 24-hour design storm resulted in higher peak discharges and larger volumes of runoff than the shorter duration storms.

It should be expected that increased costs will be incurred for the construction of drainage facilities in Clark County if longer duration design storms are used for the design of those facilities.

References

James M. Montgomery, Consulting Engineers, Inc. (1986). Flood Control Master Plan for Clark County Regional Flood Control District.

U.S. Army Corps of Engineers (1988). Hydrologic Documentation for Feasibility Study, Las Vegas Wash and Tributaries, Clark County, Nevada.

U.S. Army Corps of Engineers (1988). Special Flood Hazard Study, Las Vegas Wash and Tributaries, Clark County, Nevada.

A VALUE ENGINEERING / RISK ANALYSIS APPROACH TO OPERATION AND MAINTENANCE OF HYDRAULIC STRUCTURES

David W. Eckhoff[1], M. ASCE, and **Jeffrey R. Keaton**[2], M. ASCE

Abstract

Value engineering is an objective, systematic method for minimizing cost of a system. Risk analysis is a method of quantifying uncertainties or probabilities of possible economic loss, physical damage, or personal injury associated with a particular natural (geologic) setting or physical system condition. Together, the two methods provide a perspective for identifying and evaluating in comparable terms all hazards and processes and a means for intelligent decision-making.

A multi-disciplinary team of technical specialists is well suited to identify hazards and processes potentially affecting a facility or system. This team can develop reliable subjective estimates of the probability that a hazard will occur at a potentially damaging intensity or that a process will lead to system component failure requiring repair or replacement. The team must be well versed in hazard identification, likely system response, and appropriate repair and improvement techniques. Capital costs associated with repair and/or improvement techniques, combined with annualized probabilities of hazard occurrence or system failure, can be used to estimate annual expected values of risk exposure. Present worth of expected values can be calculated with conventional economic approaches using inflation and discount or interest rates over a design life. Systematic assessment of alternative responses to potential damage or injury permits selection of the most economical alternative which provides an acceptable level of risk.

Introduction

A value engineering / risk analysis approach provides (1) a perspective for identifying and evaluating in comparable terms all hazards and processes which might affect a structure or facility and (2) a means for intelligent decision-making based on these evaluations. Value engineering (VE) is an objective, systematic method for minimizing the total cost of a facility or system for a specific number of years, and, therefore, is an appropriate hazard management approach. 'Total cost' means ultimate costs to construct, operate, maintain, and replace a facility or system during its design life. The VE approach is a creative effort directed toward the analysis of functions; it is concerned with eliminating or modifying those aspects which *add cost* without

[1] Chairman, Eckhoff, Watson & Preator Engineering, 1121 East 3900 South, Building C-100, Salt Lake City, Utah 84124

[2] Vice President, Sergent, Hauskins & Beckwith Engineers, 4030 South 500 West, Suite 90, Salt Lake City, Utah 84123

adding function -- or, in the case of hazards, *without reducing risk.* In this context, a *hazard* is a naturally occurring or man-induced process which has the potential to cause financial loss, damage to property, or injury to people. *Risk* is exposure of something of value to potential damage or injury due to the occurrence of a hazard.

Systematic evaluation of hazards and risks permits assessment of the costs associated with alternative responses which provide an acceptable level of risk of damage to elements of a facility or system and to the facility or system as a whole. The initial phase in assessment is recognition of the hazard; failing to recognize or ignoring hazards may lead to liability for damage caused by occurrence of the hazardous processes at damaging intensities. Hazards may be evaluated in terms of the extent of exposure, probability of occurrence of the hazards, probability of damage given the occurrence of the hazards, and potential consequence of damage. Selection of an acceptable level of risk can be very difficult and, in many cases, is a public policy issue or, in some cases, owner-specified. In many cases, an expert team collectively may estimate or recommend an appropriate level of acceptable risk. Five alternative responses to recognized hazards exist and should be assessed sequentially; these responses are shown on Figure 1, a 'Hazard and Risk Evaluation Worksheet' for hazard management.

The first response, continue current practices, is not the same as 'ignore the hazard' because the nature and extent of the hazard are understood by the time this response is considered. If the risk associated with this response is acceptable, then the cost associated with the risk is estimated; if the risk is not acceptable, then the cost need not be estimated. The second response, modify the hazard, may be appropriate for some situations in which risk can be reduced by preventing a hazard from occurring (e.g., stabilize a landslide). The third response, modify the system (e.g., place an aqueduct in a tunnel under a landslide), may be appropriate for situations where modification of a hazard is very costly or technically unfeasible. The fourth response, modify operation (e.g., place warning signs to keep the public away from dangerous places on the system), may be appropriate for situations in which the physical system is functioning acceptably under most conditions. The fifth response, avoid the risk, is a last resort option to be used only if the other responses do not result in acceptable levels of risk. Avoidance means abandoning existing facilities or eliminating from further consideration a site for new construction.

Value Engineering Approach

The value engineering approach to geologic hazard risk management described by Keaton and Eckhoff (1988) analyzes function or method by asking such questions as: What is it?; What does it do?; What must it do?; What does it cost?; What other material or method could be used to do the same job?; What would be the cost? The VE approach is ideally suited for risk management required by hydraulic structures exposed to geologic and other hazards. An appropriate model for geologic hazard risk management was presented by Keaton and Eckhoff (1990).

The quantitative result of a multi-disciplinary group in generating ideas has no parallel. The expert team interaction not only results in a large number of ideas but also improves the creative ability of the participants. Many reasons exist for the large quantity of group ideas obtained but perhaps most important is the aspect of inter-disciplinary communication (Cooper and Chapman, 1987). Many times, one member's idea motivates the associative processes of other group members. This

HAZARD AND RISK EVALUATION WORKSHEET

Project Element ____________________

Subelement ____________________

Nature of Hazard(s) ____________________

____

Risk: Consequence(s) ____________________

____

Exposure(s) (Narrative) ____________________

____

(Statistical) ____________________

____

Acceptable Risk Level(s) ____________________

____

RESPONSE ALTERNATIVES

Response	Acceptable Risk?	Capital Cost	x	Annual Exposure Probability	=	Annual Expected Value
(1) Continue Current Practices	________					
(2) Modify Hazard..........	________					
________						
________						
(3) Modify System	________					
________						
________						
(4) Modify Operations	________					
________						
________						
(5) Avoid Risk?						

COMMENTS ____________________

____

REFERENCES ____________________

____

Figure 1. Hazard and risk evaluation worksheet.

phenomena produces a chain reaction, triggering many ideas, and the cycle repeats itself.

In the analytical phase, sometimes called the evaluation and investigation phase, the team examines the alternative responses generated during the preceding phase, and then develops them into lower-cost alternative solutions. The principal tasks are to evaluate, refine, and cost-analyze the ideas and to list feasible alternatives in order of descending savings potential. During this phase, the ideas must be refined to meet the necessary environmental and operating conditions of the particular situation. Ideas which obviously do not meet these requirements are dropped from further consideration.

Case History

The authors recently directed a Value Engineering / Risk Analysis effort for the appraisal of an existing hydropower aqueduct in mountainous terrain exposed to landslide and earthquake hazards, as well as a generally increasing maintenance burden caused by increasing facility age. The multi-disciplinary team consisted of a General Civil Engineer (team leader), an Engineering Geologist, a Hydropower Specialist, a Structural Engineer, a Metallurgist, and a Tunneling Specialist. The team identified by consensus the major elements and subelements of the system during an initial field reconnaissance of the aqueduct. The primary criterion was appropriate identification of elements and subelements so that unique and independent hazards and risks could be delineated. This facilitated grouping of similar hazard-risk scenarios so that categories of response could be developed and reviewed.

Hazard identification was the responsibility of each of the technical specialists on the team. The broad categories of hazard were: (1) deterioration of structural and mechanical elements; (2) erosion leading to potential undercutting and collapse; (3) protection of public safety; (4) malfunctioning elements; (5) rockfall impacting the aqueduct; (6) talus accumulations deflecting the aqueduct; (7) landslide headscarp encroachment undermining the aqueduct; (8) landslide movement displacing the aqueduct; (9) earthquake-induced buckling of aqueduct trestles; and (10) thermal strain inducing stresses which accelerate corrosion. After hazards were identified by individual team members, the team collectively reviewed the composite list of the major elements and subelements and jointly prepared worksheets (Figure 1) for each element. The worksheets consist of the following components: (1) elements and subelements; (2) nature of hazard(s); (3) risk consequence(s) and risk exposure(s); (4) acceptable risk level(s); and (5) response alternatives.

Risk exposures for each element and subelement due to each hazard were expressed in statistical terms by the appropriate technical specialists of the team; the team as a whole reviewed the rationale for each statistical risk estimate and collectively agreed by consensus on the final numerical value which was expressed as an annual probability. Subsequently, residual risks (after presumed implementation of corrective measures) were estimated in a similar fashion. The probability estimates were made in part on the basis of the results of specific published analyses (e.g., earthquake ground motion), and in part on professional, but subjective, judgments. A variety of subjective probability estimation techniques exist (Roberds, 1990); on this project, we used a modified combination of the informal expert opinion technique and the open forum technique.

Risk costs can be considered to be Expected Values. Investment decisions are often based on the Present Worth of the expected values (PW) which is calculated on the basis of annual inflation rates (j) and discount or interest rates (i) over the period of time of interest or the design life of a facility or system (t) in the following way:

$$PW = \sum_{n=1}^{t} (CEV) \left[\frac{1+j}{1+i}\right]^{n}$$

where CEV is the annualized expected value of the risk cost and n is incremented one year at a time for the total facility life. Calculation of present worth values for the variety of possible risk reduction measures permits a systematic assessment of alternatives in like terms which can facility decision-making. Similar calculations can be made for the present worth of future operating and maintenance costs.

Conclusions

The value engineering / risk analysis approach to operation and maintenance of hydraulic structures requires an integrated multi-disciplinary effort among engineers, geologists, socio-economists, and other technical specialists, working closely with those directly responsible for the hydraulic structures. This approach provides a systematic framework for considering all possible operation and maintenance strategies and scenarios, expressing them in like terms for meaningful comparison, and selecting the most cost-effective alternative. Additional research in and experience with subjective probability assessments will enhance the utility and acceptance of this approach.

Acknowledgments

Members of the team who participated on the case history project with the authors were Arthur Strassburger, Gregory Thorpe, Charles Pitt, and Gary Brierley. Joseph W. Anderson, Esq., Assistant United States Attorney, District of Utah, authorized our use of the general concepts developed and refined on the case history project described in this paper.

Appendix A - References Cited

Cooper, D. R., and Chapman, C. B., 1987, Risk analysis for large projects -- models, methods and cases: Chichester, John Wiley & Sons, 260 p.

Keaton, J. R., and Eckhoff, D. W., 1988, A value engineering approach to geologic hazard risk management [abstract]: Kansas City, Association of Engineering Geologists Abstracts and Program, p. 49.

Keaton, J. R., and Eckhoff, D. W., 1990, Value engineering approach to geologic hazard risk management: Washington, D. C., Transportation Research Board Paper Number 890208, Preprint, 18 p.

Roberds, W. J., 1990, Methods for developing defensible subjective probability assessments: Washington, D. C., Transportation Research Board Paper Number 890787, Preprint, 22 p.

FLOOD CONTROL IMPROVEMENTS ON ALLUVIAL FANS

by James D. Schall, M. ASCE,
Douglas W. Bender[1], M. ASCE
and Frank J. Peairs[2], M. ASCE

Abstract

Floodplain management on alluvial fans has become an increasingly important issue with continued growth and urbanization in the southwest. The design of flood control facilities for new development on alluvial fans must take into account potential debris flow conditions. A debris flow event results when sediment loading from an undeveloped watershed is very large, for example, after a major fire. Under such conditions the volume of sediment can nearly equal the volume of water, creating at times a slurry type flow event that can cause tremendous damage, and possible loss of life. Damages occur during debris flow events not only from floodwaters, but also from impacts such as deposition of sediment in streets and buildings, high impact boulder transport and deposition, unpredictable flow paths and flood-surge inundation.

Alternatives for managing urban development on alluvial fans include: mapping and zoning; debris basins; debris transporting channels; and various combinations of these approaches. Selection of the appropriate management approach depends on site specific conditions and generally, no single approach is universally applicable or appropriate. In the case of the Wild Rose residential development, located on an alluvial fan near Corona, California, the best alternative was design of a debris transporting channel. The Wild Rose project is a good case study of the important factors involved in the design of flood control facilities on alluvial fans.

---

[1]Water Resource Consultant and Vice-President, respectively, Church Engineering, Inc., Newport Beach, CA

[2]Chief of Planning Division, Riverside County Flood Control and Water Conservation District, Riverside, CA

Introduction

With continued growth and urbanization in the southwest land development on alluvial fan surfaces has become increasingly more common. Alluvial fans are formed where a stream enters a flatter slope, or emerges from a confined channel and is able to deposit sediment laterally (ASCE, 1977). Flow conditions on alluvial fan surfaces are quite different than conditions in the more common alluvial valley floodplain. One primary difference is the potential transport of large quantities of sediment during debris flows.

Traditional floodplain analysis and management techniques are not universally appropriate for application to alluvial fan development. The purpose of this paper is to review some unique features of debris flows that can occur on alluvial fans, and specific floodplain management techniques that may be appropriate for alluvial fans. The design of the Wild Rose residential development provides a good case study of the important factors involved in the design of flood control facilities on alluvial fans, with the selected design alternative based on debris transporting channels.

Alluvial Fan and Debris Flow Concepts

The primary problem in designing flood control facilities on alluvial fans is managing the volume of sediment, commonly referred to as debris, delivered by mountainous areas. The common sources of debris include landslides, rill erosion, dry ravel, channel bank collapse and scour of stream channel deposits (Wieczorek, 1986). Large quantities of debris loading can create very dense flow conditions (i.e., much higher density and specific weight than water alone) that further enhances sediment transport capacity, particularly of larger particles such as boulders. Such flow conditions have been commonly referred to as debris flows, although more recently the term hyperconcentrated flows, as a descriptor of conditions ranging from very high suspended sediment concentrations in streams to landslide conditions, has become more widespread.

Observations indicate that a debris flow, or a hyperconcentrated flow, involves all sizes of sediments and furthermore, that a phenomenon known as inverse grading develops in debris flows (Takahashi, 1980). Inverse grading results in finer particles at the bottom of the channel and coarser particles at the top. Boulders also accumulate and tumble at the front of the debris wave and form a lobe, behind which follows the finer-grained more fluidic debris. One of the most striking properties of debris flows is the relatively high fluidity, as demonstrated by flows that are 80-90 percent granular by weight moving in sheets about 1 meter thick over surfaces with slopes of 5-10 percent (Johnson and Rodine, 1984).

Kumar (1986) provides a more detailed description of flow conditions stating that the initial phase of a debris flow event is a mudflow traveling as a frontal wave or pulse, carrying large boulders, which demobilizes below the fan head. The trailing flood is a mud flood that deposits additional material and/or may entrain and rework existing deposits extending the deposition zone. The rearmost part of the pulse is a turbulent water flood with low to moderate quantities of sediment that further entrains and reworks debris. This series of flow events is repeated as successive debris flow surges. Surges generally originate from the temporary damming of the channel by debris, at constrictions (vegetation, boulders, or manmade), large scale soil slips, bank caving and occasional shallow landslides.

These flow conditions create unique dangers including flood-surge inundation, debris deposition, high impact boulder transport and deposition, and unpredictable flood paths (Kumar, 1986). These dangers must be recognized in designing flood control facilities on alluvial fans.

Floodplain Management Alternatives on Alluvial Fans

There are four basic ways of managing development on alluvial fans: 1) mapping and zoning to prohibit development in the areas of likely flooding and debris flows; 2) design of confined channels that transport flood and debris flows safely through the development; 3) detention of debris above the development and transport of the relatively clear water through the development; and 4) a combination of alternatives 2 and 3 where the debris basin is sized to contain debris from most events and only during large events is debris transported through the downstream facilities. The following paragraphs discuss some of the positive and negative factors with each alternative.

Accurate mapping of debris flows is difficult and requires master planning of the entire alluvial fan to be successful, potentially utilizing large areas of potential debris flow paths as open space. In general, mapping, zoning and master planning is a valuable and desirable approach; however, the technical skills and analysis for mapping and the regulatory and enforcement problems must be further researched and developed for this approach to be broadly implemented.

Transporting both water and sediment through a development (alternative 2 above) must be carefully engineered to be successful. Pederson and Kumar (1986) report that sediment carrying channels have been constructed in some areas where debris basins were not feasible; however, the channels have not always functioned as designed and deposition is difficult to avoid which may result in overflow. Sediment carrying channels do maintain the sediment supply in downstream channels which can be important for maintaining channel stability and equilibrium conditions in any natural channels below the development. This factor has added significance in southern California where the problem of beach erosion and beach sand supply from inland watersheds has become a critical issue.

Alternative 3, the use of debris basin and downstream concrete lined channels, has been the accepted standard for managing flood flows on alluvial fans in urbanizing areas. Experience has shown that this approach is effective; however, the costs of building and maintaining such facilities is high and there is still risk involved during extreme events. These risks relate to dam safety and the passage of sediment downstream if the basin does not perform as designed, or is subjected to sediment inflows greater than anticipated.

Alternative 4 represents a combination of Alternatives 2 and 3 where debris is trapped during the small events and transported during large events. This allows reducing debris basin size with the tradeoff of designing larger downstream facilities to carry bulked flow during large events. Consequently, the problems of alternative 2 (the transport of debris) are minimized without requiring construction of large basins with capacities that are infrequently utilized. This alternative is also viable when a large debris basin is not possible.

Wild Rose Case Study

The proposed Wild Rose Development is located off the Corona Freeway (I-15) about 7 miles south of Corona, California. The development is located on the alluvial fan of three side canyons which are tributary to

Temescal Wash. A potential debris flow event from any or all three side canyons was a primary concern in the design of flood control improvements.

Development of the Wild Rose property was feasible only if flood control facility improvements were utilized. There was not adequate land available for natural floodways or open-space areas (Alternative 1) that would allow the channel to move around on the alluvial fan. Therefore, Alternative 1 was not a viable option for the proposed development.

Transport of debris through the channels (Alternative 2) was a possible option, given the steep slope and high velocity that would occur in concrete lined channels. Channel alignment in the tentative tract map provided reasonably gradual curves and confluences, particularly at major confluences, which would facilitate transport of debris. However, conditions at the outlet of the development, including passage under a freeway with limited vertical clearance and a channel outlet into an existing, active gravel mining pit in Temescal Wash, created design complications.

Topographic limitations on the Wild Rose property limited debris basin volume (Alternative 3) for conventional earthfill structures. The recommended design debris production from each canyon significantly exceeded basin storage capacity. Therefore, complete trapping of the debris design volume was not possible.

Comparing the available basin capacity to the estimated debris loading indicated that debris storage volume was only one-third to one-fourth the estimated design debris loading. To evaluate the feasibility of Alternative 4 (both trapping and transport of debris) the return period of the event that would fill the available capacity was estimated. Results indicated the debris basins would be filled during the 6- to 8-year event; therefore, debris basin maintenance could be frequent and necessary after even small floods.

Given this condition the effectiveness of debris basins was diminished and the costs of construction and maintenance marginal relative to their benefit. Therefore, considering all factors it was recommended that Alternative 2 be selected for the Wild Rose Development, with all flood control facilities designed to transport debris delivered by the canyons.

Final design of the channel facilities on the Wild

Rose project has been completed. Design objectives for the debris carrying channels included minimizing grade breaks and channel curvature, utilization of bulking factors approaching 2.0 and maintenance of relatively high channel velocities to minimize potential deposition of debris. The channel bottom and the lower side slopes were designed with extra concrete thickness to provide a sacrificial wearing surface and reinforcement to provide durability and strength during a debris flow event.

Conclusions

Successful floodplain management on alluvial fans must consider the high sediment loading that can occur during a debris flow event. Four basic alternatives for alluvial fan floodplain management have been reviewed in this paper. Selection of the preferred alternative must be completed considering the specific limitations and constraints of each project. Generally, no single approach will be universally applicable or appropriate.

Appendix: References

American Society of Civil Engineers, 1977, Sedimentation Engineering, edited by Vito A. Vanoni.

Johnson, A.M. and J.R. Rodine, 1984, Debris Flow, Chapter 8 in Slope Instability, edited by D. Brunsden and D.B. Prior, John Wiley and Sons ltd.

Kumar, S., 1986, Engineering Methodology for Delineating Debris Flow Hazards in Los Angeles County, American Society of Civil Engineers, Proc. of Water Forum 86, Long Beach, CA, August 4-6.

Pederson G.J. and S. Kumar, 1986, Engineering Design and Other Mitigation Measures for Mudflow and Mudflood Areas, Proc. of Improving the Effectiveness of Floodplain Management in Arid and Semi-Arid Regions, Association of State Floodplain Managers, Inc., Las Vegas, Nev., March 24-26.

Takahashi, T., 1980, Debris Flow on Prismatic Open Channel, ASCE Jour. of Hyd. Div., Vol 106, HY3, March.

Wieczorek, G.F., 1986, Debris Flows and Hyperconcentrated Streamflows, American Society of Civil Engineers, Proc. of Water Forum 86, Long Beach, CA, August 4-6.

Risk Analysis as a Tool to Determine Spillway Design Capacities

Douglas Toy[1] and Dan Lawrence[2]

INTRODUCTION

The safety of 197 non federal dams in Arizona is regulated by the Arizona Department of Water Resources (ADWR). Sixteen of 71 high hazard dams in jurisdiction do not currently meet ADWR guidelines for spillway adequacy.

One of these is Lyman Dam located in northeast Arizona. Lyman Dam is an earth and rockfill structure located on the Little Colorado River approximately 11 miles south and upstream of St. Johns, Arizona. The dam is owned and operated by Lyman Water Company. The water is used to irrigate crops in the St. Johns area. Lyman Dam was originally constructed in 1912. It failed (killing eight people) and was rebuilt during the years 1915-1923. Additions were made in 1930, 1934 and 1948. The main dam is 53 feet high, 800 feet in length and 12 feet wide at the crest. The reservoir the dam creates has an active storage capacity of 31,150 acre-feet.

The ADWR has classified the dam as unsafe because of inadequate spillway capacity, seepage at the downstream toe, steep downstream slopes and an unusually high rate of vertical settlement of the dam itself.

Several studies examining the safety and possible repairs have been completed on the dam. In 1989, a feasibility study indicated that approximately 12.5 million dollars would be needed to reconstruct the dam to safely pass the Probable Maximum Flood (PMF). Since that time, funding from state and county sources has amounted to approximately 5.0 million dollars. However, this funding will take a period of eight years to generate. Total funding presently available from all sources is insufficient to reconstruct the dam to pass the PMF. The

1 Douglas Toy - Member ASCE; Deputy Director, Office of Engineering, ADWR, Phoenix, AZ

2 Dan Lawrence - Chief Engineer, ADWR, Phoenix, AZ

lack of funding has heightened awareness of the need to carefully allocate resources. Most recently, ADWR, Apache County Flood Control District and the consulting firm of Dames and Moore (1990) completed a risk assessment of the potential for overtopping failure under several repair options. The study did not include a direct assessment of repairs needed to strengthen the downstream slope. This paper is a report of that study.

RISK ASSESSMENT

The methodology used for the risk analysis of Lyman Dam included the identification of both economic and non-dollar-denominated consequences of dam failure for different rehabilitation options. The five rehabilitation options evaluated are briefly described as follows:

1. The existing dam with minor repairs to the outlet works and repairs to stabilize the dam embankment with no change in the conservation storage of 31,150 acre-ft.
2. The existing dam upgraded to safely pass 0.25 PMF with a new spillway with its crest at El. 5979.5 feet (an increase of 1.5 feet) and a conservation storage of 33,700 acre-ft.
3. The existing dam upgraded to safely pass 0.50 PMF with a new spillway with its crest at El. 5979.5 feet and a conservation storage of 33,700 acre-ft.
4. The existing dam upgraded to safely pass 0.75 PMF with a new spillway with its crest at El. 5979.5 feet and conservation storage of 33,700 acre-ft.
5. The existing dam upgraded to safely pass full PMF with a new spillway with its crest at El. 5979.5 feet and conservation storage of 33,700 acre-ft.

Each of the options were assigned different failure probabilities. Using these probabilities and the expected damages for each case, the costs to indemnify the owner against flood damages including the cost of rebuilding the dam were estimated (ASCE, 1988). It was assumed that the new dam, if required following the failure of the upgraded dam, will be rebuilt to safely pass the full PMF.

ADWR staff estimated flood damages downstream of Lyman Dam corresponding to flood elevations near the U.S. Highway 666 Bridge at St. Johns, Arizona. Those damages ranged from $2.34 million to $6.02 million. This included costs of emergency activities, etc.

Damage assessment for the five options investigated in the study included the routing of successively larger storm runoff hydrographs through the respective spillways up to the one which resulted in wave overtopping and a potential breach. It was assumed that a breach occurred when the maximum reservoir water surface elevation was reached within 2.5 feet of the top of the dam. It was assumed that the dam would not fail due to overtopping if it is upgraded to safely pass the full PMF.

To estimate the damages resulting from floods smaller than or equal to the design basis flood for each case, the respective storm runoff hydrograph was routed through the reservoir with the proposed spillway. The resulting outflow hydrograph was routed through the downstream river channel to estimate the maximum water surface elevation near U.S. Highway 666 Bridge at St. Johns (approximate mile 14.1 downstream of Lyman Dam). This water surface elevation was then used to obtain the corresponding flood damages using the depth vs. damage information developed by ADWR. Floods smaller than the 100-year peak are non-damaging.

To estimate the dollar-denominated consequences of each rehabilitation option, it was assumed that the residual life of the dam for all cases was 100 years, i.e., n = 100. The other variables required for the analysis are as follows:

i = (real discount rate) = 0.06 per year per dollar,
r =(nominal borrowing cost) = 0.09 per year per dollar,
s =(return on temporary investment) = 0.085 per year per dollar,
B =(total consequences of dam failure in dollars)

RESULTS

Using this data, the capital costs developed and the flood damages estimated, the dollar-denominated risks associated with each option were evaluated using the Expected Value Approach and the Indemnification Costs Approach (ASCE, 1988). The results of the expected value approach are abstracted below as are those of the idemnification cost approach.

The results presented indicate that if minimizing the dollar-denominated consequences is the only consideration, then the minimal option is preferable to the other rehabilitation options. In fact, even if the existing structure is strengthened/modified by investing a sum of up to 3.0 million dollars, this option will still remain preferable to the others.

RISK EVALUATION USING EXPECTED VALUES

Item	1 Minimal Action	2 0.25PMF Design	3 0.50 PMF Design	4 0.75 PMF Design	5 Full PMF Design
1. Probability of failure	$1x10^{-2}$	$2x10^{-3}$	$6x10^{-5}$	$2.5x10^{-5}$	$1x10^{-5}$
2. Expected annual damages (million dollars)	0.231	0.059	0.0205	0.0203	0.0202
3. Capitalized expected annual damages (million dollars)	3.84	0.98	0.34	0.34	0.34
4. Capital cost or repair (million dollars)	0.28	6.22	7.86	9.04	11.52
5. Sum of capitalized expected annual damages and capital cost (million dollars)	4.12	7.20	8.20	9.38	11.86

RISK EVALUATION USING INDEMNIFICATION COSTS

	1 Minimal Action	2 0.25 PMF Design	3 0.50 PMF Design	4 0.75 PMF Design	5 Full PMF Design
1. Probability of failure	$1x10^{-2}$	$2x10^{-3}$	$6x10^{-5}$	$2.5x10^{-5}$	$1x10^{-5}$
2. Dam failure consequences (million dollars)	3.82	4.97	5.94	6.83	0
3. Cost of reconstruction of dam to PMF (million dollars)	17.70	17.70	17.70	17.70	0
4. Total consequences of dam failure (million dollars)	21.52	22.67	23.64	24.53	0
5. Present worth of indemnification (B) (million dollars) (100 yr life)	4.61	2.55	1.99	2.05	0
6. Capital cost (million dollars)	0.28	6.22	7.86	9.04	11.52
7. Sum of present worth of indemnification and capital cost (million dollars)	4.89	8.77	9.85	11.09	11.52

The evaluations of economic risks presented indicate that so far as minimization of the dollar-denominated consequences is concerned, the minimal option is preferable. This option includes remedial repairs to the gate control tower and outlet tunnel, stabilization of existing spillway, underwater inspection of submerged structures, and strengthening the stability of the main dam.

The study indicates that the expected annual damages from the failure of Lyman Dam for various options get lower for each increase in spillway capacity. The risk assessment indicated that there were economic benefits to increasing the spillway capacity all the way to the PMF. However, the study suggests that the incremental economic benefits are not commensurate with the incremental construction costs. The clear cut economic solution is the minimal action option.

The selection of a safe rehabilitation option should be based on the evaluation of both dollar-denominated and non-dollar-denominated consequences presented.

In any case, installation of an adequate flood warning system should be coupled with the selected rehabilitation option unless the dam is upgraded to safely pass the full PMF.

The ADWR investigated the potential for loss of life for each option using the U.S. Bureau of Reclamation methods. These computations indicated that no loss of life would occur under any one of the options since the warning time is at least three and one-half hours.

The no loss of life and a decreasing incremental economic benefits to construction cost ratio are the site specific conditions that make risk assessments appropriate to this Dam.

FUTURE ACTIONS

Since a dam cannot be designed to completely eliminate the probability of failure after deciding upon a minimum acceptable level, the most realistic design criteria is to maximize benefits while minimizing risks within the funding constraints. As indicated previously, current information indicates that the existing dam has many inadequacies. Some have suggested that it may be a wise use of public monies to repair a dam to pass a lessor flood than the PMF rather than wait until sufficient funding to repair to that level can be obtained. In today's economic climate, funding for the dam to pass the PMF may never occur.

To broaden the scope of the risk assessment, ADWR is considering other non-economic issues. Clearly these other non-economic issues may have more of an impact on the public living in the flood zone below the dam than the pure economic considerations.

The ADWR is proposing to solicit input through a series of meetings with the potentially affected public. These meetings will be designed to heighten the public awareness of the risk associated with dams and to obtain their opinions on how much risk they are willing to assume.

The public will be given ADWR minimum spillway sizing criteria for Lyman Dam. They will also be given construction costs, expected damage impacts and costs, probability of the dam failing, available funding, and impacts of future tax rates (to obtain additional funding) on various spillway size alternatives. The ADWR will take the public input, along with any other pertinent information and determine minimum spillway size. The Dam owners may then be free to choose any spillway size equal to or greater than ADWR's spillway size.

Reference

Dames & Moore, 1990, Risk Assessment for Lyman Dam, Phoenix, Arizona.

ESTIMATION OF EXPECTED DAMAGES, INDEMNIFICATION COSTS AND JOINT PROBABILITIES OF DAM FAILURES

Anand Prakash[1], Fellow, ASCE

Abstract

This paper describes three methods to evaluate structural options for the rehabilitation of an existing dam using the dollar-denominated risk and capital cost associated with each option as the basis of comparison. Evaluation of risk requires estimation of joint probabilities of failures for different rehabilitation options involving different combinations of spillway sizes and dam heights. These probabilities are used to perform risk analysis using expected damages. The second approach evaluates alternative design concepts using indemnification costs based on failure probabilities within the residual life of the dam. The last approach uses failure probability within the residual life to perform conventional risk analysis. The methods of computations and results of the three approaches are illustrated by examples and the salient points of each are identified.

Introduction

The inventory of national non-federal and federal dams prepared by the US Army Corps of Engineers has identified over 66,000 non-federal and 3,000 federal dams with heights greater than 25 feet and storage volumes in excess of 50 acre-feet. Many of these dams are classified as high hazard structures and may cause severe damage to life and property in the event of failure. During the period 1960-1983, as many as 42 notable dam failures were reported within the continental United States. Since a dam is a man-made structure, its failure is less likely to be attributed to an act of God than to inadequate design, operation, or maintenance. Thus the owners and operators are often enjoined to bear the economic and environmental consequences resulting from its operation or failure. Therefore, an important consideration in selecting the design basis for a proposed dam or for the rehabilitation of an existing dam is the assessment of tangible and intangible damages attributable to the functioning or malfunctioning of this structure. This paper describes procedures to assess the flood damages associated with different hydrologic design bases for the rehabilitation of earth dams. Measures to rehabilitate an existing dam, e.g., widening the spillway, may result in increasing the outflows from the dam during all floods equal to or lower than the design basis flood. The flood damages resulting from these increased outflows along with those from an overtopping failure must be included in estimating the overall cost associated with a particular rehabilitation option.

Evaluation of Expected Damages and Indemnification Costs

A common approach to estimate flood damages associated with a particular rehabilitation option is to compute the expected value which is defined as follows (Prakash, 1985; USBR, 1986; and Yevjevich, 1972):

$$E(D) = \sum_{m=1}^{j} D(q_m) \quad DF(q_m) + L \sum_{m=j}^{k} \quad DF\,(q_m) \tag{1}$$

---

in which E (D) = expected dollar value of damages; m = 1 refers to q_o which is the non-damaging outflow from the dam in cfs; m = j refers to q_c which is the outflow from the dam in cfs at which overtopping failure of the dam may ensue; m=k refers to q_l which is peak outflow from the dam in cfs during a PMF event which is assumed to be the upper limit of probable floods; m = 1, j pertain to successively increasing values of outflow used to discretize the spectrum between q_o and q_c and m=j, k to those used to discretize the spectrum between q_c and q_m; $D(q_m)$ = dollar - denominated flood damages due to an outflow peak q_m; L = dollar - denominated flood damages resulting from dam failure including the cost of rebuilding the dam to safely pass the PMF; $DF(q_m) = F(q_{m-1/2}) - F(q_{m+1/2})$; and $F(q_m)$ = cumulative probability of q_m. In this method, all flood damages attributable to a particular rehabilitation option that are likely to occur throughout the full range of probable floods are accounted for. However, it uses the probability-weighted annual value of flood damages and under-rates the catastrophic nature of damages which may result from dam failure. In many situations, selection of the design basis may be greatly influenced by the enormity of the likely damages rather than their expected values.

The second method for the evaluation of flood damages uses the indemnification cost which is the expected present value of the opportunity cost of possible future compensation to fully cover the estimated conomic consequences of dam failure (ASCE, 1988). This indemnification cost is given by the following equation:

$$\frac{TC}{B} = \frac{r-s}{i}\,[1-(1-P)^n] + P\,\frac{i-(r-s)}{i\,(1+i)}\;\frac{1-\left(\frac{1-P}{1+i}\right)^n}{1-\frac{1-P}{1+i}} + (1-P)^n\,\frac{r-s}{i}\,\left(1-\left(\frac{1}{1+i}\right)^n\right) \tag{2}$$

in which TC=expected present value of cost to maintain a fund and to compensate for flood damages; B= monetary value of failure consequences; i=discount rate; r=nominal borrowing rate; s=rate of return on temporary investment; n=life of structure in years; and P=annual probability of failure. This approach does not include damages associated with rehabilitation which may recur during events smaller than the design basis flood for the dam due to spillway enlargement or other modifications to the dam. Therefore, it may under-estimate the damages which may require compensation. Note that these recurring damages are more real compared to those attributable to a postulated dam failure which is a low-probability event and may or may not occur sometime in the distant future.

Theoretically, risk is defined as the probability or percent chance of the exceedance or occurrence of an event, failure, or damage during the design life of a structure (Haan, 1977; Kite, 1978). Assuming that r=s, i.e., the nominal borrowing cost is equal to the rate of return on temporary investment and thereafter setting the real discount rate i=o, Eq 2 reduces to,

$$TC = B\,[1 - (1-P)^n] \qquad (3)$$

This is the value of risk as defined previously or the value of estimated damages weighted by the probability of exceedance of the design basis flood during the residual life of the rehabilitated dam. The value, TC, in Eq 3 may be called a modified expected cost. Using the probability of exceedance during the residual life of the dam as the weighting probability, the modified expected damage, $E^1(D)$, may be defined as follows:

$$E^1(D) = \sum_{m=1}^{j} D\,(q_m) \quad DF^1(q_m) + L \sum_{m=j}^{k} DF^1(q_m) \qquad (4)$$

in which $F'(q_m)$ = probability of at least one occurrence of q_m during the residual life of the dam. This approach considers flood damages throughout the full range of probable floods as also the weighting probability implied in the indemnification approach.

Joint Probabilities

The probability of failure of an earth dam due to overtopping is the joint probability of the occurrence of a flood which may cause overtopping of the dam and of the embankment material getting eroded by that overtopping so as to result in the development of a breach. Estimation of each of these probabilities requires judgement and has to be subjective. Since the results of risk analysis are sensitive to the estimated probabilities of failure, it is advisable to perform a sensitivity analysis using alternative estimates of probabilities.

Applications

The results of risk analysis using the aforementioned approaches for the rehabilitation of an existing earth dam are presented in Table 1.

TABLE 1

Results of Risk Analysis for Rehabilitation of an Existing Dam

Item	No Action	0.25 PMF	0.50 PMF	0.75 PMF	PMF
		Design Basis Flood for Rehabilitation Options			
Joint probability of failure	10^{-2}	$2\text{x}10^{-3}$	$6\text{x}10^{-5}$	$2.5\text{x}10^{-5}$	0
(i) Expected Value Approach					
Expected Annual damage	0.58	0.15	0.05	0.05	0.05
Capitalized expected annual damage	9.60	2.45	0.85	0.85	0.85
Capital cost of rehabilitation	0.28	6.22	7.86	9.04	11.52
Sum of rehab. cost and damages	9.88	8.67	8.71	9.89	12.37
(ii) Indemnification Approach					
Dam failure consequences	9.55	12.43	14.85	17.08	0
Cost of reconstruction of dam	44.25	44.25	44.25	44.25	0
Indemnification cost	11.53	6.38	4.98	5.13	0
Sum of indem. and rehab. costs	11.81	12.60	12.84	14.17	11.52
(iii) Use of Failure Probability During Residual Life of Dam					
Probability of failure	0.395	0.095	0.003	0.0012	0
Expected annual damage	22.84	6.67	2.08	2.00	2.02
Cap. expected annual damage	379.56	110.87	34.48	33.26	33.59
Sum of rehab. cost and damages	379.84	117.07	42.34	42.26	45.09

Note: i= 0.06, r=0.09, s=0.085, n=100 years.
All damages or costs are in million dollars.

Note that in this case, the expected value approach tends to under-rate the catastrophic nature of damages resulting from dam failure and indicates that the most cost-effective solution is to rehabilitate the dam to safely pass 0.25 PMF. The indemnification cost approach, on the other hand, gives a high weightage to the catastrophic damage resulting from dam failure and indicates that the dam be rehabilitated to safely pass the PMF. The computations using probability of failure within the residual life of the dam suggest that the dam should be upgraded to safely pass 0.75 PMF.

Conclusion

Three different approaches are presented to evaluate the risks associated with overtopping failure of an existing earth dam rehabilitated for different design-basis floods. These approaches include the expected value method, indemnification cost approach, and one based on probability of failure within the residual life of the rehabilitated dam. It is found that the three aproaches identified three different preferred rehabilitation alternatives based on economic considerations alone. The results of each approach are sensitive to the assumed probabilities of extreme flood events, joint probability of dam failure, and the relative costs of rebuilding the dam, rehabilitation of the dam, and associated flood damages. It may be emphasized that non-dollar-denominated consequences of flooding, though not included herein, may govern the final decision in many situations and must be evaluated separately.

References

1. American Society of Civil Engineers (ASCE), 1988, Evaluation Procedures for Hydrologic Safety of Dams.

2. Haan, C.T., 1977, Statistical Methods in Hydrology, The Iowa State University Press, Ames, Iowa.

3. Kites, G.W., 1978, Frequency and Risk Analyses in Hydrology, Water Resources Publications, Fort Collins, Colorado.

4. National Research Council (NRC), 1985, Safety of Dams, Floods and Earthquake Criteria, National Academy Press, Washington, D.C.

5. Prakash, A., 1985, Impacts of Risk-Based Analysis on Current Design Practices, Engineering Foundation Conference on Risk-Based Decision Making in Water Resources, Santa Barbara, California.

6. US Bureau of Reclamation (USBR), 1986, Guidelines to Decision Analysis, Denver, Colorado.

7. Yevjevich, V., 1972, Probability and Statistics in Hydrology, Water Resources Publications, Fort Collins, Colorado.

Future Transfers of Water and Conveyance Capacity Between the CVP and the California SWP

John Burke[1]

Abstract

On November 24, 1986, the Bureau of Reclamation (USBR) and the California Department of Water Resources (DWR) executed an agreement known as the Coordinated Operations Agreement or COA. The signing of the COA was the culmination of years of negotiations between the USBR and the DWR who are, respectively, the operators of the Central Valley Project and the State Water Project. The agreement identified the project facilities and water supplies of each party subject to the agreement. It outlined the rights and responsibilities of the parties in meeting their mutual objectives, including a method of numerically computing each Project's share of water available for export from the Sacramento-San Joaquin Delta. One of the primary objectives of the COA is addressed in its "Article 10h" which provides the framework for the negotiation of a separate contract to "assist each party in making more efficient use of the water project facilities and water supplies contemplated in this agreement". In other words, Article 10h contemplates bringing together the CVP's water supplies, which exceed its conveyance capability, with the SWP's conveyance capacity, which exceeds its water supply. The contract will be subject to Congressional ratification.

Introduction

The process of negotiating the "Wheeling, Purchase, Exchange Contract" (herein referred to as "10h" for brevity) was initiated in May 1987 at a public workshop sponsored by the Bureau of Reclamation (USBR) and the

---

[1]Hydraulic Engineer, Central Valley Operations Coordinating Office, Mid-Pacific Region, Bureau of Reclamation, 2800 Cottage Way, Sacramento, CA 95825.

California Department of Water Resources (DWR). The negotiating sessions have been conducted as open meetings during which interested parties are invited to provide comments, suggestions, or ask questions. Besides the two principle agencies, there has been significant participation by the Fish and Wildlife Service (USFWS), the California Department of Fish and Game (DFG), the State Water Resources Control Board (SWRCB), both the State and Federal water contractors, and the Natural Resources Defense Council (NRDC).

By means of the "10h" contract, DWR hopes to purchase a portion of the "firm" water supply of the CVP which can be made available for about 30-40 years prior to its withdrawal for use by other CVP contractors. The CVP water purchased by DWR will supplement SWP water supplies to meet the anticipated growth in its contractors requests. The USBR will obtain, via "10h", an amount of conveyance in the SWP's California Aqueduct, with priority equal to that of other SWP contractors. The lack of conveyance capacity South of the Delta has limited the full development of the CVP's water supplies.

The end result of these negotiations, if successful, will be a water transfer of unprecedented magnitude in California, without the need for any new physical facilities. The negotiations are now considering a total combined transfer (including both the purchase and conveyance amounts) of 300,000 to 500,000 acre-feet per year.

<u>Status of Negotiations Process</u>

As of March 1990 the DWR and USBR have accomplished the following in the "10h" negotiating process:

1. sustained significant involvement of many "interested parties", by conducting open negotiating sessions and a number of workshops on matters relating to the negotiations.
2. produced a first draft of the contract for review and comment.
3. produced a "scoping report" for the joint EIR/EIS.
4. completed several operations studies which have begun to clarify the scope and the operational consequences of the water transfer.

Contracting Principles

Some of the major contracting principles set forth in June 1987 to help guide the negotiations are:

1. Conveyance for USBR will have equal priority with SWP contractors on a "bucket for bucket" basis (a quantity of conveyance for USBR in exchange for an equal of water transferred to DWR).
2. Water sold to DWR must be in full compliance with state and federal legal requirements including Reclamation Reform Act.
3. Accounting and pricing procedures to be addressed
4. New or amended permits from SWRCB.
5. Withdrawal of water supply by USBR will require notice period (3-5 years) and only under specified conditions.
6. DWR will be treated consistent with other USBR long-term contractors when apportioning shortages.

Service Areas

CVP: Central Valley of California from Redding to Bakersfield area; also portions of the Bay Area. 5-10% M&I, over 90% Agriculture.

SWP: Feather River, portions of the Bay Area, Central Coast, portions of the Southern Central Valley, and Southern California. 60% M&I, 40% Agriculture.

CVP/SWP Project Capabilities

(maf= million acre feet)	CVP	SWP
Storage:	11.6 maf	4.6 maf
Annual Water Delivery:	8.0 maf	3.7 maf
Annual Diversion (at Delta):	3.3 maf	2.7 maf
Pumping Capability (at Delta):	3.3 maf	4.0 maf
(maximum rate):	4,600 ft3/s	6,400 ft3/s
Canal Capacity (at Delta):	4,600 ft3/s	10,300 ft3/s

Coordination Between the Projects

The CVP and SWP share responsibility for meeting Sacramento Basin water demands and water releases for salinity control and flow requirements in the Delta. The accounting of and coordination of this shared responsibility is covered in the COA. At present the CVP and SWP do exchange services to a limited degree. The DWR is permitted to wheel an annual amount of up to 194,000 acre-feet for the USBR to replace pumping foregone to protect striped bass in the Delta. The USBR's Cross Valley contractors (up to 128,000 acre-feet per year) and a few others receive non-priority pumping and conveyance through SWP facilities.

Water Rights and Permits

The SWRCB oversees water rights in California. It has the authority to issue, amend, and adjudicate them. SWRCB staff have participated in an advisory capacity in many of the "10h" negotiating sessions. This has been helpful to the negotiators in understanding SWRCB policy and procedures, and in establishing the timetable for acquisition of necessary permits from SWRCB which must consider the interaction of the "10h" process with other ongoing water rights proceedings. There are several proceedings underway in California that could have a bearing on both the DWR and USBR water rights, and could also bear directly on the implementation of the "10h" contract. Foremost among them is the Bay/Delta Estuary proceedings.

The Bay/Delta Estuary proceeding will re-evaluate the SWRCB's plan for protection of beneficial uses of water in the Sacramento-San Joaquin Delta. Decision 1485, the existing plan for Delta protection, was issued in 1978, and was incorporated into the COA by the USBR and DWR in 1986. The outcome of the Bay/Delta proceedings is expected to have a major impact on the direction of California's future water resource management. As part of the proceedings the SWRCB will consider modifying USBR permitted "points of diversion" to allow CVP water to be diverted at the SWP's Delta pumping plant. This permission is central to the "10h" wheeling-purchase concept.

Reclamation Law

Since the start of the "10h" negotiations, it has been acknowledged that a demonstration of compliance with Reclamation Law will be required in order to obtain Congressional ratification. The Reclamation Reform Act of 1982 (RRA) restricted the delivery of Reclamation water to farms with less than 960 acres. In 1988, with the help of the Kern County Water Agency, the "10h" negotiators were able to identify 4800 eligible landowners with 340,000 acres of eligible lands within Kern County alone. Under "commingling" provisions of the RRA, the SWP could "co-mmingle" Reclamation water with other SWP supplies for delivery throughout the SWP service area, as long as there are sufficient "eligible" farmers to receive the amounts of Reclamation water sold to the SWP.

Although the negotiators felt it was important to conceptually demonstrate compliance with acreage limitation provisions of RRA, it may be preferable that "10h" water sold to DWR not used for agriculture. Besides the issue of acreage limitations, an additional "place of use" permit would need to be acquired by the USBR from the California SWRCB. Also, any lands within the SWP service area receiving USBR water could need to be "classified" as to agricultural suitability. This would be an extensive and time consuming exercise.

The concept of the transfer of water from USBR to DWR for use as Delta water quality control has gained acceptance with the negotiators. DWR's intended use of the water for that purpose is expressed in the present draft contract. The use of water for Delta water quality control is an authorized use of water for both the CVP and SWP. Under the COA, the two projects share the responsibility to provide sufficient water to meet flow and salinity standards in the Delta. The COA makes it an operationally simple way to implement the "10h" transfer in practice. It is not thought to require new permits or water rights amendments. Finally, it will free up a like amount of SWP water supplies for delivery, thus accomplishing the same end as a direct transfer to SWP contractors.

Environmental Documentation

After considering several options for proceeding separately, the USBR and DWR agreed in 1989 to undertake a joint EIR/EIS to analyze the impacts of the "10h" contract. A notice of intent to proceed with this documentation was published in the federal register 8/17/89. Some of the stated probable environmental effects of the proposed contract are:

1. increased export pumping could impact fish survival, also endangered species.
2. changes in flow patterns in the Delta possibly affecting resident and migrating fish
3. increased drawdown of CVP reservoirs, possibly affecting fish, water quality, and recreation.
4. higher flow rates in the Sacramento River may impact riparian vegetation, water quality, recreation, and fish.
5. increased return flows from irrigated land in San Joaquin Valley could impact water quality, fish and wildlife.

A series of public scoping sessions was held in September 1989 to solicit public input. A scoping report has been drafted and will be finalized by May 1990.

The joint EIR/EIS will address the review and consultation requirements of NEPA and CEQA. Also the process will address applicable requirements of the Clean Water Act, Fish and Wildlife Coordination Act, the State and Federal Endangered Species Act, etc.

CVP/SWP Operations Studies

A technical subcommittee of the "10h" negotiating committee has the responsibility for carrying out operations studies of the coordinated operation of the SWP and CVP. The purpose of these studies are to develop and test the possible scope of the "10h" water transfer, to determine the operational impacts, to test operating strategies and to assist in determining the environmental impacts.

The operations studies are based on DWR's Planning Simulation Model of the CVP/SWP. This model simulates the operation of the projects on a monthly time increment. Features simulated include: reservoirs, pumping plants, canals, and operations for Delta water quality control. Hydrologic inputs to the model are taken from a historical data base for the years 1922 - 1978. Each operations study focuses on a particular level of development, (eg. 1990 or 2000 level of development).

The operations studies performed thus far seem to indicate that a "priority" wheeling/purchase quantity of 250,000 acre feet for each party may approximate an upper limit to the "10h" transfer. Future studies may attempt to refine purchase/exchange assumptions and explore "intermittent" purchases and exchanges between the two projects.

Future Developments

A recommended agreement should be submitted to the Directors of the USBR and DWR by the end of 1990. Completion of the EIR/EIS will take considerably longer, because of its interrelationship with the USBR's water contracting program, which is undergoing a re-evaluation. When the EIS/EIR is completed, and the necessary water rights permits have been obtained from the SWRCB, then a final agreement will be submitted to Congress for ratification.

Units of Measure

1 acre foot	=	1.233 cubic dekameters
1 acre	=	0.40469 hectares
1 ft3/s	=	0.028317 cubic meters/second

Central Valley Project Operations

by Michael Jackson [1]

Abstract

The Federal Central Valley Project (CVP) utilizes the water supply and storage regulated by the CVP's four major northern California reservoirs, Clair Engle, Whiskeytown, Shasta, and Folsom Lakes, to meet the majority of the CVP's needs and obligations. These northern reservoirs have a combined storage capacity of 8.25 million acre-feet. This storage is primarily regulated by making releases through the 6 powerplants and 2 power tunnels that are inter-related with the operation of these northern reservoirs. Water transfers from these northern facilities to the San Joaquin Valley are accomplished by providing sufficient flows to meet the instream needs of water rights holders, fishery, water quality standards in the Sacramento-San Joaquin Delta Estuary, and Delta exports. Energy generated by these northern facilities is used to help meet the power needs of the CVP's export and pumping facilities south of the Delta. The major southern facilities of the CVP include the Delta Mendota Canal, San Luis Canal, and the Federal portion of San Luis Reservoir in central California. These southern facilities are operated to transfer the water south of the Delta.

All of the CVP's needs and obligations are met throughout the year while the CVP operators do their best to balance the refill potentials of the major northern California storage reservoirs.

[1] Hydraulic Engineer, Central Valley Operations Coordinating Office, Mid-Pacific Region, U.S. Bureau of Reclamation, Sacramento California.

This paper is intended to give the reader an overview of how the CVP operators utilize the capabilities of the CVP facilities to transfer northern California water supplies to users throughout the Central Valley.

Introduction

The Central Valley Project (CVP) is one of the nation's major water conservation developments, and is the largest water project in California. The initial features of the CVP were authorized by President Roosevelt in 1935 for construction by the Bureau of Reclamation.

The CVP lies in the Central Valley Basin. This Basin includes California's two major watersheds. Namely the Sacramento River watershed in the northern portion of California and the San Joaquin River watershed in the southern portion of California. The primary impetus for the CVP may be attributed to the unequal distribution of precipitation versus agricultural lands within the Basin. The Sacramento Valley has one-third of the land and two-thirds of the rain, while the San Joaquin Valley has two-thirds of the land, but receives only one-third of the rain. The Sacramento River flows north to south about 300 river miles (480 km) and the San Joaquin River flows south to north about 200 river miles (320 km). Both of these Rivers converge in the Sacramento-San Joaquin Bay Delta Estuary (Delta).

The Delta encompasses an area of about 735,000 acres (300,000 ha). Delta lands consists largely of very fertile peat soils, which naturally makes for significant farming activities. The Delta is also home to many species of wildlife, fish, and plants. Export facilities in the Delta literally connect surplus northern California water supplies to deficient southern California water supplies. For these reasons the Delta possess many powerful water quality standards.

In extreme northern California the CVP has three major reservoirs on three separate rivers. The names of these reservoirs are Clair Engle Lake, Whiskeytown Lake, and Shasta Lake. Clair Engle resides on the Trinity River and has a storage capacity of 2,450,000 acre-feet (3020 dam3). Whiskeytown resides on Clear Creek and has a storage capacity of 240,000 acre-feet (297 dam3). Shasta resides on the Sacramento River and has a storage capacity of 4,550,000 acre-feet (5615 dam3).

These three reservoirs are connected by two tunnels with powerplants which can be operated to effect and regulate Sacramento River flows.

About 250 miles (400 km) downstream from Shasta, the American River forms a confluence with the Sacramento River. The CVP's Folsom Lake resides on the American River and has a storage capacity of 1,000,000 acre-feet (1245 dam^3).

The major Delta begins about 30 miles (50 km) downstream of the American River and Sacramento River confluence. The Tracy Pumping Plant (TPP) is about 60 miles (100 km) downstream of this confluence. The TPP is the second largest water export facility in the Delta with a pumping capacity of 4600 ft3/s (130 m3/s). It consist of an inlet channel, fish screen, six pumping units and three discharge pipes that lead into the headworks of the Delta Mendota Canal (DMC). Each of the six pumps is powered by a 22,500 hp (16,800 kW) electric motor and is capable of pumping about 800 ft3/s (23 m3/s). Power to run the pumps is supplied by CVP powerplants. TPP water is lifted about 200 ft (60 m) and travels through the three 15 ft diameter (4.6 m) discharge pipes up an inclined grade for about 1 mile (1.5 km) to the head of the DMC.

The DMC carries water southeasterly from the TPP along the west side of the San Joaquin Valley and is used primarily for meeting irrigation demands. The DMC is about 115 miles (185 km) long and terminates at Mendota Pool. The initial diversion capacity of the DMC is 4600 ft3/s (130 m3/s) and it gradually decreases to about 3200 ft3/s at Mendota Pool.

One of the major facilities on the DMC is the intake channel connecting the DMC to the O'Neill Pumping-Generating Plant. The O'Neill Plant is about 70 miles (110 km) downstream of the TPP.
The O'Neill Plant consist of the intake channel, six pump-generating units, and O'Neill Forebay. The O'Neill Plant is operated as a pumping plant whenever the DMC demands are less than TPP deliveries and is operated as a generating plant whenever the converse is true. During the fall, winter, and early spring months when irrigation demands are relatively low, these units act as pumps to lift DMC water about 50 ft (15 m) into O'Neill Forebay. And during the summer months when irrigation demands are high these units act as generators. Each of these units has both a pumping and generating capacity of about 700 ft3/s (20 m3/s).

O'Neill Forebay marks the beginning of several facilities that are jointly owned and used by the Federal CVP and the California State Water Project (SWP). The major joint-use facilities include O'Neill Forebay, William R. Gianelli Pumping Plant, San Luis Reservoir, Dos Amigos

Pumping Plant, and the San Luis Canal. The Forebay has a storage capacity of 55,000 acre-feet (70 dam3), and is utilized as a hydraulic junction point for Federal and State waters. It connects the State owned California Aqueduct to the jointly owned San Luis Canal and acts as a reregulating bay for the jointly owned San Luis Reservoir.

San Luis Reservoir has a storage capacity of about 2,020,000 acre-feet (2500 dam3). The CVP share of this storage is 965,000 acre-feet (1190 dam3) and the SWP share is 1,060,000 acre-feet (1310 dam3). It is connected to O'Neill Forebay by the William R. Gianelli Pumping-Generating Plant. The Gianelli Plant consist of eight pump-generating units. When operating as pumps these units can lift 1375 ft3/s (40 m3/s) at 290 feet (90 m) of total head. And when operating as generators these units have a capacity of 1640 ft3/s (45 m3/s) at the same head. Much like the O'Neill Plant, the Gianelli Plant is in a pumping mode when downstream demands are low, and in a generating mode whenever the converse it true. Water that is not pumped into San Luis Reservoir or generated to the DMC, travels 15 miles downstream to the Dos Amigos Pumping Plant. Dos Amigos consist of 6 pumping units with each capable of delivering 2200 ft3/s (60 m3/s) at 125 ft (40 m) of head. From Dos Amigos CVP water travels down the 100 miles (160 km) of the San Luis Canal with diversions occurring at numerous irrigation turnouts along the way.

Releases from CVP reservoirs are generally governed by flood control, water supply, water quality, irrigation demands, fishery, and power. Releases that fall under the flood control category are made for the immediate protection of public welfare, and therefore operator options are very limited. However, CVP operators generally have a myriad of options available to them for making releases under the other categories. The key to satisfying the particular need is making the **most efficient use of the water.** Choosing the reservoir which will make the most efficient use of the water is not always easy. CVP operators must consider both probable and possible impacts to all beneficial uses before the final decision is made.

Runoff and operational forecast often play a key role in helping the operators make their operational decision. Runoff forecast are made from February through May and are contractually required for much of the CVP service area. This is because many of the contracts have language which bases their water supply on the runoff forecast. Monthly inflows are derived from the runoff forecast for use in the operational forecast. The operational forecast models CVP monthly operations and gives a clue as to what flexibility, constraints, and conditions the operators will face in the near term future.

The Trinity and Shasta Divisions account for about 70% of the total CVP generation. The capacity and configuration of the powerplants and power-tunnels contained in these two divisions demonstrate why this is the case. These Divisions have a combined electrical capacity of 3950 megawatts and annually generate 3500 megawatt-hours of energy. This makes it easy to see why CVP operators first look to these Divisions whenever energy is a concern.

As mentioned earlier the Delta possess many powerful water quality standards. For most of these standards compliance is maintained by targeting a Delta Outflow quantity which is expected to provide a **reasonable** level of protection against non-compliance. The annual Delta Outflow requirement for maintaining compliance ranges from 3.0 million acre-feet (3.7 dam3) to 5.5 million acre-feet (6.8 dam3). Delta Outflow is a calculated value which represents the net flow leaving the Central Valley Basin and passing through the Delta towards the Pacific Ocean. In simplistic terms it is calculated by taking the difference between accretions, such as reservoir releases entering the Delta and depletions, such as export pumping leaving the Delta.

Achieving the targeted Delta Outflow during low flow conditions requires that the CVP and SWP coordinate their operations. This typically occurs between April and November. The contribution that each Project makes towards satisfying Delta needs is accounted for daily through the Coordinated Operations Agreement (COA).

The COA calculates contributions by crediting the appropriate Project for reservoir storage withdrawals, and deducting that Project's Delta export, and proportionate share of in-Basin use. As an example if the CVP was negative in the COA account, then the burden of sustaining the targeted Delta Outflow rest with the CVP. The CVP could sustain the Delta Outflow by either increasing CVP releases or reducing CVP Delta exports. If the operator elects to increase releases, then consideration must be given to the previously mentioned categories which govern releases as well as the current conditions at each reservoir. The operators must be careful not to draw to heavily on the chosen reservoir(s) so that a relative balance in the distribution of reservoir storage is maintained. The balance of reservoir storage is based on refill potential and the capability of the particular reservoir to meet local and downstream demands.

If the CVP operator chooses to reduce Delta exports he must be sure that the San Luis Reservoir storage and the remaining pumping capacity at the TPP can still accommodate CVP obligations south of the Delta.

In general, the CVP operators **strive** to meet all of the CVP's needs and obligations while making the most efficient use of the available water supply.

Sacramento-San Joaquin Delta
The Key to Water Transfer

Karl P. Winkler[1]

Abstract

The Delta is a key feature in California's water resources picture. This ecologically sensitive area is currently experiencing adverse effects from water diversions and needed solutions must satisfy today's regulatory and media-focused environment.

Introduction

Probably no water problems in California have involved more investigations or generated more controversy than those involving the Delta of the Sacramento and San Joaquin Rivers. The maze of islands and channels lying at the confluence of these two large rivers has become the focal point for a wide variety of water-related issues. Many different interests have a vital stake in the Delta: farmers, fish and wildlife, environmentalists, boaters, navigation, railroads, highways, and the people and industries that receive their water from the two large export systems, the Central Valley Project and the State Water Project.

The Sacramento-San Joaquin Delta, an area of 700,000 acres, was once a tule marsh fed by winter floodwaters, snowmelt, and tidal flows entering through San Francisco Bay. During flood season, the Delta became a great inland lake; when the floodwater receded, the network of sloughs and channels reappeared throughout the marsh.

Reclamation of the Delta began in the 1850s. By 1930, virtually all the marsh had vanished, to be replaced by farms growing barley, corn, pears, asparagus, and tomatoes. Many miles of entirely new channels had been dredged, and farmlands, small communities, highways, and utilities were protected -- often tenuously -- by 1,100 miles of levees, many of them built on peat soils. The present Delta is shown on Figure 1.

Both the CVP and SWP use the natural channels in the Sacramento-San Joaquin

1 Chief, Delta Planning Branch, State of California Department of Water Resources, 1416 Ninth Street, Sacramento, CA 95814.

Figure 1

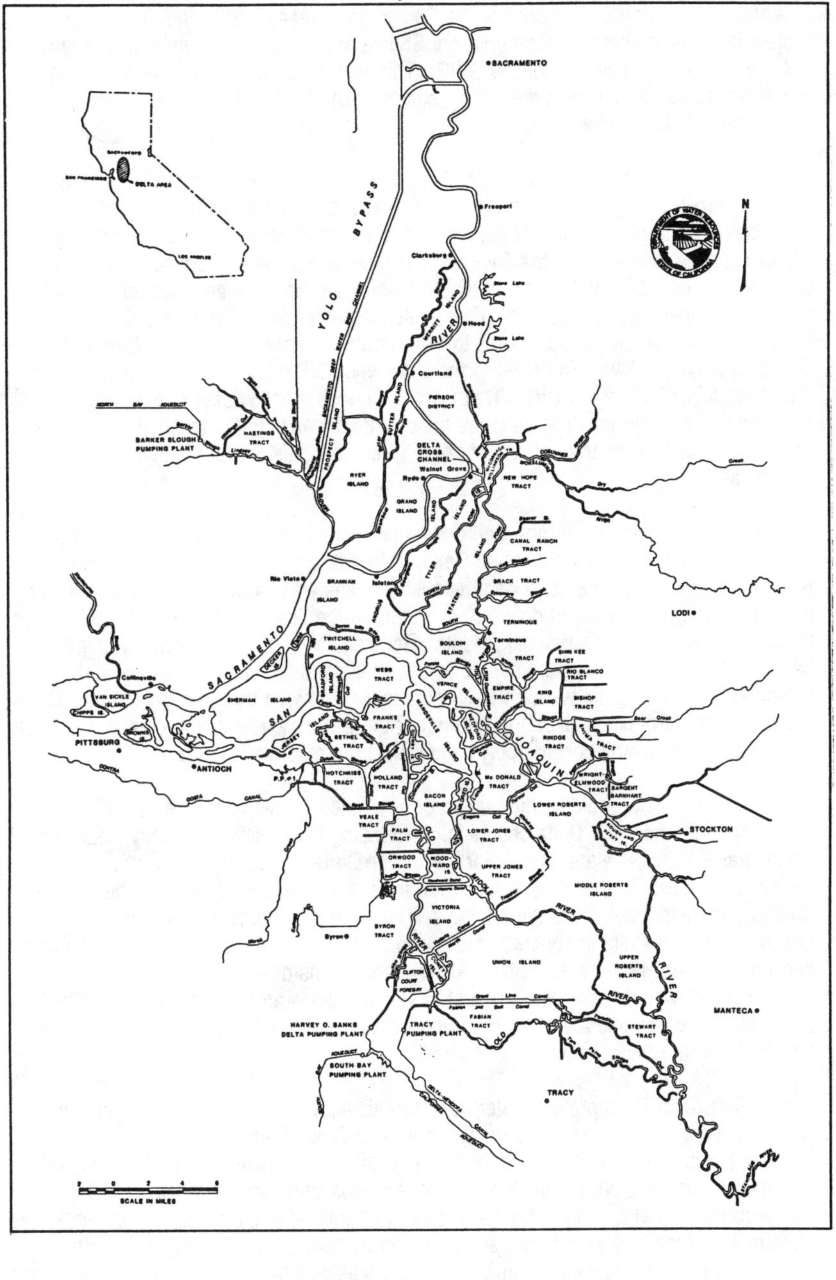
SACRAMENTO
Freeport
Clarksburg
Hood
Courtland
YOLO
BYPASS
SACRAMENTO DEEP WATER SHIP CHANNEL
RIVER
MERRITT ISLAND
SUTTER ISLAND
PIERSON DISTRICT
DELTA CROSS CHANNEL
Walnut Grove
Ryde
RYER ISLAND
GRAND ISLAND
PROSPECT ISLAND
HASTINGS TRACT
BARKER SLOUGH PUMPING PLANT
NORTH BAY AQUEDUCT
NEW HOPE TRACT
CANAL RANCH TRACT
BRACK TRACT
TERMINOUS
Terminous
LODI
Rio Vista
Isleton
BRANNAN ISLAND
TYLER ISLAND
TWITCHELL ISLAND
BOULDIN ISLAND
SHIN KEE TRACT
RIO BLANCO TRACT
KING ISLAND
BISHOP TRACT
EMPIRE TRACT
VENICE ISLAND
WEBB TRACT
BRADFORD ISLAND
SACRAMENTO
Collinsville
VAN SICKLE ISLAND
CHIPPS IS.
SHERMAN ISLAND
SAN
JOAQUIN
PITTSBURG
ANTIOCH
JERSEY ISLAND
BETHEL TRACT
FRANKS TRACT
MANDEVILLE ISLAND
RINDGE TRACT
WRIGHT-ELMWOOD TRACT
SARGENT BARNHART TRACT
HOTCHKISS TRACT
HOLLAND TRACT
BACON ISLAND
Mc DONALD TRACT
LOWER ROBERTS ISLAND
STOCKTON
P.P. #1
CONTRA COSTA CANAL
VEALE TRACT
PALM TRACT
LOWER JONES TRACT
ORWOOD TRACT
WOODWARD IS.
UPPER JONES TRACT
MIDDLE ROBERTS ISLAND
MIDDLE
VICTORIA ISLAND
Byron
BYRON TRACT
RIVER
UNION ISLAND
UPPER ROBERTS ISLAND
CLIFTON COURT FOREBAY
FABIAN TRACT
MANTECA
HARVEY O. BANKS DELTA PUMPING PLANT
TRACY PUMPING PLANT
STEWART TRACT
OLD
SOUTH BAY PUMPING PLANT
TRACY
DELTA-MENDOTA CANAL
CALIFORNIA AQUEDUCT
DELTA AREA
SAN FRANCISCO
LOS ANGELES
N
SCALE IN MILES
0 2 4 6

Delta and river systems in Northern California to transport stored water from reservoirs to the agricultural area of the Sacramento Valley and Delta. The major project facilities in Northern California are Shasta and FolsomReservoirs for the federal CVP and Oroville Reservoir for the SWP. This use of natural waterways results in regulated, stable flows during the spring, summer, and fall benefiting the water user, recreation, and fish and wildlife.

A major problem that occurs from using the natural channels in the Delta to transport project supplies are "reverse flows". The expression "reverse flows" describes a Delta problem that is caused from the lack of capacity in certain channels. Water supplies for export by the CVP and the SWP are obtained from surplus Delta flows, when available, and from upstream reservoir releases, when Delta inflow is low and surplus flows are unavailable. These releases enter the Delta via the Sacramento River and then flow by various routes to the pumps in the southern Delta. Some of these Sacramento River flows are drawn to the SWP and CVP pumps through interior Delta channels, facilitated by the CVP's Delta Cross Channel near Walnut Grove in the northern Delta. Unfortunately, because the channels aren't large enough, insufficient amounts of water enter through the northern Delta channels to meet both local and export needs.

The remaining water flows on down the Sacramento River to its confluence with the San Joaquin River in the western Delta. When fresh-water outflow is low, water in the western Delta becomes brackish because it mixes with saltier ocean water entering as tidal inflow. This water is drawn upstream (reverse flow) into the San Joaquin River and other channels by the pumping plants. Reverse flow disorients migratory striped bass, salmon, and steelhead. Reverse flow further increases the impacts on fish by pulling small fish from the western Delta nursery area toward the pumping plants. The massive amount of water driven in and out of the Delta by tidal action dwarfs the actual fresh-water outflow and considerably complicates the reverse-flow issue.

Reverse flow could be moderated or eliminated by in-creasing the transfer efficiency of the northern Delta channels. In-creasing the transfer efficiency would also considerably increase water supply for the SWP. Currently, during the operational periods that cause reverse flow, more water than is needed for export must be released from project reservoirs to repel intruding sea water and to maintain required water quality in western Delta channels to meet local and export standards. The amount of extra outflow required is substantial. An efficient means of transfer through the northern Delta would make better use of upstream fresh-water storage, and the SWP could gain up to 400,000 acre-feet more per year in depend-able supply. Delta fisheries and Delta water quality would also benefit.

The need for improved water transfer efficiency across the Delta resulted in the U. S. Bureau of Reclamation constructing the Delta Cross Channel between the Sacramento and Mokelumne rivers in 1951 to protect the quality of its Delta-Mendota exports. When the State Water Project's Delta pumps came on line in the late 1960s, it was recognized that facilities would eventually be required in the Delta to improve water transfer efficiency and to control salinity caused by tidal inflow entering the western Delta. The need and authorization for these facilities was recognized in the Burns-Porter Act, approved by the voters in 1960.

However, specific proposals to accomplish these objectives have generated much controversy, and agreement has not been reached upon the best approach to mitigating deteriorating conditions in the Delta. Fisheries declines are well documented, although the causes are not yet fully understood. Water quality continues to be a major operational problem. And Delta levees continue to fail at an accelerating rate. No one seems satisfied with today's conditions, and a consensus appears to be evolving that some form of channel improvements is needed.

The development and use of water in California is governed by a complex system of State and federal laws. This system is not fixed, but evolves year by year as new issues are raised that require changes and interpretations. The current environmental regulatory process provides a useful forum for discussions which can lead to projects that benefit all users, including instream uses. The process encourages step-by-step negotiations. The Department of Water Resources has been successful in using this approach to identify concerns, interests, and alternative solutions and to move forward with water projects.

DWR is moving ahead with environmental impact evaluations for alternative improvements in these areas; the southern, western, and northern Delta. The planning and environmental documentation process for three separate Delta Water Management Programs has started with scoping meetings and will take approximately three to four years to complete. Using the environmental documentation process during planning promotes public input and allows flexibility and adjustment in formulating project alternatives and related mitigation. This process also facilitates communication, which provides better coordination of related local, State and Federal planning efforts.

The environmental documentation process provides the information necessary for Federal and State regulatory permits and agreements. Federal regulatory permits are required to authorize implementation of any action or selected alternative. A major permit consideration is Section 404 of the Clean Water Act that regulates the discharge of dredge and fill material into waters of the United States, and establishes a permit program to ensure that such discharges comply with environmental requirements. The Section 404 program is administered at the federal level by the U. S. Army Corps of Engineers and with oversight by the U. S. Environmental Protection Agency. The U. S. Fish and Wildlife Service and the National Marine Fisheries Service have important advisory roles.

The Section 404 program is broadly recognized as the most significant federal regulatory program affecting wetlands. As defined in Section 404 program regulations, wetlands are "those areas that are inundated or saturated with surface or groundwater at a frequency and duration sufficient to support, and that under normal circumstances do support, a prevalence of vegetation typically adapted for life in saturated soil conditions". In applying this definition, indicators of vegetation, soil and hydrology are used to identify wetlands.

Efforts are also being focused on Delta levee rehabilitation. DWR, the Corps, and local interests are working to develop a long-range answer to the levee problem. Both the Federal Emergency Management Agency and the State Office of Emergency Services are reluctant to spend more money on disaster relief in the Delta without a comprehensive plan and commitment by the State. In developing a plan, it is appropriate to consider alternative approaches to dealing with the levees. The problem of subsidence is of particular concern in some Delta areas.

South Delta Water Management - In April 1987, DWR and the U. S. Bureau of Reclamation conducted public meetings to discuss southern Delta water management issues. This planning activity was initiated under an agreement among the Bureau, DWR, and the South Delta Water Agency that committed all three parties to work together to develop mutually acceptable, long-term solutions to the water supply problems of water users in the southern Delta. Objectives of the agreement are to improve and maintain water levels, circulation patterns, and water quality.

Evaluation of alternatives to meet these objectives also take into account broader objectives that recognize the many interests and beneficial uses of the Delta's water supply concerning fisheries, overall reliability and operational flexibility of the SWP and CVP, navigation, and flood protection. Channel flow control structures are proposed to use the natural tidal activity to trap higher water levels in south Delta channels to maintain water levels and control circulation to dilute salt concentrations. These structures will also be used to improve flow patterns for migrating salmon. To improve operational flexibility, the Department is also proposing enlarging the existing SWP forebay with new inlet gates to allow increased winter exports and subsequent banking of this water into storage south of the Delta. This banking concept increases SWP reliability while favorably shifting exports away from months of high fish abundance. Later, during dry conditions, this storage can be used to meet SWP demands without affecting the Delta's environment.

In addition, DWR is investigating a conjunctive use program with local interests for use of good quality water supplies stored in an upstream reservoir. Under this program, local distribution systems and wells would be used to enhance local ground water supplies that could be used during dry periods allowing added releases of good quality reservoir water to the San Joaquin River, a tributary to the southern Delta. These releases would provide instream fishery benefits, water quality enhancement in the Delta and tributary, and an added water supply for the SWP.

North Delta Water Management - Today's planning effort in the northern Delta is proceeding about six months behind southern Delta planning. Public involvement began in August 1987. Northern Delta planning focuses on providing flood protection for islands along the lower Mokelumne River in the north Delta, reducing fisheries impacts, and improving transfer efficiency of Federal and State project water across the Delta.

One promising possibility for the northern Delta is a phased program that would start with enlargement of the South Fork of the Mokelumne River. Enlargement of this channel would also help correct unfavorable flow patterns in the Delta caused by State and Federal project operations. This would help reduce fishery impacts, improve water quality and add to project water supply reliability.

Western Delta Water Management - The Department of Water Resources and Department of Fish and Game are investigating West Delta Water Management needs. One alternative is a wildlife management plan for Sherman Island in the western Delta.

The objectives of Western Delta Water Management Planning include minimizing oxidation and subsidence, improving levees and protecting existing highwaysand utilities, providing habitat for waterfowl, improving recreation, reducing fish impacts by screening, and protecting Delta water quality.

Sherman Island can be managed to provide 10,000 acres of habitat to support a diversity of species including waterfowl. Changing from current cultivation practices to managed wildlife habitat can significantly reduce subsidence. This reduced subsidence plus levee improvements will significantly reduce the probability of future flooding.

In addition to protecting important highways and utilities, flood control measures for Sherman Island will also protect Delta water quality. If Sherman Island was permanently flooded, it would increase the rate of salt and fresh water mixing in the Delta estuary and jeopardize all beneficial uses. Included would be the reliability of State and Federal projects.

Delta Flood Protection Act of 1988 - The Delta Flood Protection Act of 1988 overwhelmingly passed both houses of the State Legislature and was signed into law by the Governor. This bill creates the Delta Flood Protection Fund to make $12 million a year available for appropriation by the Legislature for 10 years for flood protection in the Sacramento-San Joaquin Delta.

It makes available for allocation $6 million annually from that fund for local assistance under the Delta Levee Maintenance Subventions Program. The remaining $6 million is for special flood control projects for eight western Delta islands and the towns of Walnut Grove and Thornton in the north Delta area. A priority list for the western islands, based on public benefit and needed flood protection work, has been developed.

The Delta Flood Protection Act of 1988 also requires investigation of other flood control measures, such as provisions to acquire easements up to 400 feet wide along levees to minimize tillage and modification of land management practices. DWR is directed to seek appropriate cost sharing for flood control plans. Provisions forprotection of fish and wildlife habitat, as determined by the Department of Fish and Game, are to be included in these plans. Finally, the Delta Master Recreation Plan also needs to be considered.

The Department is working with the U. S. Corps of Engineers and local agencies to administer this program and identify cost sharing opportunities. We hope to use the expenditures from this program as the nonfederal contribution for a future federally authorized flood control project for the Delta.

Future Expansion of the State Water Project

by John J. Silveira[1]

Abstract

Measures currently underway by the California Department of Water Resources will result in water transfer enhancements to the State Water Project. Water banking facilities, in the form of surface and underground reservoirs located south of the Delta, will allow for diversion and storage of excess winter flows for use during periods of critical need. Concurrently, plans are underway for completion of the State Water Project delivery system including the Coastal Aqueduct. As an integral part of these combined efforts, environmental issues must be addressed and optimum solutions arrived if these programs are to be successful.

Introduction

The continuing debate over California water development reveals one point on which nearly all interests agree: more emphasis should be placed on diverting water from the Delta during the less environmentally sensitive winter flow periods, and banking that water south of the Delta for use later, during times of need. In recognition of this general consensus, the California Department of Water Resources, in addition to current measures being taken to provide improved water transfer conditions in the Delta, has in the planning and implementation stages, water banking facilities which will contribute increased yield and reliability to the State Water Project. These facilities will make use of additional Delta export capacity to increase surface and underground storage capability south of the Delta.

---

[1] Chief, State Water Project Planning Branch, California Department of Water Resources, P. O. Box 942836, Sacramento, CA 94236-0001.

Concurrently, with the planning for water banking, the Department is proceeding with advanced planning studies for the implementation of facilities required to provide urban water supplies to San Luis Obispo and Santa Barbara Counties in the central coast region.

The Harvey O. Banks Delta Pumping Plant

A major flow restriction in the State Water Project system is the Banks Delta Pumping Plant. Originally planned and designed for staged development, the plant was constructed in the 1960s at 60 percent of ultimate capacity. Consequently, the plant, in its current configuration, is unable to divert adequate excess Delta flows necessary to fill proposed water banking facilities south of the Delta. To overcome this operational constraint, construction is currently underway which, when completed in 1991, will bring the plant up to its planned capacity of 10,300 cfs. With the Delta plant capable of operating at full capacity, and south Delta Management Programs in place, as described earlier in a companion paper, it will be possible to move excess winter flows to water banking facilities south of the Delta. Two such facilities currently in the advanced stages of planning are the Los Banos Grandes Reservoir and the Kern Water Bank.

Los Banos Grandes Reservoir

The Los Banos Grandes Reservoir is a 1.73 million acre-foot surface storage facility located on Los Banos Creek in western Merced County, about six miles south of the existing joint State/Federal San Luis Reservoir complex, about 80 miles south of the Delta. Delta water will be pumped into the Reservoir from the California Aqueduct for later release, as required, to meet project demands. Since the facility will provide storage capacity south of the Delta, it will be able to store water during periods of high winter flow when impacts of Delta diversion are minimal. Because the limited natural inflow generated by the Los Banos Creek basin is already committed to local uses, the reservoir will function entirely as an offstream storage facility.

Two pumping plants will be used. The first will lift water a maximum of 136 feet into an existing flood detention reservoir located some two miles upstream from the aqueduct. A second pumping plant, at the headwaters of the detention reservoir, will lift the water a maximum of 435 feet into the Los Banos Grandes Reservoir, located about six miles from the aqueduct. Total static lift at normal pool will be approximately 542 feet. Both plants will contain reversible pump-turbine units, thus allowing for generation of electrical energy during periods when water is being released from the reservoirs. When completed, this complex will add about 260,000 acre-feet of dry period yield and up to 350,000 acre-feet of average annual yield to the State Water Project.

The Los Banos Grandes Dam is planned as a zoned earthfill embankment with a height of 414 feet above streambed and a crest length of 1,760 feet at an elevation of 783 feet. In addition to the main dam, three saddle dams will be required. At the normal maximum operating pool elevation of 763 feet, the reservoir surface will encompass 12,870 acres. Maximum design flow in the pump mode at both the upper and lower sites is 3,500 cfs, with maximum design flow in the generation mode amounting to 4,650 cfs. Installed generating capacity is 135 megawatts at the upper site and 40 megawatts at the lower. Total cost is currently estimated to be on the order of $700 million plus environmental mitigation which may be substantial.

At present, lands that will be inundated by the Los Banos Grandes Reservoir are relatively undeveloped. The area is predominantly privately owned grassland, used primarily for cattle grazing. There are no permanent residents, no irrigated agriculture, and no commercial electric power service. With engineering feasibility of the Los Banos Grandes Reservoir established, study efforts are being focused on the resolution of remaining environmental issues. To the casual observer, the potential environmental impacts appear relatively minor, especially in comparison to many of the conventional reservoirs that were considered as alternatives in earlier studies. At present, three issues have emerged as the most crucial to project viability: wetlands, sycamore woodlands, and threatened and endangered species. Each of these is the subject of a specific federal or State law or regulation. Environmental studies for a final mitigation plan are nearing completion for inclusion in the draft Environmental Impact Report scheduled for release in late summer of this year. Under the most optimistic schedule for receipt of all approvals, prerequisite to project authorization, construction could begin as early as 1995, resulting in initial water storage shortly after the turn of the century.

<u>Kern Water Bank</u>

Located west of Bakersfield in Kern County, at the southern end of California's Central Valley, the Kern Water Bank will be an underground storage facility with a potential reservoir capacity of three to five million acre-feet. Water will be stored in wet years, during periods of high Delta flow, then pumped out in dry years for use by project contractors. As in the case with the Los Banos Grandes facility, the Kern Water Bank, being a south-of-Delta water storage facility, will help increase dependable water supplies and optimize project operation, while at the same time assuring minimal impacts from increased Delta diversions.

The Kern Water Bank will consist of a number of "elements". The Kern Fan Element will be a direct recharge and extraction facility located on 20,000 acres of land purchased by the Department in August of 1988. Additional "local elements" will include cooperative programs with surrounding local water districts involving new in-lieu and direct recharge activities, as well as expansion of several existing district programs. All elements will be operationally integrated.

Recently, there has been strong interest in accelerated development of the Kern Water Bank. As a result, the Department is pursuing a staged development program for the Kern Fan Element with the first stage being a scaled down operation of the ultimate full-sized project.

The first stage Kern Fan Element is planned for a maximum storage capacity of 300,000 acre-feet. Facilities will consist of 800 to 1,000 acres of recharge basins incorporating 20 to 25 existing irrigation wells on the Kern Fan Element property. In this regard, the first stage will rely on the use of existing facilities to the maximum extent possible, including existing recharge basins in the City of Bakersfield's 2800-acre recharge facility, which will be used on an as-available basis. Water for recharge will be conveyed from the California Aqueduct through the existing Cross Valley Canal. Extraction will consist of pumping and conveying water to local canals, from which much of the water will be exchanged with local water users for SWP entitlement water that would otherwise have been delivered from the California Aqueduct. A certain quantity of water will also be pumped directly back to the aqueduct during periods when exchange with local water purveyors is not feasible.

The planned maximum annual recharge for the first stage facilities is 90,000 acre-feet. Recharge capability on the initial first stage recharge facilities, plus the City of Bakersfield's 2800-acre facility, is expected to be adequate to absorb 15,000 acre-feet per month. The planned maximum extraction rate for the first stage is 72,000 acre-feet per year. At full development, dry year extraction from the element is expected to reach 360,000 acre-feet with increased long term dry period project yields amounting to 145,000 acre-feet per year.

It is the Department's intention that the first stage provide information and experience necessary to define potential problems and optimize future operations. Thus, implementation of the full-sized Kern Fan Element will, to a large extent, be based upon experience gained during design, construction and operation of the first stage.

Coastal Aqueduct and Enlarged Cachuma Reservoir Projects

The Department of Water Resources is currently conducting advanced planning studies and preparing Environmental Impact Report documentation for the second phase of the Coastal Aqueduct and a proposal for the Enlarged Cachuma Reservoir Project. The aqueduct, located in Kern, San Luis Obispo and Santa Barbara counties, will deliver a maximum of 70,586 acre-feet of entitlement water per year to San Luis Obispo and Santa Barbara counties. Concurrently, the Department is studying the possible enlargement of Cachuma Reservoir on the Santa Ynez River as a local project alternative. This project would supply 17,000 acre-feet of water per year to Santa Barbara County, with an equivalent reduction in entitlement delivery through the Coastal Aqueduct.

Based on a request of the counties, construction of the aqueduct was divided into two phases in 1964. The first phase consisted of a 15-mile canal extending from the California Aqueduct near Avenal Gap to Devils Den. Completed in 1968, the first phase of the Coastal Aqueduct delivers water to the Devils Den and Berrenda Mesa Water Districts. Implementation of the second phase of the aqueduct was deferred by the counties, pending a decision that the additional water supplies were needed.

In 1983, the Department, and the Flood Control and Water Conservation Districts of both San Luis Obispo and Santa Barbara counties, initiated joint studies to determine the status of water needs of the two counties. The results of the studies demonstrated that shortages of dependable water supplies existed in both counties, resulting in serious ground water overdraft conditions. The studies further demonstrated that for Santa Barbara County, importation of SWP water was the most economical way to alleviate the condition in the northern portion of the county, with the enlargement of Lake Cachuma Reservoir being the most economical alternative for the south county. For San Luis Obispo County, the study concluded that provision of SWP water was the least expensive alternative, with or without participation of Santa Barbara County. Based on these results, the Department was requested to proceed with advanced planning studies and preparation of environmental documentation for the possible enlargement of Cachuma Reservoir and construction of the Coastal Aqueduct, including the Santa Barbara Extension (Mission Hills Pipeline).

The proposed second phase of the Coastal Aqueduct would deliver a maximum of 70,586 acre-feet per year via a 60-inch diameter subsurface pipeline extending from near Devils Den in Kern County to a terminus near the city of Santa Maria. With 25,000 acre-feet per year being delivered within San Luis Obispo County, the remaining 45,586 acre-feet of annual entitlement would be delivered to Santa Barbara County via the proposed Mission Hills Pipeline, traversing between Santa Maria and its terminus near Lompoc.

Based upon information gathered at public workshops, input from design staff, geologists and environmental specialists, including wildlife biologists, botanists and archaeologists, an alignment has been chosen and plant facilities sited. The draft Environmental Impact Report was released for public review in April of this year and the final EIR is expected to be on the street by year end. With final design commencing in early 1991, the aqueduct could be operational sometime in 1995.

Similarly, the draft environmental documentation report, prepared jointly with the U.S. Bureau of Reclamation for the Enlarged Cachuma Reservoir, was released in April of this year, with the final document expected to be ready in late 1990.

Summary

The development and management of California's available water supplies will continue to challenge the imagination of planners well into the 21st century. As we move forward in our long-range planning efforts, the planning process, too, must continue to be responsive to the needs of the times. Today, complying with environmental regulations plays a greatly expanded role in the planning process and, in fact, in many instances _is_ the process. Hopefully, with concerted efforts aimed at balancing the needs of the environment with those of the water development community, optimum solutions will be realized.

The Need for Water Transfers and the State Water Project

Edward F. Huntley[1]

Abstract

California's water resources do not all occur when and where they are needed to meet growing urban and agricultural needs. Extensive water transfer projects have been constructed to redistribute these resources.

Introduction

California is often described in superlatives. To name a few, it has the highest elevation in the United States outside of Alaska, and the lowest elevation; it has the oldest trees and tallest trees in the world; it has highest temperature ever recorded in the United States; the greatest one month snowfall; it is the most populous state, has the most irrigated acres, and is first in agricultural income.

An appealing climate and natural resources have drawn people to California since the gold rush of 1849. With its burgeoning population and extensive agricultural development has come an increasing thirst for water. Fortunately, through a combination of foresight, fortuitous timing, and political will, California has so far been able to develop enough of its water resources to keep pace with ever-increasing needs.

By far the largest user of water in California is agriculture. Unlike the part of the U.S. from the Rocky Mountains eastward, precipitation in California does not occur during the growing season for most crops. The

---

[1] Chief, Division of Planning, California Department of Water Resources, P. O. Box 942836, Sacramento, CA 94236-0001.

answer to this has, of course, been irrigation. With an otherwise ideal climate and extensive fertile valleys, California leads the nation in value of annual crop production. Today, over 9 million acres are being irrigated, using nearly 80 percent of the State's developed water supply. About 40 percent of the irrigation supply comes from ground water, with the balance coming from local surface water sources or importation via federal, State, or local aqueducts. Ground water usage exceeds replenishment by about 25 percent, a matter of continuing concern.

While agriculture is the largest user, urban water use is presently growing the fastest. From 100,000 settlers in 1850, California's population has increased to 29 million people today. One out of nine Americans now lives in California. Half of these reside in what is commonly called Southern California or the South Coast region, which stretches from Ventura County south to the Mexican border, and inland to the Mojave Desert.

California is over 800 miles long, and the climate varies tremendously from one end to the other. Most of the precipitation falls on the northern third of the State, remote from areas of greatest need. Great distances and rugged mountains intervene between sources and demand. To deal with these disparities, local, regional, State, and federal agencies have constructed reservoirs and aqueducts throughout the State, giving California the most extensive plumbing system in the world. (See map)

Early Water Transfers

Southern California long ago outstripped its local water supplies and reached out first to the Owens Valley in 1913, when Los Angeles constructed the Los Angeles Aqueduct. Next, The Metropolitan Water District of Southern California tapped the Colorado River, completing the Colorado River Aqueduct in 1941. Later, Los Angeles reached further north to the Mono Lake basin to augment its Owens Valley supply. The first State Water Project water arrived in 1972. Southern California now imports over 60 percent of its water supply.

The San Francisco Bay area, now with over 20 percent of the State's population, likewise imports 60 percent of its water supplies, drawing first on Sierra Nevada rivers across the San Joaquin Valley to the east, and later State and federal project supplies. In 1914, San Francisco began construction of the Hetch Hetchy Aqueduct from the Tuolumne River, but continuing disputes

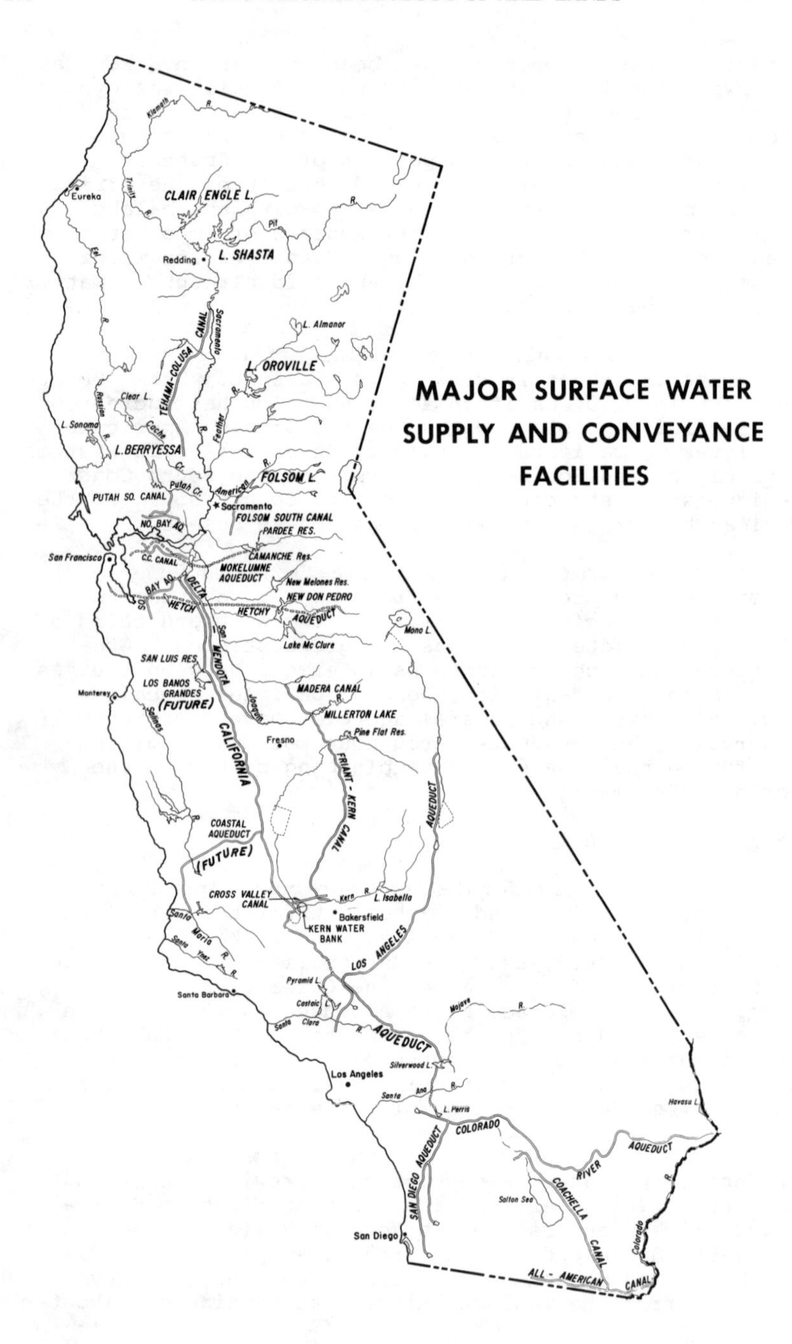
MAJOR SURFACE WATER SUPPLY AND CONVEYANCE FACILITIES
Klamath R.
Eureka
Trinity R.
CLAIR ENGLE L.
Pit
R.
Redding
L. SHASTA
Eel
R.
TEHAMA-COLUSA CANAL
Sacramento
L. Almanor
L. OROVILLE
Clear L.
Russian R.
L. Sonoma
Cache
Feather R.
L. BERRYESSA
Cr.
R.
American
FOLSOM L.
Putah Cr.
PUTAH SO. CANAL
Sacramento
FOLSOM SOUTH CANAL
NO. BAY AQ.
PARDEE RES.
San Francisco
C.C. CANAL
CAMANCHE Res.
MOKELUMNE AQUEDUCT
SO. BAY AQ.
DELTA
New Melones Res.
NEW DON PEDRO
HETCH
HETCHY
AQUEDUCT
Mono L.
San
Lake Mc Clure
SAN LUIS RES.
MENDOTA
LOS BANOS GRANDES (FUTURE)
Monterey
Joaquin
MADERA CANAL
R.
MILLERTON LAKE
Salinas
Pine Flat Res.
Fresno
CALIFORNIA
FRIANT - KERN CANAL
AQUEDUCT
R.
COASTAL AQUEDUCT (FUTURE)
CROSS VALLEY CANAL
Kern R.
L. Isabella
Bakersfield
KERN WATER BANK
Santa Maria R.
Santa Ynez R.
LOS ANGELES
Pyramid L.
Santa Barbara
Castaic L.
Mojave R.
Santa Clara R.
AQUEDUCT
Silverwood L.
Los Angeles
Santa Ana R.
L. Perris
Havasu L.
COLORADO
AQUEDUCT
SAN DIEGO AQUEDUCT
RIVER
R.
COACHELLA
Salton Sea
CANAL
Colorado
San Diego
ALL - AMERICAN CANAL

delayed delivery of the first water to the city until 1934. East Bay Municipal Utility District first brought water from the Mokelumne River to Oakland and other Easy Bay cities in 1929.

By 1920, irrigated agriculture in California already exceeded 4 million acres, much of it overlying the vast Central Valley ground water supply, which augmented local surface water supplies from nearby Sierra Nevada rivers. But, the extensive irrigation development and California's recurring droughts were outstripping available local supplies, resulting in rapidly falling ground water levels in the southern Central Valley. In 1921, recognizing this and other developing problems, the Legislature directed the State Engineer to make a comprehensive statewide investigation of California's water resources. The resulting "Report to the Legislature of 1931 on The State Water Plan", outlined a proposal that included transferring surplus water from Northern California to the southern part of the Central Valley.

In 1933, the Legislature passed the State Central Valley Project Act to implement the CVP. State funds to begin construction could not be obtained, however, because the nationwide economic depression made the revenue bonds unmarketable. Consequently, arrangements were made for federal authorization and financing. Congress authorized the project for construction by the U.S. Bureau of Reclamation. The USBR operates the CVP principally to transport water from the Sacramento, Trinity, American, and San Joaquin rivers to the water-deficient areas of the Central Valley. Further discussion of the federal CVP is presented in a separate paper.

In the late 1940s, it was apparent that while the existing and planned features of the federal Central Valley Project would provide for much of the supplemental agricultural water needs in the Central Valley, additional interbasin transfers would be needed. People were flocking to California to live. About half of the population increases were (and continue to be) non-natives. It was clear the major urban areas of the State would need more water to meet projected needs, and the increasing ground water overdraft in the southern Central Valley resulting from continued agricultural expansion was threatening the economy of that region.

The State Water Project

In 1947, the California Legislature authorized a statewide water resources investigation, which resulted in the California Water Plan, published in 1957. The Plan is a comprehensive and flexible guideline for the control, protection, conservation, distribution, and utilization of the State's waters for present and future beneficial uses. The Plan was designed to supplement existing transfer systems such as the federal Central Valley Project and those serving the San Francisco Bay area and Southern California. The Plan, in addition to identifying the need for local projects, pointed out the need for added transfers of water from the areas of surplus in Northern California to areas of need throughout the rest of the State. Studies done at the time resulted in the refinement of earlier conceptual interbasin water transfer plans into the selected Feather River Project.

In 1951 the State Legislature authorized the Feather River Project, now known as the State Water Project. The Burns-Porter Act, passed by the Legislature in 1959, authorized the Project's initial facilities. In 1960, California voters approved a $1.75 billion general obligation bond issue under the Burns-Porter Act to begin building the Project's dams, pumping plants, and aqueducts. For the first time, the State became a water development agency.

Thirty public agencies have long-term water supply contracts with the Department of Water Resources for an ultimate total of slightly over 4.2 million acre-feet a year. About 70 percent will go to urban areas, and 30 percent to farms. For most contracts, the amounts increase yearly up to the maximum annual entitlement. The calculated firm yield from existing Project facilities is about 2.3 million acre-feet per year.

The State Water Project is being built in stages. Today, the SWP consists of 14 dams and reservoirs, 17 pumping plants, 10 power plants, and a 560-mile aqueduct system. The basic operation involves storage in Oroville Reservoir for later release using the Feather River, Sacramento River, and Delta channels to convey water to the California and North Bay Aqueduct intakes in the Delta. From there, the Oroville releases and other surplus flows in the Delta are transferred to the San Francisco Bay area, to the southern Central Valley, and to Southern California. Some of the water, particularly the surplus winter Delta flows is temporarily stored in San Luis Reservoir for release to the service areas in the summer and fall.

In 1962, the first water deliveries were made from the partially completed South Bay Aqueduct, and work started on Oroville Reservoir, the key storage facility on the Feather River, and the joint-use San Luis Reservoir, an offstream storage facility adjacent to the Delta-Mendota Canal (CVP) and California Aqueduct (SWP). In 1963, work began on the California Aqueduct, and by 1968, the Project was able to deliver water in the southern Central Valley. By 1973, the initial facilities were completed, allowing water deliveries to Lake Perris, the Project's southernmost terminal reservoir.

The South Bay Aqueduct, now connected to the California Aqueduct, was completed in 1965. The North Bay Aqueduct began deliveries in 1968 through interim facilities, and was completed in 1988.

Currently, the SWP provides 20 percent of the South Coast region's water supply, and 10 percent of the Bay area's water supply. About 90 percent of SWP agricultural water entitlements are with contractors in the southern end of the Central Valley. The present level of deliveries comprise over 15 percent of the agricultural water use in the area.

The most recent update of the California Water Plan, in 1987, projected that California would grow by another 8 million people by 2010, bringing the total population to 36 million. The projection may have been too conservative, since unofficial estimates indicate the current growth at almost 700,000 people per year. The 1987 update also assumed that irrigated agriculture would level off at 9-10 million acres. The supplemental water needs for 2010 were estimated at 1.4 million acre-feet per year. Nearly all of these supplemental needs will be for urban uses, most of which will occur in service areas of the State Water Project. Thus, the Project must be expanded, including increased water transfer capability, to meet future needs.

In all, interbasin transfers of water have been the solution to many of California's water problems. The challenge for the future is to meet the State's increasing water needs in a manner that will continue to protect the valuable environmental assets that make California such a desirable place to live.

MUDFLOW RHEOLOGY IN A VERTICALLY ROTATING FLUME

Robert R. Holmes, Jr. [1], Associate Member, American Society of Civil Engineers
Jerome A.Westphal [2], Fellow, American Society of Civil Engineers, and
Harvey E. Jobson [3], Member, American Society of Civil Engineers

Abstract

Joint research by the U.S. Geological Survey and the University of Missouri-Rolla currently (1990) is being conducted on a 3.05 meters in diameter vertically rotating flume used to simulate mudflows under steady-state conditions. Observed mudflow simulations indicate flow patterns in the flume are similar to those occurring in natural mudflows. Variables such as mean and surface velocity, depth, and average boundary shear stress can be measured in this flume more easily than in the field or in a traditional tilting flume. Sensitive variables such as sediment concentration, grain-size distribution, and Atterberg limits also can be precisely and easily controlled.

A known Newtonian fluid, SAE 30 motor oil, was tested in the flume and the computed value for viscosity was within 12.5 percent of the stated viscosity. This provided support that the data from the flume can be used to determine the rheological properties of fluids such as mud.

Measurements on mud slurries indicate that flows with sediment concentrations ranging from 81 to 87 percent sediment by weight can be approximated as Bingham plastic for strain rates greater than 1 per second. In this approximation, the yield stress and Bingham viscosity were extremely sensitive to sediment concentration. Generally, the magnitude of the yield stress was large relative to the change in shear stress with increasing mudflow velocity.

Flume Wheel as a Rheologic Tool

Rheology is the study of the relation between "stress" and corresponding "strain" in a non-rigid substance (Harris, 1977). The rheology of laminar water flow is understood to be Newtonian, whereby the relation of shear stress to strain rate is linear with zero being the y-intercept or, in other words, zero yield stress (figure 1). In contrast, the rheology of mudflows is not fully understood because of the complex non-homogeneous makeup of the material. Mudflows could act as a pseudoplastic, non-linear viscoplastic, or Bingham plastic (linear viscoplastic). An understanding of mudflow rheology is an important first step to characterizing the mudflow phenomena.

---

[1] Civil Engineer, U.S. Geological Survey, 1400 Independence Rd., Mail Stop 200, Rolla, MO 65401.
[2] Professor, Department of Civil Engineering, University of Missouri-Rolla, Rolla, MO 65401.
[3] Hydrologist, U.S. Geological Survey, 12201 Sunrise Valley Drive, Mail Stop 415, Reston, VA 22092.

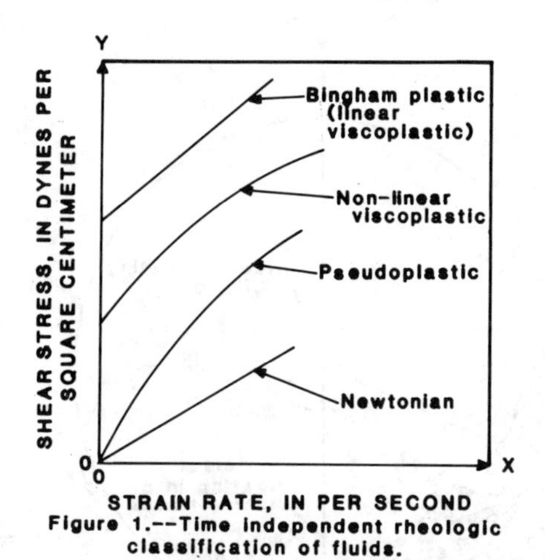

Figure 1.--Time independent rheologic classification of fluids.

Joint research by the U.S. Geological Survey and the University of Missouri-Rolla currently (1990) is being conducted on a wheel-shaped rotating flume (figure 2) used to simulate mudflows under steady-state conditions. The flume ('A' on figure 2) is 3.05 meters in diameter, 0.61 meter wide, and has walls 0.43 meter deep. A mass of material with known and controlled properties is placed inside the flume. A hydraulic, variable-speed motor ('B' in figure 2)is used to rotate the flume. The rotation initiates a relative deformation of the mass of material, inducing flow (figure 3). By rotating the flume at a constant speed, a steady-state condition is established; the material remains stationary relative to the laboratory floor but continues to flow relative to the flume bottom. When the steady-state system is established in the rotating flume, direct observation of the flow is possible for an extended period of time.

Figure 2.--Rotating flume.

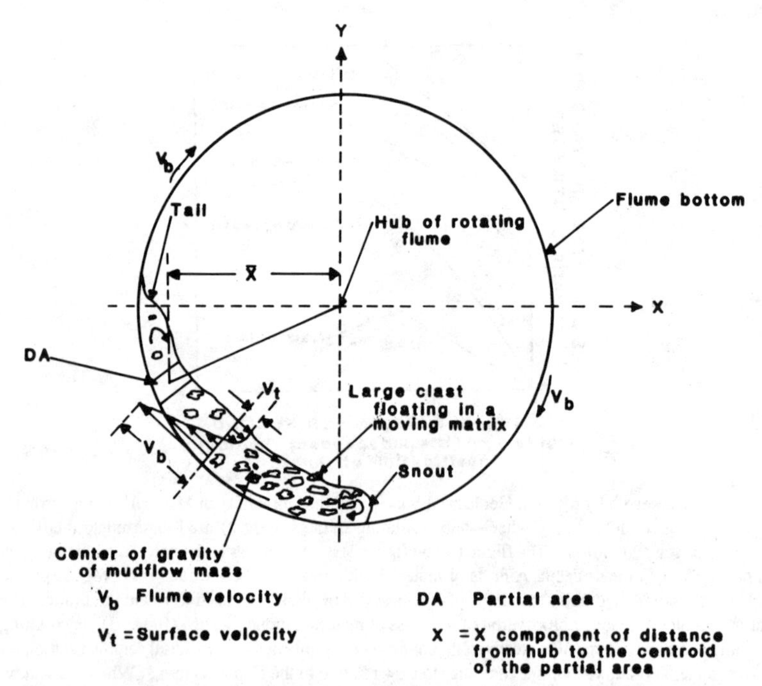

Figure 3.--Cutaway of rotating flume and mudflow, showing flow characteristics, velocity distribution and force balance analysis of the flow.

Mudflows, as have been witnessed in the field, vary in sediment concentration, grain size distribution, clay content, and clay type from one occurrence to the next. In the rotating flume, the above mentioned variables can be controlled and measurements of shear stress and strain rate can be made. From these measurements, rheologic properties can be determined and the effects of the variables (sediment concentration, etc.) can be quantified.

Boundary shear stress for flows in the flume can be calculated from a force balance analysis of the mud mass if the surface profile is known. The instantaneous surface profile of each flow is determined by a photogrammetric technique which is explained later. As the flume rotates, the center of gravity of the mudflow shifts to the left of the center of rotation of the flume (figure 3) causing an unbalanced weight moment. The unbalanced weight moment of the mud must be balanced by the moment of shearing forces acting on the bottom and sides of the flume. The weight moment was computed by dividing the profile into small incremental areas (DA, figure 3). The partial areas are converted into a partial volume by multiplying the area by the width of the flume (0.61 meter), and the moment of each partial volume about the hub is computed as $\bar{x}$ (figure 3) times the partial volume. Each partial moment is multiplied by the unit weight of the material (varying from 19800 to 21300 Newtons per cubic meter depending on the sediment concentration) and then summed to determine the total weight moment about the hub. The shearing stress times the contact area times the moment arm of the contact area must equal the unbalanced weight moment. The contact area and moment arms are determined from the surface profile.

Strain rate, which is the change of velocity over some distance, is approximated by knowing the surface velocity of the mudflow (V_t in figure 3), the flume speed (V_b in figure 3), and the maximum depth of flow. Surface velocity is the distance a wooden peg travels on the surface of the flow divided by the time of travel. The flume speed is determined as the circumference of the flume bottom divided by the time of one flume revolution. The maximum depth of flow is determined from the surface profile. Strain rate is then approximated by adding the surface velocity, and the wheel speed, and dividing by the maximum depth. It is recognized that this is a rather crude approximation of strain rate because 1) slippage velocity between the flume surface and material could occur and 2) the velocity distribution might be non-linear; figure 3. It is believed that this approximation for strain rate is adequate for the preliminary analysis presented here.

Photogrammetric Technique

Instantaneous surface profiles of the mudflow were obtained photogrammetrically. A 35-millimeter camera was rigidly mounted to the ceiling of the laboratory at an oblique angle to the flume such that the intersection of the entire surface profile of the mudflow with the back wall of the flume ('C' in figure 2) was visible in the photographs.

Any set of points on a fixed plane in space will have unique relative locations on the film of a camera which is also fixed in space. The intersection of the mud surface with the back wall of the flume defines a set of points on such a fixed plane in space. A grid, of three inch squares, with a known location and orientation relative to the flume was attached to the back wall of the flume and photographed before and after each series of experiments. In addition four targets ('D' in figure 2) were positioned at fixed locations in space and in the same plane as the back wall of the flume. These targets were visible in every photograph. The targets were used to determine the scale and orientation of each photograph.

The photographs of the grid were used to calibrate the film in that the position of known points in space were located on the film. Then, the location of points where the mud surface intersected the back wall of the flume could be obtained by comparison of the position of these points on the film to the position of the nearest four grid points. The calibration procedure and calculation of the profile points were carried out by digitizing the photographs and entering the digitized data into a computer. The accuracy of the photogrammetric technique was checked by marking several points on the back wall of the flume and then determining their locations both physically and photogrammetricly. The photogrammetric technique was accurate to within 0.003 meter.

Laboratory Procedure

For each test, 99.8 kilograms of gravel, 49.9 kilograms of sand, 24.9 kilograms of non-plastic fines, (non-plastic description determined by Atterberg limits tests to contain very little clay; see Peck and others, 1973, for discussion of Atterberg limits) and a known quantity of water are placed in the flume while it is rotating to ensure it is well mixed. The sediment concentration is determined by dividing the weight of solids by the weight of solids plus water. Grain size distribution, as well as the Atterberg limits, are carefully controlled. Sediment concentration is varied by changing the water content to determine various influences on rheologic parameters such as yield stress and viscosity.

The flume speed typically was varied from 0.06 meter per second to 2.13 meter per second. For each concentration, seven photographs of the mudflow profile are taken at each speed to determine the average shear stress and average strain rate. The surface velocity of the mud, relative to the laboratory floor, was measured several times for each flume speed by measuring the time required for a wooden peg, placed on the surface of the flow to travel a known distance. Individual measurements of surface velocity seldom differed from the mean by more than 10 percent and typical surface velocities ranged from 0.015 meter per second to 0.152 meter per second.

Flume Verification

For an experimental device to be used as a tool of understanding, the device must first be tested with materials of known behavior. A Newtonian fluid, SAE 30 motor oil, was tested in the flume and its viscosity and yield stress were determined by the previously mentioned methods. The computed viscosity differed from the theoretical value by 12.5 percent and the yield stress was determined to be 19.15 dynes per square centimeter, which is close to the theoretical yield stress of SAE 30 oil, 0 dynes per square centimeter (figure 4).

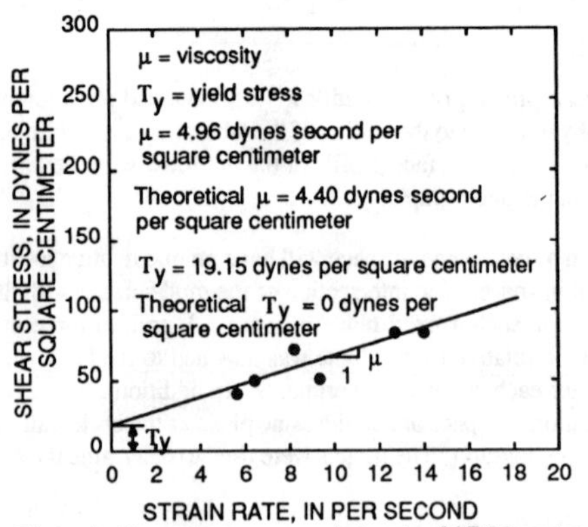

Figure 4.--Shear stress versus strain rate--SAE 30 motor oil.

For laboratory data derived from the rotating flume to be relevant to actual mudflows, the flow characteristics of actual and flume mudflows must be similar. A videotape of actual mudflows (Costa and Williams, 1984) indicates that the external characteristics of mudflows consist of: 1) A snout containing coarse particles, 2) tail containing finer particles, and 3) large clasts floating in a moving matrix. The mudflows simulated in the flume seem to have all the aforementioned flow characteristics (figure 3), which would tend to make the rheologic phenomena noted in the flume more believable in terms of applying the knowledge gained to actual mudflows.

Rheology Of Simulated Flows

Mudflows with differing sediment concentration were simulated in the rotating flume. External shear stress and strain rate were calculated for each flow. The average external shear stress and the average strain rate for each velocity with a sediment concentration of 87 percent by weight are shown in figure 5. Figure 5 is typical of the shape of all data plots. The Bingham plastic model of the stress-strain relationship is by far the easiest type of model to use in mathematical representation of the flow and all the data obtained thus far indicate that this model is a good approximation. It could be argued on the other hand, that the data could be represented slightly better by either a pseudoplastic or non-linear viscoplastic model (figure 5).

Yield stress (y-intercept) and viscosity (slope) decrease with decreasing sediment concentration (figure 6), demonstrating that sediment concentration is significant in mudflow rheology.

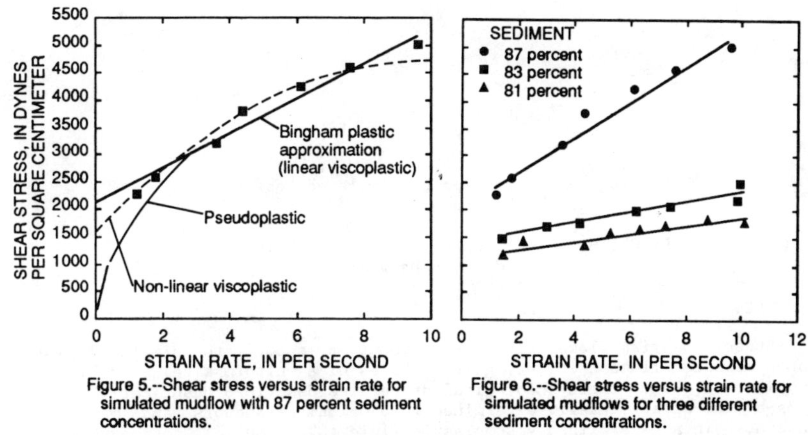

Figure 5.--Shear stress versus strain rate for simulated mudflow with 87 percent sediment concentrations.

Figure 6.--Shear stress versus strain rate for simulated mudflows for three different sediment concentrations.

Future experiments will involve varying Atterberg limits and grain size distribution. The rheologic effects of these characteristics will be studied and incorporated with continuing work on sediment concentration effects on mudflow rheology.

Appendix

Costa, J.E. and Williams, G.P., 1984, Debris-flow dynamics: U.S. Geological Survey Open-File Report videotape 84-606, 22 minutes.

Harris, John, 1977, Rheology and non-Newtonian flow: Longman, London, 333 p.

Peck, R.B., Hanson, W.E., and Thornburn, T.H., 1973, Foundation Engineering, Second Edition: John Wiley and sons, New York, 513 p.

Rheological Properties of Simulated Debris Flows in the Laboratory Environment

By Chi-Hai Ling[1], Cheng-lung Chen[1], Member, ASCE and Chyan-Deng Jan[2]

Abstract

Steady debris flows with or without a snout are simulated in a "conveyor-belt" flume using dry glass spheres of a uniform size, 5 or 14 mm in diameter, and their rheological properties described quantitatively in terms of the experimentally-determined rheological parameters and material constants in a generalized viscoplastic fluid (GVF) model. Close agreement of the measured velocity profiles with the theoretical ones obtained from the GVF model strongly supports the validity of a GVF model based on the continuum-mechanics approach. Further comparisons of the measured and theoretical velocity profiles along with empirical relations among the shear stress, the normal stress, and the shear rate developed from the "ring-shear" apparatus determine the values of the rheological parameters in the GVF model, namely the flow-behavior index, the consistency index, and the cross-consistency index. Critical issues in the evaluation of such rheological parameters using the conveyor-belt flume and the ring-shear apparatus are thus addressd in this study.

Introduction

Accurate evaluation of the rheological parameters in the generalized viscoplastic fluid (GVF) model is an important step toward applying the GVF model to solving practical debris flow problems. It requires that a steady debris flow be simulated in the laboratory environment and be readily measurable during its movement. This is accomplished by use of a "conveyor-belt" flume. The conveyor-belt flume is a tilting flume with a variable-speed conveyor-belt bed designed to move in the opposite direction of flow. By adjusting the angle of inclination and the speed of the belt, a simulated debris flow in the flume can be maintained at the steady state, thereby allowing the measurements of particle movements at different distances from the bed through transparent side walls. This experimental setup enables one to plot measured velocity profiles for various simulated debris flows in the flume. This paper describes how the rheological parameters are evaluated through comparisons between measured and theoretical velocity profiles.

Theoretical Analysis

A GVF model developed by Chen (1989) is used for expressing the stress-strain relation for flow of beads. Using the GVF model, Chen (1988a,b) expressed the relevant stresses for two-dimensional uniform debris flow in wide open channels, and Ling and Chen (1989) extended the theory to that for slightly non-uniform flow. The expressions of the stresses for slightly non-uniform flow were found to be identical to those for uniform flow (Ling

[1]Hydrologist, U.S. Geological Survey, Water Resources Division, 345 Middlefield Road, MS-496, Menlo Park, CA 94025
[2]Graduate Research Assistant, Dept. of Civil Eng., U.C. Berkeley, CA 94720

and Chen 1989) as follows:

$$\tau_{\xi z}\ (=\tau_{12}) = (c\ \cos\phi + p\ \sin\phi) + \mu_1(\partial u/\partial z)^{\eta} \quad (1)$$

$$\tau_{zz}\ (=\tau_{22}) = -\ p + \mu_2(\partial u/\partial z)^{\eta} \quad (2)$$

in which $\tau_{\xi z}$ and τ_{zz} are the shear and normal stresses, respectively; c is the cohesion; ϕ is the angle of internal friction; p is the mean normal pressure; μ_1 is the consistency index; η is the flow behavior index; μ_2 is the cross-consistency index; u is the velocity component in the ξ direction, which is the coordinate moving at the speed of the snout parallel to the bed, positive going down the slope; and z is another coordinate perpendicular to ξ, positive toward the surface of the flowing beads (see Fig. 1).

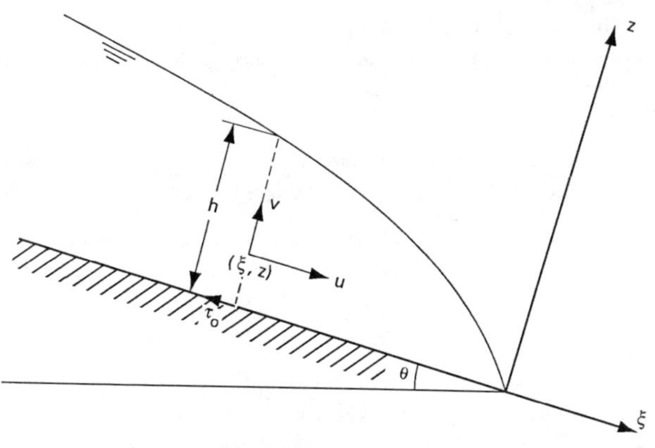

Fig. 1 Definition sketch

If the concentration of beads is assumed to be constant everywhere within the debris flow and inertial terms are neglected, Eqs. 1 and 2 become

$$\rho g(h-z)\sin\theta - \rho g(\partial h/\partial\xi)(h-z)\cos\theta = c\ \cos\phi + p\ \sin\phi + \mu_1(\partial u/\partial z)^{\eta} \quad (3)$$

$$-\ \rho g(h-z)\cos\theta = -\ p + \mu_2(\partial u/\partial z)^{\eta} \quad (4)$$

in which ρ is the bulk density of beads, treated herein as a constant; g is the gravitational acceleration; h is the depth of flow; and θ is the angle of inclination of the bed from the horizontal.

Solving Eqs. 3 and 4 yields

$$\partial u/\partial z = C_1(h-z)^{1/\eta} \quad (5)$$

where $C_1 = \{\rho g[\sin\theta - (\partial h/\partial\xi)\cos\theta - \cos\theta\sin\phi]/(\mu_1 + \mu_2\sin\phi)\}^{1/\eta}$.

Integrating Eq. 5 gives

$$u = u_0 + C_1[\eta/(\eta+1)]\{h^{1+1/\eta} - (h-z)^{1+1/\eta}\} \quad (6)$$

where u_0 is the velocity of beads at the bed (z = 0), which is different from the conveyor-belt velocity, u_b. Thus, there exists a slip velocity, u_s, equal to $u_0 - u_b$. Because the average velocity across h at any ξ in the conveyor-belt flume is zero, integrating u (Eq. 6) over h yields

$$u_0 = - C_1[\eta/(2\eta+1)]h^{1+1/\eta} \quad (7)$$

$$u = - C_1[\eta/(2\eta+1)]h^{1+1/\eta} + C_1[\eta/(\eta+1)][h^{1+1/\eta} - (h-z)^{1+1/\eta}] \quad (8)$$

For uniform-flow experiments performed in this study, $\partial h/\partial\xi = 0$ and C_1 reduces to $\rho^* g \sin\theta/\mu_1$ obtained by Chen (1988a), where $\rho^* = [1 - \text{ctn}\theta \sin\phi]\rho/[1 + (\mu_2/\mu_1) \sin\phi]$.

Normalizing Eqs. 7 and 8 by the surface velocity, u_{sf}, and h, and denoting $u_* = u/u_{sf}$, $z_* = z/h$, $u_{s*} = u_s/u_{sf}$, and $u_{b*} = u_b/u_{sf}$ yields

$$u_* = 1 - (2+1/\eta)(1-z_*)^{1+1/\eta} \quad (9)$$

$$u_{s*} = - [\eta/(2\eta+1)](\partial u_*/\partial z_*)_0 - u_{b*} \quad (10)$$

$$u_{sf} = C_1 h^{1+1/\eta}\{[\eta^2/(2\eta+1)][1/(\eta+1)]\} \quad (11)$$

For determining the η value of flowing beads under a given set of experimental conditions, setting $u_* = 0$ in Eq. 9 yields

$$z_{m*} = 1 - [\eta/(2\eta+1)]^{\eta/(\eta+1)} \quad (12)$$

Therefore, the vertical location at any z (or z_{m*}), where u (or u_*) is equal to zero, determines the value of η. For illustration, Eqs. 9 and 12 are shown in Figs. 2 and 3, respectively.

Eq. 11 upon substitution of the C_1 expression, along with the relation between μ_1 and μ_2 from Chen (1988a),

$$\mu_2 = - \mu_1/\tan\phi_d \quad (13)$$

in which ϕ_d is the "dynamic" angle of internal friction, yields

$$\mu_1 = \rho^* gh \sin\theta \; [\eta^2 h/(\eta+1)(2\eta+1)u_{sf}]^\eta \quad (14)$$

When the values of η, ϕ, ϕ_d, and θ are known, the measurement of u_{sf} and h will determine μ_1. Eq. 13 upon substitution of μ_1 so determined would then give μ_2.

Comparison with Experiment

The experimental apparatus consists of a rectangular flume with bed driven by a conveyor belt in the opposite direction of flow. It is 30 cm wide, 50 cm high, and 4 meters long, and it can be tilted from zero to 45 degrees. A rough belt was used in this set of experiments, but can be changed to a smooth one. Dry beads of a uniform size, 5 mm or 14 mm in diameter (D), were used to simulate the debris flow. A divider parallel to the side walls can be inserted in the flume to reduce its width so as to increase the depth of flowing beads under a given maximum allowable load on the driving motor without noticeable change in the flow behavior of beads. Experiments with the 14 mm beads were run both with and without the divider set at 15 cm (i.e., half of the total flume width). However, experiments using the 5 mm beads could not be run with the divider because of leakage of beads under the divider.

Movement of marked particles at different distances from the bed was measured photographically using a videocamera through a transparent side wall. Measured velocity distributions of uniform flow for 14 mm beads are plotted against the theoretical results in Fig. 4. For 5 mm beads, they are shown in Fig. 5

The value of η, as shown in Table 1, is determined from Eq. 12 (Fig. 3) upon substitution of measured z_{m*} value. For lack of an independent

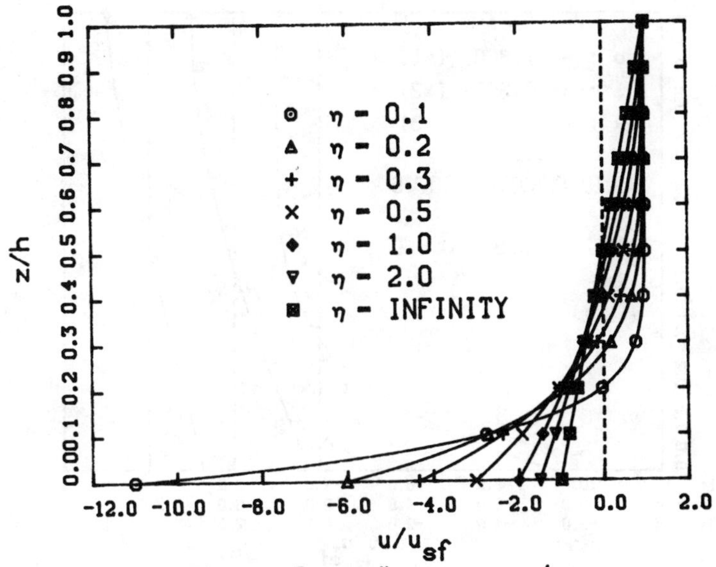

Figure 2. z/h versus u/u_{sf}

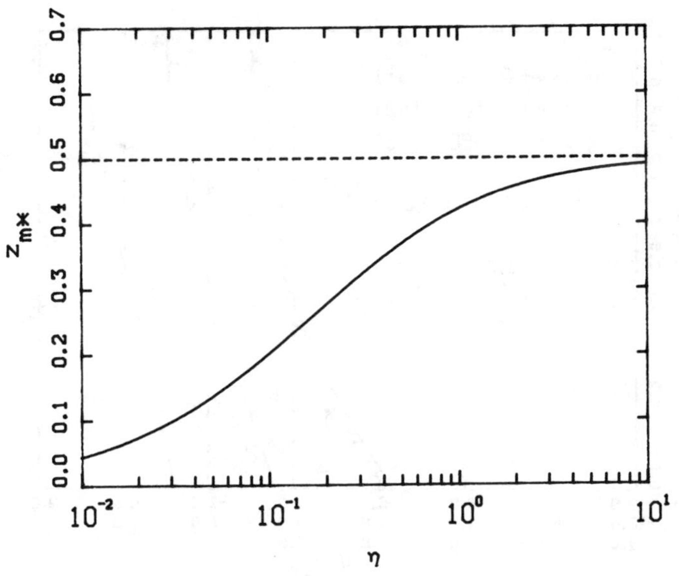

Figure 3. z_{m*} versus η

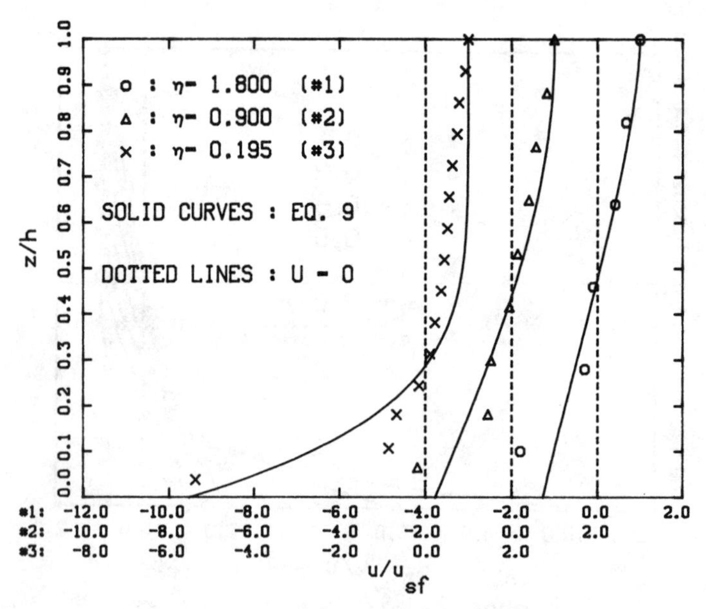

Figure 4. Velocity profiles for 14 mm beads

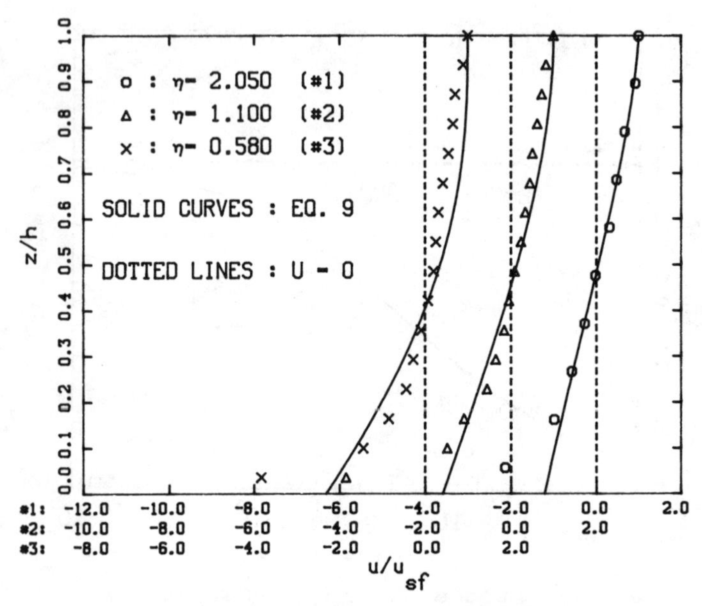

Figure 5. Velocity profiles for 5 mm beads

description of ϕ in terms of flow conditions, the values of ϕ are assumed to be zero. This is justified on the physical ground that once the beads move, the yield stress, s (= c cosϕ + p sinϕ), is negligibly small. The values of ϕ_d are obtained by use of a ring-shear apparatus designed by Professor Kyoji Sassa of Kyoto University, Japan (Sassa, et al. 1984). The ring-shear apparatus has an annular shear cell holding granular materials inside the inner and outer rings of 30 cm and 48 cm in diameter, respectively. The sample space has a height of 9 cm, which consists of identical upper and lower rings. The upper ring remains stationary, while the lower ring can be rotated at a linear speed between zero and 1 m/s of the center diameter of the inner and outer rings. For 5 mm beads, ϕ_d is measured at 13^0 from experiments, whereas 14 mm beads are too large to be treated as a continuum in the ring-shear apparatus. For consistency in data analysis, therefore, the value of ϕ_d for 14 mm beads is assumed to be also 13^0. From the measured values of u_{sf} and h, as shown in Table 1, μ_1 and μ_2 are calculated from Eqs. 14 and 13, respectively. The last two columns in Table 1 show the results of the calculation.

Results

The theoretical velocity profiles (Eq. 9) compare reasonably well with the measured velocities. The theory, however, seems to over-predict the velocity of the bottom layer of beads, but under-predict the velocity of the next layer above. This is perhaps due to the fact that with the present motor, enough layers of the 14 mm beads cannot be established in the flume to be treated as a continuum. It is believed that smaller beads will show better comparisons between the theory and measurements. Unfortunately, this belief has not yet been substantiated in this study due to the small amount of data available from experiments using 5 mm beads. (There are only Runs 18, 19, and 20 for 5 mm beads, as shown in Table 1.) Apparent errors in the measurements of the velocities of the fast moving beads at or close to the bottom layer as well as the difficulty in defining exactly the bed (z = 0) may also cause a problem in data analysis. Efforts are being made to reduce the errors in such measurements and data analysis. Furthermore, because the present theory is developed for two-dimensional flow, any side-wall effects would, of course, violate the assumption of continuity based on the theoretical two-dimensional flow scheme (Eqs. 1 and 2). Slight violation of the two-dimensional continuity is evidently reflected in measured velocity profiles, as shown in Figs. 4 and 5.

The computed consistency index μ_1 shows a wide range (Table 1). This is conceivable because a close inspection of the computed μ_1 values reveals that larger μ_1 values correspond to higher h and lower u_b values or vice versa. In either case of higher h and lower u_b values, the larger μ_1 values may be attributed to the higher concentrations of flowing beads. Physically, the higher concentration of beads does indeed give the higher value of μ_1 according to the empirical μ_1 versus concentration relation formulated by Chen (1988b). By the same token, the value of $-\mu_2$ varies accordingly with h and u_b and hence, with the concentration. In the present state of our knowledge, the variation of μ_1 and $-\mu_2$ values with the concentration must be determined by experiment. The η value determined from Eq. 12, as listed in Table 1, appears to vary with the concentration too, but their correlation is found to be less obvious than that for μ_1 and μ_2 with the concentration.

Conclusions

A preliminary comparison of simulated debris flows in the laboratory with the theoretical results has indicated that the GVF model is reasonably accurate and practical for debris flow simulation. The GVF model with the assumption of the constant concentration will predict the velocity distribution reasonably well. An analysis of computed μ_1 and μ_2 values shows that both indices are highly dependent on the concentration, as expected from the GVF model. Use of the GVF model will thus require that both indices be expressed in terms of the concentration.

References

Chen, C. L. (1988a). "Generalized viscoplastic modeling of debris flow." J. of Hydr. Engrg., ASCE, 114(3), 237-258.

Chen, C. L. (1988b). "General solutions for viscoplastic debris flow." J. of Hydr. Engrg., ASCE, 114(3), 259-282.

Chen, C. L. (1989). "Issues in debris flow research: personal views." U. S. Geological Survey Water-Supply Paper Series "Selected Papers in the Hydrologic Sciences," WSP 2340. (in press)

Ling, C. H. and Chen, C. L. (1989). "Idealized debris flow in flume with bed driven by a conveyor belt." Proceedings, National Conference on Hydraulic Engineering/HY Div/ASCE, New Orleans, Louisiana, 1144-1149.

Sassa, K., Shima, M., Hiura, M., Nakagawa, A., and Suemine, A. (1984). "Development of ring shear type debris flow apparatus." Report of Grant-in-Aid for Scientific Research by Japanese Ministry of Education, Science and Culture No.57860028, Disaster Prevention Research Institute, Kyoto University, Kyoto, Japan.

TABLE 1. Experimental Results and Evaluation of Rheological Parameters

RUN	Measurements							Calculation		
	D	θ	u_{sf}	u_b	u_0	h	z_{m*}	η	μ_1	$-\mu_2$
	(mm)	(°)	(cm/s)	(cm/s)	(cm/s)	(cm)			(dyne/cm²)(s)$^\eta$	
1	14	10	9.3	-50	-16.7	7.7	0.452	1.8	121	544
2	14	10	14.4	-100	-28.9	8.96	0.413	0.87	298	1336
3	14	10	18.6	-150	-41.5	8.96	0.416	0.9	226	1015
4	14	10	4.1	-30	-10.9	8.96	0.430	1.15	847	3795
5	14	10	7.7	-50	-18.5	10.22	0.409	0.83	677	3035
6	14	11	9.1	-70	-23.9	10.22	0.438	1.31	436	1954
7	14	11	15.3	-100	-33.3	11.48	0.416	0.9	475	2130
8	14	11	16.7	-120	-40.0	11.48	0.424	1.02	374	1675
9	14	11	17.4	-140	-44.1	11.48	0.420	0.96	389	1744
10	14	11	18.3	-160	-49.5	11.48	0.429	1.12	296	1326
11	14	11	19.6	-180	-55.8	11.48	0.429	1.12	274	1228
12	14	11	21.5	-200	-58.3	11.48	0.429	1.12	247	1107
13	14	12.5	2.7	-40	-14.5	19.04	0.273	0.195	4415	19790
14	14	12.5	4.9	-60	-22.0	19.04	0.301	0.25	3780	16940
15	14	12.5	6.9	-80	-27.2	19.04	0.313	0.28	3381	15160
16	14	12.5	6.7	-100	-35.6	19.04	0.312	0.275	3423	15340
17	14	12.5	9.1	-120	-39.2	19.04	0.318	0.293	3087	13830
18	5	12.5	6.0	-40	-23.3	7.4	0.382	0.58	721	3233
19	5	15	16.0	-50	-34.0	4.64	0.458	2.05	9.9	44
20	5	13	10.1	-60	-38.8	7.4	0.428	1.1	274	1229

RHEOLOGICAL ANALYSIS OF FINE-GRAINED NATURAL DEBRIS-FLOW MATERIAL

Jon J. Major and Thomas C. Pierson[1]

ABSTRACT

Experiments were conducted on large samples of fine-grained material (≤2mm) from a natural debris flow using a wide-gap concentric-cylinder viscometer. The rheological behavior of this material is compatible with a Bingham model at shear rates in excess of 5 sec^{-1}. At lesser shear rates, rheological behavior of the material deviates from the Bingham model, and when sand concentration of the slurry exceeds 20 percent by volume, particle interaction between sand grains dominates the mechanical behavior. Yield strength and plastic viscosity are extremely sensitive to sediment concentration.

INTRODUCTION

Recent empirical and theoretical analyses suggest that coarse-particle interaction plays a significant role in the rheological behavior of debris flows (e.g. Iverson and Denlinger, 1987) and much effort has been expended in the development of theoretical models and discussion of their relative merits (e.g. Johnson, 1970; Takahashi, 1981; Iverson and Denlinger, 1987; Chen, 1988; Campbell, 1989). However, relevant field and laboratory data that can be used to test models generally are lacking (Phillips and Davies, 1989b). This void can be filled, in part, by data collected under controlled experimental conditions.

Most previous investigators have measured rheological parameters of small-volume mixtures ranging from artificially mixed dilute suspensions of clay and water to dense suspensions of solid-liquid mixtures. Recent work has focused on testing of materials that compose debris flows at realistic sediment-water ratios and at realistic shear rates (e.g. O'Brien and Julien, 1988). These analyses have been performed on samples of limited volume and grain size; most tested slurries had volumes of only a few hundreds of cm^3 and few contained particles as large as 0.5 mm. Coarser slurries having volumes to 10^6 cm^3 and containing particles as large as 120 mm have been tested recently (Phillips, 1989; Phillips and Davies, 1989a) with a large inverted cone-and-plate rheometer.

This paper discusses results obtained from tests of large samples of natural debris flow material with a specially constructed rotating-cylinder viscometer. Specifically, we address the issue of the influence of sediment concentration and

---

[1]U.S. Geological Survey, 5400 MacArthur Blvd., Vancouver, WA 98661

the proportion of sand on the bulk rheological behavior of fine-grained material (≤2 mm) from a natural debris flow. The effect of particles coarser than 2 mm is the subject of experiments in progress and is not reported here.

VISCOMETER DESIGN, OPERATION, AND CALIBRATION

We have designed and constructed a large, coaxial, rate-controlled, wide-gap rotational viscometer capable of containing as much as 0.3 m^3 of slurry (Fig. 1). The drum that holds the slurry is driven by a 5 hp hydraulic motor and is designed to rotate and tilt so that debris and water may be thoroughly mixed. Removable fins attached to the inside of the drum facilitate slurry mixing. The rotating inner plastic cylinder, or rotor, was roughened by attaching a 13x13-mm-grid wire mesh to the surface to minimize slippage at the fluid interface. The 300-mm gap between the rotor surface and drum wall permits testing of slurries having a maximum particle diameter of about 30 mm, which enables the mixture to be treated as a continuum (Van Wazer et. al., 1963). For the tests described here, however, slurries were confined to a smaller bucket placed within the drum, which reduced the annular gap to 31 mm. Because the largest particle in the material tested was less than 2 mm, use of the smaller bucket does not violate the continuum approximation.

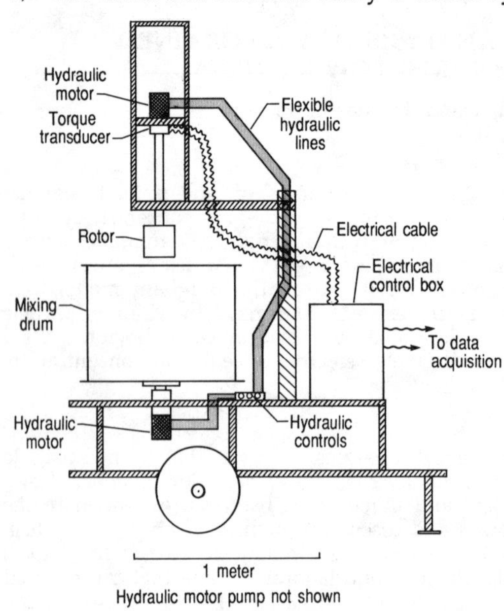

Fig. 1. *Schematic diagram of concentric-cylinder viscometer.*

The rotor is inserted into a slurry and the torque that resists rotation is measured at rotation rates of about 0.5 to 30 revolutions per minute (rpm). Torque is measured using a commercially produced transducer mounted between the rotor shaft and the hydraulic motor that drives the rotor (Fig. 1). Velocity of the rotor is measured by a small generator in contact with the rotor shaft. Rate of rotation and torque are logged automatically during each test on a personal computer at sampling rates of 25 Hz. Except for the initial run in each suite of tests, which begins from a static start, test data are collected after the rotor equilibrates to the new rate of rotation for several tens of seconds.

The viscometer was calibrated using MC-250 medium-curing cutback asphalt, a homogenous single-phase Newtonian fluid despite being a hydrocarbon mixture. The asphalt was tested at temperatures ranging from 16°C to 34°C. Independent verification of asphalt viscosity versus temperature was provided by the US Army Corps of Engineers North Pacific Division Materials Laboratory (J. Paxton, written communication, 1988).

DATA REDUCTION

Shear rate, which cannot be directly measured for wide-gap viscometers, was computed using methods devised by Kreiger and coworkers (Kreiger and Maron, 1952; Kreiger and Elrod, 1953; Yang and Kreiger, 1978), which Jacobsen (1974) extended to non-Newtonian fluids having a yield strength. Rotor end effect, which was computed to be as much as 8 percent of the rotor length (Van Wazer et. al., 1963; Spera et. al., 1988), was considered in the calculation of applied shear stress. Error-analysis theory applied to the reduction of noisy data collected from wide-gap concentric-cylinder systems (Borgia and Spera, 1990) was utilized to eliminate questionable data.

EXPERIMENTAL TECHNIQUE AND RESULTS

Rheological testing was conducted on samples of the large 1980 North Fork Toutle River volcanic debris flow collected 55 km downstream from Mount St. Helens, Washington. The debris flow deposit is very poorly sorted (Fig. 2) and is composed of sediment ranging in size from clay to boulders several meters in diameter. The clay fraction of this debris is composed predominantly of smectite; larger particles are primarily angular fragments of volcanic rock. Approximately 0.3 m^3 of material was collected, from which particles greater than 32 mm were removed. The sample was sieved and separated into constituent components of fines (<63 µm), sand (63 µm-2 mm), and coarse clasts (>2 mm). The influence of sediment concentration and sand proportion on the bulk rheological behavior of the fine-grained debris was evaluated systematically by varying the proportions of fines, sand, and water. Volume concentration of sediment in the wetted mixtures ranged from 0.44-0.66. Ratios of fines to sand ranged from 11:1 to about 1:5, the latter being slightly greater than the natural ratio of fines to sand in the original sample.

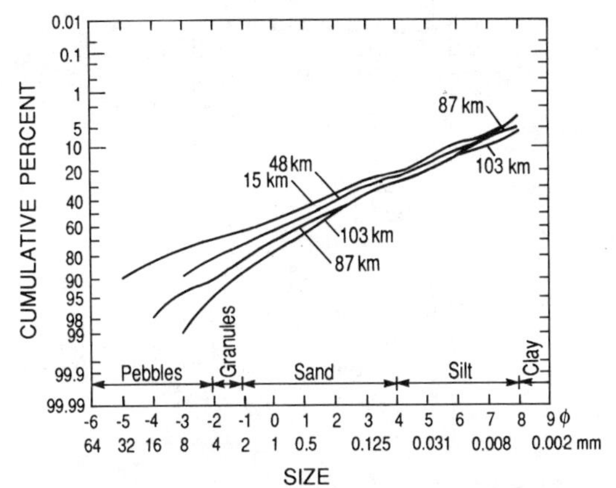

Fig. 2. *Cumulative curves of particle sizes in the North Fork Toutle River debris flow. Channel distance is in kilometers from Mount St. Helens crater (from Scott, 1988).*

The experiments indicate that the sediment concentrations at which a mixture behaves as a homogeneous slurry are strongly related to the sand concentration. As sand concentration increases, greater sediment concentrations are needed to maintain the uniform integrity of the slurry. If sediment concentration is too low, slurry integrity is degraded by particle settling and the mixture reduces to two independent phases, sediment in water. For mixtures containing only fines, homogeneity was maintained easily to a sediment concentration as low as 0.44, the lowest sediment concentration at which we could readily detect shear-stress changes by increasing rotation rate. For mixtures whose ratio of fines to sand was the same as the debris' natural state (about 1:5), sediment concentration was as great as 0.66, a limiting value that represents a mixture that could be thoroughly mixed by hand.

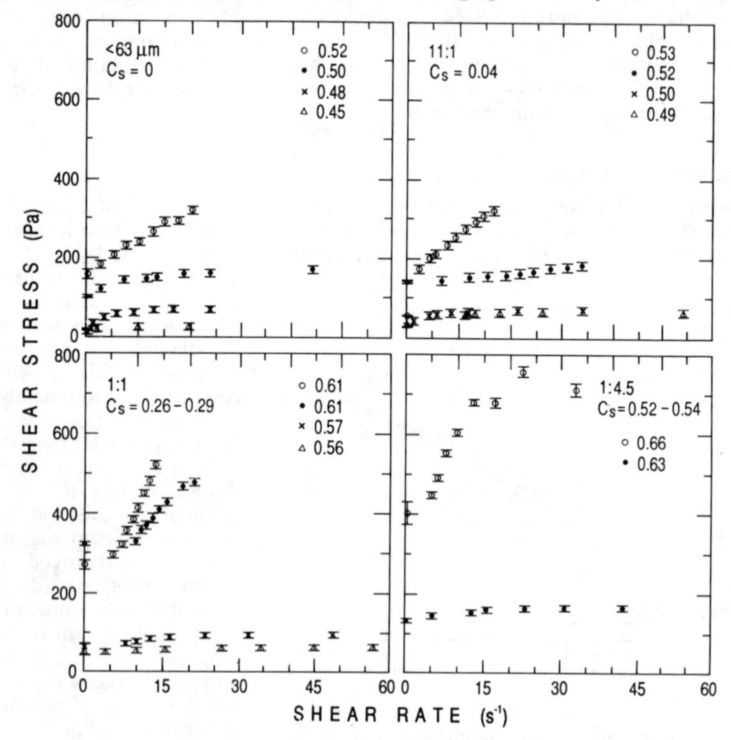

Fig. 3. *Shear stress versus shear rate. Each data point represents a time-averaged value from a single test. 1σ error is shown. Ratios (e.g. 11:1) indicate proportion of fines to sand; symbols indicate sediment concentration by volume.* C_s= *sand concentration.*

Shear stress generally increases as shear rate increases for all ratios of fines to sand and at all sediment concentrations (Fig. 3). Furthermore, as sediment concentration increases, apparent yield strength and viscosity also increase. The dependence of shear stress on shear rate is much stronger at higher values of sediment concentration for each ratio of fines to sand. At various lower threshold values of sediment concentration, which differ for each ratio of fines to sand, several of the stress-strain curves are nearly flat and shear stress is only weakly dependent on shear rate. The tested slurries exhibit variable response to increasing shear rate; several curves exhibit apparent shear-thinning whereas others exhibit weak shear-thickening or are linear. At shear rates greater than about 5 sec^{-1}, however, the curves are approximately linear and are compatible with a Bingham, or plastic, rheological model.

No systematic change is observed in relations between shear stress and shear rate at rates greater than 5 sec^{-1} as sand concentration increases. At lesser shear rates grain size and distribution appear to significantly influence slurry behavior. This behavior is best observed in relations between torque and time at the lowest applied rate of rotation (<1 rpm) and in hysteresis tests where rate of rotation was continuously increased to the maximum rate then continuously decreased to zero. For slurries having a sand concentration less than about 0.2 and for slurries having a relatively low sediment concentration, measured peak torque, once achieved, was maintained for the duration of the test and the signal emitted was relatively smooth (Fig. 4a). When sand concentration of the slurries exceeded 0.2 and sediment concentration exceeded about 0.56, measured peak values of torque were achieved rapidly but subsequently decayed to lesser, nearly constant residual values and the emitted signal sometimes fluctuated by several percent (Fig. 4b). Furthermore, in tests where rate of rotation was increased continuously, measured torque values commonly *decreased* slightly before they increased. This behavior became more pronounced as sand concentration and sediment concentration increased. Complementary behavior was observed in some hysteresis tests; as the rate of rotation continuously decreased measured torque generally decreased to some minimum value but then increased before zero rate of rotation was reached. Similar behavior was observed at very low rates of shear in the experiments reported by Phillips and Davies (1989a,b).

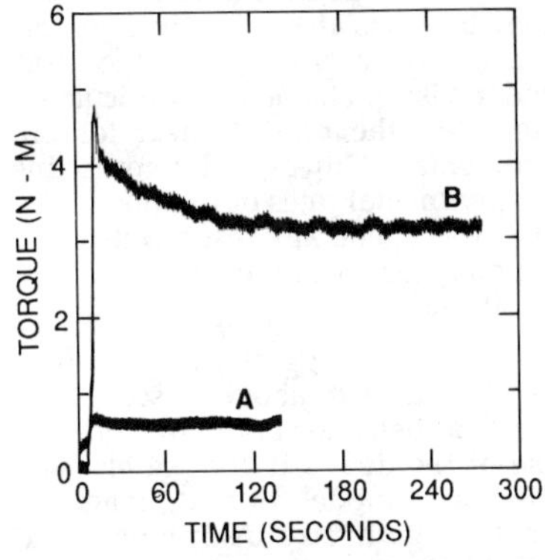

Fig. 4. *Torque versus time. **A**. Sand concentration = 0 (<63 μm); sediment concentration = 0.51. **B**. Sand concentration = 0.54 (fines:sand ≈1:5); sediment concentration = 0.66.*

The extreme sensitivity of plastic viscosity and yield strength of the fine-grained material of the debris flow to sediment concentration, particularly at higher concentrations and at higher ratios of sand to fines, is illustrated in Figures 3 and 5. Changes in sediment concentration of as little as 2 to 4 percent produce nearly order-of-magnitude changes in both plastic viscosity and yield strength (Fig. 5). Increasing the sand concentration of the slurries generally causes an increase in yield strength and plastic viscosity, primarily due to the concomitant increase in total sediment concentration needed to maintain slurry integrity. At a given sediment concentration, however, increased sand concentration in the slurry causes a *decrease* in the strength and plastic viscosity of the matrix material (Fig. 5). Trask (1959) observed a similar decrease of strength at a given sediment concentration as the ratio of sand to fines increased. He attributed this behavior to the relative proportions of bound and free water in the system. The surface area of particles per unit volume in the system decreases as the concentration of sand relative to fines increases. As a result, the relative proportion of water bound to the fines decreases and the content of free water in the system increases. Consequently the strength of the mixture decreases.

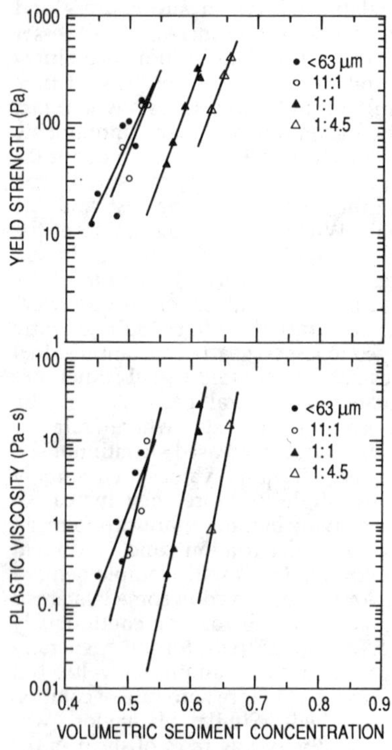

Fig. 5. *Yield strength, plastic viscosity versus volumetric sediment concentration. Symbols indicate proportion of fines to sand.*

INTERPRETATIONS AND CONCLUSIONS

Rates of shear in natural open-channel debris flows rarely exceed 50 sec^{-1} and perhaps are more commonly ≤ 10 sec^{-1} (O'Brien and Julien, 1988). Therefore experiments designed to directly measure the rheological behavior of natural debris flow materials should be conducted over a generally low range of shear rate. Because rate of rotation cannot be translated directly into rate of shear for a wide-gap concentric-cylinder viscometer it was difficult for us to select the rate of shear for our experiments. However, by restricting the experimental rates of rotation to ≤ 30 rpm, we generally succeeded in containing the maximum rate of shear near 50 sec^{-1}.

Our results suggest that for shear rates that exceed about 5 sec^{-1}, bulk rheological behavior of the fine-grained phase of the debris is compatible with the Bingham model for wide ranges in grain size, grain-size distribution, and sediment concentration. At these shear rates the behavior of the fine-grained natural debris flow material appears to be dominated by the rheology of the fluid phase, clay, silt and water, and not by the suspended sand grains. At shear rates less than about 5 sec^{-1}, the rheologic behavior of the slurry tends to deviate from the Bingham model and yield strength and plastic viscosity are extremely sensitive to sediment concentration and the proportion of sand. At these low rates of shear, when the concentration of sand exceeds about 20 percent by volume, we infer that particle interaction between sand grains (rubbing, grinding, and interlocking) dominates the mechanical behavior of the slurry.

Our results are compatible with those reported by Phillips and Davies (1989a,b) and O'Brien and Julien (1988). Phillips and Davies reported that debris flow mixtures containing particles smaller than 35 mm in diameter tended to exhibit plastic or viscoplastic behavior, but for rates of shear less than about 5 sec^{-1} they measured highly variable instantaneous shear stresses. This variability appeared to reflect significant interaction between the coarser particles, which dominated the slurry behavior at these low rates of shear.

O'Brien and Julien (1988) concluded that at rates of shear less than about 10 sec^{-1} the relation between shear stress and shear rate of debris flow material composed of fines was compatible with a Bingham rheological model. Also, they demonstrated that yield strength and plastic viscosity were very sensitive to sediment concentration. In their work, yield strength and viscosity increased by about three orders of magnitude with an increase in sediment concentration from 10 to 40 percent by volume.

APPENDIX 1. REFERENCES

Borgia, A., and Spera, F.J. (1990). "Error analysis for reducing noisy wide-gap concentric cylinder rheometric data for nonlinear fluids: Theory and applications." *J. Rheology* 34(1), 117-136.

Campbell, C.S. (1989). "Self-lubrication for long runout landslides." *J. Geology* 97, 653-665.

Chen, Cheng-lung (1988). "Generalized viscoplastic modeling of debris flow." *ASCE J. Hydr. Engineering* 114, 237-258.

Iverson, R.M., and Denlinger, R.P. (1987). "The physics of debris flows--a conceptual assessment." *Erosion and Sedimentation in the Pacific Rim*, Proceedings of IAHS Corvallis Symposium, Corvallis, OR, publ. no. 165, 155-165.

Jacobsen, R.T. (1974). "The determination of the flow curve of a plastic medium in a wide gap rotational viscometer." *J. Colloid and Interface Science* 48, 437-441.

Johnson, A.M. (1970). *Physical Processes in Geology*. Freeman, Cooper and Co.

Kreiger, I.M., and Elrod, H. (1953). "Direct determination of the flow curves of non-Newtonian fluids. II. Shearing rate in concentric cylinder viscometer." *J. Applied Physics* 24, 134-136.

Kreiger, I.M., and Maron, S.H. (1952). "Direct determination of the flow curves of non-Newtonian fluids." *J. Applied Physics* 23, 147-149.

O'Brien, J.S., and Julien, P.Y. (1988). "Laboratory anaylsis of mudflow properties." *ASCE J. Hydr. Engineering* 114, 877-887.

Phillips, C.J. (1989). "Rheology of debris flow material." PhD thesis, Lincoln College, University of Canterbury, Canterbury, New Zealand.

Phillips, C.J., and Davies, T.R.H. (1989a). "Debris flow material rheology--direct measurement." *Proceedings of International Symposium on Erosion and Volcanic Debris Flow Technology*, Yogyakarta, Indonesia, V13-1 - V13-13.

Phillips, C.J., and Davies, T.R.H. (1989b). "Generalized viscoplastic modeling of debris flow-discussion." *ASCE J. Hydr. Engineering* 115, 1160-1162.

Scott, K.M. (1988). "Origins, behavior, and sedimentology of lahars and lahar-runout flows in the Toutle-Cowlitz River system." *USGS Prof. Paper 1447-A*, 74 p.

Spera, F.J., Borgia, A., Strimple, J., and Feigenson, M. (1988). "Rheology of melts and magmatic suspensions 1. Design and calibration of concentric cylinder viscometer with application to rhyolitic magma." *J. Geophys. Research* 93(B9), 10,273-10,294.

Takahashi, T. (1981). "Debris flow." *Ann. Rev. Fluid Mech.* 13, 57-77.

Trask, P.D. (1959). "Effect of grain size on strength of mixtures of clay, sand, and water." *Geol. Soc. Am. Bulletin* 70, 569-580.

Van Wazer, J.R., Lyons, J.W., Kim, K.Y., and Colwell, R.E. (1963). *Viscosity and Flow Measurement*. Interscience.

Yang, T.M.T., and Kreiger, I.M. (1978). "Comparison of methods for calculating shear rates in coaxial viscometers." *J. Rheology* 22, 413-321.

THE BEARING CAPACITY OF DEBRIS FLOWS

By

Luis E. Vallejo*, M. ASCE

and

Hankyu Yoo*, A.M. ASCE

Abstract

Debris flows are made of a cohesive muddy matrix in which large and heavy particles exist in suspension. To explain how these particles are maintained in suspension, several mechanisms have been put forward. These mechanisms consider the effect of the cohesive strength and the buoyancy offered by the cohesive matrix, the pore water pressures in the void spaces of the muddy matrix, the dispersive pressure between the large grains, and the effect of turbulence. In debris flows, the large particles in suspension are subjected to random and violent movement during the flows. Thus, the large particles will exert loads on the muddy matrix at high strain rates. It is known that high rates of application of loads to solid samples of clay result in an increase in their compressive strength. Therefore, the objective of this study was to investigate if high rates of loading of muds by a sphere (that simulates a large particle in a debris flow) had an influence on the bearing capacity of muds forming part of debris flows. From a laboratory investigation, it was found that the higher the rate of loading of the mud by a sphere, the higher is the bearing capacity of the mud. Thus, the rate of loading of the muddy matrix by the particles in debris flows seems to be an additional mechanism that needs to be taken into consideration when explaining the capacity of debris flows to transport large particles in suspension.

---

*Associate Professor and Graduate Student, Dept. of Civil Eng., Univ. of Pittsburgh, Pittsburgh, PA 15261

Introduction

Many investigators have described the transport of large particles, some several meters in diameter, by debris flows (Johnson, 1970; Fisher, 1971). Several theories have been put forward to explain the mechanisms by which debris flows support and transport large particles. A brief summary of these theories is given next.

The first mechanism that has been suggested to explain the ability of debris flows to transport large particles is based on the cohesive strength of the muddy matrix surrounding the large particles. This cohesive strength of the matrix gives to it bearing capacity that will help to maintain the particles in suspension (Johnson, 1970; Hampton, 1975).

The second mechanism deals with the buoyancy provided by the muddy matrix. Buoyancy is determined by the difference in density between the submerged soils and the fluid matrix. According to Johnson (1970) and Hampton (1975,1979), the buoyant force acting on a boulder in a debris flow is equal to the weight of all the displaced material (fines and water) and helps with the aid of the cohesive strength of the matrix to maintain the boulder in suspension. Also, according to Hampton (1979) and Pierson (1981), buoyancy is enhanced in the muddy matrix by excess pore water pressures caused by the partial transfer of the weight of the large particles to the muddy matrix.

The third mechanism was put forward by Bagnold (1954) and is called the "dispersive pressure" mechanism. Bagnold (1954) demonstrated that when a mixture of poorly sorted grains is subjected to shear during its flow down a slope, for example, the larger particles drift toward the surface of the flow. The drift force is the result of stresses transmitted to the particles during their collision. This dispersive pressure is directly related to the square of the diameter of the particles. Since dispersive pressures will affect more large particles, they are moved toward the surface of the shear flow and are maintained there in suspension by the smaller particles.

The fourth mechanism inferred to be operative in the support of large particles by debris flows is turbulence. Enos (1977) has suggested that debris flows are turbulent. Within turbulent flows, the variation in direction and magnitude of velocity and force vectors with time could provide a mechanism contributing part of the lift force required to support the large particles within the debris flows.

The Effect of Rate of Loading on Cohesive Strength

In debris flows, the large particles in suspension are subjected to random and violent movement during the flows. Thus, the large particles will exert loads on the muddy matrix at high strain rates. From compression tests on solid samples of clay, Casagrande and Shannon (1949), and Casagrande and Wilson (1951) found that high rates of application of compressive loads to the clay result in an increase in its compressive strength, and thus in its bearing capacity. Thus, the present study involves the laboratory investigation of the braring capacity of a clay-water slurry loaded at varying strain rates by a sphere that simulates a large particle in a debris flow. If the rate of loading of the mud by the sphere increases its bearing capacity, the rate of loading will be an additional mechanism to explain why debris flows transport large particles in suspension.

For the laboratory testing, a mixture of 50% kaolinite clay by weight, 10% calgon (hexamethaphosphate, used as a dispersive agent), and 40% distilled water was used as mud. This mixture was placed in a rectangular container, 31 cm in length, 31 cm in width, and 43 cm in depth. A wooden spherical ball, 7.68 cm in diameter was used for testing the mud. The ball had an aluminum bar attached to it and to a proving ring that formed part of a Wykeham Farrance compression testing machine. The ball was fully embedded in the mud before the penetration testing started. The distance between the surface of the mud and the top of the sphere was equal to 2.5 times the diameter of the ball. The penetration of the mud by the sphere was accomplished by the upward movement of the base of the compression machine (Figure 1).

After the sphere was embedded in the mud and the dial of the proving ring that measures the reaction of the mud to penetration by the sphere was set to zero, the test was started. Three rates of penetration of the mud by the sphere were used. These strain rates were equal to 0.0508 cm/min, 0.1905 cm/min, and 0.762 cm/min.

The force needed to move the sphere through the mud at these different strain rates as a function of the depth of penetraition is shown in Fig. 2. An analysis of Fig. 2 indicates that the rate of loading of the mud by the sphere has a marked influence in the bearing capacity of the mud which is reflected in the resistance measured during testing. Thus, the increase in bearing capacity as a result of an increase in the rate of loading of the muddy matrix by particles forming part of debris flows seems to be an additional mechanism that helps to explain why large particles are kept in suspension.

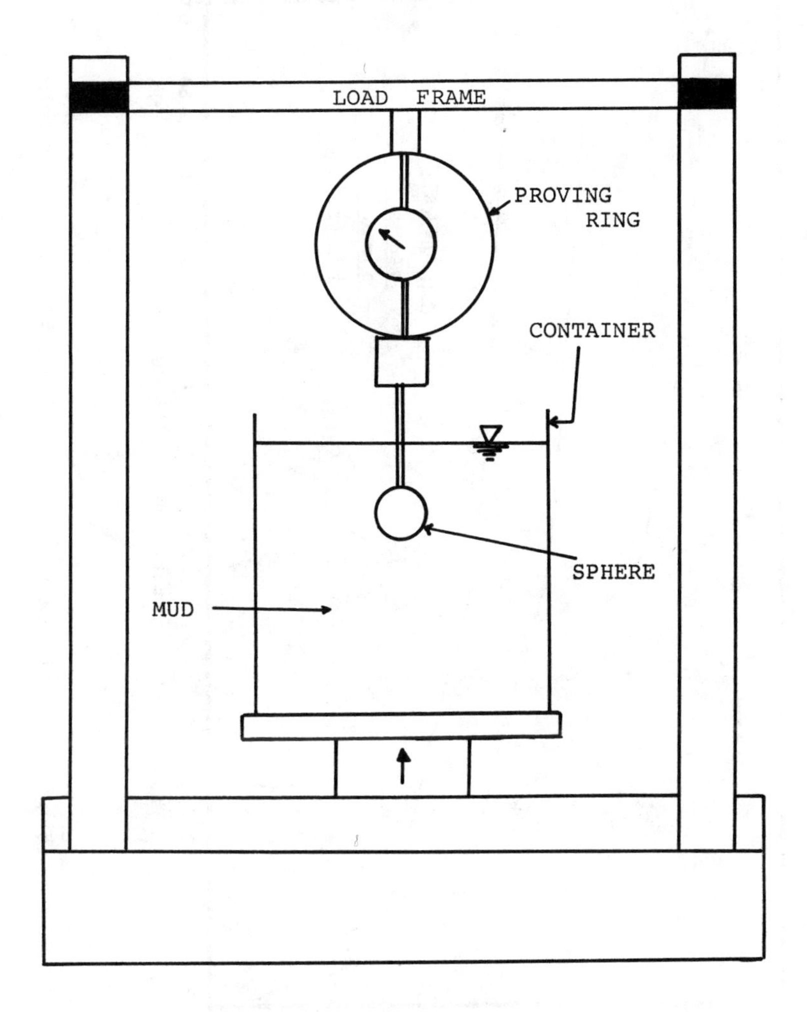

Figure 1. Laboratory Equipment Used For Penetration Tests.

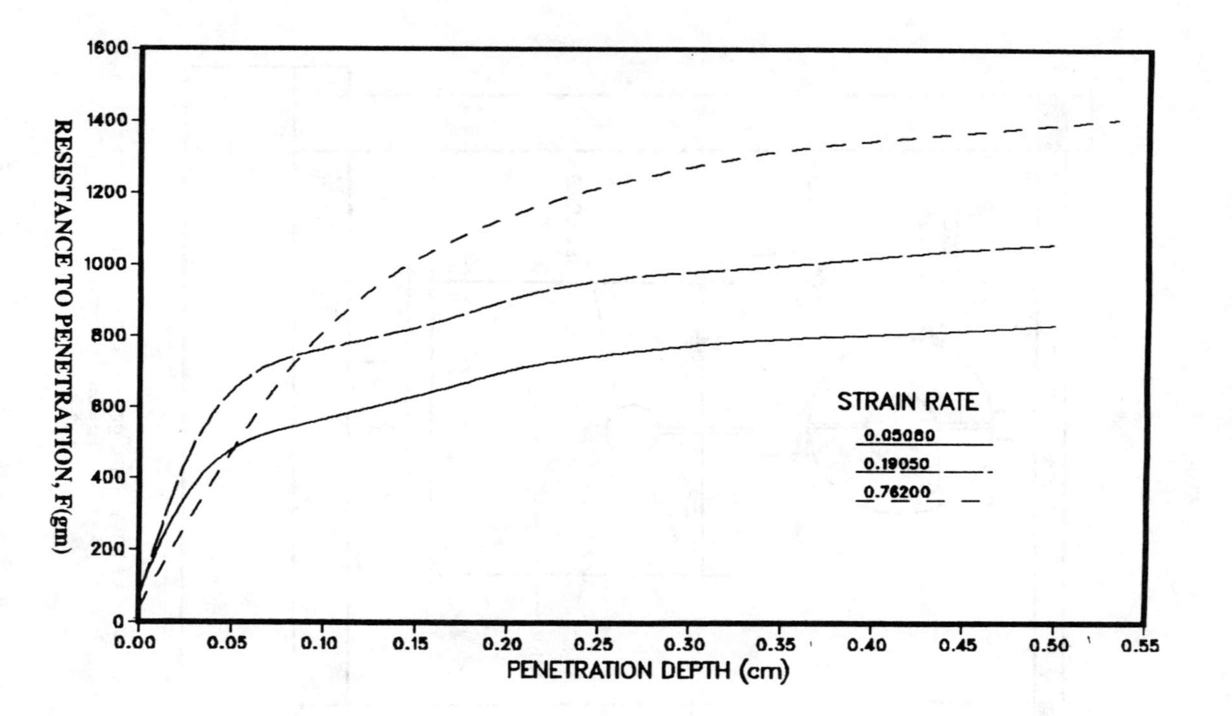

Figure 2. Resistance to Penetration (Bearing Capacity) Versus Depth of Penetration at Three Different Strain Rates (0.0508, 0.19050, and 0.762 cm/min).

Conclusion

From a laboratory study that involved the measurement of the resistance offered by a mud sample to the penetration of a wooden sphere (that simulates a large particle within a debris flow), the following can be said:

The resistance (bearing capacity) offered by the mud to the penetration of the sphere increased as the rate of this penetration increased. This increase in the bearing capacity of the mud as a result of an increase in the rate of loading by the sphere seems to represent an additional mechanism that helps to explain why debris slows suspend and move large particles.

Appendix - References

Bagnold, R.A. (1954). "Experiments on a gravity free dispersion of large solid spheres in a Newtonian fluid under shear." Proceedings Royal Society of London, Series A, Vol. 225, pp. 49-63.

Casagrande, A., and W.L. Shannon (1949). "Strength of soils under dynamic loads." Transactions ASCE, pp. 755-772.

Casagrande, A., and S.D. Wilson (1951). "Effect of rate of loading on the strength of clays and shales at constant water content." Geotechnique, Vol. 2, pp. 251-263.

Enos, P. (1977). "Flow regimes in debris flows." Sedimentology, Vol. 24, pp. 133-142.

Fisher, R.V. (1971). "Features of coarse-grained, high concentrations fluids and their deposits." Journal of Sedimentary Petrology, Vol. 41, pp. 916-927.

Hampton, M.A. (1975). "Competence of fine grained debris flows." Journal of Sed. Petrology, Vol. 45, pp. 834-44.

Hampton, M.A. (1979). "Buoyancy in debris flows." Journal of Sedimentary Petrology, Vol. 49, pp. 753-758.

Johnson, A.M. (1970). Physical Processes in Geology. Freeman and Cooper, San Francisco, 577 p.

Pierson, T.C. (1981). "Dominant particle support mechanisms in debris flows at Mount Thomas, New Zealand, and implications for flow mobility. Sedimentology, Vol. 28, pp. 49-60.

FLUME EXPERIMENTS ON DEBRIS FLOW

Lixin Wang[1]

Abstract

With the different characteristics of grain motion in debris flow, three layers are vertically divided, i.e., viscous collision layer, inertial collision layer and suspension layer. Both velocity and concentration distributions are systematically measured and the three-layer model is used to analyse the experimental data.

Introduction

Since the concept of dispersive stress was proposed by Bagnold (1954), it has been applied to many fields, e.g., debris flow, granular flow and laminated (or bed) load motion. Most work is focused on the improvement of the expression of dispersive stress. Generally the results have essentially the same form with Bagnold's.

In debris flow, the effective weight of grains is balanced by the dispersive pressure but intergranular turbulence, which has been universally accepted with the evidence that the dispersive stress has been used in nearly all models of debris flow. Although great efforts on experiments and theoretical models have been made, (e.g., Bagnold 1955, 1956; Takahashi 1978, 1980; Tsubaki 1983; Chen 1986, 1988), experimental data, especially those for concentration, are still rather rare due to the difficulties in measuring. This paper will present flume experiments on debris flow.

Experimental Facility and Measuring Methods

The experiment was made in a 6.9 m long transparent flume with a rectangular cross-section 10x10 cm and adjustable bottom slope between -4 ‰ and 13 ‰ . Two kinds of plastic grains were used and their characteristics are listed in Table 1.

The velocity was measured by tracing the marked grains in the lower part near the bottom and by a specifically designed Pitot tube in the higher part near the surface. The concentration was obtained by a radioisotope concentration meter. This meter is able to measure the points 1 mm from the bottom and has the advantage of high accuracy and disturbance-free to the flow field.

---

1 Ph. D., Dept. of Hydraulic Engineering, Tsinghua University, Beijing 100084, P.R. of China.

Table 1 Characteristics of Grains

Grain	ρ_s	D_{50} mm	C_o	ω_o cm/s
I	1.046	0.95	0.62	1.6
II	1.277	0.74	0.44	2.0

Table 2 Experimental Parameters

	$\bar{C}/C_o$ %	h cm	J_o ‰	J ‰	$\bar{U}$ cm/s	$F_r = \bar{U}/\sqrt{gh}$	U_m cm/s	T °C
RCP1	0.635	3.1	5.45	6.25	72.3	1.29		31
RCP3	0.529	3.7	4.85	5.66	71.8	1.20		27
RCP7	0.339	3.4	4.04	5.07	71.5	1.24		17
OP1	0.456	6.2	8.10	10.33	28.5	0.366	46.0	27
OP5	0.533	6.0	8.10	8.87	30.3	0.395	46.4	28
OP13	0.332	6.1	4.31	11.76	27.9	0.360	50.5	14
OP15	0.229	4.6	5.05	7.19	40.5	0.601	55.0	14

Experimental Results

A total of 21 sets of experiments was made with two kinds of grains, 14 sets for velocity distributions and 7 sets for concentration distributions (Wang 1986). The velocity and concentration were measured in different series of experiments just because the radioisotope concentration meter was more sensitive to the grain with larger density and the velocities of grains were easier to be obtained in the debris flow formed by the grain with less density. Due to the space limitation of this paper, only 7 sets which are so chosen that the mean concentration of each set is different are presented. Fig. 1 and Fig. 2 are 3 sets of concentration distributions and 4 sets of velocity distributions respectively. The parameters for each set are given in Table 2.

Three-layer Model

The bottom slope of the flume was so adjusted that there was deposit at the bottom. It was observed that there existed three layers with different characteristics of grain motion.

(1) Viscous Collision Layer Near the deposit, there is a shear layer with the slow velocity of grains, low shear rate and hyperconcentration nearly equal to that of deposit. The moving grains keep contact with each other. The grains of upper layers collide with those of lower layers and then roll on their surfaces. The exchange of grains between layers is quite rare. The intergranular fluid flow is laminar.

(2) Inertial Collision Layer Compared with the viscous collision layer, the inertial collision layer above it has the characteristics of the greater velocity of grains, higher shear rate and lower but still considerable concentration. The frequency of collision between grains is quite high and the contact time is very short. It seems that the turbulence of intergranular fluid remains very weak.

(3) Suspension Layer The layer near the surface is nearly pure water. There are a few number of grains in saltation. The

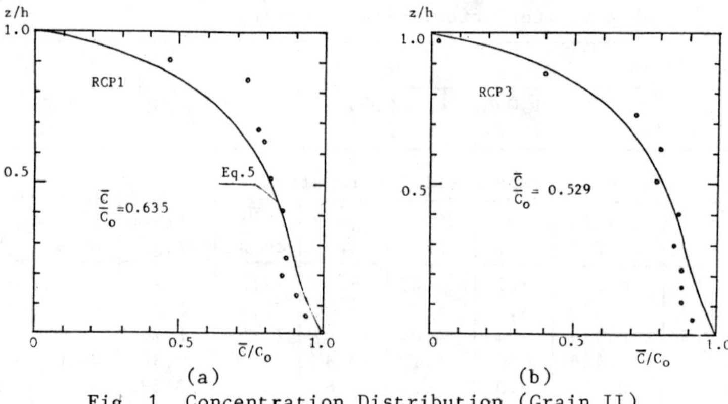

Fig. 1 Concentration Distribution (Grain II)

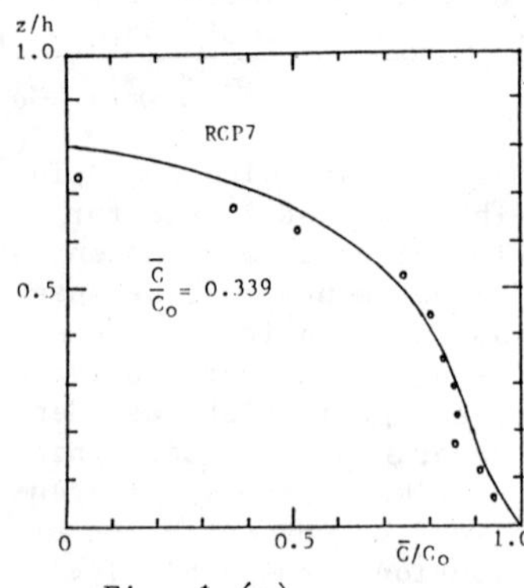

Fig. 1 (c)

flow is obviously turbulent.

The three different characteristics of grain motion mentioned above are quite distinct. However, it is very hard to exactly determine the bed of deposit and the positions of three layers. They are estimated from ovservations in the experiment.

In the viscous collision layer, the shear stress of flow τ is balanced by the dispersive shear stress τ_{zx} and the frictional shear stress τ_f between grains. In the inertial collision layer, the friction disappears due to the change of collision manner of grains. So τ is balanced only by τ_{zx}. In both layers the resistance of intergranular fluid is negligible. In the suspension layer, the logarithmic law remains valid.

From the expression of τ_{zx} by Z.Y. Wang and N. Chien (1984), the velocity distributions of debris flow in three layers can be expressed by (Wang 1986; Chien 1989)

for viscous collision layer

$$\frac{u}{u_{s1}} = \left(\frac{z}{h_1}\right)^{3/2} , \tag{1}$$

for inertial collision layer

$$\frac{u^2-u_{s1}^2}{u_{s2}^2-u_{s1}^2} = \frac{\eta^{3/2} - \eta_{s1}^{3/2}}{\eta_{s2}^{3/2} - \eta_{s1}^{3/2}} \tag{2}$$

and for suspension layer

$$\frac{u - u_{s2}}{u_m - u_{s2}} = \frac{\ln (z/h_2)}{\ln (h/h_2)} \tag{3}$$

When the viscous collision layer does not exist and the suspension layer is negiligible, the velocity curve can be simplified from Eq. 2 as follows (Wang & Chien 1984)

$$\frac{u}{u_m} = \left(\frac{z}{h}\right)^{3/4} \tag{4}$$

The availability for Eq. 4 will be discussed in the following part.

The concentration distribution in Fig. 1 can be expressed by a semi-empirical equation, i.e.

$$\frac{\lambda^2}{1+\lambda} = 18\ \frac{1-\eta'}{\sqrt{\eta'}} \tag{5}$$

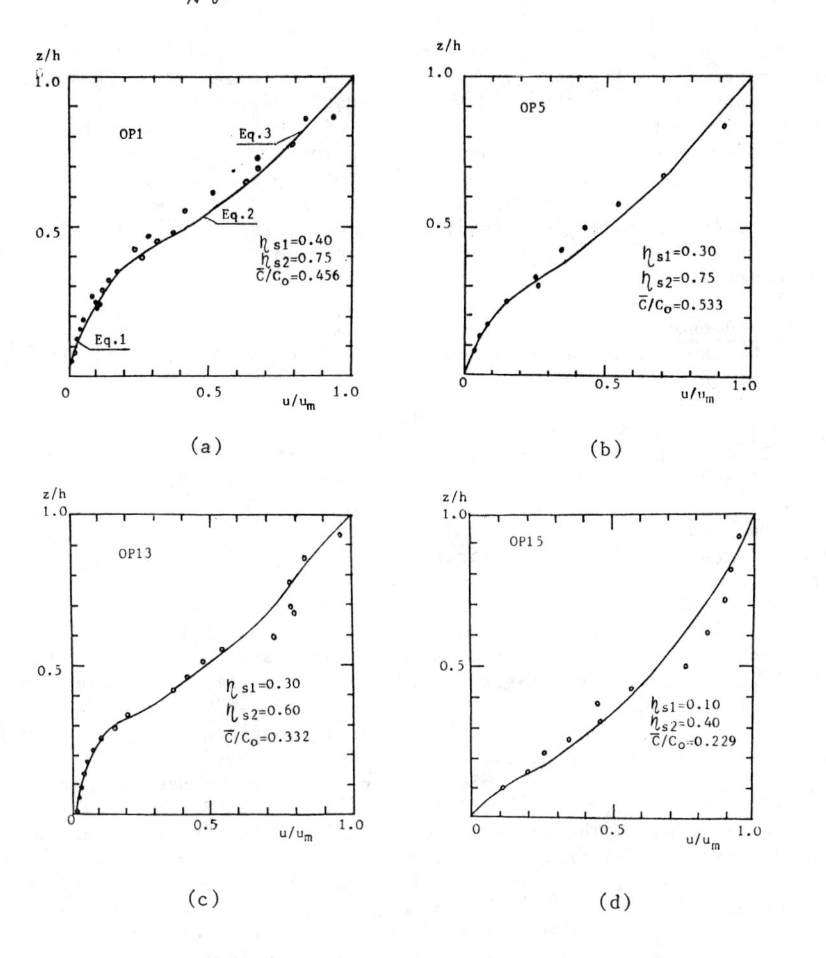

Fig. 2 Velocity Distribution (Grain I)

Discussion and Conclusions

The experiments described in this paper provided plenty of data. It will undoutedly very helpful to the studies on debris flow. Considering the difficulties in measuring, a slight adjustment of the position of deposit bed is permissive when one makes use of the velocity data.

For lack of data, the concentration distribution of debris flow was hypothesized to be vertically uniform before. The data presented is possibly available to establish more reasonable models of debris flow.

From Fig. 2, it can be seen that the velocity profile approaches to that of common hyperconcentration flow with the decrease of the mean dimensionless concentration $\overline{C}/C_0$ from 0.533 to 0.229.

It should be noted that the velocity curve has a deflection point. In contrast to general concept, the second derivative in the viscous collision layer is positive but negative. The same phenomenon is also observed in other data (Savage 1979, Takahashi 1980 (a); Tsubaki 1983). It could be of great importance to the resistance of debris flow and bed load motion. The viscous collision layer will disappear in certain conditions, say, high velocity and lack of enough movable deposit. That is why the deflection point does not exist in some data (Takahashi 1978, 1980 (b); Wang & Chien 1984). For this case, the experimental data are in good agreement with Eq. 4 (Wang 1986).

Acknowledgement

The experiment was made under the direction of the late Prof. N. Chien in 1986.

Appendix I. References

Bagnold, R.A., "Experiments on a gravity-free dispersion of water flow", Proc. Royal Soc. London, Ser A, Vol. 255, PP. 49-63, 1954.

Bagnold, R.A., "Some flume experiments on large grains but little denser than the transporting fluid, and their implications", Proc. Inst. Civil Engrgs., PP. 174-205, 1955.

Bagnold, R.A., "The flow of cohesionless grain in fluids", Philo. Trans., Royal Soc. London, Ser A, Vol.249, PP.235-297, 1956.

Chen, C.L., "Bingham plastic or Bagnold's dilatant fluid as a rheological model of debris flow?", Proc. of Third Int. Symp. on River Sedimentation, Jackson, Mississippi, 1624-1636, 1986.

Chen, C.L., "Generalized viscoplastic modeling of debris flow" J. of Hydr. Engrg., ASCE, 114 (3), PP. 237-258, 1988.

Chen, C.L., "General solution for viscoplastic debris flow", J. of Hydr. Engrg., ASCE, 114 (3), PP. 259-282, 1988.

Chien, N., "The movement of the flow with hyperconcentration", Tsinghua University Press, Beijing, PP. 141-149, 1989.

Savage, S.B., "Gravity flow of cohesionless granular materials in chutes and channels", J. Fluid Mech., Vol.92, 53-96, 1979.

Takahashi, T., "Mechanical characteristics of debris flow", J. of Hydr. Div., ASCE, 104 (8), PP. 1153-1169, 1978.

Takahashi, T., "Debris flow on prismatic open channel", J. of Hydr. Div., ASCE, 106 (3), PP. 381-396, 1980.

Tsubaki, T., Hashimoto, H. & Suetsugi, T., "Interparticle stresses and characteristics of debris flow", J. Hydroscience and Hydraulic Engrg., Vol.1, No.2, PP. 67-82, Nov., 1983.

Wang, L.X., "Experimental study on constitutive laws of the movement of laminated load", M. S. Thesis, Tsinghua University, Beijing, 1986.

Wang, Z.Y. & Chien, N., "Experimental studies on laminated load", Chinese Science, Ser A, PP. 863-870, No.9, 1984.

Appendix II. Notation

The following symbols are used in this paper.

C = concentration of debris flow by volume;
$\overline{C}$ = mean concentration by volume;
C_o = deposit concentration at the bed;
d_{50} = grain diameter by 50% weight of sample;
F_r = $\overline{U}/\sqrt{gh}$ Froude number;
g = gravitational acceleration;
h = depth of debris flow;
h_1, h_2 = depths at the tops of the viscous collision layer and inertial collision layer, respectively;
h' = depth at the zero concentration point;
J = bottom slope of flume;
J_o = energy slope of debris flow;
T = temperature of flow;
u = velocity;
u_{s1}, u_{s2} = grain velocities at h_1 and h_2, respectively;
u_m = velocity at the free surface of flow;
$\overline{U}$ = mean velocity of flow;
x = coordinate in the longitudinal direction of flow;
z = coordinate in the direction normal to the bed, positive upward;
λ = linear concentration defined by Bagnold (1954);
ρ_s = grain density;
ω_o = grain settling velocity in pure water;
τ = shear stress of flow;
τ_{zx} = dispersive shear stress;
τ_f = frictional shear stress between grains;
η = z/h dimensionless depth;
η' = z/h';
η_{s1} = h_1/h;
η_{s2} = h_2/h.

Experimental study on the rheologic behavior of the Slurry of viscous debris flow

Zhao Huiling[1]

Abstract

The slurry of viscous debris flow is a kind of dense solid–liquid mixture. It has very complicated rheologic behavior. As shown in the experiment described in this paper, the slurry exhibits time–dependent flow effect. Its viscosity varies with shearing time. The rheologic phenomena observed in the experiment are briefly introduced and discussed.

Introduction

The slurry of viscous debris flow is a kind of dense solid–liquid mixture. It always contains a certain amount of clay (d<0.005mm), the content of which may reach 3–15% of the total amount of solid particles (Kang and Zhang, 1980). Through floculation, The slurry with fine particles including clay can form a net structure. It possible to bring a lot of coarser particles in a suspension. Fine particles and suspended coarser particles constitute the slurry of viscous debris flow.

Viscous slurry is highly concentrated. for example, the volume concentration of the slurry (d<1.0mm) of viscous debris flow at Jiangia Ravine can be as high as about 35–50%, and the slurry with fine particles (d<0.05mm) about 25–35%.

Viscous slurry exhibits very complicated physical and chemical properties which exercise an important influence on rheologic behavior of viscous debris flow. Therefore, our test started with observing and measuring the slurry.

In the past few years, we conducted some rheologic tests

1Engineer, Chengdu Institute of Mountain Disasters and Environment, Chinese Academy of Sciences. Chengdu P.O. Box 417, Sichuan 610015.

on viscous slurry with a viscometer RV12, whose rotor speed can change continuously, if required (zhao and others,1989). These tests have shown that the rheologic parameters of the slurry will not be constants, but will vary with shearing time, when the sample reaches a certain degree of concentration. This paper will only introduce some of these tests, due to the limited space.

Two samples concerned in the paper were taken from the typical depositions of viscous and sub–viscous debris flow in Sichuan Province. The content of the fine particles (d<1.0mm) of sample A is 25.4%, sample B is 13.0%. After the particles of d>1.0mm had been removed from the sample, the remainder was compounded into several slurry samples of different concentrations with distilled water.

Experiment

The rheologic phenomena related to time is very complex, so far they cannot be exactly described mathematically. Therefore, we made observations on the viscosity–time relationship of viscous slurry through the following various tests.

Shear test

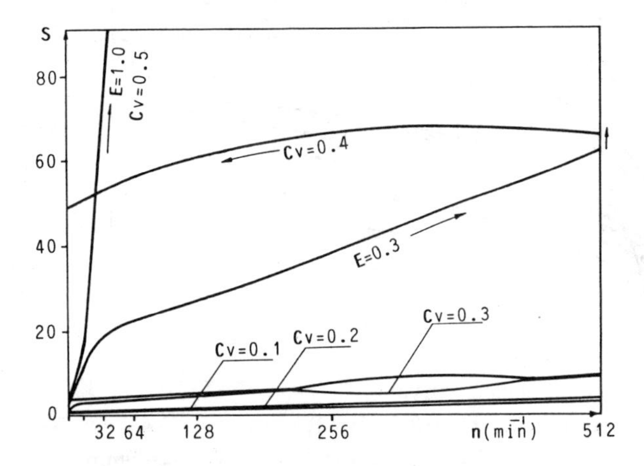

Fig1 Actual measured flow curve

sampleA: $t_1=0$, $t_2=3$min, $t_3=0.3$min.

The compounded samples whose volume concentration Cv are from 10% to 50% were respectively measured in terms of the flow curves (up– and down–curve). In our tests, we use dif–

ferent program time t_1 , t_2 and t_3 , where t_1 is waiting time, t_2 is shearing time t_3 is the period held at n_{max}, S is meauring value (scale grade) of shear stress τ , n is rotor speed. (See Fig1)

As shown in Fig1, when Cv is low, the flow curve is approximate to a straight line, up and down curves were coincident; when Cv=0.3, loop appears and slurry begins to show time–dependent effect , the area of loops increases rapidly with Cv. For instance, the size of the loop, when Cv=0.4, is much larger than when Cv=0.3, although their shear condition is the same. In a word, only if slurry has a certain concentration, can it show time–dependent effect. In other words, the slurry of non–viscous debris flow cannot have such rheologic behavior.

Cyclic shear test.

When shear rate is repeated cyclicly from zero to a preset value, then back to zero, a series of loops is obtained by shearing (Fig 2). The numbers and arrows in Fig2 express the measured sequence and the flow direction of shear curve.

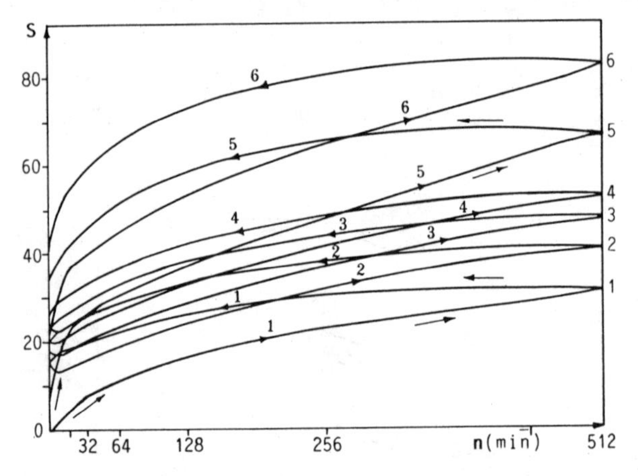

Fig2 cyclic shear curve

sample B: Cv=0.5 E=1, loop 1,2,3,4: $t_1=0, t_2=3$min, $t_3=0$, loop 5 , 6: $t_1=0$ $t_2=9$min, $t_3=0$.

The following experimental phenomena can be observed from Fig2:

First, flow curve spirals up and a down–curve is located above its up–curve. This characterastic of the loop (the sample is not pre–sheared) is difffrent from the hysteresis loop of thixotropy which spirals down and shifts towards the shear rate axis. Besides, the down–curve of thixotropy is beneath its corresponding up–curve (Cheng, 1987). Obviously, these two kinds of different loops imply different time–dependent effects.

Secondly, from Fig2, it is evident that the points at which the flow curve and the shear stress axis intersect go up along the axis, during the period when the shear rate cyclicly repeated. It is well known that the value of the shear stress at zero shear rate is yield stretss τ_y . Consequently, the result of this test indicates that yield stress τ_y increases with shearing time. That is, the structure of viscous slurry becomes stronger and stronger with time.

Thirdly, from Fig2, it is seen that apparent viscosity η_a decreases with shear rate, but at a constant shear rate it also increases with time. As a result, the viscosity also exhibits time–dependent effect. The viscous slurry thickens with time.

Cyclic shear test at low shear rate.

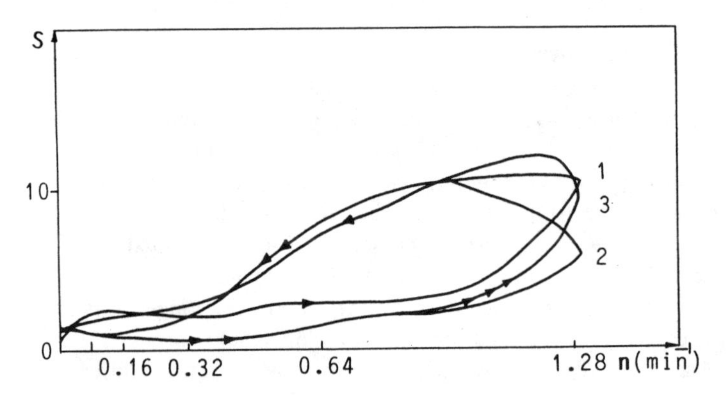

Fig3 Cyclic flow curve at low shear rate.
sample B: Cv=0.4, E=0.3, $t_1=0, t_2=3$min, $t_3=0$

The Cyclic flow curve measured at low shear rate (0–1.15 S^{-1}) is rather irreqular. There are still loops in Fig3, but the curve does not shift away obviously from the shear rate axis. It seems that shear stress varies repeatly only in a certain range. There are some differences in curve betwecn those measured at a low shear rate and those at a high one. (See Fig2). But, how to explain the differences? What shear rate will be counted as low for viscous slurry? At present, we know little about the rheologic behavior of the slurry at low shear rate. It is necessary to do more test.

Recovery test

In recovery test, shearing at a constant rate alternates with resting of different intervals. In Fig4, t_1 is resting time, t_3 is shearing time, dotted line is the reference line.

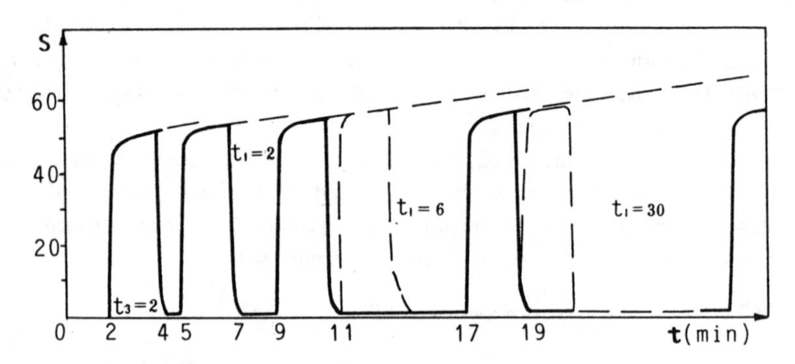

Fig4 Recovery curve

sample B: Cv=0.4, E=0.3, D=460.8 S^{-1}

As con be seen from Fig4, the heights of the five graphs increase with time. If shearing does not stop, the top of the graph will apporoximately extend along the dotted line. If a latter graph be transferred to the side of the former one, then, the top of the latter will coincide with the dotted line of the former. Only the fifth one is slightly lower than its reference, becouse its resting time (t_1=30min) is much longer than others. This result indicates that it is shearing that causes shear stress (or viscosity) to increase with time. This is quite different from the thixotropic recovery curve. The difference also brings light to the fact that viscous slurry is different from thixotropic fluid in their internal structures. The strueture of thixotropic fluid is broken down by shear and recovers at rest (Cheng, 1987). The longer the resting time, the stronger the structure. However, the structure of viscous slurry is strenthened by shear, but may be broken down or partially broken down by rest, due to settling of sediment particles. The particles which form the structure are difficult to settle, so the disruption process of the structure of slurry is extremly slow (wang and Qian, 1984)

Priliminary conclusion

The materials related to time fall into two classes: thixo tropy and rheopexy. They have contrary rheologic behavior (Van Wazer and others, 1963). Therefore, rheopexy is also called anti–thixotropy or negative–thixotropy. As shown in the above tests, the slurry of viscous debris flow exhibits time–dependent flow effect. Its viscosity increases with shearing time. That is to say, viscous slurry exhibits anti–thixotropic rheologic behavior.

The experiments also slow that the structure of viscous slurry is built up by shear at high shear rate. Hence, the view held by some scientists that shear can only disrupt structure is not very consistent with our experimental results. In fact, shear can break down structure in thixotropic materials, but, in anti–thixotropic materials, it probably has a functon, besides that of disrupting structure, of building up structure (Cheng, 1973, 1987). The two different kinds of functions depend mainly on the physical and chemical properties of materials.

These tests described in the paper have not been able to expose the complicated rheologic behavior of viscous slurry in all respects. Therefore, further tests and observation on viscous slurry have yet to be made.

References.

Cheng, D. C–H, 1973, Some measurements on a Negative–thixotropic fluid, Expanded version of article published in nature, v.245, P. 93–95.

Cheng, D. C–H, 1987, Thixotropy, International Journal of Cosmetic Science. 9. P151–191.

Kang, Z,and Zhang, S, 1980, A Priliminary analysis of the Characteristics of debris flow, in Proceedings of the International Symposium on River Sedimentation, P.225–226.

Van Wazer, J.R, Lyons, J.W, Kim, K.y, and Colwell, R.E, 1963,Viscosity and flow measurement: New York, Interscience, P.20–78.

Wang, Z, and Qian, N, 1984, Experimental Study on the Physical Properties of Sediment suspensions with Hyperconcentration, Journal of hydroulic engineering, P1–10.

Zhao, H, and others, 1989, Size properties and rheologic behavior of Debris flow, Research and prevention of debris flow, Sichuan Science and technology publisher, P165–192.

EXPERIMENTAL STUDY ON THE FLOW PROPERTIES OF DEBRIS FLOW

Fei Xiangjun[1] Fu Renshou[2]

Abstract

A new flow property model for viscous debris flow was presented in this paper. The results of the flow property model were in good agreement with the results of specific experiments. The model is also suitable for hyperconcentrated flow with wide size distribution.

Introduction

Viscous and subviscous debris flows take place frequently in the Southwest of China. The density of these kind of flows can reach 2.0 - 2.35 T/M^3, corresponding to the concentration by volume S_V=0.6-0.8. The particles size distribution of the flow is very wide and distribution of concentration in vertical is quite uniform with coarse particles keeping in suspension. It is difficult to measure the flow property by rotational or capiliary viscometer due to the sorting of coarse grains. For overcoming the difficulty in measurement and obtaining the viscosity of debris flow a new flow property model was suggested.

A new model of flow property for viscous debris flow

It is assumed that the mixture of debris flow could divided into two parts. The first part is the slurry of water-fine particles ($D < 0.1$ mm) whose flow property of the slurry can be measured by ordinary viscometer directly and the second part is the coarse particles whose diameter is more than 0.1mm. A special experimental study on the influence of coarse particles to the slurry of water-fine particles has been carried

---

1. Professor, Sediment Research Laboratory. Tsinghua University, Beijing, 100084, People's Republic of China.
2. Associate Prof., Sediment Research Lab. Tsinghua University, Beijing, 100084, People's Republic of China.

out to establish a model of flow property of viscous debris flow with wide size distribution.

There were many research results about apparent viscosity of water-sand mixture with low concentration of sediment. But when the concentration was very high, the sediment particles not only rises the shear rate of the fluid, but also leads to extra shear stress due to the friction force between particles. Both effects made apparent viscosity increase rapidly. According to this physical picture, shear stress equation along flow direction for the mixture was obtained by Shen Shouchang [1]

$$\tau = \eta' du/dy$$

$$\eta_r = \eta'/\eta = 1 + \pi \lambda^3/4(1+\lambda)^2 + K\lambda^n \quad (1)$$

where η', η, are apparent viscosity of mixture and clear water. On the right side of Eq.1, second term is the relative viscosity effected by the increase of shear rate of the fluid and third term is the relative viscosity effected by the friction force between particles. Parameters k, n are equal to 0.09 and 2.0 respectively according to the experiments. λ is the linear concentration which was defined by Bagnold.

$$\lambda = [(S_{vm}/S_v)^{1/3} - 1]^{-1} \quad (2)$$

in where, S_v, S_{vm} are the concentration and packing density of particles by volume in fluid respectively. By using the experimental viscosity data of water-sand mixture with noncohesion particles (d=0.1-0.15 mm) which was obtained by Wang Xinsheng [2], another relative viscosity equation was obtained as follows:

$$\eta_r = (1 - S_v/S_{vm})^{-2} \quad (3)$$

The experimental data were plotted as shown in Fig. 1. Replacing Eq. 3 by Eq. 2, Eq. 4 was obtained

$$\eta_r = [1 - (\lambda/(1+\lambda))^3]^{-2} \quad (4)$$

It should be pointed out that the results computed by Eq.1 and Eq.4 are almost the same under different concentration. There were many researcher to study apparent viscosity of sand-water mixture, but the results were not suitable for hyperconcentrated flow ($\lambda > 2$), because these equations did not consider the effect of friction force between particles. For example, Thomas [3] analysed the data published and obtained a

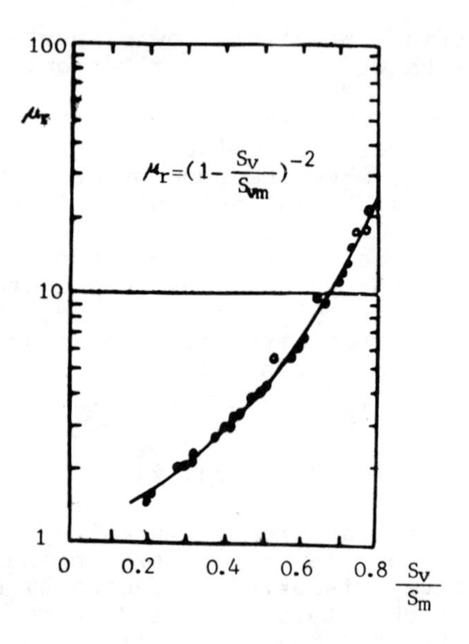

Fig. 1 μ_r vs. S_v/S_{vm} for slurries without fine particles

well-known equation as follows:

$$\eta_r=1+2.5S_v=10.05S_v^2+0.00273\exp(16.6S_v) \qquad (5)$$

The results of Eq. 5 are as almost same as the results of Eq. 6 which omits the third term of right side in Eq. 1. The results of Eq.5 and Eq.6 are compared in table 1 and the equation 6 is expressed as follows.

$$\eta_r=1+\pi\lambda^3/4(1+\lambda)^2 \qquad (6)$$

According to Eq. 3 if the water regarded as the carrier of coarse particles is replaced by the slurry of water-fine particles, the relative apparent viscosity of debris flow can be written as follows:

$$\eta_r=\eta_r'(1-S_{vc}/S_{vmc})^{-2} \qquad (7)$$

where η_r is the relative viscosity of the slurry of water-fine particles which can be measured by ordinary viscometer. S_{vc} and S_{vmc} are volumtric concentration and packing density of coarse particles respectively. V_c, V_f and V_o are the volumes of coarse, fine particles and water respectively. Thus, the concentrations of slurry S_{vf} and of the coarse particles S_{vc} are $S_{vc} = V_o/(V_o+V_f+V_o)$, $S_{vf}=V_f/(V_f+V_o)$ respectively. Total concentration of debris flow can be obtained as Eq.8.

Table 1 Comparision of Eq. 5 and Eq. 6 (S_{vm}=0.58)

S_v		0.1	0.15	0.20	0.25	0.30	0.35	0.40	0.45	0.50
λ		1.26	1.76	2.35	3.09	4.07	5.46	7.59	11.74	19.74
η_r	Eq. 5	1.37	1.63	1.98	2.43	3.05	4.02	5.70	8.95	15.75
	Eq. 6	1.31	1.56	1.91	2.39	3.06	4.06	5.67	8.52	15.04

$$S_{vT} = S_{vc}+S_{vf}(1-S_{vc}) \quad (8)$$

$$S_{vf} = S_{vT}(1-x)/(1-xS_{vT}) \quad (9)$$

where, $x=S_{vc}/S_{vT}$. When S_{vT}, x and $\eta_r' = f(S_{vf})$ have known, η_r in Eq. 7 can be obtained.

The slurry of fine particles under certain concentration has the property of a Bingham fluid. Yeild stress τ_B of debris flow also can divided into two parts, the first part is formed by flocculation of fine particles and the other is formed by friction between particles, of course, which is also related to the concentration. So far we can not distinguish these two kinds of effect in experiments, but a lot of experiments indicated that when coarse particles are put into the slurry of fine particles, the τ_B would increase to a higher value than that of slurry without coarse particles τ_B' and increase rapidly as the fraction of coarse particles increases. The value of τ_B and λ also can be expressed as follows,

$$\tau_B = \tau_B'(1+\alpha\lambda^{\beta}) \quad (10)$$

where α, β are coefficient and exponential, $\tau_B'=f(S_{vf})$.

Experimental verification

A special flow property experiment was carried out for verifying reliability of Eq. 7 and determining parameters α, β in Eq.10. Capillary viscometer of pipe diameter D=0.582 cm was adopted. At first, rheological experiments for water-fine particles ($D < 0.1$mm) slurry was carried out to obtain $\eta_r'=f(S_{vf})$ and $\tau_B'= f(S_{vf})$. Then, coarse sediment ($D > 0.25$–0.4 mm, $S_{vmc}=0.56$) was put into the water-fine particles slurry, a series of experiments for flow property was conducted with coarse particle ratios X of 0.125, 0.3, 0.48, 0.64 and 0.72

respectively. Corresponding values of τ_B and η_r were measured and the relationship between τ_B/τ_B' and $\eta_r/\eta_r' - \lambda$ were plotted as Fig.2. It can be seen that the data of experiments and the results computed by Eq. 11 and Eq. 12 were in good agreement.

$$\eta_r/\eta_r' = [1-(\lambda/(1+\lambda))^3]^{-2} \quad (11)$$

$$\tau_B/\tau_B' = (1+0.22\lambda^2) \quad (12)$$

where, $\alpha=0.22$, $\beta=2.0$ could be obtained and Eq.7 was verified under experimental concentration range.

Coarse particles with D=0.25-0.4 mm which adopted in above mentioned experiments, may be typical in the hyperconcentrated flow in Yollow River, but for viscous or subviscous debris flow taking place in the Southwest of China, the diameter of coarse sediment is not coarse enough. So other rheolgical experiments were carried out

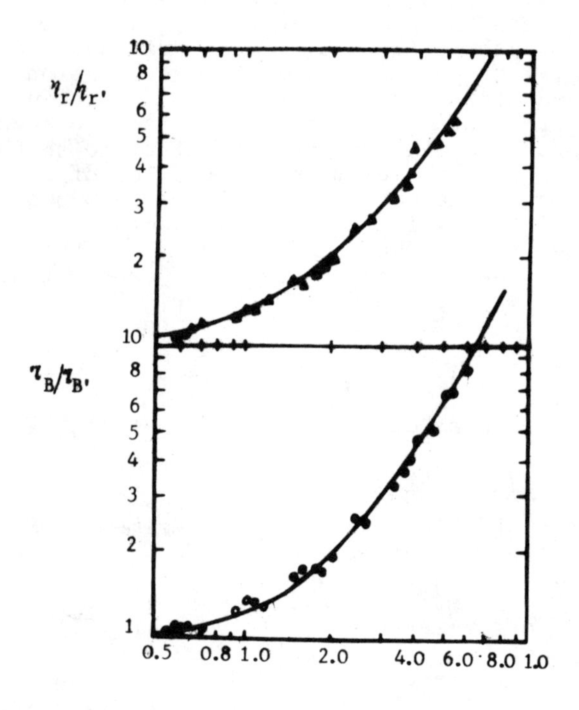

Fig.2 η_r/η_r' and τ_B/τ_B' for viscous debris flow

by large capillary viscometer. The mixture was consisted of coarse particles of D = 1.6 - 2.2 mm which made up of plastic grains with densisy r_s of 1.32 - 1.34 T/M^3 and fine particles (D < 0.1 mm)-water slurry with concentration S_{vf} = 0.18 (r_m=1.3 t/m^3), S_{vmc}= 0.64. The slurry could keep coarse particles in suspension. The results of experiments η_r and the results of computation according to Eq. 7 were listed in table 2. It can be seen that both results were in good agreement.

Table 2 Comparision of η_r values from experiments and computations

S_{vf}	η_r'	x	S_{vT}	S_{vc}	η_r(compu.)	η_r(experi.)
0.18	5.1	0.24	0.224	0.054	6.08	6.15
0.18	5.1	0.394	0.256	0.105	7.30	7.35
0.18	5.1	0.538	0.322	0.173	9.57	9.31
0.18	5.1	0.570	0.338	0.193	10.45	10.57
0.18	5.1	0.642	0.380	0.244	13.32	12.92

Acknowledgements

The research on which the work reported in this paper is based was supported by the Chinese National Science Foundation.

Reference

1. Shen Shouchang, Xie Shenliang "Mode of Structure of Debris Flow and the Effect of Coarse Characteristics of Slurry" Journal of Sediment Research 1983, No. 3.

2. Wang Xinsheng "Study on the Flow Properties of sand-water mixture without fine particles" 1981 the Thsis for Master Degree in Tsinghua University.

3. Thomas, D. G., Transport Characteristics of suspensions Part vill. "A note on the viscosity of Newtonian suspensions of uniform spherical particles" J. Colloid Sci., 20. 267 (1965).

Key Words

debris flow, hyperconcentration, rheological property, coarse particles, apparent viscosity.

TWO-DIMENSIONAL MODELING OF OVERLAND FLOW

Khalid B. Marcus[1] and Pierre Y. Julien[1], M.ASCE

Abstract

Two-dimensional distributed finite element (F-E) and finite difference (F-D) models are developed to simulate storm events on arid watersheds. The F-E model uses the kinematic wave approximation, while the F-D model uses the diffusive wave approximation. Both formulations offer tremendous flexibility in the analysis of surface runoff generated from both stationary and moving storms. In general, F-D models offer more stability than F-E models. Input data can be obtained from a network of raingages or from radar measurements. Several raingages provide ground truth information on precipitation, while radar data describes spatial variability.

Introduction

Surface runoff from natural watersheds is characterized by the spatial and temporal variability of overland flow. The physical and hydrological properties change from one point to another, which influences the runoff characteristics. The variations in geometry, slope, roughness coefficient, and excess rainfall intensity can have significant effects on the surface runoff from arid lands. In order to predict these variations, physically based distributed models are needed.

Earlier models include those of Chen and Chow (1971), Chow and Zvi (1973), Taylor et al. (1974), Constantinides et al. (1981), while more recent contributions include those of Hromadka et al. (1987), and Zhang and Cundy (1989).

---

[1] Colorado State University, Fort Collins, CO 80523

The one-dimensional kinematic wave approximation, and the diffusive wave approximation have been successfully used to simulate runoff from upland areas. It seems appropriate to develop two-dimensional models which can accurately depict the spatial variability of surface runoff. Since the F-D method has been used by other researchers to model the kinematic and the diffusive wave approximation, we will only focus on the F-E model.

Formulation of the Governing Equations

The two governing equations describing overland flow consist of the continuity and the momentum equations. A suitable resistance relationship that depends on the type of flow (laminar or turbulent) is also needed to solve the friction slope of the momentum equation.

The continuity equation in two-dimensions can be written as:

$$\frac{\partial h}{\partial t} + \frac{\partial}{\partial x}(q_x) + \frac{\partial}{\partial y}(q_y) = E \tag{1}$$

where:
- q_x = unit flow in the x direction
- q_y = unit flow in the y direction
- E = excess rainfall intensity
- t = time
- x-y = x and y coordinates

After considering the kinematic wave approximation, which substitutes the friction slope by the surface slope, a general resistance relationship can be written as:

$$q_z = C h^M S_z S^{L-1} \tag{2}$$

where z denotes either x or y, S is the total slope and S_z is the slope component along z. The parameters C, M, and L take the following values for the three types of resistance formula.

	C	M	L
turbulent (Manning)	1.49/n	5/3	0.5
turbulent (Chezy)	C	3/2	0.5
laminar (Darcy-Weisbach)	$8g/k\nu$	3	1

where:
- ν = kinematic viscosity
- n = Manning coefficient
- C = coefficient in the Chezy equation
- k = resistance coefficient

By substituting Equation 2 in Equation 1 we get:

$$\frac{\partial h}{\partial t} + \frac{\partial}{\partial x}(C\, h^M S_x S^{L-1}) + \frac{\partial}{\partial y}(C\, h^M S_y S^{L-1}) - E \qquad (3)$$

Equation 3 is solved using the isoparametric finite element method for the space domain and a finite difference for the time domain. A brief description of the finite element formulated to Equation 3 is described below.

The space domain is subdivided into quadrilateral elements. A trial solution is written for the unknown variable, h, as:

$$h\,(x,y,t) \simeq h^*\,(x,y,t) - \sum_{i-1}^{m} h_i(t)\, N_i(x,y) \qquad (4)$$

where: N_i = interpolation function
h^* = trial solution
m = number of nodes per element

The spatial variability of C, S_x, S_y, and E is described with interpolation functions. For instance, S_x is given by:

$$S_z\,(x,y) - \sum_{i-1}^{m} S_z\,(x,y)\, N_i\,(x,y) \qquad (5)$$

Equations 4 and 5, and a discretization of E, are substituted into Equation 3 to get the residual R. Using the Galerkin method, the integral equation is written as:

$$\int_D R\,(x,y,t)\, N_i\, dD - o \qquad (6)$$

where: D = solution domain

Equation 6 can be used for each element and the solution is summed over the entire domain. The assembly of the elements results in a set of global, non-linear equations which must be solved iteratively. For each time step, the iterative substitution method is used in solving the non-linear set of equations until convergence. In this way, the non-linear form of the equations is conserved.

The final form of the finite element formulation of Equation 3 can be written in matrix form as:

$$[KK]\ \{\Delta h\}^{i+1}_{t+\Delta t} = \{RR\} \tag{7}$$

where:

$$[KK] = \left([A] + [K(h^{\,i}_{t+\Delta t})\,] \right)$$

$$\{RR\} = \Delta t \{F_{t+\Delta t}\} - \Delta t [K(h^{\,i}_{t+\Delta t})]\ \{h\}^{i}_{t+\Delta t} - [A] \left(\{h\}_t - \{h\}^{i}_{t+\Delta t} \right)$$

in which {F} is the excess rainfall vector, [A] is the water storage matrix, and K(h) is the water conveyance matrix. We have:

$$h^{\,i+1}_{t+\Delta t} = h^{\,i}_{t+\Delta t} + \Delta h^{i+1}_{t+\Delta t}$$

where i refers to iteration level.

Model Testing and Application

The two-dimensional finite element model is applied to a converging section with convex surface Figure 1-a. The finite element discretization, shown in Figure 1-a, retains the actual representation of the geometry of the converging section compared to the finite difference representation. This leads to a better approximation of the outflow hydrograph, Figure 1-b, as compared to the results obtained by Kibler and Woolhiser (1970) with a 1-D finite difference model. Figure 1-c shows ponding and recession of water surface profile for part of the converging surface. Rainfall ceases at 80 seconds, after which the water surface profile still advances downstream while receding near the upstream boundary. The outflow hydrograph at the downstream end is a case of partial hydrograph.

The application of two-dimensional F-E models to natural watersheds is quite complex because of the distortion caused by the watershed topography. Comparison between different schemes shows that F-D schemes are generally easier to work with at the basin scale.

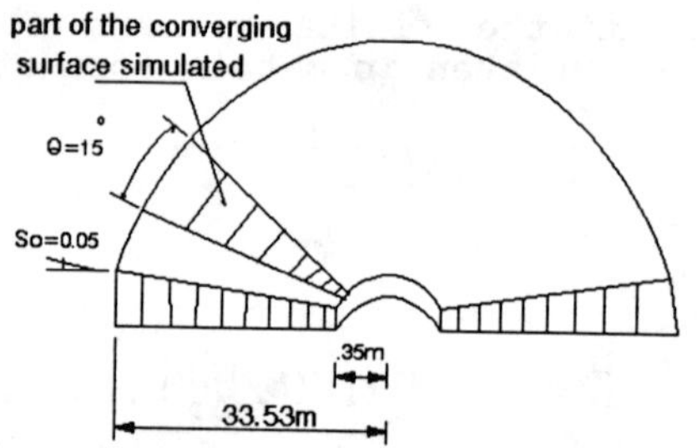

Figure 1-a Geometry of converging section

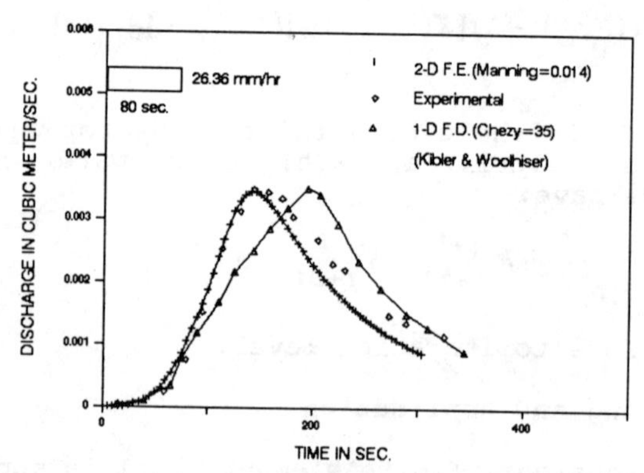

Figure 1-b Outflow hydrograph

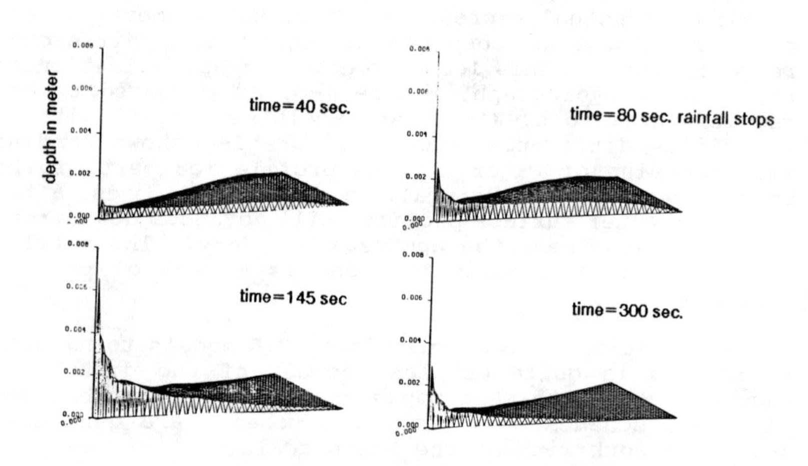

Figure 1-c Building up and recession of water surface profile for part of the converging surface

The response of natural basins to spatially varied rainfall intensity on Macks Creek, Idaho shows that surface runoff in arid lands originates from only partial areas of the basin. Although the total infiltration during one storm is more uniform over the basin, the surface runoff depends not only on the total volume of rainfall, but principally on the timing and the intensity of rainfall.

It is therefore becoming increasingly important to obtain rainfall precipitation data which reflects the local and complete nature of rainstorms in arid lands. Such information can be appropriately described by either a set of 5-10 raingages, like in the case of Macks Creek, or with more sophisticated radar data. The advantage of raingages being of collecting ground truth information while radar data describes the spatial variability.

References

Chen, C.L., and Chow, V.T., (1971), Formulation of Mathematical Watershed-Flow Model, Journal of the Engr. Mechanics Div., Proceedings of the Amer. Society of Civil Engineers, 97 (EM3), pp. 809-828.

Chow, J.T., and Ben-Zvi, A., (1973), Hydrodynamic Modeling of Two-Dimensional Watershed Flow, Jour. of the Hydr. Div., ASCE, 99 (HY11), pp. 2023-2040.

Constantinides, A.C., and Stephenson, D., (1981), Two-Dimensional Kinematic Overland Flow Modelling, Second International Conf. on Urban Storm Drainage, Urbana, Ill., U.S.A.

Hromadka II, T.V., McCuen, R.H., and Yen, C.C., (1987), Comparison of Overland Flow Hydrograph Models, Jour. of Hydr. Div., ASCE, 113 (HY11), pp. 1422-1440.

Kibler, D.F., and Woolhiser, D.A. (1970). The Kinematic Cascade as a Hydrologic Model, Colorado State University, Paper 39, Fort Collins, CO.

Taylor, C., Al-Mashidani, G., and Davis, J.M., (1974), A Finite Element Approach to Watershed Runoff, Journal of Hydrology, 21, pp. 231-246.

Zhang, W, and Cundy, T.W., (1989), Modeling of Two-Dimensional Overland Flow, Water Resources Research, Vol. 25, No. 9, pp. 2019-2035.

TWO-DIMENSIONAL MODELING OF ALLUVIAL FAN FLOWS

J. S. O'Brien[1], M. ASCE

W. T. Fullerton[2], M. ASCE

Abstract

A two-dimensional, flood routing model FLO-2D has been developed to route water and mudflows over unconfined surfaces. It predicts flow velocities and depths at discrete points for channel or overland flow. The model accounts for the effects of buildings or other obstructions that limit storage and constrict flow widths. Mudflow hydraulics which vary spatially and temporally on alluvial fans are simulated as a function of sediment concentration.

Introduction

The focus of modeling alluvial fan and floodplain flows is to predict a general range of flow hydraulics, such as velocity, depth and the area of inundation. FLO-2D was initially designed for simulating alluvial fan floods. It has been used to delineate flood hazard areas for FEMA flood insurance maps and to design structural mitigation on a variety of engineering projects such as: flood delineation on coalescing alluvial fans; design of a flood containment wall; water flooding on an urban floodplain; channel obstruction at the apex of an alluvial fan; and mudflow inundation of an urbanized alluvial fan.

Description of FLO-2D

The FLO-2D model has several important features including: flows can be routed over complex topography and roughness in two dimensions; the routing procedure

---

[1]Hydr. Engr., Lenzotti & Fullerton Consulting Engrs., Inc., Breckenridge, CO 80424.
[2]Pres., Lenzotti & Fullerton Consulting Engrs., Inc., Breckenridge, CO, 80424.

predicts peak discharge attenuation and bulking of the frontal wave; viscous mudflows can be simulated; channel flow, street flow, the interface with overland flow and return flow to the channel can be simulated; and the effects of flow obstructions such as buildings or walls which limit storage or constrict flow paths can be modeled. Buildings, levees or other flow obstructions are simulated by specifying area and width reduction factors for each grid element. Area reduction factors are defined by the amount of storage area in each grid element that is lost to buildings or natural features such as hills.

The accuracy of the FLO-2D model depends on the size and number of grid elements chosen. The number of grid elements must be balanced against the desired computational time. FLO-2D is linked to the AutoCAD-DCA software program which creates the grid element system and extracts their grid node elevations. This has significantly reduced the time required for a data file to be formulated. A typical data file of 1000 grid elements can be prepared from maps in a few days. Data revisions, changes in grid element size, alternate flow scenarios are very simple and very fast with AutoCAD-DCA.

The model has several other important attributes. Flow over adverse slope and backwater effects from constrictions or bridges can be simulated. The flow regime can vary between supercritical or subcritical. Outflow from bridges and culverts can be modeled. Channels can be represented by variable geometry cross sections. Street flow can be simulated along two perpendicular axes. For hyperconcentrated sediment flows, the flow hydraulics are governed, in part, by the sediment concentration and continuity is preserved for both the water and sediment. Sediment exchange between the flow and bed is not simulated by the model. FLO-2D is a rigid boundary model, but does account for mudflow deposition when the flow ceases.

Theoretical Considerations

FLO-2D predicts flow depth and velocity by solving the continuity equation and the two-dimensional momentum equations. The momentum equations are approximated by the diffusive wave equation. The model uses a finite, central difference routing scheme and uniform grid elements. The core of the routing algorithm was a finite difference model designed by T. V. Hromadka II and C. C. Yen and published by the USGS (Hromadka and Yen, 1987).

Routing hyperconcentrated sediment flows requires supplemental equations to define flow motion. Mudflows are nonhomogeneous, nonNewtonian, transient flood events whose fluid properties can change dramatically. They are characterized by surging, abrupt flow stoppage, and channel avulsion. Due to their transient nature, the most practical approach is to treat the flow as a continuum, combining the water and sediment components.

O'Brien and Julien (1985) reported that the total shear stress τ in hyperconcentrated sediment flows can be calculated from the summation of four shear stress components: the yield stress τ_y, the viscous shear stress, the turbulent shear stress, and the dispersive shear stress. The following rheological model has been proposed written in a quadratic form of the shear rate (dv/dy):

$$\tau = \tau_y + \eta \left(\frac{dv}{dy}\right) + C \left(\frac{dv}{dy}\right)^2 \qquad (1)$$

where C denotes the inertial shear stress coefficient.

The first two terms in Eqn. 1 are internal resistance stresses referred to as the Bingham shear stresses. The last term represents inertial stresses and is the sum of the dispersive and turbulent shear stresses. The yield stress τ_y and the viscosity η vary principally with sediment concentration. Their relationships were determined by laboratory experiment (O'Brien and Julien, 1988). For a complete solution to the equations of motion refer to O'Brien and Fullerton (1989). FLO-2D predicts mudflow hydraulics with a variable sediment concentration flow hydrograph.

Model Verification

Results from the model have been compared to the Army Corps of Engineers (COE) HEC-2 model for channel routing. A river overbank flood study was conducted and the results compared to the area of inundation predicted by the COE HEC-2 model. These comparisons confirmed the model's accuracy for water flow. For a verification of the mudflow component, the 1983 mudflow event at Rudd Creek, Davis County, Utah was simulated. This is the best available mudflow field data in the literature.

The simulated flood hydrograph was developed by the U. S. Army Corps of Engineers (COE, 1988). The field data to be replicated included: (1) The area of inundation indicated by aerial photography; (2) A surveyed volume of the mudflow deposit of approximately 84,000 yd^3; (3) A mudflow frontal velocity on the alluvial fan of approximately the speed that a man could walk (eyewitness account); (4) Observed mudflow depths that ranged from approximately 12 feet at the apex of the alluvial fan to approximately 2 or 3 feet at the debris front.

The mudflow was initiated by a landslide and a uniform sediment concentration of 45% by volume was chosen for the simulation. The maximum computed depth of 12.3 ft. near the fan apex correlated well with the 12 foot observed depth (**Fig. 1**). Mudflow velocities on the fan ranged from one to four feet per second (about walking speed). The maximum flow velocity at the fan apex was computed to be about 20 fps. Frontal lobe depths ranged from 1 to 4 feet depending on the spatial distribution of the flow on the fan. The area of inundation was accurately reproduced.

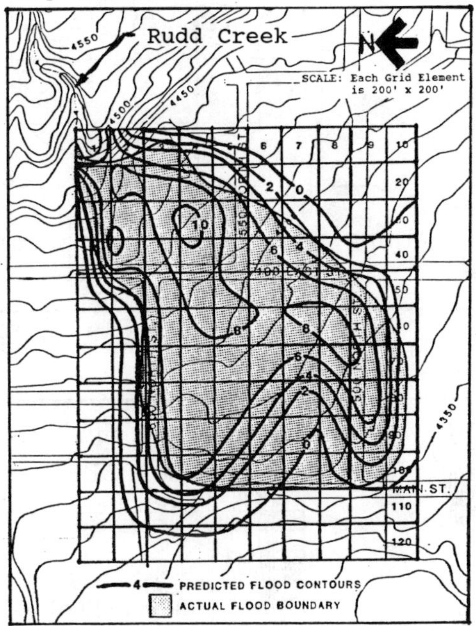

Figure 1. Rudd Creek 1983 Mudflow Predicted Flow Depth

Application of the FLO-2D Model - Case Study

FLO-2d was applied for the prediction of flood hydraulic in an alluvial wash. The results were compared with the hydraulics predicted with the FEMA methodology. A 8.0 hour, 100-year return period storm with peak flow of 11,100 cfs was simulated for Hiko Springs Wash in Nevada. The grid system consisted of 624 two-hundred foot, square grid elements within the confines of the wash walls (**Fig. 2**). The purpose of the project was to determine the hydraulics of the 100-year flood in vicinity of a proposed flood wall in the downstream portion of the wash. **Fig. 3** illustrates the contours representing maximum flow depths predicted by the FLO-2D model. A maximum flow depth of

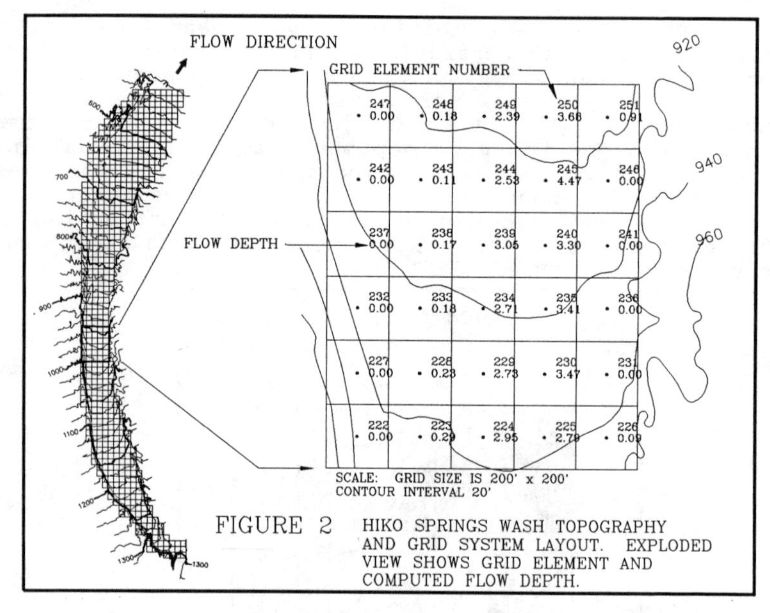

Figure 2. Hiko Springs Wash Topography and Grid System

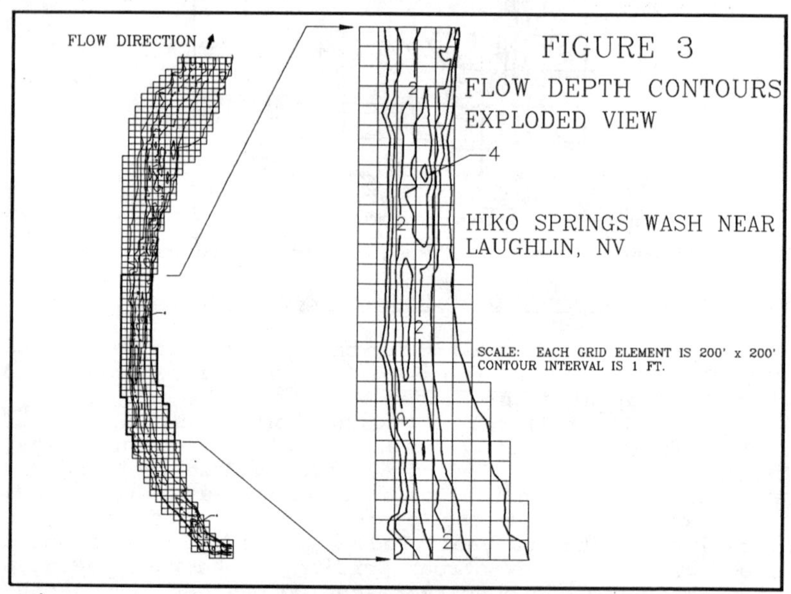

Figure 3. Hiko Springs Wash Predicted Flow Depth

4.5 ft. and a maximum velocity of 18.0 fps were predicted near the middle of the wash . The predicted maximum flow depth and velocity at the wall were 2.7 ft. and 11.0 fps respectively (O'Brien & Fullerton, 1989).

For comparison, the FEMA method for predicting alluvial fan flow hydraulics was applied to the Hiko Springs Wash. The method determines average velocity and depth using both the single channel and multi-channel methods. It predicted that the highest velocity (10 fps) and depth (3.5 ft.) would occur near the fan apex. Depths of 2 ft. and 1.5 ft. and velocities of 8 fps and 6 fps were predicted in the vicinity of the wall. These values are significantly less that the maximum values predicted by FLO-2D and it was concluded that hydraulics predicted by the FEMA methodology were not conservative by comparison and were inappropriate for design purposes.

Conclusions

FLO-2D is a valuable tool to predict the maximum flow velocity, depth and area of inundation. These predicted hydraulics can be used in the design of flood containment walls or other structural mitigation. The results illustrate FLO-2D's capability to delineate flooding hazards including shallow water flooding, mudflows, alluvial fan hydraulics, channel overbank flooding and flow obstruction in urban areas.

References

Hromadka, T. V.,II and Yen, C. C. (1987). "Diffusive Hydrodynamic Model," USGS Water Resources Investigations Report 87-4137, Denver Federal Center, CO.

O'Brien, J. S. and Fullerton, W. T. (1989). "Hydraulic and Sediment Transport of Crown Pointe Flood Wall, Hiko Springs Wash, Laughlin, Nevada." Lenzotti & Fullerton Consulting Engineers, Inc., Breckenridge, CO.

O'Brien, J. S. and Julien, P. Y. (1985). "Physical Processes of Hyperconcentrated Sediment Flows," Proc. of the ASCE Spec. Conf. on the Delineation of Landslides, Floods, and Debris Flow Hazards in Utah, Utah Water Research Laboratory, Series UWRL/g-85/03, 260-279.

O'Brien, J. S. and Julien, P. Y. (1988). "Laboratory Analysis of Mudflow Properties." J. Hydr. Div., ASCE, 114 (8), 877-887.

U.S. Army Corps of Engineers (1988). "Mud Flow Modeling, One- and Two-Dimensional, Davis County, Utah," Omaha District, Omaha, NE.

FLOOD HAZARD DELINEATION ON ALLUVIAL FANS

Thomas R. Grindeland, P.E., M. ASCE[1]
J.S. O'Brien, P.E., M. ASCE[2]
Ruh-Ming Li, P.E., M. ASCE[3]

Abstract

Alluvial fans are significant geomorphic features of arid lands. The unconfined and dynamic nature of alluvial fans makes standard methods of delineating flood hazards unsuitable. The probabilistic method suggested by the Federal Emergency Management Agency (FEMA) is currently the recognized means of assessing the flood-hazard potential of alluvial fans. The FEMA method, however, has limitations relative to its inherent assumptions. An alternative approach to delineating the flood hazards on alluvial fans utilizing two-dimensional numerical modeling has been developed and applied. Unlike the probabilistic method, the effect of existing developments and flood-control facilities on flow characteristics can be directly analyzed using this approach. The subjective interpretation of the effect of topographic or structural features on the fan is minimized.

Introduction

Currently, the National Flood Insurance Program (NFIP) administered by FEMA suggests a probabilistic-based methodology for assessing the flood-hazard potential of alluvial fans. The existing methodology, however, has inherent limitations which ignore important physical

---

[1]Senior Engineer; Simons, Li & Associates, Inc.; P.O. Box 1816, Fort Collins, CO 80522

[2]Senior Hydr. Eng.; Lenzotti & Fullerton Consulting Engineers, Inc.; P.O. Box 491, Breckenridge, CO 80424

[3]Principal Engineer; Simons, Li & Associates, Inc.; 3636 Birch St., Suite 290, Newport Beach, CA 92660

processes of flood hydraulics on alluvial fans and requires subjective interpretation of flood-hazard boundaries. Further, the currently accepted methodology is not suited to evaluating the impact of developments on the hydraulics of alluvial fans, or the adequacy of flood-control measures. Alternative methods are required to effectively assess the flood-hazard potential on alluvial fans. Efforts to develop improved means of assessing alluvial fan flood hazards have been undertaken by FEMA and others. Recently, an alternative approach utilizing two-dimensional numerical modeling to delineate the flood hazards on alluvial fans has been developed and applied.

Deficiencies of Existing FEMA Methodology

The existing FEMA methodology for delineating flood-hazard areas on alluvial fans resulted from the need for a standardized procedure to facilitate the administration of the NFIP for these areas. It incorporated a method suggested by Dawdy (1979) for estimating alluvial fan hydraulics. The FEMA method represented a pioneering step in the development of a practical means of evaluating the flood-hazard potential of alluvial fans. The method was later modified to account for multiple channel conditions (DMA, 1985). FEMA's method for the prediction of flow hydraulics on alluvial fans is based on numerous assumptions. A list of relevant assumptions and a brief discussion of their implications is presented:

1. There is an equal probability and random distribution of possible channel locations across fan contours.

This assumes that alluvial fans are of uniform elevation at locations of equal radial distance from the apex of the fan. In fact, the topography of most alluvial fans is not uniform. Local relief on alluvial fans is typically on the order of 5 to 10 feet, which is generally much greater than potential flow depths. The influence of local relief on the direction and extent of flooding on alluvial fans can be significant. Additionally, fanhead channels have been observed to be entrenched up to 50 feet. The influence of entrenchment on the direction and magnitude of downfan flooding is unaccounted for in the present methodology.

2. The alluvial fan is undeveloped.

The existing methodology does not address the effects of existing or proposed development in the flood-hazard delineation process. The influence of urbanization and flood- or sediment-control measures on occurrence probabilities and flow paths can only be subjectively assessed with the existing method. The method is therefore inappropriate for estimation of flow hydraulics for the design of flood- or sediment-control measures since it provides only estimates of flow depth and velocity. These estimates are largely dependent on the judgment of the individual practitioner. This limitation is important considering the increasing pressure to develop on alluvial fans.

3. The hydraulics of flow are determined for channel geometry [depth (D) and width (W)] such that

$$dD/dW = -0.005$$

This empirical width-depth relationship was developed from unpublished "field evidence" (Dawdy, 1979). The central equations of the FEMA methodology for predicting flow depth and velocity on an alluvial fan are derived from this single, empirically-based relationship. Since channel geometry on alluvial fans is influenced by site specific topographic, geologic, hydrologic, and hydraulic conditions, the existing methodology can only be expected to yield approximate depth and velocity estimates. Furthermore, these estimates are only valid within the range of the unknown "field evidence." Approximate values of flow depth and velocity are inappropriate for the design or evaluation of water- and sediment-control structures or for the delineation of flood-hazard zones.

4. There is no attenuation or bulking of the peak discharge as the flow progresses over the fan.

Attenuation or bulking of the peak flood discharge as it crosses an alluvial fan in not considered in the present FEMA methodology. Influences such as rainfall or hyperconcentrations of sediment can significantly affect the peak discharge and the associated hydraulics of flood events. Similarly, the influence of flood- and sediment-control structures on the peak discharge is ignored by the existing methodology.

5. A log Pearson III distribution applies to flood flows at the fan apex.

Use of a log Pearson III distribution in the FEMA methodology is consistent with the requirements of Water Resource Council procedures. Flood hazards on alluvial fans are, however, caused by both water and water-sediment mixtures. The applicability of a log Pearson III distribution to the range of potential water-sediment mixtures is unconfirmed at this time.

Alluvial Fan Flood-Hazard Delineation By Two-Dimensional Flood Routing

Alluvial fan flooding may encompass water floods, mud floods, and mud/debris flows. Mud and debris flows are of particular concern in delineating the flood-hazard potential of an alluvial fan because of the unconfined nature of the floodplain and the potential for avulsion from established stream channels. Delineation of the flood-hazard areas on an alluvial fan must consider the implications of all potential flood conditions. Computer modeling permits the efficient analysis of multiple flow scenarios, allows consideration of site specific physical influences, and can be calibrated to available historical flood data.

Two-dimensional flow routing by computer modeling has numerous advantages. These include the ability to consider the effect of complex topography and variable hydraulic roughness over the alluvial fan; attenuation or bulking of the hydrograph across the fan; the effect of varying sediment concentration on the hydraulics of flow; the hydraulic capacity of channels and overland areas; and the effects of natural or man-made flow obstructions which limit storage or constrict flow paths on the floodplain.

A computer model, FLO-2D (originally named MUDFLOW), was designed to route water floods or hyperconcentrated water-sediment flows in two dimensions (SLA and O'Brien, 1989). The model uses a central, finite difference routing scheme and uniform grid elements to apply the continuity and momentum equations. FLO-2D simulates the primary hydraulic processes on the alluvial fan. For hyperconcentrated flows, fluid flow properties are governed, in part, by sediment concentration. Sediment exchange between the flow and boundary is not simulated by the model. FLO-2D is a rigid boundary model.

The FLO-2D model has been applied to simulate water floods and mud flows over an urbanized alluvial fan in Telluride, Colorado as part of a flood insurance study (SLA and O'Brien, 1989). The results of that study have been submitted to FEMA. The model has also been used to evaluate the flood hazards for existing and proposed conditions on alluvial fans in Nevada and California. A suggested general approach to the application of the FLO-2D model is as follows:

1. Establish the clear water flood hydrograph for the return period of interest. If warranted, establish a potential hyperconcentrated water-sediment hydrograph by bulking the clear water hydrograph.

The model can be used to develop the flood hydrograph from the upstream watershed and route it to the fan apex or floodplain study area. The magnitude of a peak discharge at the fan apex can be determined through repeated application of the FLO-2D model and assumed sediment bulking scenarios. Overland and channel roughness values, excess rainfall, and sediment concentration can be varied to establish the discharge hydrograph from the upstream watershed.

2. Establish a grid system encompassing the study area. Grid elements must be chosen to present a reasonable representation of the topographic and hydraulic roughness characteristics of the study area.

3. Use the FLO-2D model to route the flow in the channel through the study area or over the unconfined floodplain.

4. Evaluate potential scenarios for flooding on the fan. These may include: failure of the existing primary channel along selected points of the water course; obstruction of the channel; relocation of the inflow hydrograph to a different portion of the grid system; obstructing or guiding the flow path to different floodplain areas; and analysis of various proposed development scenarios such as alternative locations of levees, bridges, or buildings.

   The various flood scenarios should consider the type and source of flooding such as mud flow, snowmelt, landslide, general storm or thunderstorm; and type and location of channel or levee failure or overtopping. A major advantage of computer modeling is the efficient evaluation of alternate flooding scenarios. By artificially blocking or obstructing selected grid elements the flow path of a flood event can be redirected. In this way, the effect of channel avulsion can be simulated. If blockage is simulated near the fan apex or other location, the magnitude of potential flooding over the entire fan may be evaluated. Water flooding will generally result in inundation of a broad area with shallow flooding, whereas mud flow simulations result in higher flow depths and smaller area of inundation.

5. Delineate the total hazard area for a fan by superimposing the results of water flood and mud flow simulations to describe the maximum extent, depth, and velocity of flooding.

Conclusions and Recommendations

The existing FEMA methodology for the prediction of alluvial fan hydraulics and area of inundation is based on "geomorphological principles" (Dawdy, 1979). Several major assumptions inherent to the method limit its applicability and utility. An alternative to the existing FEMA method for delineating flood hazards on alluvial fans has been presented. It is based on the hydraulic principles of continuity and momentum.

Two-dimensional modeling of alluvial fans permits the evaluation of site-specific physical influences on the magnitude and extent of flooding. Potential flood scenarios can be efficiently analyzed by computer modeling. A computer model, FLO-2D, has been developed. It has proven to be a useful tool for identifying alluvial fan flood hazards and designing flood-control structures. A general approach to delineating flood-hazard areas on alluvial fans through two-dimensional computer modeling was presented.

Additional research should be conducted to formalize a methodology for delineating flood-hazard areas on alluvial fans by two-dimensional computer modeling. Adoption of formal procedural guidelines and approval of a computer model such as FLO-2D by FEMA is suggested.

Acknowledgements

The authors wish to acknowledge FEMA for sponsoring the application of the FLO-2D model in the Telluride, Colorado Flood Insurance Study. The encouragement of Dr. John Liou, FEMA Region VIII, is also greatly appreciated.

References

Dawdy, D. R., 1979. "Flood Frequency Estimates on Alluvial Fans." Journal of the Hydraulics Division, ASCE, Vol. 105, pp. 1407-1413.

DMA Consulting Engineers, 1985. "Alluvial Fan Flooding Methodology an Analysis." Prepared for FEMA by DMA, Rey, CA, October.

Simons, Li & Associates, Inc. and Dr. J. S. O'Brien, 1989. "Flood-Hazard Delineation for Cornet Creek, Telluride, Colorado." Submitted to FEMA, March.

APPLICATION OF THE U.S.G.S. DHM FOR FLOODPLAIN ANALYSIS

T.V. Hromadka II[1], M.ASCE and J.J. DeVries[2], M.ASCE

Abstract

The two-dimensional Diffusion Hydrodynamic Model, DHM, is applied to the evaluation of floodplain depths resulting from an overflow of a leveed river. The environmental concerns of flood protection and flow velocities can be studied with the two-dimensional DHM flow model. Although the DHM generates considerable information, it is easy to use and does not require expertise beyond that required for use of the one-dimensional approaches.

Introduction

The main objective of this paper is to review the findings of a detailed study of the Santa Ana River 100-year event floodplain in the City of Garden Grove, California, using the two-dimensional Diffusion Hydrodynamic Model (DHM) (Hromadka, 1985, Hromadka et al, 1985, Guyman and Hromadka, 1986, Hromadka and Durbin, 1986, Hromadka and Nestlinger, 1985, Hromadka and Yen, 1986, Hromadka and Yen, 1987).

The local terrain slopes southwesterly at a mild gradient (i.e., 0.4%) and is fully developed with mixed residential and commercial developments. Freeways form barriers through the study site so that all flows are laterally constrained with outlets at railroads and major streets crossing under the freeways. Because of the flood flow conveyed through the floodplain and the mild cross-sectional terrian, the floodplain analysis needs to include two-dimensional unsteady flow effects.

Description of the DHM

The DHM provides the capability to model two-dimensional unsteady flow where storage effects and diverging flow paths are important, and hence, the steady state one-dimensional flow approach may be inappropriate.

---

[1]Director of Water Resources Engineering, Williamson and Schmid, 17782 Sky Park Boulevard, Irvine, CA 92714, and Associate Professor, Mathematics Department, California State University, Fullerton, CA 92634

[2]Associate Director, Water Resources Center, University of California, Davis, CA 95616

The two-dimensional unsteady flow equations consist of the equation of continuity

$$\frac{\partial q_x}{\partial x} + \frac{\partial q_y}{\partial y} + \frac{\partial z}{\partial t} = 0 \tag{1}$$

and two equations of motion

$$\frac{\partial Q_x}{\partial t} + \frac{\partial}{\partial x}\left(\frac{Q_x^2}{Ax}\right) + \frac{\partial}{\partial y}\left(\frac{Q_x Q_y}{Ax}\right) + gAx\left[S_{fx} + \frac{\partial h}{\partial x}\right] = 0 \tag{2a}$$

$$\frac{\partial Q_y}{\partial t} + \frac{\partial}{\partial y}\left(\frac{Q_y^2}{Ay}\right) + \frac{\partial}{\partial x}\left(\frac{Q_x Q_y}{Ay}\right) + gAy\left[S_{fy} + \frac{\partial h}{\partial y}\right] = 0 \tag{2b}$$

in which t is time, x and y (and the subscripts) are the orthogonal directions in the horizontal plane; q_x and q_y are the flow rates per unit width in the x and y-directions; z is the depth of water; Q_x and Q_y are the flow rates in the x and y-directions, respectively; h is the water surface elevation measured vertically from a horizontal datum; g is the acceleration of gravity; Ax and Ay are the cross-sectional areas; and S_{fx} and S_{fy} are the friction slopes in the x,y-directions. The DHM utilizes the uniform grid element to model the two-dimensional unsteady flow, therefore, Ax and Ay are defined as the length of uniform grid element times the depth of water.

The friction slopes S_{fx} and S_{fy} can be estimated by using Manning's formula

$$S_{fx} = \frac{n^2 Q_x^2}{C^2 Ax^2 R_x^{4/3}} \tag{3a}$$

and

$$S_{fy} = \frac{n^2 Q_y^2}{C^2 Ay^2 R_y^{4/3}} \tag{3b}$$

in which n is the Manning's roughness factor; R_x, R_y are the hydraulic radiuses in x,y-directions; and the constant C=1 for SI units and 1.486 for U.S. Customary units.

In the DHM, the local and convective acceleration terms in the momentum equation (i.e., the first three terms of Eq. 2) are neglected (Akan and Yen, 1981). Thus, Eq. (2) is simplified as

$$S_{fx} = -\frac{\partial h}{\partial x} \tag{4a}$$

and

$$S_{fy} = -\frac{\partial h}{\partial y} \tag{4b}$$

Combining Eqs. (3) and (4) yields

$$Q_x = \frac{C}{n} Ax\, R_x^{2/3} \frac{\left(-\frac{\partial h}{\partial x}\right)}{\left|\frac{\partial h}{\partial x}\right|^{1/2}} \tag{5a}$$

$$Q_y = \frac{C}{n} Ay\, R_y^{2/3} \frac{\left(-\frac{\partial h}{\partial y}\right)}{\left|\frac{\partial h}{\partial y}\right|^{1/2}} \tag{5b}$$

which may account for flows in both positive and negative x and y-directions. The flow rates per unit width in the x and y-directions can be obtained from Eq. (5) as

$$q_x = \frac{C}{n} Z\, R_x^{2/3} \frac{\left(-\frac{\partial h}{\partial x}\right)}{\left|\frac{\partial h}{\partial x}\right|^{1/2}} \tag{6a}$$

$$q_y = \frac{C}{n} Z\, R_y^{2/3} \frac{\left(-\frac{\partial h}{\partial x}\right)}{\left|\frac{\partial h}{\partial x}\right|^{1/2}} \tag{6b}$$

Substituting Eq. (6) into Eq. (1), gives

$$\frac{\partial}{\partial x}\left[\frac{C}{n} Z\, R_x^{2/3} \frac{\left(-\frac{\partial h}{\partial x}\right)}{\left|\frac{\partial h}{\partial x}\right|^{1/2}}\right] + \frac{\partial}{\partial y}\left[\frac{C}{n} Z\, R_x^{2/3} \frac{\left(-\frac{\partial h}{\partial y}\right)}{\left|\frac{\partial h}{\partial y}\right|^{1/2}}\right] + \frac{\partial h}{\partial t} = 0$$

or

$$\frac{\partial}{\partial x}\left[K_x \frac{\partial h}{\partial x}\right] + \frac{\partial}{\partial y}\left[K_y \frac{\partial h}{\partial y}\right] = \frac{\partial h}{\partial t} \tag{7}$$

where

$$K_x = \frac{C}{n} Z\, R_x^{2/3} \Big/ \left|\frac{\partial h}{\partial x}\right|^{1/2}$$

and

$$K_y = \frac{C}{n} Z R_y^{2/3} \bigg/ \left| \frac{\partial h}{\partial y} \right|^{1/2}$$

The numerical algorithms used for solving Eq. (7) are fully discussed by Guymon and Hromadka (1986) and in the U.S.G.S. Water Resources Investigation Report, 87-4137 (Hromadka and Yen, 1987). The data preparation needs for a floodplain analysis is also discussed in the U.S.G.S. Water Resources Investigation Report (Hromadka and Yen, 1987).

Application of the DHM to the Study Area

A global Manning's roughness coefficient of n = 0.045 was initially used in this study, except at major obstructions, such as freeways. Roughness coefficients for freeway undercrossings were assumed to be n = 0.020. Effective grid areas (Fig. 1) were also assigned to the elements that are adjacent to the freeways. This decreased the available storage of the particular grid. The net effects of using the flow path reduction factor and the decreased available storage is to achieve more realistic results for the floodplain analysis.

Based upon an aerial photograph and a field investigation of the study area, it was assumed that the flood flows will mostly be contained within flow-paths in which streets exist. On the average, widths of these flow-paths comprise one-fifth of a typical cross-section, i.e, the flow-path reduction factor is 0.8. An average effective grid area was also found from the aerial photograph. Buildings occupied thirty-five percent of the photographed area. In this study all buildings were assumed to be excluded from available storage; therefore a global effective area factor of 0.65 was applied.

A 100-year frequency runoff hydrograph of the Santa Ana River at Imperial Highway (see Figure 1) was generated by the U.S. Army Corps of Engineers, Los Angeles District (1987). This hydrograph was used in the subject DHM model, by dividing the runoff hydrograph into segments (see Figure 2) according to the peak breakout flowrates estimated in the referenced Corps of Engineers' study. The peak 5000 cfs was applied at Katella Avenue; the next 19,000 cfs breaks out just north of the Garden Grove Freeway. Immediately south of the Garden Grove Freeway, 1000 cfs breaks out on the west and east banks. Only the west bank overflow was applied to the model. An underlying assumption in the Corps' breakout analysis was that the eastern overflows return to the river downstream from the study site, so the overflow is ignored in this model. Finally, 3000 cfs overflows the west bank at Fairview Street.

The maximum flood depths calculated using the DHM are shown in Figure 3. These depths occur at various times throughout the total simulation time of 24 hours, although depths close to the maximum depths will remain for hours before and after the peaks. The floodplain boundary (see Figure 2) is derived from the maximum flood depths and the ground elevations.

In general, the DHM is used for floodplain analysis because this approach is capable of handling unsteady backwater effects in overland flow, unsteady overland flow due to constrictions, such as culverts, bridges, freeway underpasses, and so forth, unsteady flow overland flow across watershed boundaries due to backwater and ponding flow effects. In general, several important types of information can be generated from the DHM analysis. These include (1) the time versus flood depth relationship; (2) the flood wave arrival time; (3) the maximum flood depth arrival time; (4) the direction and magnitude of the flood wave; (5) the stage versus discharge relationship; and (6) the outflow hydrograph at any specified grid element within the study area.

Conclusions

The DHM (1987), which provides another tool for floodplain management, was published by the U.S. Geological Survey as a Water Resources Investigations Report (87-4137). The flow-path reduction factor and the effective grid area were added to the DHM (1987) for a more realistic representation of the field conditions.

Because the DHM provides a two-dimensional hydrodynamic response, use of the model eliminates the uncertainty in predicted flood depths due to the variability in the choice of cross-sections used in the one-dimensional models. That is, model users might select a cross-section perpendicular to the direction of flow, but on urban area the selection becomes somewhat arbitrary. Additionally, the DHM accommodates both backwater effects and unsteady flow, which are typically neglected in HEC-2 (1973) floodplain analysis.

NOTICE

The computational results shown in this paper are to be used for research purposes only. No governmental approval of the results shown are to be construed nor implied.

References

1. Akan, A.O., and Yen, B.C., Diffusion-Wave Flood Routing in Channel Networks, ASCE Journal of Hydraulics Div., Vol. 107, No. HY6, 1981, 719-732.
2. Guymon, G.L., and Hromadka II, T.V., Two-Dimensional Diffusion Probabilistic Model of a Slow Dam Break: Water Resources Bulletin, 1986, 22, 2, 257-265.
3. Hromadka II, T.V., Predicting Dam-Break Flood Depths Using a One-Dimensional Diffusion Model: Microsoftware for Engineers, 1985, 1, 1.
4. Hromadka II, T.V., Berenbrock, C.E., Freckleton, J.R., and Guymon, G.L., A Two-Dimensional Dam-Break Floodplain Model: Advances in Water Resources, 1985, 8, 1, 7-14.
5. Hromadka II, T.V., and Durbin, T.J., Two-Dimensional Dam-Break Flood Flow Analysis for Orange County Reservoir; Water Resources Bulletin, 1986, 22, 2, 249-255.

6. Hromadka II, T.V., and Nestlinger, A.J., Using a Two-Dimensional Diffusion Dam-Break Model in Engineering Planning, Proceedings: ASCE Workship on Urban Hydrology and Stormwater Management, Los Angeles County Flood Control District Office, Los Angeles, California, May 1985.
7. Hromadka II, T.V., and Yen, C.C., A Diffusion Hydrodynamic Model (DHM), Advances in Water Resources, 1986, 9, 3, 118-170.
8. Hromadka II, T.V., and Yen, C.C., A Diffusion Hydrodynamic Model, Water Resources Investigations Report, 87-4137, U.S. Geological Survey, 1987.
9. U.S. Army Corps of Engineers, Los Angeles District, Overflow Analysis for Lower Santa Ana River, Orange County, California, Proceedings: Rigid Boundary Hydraulic Problems Workshop, April 1987.

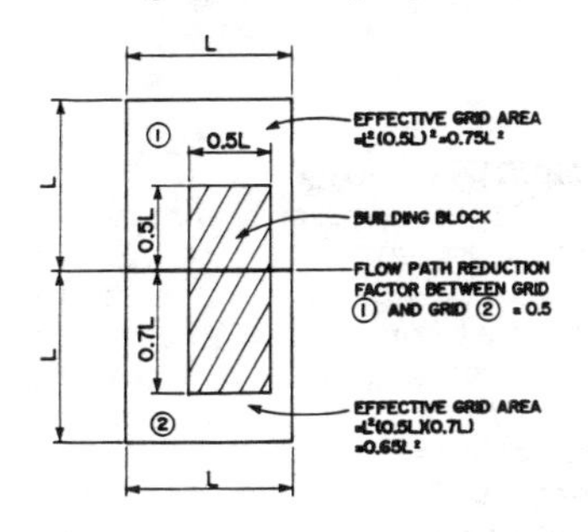

Figure 1. Flow Path Reduction Factor and Effective Grid Area Features.

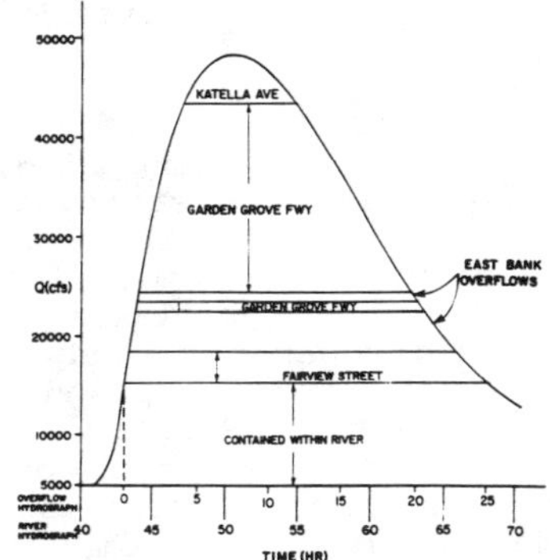

Figure 2. Segmented Santa Ana River 100-Year Runoff Hydrograph at Imperial Highway.

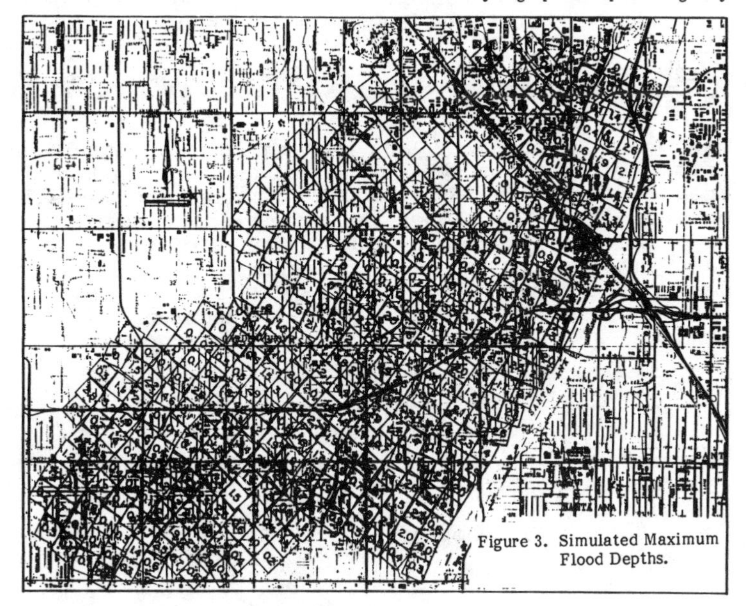

Figure 3. Simulated Maximum Flood Depths.

Phoenix Flood Hydrology for Price Expressway

Jerome J. Zovne, Ph.D, P.E.,M.ASCE[1] and
L. Steven Miller, P.E., M.ASCE[2]

Abstract

Recently HDR Engineering, Inc. (HDR) completed a study of the off-site stormwater volumes and peak flow rates for the Price Expressway and Santan Freeway in suburban Phoenix. HDR is the General Engineering Consultant to the Arizona Department of Transportation (ADOT) and is preparing general roadway plans for the six mile length of Price Expressway and an initial five mile portion of the Santan Freeway. The roadways offer unique stormwater challenges because they will be constructed as depressed roadway sections through rapidly growing Phoenix suburban communities of Chandler, Mesa, Tempe and Gilbert. This paper summarizes the stormwater modeling approach for the complex 58 square mile drainage basin intercepted by these depressed roadways (see Figure 1).

Introduction

The Price Expressway and Santan Freeway roadways are primarily depressed below existing grades in the project area. Stormwater sheet flows originating off-site and entering the ROW cannot be conveniently passed through or under the roadway. Thus, the preferred method of handling the off-site stormwater is to contain the design storm volume (the 100-year 24-hour event) in a series of detention basins on the "upstream" side of the roadway.

In the project area there is no natural stormwater outlet. The only available outlet is the Carriage Lane

[1]Project Manager, HDR Engineering, Inc., 3000 S. IH-35, Suite 400, Austin TX 78704
[2]Project Engineer, HDR Engineering, Inc., 5353 N. 16th St., Suite 205, Phoenix, AZ 85016

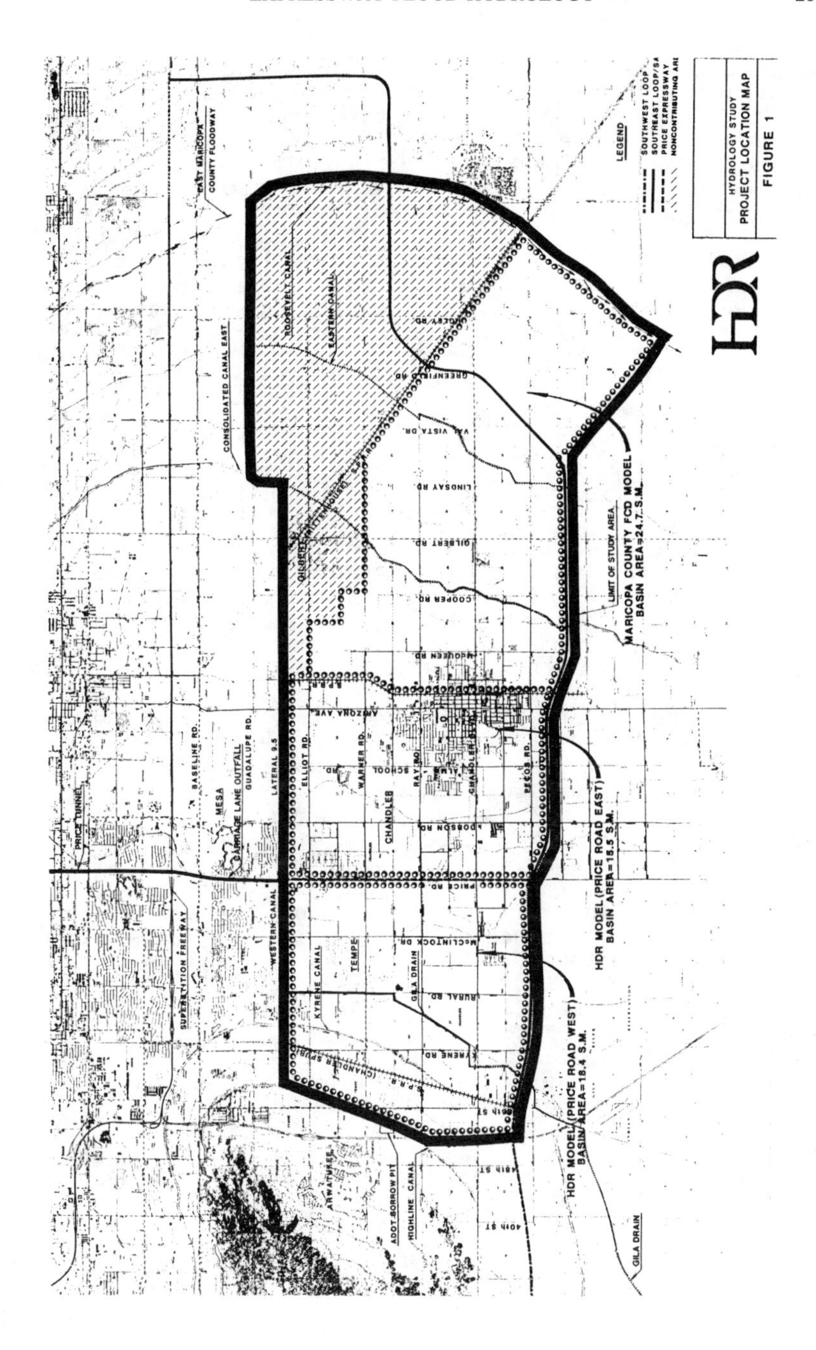
HYDROLOGY STUDY
PROJECT LOCATION MAP
FIGURE 1
LEGEND
SOUTHWEST LOOP
PRICE EXPRESSWAY
HDR
LIMIT OF STUDY AREA
MARICOPA COUNTY FCD MODEL
BASIN AREA=24.7 S.M.
HDR MODEL (PRICE ROAD EAST)
BASIN AREA=15.5 S.M.
HDR MODEL (PRICE ROAD WEST)
BASIN AREA=18.4 S.M.
EAST MARICOPA
COUNTY FLOODWAY
CONSOLIDATED CANAL EAST
ROOSEVELT CANAL
EASTERN CANAL
BASELINE RD.
MESA
CARRIAGE LANE OUTFALL
GUADALUPE RD.
LATERAL 9.5
ELLIOT RD.
WARNER RD.
CHANDLER
PECOS RD.
WESTERN CANAL
KYRENE CANAL
TEMPE
GILA DRAIN
SUPERSTITION FREEWAY
PRICE TUNNEL
ARWATUKEE
ADOT BORROW PIT
HIGHLINE CANAL
GILA DRAIN
GREENFIELD RD.
VAL VISTA DR.
LINDSAY RD.
GILBERT RD.
COOPER RD.
McQUEEN RD.
ARIZONA AVE.
DOBSON RD.
PRICE RD.
McCLINTOCK DR.
RURAL RD.
KYRENE RD.
40th ST
48th ST

Outfall and Price Road Tunnel system which is under construction and will soon provide an outlet north to the Salt River. A constraint to pumping into this system is that the flow will be restricted to 150 cfs during design peak flow conditions, per an intergovernmental agreement between ADOT, Chandler, Mesa, and Maricopa County. (ADOT, 1988)

The drainage area contributing off-site flows to the project is approximately 58 square miles and complicated by mixed land use, variable development criteria, and the presence of various significant barriers to flow. This required the development of a detailed hydrologic model which is the primary focus of this paper.

Drainage Area

An overall view of the project setting is shown on Figure 1.

The drainage area boundaries for the offsite drainage analysis is the Western Canal/Lateral 9.5 on the north, the Santan Freeway alignment on the south, Interstate 10 on the west, and the Roosevelt Water Conservation District (RWCD) Canal on the east. The general drainage pattern from east to west across the study area is modified by three significant barriers, including the East Maricopa County Flood Channel constructed along the east side of the RWCD Canal, the Western Canal and Lateral 9.5, and the Rittenhouse Road branch of the Southern Pacific Railroad (SPRR).

During the study it was also found that the Rittenhouse SPRR line, which runs diagonally across the study area, is also a significant barrier to east/west crossflow. Flood flows that reach the east side of the railroad embankment are impounded or flow northwesterly toward Gilbert (FCDMC. 1989).

Less significant barriers to east/west crossflow are the Consolidated and Eastern Canals and the Chandler SPRR embankment. Considering the barriers, the effective combined drainage area contributing offsite runoff to the Price and Santan alignments is approximately 58 sq. mi., of which 40 sq. mi. flows almost straight westerly to the Price alignment, and the remainder west of Price Road generally flows southwesterly toward the Santan alignment.

Hec-1 Model

The U.S. Army Corps of Engineers Flood Hydrograph Package, HEC-1, (COE, 1987) was utilized to model the flood hydrology of drainage areas contributing to the Price Expressway and Santan Freeway. The soil Conservation Service (SCS) runoff and unit hydrograph options were used to generate design flood hydrographs for all land use types. Combined hydrographs were routed downstream using the kinematic wave routing option. When storage structures were encountered, the modified Puls routing option was used.

As shown in Figure 1, the area east of the Chandler branch of the Southern Pacific Railroad (SPRR), one-quarter to one-half mile east of Arizona Avenue, was modeled by Franzoy Corey Engineering Company for the Flood Control Doistrict of Maricopa County (FCDMC) in a study to delineate floodprone areas upstream of railroad and irrigation canal embankments. The easternmost basin boundary of the FCDMC study was the Roosevelt Water Conservation District (RWCD) irrigation canal and East Maricopa County Flood Channel. The FCDMC model results were used as input into the HDR hydrologic model.

Contributing drainage areas were delineated from 7-1/2 minute USGS quadrangle maps, then subjected to a field inspection to verify general accuracy of delineation. The overall area was divided into subbasins ranging in size from one-half square mile to one and one-half square miles and an individual HEC-1 model created for each basin. Within a basin model, the basin area was further subdivided into sub-basins which shared a known orassumed common outfall point. Many subbasins have varied land use characteristics due to the sporadic development of land. When practical, the subbasins were delineated with a preference for size uniformity and homogeneous land use.

Key assumptions in HDR's modeling approach involve the initial abstractions and loss rates, and the manner in which the very large number of individual subbasins were linked and routed. The initial abstractions (IA) and loss rate (LR) options in the SCS method were used to account for detention in the developed areas and irrigated areas in lieu of attempting to model every on-site detention basin in the area.

The linking and routing approach was a practical consideration in which one-square mile subbasins were modelled in detail and the output hydrograph optimizing as a lumped one-square mile subbasin. The one-square mile basins were then linked and routed to the outfall.

Thus, extremely long run times (on the order of hours) were avoided and, a very complex model to interpret was also avoided.

Initial Abstractions

HEC-1 automatically calculates both the initial abstration and infiltration for a given curve number as per SCS guidelines (SCS, 1971). However, HEC-1 has an option to arbitrarily change the IA without affecting the amount of infiltration. This option was used to account for storage in residential commercial areas where detention policies were in effect, and for agricultural areas and ranchettes which were constructed to retain water. This approach was a key to HDR's modeling strategy, which was to construct models which can be regarded as site specifc yet avoiding the necessity to model details such as the characteristics of each detention pond in a developed area.

Agricultural areas were modelled as irrigated row crops on flat slopes. Most fields have berms to retain and conserve irrigation water. These are assumed to have a storage effect for the design storm. A separate study was done to determine the required increase in the IA to account for the storage effect. The study concluded that an average field can retain approximately 2.5 inches of rainfall, which can be duplicated by using an IA=1.5 inches in HEC-1.

A number of residential neighborhoods in the study area are comprised mainly of "horse acreages," in which individual lots are graded as sumps and flood irrigated. A three-inch IA was used for these areas which effectively eliminated excess stormwater runoff from these areas.

Weighted average Curve Numbers were estimated for each subbasin considering factors such as lot size, percentage of impervious area, and soil type.

Assigning an IA to account for detention policy was somewhat complicated for the project area east of Price Road because development occurred throughout the span of time during which the detention policies were changing. Therefore, available drainage reports were used to determine the actual storage provided.

West of Price Road development has primarily occurred in the 1980's when uniform and consistent policies were in effect for both Chandler and Tempe. Therefore, individual subdivision records were not investigated and uniform IA's were applied to developed subbasins.

Stormwater in excess of the provided retention volumes, typically must surcharge the basins to reach the subdivision overflow elevation. The excess storage volume between the jurisdictional retention and the overflow is not accounted in the analysis, but is potentially significant.

In developing the models, the watershed was initially subdivided into subbasin and then subareas within subbasins. Each subbasin is therefore a substantial model by itself. If all the subbasins were to be linked in a single model, the resulting combined model would have been very cumbersome to use because of long run times, large print-outs, and the amount of detail in the print-out.

To overcome the inefficiencies of this approach, the subbasin output hydrographs were duplicated by using a lumped model of the subbasin and the HEC-1 calibration option. Thus, the lumped basin hydrograph was optimized by varying the parameters of the SCS method for a single basin until the outflow hydrograph closely matched the linked subarea hydrograph.

Suitable optimized hydrographs were obtained in two or three trials. The effort was directed toward closely matching volumes rather than peaks. However, differences in peaks between the lumped and detailed models were normally insignificant after routing to the next combination point downstream. The routing functions essentially smooth the sharper peaks (and sometimes multiple peaks) of the detailed subarea models.

Results

The models were used to calculate total stormwater volumes and peak flows for design of a multiple basin and pump station stormwater management facility. This system of large collector pipes, five detention cells, and three pump stations is projected to cost in excess of $60 million. The system will control 1800 acre-feet of stormwater for the design event.

Acknowledgements

The advice and assistance of the ADOT Urban Highways Section staff; Ray Jordan, Drainage Engineer, George Wallace, Corridor Engineer and Steve Martin, Design Engineer, is greatly appreciated. Also, we would like to thank the Flood Control District of Maricopa County for allowing HDR to utilize the Franzoy Corey HEC-1 model, and Franzoy Corey Engineers for their assistance.

OPERATIONAL REMOTE SENSING OF SNOW COVER IN THE U.S. AND CANADA

by

Thomas R. Carroll[1] and Edmond W. Holroyd, III[2]

1 ABSTRACT

The Office of Hydrology of the National Weather Service maintains a cooperative National Operational Hydrologic Remote Sensing Center, based in Minneapolis, to generate remotely sensed hydrology products. The Center uses terrestrial gamma radiation sensed from low-flying aircraft to infer snow water equivalent over a network of more than 1500 flight lines covering portions of 25 states and 7 Canadian provinces. Additionally, Advanced Very High Resolution Radiometer (AVHRR) data and Geostationary Operational Environmental Satellite (GOES) data are used to digitally map areal extent of snow cover over regions covering two-thirds of the U.S. and southern Canada where snow cover is a significant hydrologic variable. This paper reviews the techniques to: (1) make airborne snow water equivalent measurements using terrestrial gamma radiation data, (2) make satellite areal extent of snow cover measurements, and (3) reduce and distribute, in near real-time, both alphanumeric and graphic products to end-users in the U.S. and Canada.

2 INTRODUCTION

The National Weather Service maintains a cooperative National Operational Hydrologic Remote Sensing Center, based in Minneapolis, to generate remotely sensed hydrology products. Real-time, airborne snow water equivalent data and satellite areal extent of snow cover data are used operationally by the National Weather Service, the U.S. Army Corps of Engineers and other Federal, state, and private

---

[1] Office of Hydrology, National Weather Service, NOAA, 6301-34th Avenue South, Minneapolis, Minnesota 55450.
[2] U.S. Bureau of Reclamation, P.O. Box 25007, D-3744, Denver, Colorado 80225-0007

agencies when issuing spring flood outlooks, water supply outlooks, river and flood forecasts, and reservoir inflow forecasts. The airborne and satellite data are ingested and processed by hydrologists in the Minneapolis office and distributed electronically, in near real-time, to NWS and non-NWS end-users in both alphanumeric and graphic format.

3 AIRBORNE SNOW WATER EQUIVALENT MEASUREMENTS

The ability to make reliable, airborne gamma radiation snow water equivalent measurements is based on the fact that natural terrestrial gamma radiation is emitted from the potassium, uranium, and thorium radioisotopes in the upper 20 cm of soil. The radiation is sensed from a low-flying aircraft flying 150 m above the ground. Water mass in the snow cover attenuates, or blocks, the terrestrial radiation signal. Consequently, the difference between airborne radiation measurements made over bare ground and snow covered ground can be used to calculate a mean areal snow water equivalent value with a root mean square error of less than one cm. The techniques used to make airborne snow water equivalent measurements in a prairie, forest, and mountain snowpack environment have been reported in detail by Carroll and Carroll (1989A, 1989B, 1990).

Airborne snow surveys are typically conducted from January through March each year using two aircraft simultaneously to provide data to support spring snowmelt flood outlooks and water supply forecasts. Maps and a user's guide are available which give the current airborne flight line network and the details of the airborne snow water equivalent measurement technique and procedures to access electronically the snow water equivalent data in real-time.

4 SATELLITE AREAL EXTENT OF SNOW COVER MEASUREMENTS

In the operational satellite snow cover mapping program, AVHRR data are ingested, radiometrically calibrated, and geographically registered to one of 16 windows. A snow/no-snow/cloud cover byte plane image classification is generated on a digital image processing system and exported to a geographic information system where digital elevation model (DEM) and hydrologic basin boundary maps reside. DEM and basin boundary data sets have been prepared for 6 windows, in the West, each of which are approximately 1000 km by 1000 km. Percent areal extent of snow cover statistics are calculated for each of approximately 5 elevation zones in each of approximately 400 major river basins in the western U.S. Ten additional windows contain basin boundary data sets and are used to map snow cover for the Upper Midwest, the Great Lakes, New England, and southern Canada. During the 1990 snow mapping season, 2,113 river basins in North America were mapped on multiple occasions using AVHRR and GOES satellite data. Additional basin boundaries are currently being added to the western U.S. and for Alberta, Saskatchewan, Manitoba, and Ontario.

Research was recently completed to evaluate different snow cover mapping techniques using GOES, AVHRR, and Landsat Thematic Mapper (TM) data over the San Juan Mountains in southwestern Colorado (Baglio and Holroyd, 1989). Change detection and multispectral space classification techniques were developed to evaluate snow cover mapping techniques using GOES and AVHRR data. Procedures were developed to screen cloud cover and to map snow cover under forest canopy using AVHRR data in the West. Landsat TM data sets were used as a base with which to compare the GOES and AVHRR snow cover mapping techniques. Digital elevation model (DEM) data were used to develop techniques to insure temporal and spatial solar normalization for the multiple 1987 and 1988 satellite image data sets (Szeliga, et al., 1990).

4.1 Refinements in Snow Cover Assessment

Techniques used to map areal extent of snow cover using AVHRR data are described by Holroyd and Carroll (1990) and Holroyd, et al. (1989) and will not be reviewed here. The refinements to the AVHRR snow mapping techniques have been described and incorporated into the NWS operational snow mapping procedures. The major refinements include: (1) scaling of AVHRR data in bands 1, 2, and synthetic band 6, (2) terrain corrections, and (3) change detection.

Data scaling of bands 1 and 2 within the 0-255 byte data range was originally giving an albedo resolution of 0.5 percent. The comparatively low resolution represented only a 5-bit data range and did not take advantage of the full byte range available for electronic transmission or the 10-bit resolution of the spacecraft data stream. Consequently, bands 1 and 2 have been rescaled to an albedo resolution of 0.25 percent.

The original data scaling of band 6 (i.e., band 3 minus band 4) within the 0-255 byte data range produced a saturation of band 3 and occasionally band 4 at radiative temperatures of 46 degrees Celsius or higher. Modifications were made to arbitrarily classify all pixels as snow-free having a band 3 value greater than 245 (40 degrees Celsius or higher) which is physically reasonable.

Classified images from the early morning NOAA-10 satellite accentuate the snow on the east-facing slopes and under represent snow on the west-facing slopes. Similarly, images from NOAA-11 in the early afternoon and NOAA-9 in the late afternoon have snow cover apparently enhanced on the sunlit slopes and diminished on the shaded slopes.

In order to correct the images in rugged terrain, 30 arcsecond digital terrain data were resampled to the 901 m pixel size and projection of the AVHRR imagery. Slope and aspect of the terrain were then calculated. The solar incidence angel, i, was then calculated by:

$$\cos(i) = \cos(z)\cos(s) + \sin(z)\sin(s)\cos(A-a)$$

where z is the solar zenith angel, s is the terrain slope angel, a is the terrain aspect angel, and A is the solar azimuth angle. Band 1 values are divided by cos(i) to give the brightness values appropriate to flat terrain and a vertical sun. These terrain corrections reduce the solar azimuth bias associated with the areal extent of snow cover. Of course, the corrections are not perfect because substantial variability of solar incidence angle exists within each pixel as a result of terrain variability.

5 AIRBORNE AND SATELLITE END-USER SNOW COVER PRODUCTS

The alphanumeric and graphic airborne and satellite snow cover data are distributed over both the National Weather Service computer network and the commercial MCI Mail electronic mail network 4 hours after the snow survey aircraft land anywhere in the country and 36 hours after the morning overpass of NOAA 10 (10 AM) or the afternoon overpass of NOAA 11 (2 PM). Areal extent of snow cover maps derived from AVHRR satellite data are generated for all 16 windows approximately once a week, the effect of cloud contamination notwithstanding. Figure 1 shows the 2,113 NWS basins where snow cover was mapped in 1990. Additional basins are being added in the West and southern Canada for use in future snow cover mapping.

The airborne and satellite alphanumeric and digital image snow cover data sets are available electronically to end-users in near real-time for use in operational and research hydrology programs in the U.S. and Canada. Snow cover maps for each of the 16 windows are faxed, in real-time, to end-users upon request. IBM-PC compatible PCX image snow cover maps for each of the windows covering North America are distributed in real-time over MCI Mail to end-users upon request. The digital, georeferenced, snow/no-snow/ cloud images for each of the windows are also available to end-users for use on digital image processing systems or geographic information systems.

6 SUMMARY

Techniques are currently being developed to incorporate DEM data and forest canopy cover data into the snow cover classification procedures to better estimate snow cover in areas where the snow surface is obscured from view by: (1) cloud cover, or (2) dense forest canopy. Research is also continuing toward improvement of normalization procedures to include corrections for within image

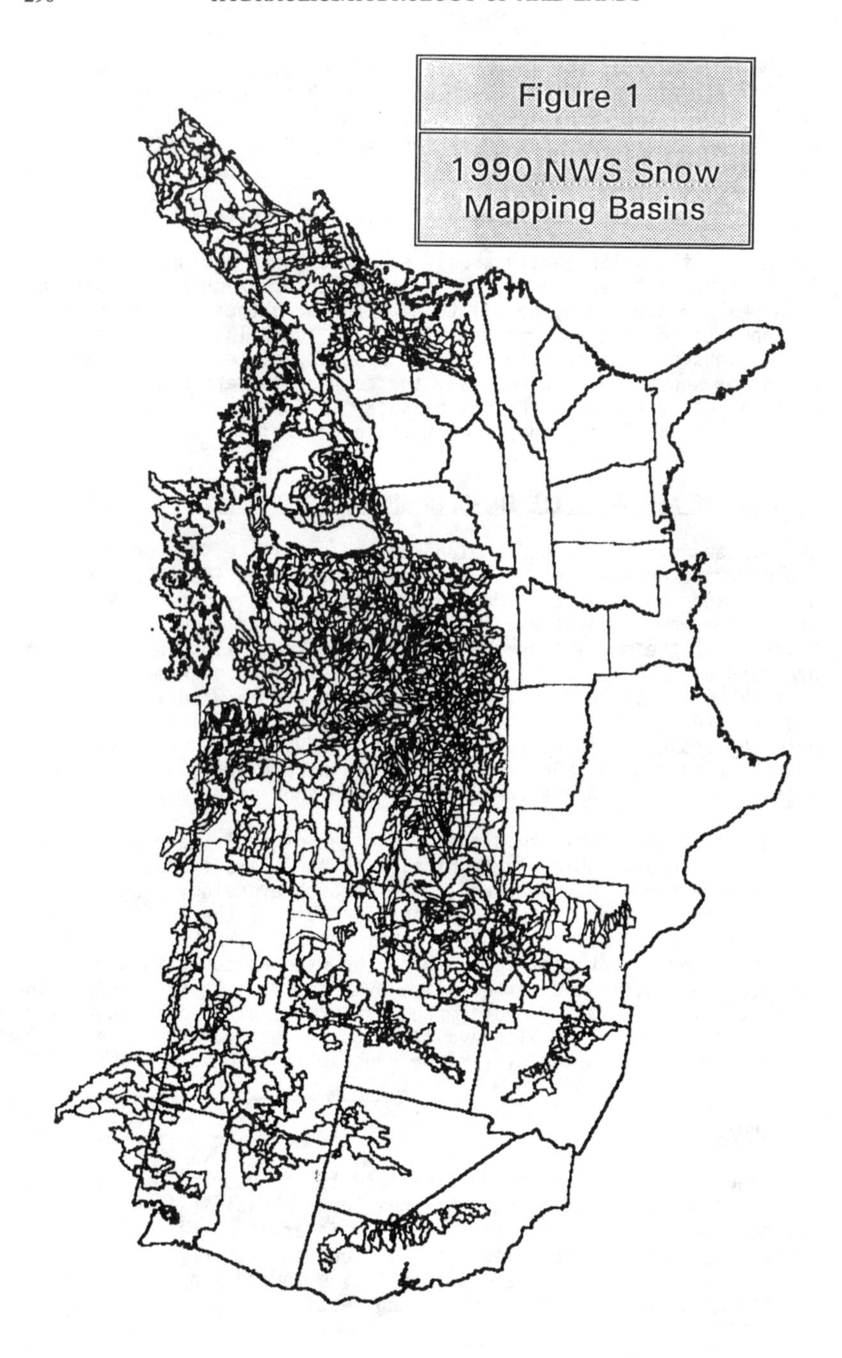
Figure 1
1990 NWS Snow
Mapping Basins

effects of terrain and for effects of atmospheric scattering and absorption. Techniques and procedures are currently being developed to generate, in near real-time, a grid-cell, or raster based, snow water equivalent data set using: (1) ground-based point snow water equivalent data, (2) airborne line snow water equivalent data, (3) satellite areal extent of snow cover data, (4) digital elevation data, and (5) forest canopy cover data sets.

7 REFERENCES

Baglio, J.V. and Holroyd, E.W. (1989) Methods for operational snow cover area mapping using the Advanced Very High Resolution Radiometer - San Juan Mountain test study. Research technical report, USGS EROS Data Center, Sioux Falls, South Dakota, March, 1989, pp. 82.

Carroll, S.S. and Carroll, T.R. (1989A) Effect of uneven snow cover on airborne snow water equivalent estimates obtained by measuring terrestrial gamma radiation. Water Resources Research. 25 (7), 1505-1510.

Carroll, S.S. and Carroll, T.R. (1989B) Effect of forest biomass on airborne snow water equivalent estimates obtained by measuring terrestrial gamma radiation. Remote Sensing of Environment. 27 (3), 313-319.

Carroll, S.S. and Carroll, T.R. (1990) Simulation of airborne snow water equivalent measurement errors in extreme environments. Nordic Hydrology (in press)

Holroyd, E.W. and Carroll, T.R. (1990) Further refinements in the remote sensing of snow-covered areas. Presented at the American Society of Photogrammetry and Remote Sensing annual meeting; Denver, CO; 1990 March.

Holroyd, E.W.,III, Verdin, J.P., and Carroll, T.R. (1989) Mapping snow cover with satellite imagery: comparison of results from three sensor systems. Proceedings of the 57th Western Snow Conference; Fort Collins, CO; April 18-20; pp.10.

Szeliga, T.L., Allen, M.W., Baglio, J.V., and Carroll, T.R. (1990) Comparison of methods of solar normalization of AVHRR data and their effects on snow cover classification. Presented at the Second North American NOAA Polar Orbiter's User's Group Meeting; New Carrollton, Maryland; 1990 May 24-25.

MULTI-DATE IMAGE ANALYSES USED FOR DETERMINING FLOOD AREA IMPACTS IN THE SAGINAW RIVER BASIN, MICHIGAN

Roger L. Gauthier[1]
William J. Kempisty[2]

ABSTRACT

During the months of September and October 1986, a series of storms swept across the Lower Peninsula of Michigan. These heavy downpours caused $400 million worth of flood damages. Significant rainfall occurred over a 26,000 square mile area on 26 consecutive days, with total rainfall recorded at 14 to 20 inches.

The problem of determining the extent of flooding over a large area is one ideally suited for the use of satellite imagery. The flood extent that occurred in the Saginaw River basin in 1986 was determined by overlays of water coverages from multidate Landsat Thematic Mapper (TM) and NOAA Advanced Very High Resolution Radiometer (AVHRR) imagery. Landsat-TM imagery collected in 1984 was used to derive land use/cover classifications. Landsat-TM and NOAA-AVHRR imagery, collected in 1986 were used to determine flood area extent.

Acreage estimates of flooding, by land use/cover type, were generated and mapped. Classification accuracies are discussed.

INTRODUCTION

The Saginaw River basin encompasses approximately 6,220 square miles in central Lower Michigan, being included within twenty-one counties. The basin is in an area that was once a glacial lake bed; topography is extremely flat, broken only by occasional low ridges.

----

[1]Supervisory Hydrologist, [2]Physical Scientist, U.S. Army Engineer District, Detroit, P.O. Box 1027, Detroit, MI

The Saginaw River has four major tributaries, these being the Shiawassee, the Cass, the Tittabawasee, and the Flint Rivers. The four tributaries meet within Saginaw County, discharging into the Saginaw River, which flows into Saginaw Bay, Lake Huron.

The September-October 1986 deluge caused widespread flood damage to homes, businesses, roads, bridges, dams, and agriculture across central Lower Michigan. Eleven dams failed or were breached intentionally within Michigan due to heavy water supplies, with up to twenty other dams being threatened with failure. Some 1,500 people were evacuated downstream of the dams. Five storm-related deaths were reported. Twenty-two counties in Michigan were declared as federal disaster areas. The additional precipitation caused Lakes Michigan, Huron, St. Clair, and Erie to rise by up to 9 inches, causing the lakes to set new historic maximums.

Areas within the Saginaw River basin received rainfall ranging from 8 to 11 inches during the initial 48 hours on September 10-11, 1986. Over the next 24 days, upwards of 6-9 inches of more rainfall were received. The Saginaw River and its tributaries experienced considerable flooding, exceeding all previous recorded flood stages, except in March 1904. Figure 1 shows a plot of daily precipitation and stage of the Saginaw River, as recorded at Saginaw, Michigan.

FIGURE 1

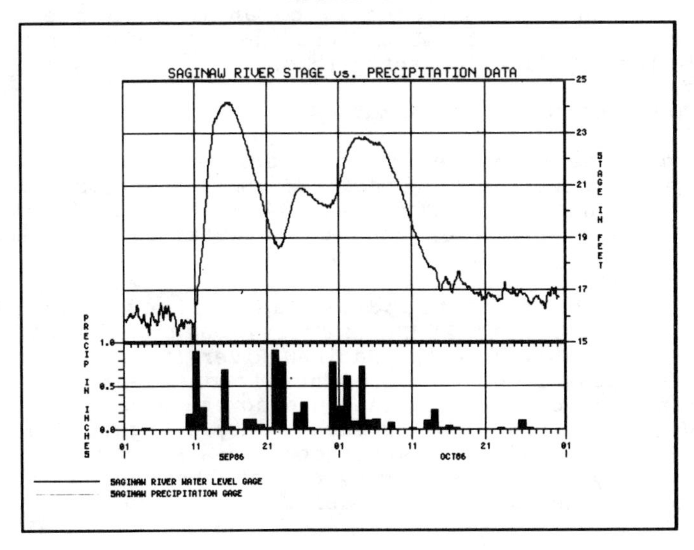

SATELLITE DATA COLLECTION

A wide variety of products can be created using data collected by the Landsat earth resources satellites and the NOAA weather satellites. Landsat and NOAA imagery have been used for delineations of flooding by use of multidate change detection, spectral composites from multiple-band combinations, or enhancements of single-bands (Deutch and Ruggles, 1978).

Among the sensor arrays aboard the Landsat-4 and 5 satellites are the multispectral TM's. The TM sensor records visible, infrared, and thermal reflective information in seven electromagnetic wavelength ranges, or channels. The ground resolution of Landsat-TM data is 30 x 30 meters, or .22 acres. The Landsat-TM imagery covers a swath width of 185 kilometers, with a revisit frequency of 16 days.

Each of the NOAA satellites carry an AVHRR, which records visible, infrared, and thermal reflective data in at least four channels. The ground resolution of AVHRR data is approximately 1100 meters on a side, or 1.1 kilometer. The swath width of an AVHRR overpass is 2400 kilometers with a repeat cycle of 12 hours.

With appropriate computer capabilities, satellite image data can be processed to produce a great variety of computer and photographic outputs. The products described herein were generated by the Detroit District, U.S. Army Corps of Engineers, on an image processing system consisting of a DEC MicroVax-II computer, running ERDAS, Inc. application software.

COMPUTER PROCESSING AND ANALYSES

Baseline land use/cover classifications and pre-flood water coverage were derived from Landsat-TM imagery collected on either July 29, 1984 or August 21, 1984, when the Saginaw River and its tributaries were within bank or "normal channel" conditions.

The extent of flooding was derived from 1986 Landsat-TM and NOAA-AVHRR imagery, both collected on October 7, 1986, when the Saginaw River system was at or near it's peak flood stage. The flood extent was then superimposed on the 1984 land use/cover image to produce a 1986 flood extent map for each county. Figure 2 is an example of the different water coverages between 1984 and 1986 for **Bay County, Michigan,** one of the most severely flooded counties within the basin.

FIGURE 2

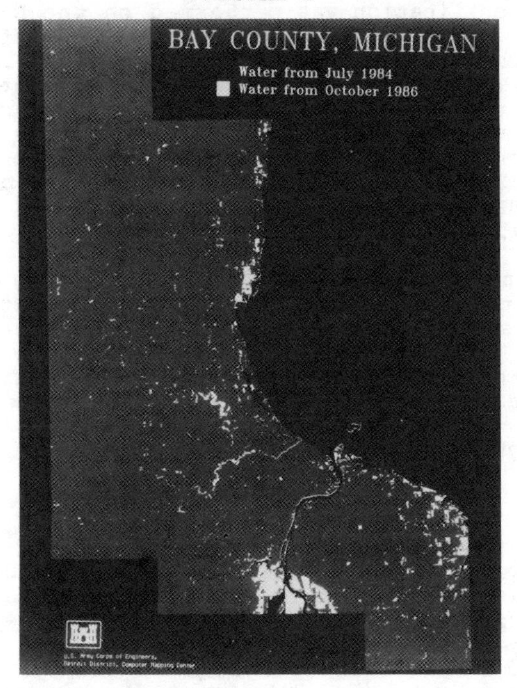

The land use/cover data for each of the twenty-one counties were generated from an unsupervised minimum distance classification of Landsat-TM channels 2, 3, 4, and 7, corresponding to the .52-.60, .63-.69, .76-.90, and 2.08-2.35 micrometer wavelengths, respectively. The spectral classes were correlated to known land use/cover by visual interpretation of aerial photographs to validate positional and categorization accuracies.

The land use/cover classes used for each county, were grouped into five general categories: 1) forested or wetland; 2) agriculture or grassland; 3) industrial or commercial; 4) residential; and 5) bare soil. These classes were chosen to cover the 64 categories of the Michigan Resources Inventory System (MIRIS) (MDNR, 1983).

Information on major transportation networks, city limits, and hydrologic basin divides for each county were digitized from USGS topographic quadrangles or other map sources, registered, and overlain onto the classified image. The county limits were digitized from USGS quadrangles to cut the image to the county borders.

For the 1986 flood extent, an unsupervised minimum distance classification was performed on the Landsat-TM channel 4. Not all of the twenty-one counties were covered by the 1986 Landsat-TM imagery, due to image size limitations and/or cloud coverage. For those areas outside of the Landsat-TM coverage, an unsupervised minimum distance classification was performed on the NOAA-AVHRR channels 1 and 2, corresponding to the .56-.68, and .72-1.0 micrometer wavelengths, respectively. The NOAA-AVHRR water extent file was resampled to a .22 acre cell size, for the subsequent overlay analyses.

All land use/cover and water extent files were rectified to the Universal Transverse Mercator (UTM) projection system using a nearest neighbor resampling algorithm. For each county, between fifteen to twenty ground control points were used for the transformation matrix, needed to register the data to a root mean square (RMS) error of less than one pixel, or within .22 acre of true geographic position.

RESULTS

These processing procedures were used to create individual products for each of the counties, including a 1984 land use/cover image and acreage report, and a 1986 flood area extent image, also with an acreage report.

The following table for **Saginaw County, Michigan,** the most severly flooded county, is an example of the acreage by land use/cover category derived from the 1984 Landsat-TM data, and the inundated acreage and percent inundation derived from the 1986 Landsat-TM data:

Saginaw County Land U/C Category	1984 Acreage	1986 Acres Inundated	Percent Inundated
Agriculture/Grassland	175,118	13,129	7.5
Forested/Wetland	181,905	11,361	6.2
Bare Soil	60,866	3,439	5.7
Residential	76,671	5,851	7.6
Industrial/Commercial	8,655	764	8.8
Total	503,215	34,544	6.9

This table indicates that there was a dramatic increase in water surface area between the two dates, with significant damages being incurred by all resource classes within **Saginaw County, Michigan.** Each of the twenty-one counties required a different interpretation of the results of these analyses, due to differing land use, precipitation intensities, and flooding extent.

ACCURACY EVALUATIONS

During all phases in the development of the flood extent mapping, efforts were made to adhere to accepted image processing procedures and standards. Land use/cover classification accuracies have not been explicitly assessed by comparisons with detailed ground-based sampling. These accuracies are expected to be within normal ranges as assessed in previous studies of this type (Gervin, et al, 1985). Detailed accuracy assessments for this study are to be conducted by digital comparisons with MIRIS current use data, derived by photo-interpretation of aerial photography (MDNR, 1987).

All acreage estimates were generated using the land use/cover classification of the 1984 satellite imagery, and the extent of flooding from the 1986 imagery. Hence, any changes in land use/cover that may have occurred between dates due to urbanization, agricultural practices, or vegetation re-growth, are not accounted for in the results.

CONCLUSIONS

Multidate image analyses of Landsat-TM, used in combination with NOAA-AVHRR imagery, proved to be a valuable tool for assessing the areal extent of flooding in 1986 over the Saginaw River basin. Multitemporal overlay analyses and mosaicking provide for both a synoptic look at flooded areas and can identify areas, by land use/cover classes, which are subject to flooding.

REFERENCES

Deutsch, M., and Ruggles, F.H., Jr., Hydrological Applications of Landsat Imagery used in the Study of the 1973 Indus River Flood, Pakistan. April 1978, AWRA Water Resources Bulletin 14 (2), pp. 261-274.

Gervin, J.C., Lu, Y.C, Gauthier, R.L., Miller, J.R., and Irish, R.R., The Effect of Thematic Mapper Spectral Properties on Land Cover Mapping for Hydrologic Modeling. October 1985, Proceedings of the U.S. Army Corps of Engineers Fifth Remote Sensing Symposium.

Michigan Department of Natural Resources, Supplemental Land Cover/Use Inventory Specifications, Exhibit A. October 1983, East Lansing, MI.

Michigan Department of Natural Resources, The Fourth Report of the Michigan Inventory Advisory Committee. November 1987, East Lansing, MI.

EVAPORATION AND SNOWMELT ESTIMATES FROM SATELLITE DATA

Woodruff Miller,[1] Member, ASCE

Abstract

Satellite imagery has been used to estimate saltwater evaporation and snowmelt runoff. Water surface temperatures can be obtained from satellite thermal infrared data. Saltwater evaporation is determined by multiplying pan evaporation by a pan coefficient and a salt concentration-dependent salt coefficient. Temperatures and evaporation values on Great Salt Lake were correlated for various time periods and locations. Results indicate that this methodology of using satellite-derived water surface temperatures along with salt concentrations can be used to estimate lake evaporation. The Martinec-Rango Snowmelt-Runoff Model (SRM) uses snowcover derived from satellite imagery to develop snow depletion curves, and along with basin parameters and meteorological data, runoff rates are simulated. The model was applied in the forecasting mode on the Sevier Basin in Utah with calibrated parameters, projected snow depletion curves, and predicted temperature and precipitation. With one to two months of calibration, the following three to four months of runoff have been forecasted with good correlation to the measured flowrates.

Evaporation Introduction

Research has been conducted using satellite thermal infrared data to estimate evaporation. Studies on freshwater Utah Lake by Miller and Rango (1985) and on Great Salt Lake by Miller and Millis (1989) have suggested that evaporation can be modeled in this manner. The objective of the study on Great Salt Lake was to evaluate the effects of salinity. Modeling evaporation by remote sensing involves correlating measured evaporation data with surface temperature data from the satellite imagery for the same time period. Because the heat transfer through the conduction layer is predominantly dependent on the surface evaporative heat transfer, there is good correlation between evaporation and surface temperature. Evaporation can be approximated by the linear function of surface temperature: $\mathbf{E = a + b \cdot T}$ where **a** and **b** are respectively the intercept and slope of the best-fit line.

Data

Saltwater lake evaporation values were obtained by multiplying pan data by salt coefficients, which reduced

[1]Professor, Department of Civil Engineering, Brigham Young University, Provo, Utah 84602.

the pan evaporation rates to that of saline water, and then multiplying by pan coefficients. The salt coefficients were calculated using the saltwater/freshwater ratio versus sodium chloride content relationship developed by Jones (1933). He made direct observations of the evaporation of Great Salt Lake water and freshwater under identical conditions. Salt coefficients are inversely related to the lake's percent salt (% TDS) in a slightly nonlinear fashion; e.g., coefficients are 0.997, 0.982, 0.952, 0.912, and 0.874 at 1, 5, 10, 15, and 20% TDS respectively.

Pan evaporation data were obtained from stations at Bear River Refuge, Saltair, Utah Lake Lehi, and Logan Experimental Farm. Salinity data were collected by the Utah Geologic and Mineral Survey (UGMS) as part of their routine sampling program. The satellite thermal infrared data came from 28 Heat Capacity Mapping Mission (HCMM) scenes and 5 Landsat Thematic Mapper (TM) scenes. Surface temperatures were derived from the satellite data using equations obtained through conversations with personnel at NASA's Goddard Space Flight Center.

Temperatures from individual lake areas were correlated with evaporation data from their nearest weather station. Also, average temperatures of the whole-lake were correlated with regional average evaporation values. The temperature for a satellite overpass day was correlated with the evaporation for the "same-day", the "day-before", and the "day-after". Temperatures were also correlated with the "two-day average" and the "three-day average" evaporation. Additionally, temperatures were averaged for the month and correlated with monthly evaporation data.

Results

The overall best short-term correlations were for the "day-before" data. Correlation coefficients **r** ranged from 0.75 to 0.85. These good results are likely because the surface temperature is mainly a function of evaporation from the day before and because the satellite passes over at midday, before much of the same day's heating-up has occurred. The total monthly pan evaporation correlated with the average monthly surface temperature yielded coefficients that ranged from .83 to .97. These are significantly higher than those for the shorter periods within the month, most likely because the monthly evaporation values and temperatures are more stable and dampen out effects of short-term variations in weather.

The average surface temperatures of the three sections of the lake and of the whole-lake were correlated with the regional average evaporation values. Table 1 shows the results of the sectional temperature and regional evaporation correlations and is an example of the many tables which were generated to display all the correlations. The correlation coefficients **r** along with **a** and **b** values for the linear equation $\mathbf{E = a + b \cdot T}$ and the number of samples **n** are given. Much better correlations were obtained by comparing regional pan averages with the sectional and whole-lake temperatures than were found by comparing an area temperature with a nearby pan. Again, the "day-before" average correlation of 0.90 was the best of the daily values, but the multiple-day correlations were also very good with an average of 0.91, and the monthly correlations had the highest average of all at 0.92.

Table 1. Corrrelation and Linear Regression Results for Great Salt Lake Sectional Satellite Temperatures versus Regional Evaporation.

Temp. Section	Same Day	Day Before	Day After	2-Day Average	3-Day Average	Total Month
Whole Lake	r= 0.86 a=-0.195 b= 0.032 n= 24	0.90 -0.244 0.036 25	0.84 -0.173 0.032 24	0.91 -0.225 0.034 25	0.89 -0.203 0.033 25	0.92 -6.191 1.048 12
South Lake	r= 0.86 a= -0.177 b= 0.032 n= 21	0.89 -0.235 0.037 22	0.86 -0.158 0.032 21	0.90 -0.211 0.034 22	0.91 -0.226 0.036 21	0.90 -6.339 1.099 10
Southeast Lake	r= 0.87 a= -0.198 b= 0.033 n= 23	0.90 -0.238 0.037 24	0.87 -0.183 0.034 23	0.91 -0.224 0.035 24	0.92 -0.235 0.036 23	0.92 -6.509 1.099 12
North Lake	r= 0.86 a= -0.183 b= 0.030 n= 20	0.91 -0.249 0.035 21	0.83 -0.153 0.029 21	0.91 -0.220 0.033 21	0.91 -0.215 0.033 21	0.95 -7.625 1.056 11

Figure 1 is an example of the many plots of evaporation versus satellite-derived temperature. This figure is for the whole-lake temperature and regional evaporation data and shows both pan correlations and correlations adjusted with pan and salt coefficients.

The results are especially interesting since 1978-86 had as great a variation in lake salinity as might ever occur. Salinity changed from 30% to 15% in the north and from 15% to 5% in the south and east areas. These results indicate that models can be developed to estimate short and long term evaporation from the whole lake and from smaller areas of the lake while taking into account the lake's salinity. Modeling evaporation with temperature and salinity should have applicability to other terminal lakes.

Snowmelt Introduction

The Snowmelt-Runoff Model (SRM) developed by Martinec and Rango (1979) uses satellite imagery as the method to determine snow extent. SRM has been applied with Landsat data by Miller and Sereno (1987) to simulate snowmelt runoff on four basins in Utah's Wasatch Mountains. Both Landsat and Advanced Very High Resolution Radiometer

(AVHRR) satellite data are being used to study the Sevier Basin above Kingston, Utah. Predicting discharge for the Sevier River is critical because the economy of the basin is dependent on irrigation water from the river. The region is semi-arid receiving only an average of 11 inches (28 cm) of precipitation annually. However, snow depths in the mountains can reach 100 inches (250 cm).

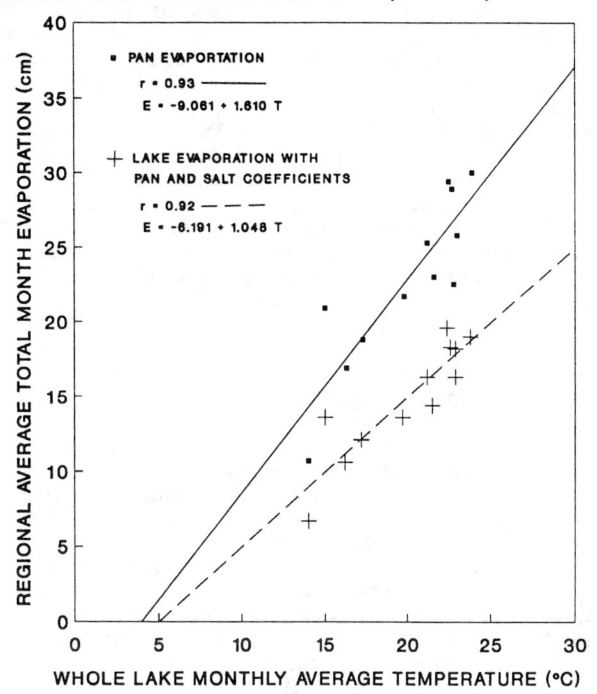

Figure 1. Whole Lake Monthly Average Temperature versus Regional Average Total Month Evaporation

SRM simulates daily streamflow in mountain basins where snowmelt is a major runoff factor. Discharge is calculated as a function of the runoff coefficients, degree-day factor, temperature lapse rate, precipitation, zonal areas, recession coefficients, and lag time, as well as percent snow covered area.

Data

The snowcover from Landsat and AVHRR images for each elevation zone in the basin is used for the snow depletion curves which depict the percentage of snow coverage as a function of time. The process involves photoscanning satellite photographs, defining basin boundaries and zones, and determining snow covered areas.

The basin was delineated and divided into four zones at intervals of about 1500 ft (460 m) on a large scale map. The map was reduced until it was the exact scale as the satellite photograph and a transparency of the map was overlaid on the photograph. The image with the attached transparency was then photoscanned in order to produce a digitized data set which was displayed as a computer image. The areas of the zones and the area of snowcover in each zone were determined by extracting the pixels representing the areas. A range of snow intensities was defined, the pixels within the selected snow intensity range were counted, and the snow areas were calculated.

Temperature data were available from Blowhard Mountain (El. 10,690 ft (3260 m)) and from Circleville (El. 6060 ft (1850 m)). The Hatch precipitation record was chosen to represent the basin. The accuracy of flow simulation is assessed by goodness-of-fit and percentage of volume difference. Both of these indexes require actual flow data which were available at Kingston and at Circleville, 10 miles (16 km) upstream of Kingston.

Results

Simulation of the 1988 streamflow was conducted using the 1988 snow depletion curves, meteorological data, and basin parameters. Calibration of the basin parameters was accomplished through optimization of the model on a monthly basis. The overall goodness-of-fit for the Upper Sevier Basin was 0.91 and the volume difference was -5%.

SRM was then used to forecast runoff in 1989 based on the simulated basin parameters from 1988. In the forecast mode, SRM needs to be updated periodically with current data. A calibration input file is created and these data are used to simulate the forecast period. Projecting the snow depletion curves into future months is also required.

Several calibration and forecasting periods were investigated. For example, March and April snowcover and streamflow simulation were used to calibrate the parameters for the May through July forecast. Temperatures and precipitation were predicted based on daily and monthly normals. Figure 2 shows the March-April calibration and May-July forecast results (goodness-of-fit at 0.98). This is an example of the several combinations of periods studied. The actual flow is also shown for comparison.

The later season forecasts were all very good. However, the April-July forecast based on the March calibration alone was quite poor. This was due to the sharp drop in flow during the first 2 weeks of April (Figure 2). A much better simulation could have been accomplished by weekly calibration rather than monthly. With good basin parameters, snow depletion curves, and weather predictions, reasonable streamflows in the Sevier River have been forecast.

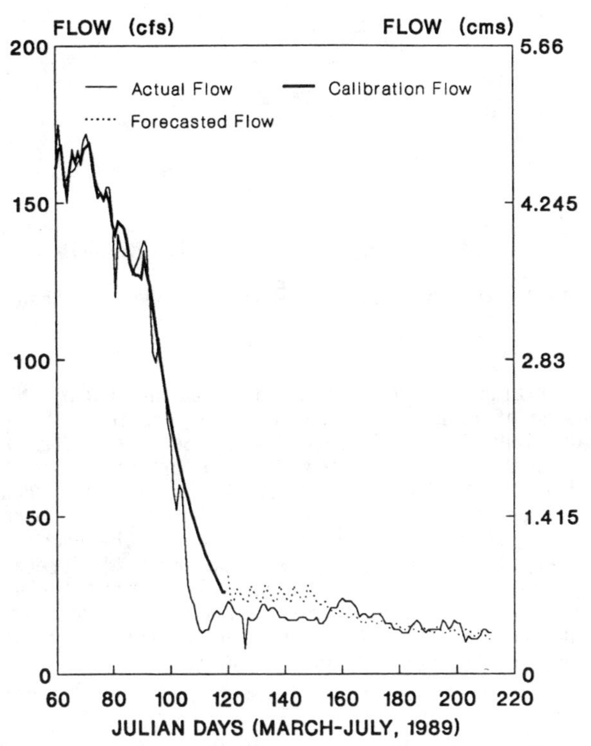

Figure 2. Sevier River Flowrate above Kingston. March-April Calibration, May-July Forecast

References

Jones, D.K. (1933). "A study of the evaporation of the water of Great Salt Lake." Unpublished Master's Thesis, University of Utah, Salt Lake City, UT.

Martinec, J., and Rango, A. (1979). "Application of a snowmelt-runoff model using Landsat data." *Nordic Hydrology*, 10(4), 225-238.

Miller, W., and Millis, E. (1989). "Estimating evaporation from Utah's Great Salt Lake using thermal infrared satellite imagery." *AWRA Water Resources Bulletin*, 25(3), 541-550.

Miller, W., and Rango, A. (1985). "Lake evaporation studies using satellite thermal infrared data." *AWRA Water Resources Bulletin*, 21(6), 1029-1036.

Miller, W., and Sereno, D.J. (1987). "Modeling snowmelt-runoff and lake evaporation using satellite data." *Computational Hydrology 87*, Proc. of 1st Intern. Conf., Anaheim, CA.

Remote Sensing of Rainfall with NEXRAD

Mark L. Walton, James A. Smith and Robert C. Shedd[1]

Abstract

In utilizing the Next Generation Weather Radar (NEXRAD) for remote sensing of rainfall, three stages of precipitation processing have been developed by the National Weather Service's (NWS) Hydrologic Research Laboratory (HRL). The Stage 1 precipitation processing occurs within the NEXRAD computer system; it produces quantitative estimates of precipitation, short-term forecasts of precipitation accumulations, and flash flood probabilities. The second and third stages of precipitation processing will occur at NWS forecast offices on systems external to NEXRAD. Stages 2 and 3 will further improve the quality of the precipitation estimates through the use of satellite imagery, additional rain gage data, and interactive quality control at the River Forecast Centers (RFC).

Introduction

One of the most important steps in the process of making river forecasts is an accurate precipitation analysis. Precipitation is one of the primary inputs to the NWS River Forecast System. Accurate precipitation analyses are also extremely important to the meteorologist faced with a flash flood situation. The NWS precipitation processing is performed in three stages. These stages of precipitation processing were developed by HRL. The first stage of precipitation processing occurs within NEXRAD. NEXRAD will provide the detailed quantitative information on the spatial and temporal variation of precipitation and the resolution necessary to capture such extreme variability as that associated with arid and semi-arid precipitation patterns. The second and third stages of precipitation processing will occur at NWS forecast offices on systems external to NEXRAD and further improve the quality of the precipitation estimates. The following sections will discuss how NEXRAD fits into the three stages of precipitation processing.

---

[1]Hydrologists, Hydrologic Research Laboratory, National Weather Service, National Oceanic and Atmospheric Administration, Silver Spring, MD 20910

Stage 1 Precipitation Processing

The Stage 1 processing within NEXRAD contains two major components: a Precipitation Processing System (PPS) and a Flash Flood Potential (FFP) System. The Stage 1 NEXRAD PPS produces quantitative estimates of 1 hour, 3 hour, and storm total precipitation. The PPS contains procedures for preprocessing and quality controlling radar signals, converting radar signals to precipitation rates and accumulations, adjusting radar precipitation estimates using observations from rain gages, and developing precipitation products. Each of these procedures will be briefly described.

The preprocessing of the radar signals consists of the development of a "sectorized hybrid scan". The sectorized hybrid scan uses the four lowest contiguous tilts of the radar volume scan to come up with the best reflectivity data at each particular radar bin, based on beam height, terrain effects, and beam blockage. Each NEXRAD radar bin is one degree in azimuth by one kilometer in range. At ranges close to the radar the higher tilts are used to reduce ground clutter. At farther ranges, to reduce problems associated with incomplete beam filling and overshooting storm systems, a combination of the two lowest tilts are used. The sectorized hybrid scan is described in detail by Shedd et al. (1989).

Quality control of the radar data occurs throughout the entire processing stream and attempts to correct for: beam blockage, isolated sample volumes, outliers, contamination of the lowest tilt by anomalous propagation or ground clutter, sudden and physically unreasonable growth or decay of precipitation, range effects, unreasonable precipitation rates and accumulations, and radar bias.

Following the construction of the sectorized hybrid scan, reflectivity data from the hybrid scan are converted to precipitation rates using the empirical relationship $Z_e = aR^b$. The equivalent reflectivity factor is Z_e, R is the precipitation rate. Constants a and b must be derived by parameter estimation procedures using archived radar and rain gage data. The precipitation rates are then used to estimate scan-to-scan and hourly accumulations.

Despite all the efforts to maintain a high level of quantitative accuracy in estimating precipitation from radar data, there will be errors in these estimates. While some of these errors will be localized, some will produce a generally uniform multiplicative bias in the radar estimated precipitation. In either case, a mean bias correction can be applied to the entire field in an attempt to insure that the estimate of total field volumetric water closely equals the true field volumetric water. The mean field radar bias is computed by comparing the hourly precipitation from precipitation gages to the associated hourly precipitation accumulations from the radar. The precipitation accumulations are then adjusted by multiplying by this bias.

The quantitative estimates of precipitation from the PPS in Stage 1 are updated every volume scan (approximately every 5 minutes) and consist of digital and graphical products. The spatial resolution of the graphical products are 2 km by 2 km. The graphical precipitation products from Stage 1 consist of 1 hour, 3 hour, and storm total precipitation accumulations. The graphical products will be used by the forecasters to evaluate precipitation distribution over a basin and assess flash flood potential. The primary digital product is a 1 hour precipitation accumulation with a resolution of approximately 4 km by 4 km. This product is non-displayable at the NEXRAD workstation and will serve as one of the inputs into Stage 2 processing. A more detailed discussion of the PPS is described by Ahnert et al. (1983, 1984).

The Stage 1 NEXRAD FFP System produces short-term forecasts of precipitation accumulations and flash flood potential. The NEXRAD FFP System consists of a precipitation projection procedure and a flash flood potential assessment procedure. The precipitation projection procedure forecasts precipitation accumulation up to 1 hour into the future. The flash flood potential assessment procedure produces an estimate of the probability that the actual precipitation for some time during the precipitation event has exceeded or will exceed the flash flood guidance value. These guidance values, generated at the RFCs using hydrologic models, are estimates of how much precipitation would be required over specified durations to produce flooding at one or more locations within a zone or county. Further details on the FFP System are available in Walton et al. (1985, 1986, 1987).

In Stage 1 the data are processed to a level of refinement that can be achieved with modest computer resources, yet provide an accuracy that make the precipitation estimates useful for local real-time applications.

Stage 2 Precipitation Processing

The Stage 2 Precipitation Processing program is used to compute hourly precipitation estimates on a 1/40th Limited Fine Mesh (LFM) grid (approximately 4 km on a side) for the area covered by a single NEXRAD, which has a radius of 230 km. Input to Stage 2 includes hourly digital precipitation data from Stage 1 processing, GOES infrared imagery, and rain gage data. The Stage 2 program will run at each NWS Weather Forecast Office (WFO) collocated with a NEXRAD. Stage 2 precipitation analyses will be used by the WFO in providing forecast guidance during periods of severe weather and as input to Stage 3 precipitation processing at RFCs.

Stage 2 precipitation processing differs from Stage 1 in several important ways. Additional quality control steps are carried out in Stage 2 processing. Satellite and rain gage data are used to detect and eliminate errors in the NEXRAD data associated with clear-air anomalous propagation or other data contamination not detected or eliminated during Stage 1 processing (Fiore et al., 1986). From the satellite data it can be determined whether clouds are contained in a 1/4 LFM grid box. If radar detects precipitation in a 1/4 LFM grid box for which satellite data indicate no clouds are present and for which no rain gages record precipitation, then the radar precipitation estimates are replaced by zero values.

Stage 2 processing will produce two estimates of the precipitation field for the area covered by a single NEXRAD. One is an estimated "multisensor" precipitation field and the other is a "gage-only" precipitation field.

In the multisensor precipitation field, radar and rain gage data are "merged" to form an optimal "multisensor" estimate of the precipitation field. Merging procedures used for multisensor estimation employ nonparametric interpolation techniques (Seo and Smith, 1990a and b). The procedures account for the strengths and weaknesses of the two measurement systems. Prior to application of the merging procedure a mean field bias correction is computed. Unlike the Stage 1 bias correction procedure, the Stage 2 procedure can operate on data from preceding hours (Smith and Krajewski, 1990). In Stage 1, the number of rain gages used in the adjustment procedure is limited due to the computational load on NEXRAD. In Stage 2, all the rain gage data available at the WFO is used. To estimate precipitation at a given location, a rain gage observation will be heavily weighted only if it is close to the location. The weight that a rain gage receives will also depend on characteristics of the precipitation field. For precipitation fields with large spatial variability, as is typically the case with convective storms, rain gage observations will generally receive lower weights than for more uniform precipitation fields, associated, for example, with stratiform precipitation.

The gage-only precipitation field produces a precipitation estimate based largely on rain gage data. In the gage-only precipitation analysis, radar and other data are used for the time distribution of rain gage data, and to delineate regions receiving no precipitation.

Quantitative estimates of precipitation from Stage 2 consist of graphical and digital products. Graphical products will allow display of Stage 2 "multisensor" and "gage-only" precipitation estimates at the WFO. Summary information, such as mean precipitation over the field and maximum point precipitation, will also be displayed by the Stage 2 program. The digital products will be forwarded onto the RFCs for incorporation into Stage 3 processing.

Stage 3 Precipitation Processing

Stage 3 precipitation processing will occur at the 13 RFCs. Within an RFC coverage area there are 8 to 20 WFOs. Data are combined from each of the WFO Stage 2 analyses within the RFC's area of responsibility to generate a mosaicked precipitation field. Stage 3 is the only stage for which the forecaster can interactively make adjustments to the precipitation field.

The forecaster has two quality control decisions that can be made within Stage 3. The first is a decision on the quality of the individual Stage 2 multisensor or gage-only fields. Stage 3 has the capability of displaying side by side on the workstation monitor the multisensor and gage-only fields generated in Stage 2. If it is believed that the multisensor field is contaminated by anomalous propagation or other errors, which will be generally apparent to the eye, the forecaster can elect to remove the multisensor field from further analysis, and the gage-only field will be included in the precipitation mosaic. This decision can be made on a radar by radar basis. The second quality control decision the forecaster can make regards the quality of the gage data used in the Stage 2 analysis. Stage 3 has an option to display the hourly gage accumulation along with corresponding radar accumulations. The forecaster at the RFC may, if it is believed that a gage report is in error, remove a gage (or multiple gages) from the database and Stage 2 runs for the affected radars will be automatically re-submitted from the RFC. Following completion of the re-submitted Stage 2 runs, Stage 3 would be restarted.

In addition to those mentioned, Stage 3 has other options to aid in the precipitation analysis. One is the capability to zoom into a small portion of the RFC coverage area to view the precipitation field in greater detail. A time lapse feature will also be available to show an animation of the previous six hours of mosaicked data. A number of map overlays will also be provided to assist in determining precise geographic location of precipitation features.

The result of these interactive quality control steps is a precipitation mosaic for the RFC area of coverage. For areas of overlapping radar coverage, the mosaicked value will consist of the average of the non-zero precipitation accumulations. Earlier studies have shown this method to be an effective mosaicking scheme (Hudlow et al., 1979).

The Stage 3 analysis will be used in several ways. First, it will be used as input to the Mean Areal Precipitation (MAP) preprocessor of the hydrologic models in use at the RFCs. The MAP preprocessor accumulates and averages hourly precipitation estimates over a basin to the time and space resolutions required for hydrologic forecasting. The MAP time step required by the RFC may vary from one to twenty-four hours. Second, it will be transmitted to the National Meteorological Center where precipitation fields may be combined from each of the RFCs to develop a national precipitation map.

Summary

Testing and evaluation of the Stage 1 precipitation processing have been conducted by HRL in collaboration with the NEXRAD Operational Support Facility (OSF) in Norman, Oklahoma, and the Prototype Regional Observing and Forecasting Service (PROFS) located in Boulder, Colorado

(O'Bannon and Ahnert, 1986; Kelsch, 1989). Since summer 1988, the PPS has undergone operational testing at the Denver forecast office using a radar system similar in many characteristics to NEXRAD. Starting spring 1990, the FFP will undergo operational testing at the Denver forecast office. Testing and evaluation of Stages 2 and 3 are underway at HRL.

The three stages of precipitation processing will allow hydrologists and meteorologists to not only issue more accurate forecasts of precipitation events and their consequences, but also routinely measure rainfall with accuracies never before achieved over large areas in real-time.

Precipitation data coming out of the various stages of precipitation processing will be available for cooperating agencies and the private sector. The quantitative precipitation estimates coming out of Stage 3 will ultimately become an extremely important data source for Geographical Information Systems (GIS) and could, in fact, revolutionize hydrologic forecasting and the way we manage our water resources.

Acknowledgments

The authors would like to thank the NEXRAD Joint System Program Office for supporting development of the Stage 1 precipitation processing software. Thanks also goes to the Denver forecast office for operational testing of the Stage 1 precipitation processing. The cooperation and support of PROFS, and the NEXRAD Operational Support Facility (OSF) are also greatly appreciated. Critical test data for the three stages of precipitation processing have been provided by the National Severe Storms Laboratory, NEXRAD OSF, and the National Center for Atmospheric Research.

References

Ahnert, P.R., Hudlow, M.D., Johnson, E.R., Greene, D.R., and Dias, M.P.R., "Proposed 'On-Site' Precipitation Processing System for NEXRAD", Preprints, 21st Conference on Radar Meteorology, AMS and Alberta Research Council, Canadian Meteorological and Oceanographic Society, Edmonton, Alberta, Canada, September 19-23, 1983, pp. 378-385.

Ahnert, P.R., Hudlow, M.D., and Johnson, E.R., "Validation of the 'On-Site' Precipitation Processing System for NEXRAD", Preprints, 22nd Conference on Radar Meteorology (AMS), Zurich, Switzerland, September 10-14, 1984, 10 pp.

Fiore, J.V., Farnsworth, R.K., and Huffman, G.J., "Quality Control of Radar-Rainfall Data with VISSR Satellite Data", Preprints of the 23rd Radar Meteorology Conference, Snowmass, Colorado, September 1986, pp. JP15-JP18.

Hudlow, M.D., Smith, J.A., Walton, M.L., and Shedd, R.C. "NEXRAD - New Era in Hydrometeorology in the United States", Proceedings of the International Symposium on Hydrological Applications of Weather Radar, Salford, England, August 1989, 12 pp.

Hudlow, M.D., Arkell, R., Patterson, V., Pytlowany, P., Richards, F., and Geotis, S., "Calibration and Intercomparison of the GATE C-Band Radars", NOAA Technical Report EDIS 31, November 1979, 98 pp.

Kelsch, M., "An Evaluation of the NEXRAD Hydrology Sequence for Different Types of Intense Convective Storms in Northeast Colorado", Preprints, 24th Conference on Radar Meteorology, AMS, Tallahassee, Fla., March 27-31, 1989, pp. 207-210.

O'Bannon, T., and Ahnert, P.R., "A study of the NEXRAD Precipitation Algorithm Package on a Winter-Type Oklahoma Rainstorm", Preprints, 23rd Conference on Radar Meteorology, AMS, Snowmass, Colorado, September 1986, pp. JP99-JP101

Seo, D.J., and Smith, J.A., "Rainfall Estimation using Rain gages and Radar - A Bayesian Approach: 1. Derivation of Estimators", Submitted to Stochastic Hydrology and Hydraulics, 1990a.

Seo, D.J., and Smith, J.A., "Rainfall Estimation using Rain gages and Radar - A Bayesian Approach: 2. An Application", Submitted to Stochastic Hydrology and Hydraulics, 1990b.

Shedd, R.C., Smith, J.A., and Walton, M.L., "Sectorized Hybrid Scan Strategy of the NEXRAD Precipitation Processing System", Proceedings of the International Symposium on Hydrological Applications of Weather Radar, Salford, England, August 1989, 9 pp.

Smith, J.A., and Krajewski, W., "Estimation of the Mean Field Bias of Radar Rainfall Estimates", Submitted to Journal of Applied Meteorology, 1990.

Walton, M.L., Johnson, E.R., and Shedd, R.C., "Validation of the On-Site Flash Flood Potential System for NEXRAD", Proceedings of the Twenty-First International Symposium on Remote Sensing of Environment, Ann Arbor, Michigan, October 26-30, 1987, 12 pp.

Walton, M.L., and Johnson, E.R., "An Improved Precipitation Projection Procedure for the NEXRAD Flash-Flood Potential System", Preprints of the 23rd Radar Meteorology Conference, AMS, Snowmass, Colorado, September 1986, pp. JP62-JP65.

Walton, M.L., Johnson, E.R., Ahnert, P.R., and Hudlow, M.D., "Proposed On-Site Flash-Flood Potential System for NEXRAD", Preprints, Sixth Conference on Hydrometeorology of the American Meteorological Society, Indianapolis, Indiana, October 29, 1985, pp. 122-129.

Towards a Rainfall Estimation Using Meteosat over Africa

Yves Arnaud[1], Michel Desbois[2], Alain Gioda[1]

Abstract

The scope of the research is to develop a method of rainfall estimation at both small space and time scales (9 pixels -225 km^2-, and single shower duration -several hours-) using Thermal Infra-Red Meteosat data. The method is based on the cloud-top radiometry. The study area is located between 13°-14°N and 2°-3°E (Niamey area). A dense recording raingauge network implanted in Niger provides the data to calibrate satellite estimations. A relationship was found to link the maximum digital count to the squall line rainfall amount. The use of area average rainfall estimates increases the relationship's reliability.

Introduction

The interest of satellite study in hydrology is important especially in areas like the Sahel where raingauge networks are sparse. Nowadays, satellite rainfall estimations over western Africa are used for time periods longer than ten days (Carn et al. 1989). Methods change according to the time and space resolution chosen for the application (Barrett and Martin 1981). The EPSAT-Niger (Estimation des Précipitations par SATellite) experiment has the specific objectives of both improving our knowledge of precipitation processes and deriving operational algorithms to estimate rainfall over the study area (Hoepffner et al. 1989). The study area is a square of about 110x110 km^2 (13-14°N/2-3°E) located in the Niamey outskirts. The so-called "square degree" is representative of the large Sahelian belt situated between annual isohyets 400 and 800 mm from N'Djamena (Chad) to Kayes (Mali). The study area has a smooth relief (maximum elevation difference of 100 m). Hence orographic effects can only be moderate. The Sahelian precipitation regime is characterized by two seasons: a

[1]ORSTOM BP 5045 34032 Montpellier Cx 01 France
[2]LMD/CNRS Ecole Polytechnique 91128 Palaiseau Cx France

dry season (October to May) and a rain season (June to September). The maximum monthly rainfall occurs generally in August. The major rain-bearing cloud system is the squall line : a convective system moving westward with a typical velocity of 15 $m.s^{-1}$. Mean annual rainfall is 546 mm at Niamey City (1905-1987). In 1988, 558 mm was recorded and 600 mm in 1989.

Raingauge Network and Meteosat Data

The high density static memory raingauge network (rainfall accuracy of 0.5 mm for a minimum of one second time interval) is presented by Hoepffner et al. (1989). The basic pattern of raingauge distribution is a regular 13x13 km^2 mesh, and a target area located at the center of the square degree receives 16 raingauges. In 1988, 37 raingauges were set up and the network was completed until October 1989 with a total of 79 raingauges installed on the square degree.

The study is based on real time window rectified image data for the 1988 and 1989 rain seasons. The Thermal Infra-Red (TIR) "window" channel (10.5-12.5 μm wavelength) is used at full space resolution (25 km^2 pixel area). The frequency of these images is 48/day (periodicity 0.5 h). We can consider Meteosat images as a snapshot of the field area because data acquisition durations last only 12 seconds. The image quality and the navigation algorithm allow a localization accuracy of ±1 TIR pixel. In the paper the Maximum Digital Count (MDC) stands for the lowest temperature reached by a pixel, which is usually represented by the minimum digital count.

1988 Preliminary Study

The first step consisted of finding significant parameters from images for estimating rainfall at a time scale adapted to the squall line (several hours). To obtain this information we compared half hourly time series of rainfall at one raingauge station and parallel modification of the digital counts at the corresponding pixel (figures 1/2).

On July 17 (see figure 1), the cloud's coldest part (< -40°C) climbs progressively for 3 h and stays 5 h on the same pixel. Rainfall intensities are high at the beginning (shower body) corresponding with the MDC arrival. After the shower body, rainfall intensities become moderate (shower train) with no notable modification of the digital count. The typical areal shape of the cloud (well defined convex western edge and badly defined eastern edge), as well as the rainfall intensity pattern correspond to Sahelian squall lines (Desbois et al. 1988).

On August 10 (see figure 2), the cloud's coldest part climbs quickly (0.5 h) and stays 7 h on the same pixel. The precipitation occurs when the coldest part is established over the station. The shower intensity pattern does not present an organized structure. It is likely that this event is not a squall line, but rather a diurnal convective event.

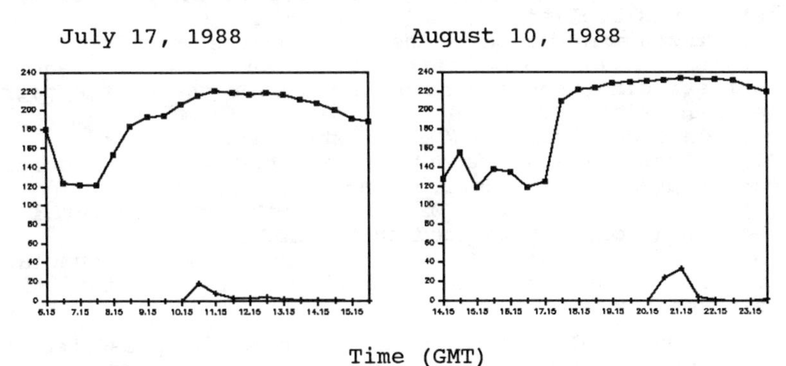

Figures 1/2: Evolution of the rain depth (+) in mm and the satellite digital count (▪). An example: Gassanamari station.

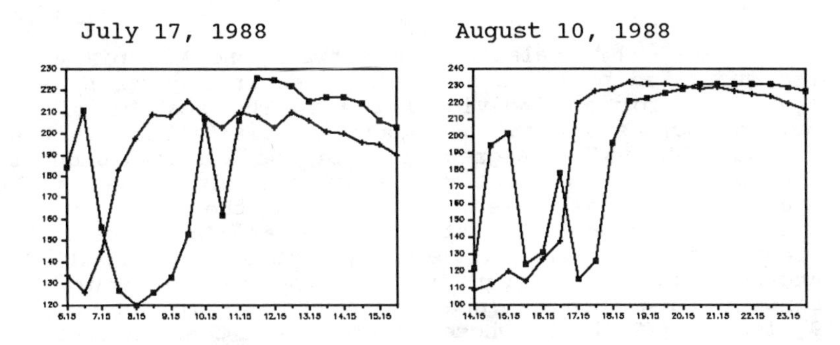

Time (GMT)

Figures 3/4: Comparison of the digital count evolution at two stations. Examples: (▪) Niamey and (+) Gamonzon

An investigation was then undertaken to find a rainfall indicator from Meteosat images for single showers. On July 17 (see figure 3) we have two stations (Niamey and Gamonzon) corresponding to two different pixels. The Niamey MDC is higher than the Gamonzon one.

On August 10 (see figure 4), no significant MDC difference exists between the two stations. On the squall line of July 17 the MDCs differ, and we could suppose that this difference is linked to the rainfall depth at the station.

1989 Elaborate Study

To verify the preliminary results, we compared systematically the rainfall and the MDCs. The following method was applied:
(1) The beginning and the end of the rainfall for a single event is detected. The image sequence corresponding with the rain period over the square degree is then selected.
(2) On each pixel the MDC is extracted
(3) A new image representing this MDC where the coldest part represents the active zone is produced.
(4) Cumulated rainfall for each station is computed and these stations are located on the MDC image.
(5) The MDC pixel and the corresponding cumulated rainfall are compared, looking for a statistical relationship.

The application of this method is presented for two squall lines : June 29 and August 4, 1989 (figures 5/6). Note that the return period of the August 4 shower (119 mm) at Niamey is about one hundred years.

Analysis

We find a relationship between one MDC pixel and the cumulated point rainfall (squares in figures 5/6) at a station for an individual event. The scatter of the points increases with the amount of rainfall as well as with the MDC value. A great part of the scatter is due to the lack of representativeness of a point value as the mean of the pixel area. The greater the area and the distance between a point and the area's center, the greater is the error involved (Flitcroft et al. 1989). Another source of error is the localization accuracy : since a point can be located anywhere within an area of 9 pixels (225 km^2) whose center is the theoretical corresponding pixel given by the conversion of the geographical coordinates into image coordinates. To overcome these observational errors we have considered an area's average rainfall and we have compared it to the corresponding satellite signal of the 9 pixels area. Following different works (Lacomba (1986), Lebel et al.(1987), Thauvin and Lebel (1989)) we have thus computed areal rainfall values using kriging. These values (squares with vertical bars in figures 5/6) were used to replace the point values formely used for studying the relationship between the MDC and the ground rainfall (vertical bars represent one standard deviation interval). As can be observed this improves the relationship.

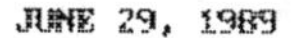

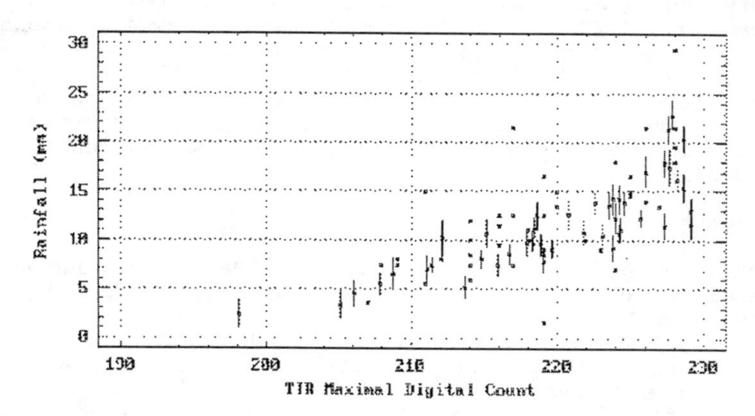

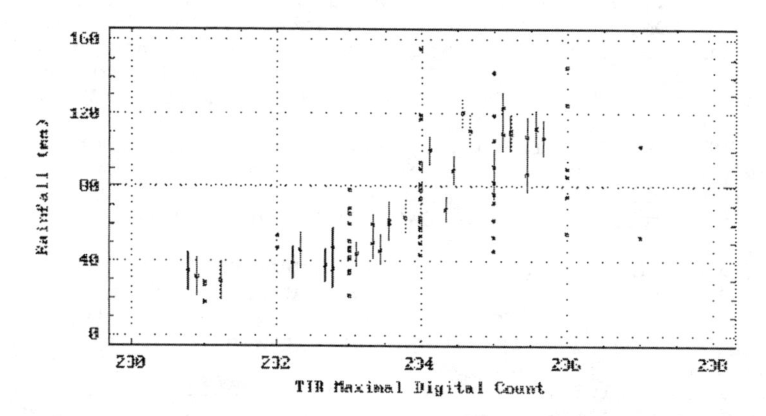

Figures 5/6 : Rainfall / TIR Maximal Digital Count relationships.(▪) point rainfall and associated single pixel value, (ɸ) area average rainfall and associated 9 pixels value where the vertical bar is the standard deviation of the areal average.

<u>Conclusion</u>

The relationship shows the existence of a link between a simple variable (MDC) and rainfall, and gives a rough estimate of rainfall depth over an area. The improvement due to the use of areal rainfall is derived from the utilization of appropriate ground based data for the calibration of satellite rainfall estimate algorithms. Thus more attention must be payed to the validation of satellite rainfall estimates. However more events must be analyzed to assess the relationship between rainfall and satellite MDC during rain seasons over a given area. This paper is the first step of a larger study which aims to assess the rainfall estimation capabilities of our method. A dense raingauge network such as the EPSAT-Niger network is of prime importance for the improvement of satellite rainfall estimation over the Sahel.

References

BARRETT, E. C., and MARTIN, D. W. (1981). The use of satellite data in rainfall monitoring. Academic Press, London, U.K.

CARN, M., DAGORNE, D., GUILLOT, B., LAHUEC, J.P. (1989). "Estimation des pluies par satellite en temps réel en Afrique sahélo-soudanienne. Essai d'utilisation d'une calibration du champ de température maximum de surface." Veille Climatique Satellitaire, 28 ,47-54.

DESBOIS, M., KAYIRANGA, T., GNAMIEN, B., GUESSOUS, S., and PICON, L. (1988). "Characterization of some elements of the Sahelian climate and their interannual variations for July 1983, 1984 and 1985 from the analysis of METEOSAT ISCCP data." J. Climate, 1 (9), 867-904.

FLITCROFT, I. D., MILFORD, J. R., and DUGDALE, G. (1989). "Relating point to area average rainfall in semiarid West Africa and the implications for rainfall estimates derived from satellite data." J. Appl. Meteor., 28, 252-266.

HOEPFFNER, M., LEBEL, T., and SAUVAGEOT, H. (1989). "EPSAT-Niger : a pilot experiment for rainfall estimation over West Africa." Proc. WMO/IAHS/ETH Workshop on Precipitation Measurement, St Moritz, Switzerland, 251-258.

LACOMBA, P. (1986). "Evaluation des précipitations par combinaison d'images satellitaires V.I.S. et I.R. et de réseaux de pluviomètres. Application à la péninsule arabique et au sud de la France." Thèse de Docteur-Ingénieur, INPG-Université Grenoble ,France.

LEBEL, T., BASTIN, G., OBLED, C., and CREUTIN, J. D. (1987). "On the accuracy of areal rainfall estimation : a case study." Water Resour. Res. , 23 (11), 2123-2134.

THAUVIN, V. and LEBEL, T. (1989). "EPSAT : study of rainfall over the Sahel at small time steps using a dense network of recording raingauges." Proc. WMO/IAHS/ETH Workshop on Precipitation Measurement, St Moritz, Switzerland, 259-266.

Droughts and Water Supply

John A. Dracup and Donald R. Kendall[1]

Abstract

The current drought in California and the western U.S. are set in the framework of drought persistence, drought forecasting and reducing the consequences of droughts. During the fifteen year period from 1975-1990, six years have been significantly below average in precipitation and runoff in California and the western U.S. Reducing drought consequences can be achieved through an expanded use of ground water recharge during wet years and expanded ground water pumpage during dry years. Legal and institutional barriers are currently delaying an increased usage of this water management scheme.

The Ongoing Four Year Drought in the Western U.S. (1986-1990)

California and much of the Western U.S. is in the fourth year of drought conditions (USDA, et al. 1990 and CDWR, 1990). Table 1 indicates that as of March 31, 1990, the snow depths at the Central Sierra Snow Laboratory, runoff to date and forecasted runoff were at 40% of average (CDWR 1990).

The USDA and USDC April 1, 1990 forecasts predict streamflows to be near to well below average for most of the western states. Reservoir storages, on the last day of March, 1990, were reported to be well below average for Nevada, Arizona and California.

[1]Civil Engineering Department, School of Engineering and Applied Science, University of California, Los Angeles, CA 90024-1593.

Table 1					
	1990	1989	1988	1977	1976
Precipitation	60	85	75	35	60
Snow Water	40	75	30	25	40
Storage	75	85	85	55	90
Runoff to date	40	75	50	20	50
April through July	40	75	35	25	40
Water Year	40	75	45	25	45

Comparison of April 1, 1990 statewide conditions in California with previous dry years. Source: CDWR, 1990.

All of the major river basins in the southwest: the Colorado, the Great Basin, the San Joaquin Basin, the Sacramento Basin, and the North Coastal Basins all reported below to much below average precipitation, snowpack storage and forecasted runoff (USDA et al. 1990).

This severe dry period raises a myriad of questions concerning climate variability, anthropologenic influences on climate, the causes of drought, drought persistence, drought forecasting and the prospects for limiting the consequences of long term drought. Obviously all of these questions cannot be adequately addressed here. The reader is referred to works by Dracup (1980 a,b, 1985, 1986), Bernan and Rodier (1985), Namias (1978), Wallis (1977), Berger, Dickinson and Kidson (1989) and Peterson (1989) and others for further study.

Drought Persistence

The causes of drought and their persistence are complex and not yet completely understood by atmospheric scientist and climatologists. Namias (1985) states:

> "... it should be made clear that there are many unsolved "mysteries" of drought. While some physical understanding has been achieved for droughts that last for a month to a season, spells of years characterized by drought are poorly understood, and thus remain on the agenda for research climatologists."

Which of the climatological processes occurs first is currently unknown. That is, does first an increase in sea surface temperature (SST) occur, which causes a temperature gradient between the warm eastern and the colder central and western Pacific Ocean, which then causes high-pressure cells to emerge in the mid-troposphere, or does the opposite sequence occur? However, it is known that these high pressure cells cause subsidence (sinking) of warm, dry air in the middle troposphere, which then causes adiabatic heating, low relative humidity and a reduction in cumulus clouds. The net result over land in the west coast of the western hemisphere is a reduction in precipitation, increased insolution, the drying of soils and increased albedo.

Once the high-pressure cells are formed and the drought is initiated, there is a self-perpetuating mechanism that continues the process. One theory holds that the land, rendered hot and dry during drought, further heats the air above it and thereby enhances the regional high pressure zone. A second theory states that drought causes an increase in fine dust particles in the air, which lead to high concentrations of very small cloud droplets whenever cumulus clouds form and thus make it more difficult for precipitation to form (Twomey and Squires, 1959). A third theory suggests that high albedo in dry areas creates a mechanism that produces warm air aloft (Charney, 1975). Whatever the mechanism, there is statistical evidence that hot, warm springs over the plains of the United States are followed by hot, dry summers and these tend to persist from one year to the next (Namias, 1960).

Drought Forecasts

The current operation drought forecasting techniques are divided into meteorological methods and hydrological methods.

Meteorological Forecasts

The Scripps Institute of Oceanography (SIO) provides the California State Department of Water Resources (CDWR) with seasonal temperature and precipitation forecasts. Namias (1984) states that these forecasts are

> "... physically based using statistical and synoptic methods which employ large scale fields of Northern Hemisphere atmospheric geopotential height and Pacific Ocean sea surface temperatures."

The forecasts are made in equally likely tercile classes, light (L), moderate (M), and heavy (H). The forecasts are called "experimental" and are part of ongoing research at SIO aimed at improving long-range-forecasting techniques. A total of four forecasts is made each year starting in September. The fall forecast is for September, October, and November. The winter forecast is for December, January, and February. Similarly, three-month forecasts are made for the spring and summer months. In addition to the precipitation forecasts, the movement of various upper level winds and storm tracks also are predicted. In addition to pictorial forecasts, an annual report is furnished (Namias, 1984). The report contains a discussion of results of the project forecasts to date and an analysis using a "skill score."

The CDWR translates these forecasts into California Water Supply Outlook report (1989) and a State Water Project Water Delivery Rule Curve and Criteria for 1989 report. These reports are mainly used to operate reservoirs throughout California mainly for irrigation supply deliveries (agricultural irrigation accounts for 87 percent of water supply deliveries in California).

These meteorological forecasts for California and the western United States are evaluated using the following skill score equation:

$$\text{Skill} = \frac{\text{Correct Forecasts - Correct Forecasts Expected by Chance}}{\text{Total Forecasts - Correct Forecasts Expected by Chance}}$$

The average of the quarterly skill from 1978 through 1989 are as follows (Roos, 1990):

Quarter		Average Skill Score
FALL	SON	0.02
WINTER	DJF	0.35
SPRING	MAM	0.19
SUMMER	JJA	-0.09

These skill score reached a peak value of 0.53 during the winter quarter of 1985 (Roos, 1990).

Limiting Drought Consequences

The U.S. population, which, if the 1900-1980 growth rate continues into the future, will reach a population of 500 million sometime in 2033 and one billion in 2088. This highly mobile population tends to favor the sun belt of the nation over the frost belt and moves to new locals fully expecting adequate water resources to be available. We have yet to encounter an individual who writes ahead to an agency to inquire whether or not water will be available for them when they arrive. Given the great difficulties encountered in expanding current facilities, the challenge for the water resource engineer is to carefully manage existing supplies to meet future demands.

The philosophy of meeting all demands by expanding supplies is being changed to that of managing the demand. Demands can be tempered via water conservation, increased use of reclaimed waste water and increased prices. Supplies can be managed via the conjunctive use of surface and ground water, the reduction of evaporation and land use management.

We believe that the most promising approach for reducing the consequences of drought are through a vigorous program of ground water recharge during wet years using excess surface water runoff. These supplies would then be available for pumpage during periods of dry years. Legal and institution problems appear to be the major stumbling block in an expansion of this approach throughout the western U.S.

References

Berge, A., R.E. Dickinson, and J.W. Kidson, 1989, "Understanding Climate Change", Geophysical Monograph 52, IUGG, Vol. 7, AGU, Wash. D.C.

Bernan, M.A. and J.A. Rodier, 1985, "Hydrologic Aspects of Drought, Studies and Reports in Hydrology Series", UNESCO/WMO Panel, Rep. 39, Paris.

California Department of Water Resources, 1989, California Water Supply Outlook, Sacramento, Calif.

California Department of Water Resources, 1989, State Water Project Water Delivery Rule Curve and Criteria for 1990, Sacramento, Calif.

California Dept. of Water Resources, 1990, Water Conditions in California, Report 3, Bulletin 120-90, Sacramento, CA.

Charney, J.G. 1975, "Droughts in the Sahara", Science, 187:435-436.

Dracup, J.A., 1985, "Causes and Occurrence of Drought", in Drought Management and its Impact on Public Water Systems, Natl. Academy Press, Wash. D.C.

Dracup, J.A., D.L. Haynes, and S.D. Abramson, 1985, "Accuracy of Hydrologic Forecasts", Proceedings of the 53rd Annual Meeting, Western Snow Conference, Boulder, CO.

Dracup, J.A., K.S. Lee, and E.G. Paulson, Jr. 1980a, "On the Definition of Droughts", Water Resour. Res., 16(2):297-302.

Dracup, J.A., K.S. Lee, and E.G. Paulson, Jr. 1980b, "On the Statistical Characteristics of Drought Events", Water Resour. Res., 16(2):289-296.

Namias, J., 1960, "Factors in the Initiation, Perpetuation and Termination of Drought", IAHS Publ. 51, Symposium on Surface Waters, Helsinki, pp. 81-94.

Namias, J., 1978a, "The Enigma of Drought -- A Challenge for Terrestrial and Extra Terrestrial Research", in Proceedings of the Symposium on Solar Terrestrial Influence on Weather and Climate, Ohio State University, Columbus.

Namias, J., 1985, "Factors Responsible for the Droughts", Chapter 3 in Hydrologic Aspects of Drought, M.A. Bernan and J.A. Rodier, rapporteurs. Rep. 39. Studies and Reports in Hydrology. UNESCO-WMO, Paris.

Namias, J., 1989, Seasonal Precipitation Forecasting 14th Annual Report, Scripps Institut. of Oceanography, La Jolla, Calif.

Peterson, D.H., 1989, "Aspects of Climate Variability in the Pacific and the Western Americas", Geophysical Monograph 55, AGU, Wash. DC.

Roos, M., 1990, Chief, Hydrology Branch, California Department of Water Resources, personal communication, May.

Twomey, S., and P. Squires, 1959, "The Influence of Cloud Nucleus Population on the Microstructure and Stability of Convective Clouds". Tellus 11:408-411.

U.S. Dept. of Agriculture and U.S. Dept. of Commerce, Soil Conservation Service, Natl. Oceanic and Atmospheric Adm., Natl. Weather Service, 1990, Water Supply Outlook for the U.S., April 1st.

Wallis, J.R., 1977, "Climate, Climate Change, and Water Supply", EOS, Trans. AGU 58(11):1012-1024.

The Development of a Hydrologic Drought Index and Termination Rate Probabilities

Donald R. Kendall[1]
John A. Dracup[2]

Abstract

The development of a hydrologic drought index (HDI) based on drought severity is presented along with a new formulation for drought termination probabilities. Results indicate that termination rate probability curves vary according to discrete levels of HDI. Data is presented for the Sacramento River near Red Bluff, California.

Drought Definition

Drought can be defined as a water shortage relative to a specified need for water in a conceptual supply and demand relationship. Dracup, Lee and Paulson (1980a) defined four decisions which are required in arriving at a viable definition of drought; the nature of the water deficit (hydrologic, agricultural,or meteorological), the basic time unit of the data, the threshold level, and the selection of a regionalization or standardization approach.

A drought may be characterized as any year or consecutive number of years during which average annual streamflow is continuously below some specified threshold level which is typically taken to be the long term mean (Yevjevich, 1967; Dracup et al., 1980a). A drought event is considered to be composed of three defining attributes: duration, D , severity which is the cumulative deficit, S , and magnitude which is the average water deficit, M , such that $S = M \times D$. These drought parameters are interrelated and only two of them are necessary and sufficient to completely define a single drought event

1 Assistant Professor, Loyola Marymount University
2 Professor, University of California, Los Angeles CA

(Dracup et al., 1980b). Duration and severity are the most highly correlated parameters, and may be considered to be the two primary parameters which depend directly on the streamflow values. In contrast, magnitude is a secondary parameter since duration and magnitude are the least correlated.

Based on this definition of drought we may write the severity of a, *d*, year drought as

$$S_d = Y_1^d + Y_2^d + \ldots + Y_d^d$$

where:

d = drought duration (years).

Y_n^d = deficit in year n of a d year drought.

Assuming that the deficits are independent and identically distributed we may write the expected value and the variance of drought severity as:

$$E\,[S_d] = E\,[Y_1^d + Y_2^d + \ldots + Y_d^d] = d \cdot E\,[Y_i^d]$$

and:

$$VAR\,[S_d] = E\,[d \cdot VAR\,[Y_i^d]] + VAR\,[d \cdot E\,[Y_i^d]]$$

These relationships hold without any additional assumptions when viewed from a computational standpoint. As will be shown later, the expected value of streamflow deficit is a function of drought duration, *d*.

The analysis of droughts and the development of a hydrologic drought index is complicated by two factors. First, the problem of sample size limitation must be addressed. For example, eight years of below normal streamflow comprise only *one* drought event. One of the critical tasks required in the HDI development is to establish reliable termination probabilities between drought events which requires a larger sample size. Secondly, duration and severity when taken as random variables come from two different probability distributions, making a drought frequency analysis more difficult. A data decimination and standardization procedure was used to increase the sample size (Kendall, 1989).

Once the data record has been extended, it becomes possible to identify probability distributions for both the high streamflow and drought duration. For this study, the discrete geometric distribution was assumed. It also becomes possible to identify probability distributions for high flow and drought deficit. The gamma distribution was utilized for deficit.

Termination of a Drought Event

For years, hydrologists and meteorologists have struggled with the task of being able to forecast the end of a hydrologic drought event, and in some cases, assess the likelihood of the initiation of a drought. The meteorologists have relied to a large extent on general circulation models of the earth and have tried to forecast from a deterministic framework. In a very small window of time, they have been largely successful. But for times greater than four months, the deterministic finite difference models become poorer forecasters. Hydrologists for the most part, rely on existing data composed primarily of precipitation records, streamflow records, tree ring data, and even sun spot data. The approach in this research effort was to study unimpaired hydrologic streamflow, viewed as a fingerprint of nature (Dracup and Kendall 1989). The key question is, by studying the information that nature gives, is there a way to assess the probability of a drought ending or continuing before it actually does so? There is a large body of work in the open literature dealing with time series analyses primarily concerned with frequency components of the series itself as well as its energy (i.e. spectral density analyses). The main emphasis presented here is a statistical approach used in the development of a hydrologic drought index. The body of the approach lies in the development of drought termination rate probabilities.

Drought Termination Probabilities

At first glance its seems straight forward to develop termination probabilities for drought events (Lee et al. 1986). However this is not the case. It must be remembered that hydrologic drought is based on three parameters, namely *duration, severity* and *magnitude*. Universally, termination probabilities have been based on duration only. However, there is a parametric dependence on severity. Termination rate probabilities could have been based on severity, or equally on magnitude. What this says basically is that the development of termination rate probabilities is largely a function of the amount of given information we choose to incorporate.

Stated mathematically:

$$P(X = t \mid X \geq t, F_t) = \frac{P(X = t, F_t)}{P(X \geq t)\, P(F_t)} = \frac{P(X = t \mid F_t)}{P(X \geq t)}$$

Some assumptions have been made regarding the choice of what probability distribution to use to describe drought deficit for a particular event. For a two year event for example, a gamma distribution was utilized. It follows then that the expected value and variance of severity, S_d, are given as:

$$E[S_d] = \frac{d \cdot r}{\lambda}$$

$$VAR[S_d] = \frac{d \cdot r}{\lambda^2}$$

An HDI was developed by studying a drought's deviation from its expected value $E[S_i]$. The following index was developed:

$$S_i = \frac{S_i - E\,[S_i\,]}{\hat{\sigma_i}}$$

Drought events that provided a value of, S_i, close to zero could be described by termination probabilities based on duration. The results using this form of an index were very surprising. Utilizing the Sacramento River unimpaired streamflow it was determined that drought events before 1940 behaved differently than events after 1940.

A drought's state can be described by any two of three parameters defined earlier, namely, duration, magnitude, or severity. Duration and severity are the most correlated parameters, while duration and magnitude are the least correlated. The calculation of drought termination probabilities is good, but does not necessarily capture additional informaton that may be given by a drought's accumulated severity. As a means of incorporating more information, the following method was developed which utilizes discrete levels of, S_i. These discrete levels can be thought of as energy deficits for particular drought events. Drought termination probabilities can then be developed for each level utilizing the equation at the top of the page. It is apparent that the behavior of drought events prior to 1940 is different from those events after 1940. Possible reasons for this will be discussed later. For purposes of this analysis, the focus will be on those events *after* 1940.

Discrete levels of, S_i, were arbitrarily divided into the following bands:

$$-1 \leq S_i \leq 0$$

$$0 < S_i \leq 1$$

$$1 < S_i \leq 2$$

$$2 < S_i$$

Utilizing the database developed from the decimination and standardization procedure mentioned earlier, drought termination probabilities were developed for the data after 1940. Because of limited space available for these proceedings, only two Figures can be shown. The first is for an HDI between 0 and 1. The other is for an HDI greater than 2. As seen in Figure 1, there is a characteristic bucket which indicates that the probability of a drought ending actually decreases before it increases. However for an HDI greater than 2 (Figure 2), no such bucket is evident. This indicates that severe droughts have a larger probability of terminating sooner, as opposed to less severe droughts.

Conclusions

Use of, S_i, as an index is useful from several viewpoints. First of all, it gives an immediate indication of the behavior of an event relative to its expected value established from the historical record. Secondly, it is useful as an *indicator* when plotted as a function time since it may also provide information regarding a particular droughts propensity to continue. When combined with drought termination probabilities, it can give a water resource manager reliable information about how to plan for the next year since what it is really providing is more information under a planning under *uncertainty* process.

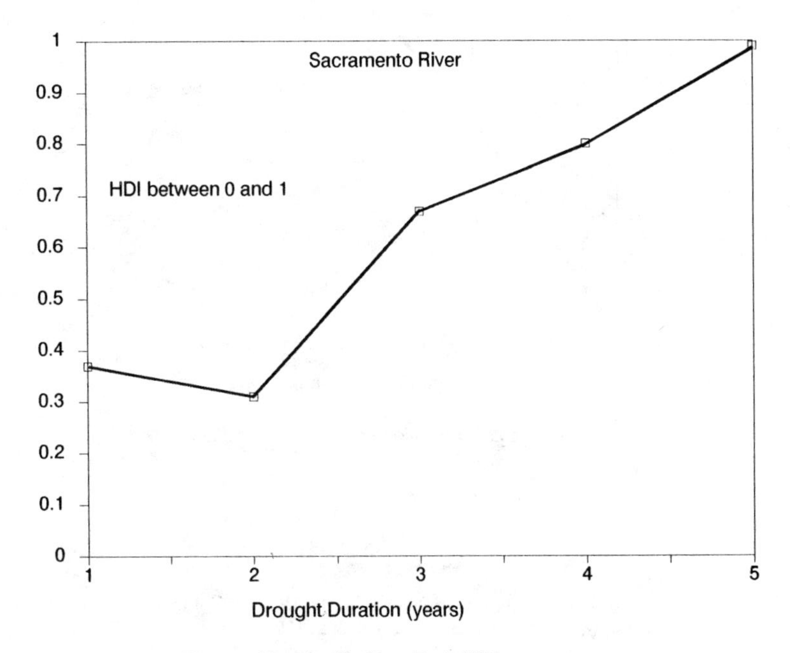

Figure 1. Termination Rate Probabilities

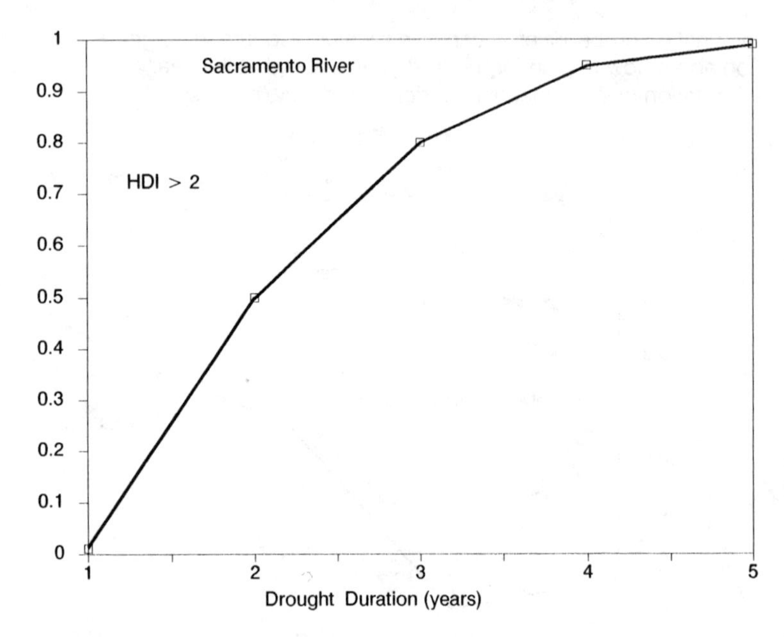

Figure 2. Termination Rate Probabilities, HDI > 2.

References

1. Dracup, J.A., K.S. Lee, E.G. Paulson, On the Definition of Droughts. *Water Resources Research*, Vol 16, Number 2, pages 297-302, 1980a.
2. Dracup, J.A., K.S. Lee, E.G. Paulson, On the Statistical Characteristics of Drought Events. *Water Resources Research*, Vol 16, Number 2, pages 289-296, 1980b.
3. Dracup, J.A., D.R. Kendall, Floods and Droughts, in ed. Waggoner, Climate Change and U.S. Water Resources, John Wiley, New York, June 1989
4. Kendall, D.R., Drought Analysis Using a Renewal-Reward Model and a Hydrologic Drought Index, *PhD. Dissertation,* University of California, Los Angeles, 1989.
5. Lee, K.S., J. Sadeghipour, J.A. Dracup, An Approach for Frequency Analysis of Multiyear Drought Durations, *Water Resources Research*, Volume 22, Number 5, 1986.
6. Yevjevich, V.M., Objective Approach to Definitions and Investigations of Continental Droughts. *Hydrology Paper 23*, Colorado State University, Fort Collins, Colorado 1967.

OWENS VALLEY GROUNDWATER BASIN

PROPOSED MANAGEMENT

Gene L. Coufal[1/] and Cecilia K. Trehuba[2/]

ABSTRACT

The City of Los Angeles imports over 60 percent of its water supply from the Owens River Watershed through a program of groundwater pumping and surface water diversions as a part of the resources used in serving over 3.5 million people. A joint long-term groundwater management plan has been proposed by Los Angeles and Inyo County to meet the water resource needs of the city while achieving the environmental goals established for the valley.

INTRODUCTION

The Owens Valley Groundwater Basin is located in east-central California along the western edge of the Great Basin Region. The watershed of the Owens Valley is located within Inyo and Mono Counties of California, about 250 miles (402 km) north of Los Angeles (see Figure 1). The valley lies between the Sierra Nevada and the Inyo-White Mountain Range about 120 miles (193 km) in length and from 15 to 30 miles (24 to 48 km) wide, with a total area of over 3,000 square miles (7,800 km^2). Approximately 30 million acre-feet (37,000 hec^3) of groundwater is stored within this basin.

The City of Los Angeles (LA) imports groundwater extractions and surface water diversions from the Owens Valley via an aqueduct system consisting of the LA Owens River Aqueduct (completed in 1913) and the Second LA Aqueduct (completed in 1970). With the completion of the Second LA Aqueduct, groundwater extractions from the Owens Valley increased, resulting in concerns regarding the environment of the Owens Valley and, in particular, vegetation on the valley floor.

1/ Hydrologic Engineer, LADWP, Los Angeles, CA

2/ Hydrologic Engineering Associate, LADWP, Los Angeles, CA

Since 1982, the USGS, Inyo County and LA, cooperatively have been performing extensive studies on the hydrogeology and plant ecology of the Owens Valley. These studies included groundwater investigations and development of a groundwater model to study the relationship between pumping, water table fluctuation, and the use of water by vegetation; vegetation survivability studies to evaluate the response of vegetation to changes in the shallow water table; and vegetation studies to provide data on plant water use and moisture stress for use in both the groundwater investigations and vegetation survivability studies[3]. These studies were performed for the purpose of providing data necessary in developing the groundwater management program for the Owens Valley, and to increase understanding of the groundwater resources and its relationship to the native vegetation.

GEOLOGY/HYDROLOGY OF THE BASIN

The Owens Valley was formed primarily by faulting along the Inyo-White Mountain Range and the Sierra Nevada. The valley is essentially a down-dropped block between these mountain ranges. Through the

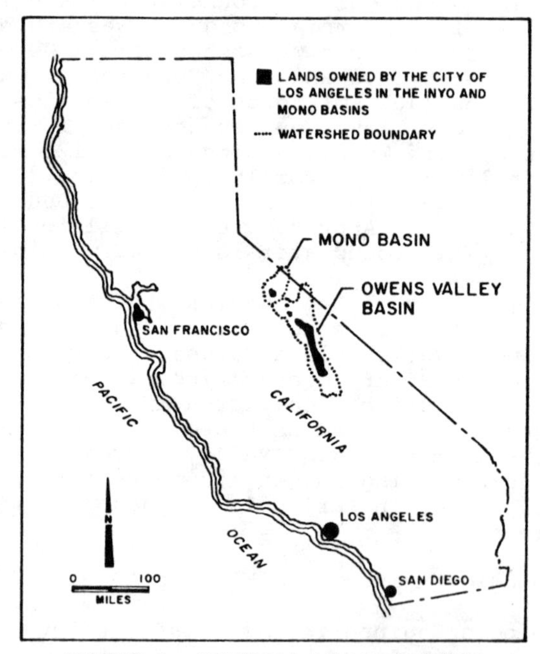

FIGURE 1. GENERAL LOCATION MAP

process of successive down-faulting and erosion, sediments from the mountains have filled in the valley. The geologic formations of the valley are divided into water-bearing (valley fill) and nonwater-bearing formations (hills and mountains). The water-bearing formations consist of alluvial deposits, lake beds, and highly fractured volcanic flows.

Among the geological conditions that influence the effect a pumping well will have on the surrounding water levels is the amount of confinement and the source of water to the well. A generalized cross-section across the Owens Valley is shown in Figure 2. Precipitation falling directly on the alluvial fans and runoff flowing within stream beds across the fans percolate downward and easterly into the upper and lower layers of the aquifer system.

In wells that penetrate the confining clay-silt layers, the water level in the well often rises several feet above the free water table or rises above the ground surface elevation. Just as confinement impedes the flow of groundwater from the lower aquifers to the upper zone, the confinement also prevents or delays the downward flow of groundwater from the upper to the deeper zones when the lower aquifers are pumped. This is evident in wells pumping from the deeper confined zones where immediate drawdowns occur in the deep wells while the shallow water table wells have responded much slower and by smaller amounts. The effect of taking groundwater from the lower confined aquifers has therefore had minimal short-term impact on the shallow water table of the Owens Valley. Evidence of confinement has been observed throughout much of the valley floor.

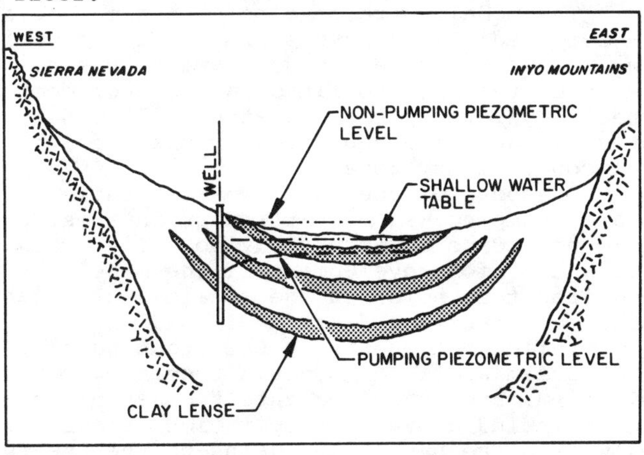

FIGURE 2. OWENS VALLEY CROSS-SECTION

The water supply to the Owens Valley consists of (1) runoff from precipitation as a result of melting snow from the surrounding hills and mountains; (2) precipitation on the alluvial fans and valley floor area; (3) flow in the Owens River, originating from the Mono Basin, Long Valley, and Round Valley areas to the north; and (4) groundwater underflow from the north.

Broad categories of major uses of the Owens Valley water supply (outflows) are: (1) export to LA (2) evapotranspiration from vegetation (including evaporation from the soil surface); (3) evaporation from water surfaces; and (4) groundwater underflow from the basin. LA obtains groundwater from the Owens Valley from natural spring flow, flowing wells, and pumping wells. About one-half of the groundwater is used in the Owens Valley, with the remainder being exported to LA.

MODELLING OF THE OWENS VALLEY GROUNDWATER BASIN

As part of the the Owens Valley groundwater studies, the USGS developed an overall valley-wide model and Inyo County and LA developed more detailed half-valley models[2,4]. The finite difference USGS MODFLOW computer code was used to simulate the response of water levels in the groundwater basin (both water table and confined) due to various natural and man-induced changes and to simulate groundwater flows in areas of interest within the valley. The models are considered tools to provide greater geohydrologic understanding of the groundwater flow system in the Owens Valley, to help identify areas where impacts due to pumping may occur, and to develop strategies for mitigating any impacts.

The models were developed to simulate an aquifer composed of two layers similar to conditions existing in the Owens Valley groundwater basin. The shallow water table (unconfined) and deeper composite (confined) aquifers were designated as layers 1 and 2, respectively, and were separated in the model by a low vertical conductivity zone.

Response of the shallow water table aquifer to groundwater pumping was of particular interest and therefore needed to be accurately modeled. The water level data used for development of the model was obtained from test holes in the shallow water table and deeper confined aquifers.

Inflow components to the groundwater system included stream and intermountain recharge, precipitation, and underflow into the basin; outflow components included evapotranspiration, flowing wells, underflow from the basin, and gains to the LA Aqueduct and Owens River.

The steady-state analysis was performed to calibrate the model. The goals for the steady-state model calibration were to produce an accurate array of aquifer heads to present as initial conditions to the transient simulation and to accurately simulate the water balance of the modelled area in an unstressed condition. The objectives of the calibration were that the model's calculated heads be within ten feet of the actual field conditions and that appropriate vertical gradients were produced by the models in addition to providing a reasonable water balance.

The results of the transient analysis were evaluated by comparing simulated runs with actual hydrograph data and evaluating the reasonableness of the responses of the system to pumping, differing recharge conditions, and vegetation activity.

PROPOSED LONG-TERM GROUNDWATER MANAGEMENT PLAN

Based on information obtained from the groundwater and vegetation studies and modelling efforts, a preliminary long-term groundwater management plan has been developed jointly by LA and Inyo County. The plan is based on defined environmental standards for the protection of Owens Valley vegetation rather than on specified numerical pumping limits. The overall goal is to manage the water resources within Inyo County to avoid certain described decreases and changes in vegetation and to cause no significant effect on the environment which cannot be acceptably mitigated while providing a reliable supply of water for export to LA and for use in Inyo County[1].

To achieve the goal of the proposed management plan, it was necessary that the management program incorporates an acceptable method for establishing a consistent and known relationship between groundwater withdrawals and environmental impacts. An inventory and classification of vegetation in Owens Valley was performed and documented on "Vegetation and Well Field Management Area" maps. Overlayed on the vegetation were boundaries generated by the two detailed groundwater flow models under a pumping scenario of all existing wells pumping during a worst case 3-year drought. The boundaries delineated the area in which drawdowns of the water table of ten feet (3m) or greater would be expected during the worst case low runoff/maximum pumping scenario, i.e., the area of greatest response to pumping and vegetation concern. Included in each management area are monitoring sites for tracking vegetation and groundwater conditions. Monitoring includes measurements of vegetation vigor, retained soil water (use of pychrometers) and water levels in shallow and deep wells. A monthly water balance for each monitoring site will be made by comparing the estimated amount of soil moisture available to vegetation at the site with the estimated required water needs of the vegetation for the growing season at each monitoring site. If as of July 1 or October 1, the projected amount of available soil water at a monitoring site is less than the estimated water needs of the vegetation for the growing season, wells linked to that monitoring site would be turned off immediately. Wells would only be turned back on if soil water recovers sufficiently to meet the estimated water needs of the vegetation at the time the well was turned off, or if it is determined that mitigation measures are effectively preventing vegetation from being harmed.

STATUS

Currently, a draft EIR on this proposed management plan is being prepared. It is expected that the Final EIR will be filed with the Third District Court of Appeal (3rd DCA) this fall. If the 3rd DCA approves the EIR, a 1972 environmental lawsuit between Inyo County and LA will be resolved. Also, a stipulated judgement between Inyo County and LA that incorporates the provision of this plan will have to be approved by the Inyo County Superior Court before it can be implemented.

CONCLUSION

The concepts and basin management procedures proposed in the preliminary long-term groundwater management plan for the Owens Valley are the result of years of extensive studies performed in the valley. These studies considered not only the geologic and hydrologic characteristics of the valley, but vegetation response to natural and man-induced changes as well. The proposed water resources management concepts will ensure that the Owens Valley groundwater basin is managed in a manner which meets both the water resources need of LA while achieving environmental goals established for the valley.

APPENDIX

1. City of Los Angeles and Inyo County, 1989, Preliminary Agreement between the County of Inyo and the City of Los Angeles and its Department of Water and Power on a Long-Term Groundwater Management Plan for the Owens Valley and Inyo County.

2. Danskin, W.R., 1988, Preliminary Evaluation of the Hydrogeologic System in Owens Valley, California, U.S. Geological Survey Open-File Report 88-4003.

3. Los Angeles Department of Water and Power, 1986, Owens Valley, California, Basin Management to Meet Both Water Resources and Environmental Goals.

4. Los Angeles Department of Water and Power, 1988, Development of a Mathematical Groundwater Flow Model of the Owens Lake Basin Area, California.

Expanding Groundwater Production in Southern California

Andrew Sienkiewich[1]

Abstract

The incentive to achieve greater conjunctive use in Southern California has been provided by the Metropolitan Water District of Southern California through a special pricing mechanism known as Seasonal Storage Service. A discussion of Metropolitan's water rate structure is presented as well as a brief description of some projects and studies that are being carried out in the service area. A substantial increase in conjunctive use practices will be seen in the 1990's.

Introduction

The Metropolitan Water District of Southern California (Metropolitan) is pursuing actions at the regional level to expand groundwater storage and production in conjunction with its imported supplies. The objective is to improve its ability to provide a dependable water supply to its 27 member public agencies. Nearly 15 million people reside in Metropolitan's 5,200 square mile service area on the south coastal plain, an inherently semi-arid region.

Water Supply

Currently, about one-third of the region's supply comes from locally managed groundwater basins (see Figure 1). Metropolitan does not have direct management responsibilities for groundwater basins. Local agencies have almost completely developed the safe yield of the region's groundwater basins which is currently about one million acre-feet per year.

---

1. Andrew Sienkiewich, M.ASCE is the manager of the Groundwater and Regional Resources Branch, Metropolitan Water District, Los Angeles, CA

FIGURE 1

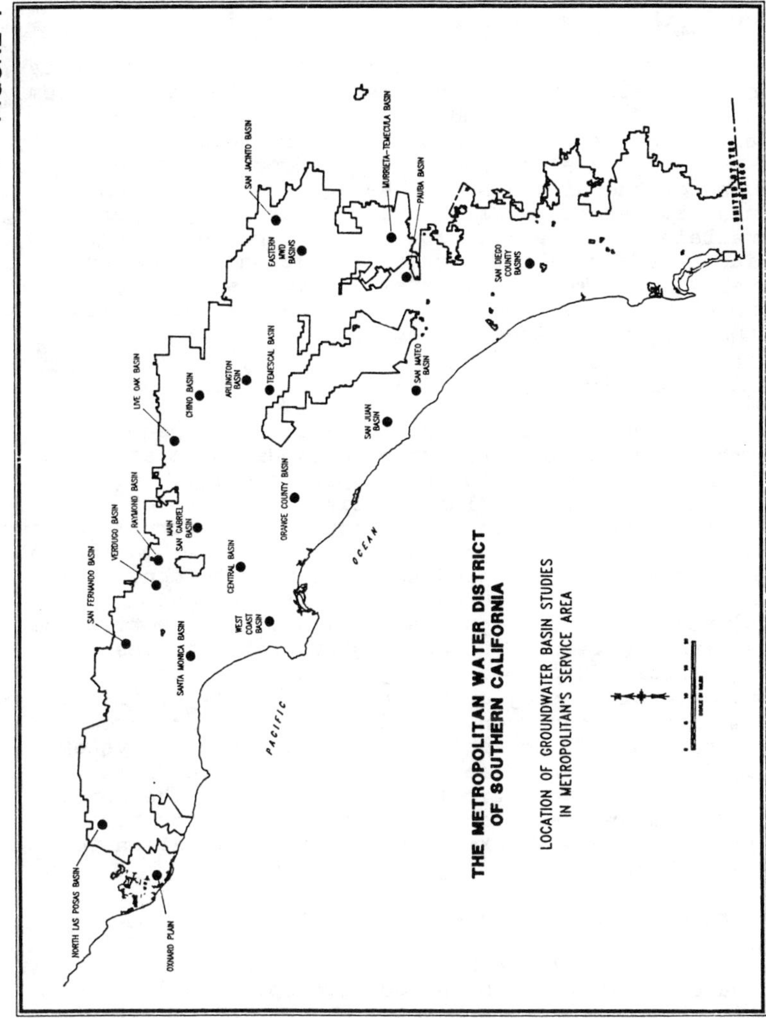
THE METROPOLITAN WATER DISTRICT
OF SOUTHERN CALIFORNIA
LOCATION OF GROUNDWATER BASIN STUDIES
IN METROPOLITAN'S SERVICE AREA
NORTH LAS POSAS BASIN
OXNARD PLAIN
SAN FERNANDO BASIN
SANTA MONICA BASIN
VERDUGO BASIN
RAYMOND BASIN
MAIN SAN GABRIEL BASIN
WEST COAST BASIN
CENTRAL BASIN
LIVE OAK BASIN
CHINO BASIN
ORANGE COUNTY BASIN
ARLINGTON BASIN
TEMESCAL BASIN
SAN JUAN BASIN
SAN MATEO BASIN
SAN JACINTO BASIN
EASTERN MWD BASINS
MURRIETA-TEMECULA BASIN
PAUBA BASIN
SAN DIEGO COUNTY BASINS
PACIFIC
OCEAN
UNITED STATES
MEXICO

Development of additional local yield is limited by the region's semi-arid climate. Annual rainfall averages 13 inches along the coast and 7 inches in the drier inland valleys. Local rivers are seasonal with virtually no flow in the summer months.

Metropolitan is a wholesaler of imported water, providing about half of the region's water supply from its Colorado River Aqueduct and from its contract for water under California's State Water Project. The City of Los Angeles also imports water from the Eastern slopes of the Sierra Nevada Range through the Los Angeles Aqueduct. Southern California is currently experiencing constraints on the availability of imported water. Arizona is now taking a greater portion of Colorado River water. The State Water Project is only half complete and environmental issues have reduced allowable flows in the Los Angeles Aqueduct.

Conjunctive Use

Member agencies principally rely upon Metropolitan to meet the water demands associated with new growth of about 300,000 people annually. However, legal and environmental constraints on Metropolitan's imported water are requiring Southern California to seek innovative and efficient approaches to water management. One approach is to expand the utility of groundwater basins to serve as storage reservoirs managed conjunctively with imported supplies. Conjunctive use means that water is stored in groundwater basins during wet periods and pumped during dry and peak demand periods. State project water is generally available in the winter months when demands in the service area are low and capacity is available in the aqueduct systems. Unused agricultural water and surplus Hoover Dam releases are also periodically available. The concept is not new and has been long practiced in Southern California, however expansion of its application would clearly help meet growing water demands.

Water Rates To Encourage Groundwater Development

Metropolitan has established different classes of water service to encourage development and conjunctive use of groundwater basins. The basic service that Metropolitan provides from its pipelines, is Noninterruptible Service and presently costs $197 per acre-foot untreated and $230 per acre-foot treated.

Discounted water rates are also available to encourage groundwater recharge and recovery through direct and in-lieu means. Direct recharge is accomplished through spreading and injection operations. In-lieu recharge is accomplished when agencies take imported water in lieu of pumping and thereby leave annual yield in storage.

Metropolitan's interruptible service may be used for in-lieu storage and costs $153 per acre-foot untreated and $186 per acre-foot treated. Following certain criteria, Metropolitan may interrupt or cease providing this service during drought and emergency periods. Agencies are then obligated to provide substitute water service by drawing on groundwater basin or surface reservoir storage. Agencies taking interruptible service receive a financial incentive in the form of discounted water rates of $44 per acre-foot to help pay the extra costs of storing and recovering water.

More recently, Metropolitan has established Seasonal Storage Service at the rate of $115 per acre-foot untreated and $135 per acre-foot treated to encourage greater groundwater production during droughts and peak demand periods. Seasonal Storage Service is normally provided during the period of October through April when unused State project water is generally available and demands are low. Agencies receive the price discount of $82 to $100 per acre-foot for placing water into groundwater storage through direct or in-lieu means by following certain qualification criteria. Use of reservoir storage may also qualify.

In-lieu qualification criteria for Seasonal Storage Service requires agencies to develop extraction facilities and use those facilities to meet peak demands each May through September. Agencies may also use Seasonal Storage Service to place water into long term storage to correct over draft or to hold in reserve for drought supply. The discounted rate helps local agencies pay for the cost of storage and extraction facilities. It also helps encourage agencies to maintain groundwater production capacity when faced with contamination issues. During severe droughts or emergencies, Seasonal Storage Service would not be available and agencies would have the option of ordering higher priced noninterruptible water or drawing on groundwater storage reserves that would be replenished at later dates using the discounted Seasonal Storage Service.

Storage Projects

While the concept of conjunctive use seems simple enough, implementation is a significant challenge. Economics of storing and pumping water is of great concern to the local purveyor. Groundwater contamination plagues almost all local groundwater basins and is getting worse. Storing too much water can raise water tables adversely affecting overlying land use. The legal right to store and retrieve water and protect it from use by others must be considered. Excessive extraction during drought periods can lower the water table and impact the operation of other nearby wells. Subsidence is also a consideration. To help implement programs, Metropolitan is providing technical assistance. Progress in selected projects are described below.

City of Oxnard

The City of Oxnard has initiated an innovative groundwater injection and recovery program that is storing and recovering about 2,000 acre-feet per year of imported water in an unconfined aquifer. Oxnard is purchasing Seasonal Storage Service from Metropolitan's member agency, Calleguas Municipal Water District which delivers the water through it distribution system to Oxnard. Calleguas passes through Metropolitan's water rate discount and also further discounts its normal water rate surcharge by an additional $12 per acre-foot. The overall cycling of 2,000 acre-feet this year will save the city about $40,000 over and above start-up costs during the one-year pilot phase of the program. Oxnard is proposing a two-phased expansion of this program which will allow for the seasonal storage of about 6,000 acre-feet per year, and will amount to a savings of about $300,000 annually.

Oxnard retrofitted existing pumping wells in order to inject water at a location about 4 miles inland from the Pacific Ocean and subject to seawater intrusion. Injection is accomplished by gravity flow in the well casing. Metropolitan provided technical assistance to help Oxnard develop the program. Evaluations showed low potential for well screen encrustation caused by mixing imported and native waters. Analysis of the plume of injected water indicated that it would be recoverable and that migration would not preclude extraction. Oxnard will begin extracting water this May using the same wells.

North Las Posas

Metropolitan and Calleguas MWD conducted a joint hydrogeologic study identifying storage potential for imported water of 300,000 acre-feet in the North Las Posas Basin, located in Ventura County. The basin has perhaps the best water quality in the County with TDS on the order of 300-500 ppm. Furthermore, being a confined aquifer, nitrate contamination from overlying agricultural practices has not been evident.

Two methods of recharge are currently being investigated. The first involves the development of traditional spreading basins in areas where the confined Fox Canyon aquifer daylights. The second involves the development of direct injection and extraction wells similar in principal to the kind being demonstrated in Oxnard.

Raymond Basin

Metropolitan and the City of Pasadena are entering into the second phase of a conjunctive-use study of the Raymond Basin. The Basin underlies 40 square miles of highly urbanized land near the City of Pasadena. Local annual demand of 60,000 acre-feet is met by using 31,000 acre-feet of safe yield plus imported water. Results indicate that about half of the 400,000 acre-feet of vacant storage may be filled for conjunctive-use operations. Four conceptual stages of basin development were identified and are under further investigation:

1. Partial reduction of summer peaking off imported water supplies by fully using existing well capacity in the summer period.

2. Exclusive use of groundwater during the summer by developing new injection and extraction wells.

3. Storage of imported water supplies in the basin to meet summer peak and drought year water demands.

4. Banking of groundwater to meet local (overlying) demands and provide export capacity to bolster regional supply during droughts.

Alternative 1 would be accomplished under in-lieu replenishment with no new construction and would therefore require minimal new construction. Accomplishment of Alternatives 2, 3 and 4 would require new recharge and extraction facilities. While there is considerable study still required to define the appropriate level of basin development, the Seasonal Storage Service rate incentive would provide significant financial support for development of these programs.

Conclusions

Conjunctive use will be a major water management tool for Southern California in the 1990's. Discounts provided through Metropolitan's water rate structure is providing the incentive to member agencies to utilize their groundwater resources in a more efficient manner.

TRANSMISSION LOSSES, FLOOD PEAKS, AND GROUNDWATER RECHARGE

Leonard J. Lane[1], M. ASCE

Abstract

Abstractions of streamflow in ephemeral stream channels from infiltration in the channel beds and banks are called transmission losses. These losses are important because water is "lost" as flood waves travel through the normally dry channel networks. Thus, local aquifers are recharged and runoff volumes and flood peaks are reduced over what they would be in the absence of transmission losses. Stream channels crossing alluvial fans transport water from mountain fronts to lower portions of the watersheds. Although these channels are unstable and variable in time and space, they retain their ephemeral character and thus transmission losses can exhibit their influence on flood peaks, water yield, and groundwater recharge as described for ephemeral stream channel networks. Recently developed procedures to estimate transmission losses for individual flow events in ephemeral stream channels are described. Parameters of the transmission-loss model are determined, by calibration, using measured inflow and outflow volumes from gaged ephemeral stream channel segments. Data from 127 hydrographs on 10 channel reaches in Arizona, Kansas, Nebraska, and Texas are used to develop parameter estimation equations and tables of parameter values for the transmission-loss model. Example applications of the transmission-loss model in predicting flood frequency curves and in estimating potential groundwater recharge from transmission losses are described.

Introduction

In arid and semiarid regions, increasing populations, urbanization, expanding industry, and irrigated agriculture are increasing demand for water resources. This demand results in increasing competition for existing water supplies and pressure to develop new sources of water. The increased demand for water resources requires better methods of assessing streamflow and assessing the interaction between streamflow, flooding,

---

[1]The author is a hydrologist, USDA-ARS 2000 E. Allen Rd., Tucson, AZ 85719.

infiltration losses in channel beds and banks, and groundwater recharge. Abstractions of streamflow in stream channel systems from infiltration in the channel beds and banks are called transmission losses.

Transmission losses are important because water is "lost" as flood waves travel through the normally dry stream channel systems or networks. Thus, runoff volumes and flood peaks are reduced over what they would be in the absence of transmission losses (e.g. see Babcock and Cushing, 1941 and Renard, 1970). Transmission losses are an important component of the water budget because surface water yields are reduced, riparian vegetation and wildlife are supported, and local aquifers are recharged (e.g. see Renard, 1970). Therefore, prediction of flood peaks and calculation of water budgets for watersheds in arid and semiarid areas require quantification of the impacts of transmission losses on components of the hydrologic cycle.

As stream channels traversing alluvial fans transport water from mountain fronts to lower portions of the watersheds, significant flow occurs in channels incised into the alluvium forming the fan (Goudie and Wilkinson, 1977). Although these channels are unstable and variable in time and space, they retain their ephemeral character and thus transmission losses can exhibit their influence on flood peaks, water yield, and groundwater recharge as described for ephemeral stream channel networks.

In terms of flood routing and transmission losses, the main differences between ephemeral stream channel networks forming the drainage pattern in watersheds and ephemeral channel segments traversing alluvial fans are due to the nature of their structure and linkage. Channel systems in watersheds tend to be dendritic in structure with main channels collecting tributary inflow in the downstream direction. Channel segments on alluvial fans tend to be singular or bifurcating in the downstream direction. Usually there is no tributary inflow but channels can split or diverge resulting in tributary outflow in the downstream direction. In spite of these differences, many of the same flow processes occur in watersheds and on alluvial fans and procedures developed to consider streamflow and transmission losses by individual stream channel segment can be applied to either system.

Overview of the Model

Procedures have been developed to estimate transmission losses for individual flow events in ephemeral stream channels (Lane, 1982 and Lane, 1985). The rate of change of runoff volume with distance downstream in an ephemeral stream channel segment subject to transmission losses is described by a first order differential equation. The differential equation assumes the volume of losses in a reach is proportional to the volume of upstream inflow, a constant or steady-state loss rate, and the rate of lateral inflow per unit length of channel.

In equation form,

$$dV(x,w)/dx = - wc - wkV(x,w) + V_L/x \quad (1)$$

where: V(x,w) is the volume of flow (acre-ft or m³) in a channel segment of length x (ft or m) and mean width w (ft or m), V_L is the volume of lateral inflow (assumed uniform along the reach) in the same units at V(x,w), and c and k are parameters. The solution to Eq. (1) is:

$$V(x,w) = a(x,w) + b(x,w)V_u + F(x,w)V_L/x \quad (2)$$

where: $V(x,w) \geq 0$ is the outflow volume in acre-ft or m³, V_u is the upstream inflow volume in the same units, and a(x,w), b(x,w), and F(x,w) are functions described below. Notice that in the absence of lateral inflow, the upstream inflow V_u must be larger than -a(x,w)/b(x,w) or all the inflow is lost in the channel segment and V(x,w) = 0. If there is lateral inflow then there will always be some outflow and V(x,w) will be greater than zero.

To calculate the volume of transmission losses in a channel segment rather than the volume of outflow, the volume of transmission losses is computed as the sum of the upstream and lateral inflow volumes minus the outflow volume. In equation form:

$$TL(x,w) = V_u + V_L - [a(x,w) + b(x,w)V_u + F(x,w)V_L/x] \quad (3)$$

where TL(x,w) is the volume of transmission losses in the segment in the same units as V(x,w).

The relationships between the functions and the parameters c and k are:

$$a(x,w) = [a/(1-b)][1 - b(x,w)] \quad (4)$$

$$b(x,w) = \exp(-kxw) \quad (5)$$

$$F(x,w) = [1 - b(x,w)]/(kw) \quad (6)$$

and

$$c = -ka/(1-b) \quad (7)$$

Values of a, k, and b have been related to the effective, steady-state hydraulic conductivity K (in/h or mm/h), the mean duration of inflow to the reach D (h), and the mean volume of inflow to the reach V (acre-ft or m³) (Lane, 1982) in English units as:

$$a = -0.00465\ KD \quad (8)$$

$$k = -1.09\mathrm{Log}_e(1 - 0.00545KD/V) \quad (9)$$

and

$$b = \exp(-k) \tag{10}$$

Earlier analyses (Murphey, and others, 1977) of data from experimental watersheds in southeastern Arizona produced a statistical estimation equation for the mean duration of flow as

$$D = C_1 A^{C2} = 2.53A^{0.2} \tag{11}$$

with $R^2 = 0.78$ and A as the watershed area in sq mi. A similar equation for the mean volume of flow is

$$V_{in} = C_3 A^{C4} = 0.05A^{-0.2} \tag{12}$$

with $R^2 = 0.61$, A as the watershed area in sq mi, and V_{in} is the mean volume of runoff in inches. Notice that V_{in} must be converted to V in acre-ft before it is used in Eq. (9).

Data Base Used for Calibration of the Model

The data base used to derive Eqs. (8) - (10) was taken from 10 gaged channel reaches in Arizona, Kansas, Nebraska, and Texas and represents 127 individual event hydrographs. Therefore, application of the transmission-loss model to streams in other areas is probably not warranted without local calibration data for a, k, b, D, and V in Eqs. (8) - (12).

The effective saturated conductivity, K, represents the steady-state conductivity of the channel bed material under field conditions of entrapped air and sediment laden flow. Therefore, it can be an order of magnitude less than conductivity estimates made with infiltrometers and clear water. Values of the effective conductivity were derived by taking the total losses from an event divided by the length and width of the segment and by the duration of flow. With proper units conversion, the result is an estimate of K in in/h for each flow event. These estimates were averaged over all flow events for a channel segment to derive an estimate of the mean effective hydraulic conductivity. Values of K for different bed material classes were tabulated by Lane (1982).

Example Application

Solutions to the differential equation for transmission losses with parameter values as described above account for empirically observed dependence of infiltration losses on rate of inflow to a channel reach and simulate reductions in flood peaks and volumes measured in ephemeral stream channel networks.

Estimated flood peaks from observed data and from applying a distributed watershed model incorporating the transmission-loss model are given in Table 1. These data represent 8 very small to small watersheds in southeastern Arizona.

Table 1. Comparison of estimated flood peaks derived from measured data and simulation results using a distributed watershed model (Lane, 1982; Lane 1985) incorporating the transmission loss model.

Watershed	Record Length (yr)	Area (sq mi)	Estimated Flood Peaks in cfs per sq mi 1,2			
			Observed		Simulated	
			2 yr	100 yr	2 yr	100 yr
Walnut Gulch, AZ						
63.103	17	.0142	620.	2960.	610.	3790.
63.104	17	.0175	710.	5160.	630.	3740.
63.111	20	.223	600.	3190.	370.	2230.
63.011	13	3.18	210.	2520.	230.	2890.
63.008	13	5.98	120.	1050.	140.	840.
Safford, AZ						
45.001	30	.81	100.	1240.	110.	1220.
Tucson, AZ						
High School Wash	8	.90	420.	1690.	300.	2150.
Big Wash	11	2.75	80.	2480.	270.	1520.

1. 1 cfs per sq mi = 0.0109 cms per sq km.
2. Log-normal probability distribution used to estimate flood frequency.

An important consequence of the transmission-loss model and simulation results summarized in Table 1 is a partial explanation of empirical observations of decreasing flood peaks and volumes with increasing drainage area on the Walnut Gulch Experimental Watershed (Keppel, 1960). Calculations with the simulation model with and without transmission losses suggest the following. For the 2 yr flood on watershed 63.103 (0.0142 sq mi), transmission losses reduced the peak discharge about 2%. But, the corresponding reduction for the 5.98 sq mi watershed 63.008 with an extensive channel system was estimated as about 30% in the peak discharge and runoff volume. It is estimated that on watershed 63.008 about 1/3 of the runoff volume from the 2 yr flood becomes transmission losses and thus potential groundwater recharge. The importance of recharge through the ephemeral stream channels on Walnut Gulch has been confirmed by increases in water levels in wells in and adjacent to the main channels following flood events (Wallace and Renard, 1967).

References

Babcock, H. M., and Cushing, E. M. (1941). "Recharge to ground water from floods in a typical desert wash, Pinal County, Arizona." Trans., AGU, 23(1):49-56.

Goudie, A., and Wilkinson, J. (1977). The warm desert environment. Cambridge Univ. Press, Cambridge, 88 pp.

Keppel, R. V. (1960). "Transmission losses on Walnut Gulch Watershed." In Proc. Joint ARS-SCS Hydrology Workshop, New Orleans, LA, pp. 21.1-21.8.

Lane, L. J. (1982). "Distributed model for small semiarid watersheds." J. Hydraulics Div., ASCE, 108(HY10):1114-1131.

Lane, L. J. (1985). "Estimating transmission losses." In Proc. Spec. Conf., Development and Management Aspects of Irrigation and Drainage Systems, Irrig. and Drain. Engr. Div., ASCE, San Antonio, TX, pp. 106-113.

Murphey, J. B. , Wallace, D. E., and Lane, L. J. (1977). "Geomorphic parameters predict hydrograph characteristics in the Southwest." Water Resources Bull., AWRA, 13:(1):25-38.

Renard, K. G. (1970). "The hydrology of semiarid rangeland watersheds." ARS 41-162, USDA, ARS, Washington, DC, 26 pp.

Wallace, D. E., and Renard, K. G. (1967). "Contribution to regional water table from transmission losses of ephemeral streambeds." Trans. ASAE, 10(6):786-789, 792.

RAINFALL INFILTRATION AND LOSS ON A BAJADA IN THE CHIHUAHUAN DESERT, NEW MEXICO

Susan Bolton[1], A.M. ASCE, Tim J. Ward[2], M. ASCE, and Walter G. Whitford[3]

ABSTRACT: Watershed runoff from thunderstorms is heavily dependent upon the rainfall intensity and the infiltration or loss rate of the watershed. Although there are techniques available for determining infiltration rates of soils, these techniques are inadequate when dealing with vegetation-soil complexes found in arid and semi-arid environments. A more representative technique is on-site measurements of runoff from natural and/or simulated rainfall. In this paper, results from natural and simulated rainfall-runoff plots located on a bajada northeast of Las Cruces, New Mexico, are presented. Analyses of the measurements indicate that: the sparse desert vegetation does not have a significant effect on runoff depth (natural plots), vegetation has a slight but significant effect on infiltration rate (simulator plots), and there is a seasonal effect on infiltration losses.

INTRODUCTION

Bajadas or alluvial fans are very hydrologically active in that they are intermediate between the steep mountain fronts and the flatter valley bottoms. In such locations, bajadas are subject to flooding from the mountains and from runoff generated on the bajada itself. It is these two sources of runoff and associated sediment and nutrients which give rise to the diversity of topography and biota found on the bajada. One of the problems an engineering hydrologist faces is how much runoff will be generated on-site.

On-site runoff production is important from the standpoint of comparing conditions before and after development. Numerous studies have investigated the hydrology of arid and semi-arid areas (e.g., Lane et

[1]Engineer I, Dept. of Civil, Agric. and Geol. Engineering, New Mexico State Univ., Las Cruces, NM 88003.
[2]Professor, Dept. of Civil, Agric. and Geol. Engineering, New Mexico State Univ., Las Cruces, NM 88003.
[3]Professor, Dept. of Biology, New Mexico State Univ., Las Cruces, NM 88003.

al. 1987; Ward and Bolin 1989; Sabol 1989). In this paper, measurements from natural and simulated rainfall-runoff plots on a desert bajada are analyzed in order to extend previous studies and to test some conventional wisdom about hydrologic processes in natural ecosystems.

METHODOLOGY

General Site Description

The Jornada Long Term Ecological Research (LTER) site is located in the Chihuahuan Desert on the New Mexico State University College Ranch, 40 km northeast of Las Cruces, New Mexico. Distinct vegetation zones occur as one descends from the mountain shrubland (1501 m) to the grassland playas at the lower elevations (1318 m). The runoff plots are in a creosote shrub zone. Wondzell, et al. (1987) indicate an average vegetation cover of 34% with 22% of that consisting of creosote.

Average annual precipitation is 23 cm. Class A pan evaporation is about ten times higher than precipitation with 52% of the precipitation occurring as rainfall between July 1 and September 30. June is the warmest month with an average maximum temperature of 36 degrees C. The average maximum temperature in January, the coolest month, is 13 degrees C (Wierenga et al. 1985).

Description of Natural Rainfall Plots

The natural runoff plots are approximately 4 square meters in area. Slopes on three of the plots are about 2.5%, but one is slightly steeper with a slope of approximately 5%. Two of the plots have a creosote bush located in the center and two plots have little or no perennial cover (Table 1). There are weighing bucket and tipping bucket rain gages at the site. Water and sediment runoff each the plot flows into a PVC trough and then into a large, calibrated collection tank.

An attempt is made to examine the plots after every rain event (but no more frequently than once every 24 hours) to determine if runoff has occurred. However, some samples from the collection tank represent more than one precipitation event.

Description of the Rainfall Simulator Plots

The simulator plots are 1m by 1m. Four of the plots were placed beneath creosote bush canopy (under shrub), four were placed beneath shrub canopy but in an area where overland flow occurs (intershrub), and four

TABLE 1.--Plot Characteristics. (Canopy cover is computed separately from the other categories. Rock includes gravel)

Plot	Cover Type	Perennial Shrubs %	Grasses and Forbs %	Litter %	Rock %	Bare Ground %
N	S	60.0	5.4	2.4	28.6	63.6
N	B	5.0	6.5	1.6	19.0	73.6
S	I	35.0	0.0	30.0	1.3	62.3
S	U	49.0	2.0	49.7	0.7	45.3
S	B	0.0	0.0	1.0	10.0	89.0

Plots with N designation are natural plots, S plots are simulator plots. Cover types are S-creosote shrub, B-plots with no creosote, I-plot with creosote canopy and overland flow processes, U-plot under creosote canopy.

were placed in open areas with no shrubs and little perennial cover. Simulator plots were rained upon in pairs. Each pair of plots was rained upon once in a "dry" or initial soil moisture condition and again approximately 24 hours after the first rain in a "wet" condition.

RESULTS AND ANALYSES

Rainfall on Natural Plots

Seventeen collection dates through 1986 were analyzed. These collection dates were characterized by runoff occurring at all four of the plots and no missing data. A more detailed description of the natural rainfall events can be found in Bolin and Ward (1987).

All of the collection periods analyzed were between June and November with 12 of the records in August, September, and October. Precipitation occurs between December and April at the site, but typically little to no runoff is generated.

The median number of rainfall events per collection period was four. The average amount of rain that fell in each collection period was 23.9 mm with a range of 3.6 to 64.0 mm. The average of the highest rainfall intensity per collection was 17.5 mm/hr compared to the overall average intensity per collection period of 8.6 mm/hr. Peak intensities ranged from 4.2 to 123.8 mm/hr.

Rainfall on Simulator Plots

Rainfall simulation using a modified Purdue simulator (Ward and Bolin, 1989) was conducted at the LTER site in May and July, 1988 and 1989. Half of the plots received rain at about 89 mm/hr and half of the plots received an applied intensity of 185 mm/hr. On average, the low intensity plots received 36 mm of rain, while the high intensity plots received 67 mm. The simulator plots received much more intense rainfall and more total rainfall than did the natural plots.

Runoff Analyses

Table 2 lists the statistics for various runoff values for the natural and simulated rainfall plots. Analyses of the natural rainfall plots indicated no significant differences ($p < 0.05$) in runoff depths or runoff to rainfall ratios between the plots with creosote bushes and plots without bushes. Although the mean runoff was higher from the bare plots, there is no consistency for individual rain events as to whether the bare plots or the shrub plots produced more runoff.

Analysis of the natural plot data indicated no difference in runoff by month for the period studied, June to November, but there are very few instances of runoff occurring outside of these months. By using artificial rainfall, infiltration responses were studied in different months.

Because the same plots were rained upon each experiment, a paired difference t-test was used to check for seasonal differences. Infiltration rates, runoff depths and runoff ratios were all significantly higher for the July simulations as compared to May simulations. Paired difference t-tests of the July data indicated that soil moisture conditions significantly affected infiltration on the bare plots (wetter conditions had decreased infiltration rates), but not on the intershrub or under shrub plots. A Mann-Whitney test indicated no differences in final infiltration rate between the high and low rainfall intensities.

Because soil moisture is important on the bare plots, further analyses divided the data into dry and wet soil moisture sets. Only the July data set is large enough to be analyzed using this division to compare bare plots and covered plots. Infiltration rates on the

TABLE 2.--Average runoff, ratio of runoff to rainfall (RO/RF), and infiltration rate by rainfall type and plot cover type. (Standard deviations in parenthesis)

Plot	N	AMC	Month	Runoff (mm)	RO/RF (percent)	Steady-State Infiltration (mm/hr)
N-C	34	NA	NA	3.54 (7.21)	0.117 (0.12)	NA
N-B	34	NA	NA	4.43 (9.20)	0.167 (0.17)	NA
S-U	4	D	MAY	14.8 (15.7)	0.198 (0.16)	85.3 (13.4)
S-U	2	W	MAY	24.2 (5.9)	0.547 (0.11)	61.1 (7.6)
S-I	4	D	MAY	15.7 (13.8)	0.246 (0.17)	87.2 (9.7)
S-I	2	W	MAY	32.0 (20.6)	0.621 (0.18)	37.0 (9.7)
S-B	4	D	MAY	20.2 (19.1)	0.290 (0.19)	70.5 (18.2)
S-B	2	W	MAY	27.7 (21.9)	0.600 (0.29)	35.5 (6.2)
S-U	7	D	JULY	29.4 (21.7)	0.478 (0.17)	60.2 (13.5)
S-U	9	W	JULY	24.4 (16.9)	0.619 (0.37)	54.2 (16.4)
S-I	7	D	JULY	30.9 (15.7)	0.588 (0.18)	47.7 (13.8)
S-I	7	W	JULY	24.5 (15.1)	0.579 (0.21)	43.5 (20.5)
S-B	7	D	JULY	35.9 (29.8)	0.595 (0.15)	42.1 (7.5)
S-B	7	W	JULY	34.3 (22.1)	0.687 (0.14)	28.6 (10.2)

N-C are natural plots with creosote; N-B are natural plots without creosote; S-U are simulator plots under creosote; S-I are simulator plots between creosote; S-B are simulator plots without creosote. AMC is antecedent soil moisture, Dry (D) or Wet (W). NA is information that is currently not available or not applicable.

bare plots were not significantly different from the intershrub plots but were significantly lower those than on the undershrub plots for dry or wet runs. Infiltration rates on the intershrub plots were not significantly different from the under shrub plots for dry or wet runs.

DISCUSSION

Many studies in arid regions that have used rainfall simulators have shown significant differences in plot responses based on vegetative and soil surface conditions. In the Pecos basin region of New Mexico, Smith and Leopold (1942) found that infiltration was positively correlated with vegetal density. Kincaid et al. (1964) found shrub cover, grass and litter cover and gravel cover to be negatively related to runoff. In contrast, some studies (e.g. Blackburn 1975, Tromble et al., 1974) found rock cover and erosion pavement to be positively related to runoff. Lane et al. (1987) found rock and gravel cover and canopy cover to be negatively correlated with runoff depth.

These studies and others that used rainfall simulators in arid and semi-arid regions with low vegetation cover have found that cover, (shrub canopy cover in particular) is an important factor in reducing runoff and erosion. Yet, studies of natural rainfall plots in semi-arid areas have found it difficult to identify systematic differences between plots with different physical (Cordery, 1983) or vegetative characteristics (Bolin and Ward 1987).

The analysis of the simulated rainfall plots indicates a difference in infiltration rates between sites with no vegetation and sites directly beneath creosote shrubs. However, the infiltration in areas away from the base mound of the creosote but still under creosote canopy are not statistically distinguishable from either the bare areas or the mound area.

The analyses presented above indicate that at a small scale (1m x 1m) there are notable differences in infiltration rates both spatially and temporally. Distinctions may be harder to see at larger plot sizes due to integration of conditions. To properly evaluate runoff production potential from bajadas, on-site studies of actual infiltration should be undertaken in a variety of soil-vegetation complexes. Point based equations are not adequate to address the spatial and temporal variability that exists.

ACKNOWLEDGMENTS: The authors wish to acknowledge the National Science Foundation, New Mexico Department of Game and Fish, the USDA-Forest Service, and the New Mexico Water Resources Research Institute for providing funds to support the research from which this paper was developed.

APPENDIX 1. REFERENCES

Bolin, S. B. and T. J. Ward. 1987. "Cover effects on runoff-erosion processes on desert lands", in, <u>USDA Forest Service, General Technical Report RM-150</u>, pp. 196-200, Fort Collins, Colorado.

Blackburn, W.H. 1975. "Factors influencing infiltration and sediment sediment production of semiarid rangelands in Nevada". <u>Water Res. Res.</u> 11,929-937.

Cordery, I., D.H. Pilgrim and D.G. Doran. 1983. "Some hydrological characteristics of arid western New South Wales". Paper presented at the Hydrology and Water Resources Symposium, Hobart, November 8-10, 1983.

Gifford, G.F. 1985. "Cover allocation in rangeland watershed management (A review)", in Watershed Management in the Eighties, edited by E.B. Jones and T.J. Ward, pp.23-31, Am. Soc. of Civil Engineers, New York.

Kincaid, D. R., J.L. Gardner, and H.A. Schreiber. 1964. "Soil and vegetation parameters affecting infiltration under semiarid conditions. Bull IAHS 65:440-453.

Lane , L.J., J.R. Simanton, T.E. Hakonson and E.M. Romney. 1987. "Large-plot infiltration studies in desert and semiarid rangeland areas of the Southwestern, USA", in Proceedings of the International Conference on Infiltration Development and Application, Honolulu, Hawaii, Jan 6-8, 1987.

Sabol, G.V. 1989. Draft Copy: Maricopa County Hydrology Manual Rainfall Losses Section. Preliminary copy provided by G.V. Sabol.

Smith, H.A. and L.B. Leopold. 1942. "Infiltration studies in the Pecos River watershed, New Mexico and Texas". Soil Science 53,195-204, 1942.

Tromble, J.M., K.G. Renard, and A.P. Thatcher. 1974. "Infiltration for three rangeland soil-vegetation complexes". J. Range. Management 27,318-321.

Ward, T.J. and S.B. Bolin. 1989. A Study of Rainfall Simulators, Runoff and Erosion Processes, and Nutrient Yields on Selected Sites in Arizona and New Mexico. N. M. Water Res. Res. Institute, Tech. Completion Report, No. 241, Las Cruces, NM

Wierenga, P.J., J. Hendrickx, M.H. Nash, J. Ludwig, and L. Daugherty. 1985. "Variation of soil and vegetation with distance along a transect in the Chihuahuan Desert". J. of Arid Environments 13:53-63.

Wondzell, S.M., G.L. Cunningham, and D. Bachelet. 1987. "A hierarchical classification of landforms: Some implications for understanding local and regional vegetation dynamics", in, USDA Forest Service, General Technical Report RM-150, pp. 15-23, Fort Collins, Colorado.

PIEDMONT-FAN FLOOD HAZARD ANALYSIS FROM GEOMORPHOLOGY AND SURFACE WATER HYDROLOGY, HUDSPETH COUNTY, TEXAS

Jeffrey R. Keaton[1], M. ASCE, **Roy J. Shlemon**[2], **Richard H. French**[3], M. ASCE, and **David R. Dawdy**[4], M. ASCE

Abstract

A 2-mi^2 (5.2 km^2) site selected by the Texas Low-Level Radioactive Waste Disposal Authority was technically evaluated for local west Texas governmental agencies by a team including the authors. This site is located on a piedmont-fan surface within the Hueco Bolson on tributaries of the Rio Grande. The 12-mi^2 (31 km^2) drainage basin includes the southern edge of the Diablo Plateau where flat-lying Cretaceous limestone is exposed in a 600-ft (183-m) high escarpment. The 100-yr flood plain is a regulatory exclusion for siting low-level radioactive waste facilities. The FEMA Flood Insurance Rate Map (FIRM) shows a narrow strip of Zone A across the site; however, the geomorphic appearance of the piedmont-fan surface indicates that the floodplain is much more extensive.

The piedmont-fan surface is characterized by creosote and scattered mesquite, low relief, and a gentle southwestern slope. A near-surface calcrete, possibly 400,000-yr old, is regionally extensive. However, trenches to about 20 ft (6 m) deep reveal that (1) the calcrete is laterally discontinuous and has been cut locally by latest Pleistocene and Holocene gravel-filled channels; (2) fluvial fine sand and silt and discontinuous intercalated buried paleosols overlie the channels and calcrete; (3) probable middle to late Holocene bars and channels locally interfinger the post-calcrete deposits; and (4) the modern surface is geomorphically active and bears only a weak cumulic (pedogenic) soil profile.

These observations indicate that the site (1) is dominated by fluvial erosion and deposition typical of proximal and medial piedmont-fan systems, (2) is geomorphically unstable, and (3) has been subjected to repeated high-energy flood events in the recent geologic past. Contemporary flood hazards were analyzed using HEC1-HEC2 and the FEMA method. The results of both analyses show that virtually the entire site will be inundated by the 100-yr storm runoff.

This case illustrates how engineering analyses of arid-land flood hazards can be corroborated by independent geomorphic and paleo-flood assessments.

---

[1] Sergent, Hauskins & Beckwith, 4030 S. 500 W., Suite 90, Salt Lake City, Utah 84123.
[2] Roy J. Shlemon & Associates, P.O. Box 3066, Newport Beach, California 92659.
[3] Consultant, 2268 East Hacienda Avenue, Las Vegas, Nevada 89119.
[4] Consultant, 3055 23rd Avenue, San Francisco, California 94132.

Introduction

A consulting team including the authors was retained by Hudspeth County, Hudspeth County Conservation and Reclamation District No. 1, Hudspeth County Underground Water Conservation District No. 1, and El Paso County, Texas to evaluate a 2 mi^2 (5.2 km^2) site selected by the Texas Low-Level Radioactive Waste Disposal Authority. The team has prepared a 3-volume report addressing technical issues at the site, including flood hazards (Sergent, Hauskins & Beckwith, 1989).

The Fort Hancock site is located near the town of Fort Hancock in southwestern Hudspeth County, about 12 mi (19 km) north of the Rio Grande and about 30 mi (48 km) east of El Paso. The Flood Insurance Rate Map (FIRM) shows narrow strips of Zone A across the site (FEMA, 1985a) and apparently was used by the Authority to define the 100-yr floodplain. A cursory reconnaissance by the authors revealed the position of the site in an alluvial-fan setting which would not have a simple 100-yr floodplain. Consequently, flood hazards were evaluated by conventional engineering hydrology techniques.

During a reconnaissance in November 1988, the senior author discovered a previously unknown earth fissure on the site. Trenches excavated at the direction of the consulting team to investigate the earth fissures exposed geologic deposits and relationships critical to interpreting flood hazards on the site. The geologic information corroborates the engineering flood hazard analyses.

Geomorphology

The site is located on a piedmont surface within the Hueco Bolson below the Diablo Plateau and drains tributaries of the Rio Grande. The rim of the Plateau is about 600 ft (183 m) high and exposes flat-lying to gently dipping limestone and local sandstone units of Cretaceous age. Using terminology of Peterson (1981), the site geomorphic surface is a fan piedmont, the largest and most extensive landform of the piedmont slope. Subsumed within the definition of a fan piedmont are alluvial fan, segmented alluvial fan, fan, and bajada. The site fan piedmont is comprised of coalescing alluvial fans emanating from canyons cut into the southern edge of the Diablo Plateau. The typical origin and general features of fan piedmonts have been described by Peterson (1981, p. 22):

> "The alluvial mantles issue individually from fanhead trenches, interfan-valley drainageways, and onfan drainageways rather than being mere undifferentiated extensions of alluvial fans.... The provenance of alluvium along any reach of a fan piedmont is the alluvial fans immediately upslope and one or only a few mountain valleys... Fan piedmonts are built of sheet-like alluvial mantles that are only a few feet thick... Fan piedmont construction can be pictured as deposition of such successive, overlapping, or imbricated, alluvial mantles during the Pleistocene and Holocene epochs. During any one deposition interval, the individual thin alluvial mantles are emplaced along broad swales on the piedmont slope. Along the central part of these broad drainageways, active arroyo cutting and filling is reflected by channel zones, some tens or hundreds of feet wide, where the basal alluvium is most gravelly and best stratified."

The Fort Hancock site surface supports common creosote and scattered mesquite bushes. Relief is about 140 ft (43 m) grading west-southwest about 70 ft per mile (s ≈ 0.013). A regionally extensive calcrete (carbonate-cemented alluvium) was assumed to extend across the site, mantled only by a thin veneer of Holocene eolian silty sand. Based on probable correlation with a surface in New Mexico about 70 mi (120 km) to the northwest, the calcrete has been estimated to be 400,000 yr old. However, trenches excavated across the earth fissure revealed that the calcrete is discontinuous, indicating that the site has not been a stable, benign geomorphic surface for hundreds of thousands of years as heretofor presupposed.

The trenches further revealed that the calcrete was cut locally by gravel-filled channels. These channels are part of a fluvial sequence that includes about 6 ft (2 m) of fine sand and silt. Discontinuous and intercalated buried paleosols have formed in the finer fluvial deposits overlying the calcrete. Gravel-filled channel and bar deposits also interfinger the finer fluvial deposits, indicating pulses of high-energy flood discharge. The modern surface at the site bears only a weak cumulic (A-C) soil profile, indicating a Holocene age probably less than a few thousand years old.

Shallow subsurface geologic conditions at the site are shown diagrammatically in a representative cross-section (Figure 1). We have reconstructed the geomorphic history of the site, based on features and relationships exposed in the trenches, as: (1) alluvial-fan sedimentation in early to mid-Pleistocene time consisting of braided gravel-filled channels and fine sand and silt overbank floodplain deposits; (2) geomorphic stability in mid- to late Pleistocene time when the calcrete formed on alluvial-fan parent material (some of the calcrete may be groundwater rather than pedogenic in origin, in which case geomorphic stability would not necessarily be indicated); and (3) return to alluvial-fan sedimentation in latest Pleistocene and Holocene time consisting of braided gravel-filled channels and fine sand and silt overbank floodplain deposits. The modern surface at the site is geomorphically active, preventing more than a weak soil profile to develop on the Holocene alluvial-fan deposits.

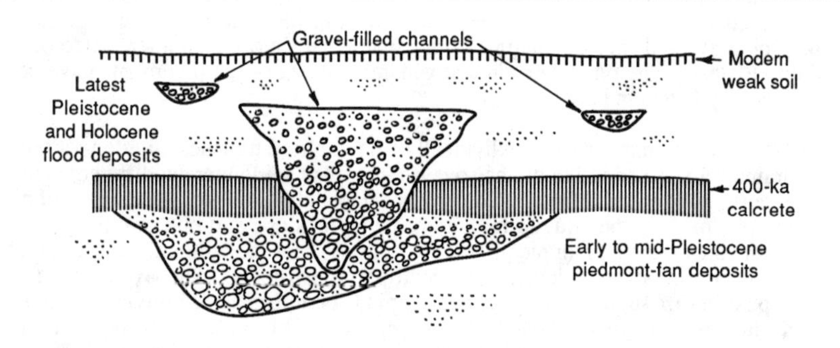

Figure 1. Representative cross-section of the site showing latest Pleistocene and Holocene gravel channels (paleo-flood indicators).

Hydrology

The 100-yr floodplain is a regulatory exclusion for siting low-level radioactive waste disposal facilities (10 CFR 61); and, thus, is a potentially critical issue at a site where the floodplain is Zone A on the FIRM. The drainage basins above the proposed site have a total area of about 12 mi^2 (31 km^2), and these basins extends up onto the Diablo Plateau. An orthotopographic map of the drainage basins above the site was prepared at the direction of the consulting team at a scale of 1:6,000 with a 2-ft (0.6-m) contour interval. This map demonstrated that flood flows are confined to incised channels from the Diablo Plateau rim to a point a short distance above the site. At the point where the flood flows cease to be confined to incised channels, the probabilistic analysis defined by FEMA (1985b) was used to define the floodplain. It is critical to note that the definition of an alluvial fan floodplain from an engineering viewpoint may differ from the strict geomorphologic definition of an alluvial fan. That is, the engineer is considering the surface from the viewpoint of hydraulic engineering and the hydraulic behavior of flows across the surface, while the geomorphologist is interested in precisely defining a landform from the viewpoint of geology. From the viewpoint of an engineer, alluvial fan areas are characterized by (1) active surfaces of deposition or erosion on an engineering time scale (in some cases, such as the radioactive waste disposal site, the design life of the facility begins to approach a geologic time scale); (2) well-defined apexes from which major channels debouch onto surfaces with no well-defined channels; (3) primary slopes in the radial direction with only minor slopes in the transverse direction; (4) poorly defined channels on the surface which are incapable of conveying peak flows associated with long return period floods; (5) flood flows cutting their own channels through the alluvial material that composes the surface; and (6) longitudinal slopes sufficiently long to support critical or near critical flows. Finally, the damage caused by floods in the arid environment is less related to the magnitude of the flood than to its quickness and ferocity.

In the alluvial fan environment, the HEC2 model (US Army Corps of Engineers, 1981a) is inappropriate because of the absence of well-defined channels. The application of such a model in the alluvial fan environment requires the definition of artificial channels based on arbitrary assumptions. Engineering judgement and common sense suggest that alluvial fan surfaces cannot develop their symmetric shapes unless the flood flows conveying sediment migrate across these surfaces. Channels cut by flood flows are done so during extreme events and are subject to migration. On a virgin fan surface (i.e., a surface unmodified by the improvements of man, such as roads and housing developments), the FEMA (1985b) method is very appropriate for defining flood hazard zones. It should be noted that the improvements of man can create channels which render the alluvial fan methodology inappropriate.

Although the proposed low-level radioactive waste disposal facility will occupy a relatively large area (approximately 300 ac [1.2 km^2] of the 2 mi^2 [5.2 km^2] site), the modification of the FEMA (1985b) alluvial-fan methodologies proposed by French and Lombardo (1984) and Mifflin (1988) for large facilities was not applied. Therefore, the analysis was not conservative from an engineering viewpoint. Even neglecting the size of the facility, only 120 contiguous acres (0.5 km^2) of the proposed site was found to be out of the 100-yr floodplain. The proposed site is only a few miles away from the Rio Grande, the border between Mexico and the United States. The unearthing and dispersal of the waste intended to be buried at this site could constitute not only a national problem, but an international incident. The floodplain analysis performed was the most favorable for the advocates of the site; however, from the viewpoint of hydrology, the proposed site could not have been located in a more dangerous environment since it is well within the 100-yr floodplain and could result in erosion of the site and dispersal of the wastes over a wide area.

Conclusions

The site geomorphology is mainly a fan piedmont surface comprised of coalescing alluvial fans. Latest Pleistocene and Holocene stratigraphy and soil horizons exposed in shallow trenches indicate that the site (1) is dominated by fluvial erosion and deposition typical of proximal and medial piedmont-fan systems, (2) is presently geomorphically unstable, and (3) has been subjected to repeated high-energy flood events in the recent geologic past. Flood hazard analyses using HEC1-HEC2 and the FEMA methods indicate that virtually the entire site will be inundated by the 100-yr storm runoff.

This case study illustrates how engineering analyses of arid-land flood hazards benefit from independent geomorphic and paleo-flood assessments.

Acknowledgments

This paper is published with the consent of El Paso County, Texas, and authorized by Darcy Frownfelter, Esq. Logistical support and general project coordination were provided by Mark Turnbough. The authors benefitted from discussions with George Beckwith, Ramon Martinez, Mary Gillam, and Tom Walker.

Appendix A - References Cited

FEMA, 1985a, Flood Insurance Rate Map, Community Panel Number 480361 0605B: Washington, D.C., Federal Emergency Management Agency, Nov. 1.

FEMA, 1985b, Flood insurance study - guidelines and specifications for study contractors: Washington, D.C., Federal Emergency Management Agency, Federal Insurance Agency Report FEMA-37.

French, R. H., and Lombardo, W. S., 1984, Assessment of the flood hazard at the Radioactive Waste Management site in Area 4 of the Nevada Test Site: Las Vegas, Water Resources Center, Desert Research Institute Report No. DOE-NV-L0162-15.

Mifflin, E. R., 1988, Design depths and velocities on alluvial fans, *in* Abt, S. R., and Gessler, J., eds., Proceedings of the 1988 National Conference on Hydraulic Engineering: New York, American Society of Civil Engineers, p. 155 - 160.

Peterson, F. F., 1981, Landforms of the Basin & Range Province defined for soil survey: Reno, Agricultural Experiment Station, University of Nevada Reno, Technical Bulletin 28, 52 p.

Sergent, Hauskins & Beckwith, 1989, Preliminary geologic and hydrologic evaluation of the Fort Hancock Site (NTP-S34), Hudspeth County, Texas, for the disposal of low-level radioactive waste: El Paso and Phoenix, unpublished consultant's report prepared for Hudspeth County, Texas, Hudspeth County Conservation and Reclamation District No. 1, Hudspeth County Underground Water Conservation District No. 1, and El Paso County, Texas, 3 volumes.

US Army Corps of Engineers, 1981a, Water Surface Profiles: Davis, CA, Hydrologic Engineering Center Computer Program HEC2.

US Army Corps of Engineers, 1981b, Flood Hydrograph Package: Davis, CA, Hydrologic Engineering Center Computer Program HEC1.

Large Floods and Climate Change in the Southwestern United States

Lisa L. Ely[1] and Victor R. Baker[2]

Abstract

Floods can have a tremendous impact on water resources but their properties may fluctuate at frequencies that defy recognition in short-term instrumented records. Paleoflood deposits provide insight into this component of the hydrological system by preserving evidence of the largest discharges that have occurred on a given river over time periods that often exceed 1000 years. A regional paleoflood analysis currently in progress in the southwestern United States incorporates study sites on more than twenty rivers. Previous studies on rivers in central Arizona and southern Utah indicate periods of flooding around 800-1000 A.D., 1400-1600 A.D., the late 1800's and the present, separated by periods of relative quiescence. These consistent variations in the long-term flood series within a region suggest a hydroclimatic cause. As the paleoflood dataset from the Southwest becomes more complete, it will indicate variations in the spatial and temporal distributions of large floods in relation to past climatic conditions. The present flooding regime, while not particularly anomalous when viewed in the context of the entire paleoflood record, nevertheless does not contradict the possibility that we are moving into a period of climatic response to general global warming.

Introduction

Water availability looms as perhaps the single most important consequence of future climate change in the rapidly developing arid and semi-arid southwestern United

---

[1]Research Associate, Dept. of Geosciences, University of Arizona, Tucson, AZ. 85721

[2]Regents Professor, Dept. of Geosciences, University of Arizona, Tucson, AZ. 85721

States. The possibility of global warming in response to increasing amounts of atmospheric "greenhouse gases" such as carbon dioxide and methane poses difficult questions in water resources management. Is imminent global warming a real threat? How would global warming affect the hydrologic cycle, particularly within the arid Southwest? These questions must be addressed before effective management strategies can be formulated. Obviously this is a tremendous task, and no single avenue of research can supply a complete answer. This paper focuses on paleoflood analysis, a virtually untapped source of information that could aid in addressing some of these issues by providing insight into the extremes of the hydrologic system in the Southwest.

In order to examine the hydrological effects of possible future climate change, we must first understand more fully both the natural variability of the system and how it responds to different environmental conditions, particularly the extremes. There are two ways to approach this task: through physical models, such as general circulation models (GCMs), or through analysis of paleohydrological data. The two approaches are complementary rather than competitive, since physical models require testing against independent field evidence in order to be considered valid predictive tools. The paleoclimatic record is the only source of independent data that covers sufficiently long time periods to encompass the range of natural variability. Modern climatic and hydrological records for the western United States are too short to reflect long-term climatic trends that fluctuate with periodicities greater than a few decades. Similarly, the limited time period of instrumented records may not adequately represent the hydrological extremes, such as catastrophic storms, floods, severe drought, or prolonged periods of increased precipitation.

Many questions remain as to the precise effects of greenhouse warming and whether those effects are beginning to strongly influence the climatic system. For example, Hansen et al. (1988) have postulated from the results of a GCM that global warming would intensify both dry and wet extremes of the hydrological cycle. Their model indicates that global warming increases the strength and intensity of tropical storms, which are one source of large floods in portions of the Southwest (Hirschboeck, 1985; Webb and Betancourt, 1990). A regional paleoflood record should provide useful information toward evaluating this particular model prediction. Paleoclimatic information allows one to place the present conditions in perspective relative to climatic variability over the last few thousand years. Many proxy paleoclimatic indicators record long-term, or in some cases annual, changes in precipitation and temperature over the late Holocene. Paleoflood data are unique in providing information on the frequency of individual, extreme precipitation events. Thus the paleoflood record provides insight into a virtually unexplored aspect of the climatic and hydrologic system. Spatial and temporal patterns in flooding within the Southwest over periods of centuries to millennia can be compared with other paleoclimatic indicators to determine how these extreme events have responded to past climatic conditions and might therefore respond to future climate change. Although floods represent only one aspect of the hydrologic system, together with other lines of evidence an appropriate paleoflood database should demonstrate whether the present is a period of exceptional hydro-climate or simply a part of a natural cycle that has precedent in the past.

Method of Paleoflood Analysis

The Arizona Laboratory for Paleoclimatological and Hydroclimatological Analysis (ALPHA) is currently analyzing paleoflood information across a broad region of the Southwest in order to understand better how flood-generating storms vary with long-term climatic trends. The study incorporates paleoflood chronologies from a large number of rivers along a wide north-south transect extending from the Colorado Plateau in southern Utah and northern Arizona, through the Basin and Range of central and western Arizona, to northern Sonora, Mexico. The semi-arid climate and abundance of bedrock-confined rivers in the Southwest make this region especially conducive to the accumulation and long-term preservation of slackwater flood deposits, which record the largest discharges that have occurred on a given river over time periods that often exceed 1000 years.

During large floods in bedrock canyons, fine-grained sand and silt settle rapidly out of suspension in areas of markedly reduced flow velocity such as back-flooded tributaries and eddies at channel irregularities (Baker and Kochel, 1988). Many sites contain multiple flood deposits that accumulate both vertically and as insets, with each deposit representing a separate flood. The deposits left by the largest floods generally lie farthest outside the effective flow boundary and are thus the best protected from subsequent erosion. For this reason the slackwater deposits selectively preserve evidence of the largest floods, while the deposits associated with smaller floods are flushed out.

Individual flood units are distinguished based on various criteria, such as laterally extensive sedimentological differences, basal or capping silt layers, incipient soil development, mudcracks and buried vegetation. Radiocarbon dating of associated organic material aids in correlating flood units between sites and provides the basis for constructing a flood chronology. Under ideal conditions the height of each flood deposit approximates the peak stage of the associated flood (Baker and Kochel, 1988). In combination with silt lines, scour lines and flood debris, the slackwater deposits delineate the water-surface elevations of a particular flood throughout a river reach. This paleostage information is transformed into discharge estimates using computerized step-backwater routines.

Results and Discussion

Recent paleoflood studies on rivers in central Arizona and southern Utah reveal a clustering of large floods through time (Ely et al., 1988; Webb et al., 1988). Periods of increased frequency of flooding around 800-1100 A.D., 1400-1600 A.D., the late 1800's, and the present are separated by periods of few large floods. These preliminary results require confirmation at an expanded number of study sites throughout the Southwest. Nevertheless, the fairly consistent variations in the long-term flood series from these sites suggest a climatic cause.

Under modern climatic conditions, a complex variety of meteorological conditions can cause large floods within the study region (Hirschboeck, 1985). However, most flood-generating storms in this area of the Southwest can be classified into three broad categories: summer monsoonal storms, winter frontal storm systems, and tropical cyclones (Webb and Betancourt, 1990). In addition, floods on rivers that head at high elevations can result from rapid snowmelt, although the snowmelt is often accentuated by spring rainfall from frontal storm systems. The predominant storm type associated with the largest floods on a river is influenced by geographical location, regional topography and drainage basin size. Large floods caused by monsoonal storms occur in most of the study area but are almost entirely restricted to rivers with relatively small drainage basin areas, as these are generally intense, localized thunderstorms. Winter frontal systems are areally extensive and generally last for several days. This storm type can cause flooding on rivers throughout Arizona and southern Utah, and is particularly important on rivers with large drainage basins, such as the Salt and Verde Rivers of central Arizona. Moisture from tropical cyclones that is drawn inland over the Southwest can also produce intense regional precipitation, and has caused the largest historical floods on many of the small and intermediate-sized drainages included in the study. The frequency of tropical storm incursion increases toward the southern end of the transect. The Colorado River drainage basin reaches well beyond the limits of the study area and is affected by factors other than those mentioned above. This river is included in the study for comparison with the regional southwestern rivers but falls into a hydroclimatic category of its own.

The ALPHA regional paleoflood study will analyze any changes in the long-term frequency of large floods in light of global and regional paleoclimatic reconstructions for the same periods. Because the study region is affected by both tropical and winter frontal storms, certain portions of the region might exhibit an increase in paleofloods during warm periods and others during cool periods. If model predictions prove correct (Hansen, 1988), rivers in areas dominated by tropical storms could show increased flooding during global warming. Enzel et al. (1989) have documented perennial lake deposits in the Silver Lake Playa of the Mojave Desert during a cool period from the late 1500's to early 1600's A.D., which they relate to an increase in the frequency of floods from North Pacific winter storms. Rivers in central Arizona and southern Utah also indicate increased flooding during this period, perhaps also due to enhanced winter storm activity. In many cases, part or all of the region might experience increased flooding under more than one set of climatic conditions. At the scale of resolution in dating paleofloods, the past 100 years is a period of relatively high flood activity compared to certain periods in the past. The present may be a peak in the natural long-term flood frequency cycle, which could be accentuated by anthropogenic effects.

The value of this type of study is that it provides information on regional hydrologic responses rather than simply the history of one specific river. A single storm can sometimes cause catastrophic flooding on one river and not on an adjacent drainages. By looking at periods of flooding on the scale of hundreds of years over a broad area, there is an opportunity for discovering hydroclimatic patterns that are not detectable by viewing a more restricted dataset. The spatial and temporal patterns that emerge from this regional study will provide useful information on the type of flood-frequency response that can be expected under future climate scenarios.

Acknowledgements: Research supported by the Engineering Directorate, Natural and Manmade Hazards Mitigation Program, National Science Foundation, Grant BCS-8901430. This report is Contribution No. 3 of the Arizona Laboratory of Paleohydrological and Climatological Analysis (ALPHA), University of Arizona.

References

1. Baker, V.R. and Kochel, R.C., "Flood Sedimentation in Bedrock Fluvial Systems," Flood Geomorphology, V.R. Baker, R.C. Kochel, and P.C. Patton, Eds., Wiley, N.Y., 1988, pp. 123-137.

2. Ely, L.L., O'Connor, J.E., and Baker, V.R., "Paleoflood hydrology of the Salt and Verde rivers, Central Arizona," Proceedings, 8th Annual USCOLD Lecture Series; Salt River Project, Tempe, Arizona, January, 1988, pp. 3.1-3.35.

3. Enzel, Y., Cayan, D.R., Anderson, R.Y., and Wells, S.G., "Atmospheric Circulation During Holocene Lake Stands in the Mojave Desert: Evidence of Regional Climate Change," Nature, Vol. 341, 1989, pp. 44-47.

4. Hansen, J. Rind, A., DelGenio, A., Lacis, A., Lebedeff, S., Prather, M., Ruedy, R., and Karl, T., "Regional Greenhouse Climate Effects,' Coping with Climate Change, Proceedings, the Second North American Conference on Preparing for climate Change, December 6-8, 1988, Climate Institute, Washington, D.C.

5. Hirschboeck, K.K., Hydroclimatology of Flow Events in the Gila River Basin, Central and Southern Arizona, Ph.D. Dissertation, University of Arizona, Tucson, 1985, 336 p.

6. Webb, R.H. an Betancourt, J.L., "Climatic Effects on Flood Frequency: An Example from Southern Arizona," Proceedings, Sixth Annual Pacific Climate (PACLIM) Workshop, J.L. Betancourt and A.M. Mackay, Eds., California Department of Water Resources, Interagency Ecological Studies Program Tech. Rpt 23, March 5-6, 1989, pp. 61-66.

7. Webb, R.H., O'Connor, J.E., and Baker, V.R. "Paleohydrologic Reconstruction of Flood Frequency on the Escalante River," Flood Geomorphology, Baker, V.R., Kochel, R.C. and Patton, P.C. Eds., Wiley, N.Y., 1988, pp. 403-418.

Misapplication of the FEMA Alluvial Fan Model: A Case History

Jonathan E. Fuller[1]

Abstract

The Federal Emergency Management Agency (FEMA) recently issued revised Flood Insurance Rate Maps (FIRM) for the Tortolita piedmont, located northwest of Tucson, Arizona. Floodplain delineations shown on the revised FIRM are based on FEMA's alluvial fan methodology. Floodplain management of the piedmont using the revised FIRM is hampered by the flawed theoretical basis of the model, by technical deficiencies in the application of the FEMA alluvial fan model to the Tortolita area, as well as by shortcomings in the flood zone designations used on the FIRM.

Introduction

Alluvial fans are geomorphologic features characterized by a cone-shaped deposits of sediment eroded from mountain slopes, and deposited on the valley floor downstream of the mountain front. Mountain watersheds which supply sediment in excess of the on-fan sediment transport rate form active, or aggrading, fans. Active fans are characterized by aggrading channels which migrate in response to local, deposition-induced slope changes. Inactive fans and pediments are characterized by well developed channel networks dominated by erosion processes.

The Tortolita piedmont, located northwest of Tucson, Arizona, (Figure 1) is an 80 mi^2 (200 km^2) pediment surface adjacent to the western slopes of the Tortolita Mountains, a range of low-relief and limited areal extent. The piedmont is drained by eight principal streams which head

[1]Principal Hydrologist, Planning Division, Pima County Flood Control District, 32 N. Stone Avenue, Suite 300, Tucson, Arizona 85716.

above the mountain front, with mountain watersheds ranging from 0.18 mi^2-6.6 mi^2 (0.46 km^2-17 km^2), and from 420 ft-1620 ft (128 m-494 m) in relief. The southern portion of the piedmont is moderately well developed, with housing densities ranging from working cattle ranches at less than 0.1 residence per acre (RAC) to 5 RAC subdivisions.

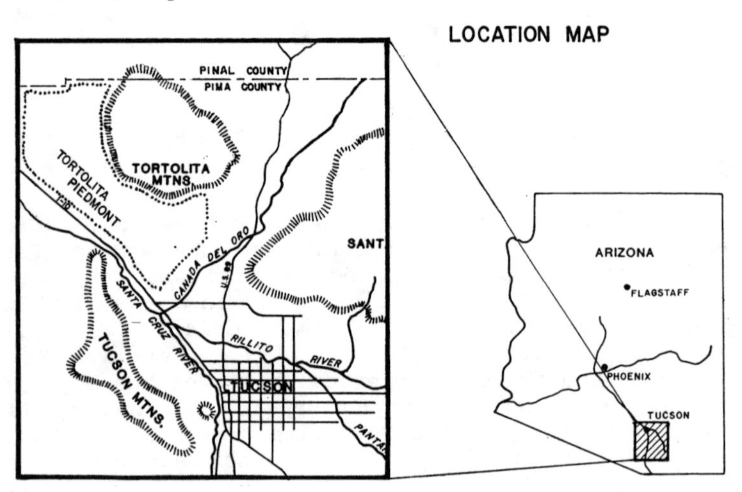

Figure 1. Location map for the Tortolita piedmont. Not to scale.

Theoretical Considerations

The FEMA alluvial fan methodology is based on procedures developed by Dawdy (1979), and is described in FEMA 37 (1985). Key assumptions of the FEMA model are that: (1) alluvial fan flooding is conveyed at critical depth in self-formed channels governed by regime equations of depth, velocity, and discharge at the apex; (2) the location of the flood channel is unpredictable; and (3) topographic relief and urbanization on the fan are minimal. These and other implied assumptions are thoroughly discussed by French (1987). Theoretical weaknesses of the FEMA model are: (1) lack of universal applicability of the regime equations used to develop depth and velocity zones (French, 1987); (2) sensitivity of Log Pearson Type III statistical parameters to the use of synthetic discharge estimates (Faltas, unpub.); and (3) failure to recognize that migration of the primary flood channel is only random in geologic time, rather than engineering time.

Application of the FEMA alluvial fan model to the Tortolita Piedmont violates the key assumptions of the FEMA model. That the flood water is conveyed in self-formed channels governed by the FEMA model's regime equations is contradicted by field measurements of flood deposits (Baker et al, 1990), as well as by the presence bedrock and caliche controlled channels within the FEMA fan boundaries. That channel locations are unpredictable is belied by geomorphic soils analysis (Baker et al, 1990) and historic photographs dating to 1936 which document remarkable channel stability within the FEMA fan boundaries. That flow depth and velocity is a function of the discharge at the apex is complicated by significant attenuation by sheet flooding. Peak discharges are further complicated by large on-fan drainages which confluence below the apices. That topographic relief and degree of urbanization are small are disproved by topographic maps, aerial photographs, and field inspection. Relief up to 15 feet (4 meters) within fan boundaries, and 5 RAC developments with major drainage structures are found within the mapped fan boundaries.

Technical Considerations

Deficiencies in basic hydrologic data further compound flaws in the application of the FEMA model to the Tortolita piedmont. 100-year discharge estimates were overestimated by approximately 50 percent (Baker et al, 1990). Peak discharge values for the two- and ten-year floods were determined using fixed ratios of the 100-year value, a procedure which has been criticized by Faltas (unpub.). Overestimation of peak discharges results in location of fan apices too far upstream, extension of FIRM depth and velocity zones too far downfan, and wider fan boundaries than are justified. In short, extensive flood-free land areas are considered flood prone.

Critical gaps in FEMA guidelines also resulted in erroneous floodplain delineations on the Tortolita FIRM. FEMA has no official methodologies for determining individual fan boundaries on coalescing alluvial fans, for determining the apex location for a fan, or for determining the location of the fan terminus. For the Tortolita piedmont, the fan apices and boundaries were defined by topographic confinement of the clear water energy grade line elevation of the 100-year flood. This procedure is internally inconsistent given the channel aggradation predicted by the FEMA model, and given that the model does not consider topographic variation within fan boundaries. While the total energy elevation criterion is easy to apply, it ignores physical evidence which precisely details fan boundary geometry (Baker et al, 1990).

Guidelines for determining the location of the fan terminus are also lacking from the FEMA model. Normally, the downstream end of a fan is defined by the axial valley stream, or the point where the model predicts depths less than 0.5 ft (0.14 m). The base of the Tortolita piedmont, however, has undergone secondary entrenchment due to base level fall of the Santa Cruz River. As individual members of the distributary flow network reach the entrenched lower area of the piedmont, they flow into well-defined channels bounded by 10 - 20 ft (3 - 6 m) ridges comprised of 10000+ year old surfaces (Baker et al, 1990). The FEMA model was applied as if these ridges did not exist, with ridges subject to the same flood hazard as the channels.

Floodplain Management Issues

Deficiencies in the theoretical basis and technical application of the FEMA alluvial fan model to the Tortolita piedmont are realized in floodplain management. Use of the FEMA alluvial fan model results in flood zone designations on FIRM panels. FIRM panels are one of the primary tools used in floodplain management. In addition, the FIRM are used by land appraisers and developers to help determine the value of undeveloped land. Therefore, it is critical for the FIRM to be accurate. For the Tortolita piedmont, theoretical inconsistencies and technical errors resulted in vast areas incorrectly identified as flood prone.

However, even if the FEMA model is correctly applied, problems with FIRM zone designations would still hinder sound floodplain management. First, because the predicted depths are a function of the probability of occurrence of the regime channel at a given point, regulatory depth decreases as fan width increases. Thus, FIRM depth zones do not indicate actual flow depths in either existing or predicted regime channels. If the fan is wide enough, the FIRM depth will be less than one foot, resulting in a B zone designation. FEMA has no elevation or design requirements for development in B zones, giving homeowners an unwarranted sense of security. Thus, the FIRM tacitly encourage development in the lower portions of even the most active fans, which in turn hinders community efforts for "whole fan" management solutions.

The existence of velocity zones on the FIRM are also problematic. If the FIRM are to be used only for insurance purposes, velocity is irrelevant, as insurance rates are based strictly on depth. If the FIRM are to be used for floodplain management, then the velocities printed on the FIRM are inadequate because they are regime velocities factored by the fan width-probability relationship, and thus cannot be used for hydraulic design. A structure designed to the FIRM depths and velocities would be damaged if the design flood were to occur at the structure.

Enforcement of uniform management standards within each FIRM zone is also difficult to justify. FEMA elevation requirements are the same for a structure built in an existing wash bottom as for a structure built on a ridge as long as both are equidistant from the apex. Clearly, the risk within an existing wash is significantly greater, particularly for low recurrence interval events. FEMA's policy of disallowing Letters of Map Revision (waives insurance requirement) based on elevation on mapped alluvial fans is also a disincentive to sound design.

A final aspect of misapplication of the FEMA model to the Tortolita piedmont which should not be overlooked is the affect on public support for the National Flood Insurance Program (NFIP) and local floodplain managements. Common sense dictates that NFIP alluvial fan regulations do not apply in the Tortolitas. Support for more rational, and much-needed, floodplain management programs has been undermined, as alluvial fan regulations are enforced where they are perceived to be inappropriate.

Conclusion

The FEMA alluvial fan model was incorrectly applied to the Tortolita piedmont. The resulting FIRM include vast land areas incorrectly designated as alluvial fans and as floodprone. The FIRM hinder effective floodplain management of the area. The economic impact on the Tortolita area is likely to be severe, with the community forced to chose between costly over-designed flood control structures, permanent flood insurance bills, or acquisition of the land designated as alluvial fan.

Misapplication of the FEMA alluvial fan methodology to an area has profound implications for the development of that area. Unlike overestimated water surface elevations in riverine flood studies, designation of an area as an alluvial fan brings the most severe design restrictions and lack of flexibility in use of alternative models to analyze the hazard. Where active alluvial fans with debris flow hazards, shifting channels, and severe erosion potential exist, strict management is justified. Where inactive fans or pediments are found, the hazard is no more severe than for other alluvial watercourses typical of the arid southwest.

Future applications of the FEMA alluvial fan model should be preceded by careful consideration of the model's assumptions, precise determination of hydrologic input, and examination of physical evidence in order to define limits of applicability.

References

Baker, V.R., et al, (1990). "Application of Geological Information to Arizona Flood Hazard Assessment." Paper presented ASCE/H2AL Conference, San Diego, California.

Dawdy, David R. (1979). "Flood Frequency Estimates on Alluvial Fans." Journal of the Hydraulics Division ASCE, Proceedings, Vol. 105, No. HYII, 1407-1413.

Faltas, M.E. (unpub.). "Evaluating Flood Hazards on Alluvial Fans." unpub., Michael Baker, Jr., Inc., Alexandria, Virginia.

Federal Emergency Management Agency, (1985). Guidelines and Specifications for Study Contractors. Washington, D.C.

French, R.H. (1987). Hydraulic Processes on Alluvial Fans. Elsevier Scientific Publishers, Amsterdam.

Conjunctive Use of Surface and Groundwater Resources in the Central Valley of California

Sushil Arora[1], M. ASCE, Sina Darabzand[2]

Abstract :

California Department of Water Resources (DWR) is evaluating the feasibility of incorporating a conjunctive use program into the State Water Project (SWP) to augment its delivery capability and reliability. This paper presents some of the conjunctive use techniques under consideration by the DWR. Also a brief analysis of the results of planning operation studies under two options of conjunctive use program is presented.

Major Features of the State and Federal Water Projects in California

Hydrologic and demographic characteristics of the State of California have made storage and conveyance systems necessary to transfer water from areas of surplus to areas of need. In addition to a few large-scale water resource development projects that are operated by local agencies, the two major systems operated by the state and federal governments are the State Water Project (SWP) and the Central Valley Project (CVP), respectively. Features of the two systems are shown in the figure below.

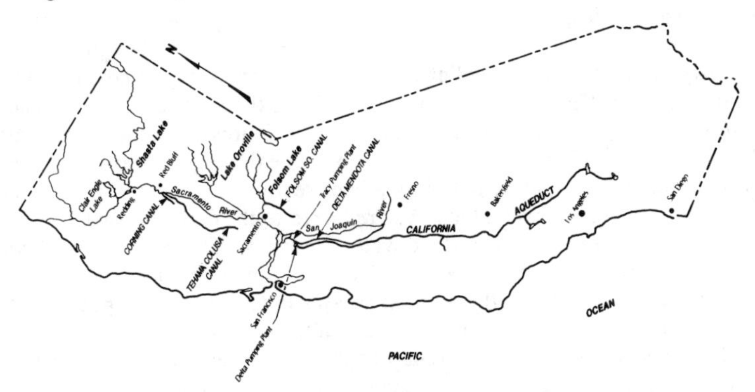

MAJOR FEATURES OF THE STATE AND FEDERAL WATER PROJECTS IN CALIFORNIA

1. Senior Engineer, Division of Planning, California Department of Water Resources, Sacramento, California.
2. Associate Engineer, Division of Planning, California Department of Water Resources, Sacramento, California.

The inter-basin transfer from Clair Engle Lake and releases from Shasta Lake, Folsom Lake, and Lake Oroville flow down the Sacramento River, forming the major component of the total inflow to the Sacramento-San Joaquin Delta. A portion of the total Delta inflow, remaining after meeting the in-basin uses and the outflow requirements of the San Francisco Bay is available to both systems for export to central and southern California. Harvey O. Banks Delta Pumping Plant with an existing capacity of 6,400 cfs, and Tracy Pumping Plant of 4,600 cfs, both located at the southern end of the Delta, divert the exportable water for the SWP and the CVP systems, respectively.

State Water Project and the Future Conditions

The existing SWP system can not provide a dependable water supply to meet projected future demands. Long-range projections by the Department of Water Resources (DWR) indicate that the demand on the SWP system will grow from the present 3.1 million acre-feet per year (MAF/yr) to more than 4.2 MAF/yr by the year 2035, with the bulk of this increase occurring by the year 2010. Along with exploring every practical means of water conservation and efficient water use, DWR planners have been considering various alternatives to expand storage and conveyance capacities of the SWP system to cope with the projected increase in demands. One alternative being considered by the DWR is the utilization of Kern County groundwater basin in the southern part of the California Central Valley. The general plan consists of various conjunctive use programs of surface and sub-surface sources of water.

Conjunctive Use Programs

Conjunctive use, as the term implies in the water resources planning investigations, is devising an operation policy for surface water facilities *in conjunction with* groundwater facilities to augment water supply of the system. The conjunctive operation could consist of the combination of the following components : direct recharge, in-lieu recharge, direct extraction, and exchange. Two specific techniques are described in more detail, in a later section. In general, the operation parameters of the groundwater facilities are linked to the delivery capability of the system, and to storage in the surface water facilities. The purpose of planning operation studies is to develop a conjunctive operation policy that leads to the best overall results. In the overall results, increases in the delivery capability of the system should be evaluated along with the impacts on the operation of surface water facilities. Changes in the long-term operation of the groundwater basins that are affected by the conjunctive use program should also need to be studied.

Kern Water Bank (KWB) is a comprehensive conjunctive use program under development by the DWR which includes various components of direct recharge and extraction program (Kern Fan Element project) and in-lieu recharge and exchange (Semitropic Water Storage District in-lieu project).

Direct Recharge-Extraction Programs

Direct recharge-extraction programs involve construction, operation, and maintenance of recharge sites and well fields. In years when surface water is available, a portion of that water could be diverted into percolation ponds. In years when the supply of surface water is inadequate the groundwater supply can be relied on to meet part of the needs, and thereby increase the delivery capability of the system. The groundwater supply capability is limited by the number of wells and potential impacts of the drawdown on the storage characteristics of the groundwater basin.

The Kern Fan Element project is a direct recharge-extraction program on a 20,000-acre parcel in the predominantly agricultural area of Kern County.(1) DWR purchased this property in August 1988, and is now in the process of developing plans for the first phase of the project. The project, at its ultimate phase, will provide the SWP with a groundwater storage capacity of approximately 1.0 million acre-feet. Planning operation studies referred to in this paper assumed 30 thousand acre-feet per month effective recharge and extraction capacities.

In-Lieu Recharge and Exchange Programs

In this type of conjunctive use program water users who rely on both surface and ground water resources would receive more than their entitled share of surface water when the system's delivery capability is high, allowing the groundwater basins to recover. In exchange, these water users would defer receiving their full share of surface water entitlement when the system's delivery capability is low. In these type of years the participant water users would rely on alternative sources, mainly their underlying groundwater basins. This type of program is specially tailored to agricultural water users who rely on a combination of imported surface water and pumping of the groundwater basins underlying their lands to meet their total needs.

Semitropic Water Storage District (SWSD), located in the north-west portion of Kern County, has long-term contracts to receive a maximum of 158 TAF/yr of water from the SWP. In addition to this imported surface water, the groundwater basin provides a major source of irrigation water to meet the remaining water demand, which amounts to about 300 TAF/yr in some years. This water district is viewed as a potential candidate for entitlement exchange program. DWR planning studies for this project are presently under way.

Simulation Model of the CVP and SWP Systems

DWR's Division of Planning has developed a statewide computer model, DWRSIM, which simulates the operation of major reservoirs and conveyance facilities of the CVP and SWP systems.(2) The model is intended for use in planning type studies wherein alternative facilities and operation rules can be simulated and the impacts on SWP-CVP system water supply capability, reservoir storage levels, and instream flow conditions determined. Normally the studies are used in the comparative, rather than absolute mode to evaluate impacts with and without a new feature. Simulation runs of the SWP-CVP system can be made on an average monthly basis over a sequence of any number of years; however, current use has been limited to a maximum of 57 water years (October 1922 through September 1978).

The simulation of the SWP-CVP system is very detailed and complex. The model has to account for an extensive array of physical constraints, as well as legal and institutional agreements and statutes. These parameters include requirements for flood control storage, instream flows for fish and navigation, allocation of storage among system reservoirs, hydroelectric power generation, pumping plant capacities, minimum outflow at the Delta for maintaining required water quality parameters, etc. A more detailed description of the DWRSIM model, as well as other operation criteria of the SWP and CVP systems, are available in various publications of the DWR and USBR.

Methodology of Operation Studies

As mentioned above, planning studies with DWRSIM are most often used to compare the performance of the SWP system under a base condition and under a condition with a proposed feature of interest, may it be an additional facility, or a change in the operation rules. A typical planning question to be investigated is expressed as : "What would be the likely contribution of an added feature to the performance of the SWP system, at a given level of future development?" Performance indices always include the reliability of the system to meet its future water supply commitments. In the case of investigating conjunctive use options the changes in the operation of the surface water facilities is also an important index of performance, because benefits such as recreation use or hydroelectric power generation are associated with surface water facilities only. Delivery capability of the SWP system as simulated in DWRSIM is a function of the total available storage in all SWP conservation facilities. This functional relationship sets the required carry-over storage at the end of every water-year. The base condition simulation run is conducted for the study period of 57 years and the performance indices of interest are compiled.

The following is a brief discussion of how the conjunctive use program is integrated into the SWP system. As mentioned before, the annual delivery of the SWP is primarily determined by the amount of conservation storage in the system. Additional water stored in the groundwater facility is treated as a part of the conservation storage of the system. Therefore, higher storages resulting form the conjunctive operation permit higher annual water deliveries. Water for the recharge of the groundwater basin is treated as additional demand on the system and is diverted at the Delta by the Banks Pumping Plant. Water extracted from the groundwater basin is treated as additional inflow to the system. The decision to recharge or extract is, primarily, based on the delivery capability of the system. For instance, recharge at the highest possible rate may be warranted only when the system is able to deliver more than 90% of the water users' annual requests. In the same manner, extraction at the highest possible rate may be warranted when the system can not deliver more than 70% of the total annual requests. Operation criteria to trigger alternative levels of recharge and extraction have been incorporated in the model. Provisions have been made in DWRSIM to control the mode of operation, on a monthly basis, based on the storage in the surface water facilities. For instance, if the initial decision (the yearly decision based on the annual delivery) is to recharge at the highest level, it may be changed to a lower recharge rate, if the storage in Lake Oroville falls below 1.5 MAF, or changed to extraction at the highest rate if the storage falls further to 1.0 MAF. Operation studies are in progress to determine the parameters that lead to the most beneficial conjunctive use program.

Results of Operation Studies

As mentioned above, operation studies are in progress to evaluate the likely impacts of conjunctive use operations on the SWP system. The following examples show typical impacts to be expected on some of the significant variables.

Recharge-Extraction Option (Kern Fan Element Project)

Figure 1 shows the frequency of annual deliveries made to the SWP contractors in a 57-year simulation run. This figure shows that the incorporation of a 1.0 MAF capacity groundwater banking project could improve the reliability of the system by approximately 21 percent, while the annual delivery in the single most critical year is increased by more than 500 TAF. Figure 2 (p. 6) shows the monthly operation of the Kern Fan Element, along with the fluctuations of the total SWP conservation storage, with and without the groundwater banking project.

Entitlement Exchange Option (SWSD Project)

Figure 3 (p. 6) shows that an entitlement exchange program applied on 74,000 irrigated acres in the SWSD service area could bring about moderate gains in the delivery capability of the system over a wide range. The contribution of this project to the delivery capability during the driest period of record (1928-34) amounted to 110 TAF/yr, while the long-term average delivery increased by approximately 105 TAF/yr.

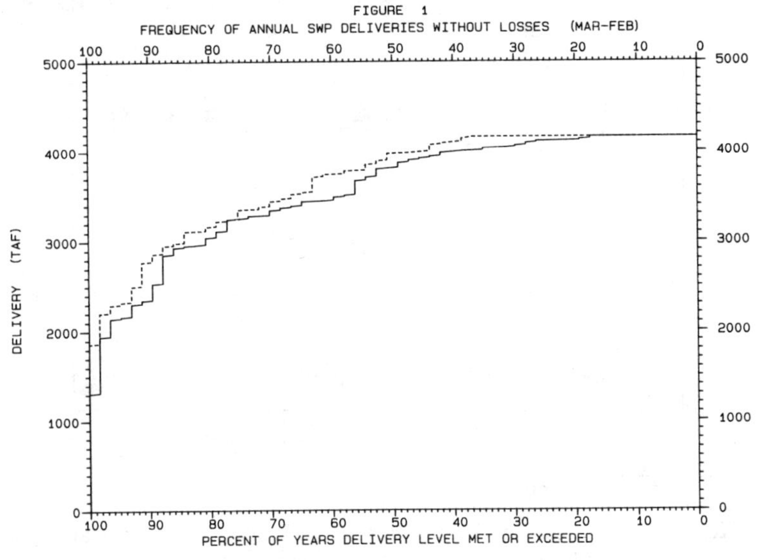

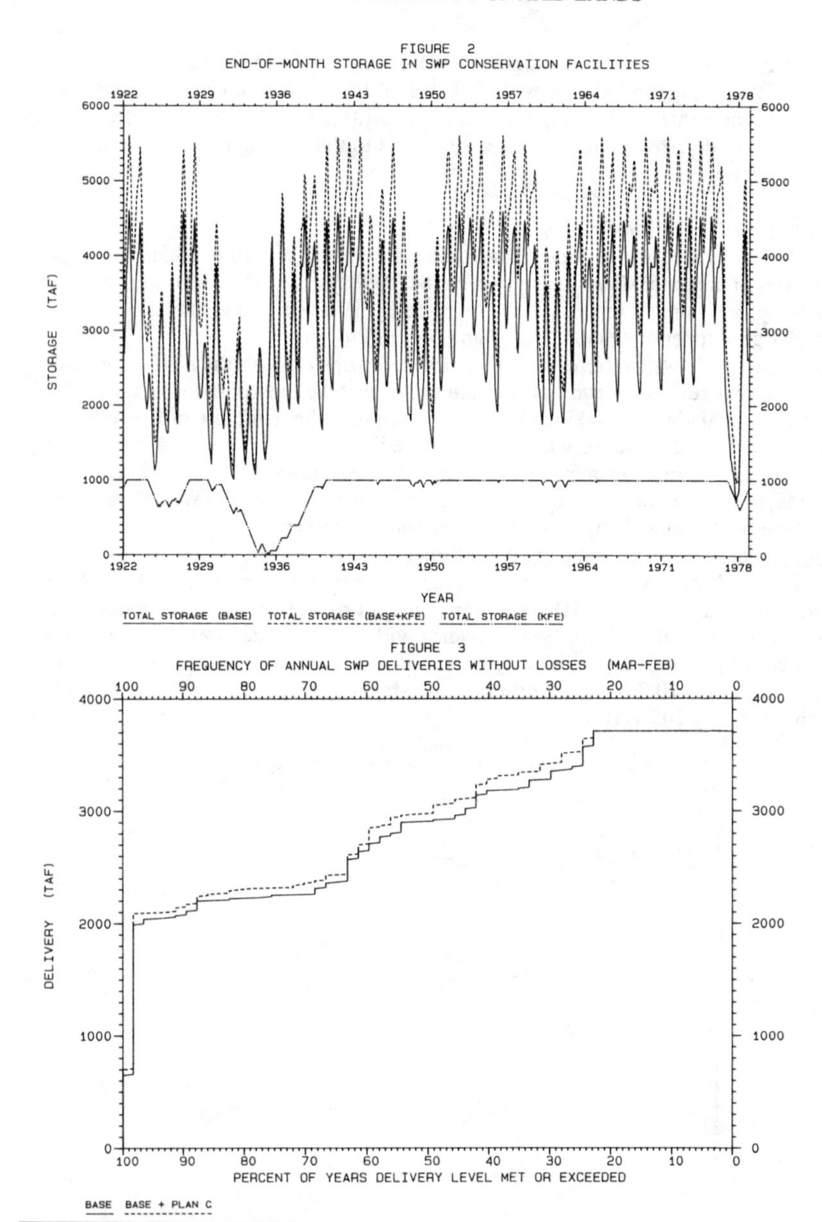

References :

(1) "Kern River Fan Element, Kern Water Bank", Preliminary Technical Report, Department of Water Resources, April 1987.

(2) "Operational Planning for California Water System", G. W. Barnes, F. I. Chung, *Journal of Water Resources Planning and Management*, 112(1), 1986.

Feasibility of Stormwater for Recharge in the Las Vegas Valley

By: Virginia Bax-Valentine, P.E.[1], Member, ASCE
Lazell Preator[2]
John Hess, PhD[3]

Abstract

Rapid growth in the Las Vegas Valley has increased the pressure on the limited water supply. The increased water demand has stimulated public interest in water conservation and the potential for stormwater recharge. Many communities in less arid regions have implemented programs to utilize stormwater runoff. In the Las Vegas Valley, which receives less than four inches of rainfall annually, the concept is new and poses several unique considerations, including extremely low volumes of runoff, aquifer depth, and storm frequency and distribution, high evaporation rates, low permeability, water quality, and geologic conditions. In this paper current stormwater recharge practices in arid regions will be discussed. The feasibility of converting detention basins into recharge facilities will be explored and the mechanics, limitations, and economic factors of stormwater recharge in the Las Vegas Valley will be evaluated.

Introduction

Las Vegas is the second fastest growing area in the Country. It is estimated that 780,000 people now live in the urban area and over 4,000 people per month move to Las Vegas. The phenomenal growth has resulted in increased water demand while the supply from the Colorado River and ground water have remained constant.

Comparisons of future water demands and the available supply have resulted in a growing interest in conservation and new sources of water. One as yet unclaimed source is stormwater runoff. Because of the low volumes and unpredictable nature of

1 General Manager, Clark County Regional Flood Control District, Nevada
2 Student Engineer, Clark County Regional Flood Control District, Nevada
3 Director, Water Resources Center, Desert Research Institute, Las Vegas, Nevada

meteorological events in the southwest artificial ground-water recharge is probably the most feasible method of utilizing stormwater. In this paper the feasibility of converting flood control detention basins to recharge facilities will be explored. Other southwest communities have used or propose to use stormwater for recharge, though to date none use stormwater exclusively. The following communities have recharge programs; Fresno, since the 60's, AAR[4] 10"; Phoenix, abandoned, AAR 9"; Los Angeles, since the 40's, AAR 12.5"; San Diego, current, AAR 15"; Maricopa County, Arizona, proposed system, AAR 8" (Randall).

In this study, the three existing Las Vegas Valley Detention basins were evaluated. Two are located on the alluvial areas surrounding the metropolitan area, and one is located in the urban area. A screening process based on water quality impacts, potential yield, and public health considerations was used to establish whether more in-depth study was warranted.

Description of Study Area

The Las Vegas Valley is a 1518mi^2 depression 35mi northwest of Lake Mead and the Colorado River in Clark County, Nevada. The Valley lies between the Great Basin section and the Mojave Desert section of the Basin and Range Physiographic Province. The Valley is surrounded by the Spring Mountains to the west, the Sheep Range and the Las Vegas Range on the north, Frenchman and Sunrise Mountains on the east, and the McCullogh Range and the River Mountains on the south. The highest point in the watershed is Charleston Peak at 12,000' above sea level and the lowest is Las Vegas Wash at 1450' above sea level (Noack).

In 1988, 268,000 acft of water was used in the metropolitan area. Of that 75%, or 201,000 acft, was supplied from the Colorado River. The remaining 25% or 67,000 acft was groundwater. The LVVWD[5] in 1987 estimated that the natural recharge rate to the aquifer was approximately 30-35,000 acft per year. Nevada's allocation of Colorado River water is 300,000 acft per year, and credits are allowed for return flow. Ground-water mining has lowered the watertable, caused the collapse of the previously saturated clays, and has resulted in subsidence of some areas.

The Las Vegas Valley is one of the more arid regions in the United States. Located in the Mojave Desert, the average annual rainfall is 4.2 in, with a temperature variation from 115°F to 16°F. The Valley floor is comprised of semi-arid coalescent alluvial aprons surrounded by sub-humid mountains. The Spring Mountains receive up to 28 inches of rain per year. This orographic feature contributes to the Valley's arid climate.

Basins and Methodology

The Meadows Detention Basin is located in the City of Las

4 Average Annual Rainfall
5 Las Vegas Valley Water District

Vegas just west of the downtown area. It has a capacity of 270 acft and a surface area of 34 acres. The service spillway has a capacity of 136 cfs and the emergency spillway has a capacity of 990 cfs. The structure is a zoned earth embankment, with a soil cement section in an area of potential erosion. The Meadows Basin lies within the LVVWD's main well field and is approximately 50' from the nearest producing well.

According to the LVVWD the principal aquifer is overlain with 200-300' of clay at the Meadows site. Recharge at this site would require direct discharge into the aquifer because of the overlying impervious soils. Due to the poor quality of runoff from the urbanized contributing watershed, and the close proximity to domestic wells, direct recharge would not be permitted at this location.

The North Las Vegas Detention Basin lies within the northern reach of the Upper Las Vegas Wash. It has a capacity of 1,650 acft and an impondment area of 114 acres. The earthen embankment has an outlet capacity of 4,500 cfs and an emergency spillway designed for 9,000 cfs.

Geotechnical investigations for a proposed future expansion of the basin indicate very fine silty soils over varied calcrete at the site. Industrial works upstream have had a small but definable impact on surface water quality at the site (Leskys, 1989).

This basin is an off-channel facility, and conversion of this facility would require allowing frequent flood events to bypass the structure. Sediment transport of the fine-grained soils in the upper watershed could further complicate the use of this site. This basin was eliminated from further consideration due to potential water quality problems, the presence of calcrete deposits, and the previously mentioned sediment considerations.

The Red Rock Detention Basin lies on the western fringe of the Las Vegas Valley in Clark County. It is a 1,700 acft facility with a surface area of 84 acres. It has a 1350 cfs capacity service spillway and a 60,000 cfs emergency spillway. The structure is a 63 ft high zoned earth embankment.

This basin is believed to be over a major portion of the regional carbonate aquifer (Noack). The water table has been estimated to be 150 ft below the surface at the apex of the fan where the basin is located (McDonnell, 1989). The alluvium is underlain by moderately hard to hard calcrete formations with a surface percolation rate of 1×10^{-3} (Leskys, 1986).

Infiltration wells were determined to be the most effective method for recharge. The model well was modified from Hannon's. The screen would begin at the static water table (150') and extend to 450' below the surface.

A TR-55 model was used to determine runoff volumes for a series of 6-hour storms. Point depths were adjusted using depth-area ratios. A curve number of 90 was used for the 53 mi^2 drainage area. It was determined that a 2-year storm was the most frequent event which would fill the basin to a level high enough to enter the recharge chamber. The openings were elevated to prevent sediment deposition from plugging the chambers. With the outlet blocked, a 25-yr storm would fill the basin.

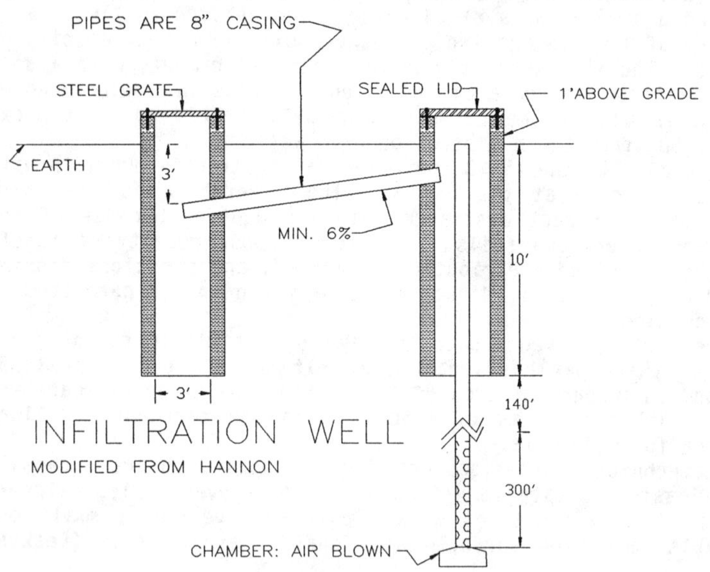

The annual potential yield was estimated by integrating the volumes between the most frequent event, the 2-year storm and the maximum event, the 25-year storm. Events with a less frequent return period would be discharged through the spillway. Less frequent events would also yield the 25-year recharge volume, but were considered sufficiently rare to be ignored for this analysis. Curvilinear regression was used to get the equation of the line $y'=a*b^x$ from $\log y'=\log a+x*\log b$. A reduced varient conversion was needed for x; x=-LN(-LN(1-Pe)).

Using the equation $y=263(1.51)^x$, the area under the line was integrated by the equation:

$$\int_{TR=2}^{TR=2.5} y = \int_{1/2}^{1/25} 263(1.51)^x dx$$

To determine the potential recharge volume, evaporative losses and losses through the basin bottom and dam were subtracted from the available volumes. A value of 6.7 acft/ac/yr was used for evaporation giving a 10 day loss of approximately 15.5 acft. Darcys Equation, Q=KiA, was chosen to evaluate basin losses. In this evaluation $K=1 \times 10^{-3}$, estimated by Converse Consultants, and i=1/1 were used. This methodology is known to be an oversimplification and Q may be less by magnitudes. For the purposes of this feasibility level study these simplifications were deemed justified. The losses for the basin and dam were approximately 1,200 acft over 10 days. The remaining 484 acft was used to determine the number of infiltration wells required.

The wells were rated using the equation:

$$Q_{gpm} = \frac{Kb(hw-H_o)}{528\log(r_o/r_w)}$$

The radius of influence (r_o) was chosen to be 200' and Kb=T=1000. Each well would recharge 142 gpm or 6.3 acft in 10 days. A 10-day recharge period was selected based on local data for vector development. It was determined that seventy-eight dry wells would be needed. Each dry well was estimated to cost $30,000. Combined with other appurtment costs, the total cost for the recharge facilities was estimated to be $2.5 million.

Using this methodology the average annual yield was estimated to be 135 acft/yr at the Red Rock Basin. Assuming a 100-year design life for the facilities and an interest rate of 7%, the annual cost was estimated to be $150,450.

Water in Las Vegas currently costs 73¢ per thousand gallons (Brothers). The cost to capture the recharged stormwater is approximately $4.00 per thousand gallons; more than 5 times the current cost. The best measure of cost-effectiveness would be to compare the cost of stormwater recharge to the cost of other new sources of water such as importation from other basins. Importa-ion of water from other areas is presently under investigation, however, no cost estimates are yet available.

Conclusions and Recommendations

Neither the Meadows nor the North Las Vegas Detention Basins should be used for stormwater recharge due to potential water quality problems. Furthermore, at this time it is not economically feasible to use this type of recharge in the Red Rock Detention Basin.

At such time as recharge is seriously considered, several factors should be addressed in more depth. Water and soil chemistry were not evaluated in this analysis but could be limiting factors in some areas. Mixing two chemically incompatible waters may produce $CaCO_3$ deposits or other clogging precipitates (Brothers).

Reactive soils, although mostly limited to the southeast portion of the Valley, may be present. Unless discharged directly into the primary aquifer recharge may aggravate perched zones which already cause foundation problems in some areas.

Public health and safety factors are also important considerations. Within a few years the area surrounding the basins will probably be developed. Mosquitoes and other insects will be a concern, therefore, the 10-day or less drawdown period must be used (Hicks). Most of the existing detention basins were designed to detain water for 32-72 hours. The hydrograph through the embankments should be examined to determine the effects of longer detention times on the structures.

Maintenance cost must also be factored into any future analysis. Both sediment and other debris must be cleaned out regularly to keep insect and rodent populations to a minimum and to protect water quality. Because the basin will retain water more frequently for longer periods of time, multiple use recreational opportunities at the sites may be lost. Additionally, infiltration wells would have to be obviously marked and maintained for safety purposes.

Water costs are certain to continue to rise, and demand is expected to exceed the supply in the next decade. The cost of importation from other areas is expected to be extremely expensive compared to current costs. At the time importation becomes necessary, it is likely that stormwater recharge will become a viable supplement to existing water supply.

Selected Bibliography

Brothers, T., and Katzer, K., (1989), "Water Banking through Artificial Recharge, Las Vegas, Clark County, Nevada," Las Vegas Valley Water District, Las Vegas, Nevada.

Brothers, K., Las Vegas Valley Water District, 1/31/90 phone

Colorado River Commission, (6/22/89), "1988 Statistics-Colorado River Water Diverted for use in Southern Nmvada," CRC, Las Vegas, Nevada.

Hannon, J. B., (1983), "Underground Disposal of Storm Water Runoff; Design Guidelines Manual," United States Department of Transportation, Washington, D.C.

Hicks, R., Clark County Public Works, Vector Control Department - 27 February phone

Leskys, A. G., (1986), "Geotechnical Investigation, Red Rock Wash RR-2 Detention Basin Site, Clark County, Nevada," Converse Consultants Southwest Inc., Las Vegas, Nevada.

Leskys, A. G., (1989), "Preliminary Geotechnical Evaluation for Facility Planning, Upper Las Vegas Wash Flood Control Projects, North Las Vegas, Nevada," Converse Consultants Southwest Inc., Las Vegas, Nevada.

McDonnel-Canan, C., (1989), "Morphology and Development of the Red Rock Canyon Alluvial Fan, Clark County, Nevada," Masters Thesis, University of Nevada, Las Vegas.

Nightingale, H. I., (August 1977), "Accumulation of As, Ni, Cu, and Pb in Retention and Recharge Basins Soils from Urban Runoff," *Water Resources Bulletin*

Noack, R. E., (1988), "Sources of Ground Water Recharging the Principal Alluvial Aquifers in Las Vegas Valley, Nevada," Masters Thesis, University of Nevada, Las Vegas.

Randall, R. A., (1988), "Groundwater Recharge Feasibility Investigation-Maricopa County, Arizona," CH2M Hill Inc./Erol L. Montgomery & Assoc.

Salo, J. E. et al, (1984), "Fresno Nationwide Urban Runoff Program Project-Final Report," Fresno Metropolitan Flood Control District, Fresno, California.

HARVEST OF EPHEMERAL RUNOFF FOR ARTIFICIAL GROUNDWATER RECHARGE IN ARID BASIN WATER MANAGEMENT

G.F. Cochran, Member ASCE[1]

ABSTRACT

Harvest of springtime flow to a playa in Lemmon Valley, north of Reno, was evaluated as a means of increasing local groundwater supplies through artificial recharge. Surface water and groundwater studies in the donor area estimated the long–term harvestable volumes. Hydrogeologic and geochemical models of the receiving area were used to determine the effects of artificial recharge. Costs for the harvest/recharge program were estimated to compare favorably with alternative water supply solutions.

INTRODUCTION

The high cost of land and housing in the Reno, Nevada area during the rapid population growth of the early 1970's resulted in residential development in nearby valleys outside of the Truckee River basin. The closest of these is the 241 km^2 Lemmon Valley which has no major perennial streams that reach the Valley floor and only a limited annual groundwater recharge. Hydrographically, Lemmon Valley is divided into two sub–basins by a low topographic divide. The two sub–basins, East Lemmon and Silver Lake, each contain playas which are partially flooded each spring.

Significant development occurred in Lemmon Valley in the form of rural, large–lot housing subdivisions. In many of these subdivisions, homes are served by individual domestic wells and septic tanks. Except for the Stead area of Lemmon Valley, all residential and commercial water comes from the alluvial aquifers. Total groundwater commitment represented by private domestic wells and water utility wells is approximately 10 times the estimated annual recharge.

Approximately 75 percent of Lemmon Valley's 15,000 residents derive their water supply from the groundwater reservoir(s) and of these, approximately half are dependent upon private wells. The other half are provided with

---

[1] Water Resources Center, Desert Research Institute, University of Nevada System, P.O. Box 60220, Reno, Nevada 89506

groundwater by water utilities from a limited number of large capacity wells. The other 25 percent of the valley's residents receive surface water imported from Truckee River through a pipeline constructed in 1940 to the Stead area. Legal constraints prohibit expansion of the pipeline capacity.

HARVEST/RECHARGE STUDY DESIGN

Exceptionally large spring runoff in 1982 and 1983 resulted in significant flooding of the Silver Lake and East Lemmon Valley playas by runoff from Peavine Mountain which rises to an elevation of 2,440 m amsl on the southeastern boundary of the valley. These large runoffs stimulated interest in the possibility of harvesting the ephemeral streamflows from Peavine Mountain for artificial recharge into the valley's aquifers. A preliminary feasibility study (Cochran et al., 1984) indicated that there was probably sufficient good quality water available on a sustained basis, and that costs of a harvest/recharge program would be favorable compared to alternative water supplies. Based on these conclusions, a more detailed investigation was implemented (Cochran et al., 1986). Because the declining water level problem is most serious in the Golden Valley area of southeastern Lemmon Valley, it was selected for detailed study.

Harvestable Water

To validate the potential amount of harvestable water available from the Silver Lake sub–basin, three representative drainages on Peavine Mountain were instrumented. A staff gauge was installed in Silver Lake and a bathymetric survey was conducted. Supplemental streamflow measurements were also made. Water samples were collected at locations ranging from the top of Peavine Mountain down to Silver Lake playa for analysis of chemical and isotopic composition. The network was monitored during water year (WY) 1984 (Sept. 1983 through Aug. 1984) and WY 1985, both of which produced less than average precipitation. The resulting data were used to develop a water balance model for Silver Lake playa and a "Discrete State Compartment" groundwater flow model of the southern portion of the Silver Lake sub–basin (Dale, 1986).

Hydrologic data analysis indicated that a substantial portion of the groundwater recharge and surface runoff from Peavine Mountain is discharged back to the atmosphere on Silver Lake playa (Cochran et al., 1986). Full development of the estimated potential yield from watersheds tributary to Silver Lake playa would, at maximum, salvage only 60 to 70 percent of the estimated Silver Lake playa discharge. Diversions from two of the drainages would probably reduce the discharge from a developed spring area and thus, adversely affect the local water supply. Elimination of natural discharge should not, however, preclude the ability to develop and maintain pumped wells to replace the lost springflow.

Golden Valley Hydrogeology

Golden Valley encompasses an area of approximately 9.3 km^2 and ranges in elevation from 1,524 m a.m.s.l. along its western margin to more than 1,783 m at the highest point on the northeastern drainage divide. Precipitation is estimated to range from 250 mm/year^{-1} on the valley floor to over 355 mm year^{-1} on the northeast peaks. There are no active perennial or ephemeral streams in Golden Valley.

The Golden Valley aquifer system is composed of two aquifers: a) a bedrock aquifer; and b) an alluvial aquifer. At various depths in the alluvial aquifer, driller logs indicate lenses of decomposed granite interlayered with coarse sand, silts, and gravels. Alluvium reaches a maximum thickness of approximately 37 m in the central part of the valley and pinches out as it approaches exposures of bedrock.

Four types of fractured crystalline rocks outcrop on the highlands and are buried beneath the valley fill. In the north and south are outcrops of granodiorite; in the east are outcrops of quartz monzonite; in the southeast are rhyolitic tuffs; and in the southwest a "finger" of metamorphic sequence occurs. These crystalline rocks have been extensively fractured and readily yield water to wells.

Conductivities of these two aquifers were estimated from bailer tests of wells screened solely within either the alluvial fill or the fractured zone. The mean hydraulic conductivity of the fractured bedrock and the alluvial aquifers were respectively calculated to be 0.7 and 0.9 m day^{-1}. No significant correlation with depth was found. Despite similarities in the conductivities, the specific yield of the fractured bedrock is significantly lower than that of the alluvial fill.

Much of the natural recharge occurs in the eastern and northeastern portions of the valley along the weathered granitic outcrops. The surficial deposits in the central portions of the valley are also decomposed granite, but contain a higher percentage of fines and thus are less permeable. Along the margins, however, there is ready access by infiltrated water to the fractured and weathered granitic aquifer. Significant additional recharge occurs throughout the valley from the septic tanks serving each of the nearly 400 homes.

Based on a water budget analysis, from 1973 to 1984 a total of approximately 1,420 x 10^3 m^3 had been pumped from Golden Valley groundwater storage. Groundwater level declines in the western portion of the valley, for the periods 1972–1975 and 1978–1980, averaged approximately 0.43 m year^{-1}. However, because of the 1976–1977 drought, the average decline from 1972 to 1982 was nearly 0.85 m year^{-1}. The extremely wet conditions during the winters of 1982–83 and 1983–84 caused a water table rise of nearly 2 m. Storage depletion is not uniformly distributed, but is concentrated near the center of pumpage and in pumped areas of low transmissivity. Observed water level declines have ranged from less than 0.3 m year^{-1} to in excess of 3 m year^{-1}. Many domestic wells have been deepened to chase the falling water table.

Golden Valley Geochemistry

Water samples for chemical analysis were collected from 59 wells during the summer of 1984. These data were combined with available historic chemical analyses to characterize the valley's water quality and to evaluate potential chemical effects related to artificial recharge with either harvested Peavine Mountain water or Truckee River water (Barry, 1985).

Groundwater in Golden Valley is generally chemically dilute with the electrical conductivity (EC) ranging from 248 to 895 μmhos cm^{-1}. Water associated with the fractured granitics and alluvium is Ca–HCO_3 water and water associated with the fractured rhyolites in the southeast corner of the valley is a Na–SO_4 water. A number of samples, especially in the fractured granitics, have high concentrations of NO_3 and Cl which indicate septic tank effluent. Waters containing low concentrations of TDS, Cl and SO_4 are found in the fractured granodiorites, reflecting natural recharge. Groundwater in the alluvium overlying fractured granitics has relatively low Cl and SO_4 concentrations. Groundwaters associated with the fractured rhyolites are significantly higher in SO_4 and range widely in Cl concentration. Some of the wells in the southwest portion of the valley that penetrate the metamorphic rocks have high SO_4 concentrations.

Mixing of Golden Valley groundwater with water from both Peavine Mountain and Truckee River was simulated with a chemical model to predict mineral phase that would be likely to precipitate or dissolve. Two mixing scenarios were developed for recharge with Peavine Mountain water and one for recharge with Truckee River water. The worst case scenario was based on mixing Peavine water with poor quality water in the southeast corner of the valley. The average case scenario was based on mixing Peavine water with water representative of the average groundwater composition in the valley.

The greatest chemical clogging potential from artificial recharge is in the southeast corner of the valley. Groundwater in this area is high in iron and sulfate and is reducing in some areas. However, given the initially low iron concentrations in most areas of the valley, clogging by iron hydroxide precipitation should not be a significant problem if the injection sites are located away from the rhyolites. Optimal recharge locations, based on water chemistry considerations, were determined to be in the fractured granitics and alluvium in the northeast portion of the valley.

Flow System Response To Artificial Recharge

A transient state, three–dimensional finite difference model was selected for analyzing the Golden Valley flow system (Cochran et al., 1986). The drainage boundary along all but the western edge of Golden Valley was defined as a no–flow boundary. Groundwater which flows out of Golden Valley along the western margin to the East Lemmon Valley sub–basin was simulated as a line of drains along that boundary. The base of the aquifer was assigned a no–flow boundary condition. Fractures were encountered by the deepest wells in the valley at a depth in excess of 91 m into the bedrock and thus, the bottom of the aquifer was set at a depth 91 m below the top of the fractured bedrock.

Specific yield was modeled as ranging from 0.01 in the fractured granitics and rhyolites, to 0.30 in the valley fill. Bedrock outcrops and shallow alluvium were modeled as having specific yield between 0.01 and 0.10. The alluvial fill, with its large variability in composition, displays a similar variability in specific yield. Calibrated specific yield of the alluvium ranges from 0.10 to 0.30. In the south–central portion of the valley, the values drop to 0.01 and 0.07 in the area of the "finger" of metamorphic rocks.

Model calibration was achieved by varying hydraulic conductivities and values of specific yield using 1971 water levels as initial conditions. The model was stepped forward in one year time steps to the spring of 1984 when another set of water levels was available. The model was verified using water levels measured in September 1984 and May 1985 and by comparison with the declining water levels measured in an observation well in the western portion of the valley. Use of average pumping rates failed to produce the observed seasonal water level fluctuations.

After verification, a water balance for the 1984 simulation was conducted. Groundwater pumpage for 1984 was modeled as approximately 450 x 10^3 m^3. Subtracting the western boundary outflow and pumpage from the volume recharged resulted in a deficit of approximately 360 x 10^3 m^3, which was removed from storage.

Projected Water Levels Without Artificial Recharge

The numerical model was used to simulate future water levels given four residential growth scenarios. These scenarios are as follows.

a) "No Growth": No further development of any kind occurs.

b) "Constant Growth": Seven percent of the presently undeveloped residential home sites are developed each year until the year 2000.

c) "Instant Growth": All of the subdivided lots are developed by 1990.

d) "Total Development": Every available space in the valley, including the presently unsubdivided lots, is developed by 1990.

Under "No Growth", water levels in Golden Valley would decline at a slow rate to in excess of 6 m by the year 2000. These declines would occur for the most part in the central portions of the valley where much of the pumping of the residential wells takes place. Under "Constant Growth", the areas of declining water levels would be greatly expanded. Declines in excess of 24 m would be experienced in the central portion of the valley. "Instant Growth" water level declines would continue to increase in magnitude such that declines in excess of 30 m would occur. "Total Development" would cause the areas of declining water levels to expand into areas where new residential development would occur and result in extensive areas with declines in excess of 30 m. All of these simulations indicated that by the year 2000, many of the valley's homeowners would be required to deepen their wells in order to secure an adequate water supply.

Effect of Artificial Recharge

Numerical simulations of several artificial recharge scenarios were conducted to determine effects of recharge operations. Possible annual recharge rates were assumed to range from 62 x 10^3 to 620 x 10^3 m^3 from March through June, followed by an eight–month rest period. All of the recharge simulations were performed assuming the "Constant Growth" scenario, with the further assumption that recharge facilities would be coming on–line in the spring of 1987. When both chemical and hydrological factors are considered, the most favorable recharge site is in the northeast area of the valley. This artificial recharge location will not, however, directly benefit the southern portion of the valley.

Through simulation, it was determined that a rate in excess of 37 x 10^3 m^3 per season per well might cause the water table to rise to within 3 m of the land surface. High water table conditions can adversely affect treatment performance of septic tank leach fields. For this reason, a series of low capacity wells should be used. Three wells were simulated across the northeastern portion of the valley, with a maximum injection rate per well of 3.2 l s^{-1} (33 x 10^3 m^3). This simulation indicated that water level declines would be 3 to 6 m less in the central and northern portion of the valley than without recharge. Thus, with this artificial recharge scenario, water levels in the central and northern portion of the valley would be stabilized at near the 1984 level. Water levels in the southern portion of the valley would continue to decline but at a lower rate, experiencing up to 3 m less decline than would occur without recharge operations.

REFERENCES

Barry, J.M., 1985. "Hydrogeochemistry of Golden Valley, Nevada and the Chemical Interactions During Artificial Recharge", unpublished M.S. thesis, University of Nevada–Reno.

Cochran, G.F., M.N. Dale and D.W. Kemp, 1984. "Peavine Mountain Water Harvest: Preliminary Feasibility Report". Water Resources Center, Desert Research Institute publication #41094. University of Nevada–Reno.

Cochran, G.F., J.M. Barry, M.W. Dale and P.R. Sones, 1986. "Water Harvest from Peavine Mountain with Artificial Recharge in Golden Valley, Nevada: Hydrologic Feasibility and Effects". Water Resources Center, Desert Research Institute publication #41103. University of Nevada–Reno.

Dale, M.W., 1986. "Hydrology of Peavine Mountain and Southern Part of Silver Lake Sub–basin, Lemmon Valley, Nevada", M.S. thesis, University of Nevada–Reno.

Management Strategies for Baseflow Augmentation

Victor M. Ponce[1] and Donna S. Lindquist[2]

Abstract

Management strategies for **baseflow augmentation** fall into the following categories: (1) livestock management, (2) upland vegetation management, (3) riparian vegetation management, (4) upland runoff detention and retention, and (5) the use of instream structures. The benefits of a watershed management strategy focused on baseflow augmentation are the following: (1) increased summer flows, (2) healthier riparian areas, (3) increased channel and bank stability, (4) decreased erosion and sediment transport, (5) improved water quality, and (6) enhanced fish and wildlife habitat.

Introduction

Baseflow augmentation refers to the temporary storage of subsurface water in floodplains, streambanks, and/or stream bottoms during the wet season, either by natural or artificial means, for later release during the dry season to increase the magnitude and permanence of low flows. The importance of streambank storage and its effect on stream hydrology, ecology, and geomorphology is becoming increasingly apparent to a broad spectrum of practitioners, including hydrologists, engineers, ecologists, and watershed managers. The temporary storage of precipitation in soils adjacent to streams, for later release during the dry summer months, can directly benefit many stream uses and users.

Hydrologic Aspects of Baseflow Augmentation

Moisture exists in the soil under either unsaturated or saturated conditions. In the unsaturated zone, the preferred path of movement of moisture is vertical, by percolation, toward the saturated zone. In the saturated zone, the preferred path of movement of moisture is horizontal, toward aquifer discharge areas. Sustainable low flows in streams are generally due to aquifer discharge. Therefore, baseflow augmentation may result in the conversion of ephemeral and intermittent streams into perennial streams.

Given the proper lithology and relief, it may be possible for unsaturated flow to contribute to streamflow. However, sustained low flows appear to be possible only in streams that remain effluent throughout the dry season. For a stream to remain effluent, the following conditions should be met:

1. The draining aquifer should be recharged seasonally with adequate amounts of moisture originating in natural and/or artificial sources.

---

[1]Professor and Chair, Department of Civil Engineering, San Diego State University, San Diego, CA 92182.

[2]Research Scientist, Pacific Gas and Electric Company, Department of Research and Development, San Ramon, CA 94583.

2. The watertable should be shallow enough to be intersected by the stream.
3. The aquifer's size and hydraulic properties should be conducive to the maintenance of flows throughout the dry season.

Generally speaking, adequate aquifer replenishment leads to shallow groundwater tables. In turn, shallow watertables lead to effluent, i.e. perennial streams. Therefore, adequate aquifer replenishment should cause streams to flow year-round. For a given climate, the larger the fraction of precipitation that is allowed to infiltrate, the more likely it is that the infiltrated water will eventually go on to replenish the local groundwater reservoirs. While aquifer replenishment is subject to management, the hydraulic properties of aquifers are largely determined by nature. Therefore, it is possible to accomplish baseflow augmentation by increasing seasonal aquifer replenishment. The aquifer's size and hydraulic properties can be used to identify those aquifers which can be readily managed for baseflow augmentation. In general, large and relatively slow-draining aquifers are good candidates for baseflow augmentation. On the other hand, small and relatively fast-draining aquifers do not show promise for baseflow augmentation.

Vegetative Aspects of Baseflow Augmentation

Riparian areas generally support a great diversity of riparian vegetation. The relationship between **baseflow augmentation and riparian vegetation** is unclear at this time, despite the many efforts to document the link between them (Heede, 1977; Stabler, 1985, Elmore and Beschta, 1987). A plausible scenario is the following: Increased amounts of subsurface moisture in streambanks, resulting from natural and/or artificial aquifer replenishment, encourage the establishment and growth of riparian vegetation and assure its survival from year to year. In turn, the riparian vegetation increases sediment deposition, soil infiltrability, soil-moisture retention capacity, and reduces stream velocities, further increasing the rate of subsurface moisture replenishment during high flows. Increased subsurface moisture replenishment then leads to saturated groundwater flow and to groundwater accretion, and raises the watertable near the streambanks. With an aquifer of the proper geometric and hydraulic properties, the rise of the water-table near the streambanks can change the character of the adjoining stream from intermittent to perennial. Moreover, the magnitude and duration of summer streamflows is a function of the aquifer properties and of the effectiveness and amount of replenishment.

Case Studies of Baseflow Augmentation

Baseflow augmentation has been accomplished in a few documented cases. The experiences of Camp Creek, Oregon; Sheep Creek Barrier Dam, Utah; and Alkali Creek and Trout Creek, Colorado, are reviewed here.

Camp Creek, Oregon

Camp Creek is a tributary of the Crooked River, in semiarid central Oregon. Before extensive settelement in the mid-nineteen century, the area drained by Camp Creek was a grassy wetland, with predominantly grassy vegetation and a shallow groundwater table maintained within root reach throughout the year. The grassy vegetation encouraged soil infiltrability and made possible an effective aquifer replenishment from year to year. Around the turn of the century, overgrazing of the meadow led to increased quantities of surface runoff and incipient gully development. Under these conditions, severe floods triggered accelerated gully erosion and led to the development of the deeply incised channel with almost vertical banks which now cuts through the Camp Creek valley. In time, the inadequate aquifer replenishment led to the draining of the meadow and the lowering of the watertable to the point where it remained permanently out of reach of most herbaceous vegetative species. Over a period of several decades, this sequence of events led to the transformation of the Camp Creek meadow from a grassy wetland into an area that could support only dryland species, with sagebrush taking over as the predominant type of vegetation. Beginning about 1968, increasing lengths of Camp Creek were fenced to exclude livestock, with fencing continuing to date (Elmore and Beschta, 1987). Over the years since the original fencing, the exclusion of livestock from the creek has allowed the development of a healthy riparian area. This has led to a buildup of the streambed and streambanks, a deeper and more stable low flow channel, and a reduction in the amount of sediment being transported by the stream. A byproduct of the Camp Creek fencing has been the substantial **increase in summer flows** within the exclosure.

Sheep Creek Barrier Dam, Utah

The Sheep Creek Resource Conservation Area Project was implemented from 1957 to 1966 to stabilize and rehabilitate the upper watershed of Sheep Creek, a tributary of the Paria river, in southern Utah. Measures included the construction of detention dams, dike water-spreader systems, gully plugs, check dams, seeding, and intensive grazing management. As part of the project, the Bureau of Reclamation built a large sediment detention dam on Sheep Creek. This dam has been very successful in trapping sediment eroded from gullies and other headwater sources. In 1961, only one year after its completion, the reservoir was completety filled with sediment to spillway crest elevation. Over the ensuing years, sediment has continued to accumulate behind the dam, creating a sediment wedge. More than 75% of the sediment accumulated behind Sheep Creek dam is now above spillway crest (Van Haveren et al, 1987). Flooding occurs on the sediment wedge for all significant runoff events, recharging the aquifer. A perennial flow at the dam has resulted from the water draining slowly from the accumulated sediment deposits. The size and recharge characteristics of the sediment wedge are apparently **sufficient to maintain perennial flow** at the Sheep Creek Barrier dam site.

Alkali Creek, Colorado

Alkali Creek is located in the White River National Forest, in western Colorado. In 1958, the Forest Service initiated the Alkali Creek Soil and Water Conservation Project, in response to increased land use pressure, as shown by accelerated gully development throughout the first half of the twentieth century. The gully growth is attributed to the combined effect of drought and overgrazing, as well as the overuse of agricultural lands below the watershed, which led to greater channel incision and lowering of the base level.

In 1963, the Forest Service constructed 133 check dams in about half of the gullies located within the project area (Heede, 1977). Seven years later, the previously ephemeral flow became perennial, although there was no noticeable change in average annual precipitation during the period. The change from ephemeral to perennial flow is attributed to the establishment of vegetation in the gullies and to the subsequent recovery of baseflow. The amount of moisture stored in streambanks and stream bottoms at or near the check dams are apparently **sufficient to maintain perennial flow** at the Alkali Creek watershed outlet.

Trout Creek, Colorado

The Trout Creek watershed is located within the San Isabel National Forest east of Buena Vista, Colorado. Following settlement in the mid-nineteenth century, the Trout Creek watershed was subjected to intensive exploitation for timber, ranching, and farming, which eventually led to the development of an extensive gully network. In 1933, the Trout Creek Civilian Conservation Corps was established to carry out rehabilitation works in Trout Creek. This activity resulted in the construction of 9 concrete dams, 53,996 temporary dams (rock, log, and brush-wire), 164 ha of gully seeding and sodding, 4 km of terracing, and 2,400 ha of tree planting, with 38% considered successful (Jauch, 1957). These varied soil and water conservation strategies aided in the restoration of the vegetative cover, stabilization of soil and water, and in raising the watertable to its original (or pre-impact) condition. Moreover, the sediment deposits accumulated at the gully-control structures have acted as artificial aquifers, storing moisture for eventual release as baseflow. Many of the Trout Creek drainages are now either completely healed or well along in the process of healing, pointing to the success of the combined structural and vegetative treatments implemented in the Trout Creek watershed.

Management Strategies for Baseflow Augmentation

Management strategies for baseflow augmentation fall under the following categories: (1) livestock management, (2) upland vegetation management, (3) riparian vegetation management, (4) upland runoff detention and retention, and (5) the use of instream structures.

Livestock Management

The hydrologic effects of livestock grazing have been recognized for several decades. Poorly managed livestock grazing affects watershed response by the removal of protective plant covering and by excessive trampling. Removal of vegetation has the following effects: (1) It increases the impact energy of raindrops, encouraging splash erosion and dislodgement of soil particles from the surface; (2) it decreases soil organic matter, decrasing soil infiltrability, and (3) it increases surface runoff, encouraging the entrainment and transport of fine suspended sediments and the eventual development of relatively thin impermeable crusts, further abetting surface runoff and sheet erosion. Excessive trampling has the following effects: (1) it destroys the protective cover of plant litter, decreasing infiltration and surface detention and increasing surface runoff, and (2) it increases the bulk density of the soil beneath the surface, decreasing soil infltration and increasing surface runoff. Throughout the years, numerous studies have shown that uncontrolled grazing of livestock in the semiarid western U.S. has a pronounced effect on soil and water conservation. Barring grazing exclusion, livestock management is currently seen as a viable alternative. Livestock management aims to control the intensity, duration, and season of use of vegetation by livestock. Poorly managed rangelands invariably lead to watershed degradation, excessive surface runoff, floods, accelerated erosion, depletion of subsurface moisture and groundwater, and loss of baseflow. Conversely, numerous examples show the multiple benefits of sound range management, including effective soil and water conservation, reduced incidence of floods, enhanced erosion control, and adequate subsurface moisture replenishment, the latter often translating into net gains in baseflow (see, for instance, Heede, 1977).

Upland Vegetation Management

The effects of upland vegetation management on baseflow augmentation are not very well defined to date. Past emphasis has been largely on the effects of vegetation conversion on water yield rather than on baseflow augmentation. In general, vegetation clearing solely for the purpose of increasing water yield appear to be misdirected, since it is likely to lead to increased erosion potential and to negatively impact water quality. There are, however, exceptions, particularly when the emphasis is shifted from water yield to baseflow augmentation. For instance, in central Oregon, juniper stands in their natural state are known to consume large amounts of water and to inhibit the growth of grassy vegetation, thereby encouraging surface runoff and erosion. Partial clearing of these forests to allow the reestablishment of grassy vegetation leads to increased infiltration, reduced surface runoff and erosion, adequate soil moisture replenishment, and baseflow augmentation to nearby creeks (Stabler, 1985).

Riparian Vegetation Management

The literature provides ample evidence in support of the statement that baseflow augmentation in small upland streams is directly related to the enhancement of riparian vegetation. Field observations indicate that riparian vegetation becomes established in streambanks where there is an ample supply of subsurface moisture year-round. Healthy stands of riparian vegetation act to reduce stream velocities, to encourage overbank flooding and sediment deposition, and to increase bank stability. In addition, they play a key role in moderating streamflows, serving as recharge areas during high flows and as discharge areas during low flows. Thus, riparian vegetation acts as an effective agent for aquifer recharge, permitting the storage of significant amounts of water in streambanks and stream bottom, for gradual release as baseflow during the dry season.

The multiple benefits of riparian vegetation to the hydrology and ecology of upland streams appear beyond doubt. However, for lowland streams, the issue of preservation of riparian vegetation vs traditional flood control objectives is very complex, with hydraulic, biological, ecological, and legal implications. For lowland streams, the benefits of riparian vegetation msut be reconciled with the associated loss of channel conveyance.

Upland Runoff Detention and Retention

The positive effects of upland runoff detention and retention on the conservation of soil and water, the replenishment of subsurface and groundwater, and the stabilization of streamflows have been recognized for some time. As early as 1937, Horton identified the following strategies for streamflow stabilization (i.e., baseflow augmentation): (1) an increase in the infiltration capacity of the soil, (2) an increase in depression storage, (3) a decrease in the rate of overland flow, and (4) the use of grasses. Increases in depression storage serve to moderate runoff, to lengthen the ponding time of surface water, to increase total infiltration, and to replenish subsurface moisture. Depression storage can be effectively increased by the construction of water-spreader dikes, which can detain runoff in slopes and meadows, encouraging infiltration. Decreases in the rate of overland flow can be accomplished by increasing surface roughness, decreasing surface slope, and/or increasing the length of overland flow. Decreases in overland flow serve to slow down the rate of water exit from soil surfaces, increasing ponding time and total infiltration. The use of grasses serve to reduce surface runoff as compared with areas under cultivation. Grasses provide a higher and better-sustained soil infiltration capacity. This is due to root, earthworm, and insect borings, and to the prevention of soil inwashing by the grass mat. Another reason for the observed low surface runoff from grass-covered areas is the greatly increased resistance to overland flow. The additional friction due to subdivision of the flow by plant stems and leaves lengthens overland flow detention time and increases total infiltration.

Use of Instream Structures

Baseflow augmentation can also be accomplished by the use of instream structures. Natural instream structures are log steps and beaver dams. In small mountain streams bordered by forests, trees and logs falling across the channel create log steps, which act as check dams and influence the hydraulic geometry. Beaver dams change the hydraulic regime of a stream, decreasing flow velocities and encouraging the growth of riparian vegetation, which aids in the replenishment of subsurface and groundwater. Manmade instream structures are either large dams, check dams, or trap dams. Large dams impact all major stream uses, including stream hydrology, sedimentology, ecology, fisheries, and aquatic biota. In addition, large dams serve to moderate streamflows, decreasing the magnitude and frequency of floods and increasing the permanence of low flows. Check dams are small structures built primarily for the purpose of controlling the base level of badly degraded streams. These dams can be either temporary (log, brush, and brush-wire), more-or-less permanent (rock dams and babions), or permanent concrete structures. Trap dams are low instream structures designed to fill with coarse sediment over a period of several years following construction. The sediment deposited behind the dam serves as an artificial aquifer for the storage of flood waters and their eventual release as streamflow.

References

Elmore, W., and R. L. Beschta. (1987). "Riparian Areas: Perceptions in Management," *Rangelands*, Vol. 9, No. 6, December, pp. 260-265.

Heede, B. H. (1977). "Case Study of a Watershed Rehabilitation Project: Alkali Creek, Colorado," *Research Paper RM-189,* USDA Forest Service, Rocky Mountain Forest and Range Experiment Station, Fort Collins, Colorado, June.

Horton, R. E. (1937). "Hydrologic Aspects of the Problem of Stabilizing Streamflows." *Journal of Forestry*, Vol. 35, No. 11, November, pp. 1015-1027.

Jauch, J. (1957). "A Guide to Watershed Rehabilitation as it Pertains to the Trout Creek Watershed," *Internal Document,* USDA Forest Service San Isabel National Forest, Salida District, Salida, Colorado, January.

Stabler, F. (1985). "Increasing Summer Flows in Small Streams Through Management of Riparian Areas and Adjacent Vegetation," in *Riparian Ecosystems and Their Management,* General Tech. Rep. RM-120, USDA Forest Service Rocky Mountain Forest and Range Experiment Station, Ft. Collins, Colo., pp. 206-210.

Van Haveren, B. P., W. L. Jackson, and G. C. Lusby. (1987). "Sediment Deposition Behind Sheep Creek Barrier Dam, Southern Utah," *Journal of Hydrology (New Zealand)*, Vol. 26, No. 2, pp. 185-196.

Unsteady Flow of Bingham Viscoplastic Material in an Open Channel

By W.Z. Savage[1], B. Amadei [2], A.M. ASCE, and S.H. Cannon[3]

Abstract: An analytic solution for transient flow of a Bingham viscoplastic material in an open channel is presented. The solution for the response to a time-varying pressure gradient predicts shear stress and velocity in a viscoplastic layer where shear stress exceeds material strength and shear stress and velocity in an overlying rigid layer where shear strength is less than material strength. Under a constant pressure gradient, the thickness of the viscoplastic layer asymptotically approaches the steady flow thickness. With removal of the pressure gradient, the thickness of the viscoplastic layer asymptotically approaches zero and flow stops.

Introduction

The concept of a viscous fluid with a finite yield strength was introduced by Bingham in 1922. Since that time very few analytic solutions for unsteady flow of such materials (Bingham viscoplastic materials) have been published. Atabek (1964) obtained solutions by Laplace Transform methods (Carslaw and Jaeger, 1959) for shear stress and velocity caused by a pressure gradient in a cylinder. Blake (1974) analyzed the deceleration of a Bingham material between two flat plates by a method of matched asymptotic expansions.

In the following, the response to a time varying pressure gradient of Bingham material in an open channel is analyzed. This analysis is a first step towards a solution for the response to accelerations and decelerations in debris flows of the type described by Johnson (1972).

---

[1] Geologist, US Geological Survey, Box 25046, MS 966, Denver Federal Ctr.,Denver, CO 80225

[2] Assoc. Prof. of Civ. Engrg., U. of Colorado, Boulder, CO 80309

[3] Geologist, Colorado Geological Survey, 1313 Sherman St., Denver, CO 80302

Rectilinear flow of Bingham material

Consider rectilinear flow in the x-direction of a Bingham material of thickness h in an infinitely wide channel. For such a flow the velocities and deviatoric stresses are functions of the y-direction and are given by v = v(y,t) and $\sigma_{xy}=\tau(y,t)$. The effect of body forces are ommited. Continuity is identically satisfied and the equations of motion reduce to

$$\rho\frac{\partial v}{\partial t}=-\frac{\partial P}{\partial x}+\frac{\partial \tau}{\partial y} \quad (1)$$

$$\frac{\partial P}{\partial y}=\frac{\partial P}{\partial z}=0 \quad (2)$$

from which it is seen that the pressure, $P=P(x,t)$.

For a Bingham material, no flow will occur and the material will behave elastically if the second invariant of the stress deviator tensor, is less than or equal to the square of a constant yield stress, K. Otherwise, flow occurs. For the rectilinear flow being considered, the Bingham constitutive equations (Bingham, 1922, Oldroyd, 1947; Prager, 1961) reduce to

$$\tau=\mu\frac{\partial v}{\partial y}+K \qquad \tau>K \qquad (3)$$

$$\frac{\partial v}{\partial y}=0 \qquad \tau\leq K \qquad (4).$$

Elastic strains in portions of the body where the strain rate vanishes are assumed infinitesimal. As the focus of this paper is on viscoplastic flow, these small strains are ignored and the elastic region is treated as an essentially rigid plug moving at a velocity equal to the velocity at the elastic-plastic interface.

At the base of the channel, y = 0, the adherence condition, v = 0, is assumed to hold. Equation 3 yields

$$v(y,t)=\frac{1}{\mu}\int_0^y \tau\,dy-\frac{K}{\mu}y \quad (5)$$

valid for $\tau>K$. Substituting equation 5 in equation 1 gives

$$\frac{\rho}{\mu}\frac{\partial}{\partial t}\int_0^y \tau\, dy = -\frac{\partial P}{\partial x} + \frac{\partial \tau}{\partial y} \quad (6)$$

which is again valid only where $\tau > K$.

From the second constitutive equation $\partial v/\partial y = 0$ when $\tau \leq K$. Then, within the rigid plug, velocity is a function of time only. Substituting this result in equation 1 yields

$$\rho\frac{\partial v(t)}{\partial t} = -\frac{\partial P}{\partial x} + \frac{\partial \tau}{\partial y} \quad (7),$$

the equation of motion for $\tau \leq K$ which is subject to the condition that shear stress vanish on the flow surface.

Taking derivatives with respect to y on both sides of equations 6 and 7 results in

$$\frac{\rho}{\mu}\frac{\partial \tau}{\partial t} = \frac{\partial^2 \tau}{\partial y^2} \quad (8)$$

for $\tau > K$ and

$$\frac{\partial^2 \tau}{\partial y^2} = 0 \quad (9)$$

valid when $\tau \leq K$.

The Bingham material in the channel remains at rest until the shear stress created by the pressure gradient exceeds the shear strength. This shear stress must vary linearly (equation 1), vanish at the surface, and will first exceed the shear strength at y = 0 when motion begins. These conditions are satisfied when the initial distribution of shear stress is given by

$$\tau = K\left(1 - \frac{y}{h}\right) \qquad 0 \leq y \leq h\ ,\ t = 0 \qquad (10).$$

To summarize; shear stresses in rectilinear flow of Bingham material in an open channel under a time-varying pressure gradient are governed by partial differential equations 8 and 9, the boundary conditions

$$\tau = 0 \qquad y = h \ , \ t>0$$

$$\frac{\partial \tau}{\partial y} = \frac{\partial P}{\partial x} \qquad y = 0 \ , \ t>0$$

and the initial condition given by equation 10. The second boundary condition results from setting y = 0 in equation 6 or velocity to zero in equation 7.

The solutions to equations 8 and 9 satisfying these boundary and initial conditions are

$$\tau = K\left(1-\frac{y}{h}\right) + \frac{2\mu}{\rho h}\sum_{n=1}^{\infty}(-1)^{n-1}\sin\left[\frac{(2n-1)\pi(y-h)}{2h}\right] \times \int_0^t \exp\left[\frac{-(2n-1)^2\pi^2\mu\lambda}{4\rho h^2}\right]\Gamma(t-\lambda)\,d\lambda \qquad (11)$$

which is obtained by the Laplace Transform method and holds where $\tau > K$ and

$$\tau = \frac{K(y-h)}{\xi(t)-h} \qquad (12)$$

valid where $\tau \le K$. The function $\Gamma(t)$ in equation 11 is

$$\Gamma(t) = \frac{\partial P}{\partial x} + \frac{K}{h} \ .$$

Equation 12 arises from the general solution of equation 9, which is

$$\tau = f_1(t)y + f_2(t) \ ,$$

and the conditions that shear stress equal yield stress at the boundary between the rigid and flowing parts of the body and shear stress vanish at the surface, y = h. The expression $\xi(t)$ represents y values where $\tau = K$ during flow.

Where $\tau > K$ velocities are obtained by substituting equation 11 in equation 5 and integrating. This yields

$$v = -\frac{4}{\rho\pi}\sum_{n=1}^{\infty}\frac{(-1)^n}{2n-1}\cos\left[\frac{(2n-1)\pi(y-h)}{2h}\right]$$

$$\times\int_0^t \exp\left[\frac{-(2n-1)^2\pi^2\mu\lambda}{4\rho h^2}\right]\Gamma(t-\lambda)\,d\lambda \;-\; \frac{Ky^2}{2\mu h} \qquad (13)$$

for velocity when $0 \le y \le \xi(t)$. Velocities where $\tau \le K$ depend on time only and are obtained by substituting $\xi(t)$ for y in equation 13.

Velocities for the case where a constant pressure gradient is maintained for a fixed time, t_1 , are shown in dimensionless form in Figure 1. The dimensionless quantities are

$v^* = \mu v/Kh$, $y^* = y/h$, $t^* = \mu t/\rho h^2$, $t_1^* = \mu t_1/\rho h^2$, and $P_o^* = \frac{h}{K}\frac{\partial P}{\partial x}$.

Here $t_1^* = 1.2$ and $P_o^* = -2$.

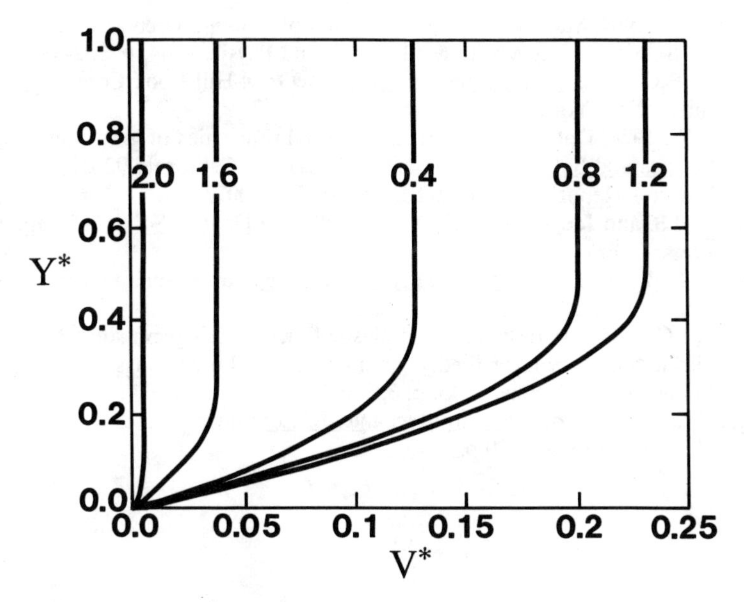

Figure 1. Velocity distribution for various times, t^* .

Values of $\xi(t)$ are obtained by setting $\tau = K$ in the integrated form of equation 11 and solving the resulting equation for $\xi(t)$ by Newton-Raphson iteration. Note the decrease in thickness of the rigid plug and increase in velocity during application of the pressure gradient ($0 \le t^* \le t_1^*$) and subsequent increase in plug thickness and decrease in velocity after removal of the pressure gradient.

Conclusions

The flow of a Bingham material in an open channel under a time varying pressure gradient depends on the magnitude of the shear stress. Where the shear stress exceeds the yield stress, the viscoplastic region, shear stresses and velocities behave in the diffusive manner typical of viscous flow. Where the shear stress is equal to or less than the yield stress, an elastic layer characterized by a linearly varying shear stress forms. This layer, which is taken to be an essentially rigid plug, moves at the velocity of the interface between it and the underlying viscoplastic region.

Appendix.--References

Atabek, H.B.,1964, Start-up flow of a Bingham plastic in a circular tube: Zeitschrift fur angewadte Mathematik und Physik, v. 44, p. 232-333.

Bingham, E.C., 1922, Fluidity and Plasticity: McGraw-Hill Book Company, Inc., New York.

Blake, T.R., 1974, Determination of the material properties of grout: Systems, Science and Software, P.O. Box 1620, LaJolla, California 92037, Progress Report for Contract DNA 001-74-C-0077.

Carslaw, H.S. and Jaeger, J.C., 1959, Conduction of Heat in Solids: Clarendon Press, Oxford, 510 p.

Johnson, A.M., 1970, Physical Processes in Geology: San Francisco, Freeman, Cooper, 577 p.

Oldroyd, J.G., 1947, Two-dimensional plastic flow of a Bingham solid-A plastic boundary layer theory for slow motion: Proceedings, Cambridge Philosophical Society, v. 43, p. 383-395.

Prager, W., 1961, Introduction to Mechanics of Continua: Ginn and Company, Boston, 230 p.

Waves in a Fluid Mud Layer Flowing Down an Incline

Ko-Fei Liu[1] and Chiang C. Mei[2]

Abstract

To model the dynamics of fluid mud, lava, or debris with a high concentration of cohesive clay particles, we consider a thin sheet of Bingham plastic fluid flowing down a slope. The Bingham fluid is characterized by a finite yield stress and high viscosity. We first assume long waves and approximate the profile of the local longitudinal velocity by a plug zone on top and a shear zone on the bottom with a parabolic velocity distribution. The instability of a uniform flow to infinitesimal sinusoidal disturbances is then examined. Numerical solutions of the nonlinear post-instability development are discussed. Specifically we present results of periodic bores and finite amplitude waves due to a fluid source at a fixed point upstream of an otherwise uniform flow. The second case resembles the pulsating flow in a muddy river with tributaries. Both can occur beneath a layer of clear water.

Introduction

Fluid mud of high concentration of cohesive clay particles are abundant on the bottom of some estuaries and coastlines, waterways and river basins. Its movement, driven by gravity, currents or waves can be important to the evolution of shorelines and river deltas, and the silting of harbors and locks. In navigable waterways it affects the movement of ships and defines the *nautical depth.*

After an intense rain, mud concentration in some mountain streams is also high from bank erosion and surface runoff. At high enough flow rates mud can carry rocks and fallen trees to form the so-called *debris flow* which can cause severe erosion and flood damage downstream. Similar problems

[1] Research assistant, [2] Member, ASCE and Professor, Department of Civil Engineering, Massachusetts Institute of Technology, Cambridge, MA, 02139.

arise in volcanic zones where volcanic ashes produces mud flows in rivers, as on the island of Sakurajima, Japan.

Very often mud and debris flows are laminar (Johnson, 1970). Their rheological behavior is of the Bingham plastic type and is characterized by a yield stress and a constant viscosity. Thus in simple shear the stress strain relation can be expressed by

$$\mu\frac{\partial u}{\partial z}=\begin{cases}\tau-\tau_o, & \text{if } \tau>\tau_o;\\ 0, & \text{if } \tau<\tau_o\end{cases} \tag{1}$$

It is known that both τ_o and μ increase with clay concentration. The transition from laminar to turbulent flows takes place when the effective Reynolds number Re_e exceeds about 2100 where

$$\frac{1}{Re_e}=\frac{1}{Re_\mu}+\frac{1}{Re_\tau} \tag{2}$$

with

$$Re_\mu=\frac{4\rho Uh}{\mu}, \qquad Re_\tau=\frac{8\rho U^2}{\tau_o} \tag{3}$$

In the preceding equation, ρ denotes the density of fluid mud, U the mean velocity, and h the hydraulic radius. It can be shown that the flow rates can be reasonably high for the mud flow to remain laminar if the concentration is sufficiently high (Mei and Liu 1987).

In natural streams it is well known that mud flows are often pulsating especially when there are inflows from small upstream tributaries. These pulsating flows have been likened to *roll waves* in a clear water. A hydraulic theory based on Chezy's formula for the bottom friction has been advanced by Dressler(1949) who modeled the roll waves as periodic bores. For a thin sheet in laminar flow Isihihara et al (1954) have also found by theory and experiment similar roll waves. More recently Savage (1989) combined Coulomb and Chezy frictions to model mud flow and examined the instability of a flowing mud layer to wavy disturbances of infinitesimal amplitude.

In this paper we report some theoretical results of mud flow instability. The analysis is based on the long wave equations for a Bingham plastic fluid. Instability of infinitesimal disturbances is first examined. Finite amplitude waves including bores due to unstable initial disturbance or to an upstream source are presented.

Governing Equations

The governing equations are derived on the basis of the following assumptions. First the longitudinal scale in the direction of flow is very much greater than that in the transverse direction normal to the flow. Hence the the velocity components are in the similar ratio and the pressure is hydrostatic. Near the bed the bottom stress exceeds the yield stress and there must be a shear layer in which the longitudinal momentum balance is described by

$$\frac{\partial u}{\partial t} + u\frac{\partial u}{\partial x} + w\frac{\partial u}{\partial z} = g \sin \theta - g \cos \theta \frac{\partial h}{\partial x} + \nu \frac{\partial^2 u}{\partial z^2}, \qquad 0 \leq z \leq h_o \tag{4}$$

where h_o is the yield surface on which $|\tau| = \tau_o$. Assuming that u has a parabolic profile in z we can integrate the preceding equation vertically from $z = 0$ to $z = h_o$ to obtain

$$\frac{2}{3}h_o\frac{\partial u_p}{\partial t} - \frac{1}{3}u_p\frac{\partial h_o}{\partial t} + \frac{2}{5}h_o u_p\frac{\partial u_p}{\partial x} - \frac{2}{15}u_p^2\frac{\partial h_o}{\partial x}$$

$$= gh_o(\sin \theta - \cos \theta \frac{\partial h}{\partial x}) - \frac{2\nu u_p}{h_o} \tag{5}$$

where the last term on the right-hand side is the bottom stress.

Near the free surface at $z = h$ the shear stress is below the yield stress so that the longitudinal velocity $u = u_p(x, t)$ is independent of the transverse coordinate z, and is governed by

$$\frac{\partial u_p}{\partial t} + u_p\frac{\partial u_p}{\partial x} = g \sin \theta - g \cos \theta \frac{\partial h}{\partial x} - \frac{\tau_o \text{sgn}(u_p)}{\rho(h - h_o)} \qquad h_o \leq z \leq h \tag{6}$$

Conservation of mass across the entire mud layer requires that

$$\frac{\partial h}{\partial t} + \frac{\partial}{\partial x}\int_0^h u\, dz = \frac{\partial h}{\partial t} + \frac{\partial}{\partial x}(h - \frac{h_o}{3})u_p = q_{in} \tag{7}$$

where the right-hand side denotes the inflow from external sources.

Equations (5) to (7) are the governing equations for four unknowns h, h_o, and u_p for a *moving* mud layer. At the threshold of motion, the magnitude of the bottom stress must exceed τ_o. Since at the incipient motion $u = 0$, the whole layer is a plug flow. The bottom stress is

$$\tau_b = -\rho g h(\sin \theta - \cos \theta \frac{\partial h}{\partial x}) \tag{8}$$

The requirement that $|\tau| > \tau_o$ gives

$$-\rho g(\sin \theta - \cos \theta \frac{\partial h}{\partial x})h \gtrless \tau_o \qquad \text{for} \quad \begin{matrix}\text{downstream}\\ \text{upstream}\end{matrix} \quad \text{motion} \tag{9}$$

which must be satisfied in order for governing equations (5) to (7) to apply.

These equations can be applied to a two-layered system where a clear water layer of negligible viscosity lies above a layer of fluid mud. One only needs to replace the gravity constant g by the effective gravity $g' = (\rho - \rho_w)/\rho$ everywhere.

To normalize z, h, and h_o we define the depth scale

$$\overline{h} = \frac{\tau_o}{\rho g \sin\theta} \tag{10}$$

We also normalize the longitudinal coordinate by $\overline{h}\cot\theta$, time by $\mu\cot\theta/\tau_o$ and velocity by $U = \tau_o\overline{h}/\mu$. A Reynolds number

$$\beta = \frac{\rho U \overline{h}}{\mu}\tan\theta \tag{11}$$

appears. In dimensionless form the uniform flow is represented by

$$h = D, \quad h_o = D - 1, \quad \text{and} \quad u_p = \frac{1}{2}(D-1)^2 \tag{12}$$

which exists only if $D > 1$.

Instability of Infinitesimal Disturbances

We now superpose small disturbances to (12), linearize the governing equations and then assume wave disturbances in the form

$$H = \hat{H}\, e^{i(kx-\omega t)}, \quad H_o = \hat{H}_o\, e^{i(kx-\omega t)}, \quad U = \hat{U}\, e^{i(kx-\omega t)} \tag{13}$$

where ˆ indicates the amplitude. An eigenvalue condition is then obtained, which is a complex polynomial of third degree for ω as a function of k with parameters β and D.

Only one among the three roots may have a positive imaginary part, which corresponds to the unstable mode. A sample plot of the growth rate is shown in Fig. 1.

Periodic Roll Waves

For the initial unstable disturbance we choose $\beta = 1, D = 3$ and $k = 1.2$ with the initial amplitude 0.03. It is numerically found that any initial amplitude less than this value does not affect the nonlinear development. A sample time series is shown in Fig. 2. The first bore appears at $t = 12$ and a steady state of periodic bores is reached at $t = 13$.

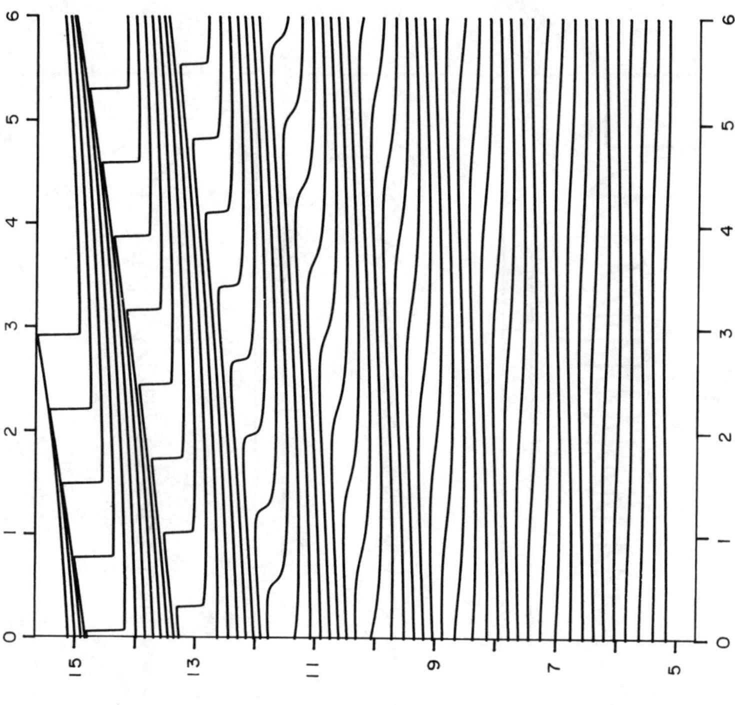

Fig. 1. Growth rate of the unstable mode for $\beta = 1$ ⇑

Fig. 2. Evolution of periodic roll waves ⟹

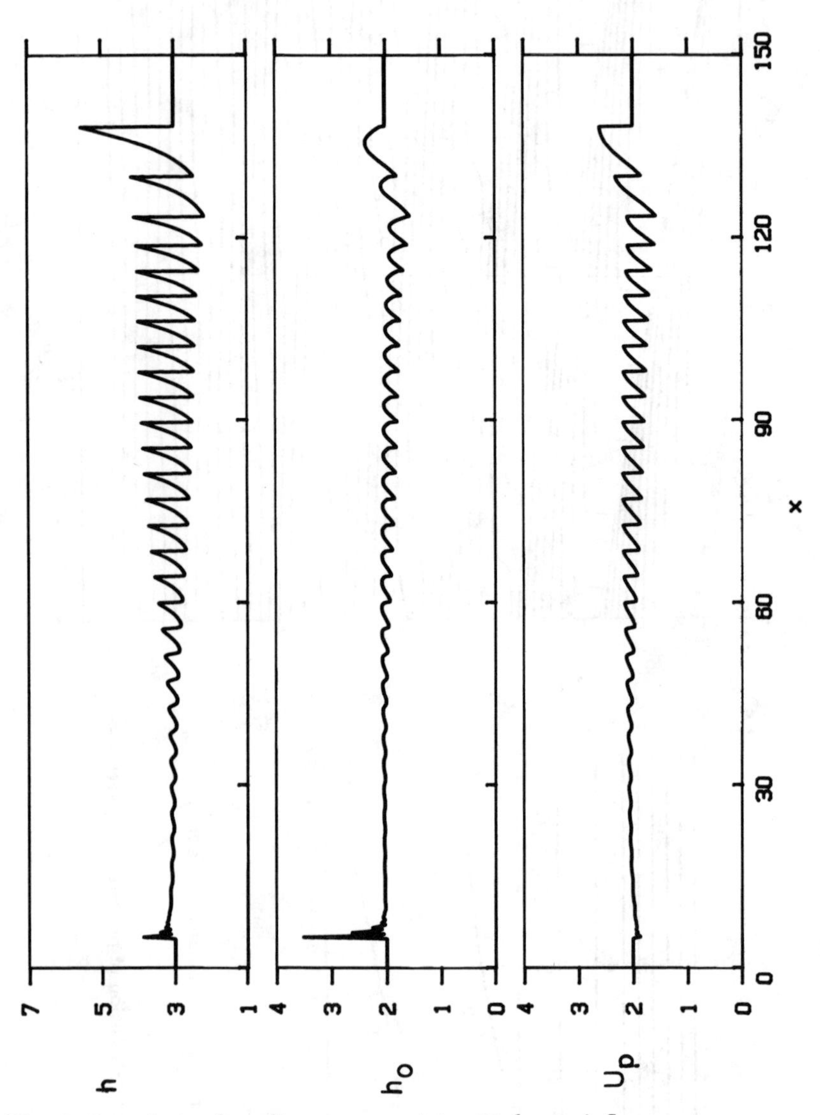

Fig. 3. Snapshots of nonlinear waves at $t = 30$ due to inflow upstream.

Nonlinear Waves Due to a Periodic Source Upstream

From field and laboratory observations (Davis, 1985), waves in a muddy river often occur at a long distance downstream from a small tributary. To simulate this phenomenon, we introduce a small periodic external input along a fixed stretch upstream as follows

$$q_{in} = \begin{cases} Q_o[1 + \sin(\frac{2\pi}{T}t - \frac{\pi}{2})], & -L < x - x_o < L; \\ 0, & \text{otherwise.} \end{cases} \quad (14)$$

We choose $L = 0.1$ to make the inflow appear localized, and a weak influx rate: $Q_o = 0.1$, with $T = \frac{\pi}{4}$, in a deep layer at relatively low Reynolds number: $D = 3$ and $\beta = 1$. At first the free surface remains flat. Much later disturbances grow and become appreciable far downstream. Then the first wave becomes a bore, whose amplitude and propagating speed increase in time. This is followed by new bores behind the front. A sample snapshot is shown in Fig 3. For higher Q_o the shear layer depth may vanish at large t and the computation has to be terminated, marking the breakdown of our theory. Mathematically other terms representing two dimensionality and dispersion must be added to allow further computation. Physically the results suggest severe scour.

Acknowledgement

This research has been supported by the Office of Naval Research, Ocean Engineering Program under Contracts N00014-83K-0550 and 89J-3128.

Appendix: References

Davis, T. R. H., (1986), Large debris flows, a macroviscous phenomenon; *Acta Mechanica*, 63, 161-178.

Dressler, R. F., (1949), Mathematical solution of the problem of roll-waves in inclined open channels; *Comm. Pure & Appl. Math.*, 2, 149-194.

Ishihara, T. , Iwagaki, Y., & Iwasa, Y., (1954), Theory of roll wave trains in laminar water flow on a steep slope surface; *Trans. Japan Soc Civil Engrs.* 19., 46-57.

Johnson, A. M., (1970), *Physical Processes in Geology*, Freeman, Cooper & Co.

Mei, C. C., & Liu, K. F., (1987) A Bingham plastic model for a muddy seabed under long waves; *J Geophy. Res*, 92, 14581-14594.

Savage, S. B. (1988) Flow of granular materials; *Theoretical and Appl. Mechan. IUTAM Symposium*, 241-266. Edited by P. Germain, M. Piau, & D. Caillerie; Elsevier.

Field Observation of Roll-waves in Debris Flow

Makoto Hikida[1]

Abstract

In 1979, the observatory of debris flow was set up in the Hase River of Mt. Sakurajima in Japan which is one of the most active volcanoes in the world. Obvious roll-waves on each debris flow have been observed through the use of a VTR system. The wave velocity, period and profile of the roll-wave, etc. are obtainable by the analysis of the video tapes. In this paper, the equipment, its use, and the measurement techniques used at the observatory are described. The values computed by the data presented here show fairly good agreement with the ones observed.

Introduction

Roll-waves in inclined open channels have been studied by Mayer(1959) and Ishihara, et al.(1960) in water flow. Takahashi, et al.(1980) reported that roll-waves in debris flow are similar to those in water flow. Davies(1985) introduced some observations of pulsing flows in small debris flows and debris slides. These field data on roll-waves in debris flows are very difficult to be obtained.

Field Observation

Figure 1 shows a bird's eye view of the debris flow observation site, which is at the Hase River. It has two sets of VTR cameras, staff gauges on the bank of the river and a wire to detect the occurrence of debris flow. The wire is stretched across the river and tied to two stones on the bed. When the front of a debris flow moves the stones, the wire will be cut and the VTR will be activated. Thus it is possible to grasp the debris flow visually.

Figure 2 shows schematic diagram of VTR system used to observe debris flow. Two series of the VTR systems

---

[1] Prof. of Civ. Eng., Kagoshima Nat'l College of Tech., 1460-1 Shinko, Hayato-cho, Kagoshima 899-51, Japan.

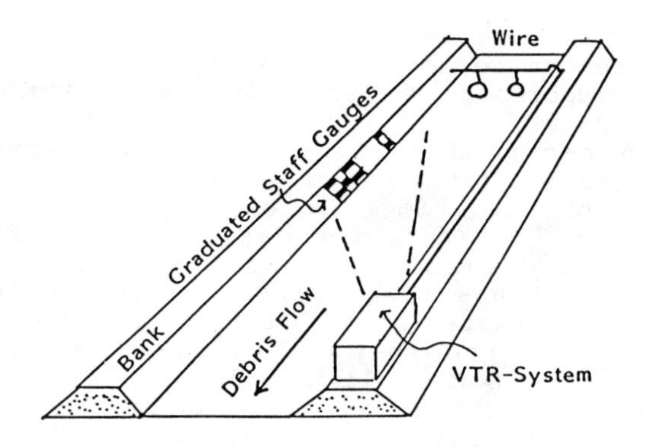

Figure 1. Sketch of the field observation system of debris flows at the Hase River.

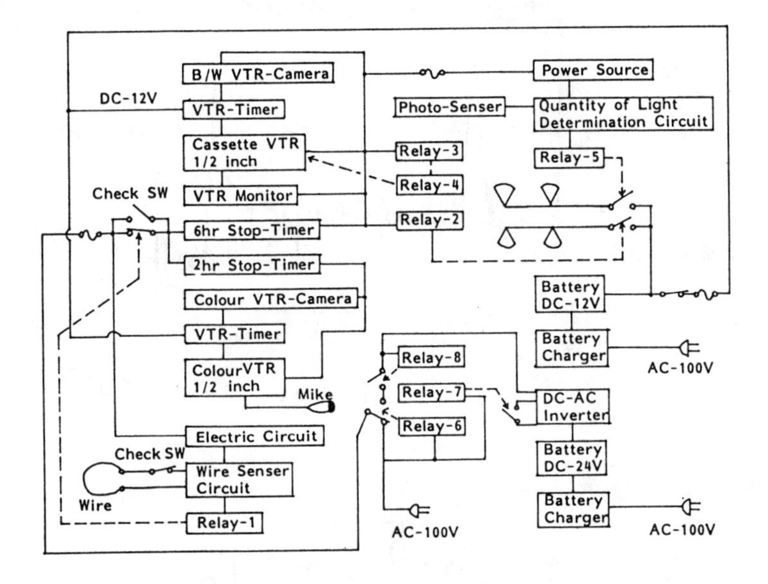

Figure 2. Schematic diagram of VTR system for the observation of debris flows.

are adopted. One is a color VTR system for daytime which can record for one hour at the most using a 2/3 inch open videotape. The other uses black and white VTR camera for night, using a 1/2 inch cassette videotape which can record for six hours at the most. These recording tapes have a time read-out system which describes absolute time i.e. day, hour, minute, second and 1/100 second. Thus, the data of debris flow have been presented clearly and precisely. Since the electric current has been often shut off during heavy rainfall, three batteries with DC-12V and 200A are used. One is set for recording absolute time into the videotape, and is used all the time. The other two are set for VTR operation and lighting by using a DC-AC invertor, converting from DC-24V to AC-100V by using a relay circuit as shown in Fig. 2.

The cross section of the stream bed forms a trapezoid and the width of the river bed is 8 m at the staff gauge. The slope of the river bed is $\tan\theta = 0.07$.

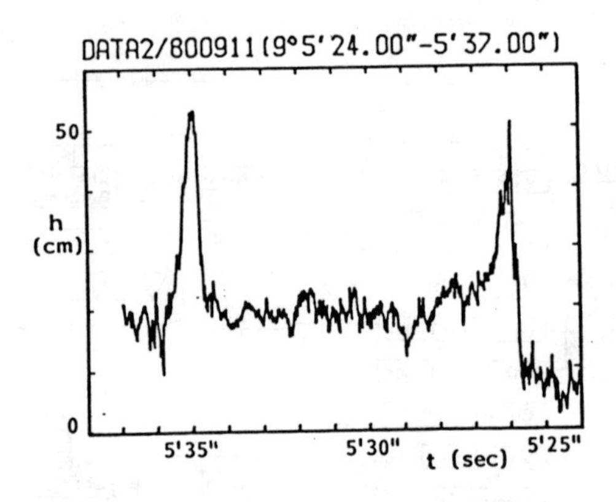

Figure 3. Typical longitudinal profile of roll-waves.

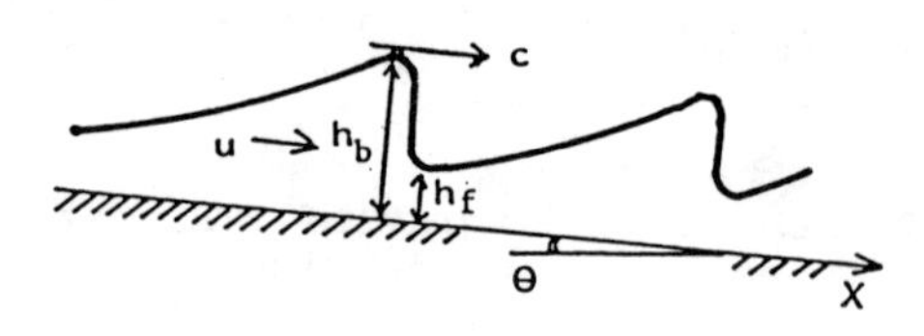

Figure 4. Diagram of longitudinal profile of roll-waves.

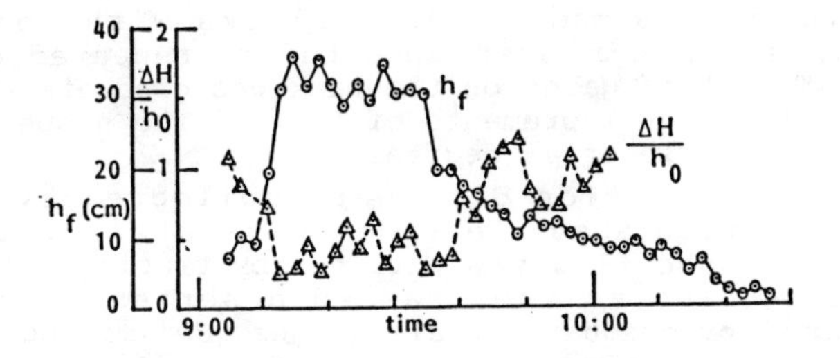

Figure 5. Series of trough depth of roll-waves h_f and relative height $\Delta H/h_0=(h_b-h_f)/h_0$; h_b and h_0 are the peak and mean depth of the roll-waves respectively.

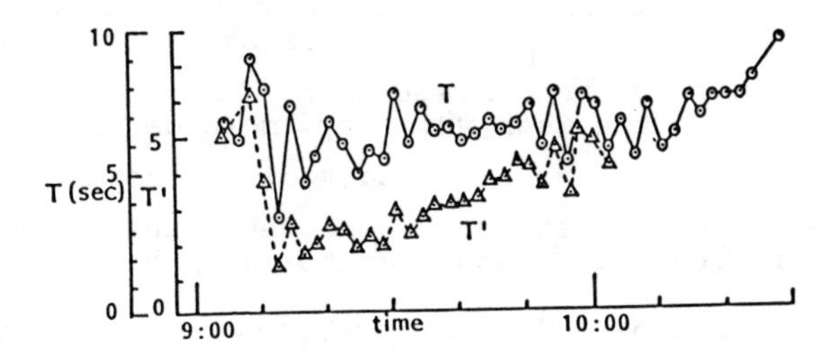

Figure 6. Series of wave period T and the dimensionless wave period $T'=T*\sqrt{g*\sin\theta*\tan\theta/h_0}$.

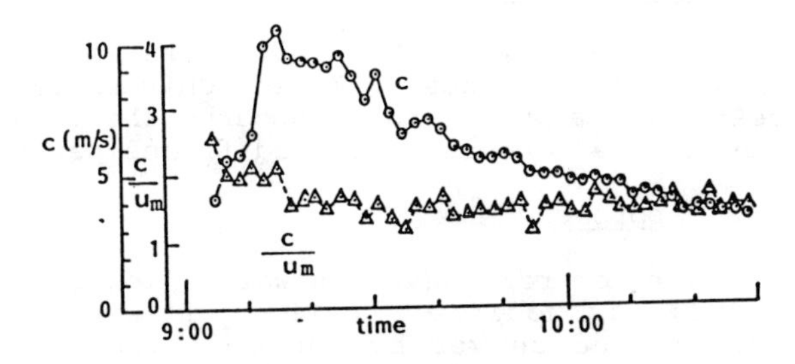

Figure 7. Series of wave velocity c and the ratio of wave velocity and mean flow velocity c/u_m.

The river bed is made of lava blocks with concrete joints, which has been smoothed by repeated debris flows. The staff gauge on the bank was graduated by the lateral stripe of increments of 50 cm. The slope of the bank has an angle of 45 degrees.

The observation data were collected by using videotapes with absolute time read-outs. The mean velocity of the flow was obtained by tracing the movement of pebbles as they passed a series of points designated by graduated staff gauges, delimiting a distance from upstream to downstream to a distance of 10 meters in length, as shown in Fig. 1. The pebbles having the diameter approximately equal to the mean depth were selected as the tracers. The distortion resulting from the effect of perspective was compensated. The water level at the staff gauge was measured by using a sonic digitizer with a micro-computer in the laboratory. Figures 3 and 4 show a typical longitudinal profile of roll-waves and the notations: h_b and h_f are the heights at the front and trough of a roll-wave respectively. c is the wave velocity, u_m is the mean velocity, mean depth is defined as $h_0 \simeq 1.1*h_f$ from the wave longitudinal profile and T is the wave period. These data were input into floppy diskettes using the micro-computer to make the calculation easy.

Figure 5 shows that the relative height of roll-waves at peak discharge time is smaller than that at non-peak discharge time. Figures 6 and 7 show that the value of wave period T, and the ratio of wave velocity and flow velocity c/u_m take a constant value respectively.

Laboratory experiments

Some flume experiments on roll-waves were performed by using an inclined flume at Kyushu University. A 16mm high speed camera was used to observe the phenomena of roll-waves at the rate of a hundred frames per second.

Theoretical approach

The theory of roll-waves in water flow by Ishihara, et al.(1960) was modified to the roll-waves in debris flow. Calculated curves by the modified theory were compared with the experimental data and field ones. In Fig. 8, the theoretical curves of dimensionless wave length λ' show good agreement with the experimental and field data of debris flow respectively where α is a momentum correction factor.

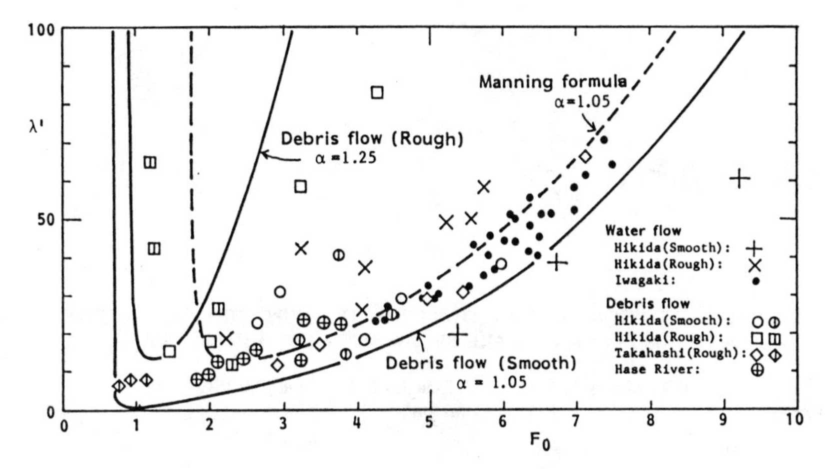

Figure 8. Dimensionless wave length $\lambda'=\lambda*\tan\theta/h_0$ versus Froude number $F_0= u_m/\sqrt{gh_0\cos\theta}$).

Acknowledgment

The author would like to express my deep appreciation for the instructive efforts made by Emeritus Prof. T. Tsubaki and Prof. M. Hirano of Kyushu University. Kind help for this observation was received from the Erosion Control Section of Kagoshima Prefecture in Japan.

References

Davies,T.R.H.(1985). "Large Debris Flows: A Macro-viscous Phenomenon", ACTA MECHANICA 63, 161-178.

Ishihara,T., Iwagaki,Y. and Iwasa,Y.(1960). Discussion of "Roll waves and slug flows in inclined open channels", J. Hydraul. Div., ASCE, 86, 45-46.

Mayer,P.G.(1959). "Roll Waves and Slug Flows in Open Channels", J. Hydraul. Div., ASCE, 85, 99-141.

Takahashi,T. and Hasegawa,S.(1980). "On the Characteristics of Debris Flow in a Fixed Uniform Flume", Proc. 35th Annu. Conf. JSCE, 2, 354-355 (in Japanese).

NUMERICAL SIMULATION OF MUDFLOWS FROM THE HYPOTHETICAL FAILURE OF A DEBRIS BLOCKAGE LAKE BELOW MOUNT ST. HELENS, WA

Robert C. MacArthur[1], M.ASCE, Douglas L. Hamilton[2], M.ASCE, and Ronald C. Mason[3], M.ASCE

Abstract

This paper evaluates the characteristics of mudflow events resulting from the hypothetical breaching of a debris blockage dam using a variety of lake levels and impounded water volumes for the initial breach conditions. A one-dimensional (Petrov- Galerkin finite element) unsteady mudflow routing model is used to simulate the movement of the dam break-induced mudflow events downvalley through the Corps of Engineers' Sediment Retention Structure.

Introduction

The May 18, 1980 eruption of Mount St. Helens, WA, produced a debris avalanche which flowed down the North Fork Toutle River damming several tributary streams. The blockage at the confluence of South Fork Castle Creek and Castle Creek produced a natural debris dam approximately 190 high. Figure 1 shows the general study area near Mount St. Helens and the location of Castle Lake. Snow melt and runoff waters captured behind the blockage quickly formed a lake. To prevent overtopping and a potentially catastrophic failure of the blockage retaining Castle Lake, the U.S. Army Corps of Engineers (COE) constructed an SPF spillway at the eastern end of the blockage to stabilize the lake at elevation 2,577 feet MSL. Recent studies by the U.S. Geological Survey (USGS) indicate that "the blockage is potentially unstable against failure from piping due to heave and internal erosion when groundwater levels are seasonally high" and that an earthquake of 6.8 or greater might initiate such a failure (Laenen and Orzol, 1987). If the Castle Lake blockage were to fail rapidly by the mechanism suggested by the USGS, approximately 18,500 acre-feet (AF) of stored water in the lake could create a mudflow flood event in the North Fork Toutle River. The USGS (Laenen and Orzol, 1987) estimates that an event of this nature could result in a peak discharge of 2,100,000 cfs at the Corps' N-1 debris retention dam twelve miles downstream from Castle Lake (see Figure 1).

In May 1982, President Reagan directed the Corps of Engineers to prepare a comprehensive

[1] Hydraulic Engineer, U.S. Army Corps of Engineers, Hydrologic Engineering Center, 609 2nd St., Davis, CA 95616, U.S.A.

[2] Principal, RIVERTECH, Inc., 23332 Mill Creek Drive, Laguna Hills, CA 92653, U.S.A.

[3] Chief, Hydrologic Section, U.S. Army Corps of Engineers, Portland District, P.O. Box 2946, Portland, OR 97208 U.S.A.

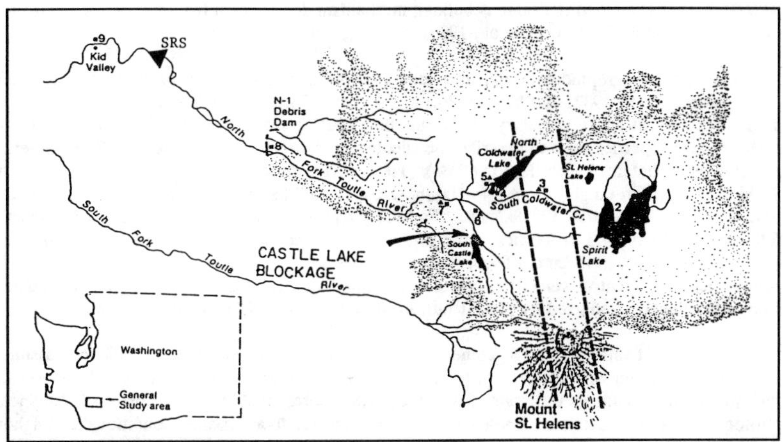

Figure 1 General Study Area

plan for long-term flood control and navigation maintenance in the wake of the Mount St. Helens eruption. A major component of the resulting plan is the recently constructed Sediment Retention Structure (SRS). The primary function of the $56.5 million dam is to trap the huge amounts of sediment expected to continue to move down the North Fork Toutle River. The SRS was designed to capture runoff-induced sediment from the blast zone, thus preventing sediment deposition and reduced flood routing and navigation capacities in the Cowlitz and Columbia Rivers. Failure of the Castle Lake blockage resulting in the possible occurrence of a mudflow event could jeopardize the safety and performance of the SRS.

The purpose of this study is to evaluate the hydraulic characteristics of mudflow events resulting from the hypothetical failure of Castle Lake and to examine the ability of the SRS to capture and pass such events through its spillway for various initial conditions at Castle Lake and in the SRS. More specifically, the study is to: (1) determine if flows will exceed the present spillway capacity at the SRS, (2) determine if the SRS will be overtopped, (3) estimate how the peak discharge downstream from the SRS will be affected by the presence of the SRS, (4) evaluate the effects on the resulting routed mudflow hydrograph due to lowering the initial Castle Lake levels at the time of breaching, and (5) evaluate the performance of the SRS during these various events when the SRS is empty with (a) "existing conditions", (b) "half full" of accumulated sediment and debris, or (c) "completely full" of sediment deposits up to the spillway crest.

Approach

The Hydrologic Engineering Center (HEC) applied the one-dimensional (Petrov- Galerkin finite element) unsteady mudflow routing model (MacArthur, et al., 1988) to route several hypothetical mudflow events from Castle Lake to the SRS. The Mudflow Model was modified to incorporate the 400 foot wide spillway at the SRS as well as the possibility of overtopping the structure. Therefore, the compute outflow hydrographs downstream from the SRS include effects due to storage (ponding) inside the SRS, flow through the spillway, and overtopping of the SRS. The upstream boundary of the modeling reach was established at the N-1 structure to correspond with the location of the USGS developed mudflow hydrographs for various initial lake elevations prior to breaching. The routing reach is approximately 6.1 miles with an average bed slope of

0.0093ft/ft. HEC applied the same debulking mechanism downstream from the N-1 structure that the USGS prescribed (Leanen, et al., 1990).

HEC used topographic data and measured cross-section information prepared by the USGS (Laenen and Orzol, 1987) and by the Portland District Corps of Engineers to depict the natural valley geometry of the mudflow routing reach from the N-1 structure to the SRS. Fluid properties used to describe the rheological characteristics of the USGS' hypothetical mudflow were obtained from Major (1984). He reported fluid viscosities from the 1980 Mount St. Helens eruption ranging from 6 to 100 lb.-sec./sq.ft., yield strengths from 2 to 31 lb./sq.ft., and unit weights from 100 to 125 lb./cu.ft. For the purposes of this investigation, HEC used the following constant fluid properties: Viscosity = 6.0 lb.- sec./sq.ft., yield strength = 2.0 lb./sq.ft., and unit weight = 110 lb./cu.ft. These fluid properties were chosen to give the most conservative results (e.g., the greatest velocities and the least attenuation). Sensitivity and model validation studies previously conducted by MacArthur, et al., (1985 and 1987) support this reasoning.

Three different valley and SRS geometry scenarios were evaluated. First, the SRS was assumed to be empty of water with the present (existing) dry reservoir bottom at elevation 870' NGVD and the valley and channel upstream from the SRS represented by the Corps of Engineers photogrametric cross-sections. These conditions are referred to as "existing conditions" throughout this paper. The second scenario assumes that the SRS is "half full" of sediment deposits and the channel bed in the North Fork Toutle River immediately upstream from the SRS is full of sediment with a bed slope of 0.006 ft/ft (approximately half the original bed slope). The third scenario assumes the SRS is "full" of sediment deposits up to the spillway crest elevation of 940' NGVD and the bed slope in the channel upstream from the reservoir is at 0.006 ft/ft.

In order to evaluate the downstream effects of lowering the initial lake levels behind the Castle Lake blockage, five different initial lake surface elevations were used to develop five different dam break hydrographs (Laenen, et al., 1990 Unpublished report). Those five dam break hydrographs were bulked up to reflect the likely entrainment of sediment and debris materials (Laenen and Orzol, 1987). Each of the five hydrographs was routed from the upstream boundary of the modeling reach at the N-1 structure, downstream to the SRS for each of the three initial SRS sedimentation conditions - (a) existing, (b) half full and (c) full. Figures 2, 3 and 4 show the five bulked mudflow hydrographs at the N- 1 resulting from each of the five starting lake elevations and the computed outflow hydrographs from the SRS for each of the three initial SRS conditions.

The mudflow model was modified to include the effects of the 400 foot wide spillway at the SRS as well as the possibility of overtopping the structure. The modelers assumed that the six rows of 3-foot diameter conduits in the SRS would clog and become inoperative during events of these magnitudes. This is a reasonable and conservative assumption. Therefore, routed mudflow hydrographs downstream from the SRS include effects due to storage (ponding) inside the SRS, flow through the spillway, and overtopping of the SRS.

Results and Discussion

Computed travel times for the leading edge of the mudflow bores are summarized in Table 1. They range from a low of 25 minutes for existing SRS conditions with 2.10 Mcfs inflow to a high of 29 minutes for a full SRS with the same inflow. Therefore, mudflows resulting from a "full lake" breach and entering an "existing" SRS, or a "full" SRS will move downvalley from the N-1 at speeds of approximately 21.5 ft/sec and 18.5 ft/sec, respectively. Lowering the initial Castle Lake elevations by 60 feet increases the computed travel times to 51 minutes and 65 minutes, respectively (flow velocities of 10.5 and 8.3 ft/sec). Table 1 and Figures 2 and 3 summarize the five different inflow and routed outflow hydrographs for the three different initial SRS conditions (existing, half full, and full), respectively. The peak discharges at the upstream model boundary

Table 1 Castle Lake Dam Breach Routing Results

Run No.	SRS Condition	(1) Initial Castle Lake Elev(ft)	(2) Peak Inflow at N-1 (cfs)	(3) Peak Outflow at SRS (cfs)	(4) Peak Stage at SRS (ft)	(5) Travel Time (min)	(6) Average Wave Velocity (ft/sec)
1	Existing	2577	2.10 M	196,000	976	25	21.5
2	Existing	2562	1.61 M	71,000	957	27	19.9
3	Existing	2547	1.18 M	0	938	31	17.3
4	Existing	2532	0.728 M	0	918	39	13.8
5	Existing	2517	0.437 M	0	902	51	10.5
6	Half	2577	2.10 M	266,000	987	26	20.6
7	Half	2562	1.61 M	189,000	975	29	18.5
8	Half	2547	1.18 M	108,000	962	33	16.3
9	Half	2532	0.728 M	0	920	39	13.8
10	Half	2517	0.437 M	0	908	51	10.5
11	Full	2577	2.10 M	600,000	1,009	29	18.5
12	Full	2562	1.61 M	411,000	1,003	32	16.8
13	Full	2547	1.18 M	315,000	995	47	11.4
14	Full	2532	0.728 M	249,000	985	49	11.0
15	Full	2517	0.437 M	230,000	982	65	8.3

(1) Starting lake water surface elevations assumed by the USGS (Laenen, et al., 1990)
(2) Peak inflow at the N-1 structure from the USGS (Laenen, et al., 1990)
(3) Mudflow routing results at the SRS, this study
(4) Computed peak stages, this study
(5) Estimated flood wave travel time, this study
(6) Average tip velocity from N-1 to SRS, this study

(at the N-1 structure) range from 2.10 Mcfs for Castle Lake full at elevation 2,577 ft. MSL to 0.437 Mcfs, assuming Castle Lake has been lowered to elevation 2,517 ft. MSL prior to breaching. The maximum computed outflow from the SRS using full Castle Lake starting conditions ranges from 196,000 cfs to 600,000 cfs for "existing" and "full" SRS conditions. Results presented in Table 1 and Figures 2 and 3 show that mudflow events for run numbers 1 through 10, 13, 14, and 15 are contained within the SRS and do not overtop the structure or exceed the designed spillway capacity of 340,000 cfs. Mudflow events exceed the capacity of the spillway and overtop the dam for runs 11 and 12 only. The peak outflows from the SRS under these conditions are 600,000 cfs and 411,000 cfs respectively.

Therefore, under "existing conditions" and the initial lake level at 2577, the SRS reduces the peak discharge in the North Fork Toutle River by 90% percent (from 2.10 Mcfs to 196,000 cfs). For "half full" conditions the peak outflow from the SRS is reduced by 87% percent and by 71% percent for "full conditions." All five mudflow hydrographs for different initial lake levels are fully contained within the SRS and Spillway without overtopping for "existing" and "half full" conditions. Overtopping occurs for initial lake levels above 2547' MSL and "full" SRS conditions. For the worst case conditions (SRS full and initial lake elevation at 2577' MSL) the peak outflow is 600,000 cfs. The SRS obviously reduces the peak discharge in the North Fork Toutle River downstream from the SRS for all three initial SRS infill conditions. Lowering the initial Castle Lake elevations at the time of the assumed breach also reduces peak flows entering and leaving the SRS. HEC incorporates into all these results the estimated volume reduction due to debulking of the flows below the N-1 as prescribed by Laenen and Orzol (1987).

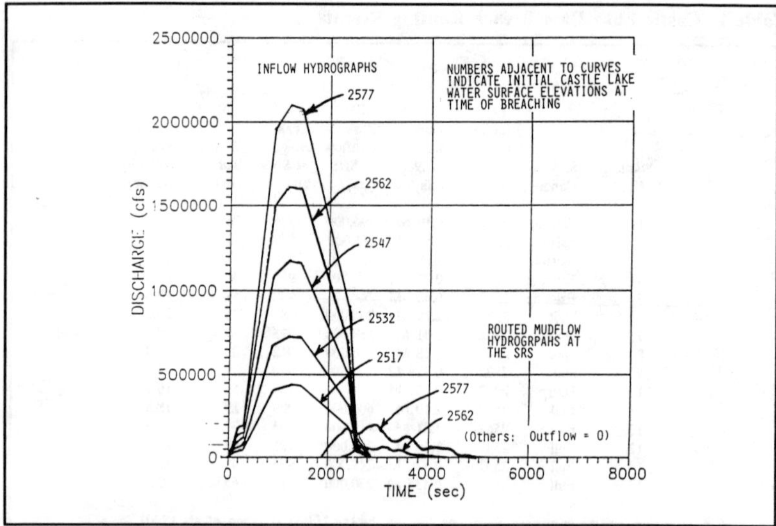

Figure 2 Inflow and Outflow Hydrographs for SRS "Existing" Condition.

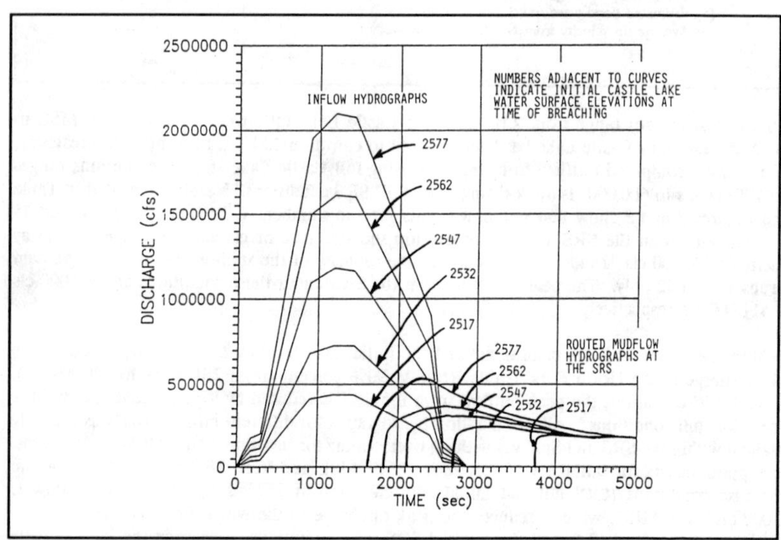

Figure 3 Inflow and Outflow Hydrographs for SRS "Full" Condition

Summary and Conclusions

The one-dimensional (Petrov-Galerkin finite element) unsteady mudflow routing model was used to route the five different hypothetical mudflow events from the Corps' N-1 structure to the SRS for three beginning-of-event geometry scenarios (SRS initial conditions). For "existing conditions" in the SRS the maximum (lake full) mudflow hydrograph is reduced by 90 percent and the SRS is not overtopped. If the SRS is "half full" of sediment deposits when the maximum (lake full) mudflow occurs, the peak mudflow hydrograph is reduced by 80 percent and the SRS is not overtopped. However, the peak stage comes within 10 feet of the top of the dam. If the SRS is initially "full" of sediment to the crest elevation of the spillway, the dam is overtopped for mudflow events where the assumed initial Castle Lake water elevations are higher than 2547 MSL. The following conclusions are made based on the results presented in this report:

1.The SRS reduces flows in the downstream reaches of the North Fork Toutle River, even for a maximum mudflow event of the magnitude estimated by the USGS (peak Q = 2,100,000 cfs). Maximum peak mudflows are reduced approximately 90%, 87% and 71% if the SRS is in its existing, half full and full condition, respectively.

2.For existing and half full conditions, no overtopping occurs at the SRS and the peak discharge into the downstream reach is reduced to 196,000 cfs and 266,000 cfs, respectively.

3.For full SRS conditions, the dam is overtopped for those mudflows that occur with initial Castle Lake elevations greater than 2547 MSL. The maximum depth of overtopping is approximately 10 feet.

4.Reduction of the initial Castle Lake levels significantly reduces the magnitude of the resulting dam breach-induced mudflow.

5.Additional economic analyses are necessary to evaluate the cost/benefit characteristics of constructing mitigative measures for reducing the initial lake level. Studies are being conducted by the U.S. Army Corps of Engineers, the U. S. Forest Service and the U.S. Geological Survey to better determine the "most likely dam breach scenario" and the most effective way to insure safe operation of the Sediment Retention Structure.

References

1.Laenen, Antonius and L. L. Orzol, 1987. "Flood Hazards Along the Toutle and Cowlitz Rivers, Washington, From a Hypothetical Failure of Castle Lake Blockage," Water Resources Investigations Report 87-4055. U. S. Geological Survey, prepared in cooperation with the State of Washington Department of Emergency Management, Portland, Oregon.

2. Laenen, Antonius, K. K. Lee and L. L. Orzol, 1990, (Unpublished Report). Computation of Flood Hydrographs at the Sediment Retention Structure for Hypothetical Failure of the Castle Lake Blockage for Several Different Starting Elevations, USGS, Seattle, WA.

3. MacArthur, Robert C., David R. Schamber, and Douglas L. Hamilton, 1985. "Toutle River Mudflow Investigation," Special Projects Report No. 85-3. Prepared for the Portland District, U.S. Army Corps of Engineers, Portland, Oregon.

4. MacArthur, Robert C., David R. Schamber, Douglas Hamilton and Mary West, 1987. "Verification of A Generalized Mudflow Model," Proceedings of the National Conference on Hydraulic Engineers, ASCE, Williamsburg, VA, August 3-7.

5.MacArthur, Robert C., Douglas L. Hamilton and David R. Schamber, 1988. "Petrov- Galerkin Finite Element Formulation of the One-Dimensional Mudflow Model," prepared by Simons, Li & Associates, Inc., for the Portland District, U. S. Army Corps of Engineers, Portland, Oregon, March.

6. Major, Jon J., 1984. "Geologic and Rheologic Characteristics of the May 18, 1980 Southwest Flank Lahars at Mount St. Helens, Washington," Master of Science Thesis, Pennsylvania State University, University Park, Pennsylvania.

Some Aspects of the Mechanism of Debris Flow

Jinren Ni[1] and Guangqian Wang[2]

Abstract

As the first step of the development of a general two-phase flow model for debris flows, a simple granular model, which is considered as a special case of the two-phase flow model, is suggested here for ordinary debris flows. Although the preliminary study shows a good fit with the experimental results, a refined two-phase flow model is still needed in order to have insights into more aspects of debris flows.

Introduction

Generally speaking, the theoretical models on the mechnisms of debris flows can be classified into three categories, or, non-Newtonian fluid model, granular flow model and two-phase flow model. Among these models, the non-Newtonian fluid model is the most widely used one and thus a lot of excellent work have been done hitherto (Bagnold, 1954; Takahashi,1980; Tsubaki etal, 1982; Chen, 1988). On the contrary, the granular fluid model and the two-phase model are not very popular owing to their complexty natures. In fact, the real improvements of the granular flow theory are finished in the recent ten years. The development of the two-phase flow model, in which both the Newtonian or non-Newtonian nature of the liquid phase and the interactions between the phases should be counted in, will be a more difficult matter. Nevertheless, such a model is greatly needed for the purpose of developing a general system in which most of the flows, including single-phase liquid flow, liquid/solid two-phase flow, granular flow and non-Newtonian flow, can be considered as a whole. As the first step, only a very simple granular model, which corresponds with the case of little liquid effect in solid/liquid two-phase flow, is introduced here for approaching the mechanisms of debris flows. In the following discussions, special attentions will be paid to the solutions of velocity distributions of debris flows with considerations

---

1 Ph. D., Department of Geography, Peking University, Beijing 100871, China.

2 Ph. D., Department of Hydraulic Engineering, Tsinghua Univ., Beijing 100084, China.

of the concentration distribution.

Development and Test of the Model

The behaviours of grannular flows are very similar to those of low-viscous debris flow, and thus exists a good bbasis for the direct analogy between them or for a revised version. Usually, the total stresses for the granular flow are expressed as

$$P_{ij} = P_{fij} + P_{dij} + P_{cij} \tag{1}$$

here P_{ij}, P_{fij}, P_{dij} and P_{cij} are components of the total shear stresses, friction stresses, dispersive stresses and collision stresses, respectively. For the simple two-dimensional shear flows with the neglect of the dispersive stresses, the relation (1) can be further written in the Cartesian coordinate system, or

$$P_{xy} = P_{fxy} + P_{cxy} \tag{2}$$

$$P_{yy} = P_{fyy} + P_{cyy} \tag{3}$$

In which, P_{fxy} and P_{fyy} have been represented by Savage(1983) and Johnson et al (1987) as

$$P_{fxy} = P \sin\phi, \quad P_{fyy} = P \tag{4}$$

where P is the normal pressure and ϕ is the internal friction angle, both of them are related to the particle concentration. According to Wang & Fei (1989), P_{cxy} and P_{cyy} can be obtained by the analogy between the motions of sediment grains and those of dense gas molecules, i.e.,

$$\begin{aligned} P_{cxy} &= \frac{4(1+e)}{15R}\sqrt{\frac{3}{\pi}}\left(1+\frac{1}{8Cg_o}\sqrt{\frac{2\pi}{3}}\right)\rho Cg_o D^2\left(\frac{du}{dy}\right)^2 \\ &= \frac{4}{15}(1+e)\sqrt{\frac{3}{\pi}}\left(1+\frac{1}{8Cg_o}\sqrt{\frac{2\pi}{3}}\right)\rho Cg_o VD\,\frac{du}{dy} \end{aligned} \tag{5}$$

and

$$\begin{aligned} P_{cyy} &= \frac{2}{3R^2}(1+e)\rho Cg_o D^2\left(\frac{du}{dy}\right)^2 \\ &= \frac{2}{3}(1+e)\rho Cg_o V^2 \end{aligned} \tag{6}$$

here, V = the mean square root of the particle fluctuating velocity, its correlation with the granular flow temperature is as such, $(3/2)T=(1/2)V^2$; C=volumetric concentration; D, particle diameter; u=temporal mean velocity in x-direction; $\rho = \rho_p C$, the density of granular flow (ρ_p is density of particles); e = a elastic restoring coefficient; $g_o(C)$ is a radial distribution function which takes the form suggested by Canahan & Starling (1969) as

$$g_o(C) = \frac{1}{1-C} + \frac{3C}{2(1-C)^2} + \frac{3C^2}{(1-C)^3} \tag{7}$$

and the parameter R is defined as

$$R = D \left| \frac{du}{dy} \right| / V \tag{8}$$

Combining all these relations with some simplifications, a new form is derived

$$r = \frac{\frac{2}{5tg\theta}\sqrt{\frac{3}{\pi}}\left(1+\frac{1}{8Cg_o}\sqrt{\frac{2\pi}{3}}\right) R-1}{1-\frac{\sin\phi}{tg\theta}} \tag{9}$$

here the bed slope θ and the parameter r, which stands for the ratio of the normal friction stresses to the normal collision stresses, are defined in the following form, respectively

$$\theta = \text{arctg}\,\frac{P_{xy}}{P_{yy}}, \qquad r = \frac{P}{\frac{2}{3}(1+e)\rho C g_o V^2} \tag{10}$$

Note that

$$P_{xy} = g\sin\theta\int_y^H \rho\,dy = \bar{\rho}\,g(H-y)\sin\theta \tag{11}$$

we obtain

$$V^2 = \frac{3g\,\bar{\rho}(H-y)\cos\theta}{2(1+r)(1+e)\rho C g_o}, \quad V_b{}^2 = \frac{3g\,\bar{\rho}_b(H-b)\cos\theta}{2(1+r_b)(1+e)\rho_b C_b g_{ob}} \tag{12}$$

Thus the ratio of them becomes

$$\frac{V^2}{V_b^2} = \frac{1+r}{1+r_b}\;\frac{\bar{\rho}}{\rho}\;\frac{\rho_b}{\bar{\rho}_b}\;\frac{C_b g_{ob}}{C g_o}\left(\frac{1-\eta}{1-\eta_b}\right) \tag{13}$$

Here, $\bar{\rho}$ =mean density of granular flow with in region of y to H (H is the flow depth); η =y/H, relative flow depth; the parameters with the footnote "b" are those at y=b from bottom.

For the rapid flow conditions, two good approximations are applicable, i.e.,

$$r \doteq 0, \quad 1+\frac{1}{8Cg_o}\sqrt{\frac{2\pi}{3}} \doteq 1 \tag{14}$$

Furthermore, in consideration of the difficulties in determining the parameter V from the energy equation, an assumed expression for V according to the observed results is used here

$$\frac{V}{V_b} = \left(\frac{\eta}{\eta_b}\right)^{\alpha}\left(\frac{1-\beta\eta}{1-\beta\eta_b}\right) \tag{15}$$

in which, α and β are two parameters which vary with the flow and boundary conditions. Then, from relation (11), (13), (14), and (15) the final relation is derived

$$\frac{du}{d\eta} = A\,\eta^{\alpha}(1-\beta\eta) \tag{16}$$

where

$$A=\frac{15\sqrt{\frac{\pi}{3}}\,gH^2\sin\theta}{4(1+e)\,D}\;\frac{\bar{\rho}_b}{\rho_b}\;\frac{1}{C_b g_{ob} V_b}\;\frac{1-\eta_b}{(1-\beta\eta_b)\,\eta_b^{\alpha}} \tag{17}$$

in which, ρ_b is the bottom density of granular flow (at y=b); $\bar{\rho}_b$, the mean density over the depth. For the sake of conveni-ence, the relation (16) is integrated and normalized in such a from

$$\frac{u}{u_o}=\frac{1}{2+\alpha-\beta(1+\alpha)}\eta^{\alpha+1}\,[2+\alpha-\beta(1+\alpha)\eta] \tag{18}$$

Here, u_o is the maximum velocity at y/H=1. Some calculations have been made by the writers with the relation (18) and the results are given in Fig. 1, from which we can see that various kinds of velocity profiles can be predicted with the granular flow model presented here. In addition, some of the measured data from Takahashi (1980) and Wang (1986) are used to test the model. As results, good agreement between the calculations and the measured results is achieved with α=1.0 and β=0.75 as shown in Fig. 2 and Fig. 3, seperatly.

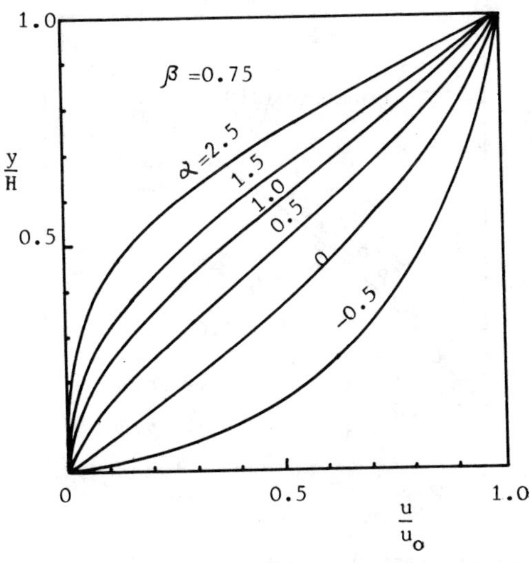

Fig. 1 Velocity profiles predicted by Eq.(18)

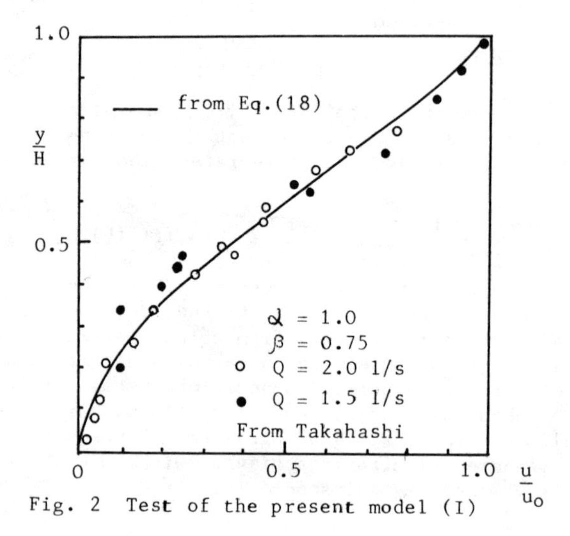

Fig. 2 Test of the present model (I)

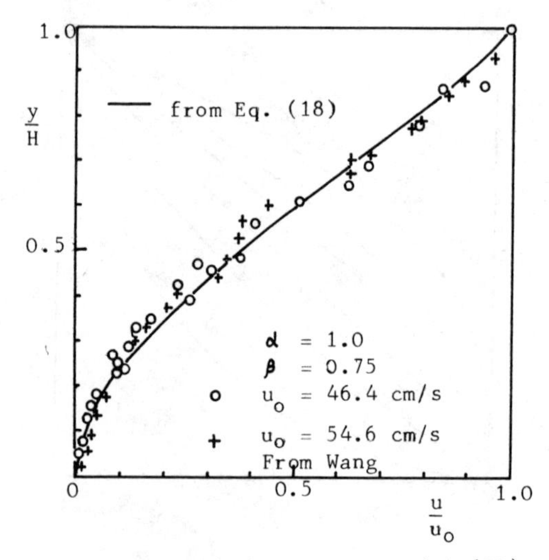

Fig. 3 Test of the present model (II)

Conclusions

From the above mentioned analyses, the following conclusions are drawn:

1. As the initial step of the development of a expected two-phase flow model for the debris flows, the simple granular flow model, which is considered as a special kind of two-phase flow model, should be well studied in advance.

2. By introducing the relevant theoretical results from the granular flow studies, a preliminary model for debris flow can be developed if the concentration distribution is properly given. The model suggested here will also present the velocity profiles which are proved to be well fitted with the measured data.

References

Bagnold, R.A. (1954). "Experiments on a gravity-free dispersion of large solid spheres in a Newtonian fluid under shear" Proc. Royal Soc. of London, Ser. A, Vol.225, PP. 49-63.

Canahan, N.F., and Starling, K.E. (1969). "Equations of state for non-attracting rigid spheres". J. Chem. Phys., Vol. 51, PP. 635-636.

Chen, C.L. (1988). "General solutions for viscoplastic debris flow". Jour. of Hydr. Engi., ASCE, Vol.114, HY3, PP. 237-258.

Johnson, P.C., and Jackson, R. (1987). "Frictional - collisional constitutive relations for granular materials, with application to plane shearing". J. Flu. Mech., Vol.167, PP. 67-93.

Savage, S.B. (1983). "Granular flows down rough inclines-review and extension". In Mechanics of Granular Materials: New Models and Constitutive Relations. Ed. by Jenkins, J. T. and Satake, M., Elsevier Science Publishers, PP. 261-282.

Takahashi, T. (1980). "Debris flow on prismatic open channel". Jour. of Hydr. Engi., ASCE, Vol.106, HY3, PP. 381-396.

Tsubaki, T., Hashimoto, H., and Suetsugi, T. (1982) "Grain stresses and flow properties of debris flow". Proc. of the Japanese Sco. of Civ. Engrs., Vol.317, PP. 79-91.

Wang, G.Q., and Fei X.J. (1989). "The kinetic model for granular flow". Proc. of the Fourth International Symposium on River Sedimentation, Beijing, PP. 1459-1466.

Wang, L.X. (1986). "Experimental studies on the laminated load motions". Master dissertation, Tsinghua University, China.

ERODIBLE CHANNEL MODELS: STATE OF THE ART REVIEW

By Subcommittee on Erodible Channels[1]
ASCE Task Committee on Flood Hazard Analysis on Alluvial Fans

INTRODUCTION

River channel behavior often needs to be studied for its natural state and response to human regulation. Studies of river hydraulics, sediment transport, and river channel changes may be through physical modeling, or mathematical modeling, or both. Physical modeling has been relied upon traditionally for river projects, but mathematical modeling is becoming more popular as its capabilities expand rapidly.

Mathematical models are developed based on convoluted physical relationships. The five most important relations are: continuity equation of flow, momentum equation of flow, continuity equation of sediment, sediment transport equation, and relations for channel deformation. These equations are applied at successive time steps to simulate the time-dependent processes. These equations may be coupled at each time step, but coupled solution is limited to rectangular channels with uniform changes in bed profile. For natural streams, the hydraulics of flow is usually uncoupled from the sediment processes.

Open channel hydraulics deals with free-surface flow, for which the water surface is the only free surface. Alluvial channels are characterized by a mobile boundary, in addition to its water-surface. Therefore, all boundaries of an alluvial channel are free surfaces. The hydraulics and fluvial processes of alluvial rivers are far more complicated than rigid channels. The study of river channel geometry and its dynamic changes presents a new challenge to the hydraulic engineering profession. This is an area for which not enough is understood, because of the complexity of fluvial processes. Therefore, this is an area for which much research and development are still needed.

This paper presents a review of existing modeling techniques, their physical and mathematical background, and applicabilities. Major areas of progress are outlined and difficulties are described to bring up-to-date recent development on mathematical modeling. An important progress in model development is the advancement from erodible-bed models to erodible-boundary models. In spite of such progress, continued research and development are still needed to expand the present modeling capabilities by including meander development and stream migration.

Predicting stream channel response to change, whether caused by nature or human, is important in river engineering, management, and environmental studies. An increasing number of problems related to stream channel changes is approached from computer-aided analyses and design. At present, mathematical models are applied to determine general scour at bridge crossings and near hydraulic structures, channel response to river control, channel design, evaluation of sand and gravel mining, dam breach analyses, reservoir sedimentation, transverse bed scour in curved channels, etc. The list of computer-aided applications is expanding rapidly.

---

[1]Members include Howard Chang (Chair), David Dawdy, Kenneth Edwards, Muna Faltas, Douglas James, Erich Korsten, Ray Lenaburg, James Slosson

The state-of-art review for erodible channel models is given herein according to models' dimensionality with regard to flow and changes in channel boundary. Stream channel changes include the following components: (1) Channel-bed scour and fill (or degradation and aggradation), (2) width variation, (3) curvature effects on bed topography, and (4) meander migration.

Mathematical models were initially developed to simulate channel-bed scour and fill. More sophisticated morphological features have gradually been added. For streams in the semi-arid west and those disturbed by human activities, width adjustment is the dominant change in channel morphology, especially visible on alluvial fans. As water and its transported sediment move downstream on a steep slope, which may well be in excess of supercritical, and yet critical flow prevails (Jarrett, 1984). Nature slows flow by adding energy dissipation by shaping irregular channel geometry, creating energy dissipation pools, and using energy to move sediment, contributing substantially to bank cutting (James, 1988). Nature slows flow also by widening whereby the large boundary resistance maintains the flow at the critical state. The channel width and roughness are two degrees of freedom among others. The roughness is flow-induced and it varies with the flow condition. This feature means that flow resistance and width variation must be modeled for such streams. The roughness formula by Brownlie (1983) provides a means of computer simulation of roughness changes with the flow condition.

ONE-DIMENSIONAL MODELS

One-dimensional models use the longitudinal coordinate along the flow direction as the spatial dimension. Cross sections are used to define the channel geometry. Such models consider changes along the flow direction; changes at a cross section are limited to scour and fill (or degradation and aggradation) of the bed without width changes. In order words, one-dimensional models are erodible-bed models. Without information on the lateral distribution of scour and fill across the width, scour and fill are assumed uniformly distributed, normally in the active bed zone. Examples include HEC-6 (Corps of Engineers, 1977), HEC2SR (Simons and Li, 1980), IALLUVIAL (Karim and Kennedy, 1982), and MOBED (Krishnappan, 1981), among others.

The one-dimensional equation for sediment continuity is as follows

$$(1 - \lambda) \frac{\partial A_b}{\partial t} + \frac{\partial Q_s}{\partial s} - q_s = 0 \qquad (1)$$

where λ is the porosity of bed material, A_b is the cross-sectional area of channel within some arbitrary frame, Q_s is the bed-material discharge, and q_s is the lateral inflow rate of sediment per unit length. According to this equation, the time change of cross-sectional area $\partial A_b/\partial t$ is related to the longitudinal gradient in sediment discharge $\partial Q_s/\partial s$ and lateral sediment inflow q_s. In the absence of q_s, longitudinal imbalance in Q_s is absorbed by channel adjustments toward establishing uniformity in Q_s.

The change in cross-sectional area ΔA_b for each section at each time step is obtained through numerical solution of Eq. 1. This area change must be applied to the active boundary area following correction techniques for channel-bed profile.

An important area of needed research, as pointed out by Dawdy and Vanoni (1986), is the allocation of scour and fill at a time step. While the total change in cross sectional area is provided by the difference in sediment inflow and outflow (Eq. 1), the adjustment of bed profile is usually not uniform. In fact, a cross section may have scour at some points while fill at other points, and vice versa. The pattern of change varies also with the changing discharge. A technique for scour and fill allocation was suggested by Chang (1988), in which the correction for scour and fill is allocated such that the resulting channel shape will expedite the movement toward uniformity in energy gradient or water-surface profile along the channel. The physical basis for this technique is that any changing alluvial stream must be adjusting toward dynamic equilibrium, under which the water-surface profile is linear. This technique brings into consideration the longitudinal profile on the cross sectional change in geometry.

The streamtube approach has been suggested by Molinas and Yang (1988) as a means of obtaining the lateral distribution of scour and fill. In this approach, the channel is divided into preselected number of tubes. Flow and sediment in each tube is computed independently to obtain the bed adjustment, so that the adjustment at each cross section may vary from tube to tube. In this approach, the change in bed elevation is based on the sediment rates computed within one tube; therefore, the lateral sediment exchange between tubes is neglected in computing streambed changes.

A second crucial problem is the sorting of bed sediment in a stream undergoing changes. While much progress has been made, such as by Borah, et al (1982), it should be pointed that in a one-dimensional model, sediment sorting is based on the cross-sectional averaged values and the lateral distribution of sediment is not considered.

The one-dimensional approach is limited to straight channels and rigid banks. Since most natural streams are sinuous, the curvature effects are important to the channel morphology but not covered in the one-dimensional models.

QUASI-TWO-DIMENSIONAL MODELS

A two-dimensional model uses a grid network for the channel and flow field definition, but a quasi-two-dimensional model uses cross sections like one-dimensional models but allows lateral changes at the cross sections. The lateral distribution of scour and fill described above may be considered as one of the quasi-two-dimensional features. A quasi-two-dimensional model also considers the geometric changes in cross section in the lateral direction, namely the channel width, in addition to scour and fill of the bed. This task has not been an easy one. While the channel boundary adjustment can be related to the sediment imbalance (Eq. 1), the total change ΔA_b needs to be applied to the bed and banks. Certain degree of success has already been achieved to allow the determination of sediment transport under the transient state of change. This development is described below.

First of all, the width changes must be for dynamic or transient conditions, but not for the long term adjustment in regime width. How the width would change has been a puzzle to researchers on river mechanics, since the only physical relationship pertaining to alluvial stream width was the following regime relation

$$B \propto Q^{1/2} \tag{2}$$

where B = width and Q = discharge. While this relationship is applicable to the regime condition or dynamic equilibrium, for which the width is a function of the discharge, it is not applicable to the situation of dynamic changes. In a transient condition, the width can be quite different under the same discharge. Such is demonstrated in Fig. 1 where a stream is undergoing transient change, by scour and fill of the bed and concomitant changes in width. The modeler is faced with two questions to resolve: the direction of width change and the rate of change. In the FLUVIAL model, the direction of width change is related to the stream channel's tendency to attain equilibrium. The dynamic equilibrium so defined consists of the two following conditions: (1) uniform sediment discharge along the channel, and (2) uniform power expenditure or uniform γQS. Since the energy gradient S may be approximated by the water-surface slope, uniform power expenditure also means a linear water-surface profile. A stream channel undergoing changes may not have a linear water-surface profile, but the changes in width are such that they will result in the linear water-surface profile subject to the physical constraints. This tendency toward uniformity in power expenditure is the physical basis for the simulation of the direction of width change. The rate of width change is then related to the flow and bank erodibility.

While the physics for width adjustment has been advanced with important accomplishment, there remains research to be done in providing more information to improve the accuracy of the rate of bank erosion and mechanism of bank failure. The effects of vegetation and bank materials are important on the rate of migration; they must be tied in with the hydraulics of flow in the river. The research development by Thorne (1982) and others on bank failure should be incorporated into the model with width adjustment. The importance on the understanding of the bank erosion processes and their influences on rivers in underlined by the special sessions on river bank erosion and deposition at the 1989 National Conference on Hydraulic Engineering and the Spring AGU meeting in 1989.

Fig. 1. Spatial variation of stream width accompanied by concurrent streambed scour and fill

Quasi-two-dimensional features also include the curvature effects on bed topography. Currently, this feature has already been incorporated into the FLUVIAL-12 model (Chang, 1985). The curvature effects are responsible for the sediment movement in the lateral direction and the transverse bed profile in curved channels. A natural extension is to incorporate bank retreat along the concave bank.

The Sedimentation Work Group (1988) of the Federal Interagency Advisory Committee on Water Data sponsored a workshop in October 1988 on mathematical models for sedimentation which was attended by over 80 participants. The models HEC2SR, FLUVIAL-12, SEDICOUP and CHARIMA, and TWODSR and RESSED were reviewed and comments were made by Nordin and Cunge. An up-to-data review of the models and their applicabilities are contained in the workshop proceedings. This publication perhaps represents the most up-to-date information on the one- and quasi-two-dimensional models.

TWO-DIMENSIONAL MODELS

Several two-dimensional models on the hydraulics of flow have been developed, examples are by Leendertse (1987), Hromadka and Yen (1987), TABS (Thomas and McNally, 1985), and many estuarine models. In such models, the two-dimensional flow equations in the x- and y-directions are couples. The model by Hromadka and Yen also considers unsteady flow. A two-dimensional model is based upon a grid system but not cross sections to define the channel geometry. The flow, sediment, and morphologic characteristics (if any) are defined at the nodal points. The two-dimensional approach is an approximation to the real flow pattern by depth-averaged flow. Because of depth-averaging, the two-dimensional approach does not include the effects of secondary flow. As rivers are seldom straight over a few channel length, it is unrealistic to neglect the effects of secondary currents. The lack of representation for secondary currents is a major reason why two-dimensional approach has been more successful for the estuarine condition rather than the riverine condition.

In those two-dimensional models involving sedimentation, the direction and quantity of sediment transport are related to the flow pattern and the sediment transport at the nodal points are computed. The channel changes due to sediment imbalance are then determined from the continuity equation for sediment, which has the following form:

$$(1 - \lambda) \frac{\partial A_b}{\partial t} + \frac{\partial (Q_s)_x}{\partial x} + \frac{\partial (Q_s)_y}{\partial y} = 0 \qquad (3)$$

The first term is the change in bed level; the second term is the spatial variation of sediment rate in the x-direction; and the third term is the variation in the y-direction. If this equation is solved using a finite-difference technique, the net storage (or depletion) of sediment is obtained from the sediment flow rates crossing the boundaries of each element. Then the change in bed elevation is the sediment storage divided by the surface area.

The two-dimensional approach provides a better definition of the flow field and sediment movement than the one or quasi-two-dimensional models. However, this approach alone does not answer the question of width variation for alluvial channels, since the mechanisms for width must be related to the turbulent eddies which are not included in the depth-averaged flow analysis. A logical extension is to incorporate the effects of secondary currents.

Computing time becomes excessive for two-dimensional model, thus preventing it from being used for long river reaches or rivers with long flow duration. A rational way to circumvent this difficulty is to integrate the two-dimensional approach with a one-dimensional approach, such as demonstrated in TWODSR (Chen in Interagency Sedimentation Work Group, 1988). The two-dimensional approach is only used in small portions of the channel in this model.

THREE-DIMENSIONAL MODELS

Three-dimensional models have been advanced basically for hydrodynamics, exemplified by those of Rodi (1980), Chiu, et al (1978), and Wang (1982). This most comprehensive approach provides the complete definition of flow including the secondary currents. The complexity of the modeling technique and computing time consumption have so far prevented them from being used for rivers of any significant length.

Most three-dimensional models only cover the hydraulics of flow but not bed sediment movement in an erodible boundary. However, they will provide the basis for the next generation of erodible boundary models. Because of the large consumption in computing time. Such models may be useful for flow determination of steady flow, but not for a long flow duration or a long river reach.

SUMMARY AND CONCLUSIONS

This review reveals that most river sedimentation models in use are either one-dimensional or quasi-two-dimensional models. In order to be useful, a model should posses the ability for lateral distribution of scour and fill, width changes, and channel morphology associated with the channel curvature. An important area of investigation is the bank erosion and failure mechanism. The work of Thorne must be coupled with the hydraulics of flow, slippage, etc. Such advances will also lead to development for predicting meander migration.

In future perspective, model development should be to expand existing capabilities. While the eventual model is a three-dimensional one, there still exit much to do even for quasi-two-dimensional models. The list of improvement may include these following areas: boundary conditions of flow and sediment; sediment transport formulas employed; bed material sorting; allocation of scour and fill; bank erodibility and width changes; curvature effects on bed topography; meander migration; braided channels, compound channels; and tributaries.

REFERENCES

Borah, D. K., Alonso, C. V., and Prasad, S. N., "Routing Graded Sediments in Streams:

Formations," *J. Hydraul. Div.*, ASCE, 102(HY12), pp. 1486-1503, December 1982.

Brownlie, W. R., "Flow Depth in Sand-Bed Channels." *Journal of Hydraulic Engineering*, ASCE, 109(7), 1983, 959-990.

Chang, H. H., "Mathematical Model for Erodible Channels," *J. Hydraul. Div.*, ASCE, 108(HY5), pp. 678-689, May 1982.

Chang, H. H., "Water and Sediment Routing through Curved Channels," *J. Hydraul. Eng.*, ASCE, 111(4), pp. 644-658, April 1985.

Chang, H. H., Osmolski, Z., and Smutzer, D., "Computer-Based Design of River Bank Protection," Proceedings of the Hydraulics Division Conference, ASCE, Orlando, Florida, August 13-16, 1985, pp. 426-431.

Chang, H. H., *Fluvial Processes in River Engineering*, John Wiley & Sons, New York, 1988, 432pp.

Chiu, C-L., Hsiung, D. E., and Lin, H-C., "Three Dimensional Open Channel Flow", *J. Hydraul. Div.*, ASCE, 104(HY8), 1978, 1119-1136.

Corps of Engineers, U. S. Army, "HEC-6, Scour and Deposition in Rivers and Reservoirs, Users Manual," Hydrologic Engineering Center, Davis, California, 1977.

Dawdy, D. R. and Vanoni, V. A., "Modeling Alluvial Channels," *Water Resour. Res.*, 22(9), pp. 71S-81S, August 1986.

Holley, F. M. Jr. and Karim, M. F., "Simulation of Missouri River Bed Degradation," *J. Hydraul. Eng.*, ASCE, 112(6), pp. 497-517, June 1986.

Hromadka, T. V. and Yen, C. C., "A Diffusion Hydrodynamic Model", USGS Water Resources Investigation Report 87-4137, NSTL, Miss.

Interagency Sedimentation Work Group, 1988, "Twelve Selected Computer Stream Sedimentation Models Developed in the U. S.", S. S. Fan, Editor, Published by Federal Energy Regulatory Commission, 353-412.

James, L. D., "Continuing Contributions of Parametric Modeling", unpublished manuscript, 1988.

Jarrett, R. D., "Hydraulics of High-Gradient Streams", *J. Hydraul. Div.*, ASCE, 110(HY11), 1984, 1519-1539.

Karim, M. F. and Kennedy, J. F., IALLUVIAL: A computer-based flow and sediment-routing model for alluvial streams and its application to the Missouri River. IIHR Report 250, Iowa Inst. of Hydraulic Research, Iowa City, 1982.

Krishnappan, B. G., "Users Manual. Unsteady, Nonuniform, Mobile Boundary Flow Model-MOBED," Hydraulics Division, National Water Research Institute, Canada Center for Inland Waters, Burlington, Ontario, Canada, 1981.

Leendertse, Jan J., "Aspects of SIMSYS2D: A System for Two-Dimensional Flow Computation," Rand Corporation, P. O. Box 2138, Santa Monica, CA, 1987, 80pp.

Molinas, A. and Yang, C. T., 1988. Applications of GSTAR model. *Hydraulic Engineering*, Proceedings of the 1988 National Conference, ASCE, 90-96.

National Academy of Sciences, "An Evaluation of Flood-Level Prediction Using Alluvial River

Models," Committee on Hydrodynamic Computer Models for Flood Insurance Studies, Advisory Board on the Built Environment, National Research Council, National Academy Press, Washington, D.C., 1983.

Rodi, W., "Turbulence Models and Their Application in Hydraulics - A State of the Art Review," University of Karlsruhe, Federal Republic of Germany, 1980, 104pp.

Simons, Li, and Associates, Inc., "Erosion, Sedimentation, and Debris Analysis of Boulder Creek, Boulder, Colorado," prepared for UPS Company, Denver, Colorado, 1980.

Thomas, W. A. and McNally, W. H., 1985. Open Channel Flow and Sedimentation, TABS-2", Instructional Report H-85-1, U. S. A. E. Waterway Experiment Station, Vicksburg, MS.

Thorne, C. R., "Processes and Mechanisms of River Bank Erosion", *Gravel-Bed Rivers*, edited by Hey, Bathurst and Thorne, John Wiley & Sons, 1982, 227-272.

Vanoni, V. A., Born, R. H., and Nouri, H. M., "Erosion and Deposition at a Sand and Gravel Mining Operation in San Juan Creek, Orange County, California," *Storms, Floods, and Debris Flows in Southern California and Arizona 1978 and 1980*, Proceedings of a Symposium, September 17-18, 1980, National Academy Press, Washington, D.C., 1982, pp. 271-289.

Wang, S. Y., 1982. Computer simulation of sedimentation processes. *Finite Elements in Water Resources*, Vol. IV, 16.35-16.47.

ARMOR LAYER EVOLUTION

Carlos E. Mosconi, A.M. ASCE

Abstract

This communication presents an analysis of celerity propagation for non-uniform materials. Based on the corresponding flow, sediment transport and sorting/armoring equations, the celerity expressions for both bed elevation and composition are derived. Good agreement is obtained when the obtained analytical expressions are compared with experimental data for the armor layer evolution phase as performed by the author at the Iowa Institute of Hydraulic Research (IIHR).

Introduction

Hydraulic engineers are often confronted with the problem of calculating the time evolution of bed elevation and composition of graded materials. Because of the complexity of this problem mathematical models are commonly used.

Several mathematical models for non-uniform materials have been formulated in the past. They must satisfy certain stability conditions which depend on both the aspect ratio for the grid size as well as on the magnitude of the physical celerity. Among other researchers, Lu (1984), Park (1986), and Mosconi (1986, 1989) pointed out the importance of the correct evaluation of the celerity propagation and the role of upstream sediment boundary conditions leading to the formation of either a partial or complete armor layer.

Celerity predictors are known and widely applied for uniform materials. Most of these analyses hinge on the pioneering work of de Vries (1965) for uniform materials and they use a power law to describe sediment transport.

On the other side, to the author's knowledge, no entirely satisfactory approach to the problem of celerity propagation for non-uniform materials prone to armor layer formation has been formulated yet. Consequently, this paper investigates the issue of celerity propagation for non-uniform materials and its interaction with armoring.

Senior Hydraulic Engineer, TAMS Consultants Inc.,
The TAMS Building, 655 Third Avenue, New York, New York 10017

Mathematical Formulation

The basic one-dimensional equations that govern flow over a deformable river bed composed of non-uniform materials undergoing a non-equilibrium process are:

Momentum Equation for Water

$$\frac{\partial U}{\partial t} + U \frac{\partial U}{\partial x} + g \frac{\partial H}{\partial x} + g \frac{\partial z}{\partial x} + g\, S_f = 0 \tag{1}$$

Continuity Equation for Water

$$\frac{\partial H}{\partial t} + U \frac{\partial H}{\partial x} + H \frac{\partial U}{\partial x} = 0 \tag{2}$$

Continuity Equation for Sediment

$$\frac{\partial z}{\partial t} + \frac{1}{(1-\lambda)} \frac{\partial G}{\partial x} = 0 \tag{3}$$

Sediment Transport Relation

$$G = G(U,H,d,g,w,S_f,\nu,\ldots) \tag{4}$$

Resistance Relation

$$f = f(U,H,G,d,w,\nu,\ldots) \tag{5}$$

Mixing Layer Relation for Sediment

$$\frac{\partial d_{50SL}}{\partial t} = (d_{50ER} - d_{50PBM}) \frac{\partial z}{\partial t} - G \frac{\partial d_{50ER}}{\partial x} \tag{6}$$

in which: d_{50SL}, d_{50ER}, d_{50PBM} are the medium-size diameters for surface layer, eroded material and parent bed material distributions, respectively; f = Darcy-Weisbach friction factor; G = sediment discharge per unit width; g = gravitational acceleration; H = mean water depth; s_f = energy slope; t = time; U = mean flow velocity; x = longitudinal coordinate measured along the channel bed; w = fall velocity of a sediment particle; z = average bed elevation with respect to an arbitrary reference level; λ = porosity of bed sediment; ν = kinematic viscosity of water and η = thickness of the mixing layer.

Mosconi (1988) correlated different percentiles of the eroded materials to the parent bed material composition by means of the dimensional analysis technique. In particular the median size of the eroded material relates to the parent bed material as:

$$d_{50ER} = [1 - (1 - \frac{U_*}{U_{*CMAX}})/\beta\ ] \quad d_{50PBM} \tag{7}$$

in which: U_* = shear velocity, U_{*CMAX} = critical maximum shear velocity associated to the coarsest armor layer that a given parent bed material can generate, and β = function of the standard deviation of the original distribution as shown on Figure 1.

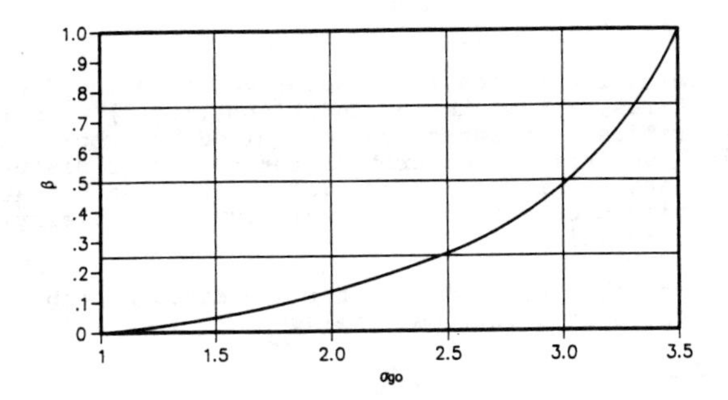

Figure 1: β **Relationship with** σ_{go} **in Equation 7.**

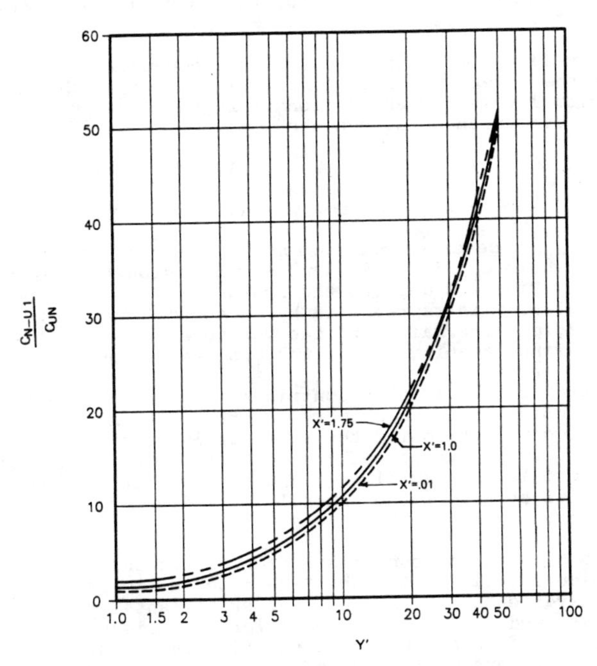

Figure 2: Normalized Bed-Elevation Celerity Variation with Y' and X' (parameter).

Analysis

On elimination of the local water velocity and depth derivative terms, the system of Equations (1)-(7) with four finite celerities is transformed to a mixed hyperbolic-parabolic system with two infinite celerities for the water-disturbance propagation and two finite celerities for the bed elevation and bed composition propagation. The validity of this simplification was further investigated.

The compatibility condition for the mixed hyperbolic-parabolic system of equations, yields

$$C^2 - C\ (C_{UN} + Y + X) + X\ C_{UN} = 0 \qquad (8)$$

in which: C = celerity expression for both bed elevation and bed composition; Y = parallel degradation with non-uniform material contribution and X = bed composition contribution. Y and X are functions of the bed surface material coarsening ratio, Froude number, ratio between acting shear velocity and maximum shear velocity, coefficients depending upon the functional form of the adopted sediment transport equation, thickness of the mixing layer and other flow parameters.

On normalization of the celerity for non-uniform materials (C_{N-U}) with the celerity for uniform materials (C_{UN}), the resulting non-dimensional celerity expressions for graded materials for the armor layer evolution phase becomes

$$C_{N-U1} = 1/2\ (Y' + X')\ \{1 + [1 - 4\ X' / (Y' + X')^2]^{.5}\} \qquad (9)$$

$$C_{N-U2} = 1/2\ (Y' + X')\ \{1 - [1 - 4\ X' / (Y' + X')^2]^{.5}\} \qquad (10)$$

in which: C_{N-U1} = celerity expression for the bed elevation propagation while C_{N-U2} = corresponding expression for the bed composition. Figure 2 depicts the non-uniform material bed-elevation celerity as represented by Equation 9 for different armoring stages. The values of the variables X' and Y' have been chosen in a range meaningful for the problem at hand.

The parallel degradation contribution to the bed-elevation celerity is observed to be of more importance than the bed composition contribution for the range of interest. For example the bed celerity for values of $Y' = 20$ magnifies the celerity for uniform materials over 20 times. The highest multiplier occurs during initial stages of the armor layer and decays as the armor layer evolves.

Figure 3 shows the comparison of observed and calculated bed-elevation celerities for the IIHR experiments performed on the armor layer evolution phase. The experimental data from Runs 3-1, 3-2, 3-3 and 3-4 have been used for comparison purposes. The comparison of Equation 9 with experimental results shows a good overall agreement in spite of the slight underestimation. The data that was compared with the analytical prediction was not included in the estimation of the numerical coefficients in the adopted bed-load transport equation.

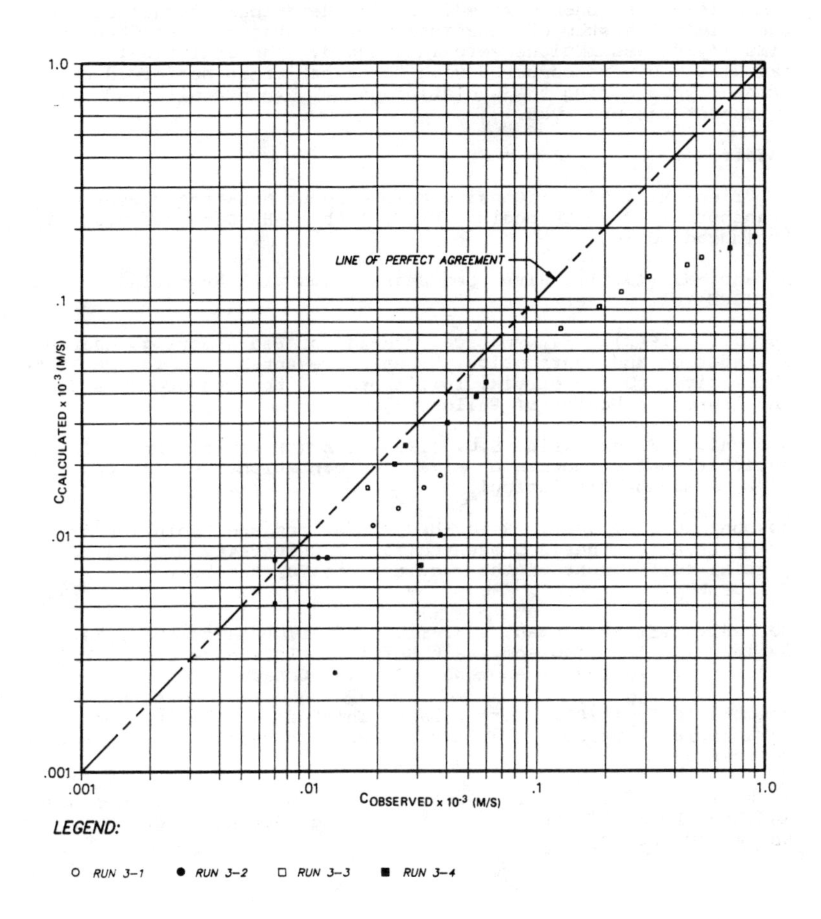

Figure 3: Comparison of Observed and Calculated Bed-Elevation Celerity during the Armor Layer Evolution Phase.

Conclusions

A methodology for computing the celerity propagations during the armor layer evolution phase for both bed elevation and composition has been derived. The mathematical characteristic of the original system of equations were maintained even when the simplifying assumptions were introduced. The analytical calculations were compared with the experiments performed on armor layers during the evolution phase. An overall good agreement was observed.

References

de Vries, M. (1965). Considerations About Non-Steady Bed-Load Transport in Open Channels. Proc. 11th. IAHR Congress, Vol. 3, Leningrad, USSR.

Hirano, M. (1971). River Bed Degradation with Armouring. Trans. of JSCE Vol. 3, Part 2.

Lu, J.Y. (1984). Mathematical Models for Bed Armoring, Channel Degradation and Aggradation. Thesis presented to Colorado State University, CO, in partial fulfillment of the requirements for the degree of Doctor of Philosophy.

Mosconi, C.E. and Jain. S.C. (1986). Armor Layer. Proc. Third International Symposium on River Sedimentation, Vol. 3, pp. 1785-1792, Jackson, Mississippi.

Mosconi, C.E. (1988). River-bed Variations and Evolution of Armor Layers. Thesis presented to The University of Iona, IA, in partial fulfillment of the requirements for the Doctor of Philosophy.

Mosconi, C.E. and Borah, D.K. (1989). Armor Layer Analysis. Seminar 4. Armoring and Grain Sorting: Influence of the Basic Concept of Bed Load Phenomena. Ottawa, Canada.

Mosconi, C.E. (1989). Armor Layer Thickness. International Workshop on Fluvial Hydraulics of Mountain Regions, Trent, Italy, October 3-6.

Park, I. and Jain, S.C. (1986). River-bed Profiles with Imposed Sediment Load. Journal of Hydraulic Engineering, ASCE, Vol. 112, No. 4, pp. 267-280.

Channel-changing processes on the Santa Cruz River, Pima County, Arizona, 1936-86

John T.C. Parker [1]

Abstract

Lateral channel change on the mainly ephemeral Santa Cruz River, Pima County, Arizona, causes damage and has spawned costly efforts to control bank erosion. Aerial photographs, historical data, and field observations are used to document the history of channel change since 1936. Variability in the nature and degree of channel change over time and space is shown. Three major channel change processes are: (1) migration by bank erosion during meander migration or initiation; (2) avulsion by overbank flooding and flood plain incision; (3) widening by erosion of low, cohesionless banks during floods and arroyo widening by undercutting and mass wasting of deeply incised vertical walls. The first process generally is a product of low to moderate flows or waning high flows; the others result mainly from higher flows, though sensitive arroyo walls may erode during relatively low flows. Channel morphology, bank resistance, and hydrology are factors determining the dominant channel-changing process on a particular reach of the river. Present river morphology reflects high flows since the1960's.

Introduction

The Santa Cruz River is a predominantly ephemeral stream with a history of lateral channel instability draining 22,200 km^2 in southeastern Arizona and northern Mexico (fig. 1). Bank erosion has caused property damage, and has spawned a costly effort to control lateral channel shifts, primarily by structural works. This paper describes the history and nature of lateral channel change on the Santa Cruz River in Pima County, Arizona, from 1936 to 1986 and evaluates the geologic, hydrologic, and hydraulic controls governing the spatial and temporal variability in lateral change. Earlier history of channel change on the Santa Cruz River is described in Betancourt and Turner (1988) and Cooke and Reeves (1976). Runoff, on the Santa Cruz River occurs in response to local thunderstorms during the summer monsoon season or to widespread winter frontal storms. Occasional tropical storms in the fall have produced record floods in the basin. Major flow events of this century include the 1977 flow, which was the flood of record until the 1983 flood, which was about 2.5 times greater (Roeske and others, 1989). In rural areas, human uses of river bottomlands are mainly agricultural. Within Tucson, residential and commercial properties occupy land adjacent to the channel.

Six reaches are defined in the study area based on morphology(fig. 1). The four upstream reaches are ephemeral. The Canoa reach has an active flood plain

---

[1] Non-member, U.S. Geological Survey, 300 W. Congress Street, Tucson, Arizona 85701

shallowly incised by a wide, sandy channel that deepens and narrows downstream. In the next three reaches, the channel flows through an arroyo 3 to 10 m deep. Most of the historic flood plain above the arroyo is inactive. The Cortaro and Marana reaches have been perennial since 1970 because of sewage effluent, and are characterized by a narrow channel flowing across an active flood plain.

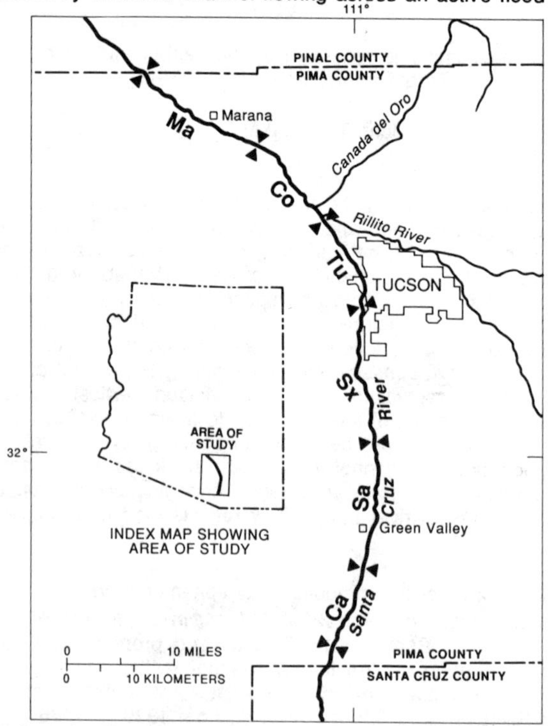

Figure 1. Study area with six reaches of the Santa Cruz River delineated in this study: **Ca**=Canoa; **Sa**=Sahuarita, **SX**=San Xavier, **Tu**=Tucson, **Co**=Cortaro, **Ma**=Marana.

Lateral channel change

Lateral channel change on the Santa Cruz River has high spatial and temporal variability. About half the channel in Pima County underwent little change from 1936 to 1986. Some reaches had continuous channel change; others changed primarily during a single flow event. Human activity has obscured the record of channel change and altered channel-changing processes. Three processes of lateral channel change were observed in the study area: (1) channel migration caused by bank erosion during meander migration or initiation; (2) channel avulsion caused by overbank flooding and flood plain incision; (3) channel widening caused by erosion of low, cohesionless banks during floods and arroyo widening, which is produced by a combination of meander migration, erosion of cohesionless materials and mass wasting of weakly indurated, deeply incised vertical walls.

Channel migration

Channel migration here refers to lateral shifts of channel centerline position associated with the inception, lateral extension, or downstream translation of meanders (Knighton, 1984). It involves continuous movement of channel position across a flood plain, rather than a discrete, abrupt channel shift caused by avulsion. Generally, channel migration increases sinuousity and lowers gradients. Primarily the product of low to moderate flows on the Santa Cruz River(Hays, 1984), channel migration apparently also occurs during recession from high flows. On ephemeral reaches, migration of confined meanders causes arroyo widening . On the Cortaro and Marana reaches, channel migration was the dominant channel-changing process from 1970, when perennial flow began, until the 1983 flood, which produced widespread avulsion. During the 1970's, the channel in those reaches narrrowed and became more sinuous in response to the low, perennial discharge (Hays, 1984). Almost all channel change on the Cortaro reach (fig. 2) during the 1966-78 interval was in response to channel migration. The migration rate was relatively low except at distances 4,500 and 6,000 m where 200 to 400 m of movement occurred.

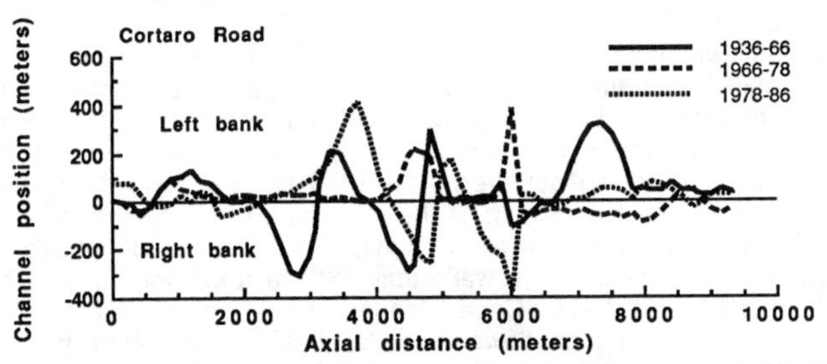

Figure 2. Migration of channel center line on part of the Cortaro reach. Upstream end is on the left. Horizontal line indicates position of channel centerline at beginning of interval.

Channel avulsion

Avulsion is an abrupt shift in channel course that occurs when overbank flow incises new channels as others are abandoned. On the Santa Cruz River, avulsion occurs mainly when high overbank flows are channelized by flood-plain topography, stripping flood-plain vegetation and eroding underlying sediment. Avulsion tends to reduce sinuousity and increase gradient. It has been a major channel-changing process on the Cortaro and Marana reaches where most channel change between 1978 and 1986 (fig. 2) resulted from avulsion during the 1983 flood. Avulsion has also occurred locally on the Canoa and Sahuarita reaches.

Avulsion occurs where the main channel is shallowly incised, the flood plain is active, and aggradation rates and sinuousity are relatively high. Low relief enables deep overbank flow to occur and incise a new channel. High sinuousity values and deposition rates enhance avulsion by lowering channel gradient and causing aggradation of the channel and flood plain thus forcing flow into a more direct, steeper course across lower flood-plain surfaces. A clear relation between flood magnitude and the amount of channel change occurring by avulsion on the Santa Cruz River is not evident. Considerable channel straightening and large shifts in channel position from avulsion took place on downstream reaches between

1936 and 1966 (Fig. 2) when the highest measured peak discharge was 481 m^3s^{-1}. Avulsion accounted for little change in the 1966-78 interval (Fig. 2) which included the1977 flood with a peak discharge of 651 m^3s^{-1}. Enhanced growth of flood-plain vegetation after 1970 may have limited avulsion by increasing resistance to flood plain erosion and channel incision. The 1,841-m^3s^{-1} 1983 flood was the most extensive avulsion event on the Santa Cruz River during the study period. In the Cortaro and Marana reaches, over 7,000 m of channel were left standing 2 m above the new channel bed, and shifts in channel position of up to 600 m occurred. In addition to the magnitude of the 1983 flood, different antecedent conditions, such as degree of pre-flood aggradation, may have produced greater channel change than occurred in the 1977 flood.

Although major shifts in channel position have occurred on the Santa Cruz River since 1936, both by avulsion and by channel migration, such changes have been restricted to a generally well-delineated, modern flood plain. Older terraces are elevated above areas affected by avulsion, and older alluvium generally is sufficiently indurated to resist bank erosion from lateral meander extension.

Channel and arroyo widening

Channel widening results primarily from high flow events that erode relatively cohesionless banks. It is a product of lateral scour (Hooke, 1979), or mass wasting of banks (Baker, 1988). Channel widening has been most profound on parts of the Canoa reach with low, sandy, sparsely vegetated banks where as much as 370 m of widening occurred from 1978 to 1986 (fig. 3). In this paper, channel widening refers to widening of the channel on an active flood plain or to widening of a low-flow channel within a high-flow channel or arroyo. Arroyo widening refers to widening of the walls that enclose a channel incised below an historic, abandoned flood plain. Once widened, arroyo boundaries tend to persist longer in the absence of high flows than do channel boundaries which rapidly narrow because of in-channel deposition, revegetation, and bank scarp degradation.

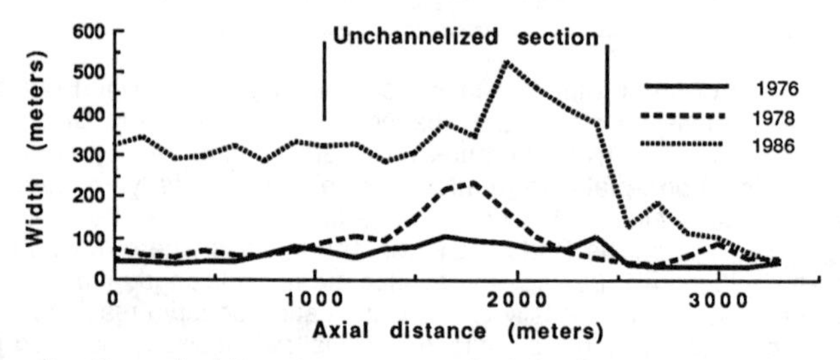

Figure 3. Channel widths at upstream end of the Canoa reach, 1976-86. Downstream direction is to the right.

Arroyo widening on the Santa Cruz River occurs when weakly indurated, oversteepened arroyo walls fail as a result of (1) undercutting of walls by flow events; (2) partial saturation and loss of cohesion of arroyo wall materials; (3) return flow of bank storage to the channel during hydrograph recession, which elevates pore pressures or liquefies basal bank sediments and promotes mass wasting. Arroyo walls may undergo dramatic retreat during extension or downstream translation of entrenched meanders, by inception of a meander within a constricted reach, or by widening of a low-flow channel until it impinges on the walls. Widening is abetted by structural weakening of arroyo walls associated with formation and enlargement of joints, tension cracks, macropores and pipes (Knighton, 1984). The most extensive arroyo widening on the Santa Cruz River has been on the San Xavier reach (fig. 4) where the channel is incised up to 10 m in Holocene age silt and sand. Over 370 m of widening occurred at some locations between 1936 and 1986, and mean arroyo width increased from 65 to 150 m. In contrast to the extreme widening seen in figure 4, arroyo widening in well-indurated alluvium of Pleistocene age at the downstream end of the Sahuarita reach is virtually imperceptible on aerial photgraphs from 1956 to 1986. Although textural variation of alluvium affects local susceptibility to erosion (Hays, 1984; Pearthree and Baker, 1987), the degree of induration, which generally increases with relative age of the material, appears to be much more significant.

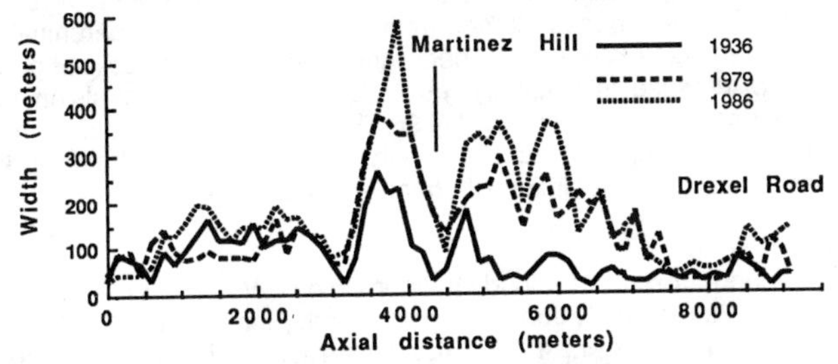

Figure 4 Arroyo widths on part of the San Xavier reach, 1936-86. Most of the change between 1979 and 1986 is a result of the 1983 flood.

Timing of major episodes of channel and arroyo widening appears directly linked to occurrence of high flows; however, other possibly significant factors, such as antecedent conditions, flow duration, and sediment discharge, have not yet been evaluated. The Canoa reach was stable from the 1950's, when large sections were channelized and armored, until the 1977 flood (fig. 3), during which widening was generally confined to unchannelized sections. The 1983 flood produced a fivefold to sixfold increase in channel width along channelized and unchannelized sections. Arroyo widening has been almost continuous on much of the San Xavier reach since 1936. The period 1936 to 1960 was dominated by relatively low peak discharges occurring during summer monsoons (Webb and Betancourt, 1990). Arroyo widening was slow but persistent and was concentrated on the outside of entrenched meanders. The rate of arroyo widening increased greatly during the 1960's and 1970's when peak discharges were higher and large fall and winter storms more frequent (Webb and Betancourt, 1990); the inner-arroyo flood plain was repeatedly scoured and arroyo wall retreat occurred on inside and outside of meanders. Finally, the October 1983 flows produced as much widening in some locations (fig. 4) as had occurred in the preceding 47 years.

CONCLUSIONS

Channel morphology, bank resistance, and hydrology appear to be most significant in determining the type of channel change to be expected at a particular location on the Santa Cruz River. Ephemeral reaches with sparsely vegetated, low sandy banks tend to adjust form by channel widening; deeply incised ephemeral channels change by arroyo widening. Avulsion may occur on shallowly incised ephemeral reaches. Narrow, perennial reaches with active flood plains change by lateral extension and downstream translation of meanders during low to moderate flows and by avulsion during extreme events.

Geomorphic and geologic controls strongly influence the general location of channel change. Shifts in position of as much as 600 m have occurred during a single flow event, but they have rarely been outside the modern flood plain. Erosion of older alluvium is minor. Confinement of lateral channel instability to the most recent flood-plain surfaces and alluvium suggests geomorphic mapping can be an important tool in designating regulatory flood plains on ephemeral streams.

The overall trend in channel morphology since 1936 has been toward greater mean width in ephemeral reaches, reflecting high flows of the 1960's through 1980's. The history of channel change on the Cortaro and Marana reaches is complicated by the human-induced change from an ephemeral to a perennial flow regime. Since 1970, those reaches have been characterized by relatively gradual lateral migration and increase in sinuosity, punctuated by a major avulsion event and associated channel straightening in 1983. The dominance of avulsion and channel- and arroyo-widening on the Santa Cruz River reflects the overriding importance of high flows in the evolution of channel morphology.

Appendix I: References

Baker, V.R., 1988, Flood erosion: *In* Flood Geomorphology, V.R. Baker, R.C. Kochel, and P.C. Patton(eds.), John Wiley and Sons, Inc., p. 81-95.

Betancourt, J.L. and Turner, R.M., 1988, Historic arroyo-cutting and subsequent channel changes at the Congress Street crossing, Santa Cruz River, Tucson, Arizona: *In* Arid lands today and tomorrow, proceedings of an international research and development conference: Boulder, Colorado, E.E. Whitehead, C.F. Hutchinson, B.N. Timmerman, and R.G. Varady (eds.), Westview Press, p.1353-1371..

Cooke, R.U., and Reeves, R.W., 1976, Arroyos and environmental change in the American Southwest, Clarendon Press, Oxford, 213 p.

Hays, M.E., 1984, Analysis of historic channel change as a method for evaluating flood hazard in the semi-arid southwest: M.S. pre-publication manuscript, University of Arizona, Tucson, Arizona, 41 p.

Hooke, J.M., 1979, An analysis of the processes of riverbank erosion: Journal of Hydrology, 42: 39-62.

Knighton, D., 1984, Fluvial forms and processes, Edward Arnold Ltd.,London, 218 p.

Pearthree, M.S., and Baker, V.R., 1987, Channel change along the Rillito Creek system of southeastern Arizona 1941 through 1983: Arizona Bureau of Geology and Mineral Technology , Special Paper 6, 58 p.

Roeske, R.H., Garrett, J.M., and Eychaner, J.H., 1989, Floods of October 1983 in southeastern Arizona: U.S. Geological Survey Water-Resources Investigations Report 85-4225-C, 77 p.

Webb, R.H., and Betancourt, J.L., 1990, Climatic effects on flood frequency: *In* Proceedings of the Sixth Annual Pacific Climate Workshop, J.L. Betancourt and A.M. MacKay (eds.), California Department of Water Resources, Ecological Studies Program Technical Report 23: 61-67.

PREDICTING STREAM WIDTH AND BANK RETREAT

Kuen D. Tsay[1], A.M. ASCE and Emmett M. Laursen[2], M. ASCE

ABSTRACT

The width of streams is dependent on the flood flow discharge which varies from year to year, the bank material which varies from stream to stream and reach to reach, the sediment load which varies from stream to stream and flood to flood, the resistance to flow which varies many ways, and probably other factors. Bank retreat can also occur because of the variability in bank shear due to the flow patterns of the non-uniform flow in natural, non-straight rivers, and because of the variation in the erodibility of the alluvially deposited bank material. Stream widening and bank retreat can be estimated to a fair degree of confidence if data from past events is supplemented by measurements that can be obtained today, and reality is interpreted through theory based on what we know about sediment transport and river behavior.

Introduction

An alluvial stream has a natural equilibrium width for every flood flow. Whether it attains this width during a given flood depends on how long the high flow persists and the erodibility of the banks. This realization is the basis for the regime equations developed for canals in India, but is equally true for rivers except that for rivers the flow and sediment characteristics are less well known or controllable. It is clear that in a bigger flood than the one which shaped the existing channel, widening can be expected with the bank retreating on one side or the other, or both.

---

[1]Civil Engineer, Pacific Gas and Electric Company, 77 Beale Street, San Francisco, CA 94106
[2]Professor Emeritus, Dept. Civil Eng. & Eng. Mechanics University of Arizona, Tucson, AZ 85721

Banks can also retreat in lesser floods because of non-uniformity of the flow. Even in a ostensibly straight flume the shear on the side walls can vary by 10 to 20 percent; the plan form of rivers can result in even greater variation in bank shear so it is to be expected that rivers will constantly shift. Added to the problem of increased bank shear due to flood flow or plan form is the variation of bank material and the consequent erodibility of the banks from cross section to cross section or, as the bank retreats, at a cross section. Great enough resistance to erosion can cause a bank to be fixed; very small resistance to erosion can result in sudden bank retreat of unexpected extent.

The processes whereby banks retreat are fairly readily understood in a qualitative sense; predicting the extent and location of the retreat is more difficult. The processes whereby banks build up again and streams narrow are less well understood, and the narrowing tends to take years rather than to occur in a single event but narrow they do, or all rivers would be as wide as they could be after the many years they have been flowing and would not widen any more. The proof that streams narrow is simply that they widen.

Because for many good reasons man likes to live and work near rivers, bank retreat is a problem he must contend with somehow. There is a need, therefore, to be able to predict bank retreat although it must be realized that any prediction must be stated on a probabilistic basis. Obviously this is the case for future floods, and unfortunately we know the probability of future floods of different magnitude less well than we would like. Shifting of the channel because of variations in flow patterns and boundary shear and of bank materials can result in the channel being anywhere in the floodplain over a long enough time span. In the next 50 or 100 years it is quite difficult to assess the probability of bank retreat of different amounts but it needs to be done. Changes in the plan form over one short reach leads to changes in the next downstream reach and the next and the next. Bank retreat in the next flood depends on the magnitude of the next flood and the flow pattern and boundary shear variation in each short reach and the bank material in each reach. What happened to the river channel in the immediate past or what could happen in the immediate future is one problem; what might happen over a long time period is another. Aerial photographs over several years can often lead to reasonable explanations of what happened, and help forecast the near future. The relationships used in this study can help project to longer time span.

This Study

This study was a M.S. thesis project (Tsay, 1983) and used an earlier Ph.D. dissertation (Silverston, 1981) as a starting point. Silverston obtained regime type equations through a computer study using the Manning equation and the Laursen total sediment load relationship (1958) and assumed values of bed material, the Manning n, and the total load sediment concentration. For each width and discharge, there was a value for the bank shear. By interpolation it was possible to find an approximate relationship between width and discharge for a constant bank shear. This relationship was

$$B = \alpha_2 \, Q^{\beta_2} \tag{1}$$

$$\alpha_2 = 0.109 \, (\tau \text{ bank max})^{-0.35}$$

$$\beta_2 = 0.853 \, \bar{c}^{\,0.0833} \, (\tau \text{ bank max})^{-0.167}$$

Where B = channel width in feet
Q = discharge in cfs
c = total sediment load in % by weight
τ bank max = maximum bank shear in pounds per square foot
= $0.76 \, \gamma ys$

Note that this equation is similar to, but more complicated than, the classic regime equation in that the exponent is a function of the sediment concentration and the permissible bank shear. Where the regime equation for width has an exponent of 0.5, for the conditions of this study the exponent of the discharge averaged 0.83. A more comprehensive study than Silverston's would probably find the coefficient and exponent would contain other variables such as Manning n, the critical tractive force of the bed material and the ratio of the shear velocity to the fall velocity of the bed material.

Aerial photographs, field inspection and measurements, and USGS flood discharge records were utilized in connection with Silverston's width equation, Laursen's total sediment load relationship, and the Manning formula. The aerial photographs were supplied by the Pima County Department of Transportation and Flood Control District. The procedure was somewhat circular in what the measured width was used to find the permissible bank shear through the Manning formula, a measured slope, an assumed n and the discharge at nearest USGS gage. The total sediment load concentration could then be calculated from Laursen's relationship and measured average bed material. The channel width calculated by Silverston's equation did not agree with the measured width, but a modified discharge estimated by adjusting for expectable tributary inflow and infiltration through the bed resulted in fair agreement between predicted and measured width. The predictions of widths could then be extrapolated to larger, rarer flood discharges. As shown in Table 1 and Figure 1, a doubling of the discharge results in an increase in width of 75 percent.

Table 1. Computed Stream Widths for Large, Rare Flows.

Reach	Stream Wash	B Measured[1] (ft)	B Computed[2] (ft)	Q Modified (cfs)	Assumed flood Q (cfs) — B Computed[3] (ft)			
					30,000	40,000	50,000	60,000
1	Pantano Wash	350	350	25,000	406	513	615	713
2	Pantano Wash	372	370	25,000	432	546	656	761
2	Tanque Verde Creek	250	283	16,000	474	601	722	839
2	Rillito Creek	697	657	45,000	469	595	714	830
3	CANADA DEL ORO	351	330	15,000	589	751	907	1,058
4	Rillito Creek	165	187	9,920	464	588	706	820
4	Santa Cruz River (N)	189	200	12,500	406	513	615	713
5	Santa Cruz River (S)	300	315	20,160	437	552	663	769

Notes: 1. Average width of reach determined from 1971 aerial photos.
2. Computed stable width using Silverston's Equations and modified discharge.
3. Computed stable width using Silverston's Equations and arbitrary flood discharge.

Figure 1. Computed Channel Widths, Reach 2, Rillito Creek.

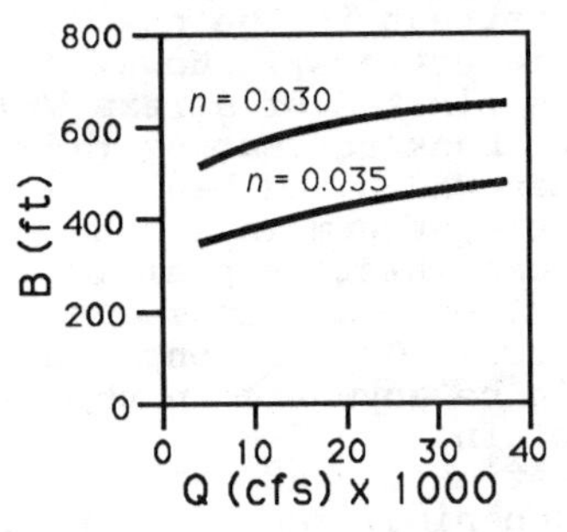

A similar widening effect can be expected due to variable flow patterns. Silverston in his measurements of boundary shear in a supposedly straight flume, found differences of 10 percent; in a river larger variations can be expected. Assuming the local shear τ_L can be 20 percent larger than the average shear τ_A, the following equivalent depth ratios, flow ratios, and width ratios can be approximated.

$$\frac{\tau_L}{\tau_A} = 1.2 = \frac{y_L}{y_A} \quad \text{(constant slope, etc.)}$$

$$\frac{Q_L}{Q_A} = \left(\frac{y_L}{y_A}\right)^{(5/3)} = 1.355 \quad \text{(constant width, etc.)}$$

$$\frac{B_L}{B_A} = \left(\frac{Q_L}{Q_A}\right)^{(0.83)} = 1.286 \quad \text{(Silverston)}$$

The y_L and Q_L values are fictitious and used to obtain the B_L/B_A ratio.

A 20 percent variation can give rise to an almost 30 percent change in width or bank retreat, for the same flow that determined the width of channel as it exists. Thus many moderate, frequent floods can add up to very significant bank retreat and shifting of the channel. In addition, if the higher than average shear happens to coincide with a bank material of less than average resistance to erosion, a very, very substantial bank retreat can occur in a single, moderate size flood.

Conclusions

Predicting bank retreat in the next flood or in the next 50 to 100 years is not easy. However, it is possible to make reasonable estimates of stream width for various discharges and of bank retreat over various periods of time. Streams can widen and banks can retreat by as much as the width of the stream in a single large, rare flood event. Streams can shift in position and banks can retreat by several stream widths during the life of man-made structures. Over a long time span the river channel can again be anywhere in the floodplain and can widen the floodplain.
Our ability to estimate what can happen, and the probability of its happening, can be improved by increasing the data base which we must work with, by measuring the widening and bank retreat in single events and over the years, by hind casting what has happened to our rivers, and by admitting that the natural behavior of rivers can effect the use of the land near the rivers.

APPENDIX I. - CONVERSION FACTOR, NON-SI TO SI (METRIC)

To convert	to	Multiply by
foot (ft)	meter (m)	0.3048
cubic foot per second (cfs)	cubic meter per second (m3/s)	0.02832
pounds per square foot (psf)	kilogram per square meter (kg/m2)	4.882

APPENDIX II. - REFERENCES

Condes de la Torre, Alberto, Streamflow in the Upper Santa Cruz River Basin, Santa Cruz and Pima Counties, Arizona, USGS Water-Supply Paper 1939-A, 1970.

Laursen, E. M., "The Total Sediment Load of Streams," Journal of the Hydraulics Division, ASCE, Vol. 84, No. HY1, February, 1958.

Silverston, E., The Stable Channel as Shape to Flow and Sediment, Ph.D. Dissertation, University of Arizona, Dept. Civil Eng. and Eng. Mechanics, 1981.

U.S.G.S. Water Resources Data for Arizona, Part I, Surface Water Record (1976 to 1981)

Tsay, K. D., The Lateral Retreat of Bank on Arizona Streams, M.S. Thesis, University of Arizona, Dept. Civil Eng. and Eng. Mechanics, 1983.

HYDRAULIC BRIDGE DESIGN FOR MOVABLE BED CONDITIONS

David Peterson[1], A.M. ASCE

ABSTRACT

This paper addresses the practical considerations of hydraulic design of a bridge under movable bed conditions. Hydraulic design tasks include pier and abutment scour calculations, design of erosion protection, and sizing of the bridge opening to avoid overtopping or adverse backwater conditions.

INTRODUCTION

Over the past 70 years, sand and gravel mining of the bed of the Santa Clara River in Ventura County, California, has contributed to changes in the morphology of the channel. Steady degradation of the channel combined with local scour caused undermining of the piers and finally the partial collapse of the State Route (SR) 118 bridge in February 1969 (Figure 1). The bridge was rebuilt following the failure, but degradation of the bed has continued, exposing pier foundation piles. Mining is now regulated to allow the channel bed to stabilize. A regulatory "red line" has been established to describe the stable invert elevations along the river's length. Mining is limited within the channel to areas where the channel invert is above the red line. At present, the bed elevation at the SR 118 bridge is approximately 8 feet below the red line elevation of 110 feet NGVD.

CH2M HILL has been hired by the California Department of Transportation to design a four-lane structure to replace the existing two-lane SR 118 bridge. Hydraulic design tasks include pier and abutment scour calculations, design of erosion protection, and sizing of the bridge opening to avoid overtopping or adverse backwater conditions.

---

[1] Civil Engineer, CH2M HILL, 3840 Rosin Court, Suite 110, Sacramento, California, 95834, (916) 920-0300

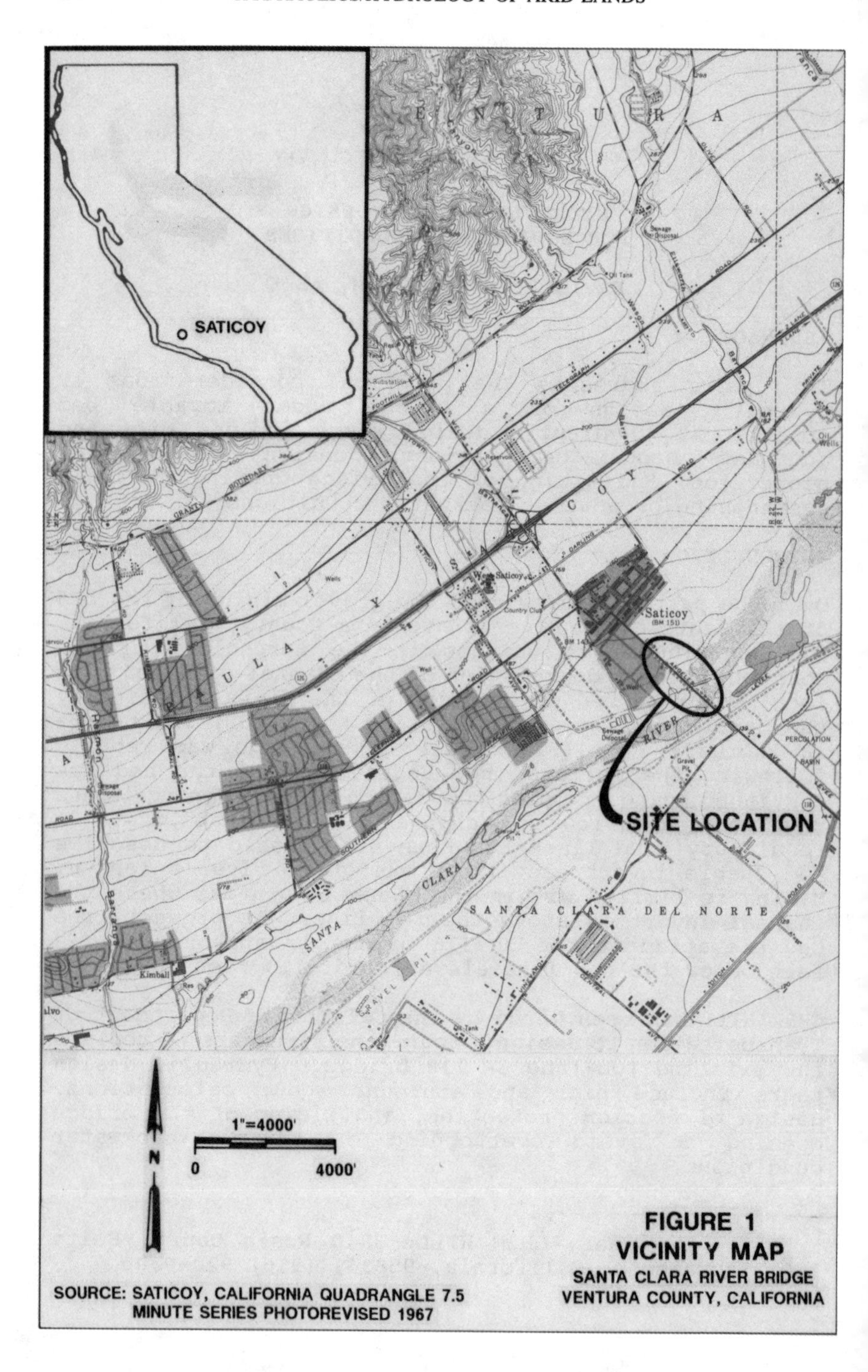

FIGURE 1
VICINITY MAP
SANTA CLARA RIVER BRIDGE
VENTURA COUNTY, CALIFORNIA

Because of the unstable bed conditions, water surface profiles for design floods can vary greatly over time. This complicates selection of the bridge low chord elevation to allow for a desired freeboard, and estimation of backwater effects of the proposed bridge during the design floods.

HYDROLOGY

Design flows for the bridge were taken from previous studies for the 1,555-square-mile drainage basin (Federal Emergency Management Agency, January 1989):

100-year flood peak discharge =	161,000 cfs
500-year flood peak discharge =	270,000 cfs
Peak historic flood (Jan. 25, 1969) =	165,000 cfs

The design flood was considered to be the 100-year flood for setting freeboard and the 500-year flood for design of scour protection.

WATER SURFACE PROFILES

The river channel is wide and deep at the bridge site, with little overbank flow during floods (Figure 2). The left (south) bank is confined by a levee constructed by the Corps of Engineers (Corps). Water surface profiles were computed for the existing and proposed bridges using HEC-2 (US Army Corps of Engineers, September 1982). Profiles were computed for three streambed conditions:

1. **Existing (1986):** Used for FEMA requirements.

2. **Pre-mining (1950s):** Defines assumed maximum potential future aggradation at the bridge. This condition controls design freeboard for the project.

3. **Red line:** Defines expected aggradation at the bridge.

Computed water surface elevations at the bridge are given in Table 1.

Since the minimum low chord elevation of the proposed bridge is 142.0 feet NGVD, freeboard will be 5.2 feet, which is sufficient to pass the debris anticipated in the river.

Table 1
WATER SURFACE ELEVATIONS
IMMEDIATELY UPSTREAM OF BRIDGE (ft NGVD)

Condition	River-bed El	Existing Bridge	Proposed Bridge
Construction Period (Existing Bed)			
100-yr	102.0	123.8	124.3
500-yr	102.0	130.7	131.2
Red line			
100-yr	109.1	127.8	127.7
500-yr	109.1	133.5	133.6
Pre-mining			
100-yr	124.0	136.2	136.8
500-yr	124.0	140.0[a]	140.9[a]

[a] The south levee and southeast bridge approach overtop at a water surface elevation of 137.4 approximately 400 feet upstream of the existing bridge. These elevations assume all of the flood is contained within the river.

SCOUR

General Scour

A 10-foot-high diversion dam (Freeman Diversion) is planned approximately 2.2 miles upstream of the bridge. This dam will intercept suspended and bedload sediments. Clearer water passing the diversion dam will tend to pick up sediments immediately downstream of the diversion dam until the water and sediment load are again in equilibrium. This process will continue until the diversion dam is full of sediment. Previous studies indicate that the scoured area will not extend to the SR 118 bridge (Simons, Li and Associates, May 1983).

Even though mining has temporarily been stopped in the reach downstream of the bridge, general scour continues at the bridge as a result of historical mining. The main head cut appears to have progressed approximately 1 mile upstream of the bridge. This is mirrored by a recovering section about 4,000 feet downstream (see 1982 and 1986 invert profiles plotted in Figure 3). The slope of the channel invert in the recovering section was projected upstream to the bridge as an estimate of the ultimate extent of general scour.

Figure 2
SANTA CLARA RIVER BRIDGE

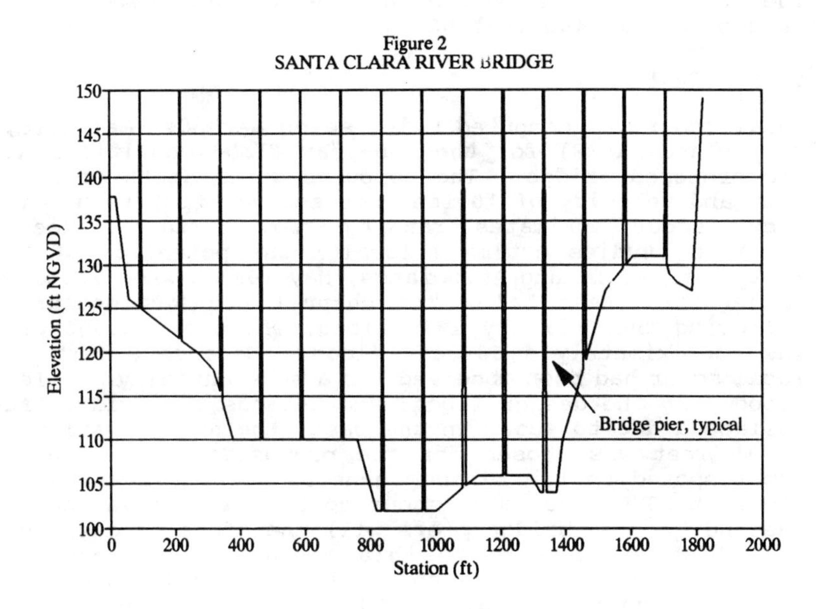

Figure 3
SANTA CLARA RIVERBED ELEVATIONS

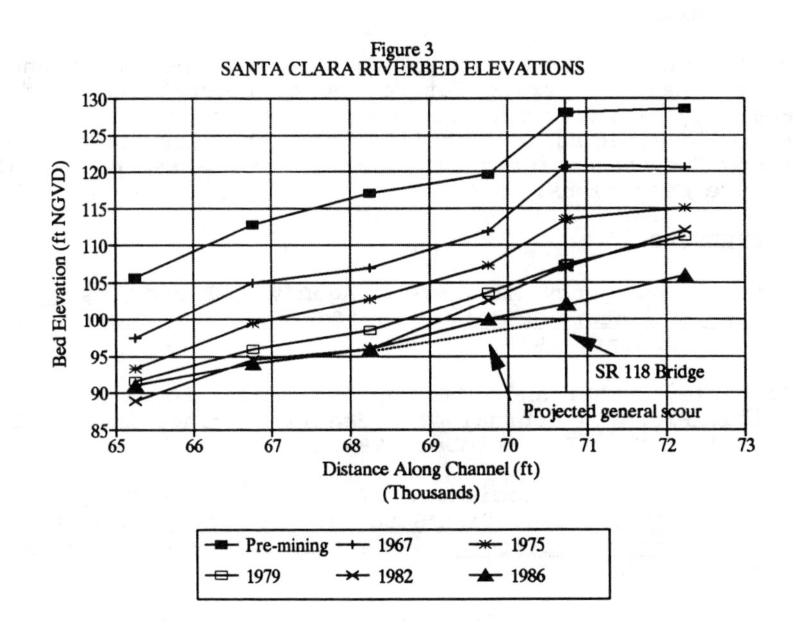

The minimum bed elevation at the bridge using these assumptions is 100 feet NGVD.

Local Scour

Local scour was computed using seven methods (Melville, Sutherland, 1988) for the 500-year flood conditions at the proposed bridge. The computed water depth of 28.3 feet and velocity of 10 feet per second (fps) result in local scour estimates ranging from 5 to 15 feet. Previous studies estimate local scour potential of 15 feet (Simons, Li and Associates, May 1983). Local scour of up to 8 feet below the channel bed was measured following the February 1978 flood, which at 98,600 cfs was approximately a 38-year flood. Because 8 feet of local scour had been observed for a substantially smaller flood than the design flood, and because the bridge had collapsed due to scour in the past, the higher estimate of 15 feet was chosen for the proposed bridge. When superimposed on the minimum general scour elevation of 100 feet NGVD, the total scour could reach elevation 85 feet NGVD. The bridge piers will therefore be designed to withstand exposure to this elevation. Embankment protection at the bridge abutments will be designed for a mean channel velocity of 10 fps, the potential for impinging flow, and scour to elevation 85 feet NGVD.

CONCLUSION

Bridge design for structures over a movable bed channel must consider varying future bed configurations in the analysis of water surface profiles and scour potential. A single water surface profile does not exist for a given flood, so the designer must attempt to bracket potential future conditions.

REFERENCES

Federal Emergency Management Agency. Flood Insurance Study - Ventura County, California - Unincorporated Areas. January 1989.

Melville, Sutherland. "Design Method for Local Scour at Bridge Piers". Journal of Hydraulic Engineering, ASCE, Vol. 114, No. 10. October 1988.

Simons, Li and Associates. Draft Report - Hydraulic, Erosion and Sedimentation Study of the Santa Clara River, Ventura County, California. Ventura County Flood Control District. May 1983.

US Army Corps of Engineers. HEC-2 Water Surface Profiles. Davis, California. September 1982.

Water Budget Analysis for Semi-Arid Alluvial Basins

Lonnie C. Roy[1], A.M., ASCE

In developing arid basins in the southwestern United States, the availability of suitable water supplies is of critical importance. Scarcity of dependable, perennial streams and rivers dictates a dependence on ground water. Additionally, spatially and temporally distributed hydrologic data in the Southwest are typically scarce, and only rough approximations are possible for basin recharge. This paper documents recharge estimates made for Lanfair Valley which is located in the eastern San Bernardino County, California.

Introduction

The United States Geological Survey (U.S.G.S.) (Freiwald, 1984) estimated recharge to Lanfair Valley as 100 to 630 acre-feet per year. This value was derived by simply equating recharge to the discharge of a single spring (Piute Spring).

This earlier work places great importance on Piute Spring as the locus of all discharge for Lanfair Valley and as the basic determinant of the water budget for Lanfair Valley. Presented herein are alternative approaches for estimating the recharge to the Valley.

Methods of Calculating Recharge

A method used extensively in Nevada by the State of Nevada Department of Conservation and Natural Resources and the U.S.G.S. is the Maxey-Eakin method. This method was developed between 1949 and 1951 by George B. Maxey and Thomas Eakin in east-central Nevada (Maxey and Eakin, 1949). The method relates recharge to precipitation excess occurring within the basin. A small percentage of the precipitation excess is assumed as recharge.

---

[1] Senior Project Engineer, The MARK Group, Engineers & Geologists, Inc., 2300 Paseo Del Prado, Suite D-108, Las Vegas, Nevada 89102.

The empirical coefficients are not specifically hydrologically based and they have been questioned in the past. They continue to be used in basins where data availability precludes the use of more sophisticated models.

Watson, et al. (1976) investigated the statistical rationale behind the Maxey-Eakin method. The authors correlated the Maxey-Eakin coefficients against independently estimated ground water recharge into 63 basins in Nevada. Their paper suggests alternate coefficients based on multiple-linear regression and on simple linear regression.

Walker and Eakin (1963) applied the Maxey-Eakin method to the Amargosa Desert in California and Nevada. The coefficient for the 8-12 inch precipitation interval was lowered to one percent of the annual precipitation. The authors suggest that the recharge estimate may be low, but could not quantify further adjustments to the coefficients used. Rush and Huxel (1966) applied the Maxey-Eakin method to Piute and Eldorado Valleys, which are adjacent to Lanfair Valley, without modification to the original coefficients.

Although the Maxey-Eakin method was developed using annual precipitation data, a number of authors (Simpson et al., 1970; Gallagher, 1979; and Mifflin, 1968) have suggested that this mountain-front recharge is a function of winter precipitation only. Reasons for this include the increased likelihood that the precipitation will occur as show; the longer duration, greater areal coverage, and lower intensity of winter precipitation events; and the lower evapotranspiration rate during the winter (Wilson et al., 1980).

<u>Precipitation in Lanfair Valley and Surrounding Areas</u>

Precipitation in semi-arid basins in the southwest is also highly variable and not well documented. Freiwald (1984) published an isohyetal map for Lanfair and Fenner Valleys. Although it is not stated in the report, it is assumed that the map was prepared by analyzing the precipitation stations shown on the map and by consideration of the topographic features of the valley.

Rush and Huxel (1966) published a precipitation vs. elevation curve for Piute Valley which adjoins the Lanfair study area. The data presented by Rush and Huxel were compared to the regression equation developed by French (1988) for the Nevada Test Site area. These curves show little difference in the 8 to 16-inch precipitation interval.

To assess variation in rainfall with season, the precipitation record for the Searchlight, Nevada gauge was examined. This gauge is close to the study area, has an elevation similar to the basin (3,540), and has a reasonably long record (1914-present).

From review of the precipitation data, the following conclusions can be made:

- The three precipitation data sources show similar trends and values for Lanfair Valley;
- A large percentage (25%) of the annual precipitation may fall as convective thunderstorms;
- Approximately 65 percent of the rainfall occurs between mid-September and mid-April.

For this study, four precipitation scenarios were assumed. They were:

- Total annual precipitation as shown by Freiwald, 1984;
- Total annual precipitation as shown by Rush and Huxel, 1966;
- 65 percent of total annual precipitation as shown by Freiwald, 1984 (winter precipitation);
- 65 percent of total annual precipitation as shown by Rush and Huxel, 1966 (winter precipitation).

Recharge Estimates

The following recharge estimates were made using the Maxey-Eakin method and the coefficients and precipitation scenarios described in the previous sections. The model scenarios are described below:

Scenario	Approach
RE1	Standard Maxey-Eakin coefficients
RE2	Multiple linear regression Maxey-Eakin coefficients (Watson, et al., 1976);
RE3	Simple linear regression Maxey-Eakin

coefficients for the 8-12 inch precipitation interval (Watson et al., 1926);

RE4 Walker and Eakin coefficients for Maxey-Eakin (Walker and Eakin, 1966).

The amount of recharge using the various combinations of coefficient and precipitation scenarios is shown below in Table 1.3.

Table 1: Calculated Recharge to Lanfair Valley

RE1--Standard Coefficients

Precipitation Scenario	Percent Recharge 8-12 in. Interval	Annual Recharge Acre-Feet
Total Freiwald	3	4,000
Total Rush & Huxel	3	3,500
Winter Freiwald	3	2,600
Winter Rush & Huxel	3	2,300

RE2--Linear Regression Coefficients

Precipitation Scenario	Percent Recharge 8-12 in. Interval	Annual Recharge Acre-Feet
Total Freiwald	4	5,300
Total Rush & Huxel	4	4,600
Winter Freiwald	4	3,400
Winter Rush & Huxel	4	2,900

RE3--Simple Regression Coefficient for 8-12 Inch Interval

Precipitation Scenario	Percent Recharge 8-12 in. Interval	Annual Recharge Acre-Feet
Total Freiwald	9	12,000
Total Rush & Huxel	9	10,500
Winter Freiwald	9	7,800
Winter Rush & Huxel	9	6,900

RE4--Walker-Eakin Coefficients

Precipitation Scenario	Percent Recharge 8-12 in. Interval	Annual Recharge Acre-Feet
Total Freiwald	1	1,300
Total Rush & Huxel	1	1,100
Winter Freiwald	1	850
Winter Rush & Huxel	1	700

In reviewing the data shown in Table 1, the following conclusions can be made:

- Precipitation reported by Freiwald shows higher recharge values by approximately 14 percent;
- Estimates using Walker-Eakin coefficients are considerably lower than others.
- The simple regression coefficients estimates are considerably higher than others. Watson, et al. (1976) report errors of +116% and +40% for Piute Mesa, Nevada, and Ash Meadow, Nevada, respectively. This may show that these coefficients over-predict for southern Nevada.

Discussion of Recharge Estimates

The recharge estimates show a possible recharge of 2,000 to 5,000 acre-feet. This amounts to 1.1 to 2.8 percent of the total rainfall on the basin. Freiwald (1984) estimated recharge on the order of 100 to 600 acre-feet solely from Piute Spring discharge. This would be only 0.05 to 0.3 percent of the total rainfall. Watson, et al. reported data for several basins in the southern part of Nevada. These basins show a percentage of total precipitation showing as discharge from 4.0 to 0.1 percent. The estimates made in this study are within the range from the Watson, et al. data, while Freiwald's recharge would be at the extreme lower end of the range.

The data and analyses represent reasonable estimates of the hydrologic cycle in Lanfair Valley. Although precise, unquestionable values of recharge and discharge are not possible, the data do suggest water may leave Lanfair Valley through other avenues as well as Piute Spring.

References

Freiwald, D.A., 1984, Ground water resources of Lanfair and Fenner Valleys and vicinity, San Bernardino County, California, Geologic Survey, Water Resources Investigations Report 83-4082.

French, R.H., 1988, Effect of the length of record on estimates of annual precipitation in Nevada, in progress.

Gallaher, Bruce M., 1979, Recharge properties of the Tucson Basin aquifer as reflected by the distribution of a stable isotope (M.S. thesis): Tucson, AZ., University of Arizona.

Linsley, R.K. and Franzini, J.B., 1984, Water Resources Engineering, McGraw-Hill Book Company, New York.

Maxey, G.B. and Eakin, T.E., 1949, Ground water in White River Valley, White Pine, Nye, and Lincoln Counties, Nevada: State of Nevada, Office of the State Engineer, Water Resources Bulletin No. 8.

Mifflin, M.D., 1968, Delineation of ground-water flow systems in Nevada: Reno, Nevada, Technical Report Series H-W, Hydrology and Water Resources Publication No. 4, University of Nevada Desert Research Institute.

Rush, F.E. and Huxel, C.J., 1966, Ground-water appraisal of the Eldorado-Piute Valley area, Nevada and California, State of Nevada, Department of Conservation and Natural Resources, Water Resources Reconnaissance Series, Report 36.

Simpson, E.S., Thorud, D.B., and Friedman, I., 1970, Distinguishing seasonal recharge to ground water by deuterium analysis in southern Arizona, in World Water Balance, Reading, 1970, Proceedings of the Reading Symposium: International Association of Scientific Hydrology.

Walker, G.E., and Eakin, T.E., 1963, Geology and ground water of Amargosa Desert, Nevada-California, State of Nevada, Department of Conservation and Natural Resources, Ground Water Reconnaissance Report 14.

Watson, P., Sinclair, P., and Waggoner, R., 1976, Quantitative evaluation of a method for estimating recharge to the desert basins of Nevada: Journal of Hydrology, v. 31, no. 3/4.

Wilson, L.G., DeCook, K.J., and Neuman, S.P., 1980, Regional recharge research for southwest alluvial basins, Water Resources Research Center, Department of Hydrology and Water Resources, University of Arizona, Tucson, Arizona.

GROUNDWATER DEPLETION AND SALINITY IN YAZD, IRAN

Zia Hosseinipour, Member, ASCE[1]
and Attaollah Ghobadian[2]

Abstract

The central Iranian plateau contains part or all of five provinces and covers more than a third of the total area of Iran (Figure 3). This area has one of the most unfavorable hydrologic conditions on earth, and as a result its water resources are very limited. With a few exceptions the only sources of usable water are the deep regional aquifers. This arid region is bordered on three sides by high mountain ranges which all but block the clouds coming from the Persian Gulf and the Caspian Sea. This study describes the results of more than two years of intensive hydrologic, erosion, salinity and water resources investigations in the central province of Yazd with an area of about 60,000 square kilometers and a population of more than half a million. The findings of this study lead the investigators to recommend drastic changes in the management of groundwaters in the region. Prime among these changes should be the abandonment of the customary agricultural and irrigation practices and limiting or changing them to save water to relieve the present stress on the aquifers.

Introduction

The city of Yazd, the capital of the province, is roughly equidistant from the Persian Gulf in the south and the Caspian Sea in the north. One can easily surmise the harshness of nature in this area by considering the fact that the average annual precipitation is about 60 mm and the average yearly evaporation can reach 3,500 mm. This imbalance has led to severe depletion of aquifers such that groundwater tables have dropped more than 50 meters in certain areas in the last 30 years. The sharp drop in groundwater levels has caused many of the Ghanats, the traditional under-

[1] Civil Engineer/Hydrologist, AScI Corporation, c/o US EPA College Station Road, Athens, GA 30613-7799.

[2] Professor of Soil Science, College of Agricultural Engineering, Jundi Shapur University, Ahwaz, Iran.

ground springs constructed and maintained over centuries, to dry up and force the depopulation of many desert rural communities. Throughout the ages these systems had maintained a natural balance between recharge and discharge from the aquifers (Fig. 2). Since these systems take years to construct, develop and use, they can not over stress the aquifers in a short.time period. Another characteristic of these systems is the role they play as drainage and indirectly prevent salinity in the semi-saline environments. Figure 1 shows the depletion of aquifers by over pumping in contrast to natural balance between the aquifer capacity and Ghanat discharge shown in Fig. 2.

Against this background and lack of basic information a study was conducted to identify problem areas such as the causes of saline and alkaline soils in irrigated lands, soil losses due to wind erosion, and water quality decline. Based on these evaluations recommendations were made as to the type of sustainable agricultural and industrial activities, suitable erosion control measures, safe yield of aquifers, level of economic growth consistent with the availability of water resources in the future, and additional studies required to better understand the natural conditions of the area to cope with the forces of progress and ecological change. This paper focuses on the hydrologic and water resources evaluations, for information on other aspects of the region's natural resources refer to Ghobadian (1982).

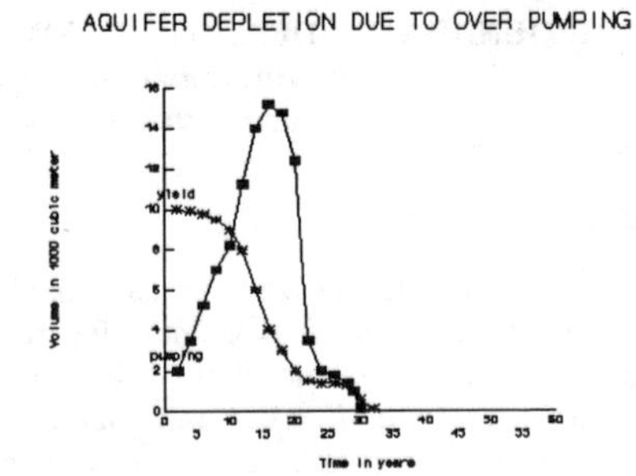

Figure 1. Depletion of Aquifer by Overpumping

Yazd's Natural Settings:

The city of Yazd one of the world's oldest is at the elevation 1222 meters above mean sea level (MSL). Its aquifers are recharged by the precipitations on the Shirkooh mountains with the peak at more than 4000 meters in the southeast direction. In general the land relief is from west to east and as such salinity and drought conditions decrease and precipitation increases. Precipitation varies throughout the province but its mean is about 60-80mm/-year. Most of the precipitation happens in the spring and often yearly amount comes down overnight. Therefore, not only the amount is low, its temporal and spatial distribution is highly irregular as well. In the mountain ranges the precipitation can exceed 250mm/year while in the flat plains of the east and northeast it can be as low as 40mm/year. The minimum, maximum and mean yearly temperatures are -20, 46 and 18 degrees centigrade respectively. Mean evaporation is in the range of 2500-

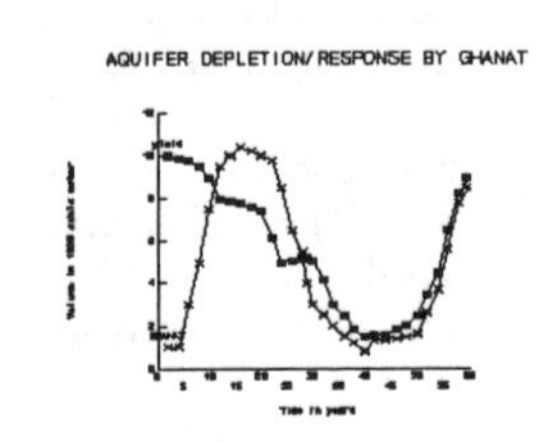

Figure 2. Aquifer Response due to Ghanat Discharge

3500mm/year in the central plains and it reaches as high as 70 times the amount of precipitation. Wind speed can reach up to 120km/hour and form sandstorms of immense proportions during which tons of soil may be moved across great distances. Based on the above data three general areas of climate have been recognized for the province; (1) Dry and cold in the high elevations of the Shirkooh mountain ranges which cover about 10% of the total area; (2) Dry, fully to semi-desert in the mid-elevations and the vicinity of the mountain ranges which cover about 20% of the area and (3) Dry, hot severe desert climate throughout the rest.

Water Resources of the Region:

With the exception of a couple of natural karst springs and a few ephemeral streams the main source of water in Yazd is groundwater. The quality and quantity of available economical groundwater sources is variable throughout the province. In recent years both the quantity and quality have diminished considerably and have caused public alarm. In the plains between the cities of Ardakan and Yazd although climatological conditions are very severe, there also exists some of the most favorable groundwater resources. The thickness of the aquifers are up to 400 meters with high yield, and if properly managed are capable of sustaining considerable human populations. These aquifers are recharged by the snow packs of the Shirkooh mountains under the favorable processes of percolation of slow melting snow and permeable geologic formations.

Figures used in the evaluation of water budgets are not definite but are approximated by field surveys and the practical judgments of investigators. To quantify the volume of recharge water the province was divided into 3 hydrologic regions:

a. The highlands in the vicinity of mountains which consist of about 10% of the total area of the province and whose elevation varies from 2500 meter to about 4075 meters above MSL. The climate of this region are classified as cold and dry from arid to semi-arid. Parts of the mountains are covered with snow throughout the year and the volume of precipitation water is calculated to be $1440X10^6$ m^3. Since the materials of the valleys are crushed rocks and gravel and therefore very porous, it is estimated that about 30% of the precipitation percolates through the top strata and moves with the natural gradient to recharge the aquifers in the Yazd and Ardakan plains. Therefore the recharge amount is $432X10^6$ m^3. There are several surface and karst springs but the amount of flows are seasonal and are not significant in the total water balance calculations. Although the potential evaporation in the region could be as high as $12X10^9$ m^3 and seemingly there is a negative water balance but due to the high permeability of the materials and the gradual melting of the snow packs a good portion of the water has a chance to percolate downward and recharge the aquifers. To maintain this situation the protection of the plant covers should have high priority since if the covers disappear the present settings will be altered and the consequences could be disastrous. Replanting efforts will be very difficult and costly if not impossible.

b. Mountain plains to semi mountain areas cover about 30% of the province area from elevations of 1500m to 2500m above MSL. In this region in spite of the bare rocky mountains and high hilly areas, there are high plains with relatively good cover plants in comparison with other areas. The climate is dry to arid and most of the range lands are located here. The area could be recharging the aquifers on a limited basis. Total volume of the precipitation is $2340X10^6$ m^3/year. If we assume that 5% of this amount is used to recharge the aquifers, total volume that percolates is $117X10^6$ m^3. The potential evaporation is high and may reach $45X10^9$ m^3 and therefore the water balance for this area is also highly negative.

c. The low lands of the province cover about 60% of the area and contain the salt flats and salt marshes. Precipitation is the lowest in this portion while potential evaporation is the highest. The only source of water is groundwater, and the aquifers are recharged by the percolating waters of the highlands. The total volume of yearly precipitation was calculated to be $2160X10^6$ m^3. A very small portion of this may be reaching the aquifers in the sand dunes areas where permeability is very high and the water percolates quickly. At best 1% of this water has a chance to recharge the aquifers and the total volume of recharge water would be $21.6X10^6$ m^3. Again evaporation is very high and there is a negative water balance.

Based on the above calculations the total yearly recharge is $57X10^7$ m^3 and sound management dictates that the discharge from the aquifers not to exceed this figure.

Erosion in Yazd:

As indicated earlier wind erosion is the dominant destructive force in Yazd and other desert regions where most of the ground cover has been eliminated. This process has been exacerbated in recent years by people who ignorant of the consequence of their actions are still plundering the aréa by cutting down the small amount of the remaining scattered desert forests. With the loss of the remaining vegetation a great desert resource which might be used to expand the wooded area could be permanently lost. Although the desert conditions are rather harsh, a lesson can be learned, since it has adopted to the biological varieties which are able to sustain in its environment. This lesson could be used to reclaim the devastated areas via ecological restoration by planting the local plants and the experience promulgated elsewhere. Hopefully with the cover plant establishment the micro-climate would change for the better and in the process while the soil erosion is reduced good top soil is generated to support light intensity full cover vegetation for the land to be also aesthetically pleasing.

Meeting Yazd Water Needs:

Yazd water needs are supplied almost entirely from groundwater. The water demands are met by about 1080 Ghanats and 2287 wells and ten or so natural springs in the mountain regions. Total water supplied from all of these sources is about 750 million cubic meters, while the maximum safe yield of all aquifers is estimated to be about 500 million cubic meters based on the amount of applicable yearly recharge. This imbalance has caused the groundwater table to drop sharply in the last three decades. Data show that 90% of the water is used in the agricultural sector for the irrigation of 60 to 70 thousand hectares of land. This is equivalent to 1000 mm per hectare per year; one can easily see that even a small improvement in the irrigation practices could save a lot of water and prevent excessive stress on the aquifers of the region. By changing the traditional flooding and furrow irrigation systems to a more efficient method (drip irrigation) it is estimated that with the present irrigation water volume three times as much land could be put into production and still save some water by reducing direct evaporation losses. This approach combined with growing varieties more natural to the desert environment could lead into even more water savings. Fortunately the people of the region are very conscious of the water shortage in their domestic uses. The per capita consumption is less than 250 liters/day and even some of this is reused for backyard gardening.

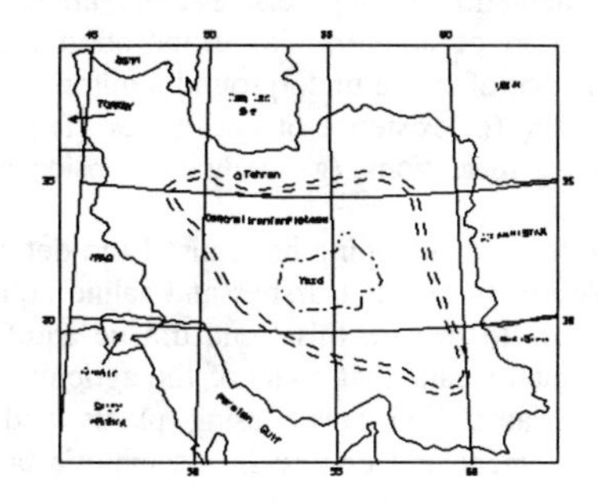

Figure 3. Map of Iran showing the Central Plateau and the study area

Salinity, Causes and Possible Remediation Procedures:

Salinity levels of soil and water are on the rise continuously in the Province. Salinity of soil and desertification are due to upward movement of water and the associated soluble salts to the soil upper layers where the evaporation process extracts the water and leaves the salts in the soil top horizons, rendering the plant function difficult and ultimately impossible. This is in part due to lack of sufficient precipitation and available soil moisture in the upper layers. Also since often the precipitation does not even infiltrate to the depth of 50 cm there is no chance for the salts to move downward to balance the upward movement. This investigation revealed that the salt movement mechanisms are of three types:

a. movement from deep zone to the surface layer by capillarity
b. horizontal movement in the direction of natural gradient to the salt flats by surface and subsurface flows
c. movement by the winds

The origins of the salts are the weathering of igneous and metamorphic rocks. Other sources are the salt domes which exist throughout the plains of central Iran and are commercially mined, as well as salt deposits around the salt marshes. The majority of the salts contain the anions CL^-, $SO4^{2-}$,$HCO3^-$ and $CO3^{2-}$ and the cations of NA^+,CA^{2+}, MG^{2+} and K^+. The soluble salts causing salinity were identified as NA2SO4, CL2CA, MGCL2, NACL, and MGSO4 as well as NAHCO3 on a smaller scale. The other group are the less soluble salts which don't cause salinity but bring about high osmotic pressure which leads to the physiological death in plants. This category of salts also form hard pressure packed layers in the crust of the earth and impede the plant growth by forming a barrier to the root expansion and basic functions. The end result is that soils will gradually accumulate high concentration of CASO4 and CAO, and prevent any kind of plant growth. These salts also fix soil nutrients and in the form of CA(HCO3)2, MG(HCO3)2 and CASO4, 2H2O may cause water hardness, the last one at solubility of about 2.6 gr/l. If sodium ion (NA^+) is present in the soil colloids the high PH can also cause super salinity.

We can safely state that the salinity of water in Yazd is in direct relationship with the depletion of aquifers. Factors affecting the salinity of water are: (a) gradual depletion of aquifers, (b) evaporation increases in salt flats and salt marshes, (c) existence of saline underground aquifers, (d) salt layers embedded between and within aquifers, (e) existence of salt domes, (f) gradual establishment of shallow saline water bearing formations, (g) gradual extension of desert salt lakes and salt marshes.

Aquifer over pumping has caused the deterioration of water quality due to salt water intrusion from salt marshes and saline aquifers where previously a hydraulic balance existed. To reverse this trend drastic aquifer management plans must be implemented which means that many of the agricultural activities must be curtailed or altered to save water. Soil cover using plants and trees which are natural to the desert environment and endure in a symbiotic fashion under severe water stress conditions must be established, and even some agricultural activities completely eliminated(i.e., alfalfa). Intensive irrigated cropping must be replaced with orchards of local trees such as pomegranates, pistachios, almonds and other low water demand high economic yield trees and plant varieties which can be sustained by normal aquifer capacity and support the farmers survival. For further information on the agricultural potential of the region the reader is referred to Clawson et al. (1971). Along with this change in agricultural activities efficient irrigation systems must be employed to use the available water wisely and expand the total area of orchards. New well construction should be restricted to the absolute minimum and even some of the existing wells put out of reservice. The abandoned Ghanats should be reclaimed and possibly new ones constructed with the application of new technologies since these systems maintain a natural balance between recharge and discharge of aquifers. An extended discussion of Ghanat construction and maintenance is given by Behnia(1988). A general historic perspective on these system is also given by Biswas(1970). In the past several years some of the abandoned Ghanats have been used for waste water disposal by the industry, mostly textiles, and others for disposal of drainage water, this practice must be prevented in order to safeguard the aquifers from chemical and salinity contamination.

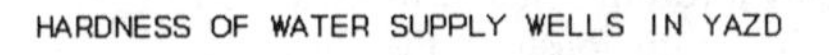

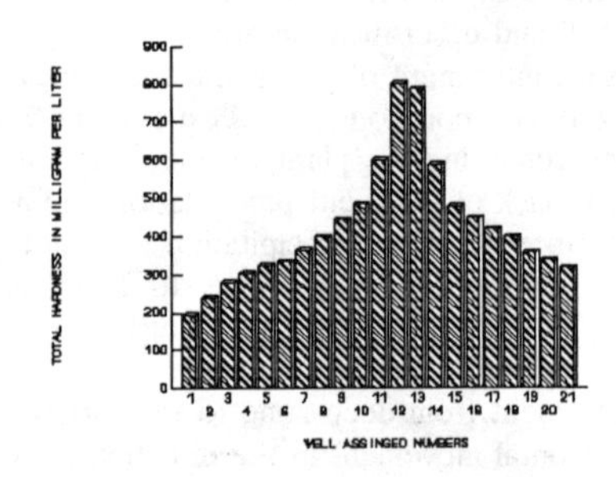

Figure 4. Total Water Hardness of Yazd Water Wells

The constant mining of the water from the aquifers has also resulted in elevated salt concentrations and decline of water quality for both domestic and industrial use. Salt concentration of water is 1 g/l on the average and can reach as high as 10 g/l with PH ranging from 7 to 8.3. With these quality parameters the water barely meets the quality standards for domestic or agricultural uses. If we consider the acceptable water hardness to be about 250 mg/l, from the following graph of the water supply wells in Yazd we see that very few of the wells meet the standards. Figure 4 shows the total hardness of the Yazd's water supply and utility wells, as seen very few meet the minimum standards.

Water Conservation and Quality Protection in Yazd:

Based on the findings of this study and the importance of water resources for survival in Yazd the following steps were recommended for the conservation and management of water resources:
a. Reducing and or preventing evaporation to the extent possible
b. preventing excessive percolation and water loss
c. research on and selection of crops with low consumptive demand
d. improving irrigation efficiency using modern technologies
e. limiting well construction and recharging aquifers artificially
f. reuse of irrigation return water and controlling saline aquifers
g. renovation of abandoned Ghanats and protection of existing ones
h. soil and water conservation through expansion of desert forests
i. limiting high water demand industrial activities
j. limit or stop grazing on the sparse desert range lands specially by destructive animals like goats in order to preserve the all valuable soil cover and encourage the range land expansions.
k. Application of groundwater flow and transport models and groundwater management models to better understand the regional flow patterns and quality changes, and study alternative management plans for aquifers.

Conclusions:

The ever increasing demand and the scarcity of water resources in the central region of Iran dictates that the available water to be managed very judiciously. Any approach for long range planning of water management in Yazd should be designed to restore the ecology and create soil cover to stop erosion and soil texture destruction. If the recommendations made is to be successfully implemented government support of the farmers in the form of technical and financial incentives is an essential part. Emergency steps must be taken to relieve the present stress on the aquifers if the region is to be saved from becoming a huge waste land in the future.

REFERENCES

Behnia, A., 1988. Construction and Maintenance of Ghanats, Jundi Shapur University, Ahwaz, Iran.

Biswas. A. K., 1970. History of Hydrology, American Elsevier, New York, N. Y. 336pp.

Clawson, M., Landsburg, H. H. and Alexander, L.T., 1971. The Agricultural Potential of the Middle East, American Elsevier, New York, N. Y., 312pp.

Ghobadian, A., 1982. Natural Resources of Yazd Province in Relation to Desert Problems, Jundi Shapur University, Ahwaz, Iran.

Van der Leeden, F., 1975. Water Resources of the World. Wiley Publishers, New York, N. Y., 568pp.

Watershed Modeling of The Western Coast Wadis of Egypt

Mostafa M. Soliman[1]

Abstract:

The main goal of this paper is to present the water resources assessment from both surface and ground-water resvoirs of the western coast watersheds in Egypt. Two submodels were used for this purpose to serve this arid and semi-arid zone.

The two submodels are; one to model the rainfall runoff relations and the other one to simulate the ground-water potentiality within the watershed boundaries. The results obtained from those submodels were calibrated and then checked with some recorded data which showed a good agreement between them.

Surface water submodels:

Two models were tested in this case which are dicussed as follows:

A- Dynamic Wave Model:

This model is used to stimulate the surface flow process for Wadi El-Ramla with the objective to produce the most reliable hydrograph with an inadequate data.

Geometrical data such as channel locations, lengths, slopes, roughness, and cross sectional dimensions are the type of data required for dynamic routing. Wadi El Ramla, together with the ohter Wadis in the northern coast of Egypt are relatively small basins. Consequently, such geometric informations can be collected in a short time span.

The surface flow model is divided into two major submodels. They are the channel flow and the lateral flow routing models. The lateral flow model routs the excess rainfall from the points where it enters the basin till it reaches a major channel. The outputs from this model are used as the spatially and temporally varied lateral flows entering the main channel system i.e. they are used as the inputs to the channel flow routing model. The later flow model routs the rainfall from the points where it enters the basin till it reaches a

----

[1] Professor of Irrigation and Hydrology, Faculty of Engineering, Ain Shams University, 24 Mohamed Mahmoud Kassim St., Heliopolis, Cairo, Egypt.

major channel. The outputs from this model are used as the spatially, and temporaly varied lateral flows entering the main channel system, i.e. they are used as the inputs to the channel flow routing model.

The numerical model is based on Saint Venant equations which are described as;

$$(B+B_o)\frac{\partial h}{\partial t} + \frac{\partial Q}{\partial x} = q \qquad (1)$$

$$\frac{\partial Q}{\partial t} + \frac{\partial (BQ^2/A)}{\partial x} + gA\left(\frac{\partial h}{\partial x} + S_f\right) = 0 \qquad (2)$$

where B, B_o are the top widths of the cross-section associated with the active section and the off-channel dead storage, respectively, q is the alteral inflow, A is the area, B is the momentum correction factor, g is the acceleration of gravity, and S_f is the friction slope computed using either Manning's or Chezy's equations.

Due to the nonlinearity of the last equations, a numerical model based on non-linear finite element formulation of these equations has been used to simulate the flow in the channel network.

For any channel with length L equations 1 and 2 can be replaced by two equivalent integrals in the following form

$$\int_L \delta h\left[(B+B_o)\frac{\partial h}{\partial t} + \frac{\partial Q}{\partial x} - q\right] dx = 0 \qquad (3)$$

$$\int_L \delta Q\left[\frac{\partial Q}{\partial t} + \frac{\partial (Q^2/A)}{\partial x} + gA\left(\frac{\partial h}{\partial x} + S_f\right) - q\frac{Q}{A}\right] dx = 0 \qquad (4)$$

where δh and δQ are arbitrary virtual elevations and discharges respectively. Finite elements were used to approximate the spatial derivatives and finite difference to approximate the time derivatives.

The lateral flow model was based on Helwa (1985) conceptual appraoch.

Since the data collected for Wadi El Ramla is almost scarce one reliable hydrograph is used for the model calibration. The results obtained from the model were compared with the available hydrograph as given in the figure.

B- SCS Modified Model:

Soil conservation unit hydrograph procedure was modified to suit the present work. The Wadi was carefully surveyed in order to determine the impervious and pervious areas beside the main channel slopes in order to obtain the representative unit hydrograph and then the designed hydrograph. Muskingum routing method was used in this respect. A computer program was formulated based on the aforementioned analysis and the resulted hydrograph compared with the recorded one, as given in Fig. (1) which one may conclude that the SCS modified model can be used with a fairly good approximation in such coastal region.

Groundwater Sumbodel:

Finite element method is used in this respect and is applied using the groundwater flow equation,

$$\frac{\partial}{\partial x}\left(Txx \frac{\partial h}{\partial x}\right) + \frac{\partial}{\partial y}\left(Tyy \frac{\partial h}{\partial y}\right) \pm Q = S \frac{\partial h}{\partial t} \qquad (5)$$

where h is the potential head; Q is a source (+ve) or sink (-ve); Txx and Tyy are the transmissivity in x and y directions, S is the specific yield; and t is time.

Triangular elements were used for its simplicity (see Fig. 2). The model was calibrated using some recorded data.

Water Resource Assessment:

The objective of this research work is to assess the water that can be yielded as surface water and groundwater storage in such kind of watersheds.

It was found from the meteorological data in this location that three to four rainy storms above 10 mm may occur during the rainy season (October to May) which are found to be effective for the surface water yield. A recorded storm of 15 mm rainful is analyzed using the two submodels analysis. From the surface water submodel a direct run-off of 5.7 mm is found to be yielded and the rest of 9.3 mm stands for the groundwater recharge and losses due to seepage and evaporation. From the groundwater submodel a ground-water outflow equivalent to groundwater recharge of 2.7 mm can be safely subtracted from the groundwater. Thus a balance of 6.6 mm is left to stand for both evaporation and seepage losses. However part of the ponded value which is lost by evaporation can be recovered by cistern construction which would undoubtedly increase the water yield. Another fact is that no water outflow from any storm having a rainfall depth less than 6.6 mm. Since three effective rainy storms can be safely assumed during the rainy season, therefore the total amount of yielded water can reach over than 50% of the total rainfal of those storms which are considered 30% of the yearly rainfall. This means that 15% of the total rainfall per year can only be recovered as surface water and ground-water discharge. This fact is considered very important to assess the water resources in this coastal zone.

Conclusion:

A representative basin in the north western coast of Egypt was selected in order to calibrate two submodels, one to find the rainfall run-off relations and the second to find the ground-water potentiality. This was essential in order to assess the water resources potentiality in this zone. It was found from this study that only 15% of the yearly rainfall on the north western coast can be recovered as surface and ground-water discharges.

References

- **Chow, V.T.** "Handbook of applied Hydrology," McGraw-Hill, New York, N.Y. 1964.

- **Cooley, R.L. and Moin, S.A.,** "Finite Element Solution of Sait-Venant Equations," Journal of the Hydraulic Division, ASCE, Vol. 102, No HYG, Proc. Paper 12213, June 1976.

- **Helwa, M.F. and Ragan, R.M.,** "A Dynamic Storage Routing Technique Adaptable to Data from Remote Sensing System" Paper presented at the International Workshop on Hydrologic Applications of Space Technology, Cocoa Beach, Florida, August 18-24, 1985.

- SCS National Engineering Handbook, Section 4, Hydrology. Soil Conservation Service, U.S. Dept. of Agriculture, 1971, V.S. Government Printing Office.

- **Soliman, M.M.,** "The Environmental Effects on Coastal Wadis in Egypt" Research Project No 842098 Presented to Supreme Council of Universities, 1987.

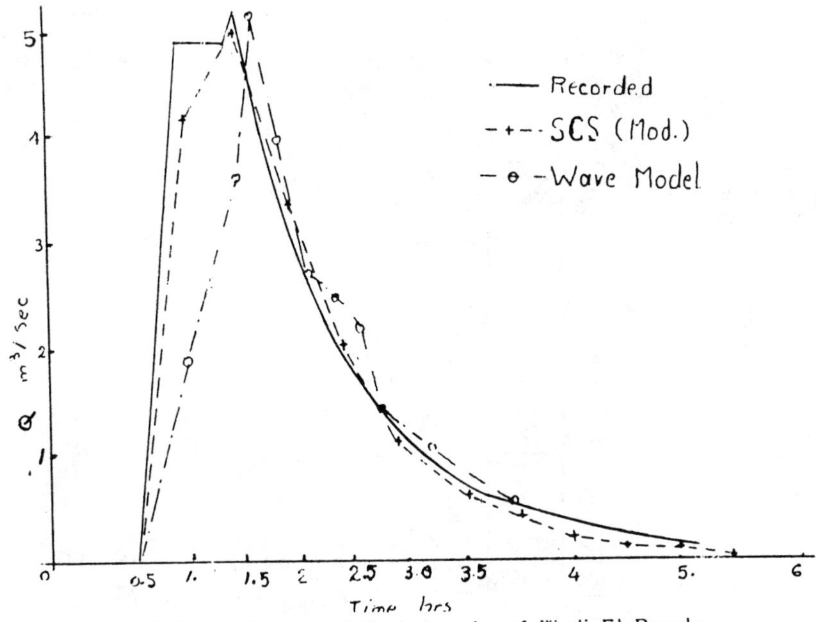

Fig. 1. Recorded and Computed Hydrographs of Wadi El Ramla.

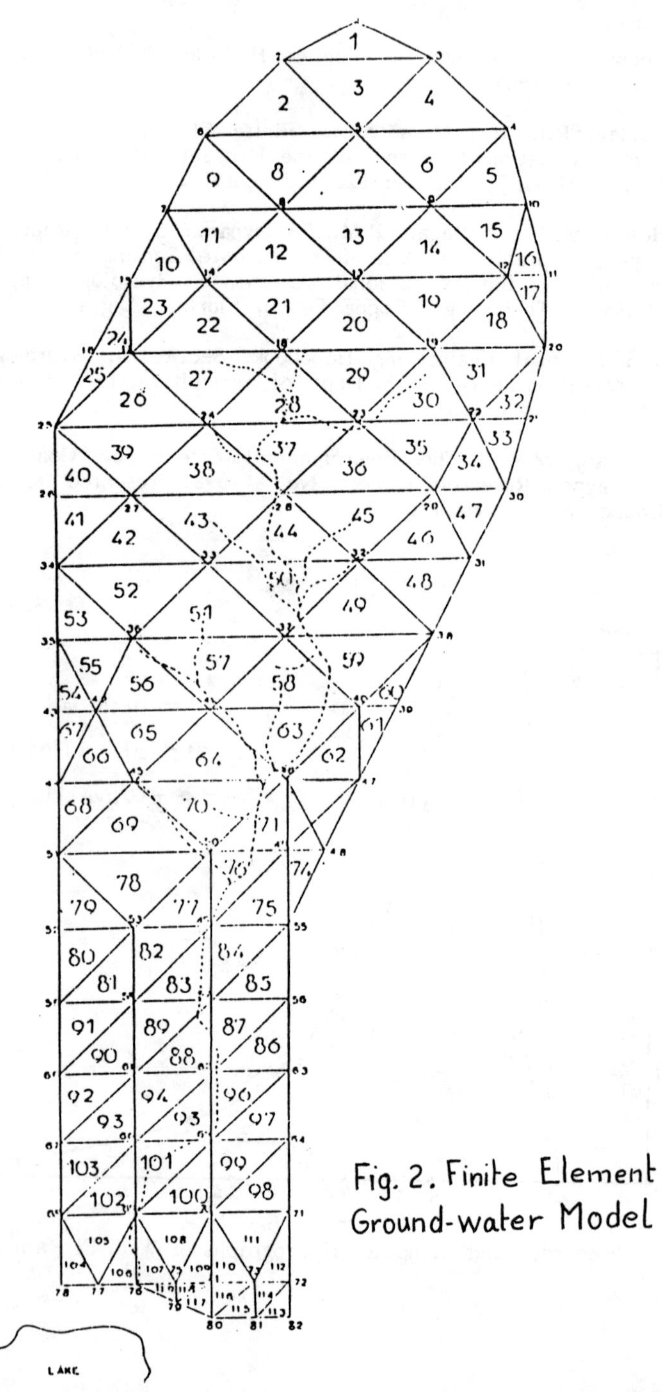

Fig. 2. Finite Element Ground-water Model

Alignment of Large Flood-Peaks on Arid Watersheds

B.M. Reich[1], K.G. Renard[2], F. ASCE and F.A. Lopez[2]

Abstract: Bridge and culvert design requires estimates for 25- to 100-year return period flood peaks (Q). Floodplain delineation calls for 500-year (Q500) estimates. Arid western flood series are shorter and more variable than eastern ones (Reich, 1969). Even if about 30 years of good data exist, various statistical models will produce widely different estimates of flood frequency. This paper shows users how variable results may be because of natural variability, mathematical peculiarities, or different flood geneses in arid regions. It also suggests how close flood estimates may be from six engineers who select best fits of the larger floods from four graphical displays.

Background and Developments: Reich and Renard (1981) used King's table (1971), as a way of portraying the forms of flood series. This table comprised a 4-by-4 array of different probability papers. If a series of observed floods appeared linear on a particular paper, say Log-Normal (LN), then that theoretical distribution could serve best for extrapolating toward larger floods. This is useful for seeking the most appropriate distribution for each flood series. It comprised plotting annual maxima on LN, Extreme Value (EV), Log Extreme Value (LEV), and Weibull papers. Observed floods that curve down from a straight, rising alignment near large floods suggest a mismatch between the paper's model and empirical data. Theoretically, only one cell in King's table can produce a straight line. Twelve other elements of that graphical aid present convex or concave curves that would appear when the wrong paper was tried on various distributions. This expedites the search for the most appropriate plotting paper.

---

[1] Consulting Engineer, 2635 E. Cerrada Adelita, Tucson, AZ 85718.

[2] Research Hydraulic Engineer and Hydrologic Technician, USDA-ARS, 2000 E. Allen Road, Tucson, AZ 85719.

Design Needs: Practitioners need estimates for return periods (RP) from 25 to 500 years. About 10 floods observed in an average 34-year history are likely to flow deep and swift enough to simulate conditions appropriate to a flood regime. The biggest observation from 34 years has a plotting position of 57 years according to Cunnane's (1978) theoretical work. The second largest recorded flood has an RP of 21 years; the following eight floods plot at RP's of 13, 9, 7, 6, 5, 4.5 and 3.6. More commonly, a flood record is only available for 20 years. It's six biggest items would have RP's of: 34, 13, 8, 6, 5, and 4 years. Search for linearity should emphasize the larger observations.

Bigger floods will usually be measured with large error. So it may be misleading to fit complex mathematical curves. The smaller half of the annual maxima do not interest design applications. In semiarid regions analysts need to indicate a preferred linear trend after testing common flood frequency papers. Positioning a straight line must recognize two possible occurrences; the chance measurement errors, and random occurrence of high or low "outliers". The selected frequency-plot should pass through about the largest 1/3 of the data and be free of systematic curvature.

An earlier test (Reich and Renard, 1981) involved seven watersheds from 2.5 to 56 square miles. They found that "straight-edge lines" were superior to statistically computed curves. Each flood series plotted into two straight lines that were separate from, and differently oriented to a large number of much smaller annual maxima. The 1982 Bulletin 17B of the Federal Inter-Agency Committee warned users that flood prediction requires more judgment than had always been used when computing flood frequencies from its 1967 Bulletin 15, Log Pearson (LP3) procedure.

Aron's multi-model PC program MINIEX (Aron, 1989) provides subroutines for implementing some of the flexibility now permitted in Bulletin 17B's text. Aron's output is limited to tabular information. However, manual plotting of floods, according to his three plotting formulae on various probability papers, is tedious and error-prone. de Roulhac's (1987) development of a PC program for plotting data and fitting classical curves was a step toward Q100. He was able to use only Cunnane's (1986) compromise plotting position formula on four probability papers in a study of short-duration rainfall intensity. He permitted us to use and modify his program.

Verification by Six Hydrologists: Four other engineering hydrologists joined the first and second authors to test our method of estimating design-sized floods from an observed series of annual maxima. These six scientists brought about 200 years of arid runoff experience into the evaluation. de Roulhac's (1987) software and automated plotting was applied to 33 yearly maxima from a 56 square mile, southeastern Arizona rangeland watershed known as the Walnut Gulch Experimental Watershed (Renard, 1970). One of their results is shown in Fig. 1. The scientists were requested to separately fit a straight line for each series on EV, Normal, LN, and LEV paper "through as many floods that each felt should influence the 25-, 100-, and 250-year estimates". They were instructed to assign a zero to a plot they felt fitted unacceptably. Four gave zero to Normal paper. The other two only graded Normal papers as 1 out of 9. In contrast, four of the six raters considered LN eye-fits the best, and the other two raters called it second best. EV eye-fits came second for five judges. Two raters called EV best.

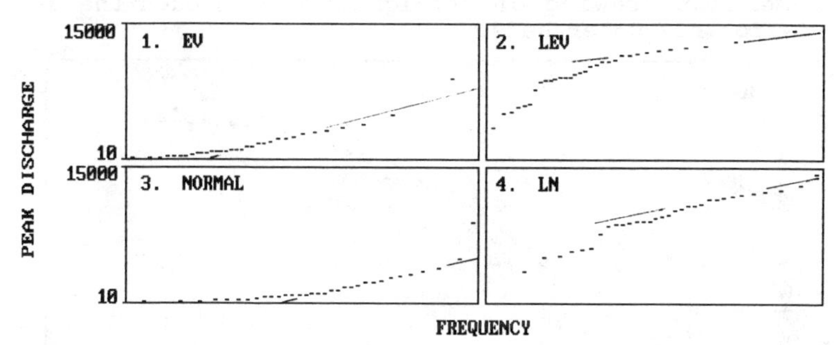

Fig. 1. Annual maximum floods plotted according to Cunnane's formula on N, LN, EV, and LEV graph papers.

The 250-year estimates from the six engineers' straight lines for LN and EV appear in Table 1. LN estimates were frequently 75% greater than EV's. Estimates by different "experts" differed by two times. Thus flood professionals may consider safety factors which are used by structural engineers.

Study of Six Western Rangeland Watersheds: Arizona and New Mexico watersheds ranging in size from 3.1 to 82 square miles were analyzed. The watersheds had record lengths of 57, 40, 33, 32, 27 and 24 years. Fig. 1 shows one station's annual maxima plotted according to Cunnane's formula on N, LN, EV, and LEV graph papers. They display "dog's legs" which seem to be characteristic of semiarid rangelands, reflecting transmission losses

Table 1. Six engineers' straight-line Q250 estimates.

	Engineer Number						Aver-age	Range
	1	2	3	4	5	6		
LN	18.0*	18.0	30.2	22.0	17.0	20.0	20.9	17.0-30.2
EV	11.5	10.5	10.2	12.6	13.0	20.0	13.0	10.2-20.0

*Discharge peak in 1000 cfs.
Computer predicted Q250: 93,000 cfs for LN; 10,900 cfs for EV.

and spatial precipitation variability. This is not common in the eastern United States. Fig. 2 shows how the solid-line Log Pearson III (LP3) flattens off for RPs greater than 10 years at less than 5,000 cfs, even though a 32-year record contained three observations from 5 to 12 thousand cfs. Fitting a straight line (dashed) through the larger half of the observations seems more satisfactory. To judge the variation between analysts in using such eye-fitted straight lines, two of us undertook independent drawing of straight lines through the data-plots for six watersheds.

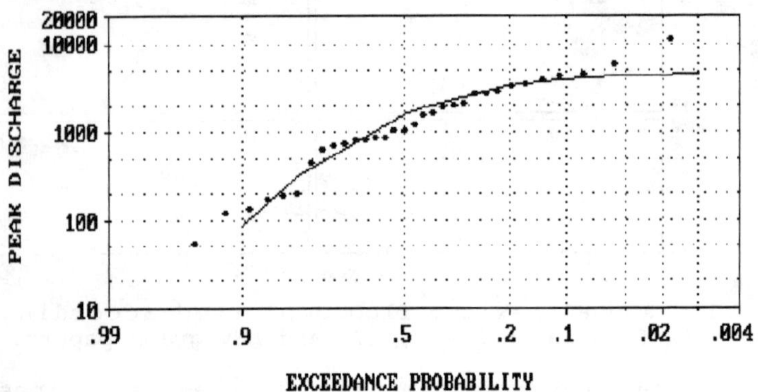

Fig. 2. LP3 math curve does not climb toward four observations between 5,000 and 12,000 cfs.

Table 2 summarizes the analyses completed by the first two authors for the six watersheds. The first three columns list some watershed information. All but Alamogordo Creek (a discontinued ARS watershed in eastern New Mexico near Santa Rosa) are located in southeastern Arizona. The Walnut Gulch watersheds, operated by ARS, have the two smaller ones as subwatersheds of the larger one. The last two on the list are operated by USGS.

Table 2. Summary of computer and straight-line frequency estimates for six southeastern Arizona "desert watersheds".

Watershed	Drainage Area sq. mi.	Record Length Years	Straight Line Q100 cfs Reich		Renard		Computer Estimate Q100 cfs LN	EV	LEV	LP3
Walnut Gulch 11	3.18	27	LN	7,500	LN	7,000	10,800	4,400	43,700	5,200
Walnut Gulch 03	3.47	32	EV	1,800	LN	2,000	2,900	1,310	10,900	1,420
Walnut Gulch 01	57.7	33	EV	9,900	LN	16,000	59,000	9,420	334,000	4,500
Alamogordo Greek (near Santa Rosa, NM)	67.0	24	LN	14,000	LN	15,000	22,400	8,700	91,000	9,770
Santa Cruz @Lochiel	82.2	40	LN	13,500	LN	13,500	23,700	10,700	75,300	5,140
Sabino Canyon	35.5	57	EV	8,000	LN	12,000	13,400	7,580	36,400	9,700

LN = log normal, EV = extreme value, LEV = log extreme value, and LP3 = log-Pearson 3.

Columns 4 to 7 show the 100-year estimates using the distribution selected by each of the two engineers as best fitting data influencing the 25- through 250-year frequency domain. The LN distribution gave the best result in most instances. One of the engineers preferred EV in 3 of the 6 watersheds.

The ratios of Renard's to Reich's Q100s were 0.93, 1.11, 1.62, 1.07, 1, and 1.5. LP3 computations produced Q100s that were only 72, 75, 35, 68, 38, and 97% of the mean of two linear straight lines through larger floods. In contrast, estimates by the two engineers were within 10% for four watersheds. Two Renard estimates are 1.6 and 1.5 times Reich's.

The last four columns show the computer estimated 100-year discharges for four models. Estimates from LN, EV, LEV, and LP3 computations deviate by several orders of magnitude. It is interesting to note that the EV computed Q100 agrees quite well with the "straight-line eye-fit" estimates for EV.

Walnut Gulch watersheds 3 and 11 which have about the same size drainage areas and are in very close proximity, show flood estimates that differ by three times. Although there are some differences in soils and vegetation, the primary explanation involves chance rainstorm position with greater storm depths and intensities on watershed 11. Thus, the two contiguous watersheds with relatively short records produce greatly different 100-year peak discharge estimates; i.e., an illustration of the variability to be expected in semiarid ephemeral streams.

Summary: Annual maximum flood series in ephemeral streams of semiarid areas comprise different populations for the less frequent larger and more frequent smaller storms. Spatial precipitation differences and transmission losses affect small floods differently from large floods.

Flood estimates such as the Q100 estimate exhibit large differences. Theoretical curves or eye-fit straight-line fitting should emphasize the larger discharge observations. The authors have argued that care must be taken in accepting computer-generated flood estimates without careful examination of the observed data. The straight-line eye-fit interpretation of the data produce consistent results.

References:

Aron, G. (1989). "Users manual." Dept. of Civil Eng., Penn. State Univ, 26 p.

Cunnane, C. (1978). "Unbiased plotting positions". J. Hydrology 37(3):205-222.

de Roulhac, D. G. (1987). "Application of computer graphics in the selection of rainfall frequency models for environmental engineering". MS Thesis in Civil Engineering, Univ. of Arizona, 106 p.

King, J. R. (1971). Probability charts for decision making. Industrial Press, New Jersey, 200 p.

Lane, L. J. (1983). "SCS national engineering handbook, Chpt. 19: Transmission losses". USDA-SCS, pp. 19-1 - 19-21.

Lane, L. J. (1982). "A distributed model for small semiarid watersheds". J. Hydraulics Div., ASCE 108 (HY10):1114-1131.

Osborn, H. B., K. G. Renard, and J. R. Simanton. (1979). "Dense networks to measure convective rainfall in the southwestern United States". Water Resources Research, AGU, 15(6):1701-1711.

Renard, K. G. (1970). "The hydrology of semiarid rangeland watersheds". USDA-ARS 41-162.

Reich, B. M. (1969). "Flood series for gaged Pennsylvania streams. Research Pub. #3, Inst. for Research on Land and Water Resources. The Pennsylvania State Univ., 81 p.

Reich, B. M., and K. G. Renard. (1981). "Application of advances in flood frequency analysis". Water Resources Bull., 17(1):67.-74.

Regional Flood-Frequency Relations for Streams with Many Years of No Flow

Hjalmar W. Hjalmarson[1], M. ASCE, and Blakemore E. Thomas[1]

Abstract

In the southwestern United States, flood-frequency relations for streams that drain small arid basins are difficult to estimate, largely because of the extreme temporal and spatial variability of floods and the many years of no flow. A method is proposed that is based on the station-year method. The new method produces regional flood-frequency relations using all available peak-discharge data. The prediction errors for the relations are directly assessed using randomly selected subsamples of the annual peak discharges.

Introduction

A standard method of defining regional flood-frequency relations at ungaged sites is to determine flood-frequency relations for gaged sites and transfer the information to ungaged sites using multiple-regression techniques (Stedinger and Tasker, 1985). In arid regions where flood-frequency relations for many stations are undefined or unreliable, regional relations estimated using the standard method also may be unreliable.

A new method of estimating regional flood-frequency information for arid regions is proposed. The method uses all available annual peak-discharge data including years of no flow at gaging stations in the area of interest. The performance of the new method was evaluated by comparison with regional relations computed using the standard method on a small group of gaging stations in Utah where the station relations could be reliably defined. The value of the method is illustrated by application to a group of 46 stations in Nevada where many of the station data have not been used in previous flood-frequency analyses.

Proposed Method and Assumptions

The proposed method, called the hybrid method, is based on the station-year method of frequency analysis that was suggested

[1]Hydrologists, U.S. Geological Survey, Water Resources Division, 300 W. Congress Street, FB-44, Tucson, Arizona 85701.

for use on floods by Fuller (1914). The station-year method is based on the assumption that independent records from a homogeneous region can be combined to form one long composite record if the peaks of the individual records can be reduced to a common base. Spatial sampling is assumed to be equivalent to time sampling if the records are independent; thus, a combination of 10 records with 10 years of record each results in a 100-year composite record. In arid or semiarid regions where most of the annual peaks are caused by local thunderstorms, the annual peaks at widely scattered gaging stations are assumed to be independent (Wahl, 1977).

The hybrid method incorporates the station-year approach with a method of standardization to produce regional flood-frequency relations based on basin and climatic parameters. The method addresses the problem of no-flow years in flood records noted by Wahl (1977), which is quite common in the southwestern United States. The hybrid method reduces flood records to a common base by dividing annual peaks by a function of common hydrologic parameters.

The equation is the familiar form used in many regional flood-frequency analyses:

$$Q_t = a\ A^b\ B^c\ C^d\ . . . , \tag{1}$$

where Q_t is the discharge Q for the t-year recurrence interval; a is the coefficient; A, B, and C are independent basin and climatic parameters; and b, c, and d are the regression exponents. Previous studies have shown that drainage area is the most significant independent variable affecting flood characteristics in the southwestern United States. Thus, drainage area is suggested for the initial standardization to determine exponent b where the other parameters, if any, are equal to 1.

The solution for equation 1 is obtained by using the following iterative technique:

1. Combine the annual peak discharges into three strata or groups of drainage area (A). Groups of equal size are preferred because a simple linear regression in log space is to be performed. For unequal numbers of peaks in the groups, it may be necessary to weight the groups to obtain unbiased relations in the regression procedure in step 5. At least 100 station-years of data with flow are suggested for each group to insure that extrapolation for the estimates of the 100-year flood is not needed.

2. Standardize the annual peaks by dividing by A^b (the initial value of b is 1.0).

3. Determine a flood-frequency relation for the standardized peaks in each drainage-area group. The log-Pearson Type III probability distribution and a conditional probability adjustment are used in this study.

4. Destandardize the computed values for the 2-, 5-, 10-, 25-, 50-, and 100-year floods for each group by multiplying by the mean of logarithm A.

5. Determine the exponent for the particular parameter in equation 1 for each recurrence interval by using linear regression methods. The regression is between the destandardized peak discharges for a given recurrence interval and the mean of the logarithms of the parameter for each group. In this example, the parameter is the logarithm of A.

6. Using the new exponent determined in step 5 for the parameter, repeat steps 2-5 until the exponent is stable. The exponent generally is stable (changes by less than 1 percent) after the third iteration. If the exponent does not become stable after the third iteration, a significant relation probably does not exist between the parameter and peak discharge, and the parameter should not be used.

Additional parameters may be included in the regional relation using the same iterative technique. Each additional parameter is separately added to the relation starting with step 1 where the new parameter is substituted for drainage area.

Performance Evaluation

The performance of the hybrid method was evaluated by comparison with regional relations determined from a standard regionalization for 12 gaging stations in the Uinta Mountains in northern Utah. These stations have well defined flood-frequency relations and similar flood characteristics and therefore the same regional relations were expected if the hybrid method is valid.

For the standard method, station flood-frequency relations were determined and the discharges for the 2-, 10-, and 100-year floods were related to drainage area using regression methods and the model defined by equation 1. The computations for the hybrid method were performed in accordance with the six-step iterative technique. Three groups of four stations each were used, and the regression was weighted for the disproportionate number of years of record in each group. The regional relations for the 2-, 10-, and 100-year floods for the two methods closely agree (table 1) and the computed errors are the same.

Application of Method

The hybrid method was applied to a homogeneous region in Nevada where most floods are caused by rainfall. Flood-frequency relations for gaging stations in this region are commonly poorly defined. Many years have no flow and the plotted annual peak discharges do not appear to define many of the computed flood-frequency relations. The homogeneous region was obtained by stratifying by geographic location, elevation, and size of drainage area. The hybrid method was applied to all gaging stations north of 37° latitude below a specified elevation with drainage areas

Table 1.--Summary of regional equations for computing peak discharge as determined from standard and hybrid methods

[Q, peak discharge, in cubic feet per second; A, drainage area, in square miles]

	Standard regionalization		Hybrid regionalization	
Recurrence interval, in years	Equation	Average standard error of estimate, in percent	Equation	Average standard error of estimate, in percent
2	$Q = 12.9\ A^{0.840}$	27	$Q = 13.2\ A^{0.835}$	27
10	$Q = 26.9\ A^{0.788}$	17	$Q = 27.5\ A^{0.798}$	17
100	$Q = 48.7\ A^{0.748}$	18	$Q = 46.9\ A^{0.767}$	18

between 1 and 100 square miles. The range of drainage area was selected because the standard deviation of the logarithms of the annual peaks for streams draining these basins was fairly uniform.

The hybrid method was applied to 46 stations that have a combined total of 833 station-years of record. Thirteen station records have more than 50 percent years with no flow and 28 records have more than 25 percent years with no flow. Drainage area for the selected stations was from 1.07 to 83 square miles, and the mean basin elevation was from 3,300 to 8,000 feet.

The characteristics of rainfall in Nevada and the geographic location of the gaging stations are persuasive arguments for the independence of annual peaks. Summer thunderstorms cause about two-thirds of the peaks and the remainder are caused by mid-latitude cyclonic storms occurring in the fall, winter, or spring. The spatial and temporal occurrence of summer thunderstorms is extremely variable, and the areal extent of a thunderstorm in Nevada typically is not greater than a few square miles. Similarly, but to a lesser degree, the rainfall of cyclonic storms also is variable. The gaging stations used in this analysis are distributed widely throughout the region and only a few of the annual peaks are on the same day. Therefore, the annual peaks are assumed to be independent.

The six-step iterative technique was used to determine a two-parameter regional flood-frequency relation of the form described by equation 1. The two independent variables are drainage area and mean basin elevation. The relation between peak discharge and drainage area was determined first, then mean basin elevation was factored into the relation.

The regional flood-frequency relations determined from the hybrid method are shown in table 2. The exponents of drainage area for the small recurrence intervals are much larger than the exponents for regional equations from studies for nearby states. One reason for the difference is that gaging stations with many

Table 2.--Regional flood-frequency relations for homogeneous region in Nevada determined from hybrid method

[Q, peak discharge, in cubic feet per second; A, drainage area, in square miles; and E, mean basin elevation, in thousands of feet]

Recurrence interval, in years	One-parameter equation	Estimated average error, in percent	Two-parameter equation	Estimated average error, in percent
2	$Q = 0.100\ A^{1.66}$	([1])	------------------------	--
5	$Q = 2.14\ A^{1.26}$	[2]30	------------------------	--
10	$Q = 11.4\ A^{1.05}$	21	------------------------	--
25	$Q = 59.3\ A^{0.820}$	21	------------------------	--
50	$Q = 157\ A^{0.683}$	22	$Q = 5{,}220\ A^{0.683}\ E^{-1.95}$	23
100	$Q = 350\ A^{0.590}$	25	$Q = 82{,}700\ A^{0.590}\ E^{-3.04}$	26

[1]Error not estimated because equation is considered a rough estimate.

[2]Estimated from graphical analysis because computed discharge was zero for several computations.

years of no flow were not included in the regional analyses for the nearby states. The cause of the large exponents in the hybrid equations probably is related to the relative size of the basins and thunderstorms. The incidence of runoff-producing rainfall is greater for large basins than for small basins, but the variability of the quantity of rainfall for a particular thunderstorm is greater for the large basins. Thus, the small basins have fewer incidents of rainfall but the average quantity of runoff-producing rainfall on the basin is large.

The significance of the parameters and the estimated error of peak discharge for the flood-frequency relations (table 2) were defined by direct assessment. Direct assessment of variability eliminates the need for complex and commonly oversimplified mathematical formulas and is feasible for this study because the laborious computations are made by digital computer. Half the 46 gaging stations were randomly selected for 30 trials. Equation 1 was solved for each of 30 subsamples using the iterative technique. The exponents and the discharges corresponding to the parameter means of the 30 subsamples were used to evaluate the significance of the parameters and the estimated error of the computed peak discharges, respectively. The significant relations are shown in table 2.

The standard error of prediction of the 100-year flood shows the effect of the large negative exponent for mean basin elevation. The standard error of prediction becomes large for mean basin elevations that are far from the average (fig. 1). The errors shown in figure 1 were estimated from several combinations of drainage area and mean basin elevation using the 30 subsamples. As

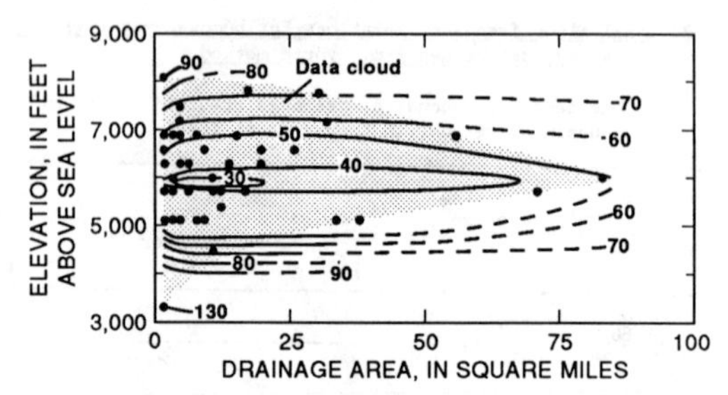

Figure 1.—Standard error of prediction of the 100-year flood magnitudes, in percent, for the two-parameter equation for homogeneous region in Nevada.

expected from the amount of the parameter exponents of the 100-year regional equation, the error is very sensitive to changes of mean basin elevation and much less sensitive to changes of drainage area. The regional equation is not reliable for combinations of area and elevation outside the data cloud because the prediction error is large (fig. 1).

Summary

A hybrid method to define regional flood-frequency relations is proposed for use in regions where methods based on flood-frequency relations from gaged sites may be unreliable. Records from all stations, including stations with many years of no flow, are used in the method. The method was applied to an arid region in Nevada where many station records have not been used in previous regional flood-frequency analyses. A two-parameter model showed that most of the variance in peak discharge is explained by drainage area and that the standard error of prediction is very sensitive to changes of mean basin elevation. A large number of peak discharges are used to define the flood-frequency relations in the hybrid method and extrapolation of the fitted relation to the 100-year flood is not needed; therefore, the results probably are less sensitive to the type of probability distribution.

Appendix.—References

Fuller, W.E., 1914, Flood flows: Transactions, American Society of Civil Engineers, 77, no. 1293, p. 564-617.

Stedinger, J.R. and Tasker, G.D., 1985, Regional hydrologic analysis—1. Ordinary, weighted, and generalized least squares compared: Water Resources Research, v. 21, no. 9, p. 1421-1432.

Wahl, K.L., 1977, Simulation of regional flood-frequency curves based on peaks of record, Proceedings of Conference on Alternative Strategies for Desert Development and Management: United Nations Institute for Training and Research (UNITAR), Sacramento, California, v. 2, 13 p.

Cyclic Streamflow Test for Validity of Randomness

By David E. Creighton, Jr., P.E., F. ASCE.[1]

Abstract

Long-term Arizona runoff records used for flood insurance frequency analysis estimates are compared with long-term precipitation station records for cyclic behavior. Runoff cumulative departure from average (CDA) displays a long-term cycle. The Verde River record shows a cycle from 1906 to 1985. A parallel charting of Flagstaff and Tucson precipitation extends the runoff chart pattern. Highest ranked discharges (HRD) fall within the rising, or high (wet) portions of the CDA curve. The period of record most used for flood insurance discharges estimates has not included HRD events of the long-term indicator stations. Runoff and precipitation CDA, when compared with HRD, call for examination of the primary assumptions of randomness and non-cyclic behavior.

Introduction

The National Flood Insurance Policy Act and subsequent legislation and regulations initially used expedient methods to guide Flood Insurance Program and floodplain management criteria. Theoretical mathematical manipulations have replaced expedients by increasing reliance on log-Pearson III. Interagency Advisory Committee on Water Data (1982) cautions "to consider all available information," and to assess the flood record for reliability by checking for cyclic and climatic trends, randomness, watershed changes, mixed populations, and reliability of flow estimates (Hirschboeck 1988).

The length of precipitation, stream runoff and discharge records is highly variable for Arizona. This generally short length of record, often truncated for

---

[1] Project Engineer, Arizona Department of Water Resources, Engineering Division, Flood Management Section, 15 South 15th Avenue, Phoenix, Arizona 85007

budget reasons, has been a major reason why there is such a strong emphasis on frequency analysis. Longer period station records are relatively few compared to the shorter period records. A numerical processing technique of annual maximum discharges, such as log-Pearson III, does not require understanding the cyclic-climatic circumstances which lead to the discharge value.

Analysis

This paper examines runoff and discharge data by chronological CDA to display cyclic tendencies. Given that runoff is dependent upon precipitation, the same analysis is applied to precipitation and tree-ring growth index records.

Analysis Method

Locating 8 annual HRD events on the longest term CDA diagram for the indicator gaging station shows the tendency for peak discharge to occur during a greater than average runoff year. At many stations used for location and regionalization frequency analysis computations (Roeske 1978; Anderson and White 1979) the long-term indicator HRD events have occurred outside the frequency analysis period of record.

Sources of proxy data in tree-rings (Schulman 1945) and paleoflood deposits (Stedinger and Baker 1987), suggest a time scale within which to estimate the record period required for a 95+/-% confidence level that a frequency magnitude event may have occurred. For this 95% level, a record period of 298 years is required for a 100-year event. Extreme events have to be considered against a possible maximum precipitation (PMP) magnitude (Stendinger and Baker 1987).

The historic record is such a small segment of the data record time line that the acceptable limits on skew of frequency diagrams and ogives for cyclic, climatic data need to be explored and clarified.

Data Adequacy Problems

Competition for public funds has severely limited acquisition and publication of runoff, discharge and precipitation data. Inadequate data collection is supported by claims that frequency analysis methodology is appropriate and adequate.

Record Availability

Ten USGS record stations with long record periods were used for this analysis. These are direct data.

Proxy data for precipitation and/or cyclic weather conditions include dendrochronology (Schulman 1945), and navigational meteorological data (Quinn et al. 1978).

Streamflow Gaging Station Records

Information on the gaging stations and watershed locations shown on Fig. 1 is given in Table 1.

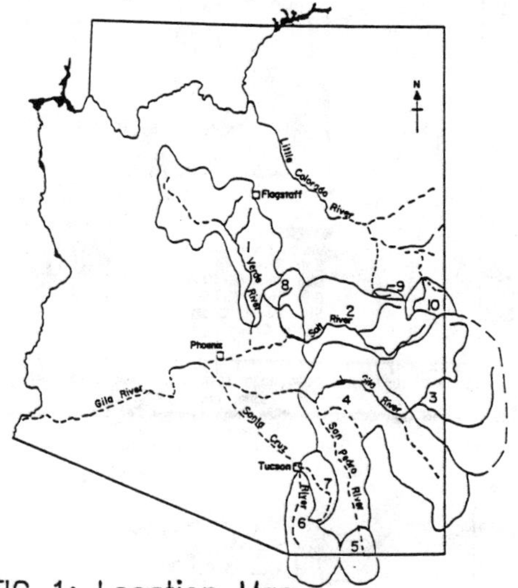

FIG 1: Location Map

Table 1. --Arizona Runoff Records Showing Cyclic Climatic Characteristics.

Line	Stream Record Station	Period of Record	Runoff Average taf/yr	(km3/yr)	Possible Length years
1.	Verde River below Bartlett Dam/nr Scottsdale	1889-->	489.5	(0.604)	99
2.	Salt River at Roosevelt	1915-->	654.8	(0.807)	74
3.	Gila River at Head of Safford Valley	1915-->	349.2	(0.431)	73
4.	Gila River at Kelvin	1912-->	358.9	(0.443)	77
5.	San Pedro River at Charleston	1919	42.9	(0.053)	76
6.	Santa Cruz River/Tucson	1906-1981	16.5	(0.020)	76
7.	Rillito Cr at Tucson	1909-1975	11.7	(0.014)	67
8.	Tonto Cr below Gun Cr	1941-->	117.0	(0.144)	47
	near Roosevelt	1914-1940	102.3	(0.126)	27
9.	Show Low Cr/Show Low	1942-->	10.0	(0.012)	46
10.	Little Colorado River above Lyman Dam	1941-->	17.3	(0.021)	47

Fig. 2 shows the CDA curves for 6 stations. Fig. 3 shows the CDA curves for 4 additional stations. The difference between river basins is interpreted as showing the role of the summertime "monsoon" as a climatic variation factor within the overall cycle.

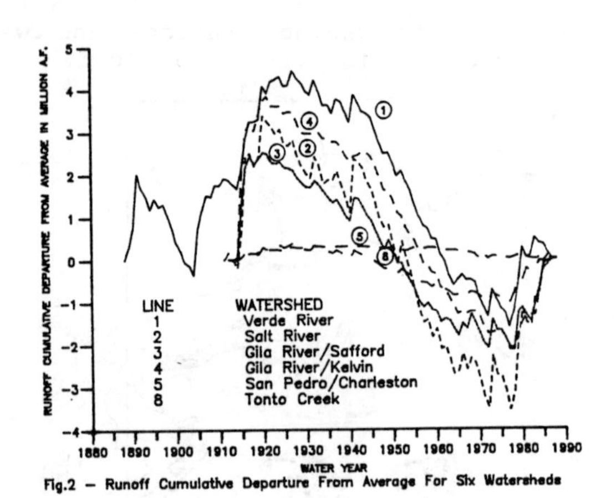

Fig.2 – Runoff Cumulative Departure From Average For Six Watersheds

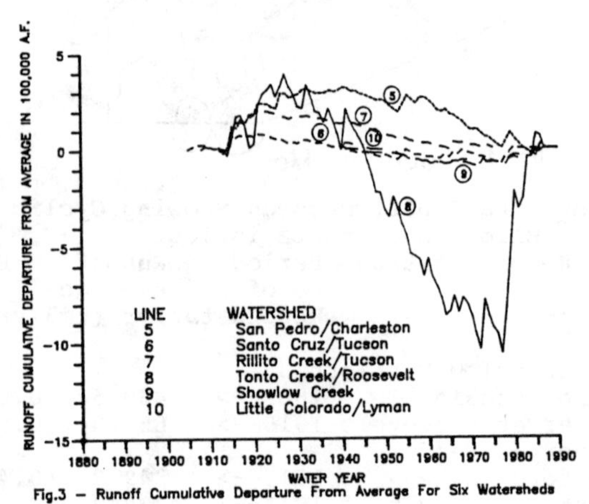

Fig.3 – Runoff Cumulative Departure From Average For Six Watersheds

Precipitation Records

The earliest precipitation records in Arizona are associated with military posts and population centers which developed with an agricultural, trading, or railroad transportation base. Flagstaff and Tucson are used as indicator stations. Their locations (Table 2) are shown on Figure 1. Fig. 4 shows the precipitation CDA for Tucson and Flagstaff as key stations located in the southern and northern parts of the drainage basins examined. The concurrent Verde River CDA with 10 HRD is added for ready comparison.

Table 2. --Indicator Precipitation Stations

Line Record Station & Stream Basin	Period of Record	Precip. Average In.	(cm.)	Record Length Years
1. Flagstaff Verde River	1988 -->	10.87	(53.01)	91
2. Tucson Santa Cruz River & Rillito Creek	1868 -->	11.40	(28.98)	121

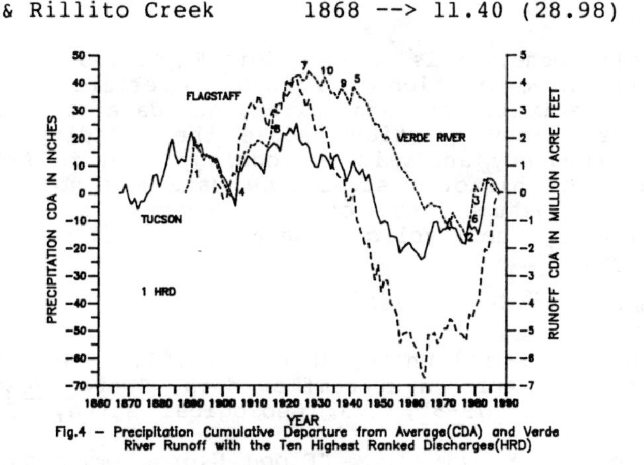

Fig.4 – Precipitation Cumulative Departure from Average(CDA) and Verde River Runoff with the Ten Highest Ranked Discharges(HRD)

Navigational Meteorological Data

Studies relating barometric pressure from the Indian Ocean to the eastern Pacific Ocean show association between the South American fisheries and the ENSO (El Niño, Southern Oscillation) (Quinn et al. 1978). Above average runoff years regionally, state- or basinwide, major eastern Pacific tropical storms with seasonal and specific named events associated with floods or recognized Presidential flood disaster declarations, need to be examined for correlation.

Tree-ring Growth Indices

Published tree-ring index values for Douglas fir in the Colorado River Basin (Schulman 1945) cover time between 1288 and 1945. Fig. 5 shows the CDA for this data. The concurrent period for precipitation in Fig. 4 and Verde River runoff (line 1 in Fig. 2) shows the same easily recognized behavior. Fig. 5 can be interpreted to suggest there may be 300+/- years between corresponding cyclic elements of above average and below average moisture.

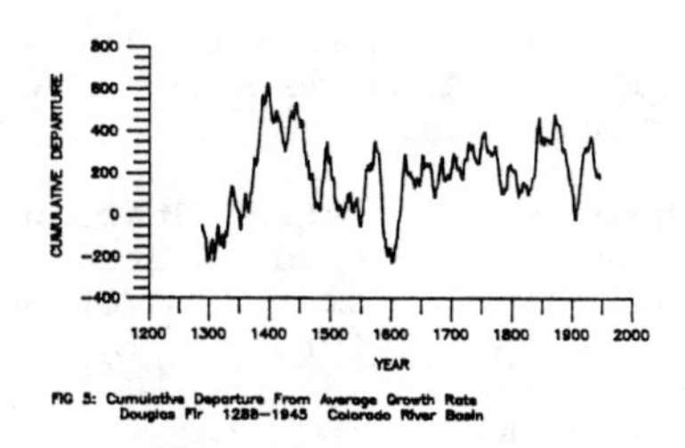

FIG 5: Cumulative Departure From Average Growth Rate
Douglas Fir 1288–1945 Colorado River Basin

Significance

Cyclic behavior is indicated in Figs. 2,3, and 4 for runoff and precipitation in Arizona. Testing the basic runoff, precipitation and discharge data for cyclic influences raises questions above the validity of the current frequency analysis methodology used in Arizona. Assessment of historic streamflow data may not support the basic assumptions of randomness, non-climatic variability, non-cyclic behavior and un-mixed populations.

REFERENCES

Anderson, T.W., and White, N.D. (1979). "Statistical summaries of Arizona streamflow data, "Water Resources Investigations 79-5", U.S. Geological Survey, Tucson, AZ

Hirschboeck, K.K. (1988). "Flood Hydroclimatology," In Flood Geomorphology, (Baker, V.R., Kochel, R.C., and Patton, P.C., ed.) John Wiley & Sons, New York, 27-49.

Quinn, W.H., Zopf, D.O., Short, K.S., & Kuo Yang, R. T.W. (1978). "Historical trends and statistics of the Southern Oscillation, El Niño, and Indonesian droughts, Fishery Bulletin. Vol. 76, 663-678.

Roeske, R.H. (1978). Methods for estimating the magnitude and frequency of floods in Arizona. Arizona Department of Transportation, ADOT-RS-15(151) Final Report, US Geological Survey, Tucson, Arizona.

Schulman, E. (1945). "Tree-Ring Hydrology for the Colorado River Basin," University of Arizona Bulletin, XVI(4), Laboratory of Tree-Ring Research, Bulletin 2.

Stedinger, J.R. and Baker, V.R. (1978), "Surface water hydrology: Historical and paleoflood information, "Rev. of Geophysics, AGU 25(2), 119-124.

Wallis, J.R., and Wood, E.F. (1985). "Relative accuracy of log Pearson III procedures," J. Hydr. Div, ASCE, III(7), 1043-1056.

Interagency Advisory Committee on Water Data. (1982). "Guidelines for Determining Flood Flow Frequency," Bulletin 17B of the Hydrology Subcommittee, US Department of the Interior, Geological Survey.

Drought Risk Analysis Based on Hydrologic Records of the River Nile

Gerold Reusing[1] and Wolfdietrich Skala[2]

Abstract

The characteristics of the parameters of hydrologic droughts are studied by analysing and modeling the maximum and "minimum" stage-level series of the River Nile from 715-1469/70 A. D.. A main factor in the statistical characteristics of these series is their long-term persistence. The results can be used to estimate values for expected drought durations, magnitudes and severities. The models are applied to the series of annual discharge sums at Aswan from 1903 to 1964. This series is not long enough to exhibit any long-term persistence. The application of models which take into account the long-term persistence lead to longer and more severe expected droughts.

Introduction

This paper deals with a probabilistic approach to assess the risk for hydrologic droughts of the River Nile. A hydrologic drought is defined by its parameters duration, severity and magnitude (Dracup et al. 1980).

The study of the probability distribution of these drought parameters is limited by the brevity of most hydrologic records. The stage levels of the River Nile, however, were measured annually during the Arabic period of Egypt at the Roda Island near Cairo. The annual maxima and "minima" (actually yearly levels measured regularly on the 3rd of July) series from 715-1469/70 A.D. represent 755 and 756 years of continuous observation, respectively. The analysis of these long historic records supplements the study and interpretation of simulated series.

[1] Research Assistant, [2] Professor, both: Institute for Geology (Mathematical Geology), Freie Universität Berlin, Malteserstr. 74-100, 1000 Berlin 46, Germany.

The simulations are based on two different modeling procedures, namely ARMA and ffGn, preserving mainly short-term and long-term persistence, respectively.

Analysis and modeling of the two stage level series

The original series have been corrected for inconsistencies as far as possible according to the information given by Popper (1951). Additionally a linear trend has been removed in both series which can be attributed mainly to the siltation of the riverbed. The residual series have been analysed with respect to their short-term and long-term persistence.

A measure for long-term persistence is the Hurst coefficient. It has been determined by range analysis and found to be H = 0.77 for the maxima and H = 0,87 for the "minima". These relatively high values indicate the existence of a strong long-term persistence in these data sets. Based on the Hurst coefficient the ffGn-generator of Chi et al. (1973) has been used for the simulation of synthetic traces.

The ARMA modeling procedure according to Box & Jenkins (1970) is concerned with fitting a model ARMA (p, q) of the form:

$$y_t = \phi_1 y_{t-1} + \ldots + \phi_p y_{t-p} - \theta_1 \epsilon_{t-1} - \ldots - \theta_q \epsilon_{t-q} + \epsilon_t.$$

As the parameters $\{\phi\}$ and $\{\theta\}$ are estimated from the autocorrelation coefficients, these models preserve mainly the short-term persistence. But it was shown that they can mimick long-term persistence as well (O'Connell 1974). The following models with their corresponding parameters have been found to fit the historical data best:

maxima ARMA (3,1)

$\phi_1 = 1.101$; $\phi_2 = -0.132$; $\phi_3 = 0.007$; $\theta_1 = 0.881$;

"minima" ARMA (2,1)

$\phi_1 = 1.210$; $\phi_2 = -0.254$; $\theta_1 = 0.795$.

The behaviour of the autocorrelation functions of original and simulated data (figs. 1 a, b) also indicates the existence of a strong long-term persistence in both data series because the correlogram decays very slowly to the zero line. This behaviour apparently is well reproduced by the ARMA-models.

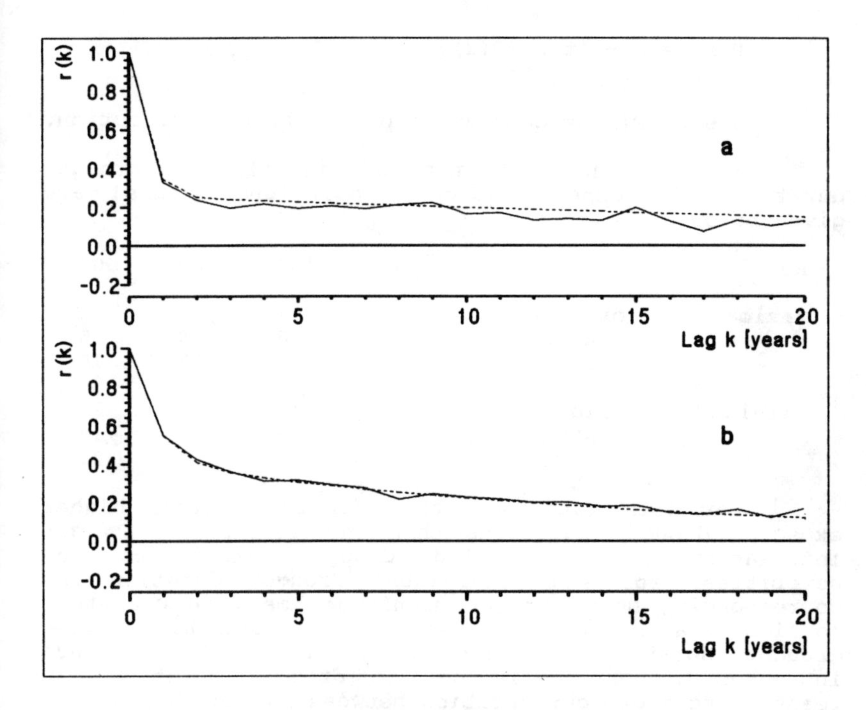

Fig. 1. Autocorrelation functions of original and simulated series of maxima (a) and "minima" (b). Solid line: original; dashed line: ARMA-model.

<u>Examination of drought parameters of original and simulated series</u>

With each model 10000 values have been simulated. The distributions of drought duration, severity and magnitude of the simulated series have been tested for goodness of fit to the distributions of the original parameters by the Kolmogorov-Smirnov test.

In case of the maxima, both modeling concepts are equally well suited for the simulation of drought parameters, in case of the "minima" only the ffGn-model reproduces all three distributions well.

The probability distributions of drought duration have been used to evaluate the expected drought durations for different return periods. The exceedance probabilities $P_e(x)$ for return periods RP of 50, 100, 200 and 500 years have been computed according to:

$$P(x) = 1 - \delta t \,/\, RP(x);$$
$$P_e(x) = 1 - P(x);$$

where: P(x) = cumulative probability distribution.

δt is given by the mean of high-flow and drought duration. The expected drought durations [years] are given below:

RP		50	100	200	500
maxima	original	5	7	10	12
	ARMA	5	8	10	15
	ffGn	6	9	12	15
"minima"	original	7	11	14	27
	ARMA	6	9	13	18
	ffGn	6	10	14	21

For both data sets the ffGn-model gives higher expected drought durations than the ARMA-model. To get information about expected drought magnitudes and severities, for each individual drought duration the corresponding mean drought magnitude has been computed. As figs. 2 a, b show, there exists a relationship between drought duration and magnitude. The mean magnitude increases proportionally to the drought duration, at least up to a drought duration between 10 and 15 years.

As the magnitudes are distributed approximately Normal and the corresponding standard deviations are given, this relationship can be used to get values for expected drought magnitudes with any desired exceedance probability. The expected drought severities can be found by multiplication of duration with magnitude.

<u>Conclusions about the risk of droughts for the Nile River</u>

If it can be assumed that the Roda maximum stage-levels are correlated with maximum discharges and also with the annual sums of discharge of the Main Nile, then the above obtained results can be applied to discharge series of the Main Nile. In this study the annual discharge sums measured at Aswan from 1903 to 1964 are used. Although this series is relatively long, it is not long enough to exhibit any long-term persistence. Significant short-term persistence can also not be found. It is a random series which can be represented by a white noise process.

The drought risk parameters have been computed from the simulated series, based on ARMA- and ffGn-models of the Roda maxima, and from a white noise process.

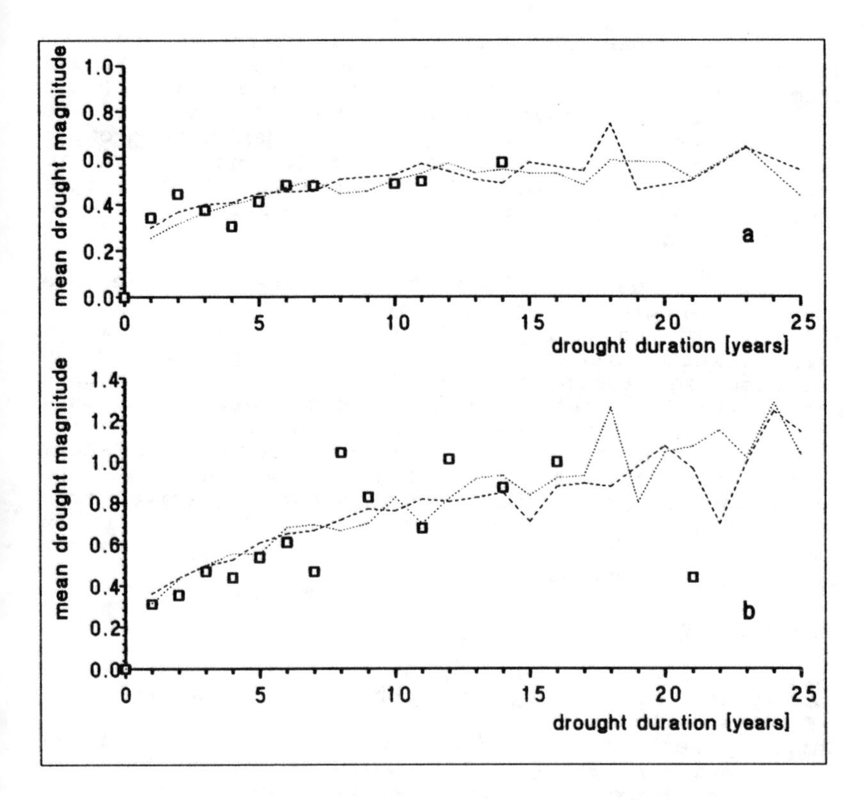

Fig. 2. Relationship between drought duration and mean magnitude [m] in original and simulated series of maxima (a) and "minima" (b). Squares: original; dashed line: ARMA-model; dotted line: ffGn-model.

The resulting expected drought durations, mean magnitudes and magnitudes with 5% exceedance probability as well as the corresponding severities are given below for a return period of 200 years:

	D [years]	M_{mean} [mio. m^3]	S_{mean} [mio. m^3]	M_{95} [mio. m^3]	S_{95} [mio. m^3]
WN	6	9671	58027	14153	84916
ARMA	10	12463	124630	17074	170740
ffGn	12	12658	151891	17601	211211

For comparison we give the parameters of the last drought period of the River Nile at Aswan. It extended from 1979 to 1988 over 9 years and had a magnitude of 9900 mio. m^3. The resulting severity is 89100 mio. m^3. If the independent model is considered, then this drought was an extraordinary one. But if a long-term model is applied, then it can be looked at as a less than average drought. Longer and more severe droughts have to be expected in future then.

It can be concluded that the choice of an adequate model is of great importance in drought risk analysis. The study of the Roda gauge-level series shows that long-term climatic trends and corresponding fluctuations in the discharge regime are important factors in the Nile basin. In this case a model which preserves long-term persistence certainly gives a more realistic estimation of drought risk than a model which preserves short-term persistence only. The problem still remains that the long-term persistence parameters cannot be derived from short time series. In risk estimation, however, it is important to know that a short-term model gives rather optimistic estimates.

Appendix: References

Box, G. E. P. and Jenkins, G. M. (1970). Time Series Analysis. Forecasting and Control. Holden Day, San Francisco etc..

Chi, M., Neal, E., and Young, G. K. (1973). "Practical Application of Fractional Brownian Motion and Noise to Synthetic Hydrology." Water Resour. Res., 9(6), 1523-1533.

Dracup, J. A., Lee, K. S. and Paulson Jr., E. G. (1980). "On the Definition of Droughts." Water Resour. Res., 16(2), 297-302.

O'Connell, P. E. (1974). "Stochastic Modelling of Long-Term Persistence in Streamflow Sequences." Ph. D. thesis presented to the Civil Engineering Department, Imperial College, London.

Popper, W. (1951). The Cairo Nilometer. Univ. of California Press, Berkeley and Los Angeles.

REGIONAL FLOOD FREQUENCY EQUATIONS FOR ANTELOPE VALLEY IN KERN COUNTY, CALIFORNIA

by
Wen C. Wang[1], M. ASCE, and David R. Dawdy[2], A.M., ASCE

Abstract

Rapid residential, commercial, and industrial development has occurred in recent years in the Antelope Valley, California. With the new development progressing at full speed, the local planning agencies face an urgent need for the development of appropriate hydrology for use in this semiarid area. Kern County Water Agency has developed regional flood frequency equations for major streams in the Antelope Valley originating in the Tehachapi Mountains (1971, 1981, 1984, 1985, 1987). For streams of small drainage area, regional flood frequency equations for the South Lahontan-Colorado Desert region developed by the U.S. Geological Survey (1977) are often used. The USGS regional equations are based heavily on the flood data of the Mojave Desert and the Imperial Valley, where streams are characterized by large flood peaks. The applicability of the regional equations to the Antelope Valley is in doubt. This paper shows that the USGS equations generally over-estimate flood peaks in the Antelope Valley, and a new set of regional equations for the Antelope Valley using local flood data are developed.

USGS Regional Equations

The existing USGS regional flood frequency equations were developed based on gaging station data for streams in the South Lahontan-Colorado Desert Region (See Fig. 1), which includes streams in southeastern California between the international boundary and Mono Lake, except the upper basins of Sierra Nevada streams draining to the Owens River and streams draining into the south San Joaquin Valley from the south and west. The region can be roughly divided into the Owens Valley in the north, Death Valley in the east, Antelope Valley in the west, Mojave Desert in the center, and Imperial Valley in the south. Also included in the region is the southern tip of the

[1] Multech, San Jose, and [2] Consulting Hydrologist, San Francisco

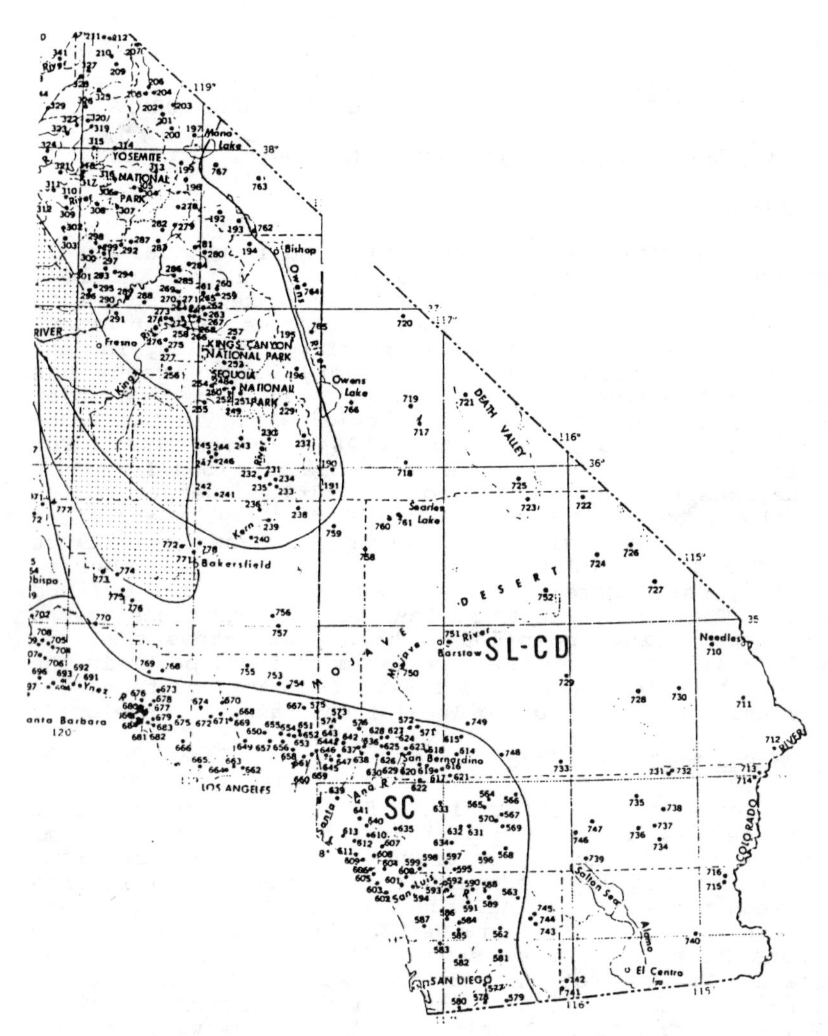

Figure 1. Location of Gaging Stations

San Joaquin Valley. The climate is semiarid. Altitudes range from 278 ft below to 14,000 ft above sea level. Floods along the western boundary of the region generally are caused by winter frontal-type storms, but annual peaks in the desert are the result of summer thunderstorms. The flood peaks are expressed as functions of drainage area in the regional equations as below:

```
Q2   =    7.3 A**0.30                    (1)
Q10  =  150   A**0.53                    (2)
Q25  =  410   A**0.63                    (3)
Q50  =  700   A**0.68                    (4)
Q100 = 1080   A**0.71                    (5)
```

To check the applicability of the equations to the streams in the Antelope Valley, flood data for the streams in the region were plotted against drainage area. A typical plot is shown in Fig. 2 for the 100-year floods. The data included those from the original study (USGS, 1977) plus additional updated data obtained from the USGS (summarized below). Numbers representing gaging station IDs are shown along with the data points. The curves representing the USGS regional equations for the 100-year flood are also shown in Figure 2. It can be seen from the figure that three separate subregions may be readily identified: the subregions for Owens Valley, Antelope Valley, and the Mojave Desert and Imperial Valley. For a given drainage area, a stream in the Mojave Desert or Imperial Valley is seen to have a greater flood peak than a stream in the Antelope Valley, while a stream in the Owens Valley is seen to have the lowest peak. Except for the 2-year flood, the distinctions among the three subregions are apparent. The distinction becomes more apparent as the return period increases. For the 2-year flood, no obvious distinction can be seen.

A reasonable explanation for this tendency is that the frequent flood events such as the 2-year flood are predominantly caused by winter frontal-type storms. These winter storms originate in the Pacific and move inland from the west and northwest, typically lasting for several days. Winter frontal-type storms cover a broad geographical area, and tend to be characterized by a rather uniform functional relationship between flood peak and drainage area over the entire region. The infrequent flood events, on the other hand, are the result of localized summer thunderstorm activity, which is affected by the climatological characteristics of the individual subregions. So the functional relationship between flood peak and drainage area varies with the subregion.

The existing USGS regional flood frequency equations were developed based heavily on the stream data collected in the Mojave Desert and Imperial Valley, which exhibited

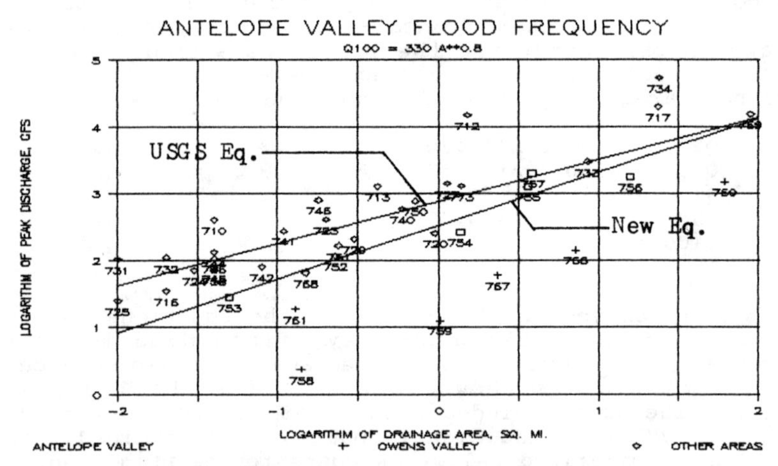

Figure 2. Peak Discharge Versus Drainage Area for the 100-Year Flood

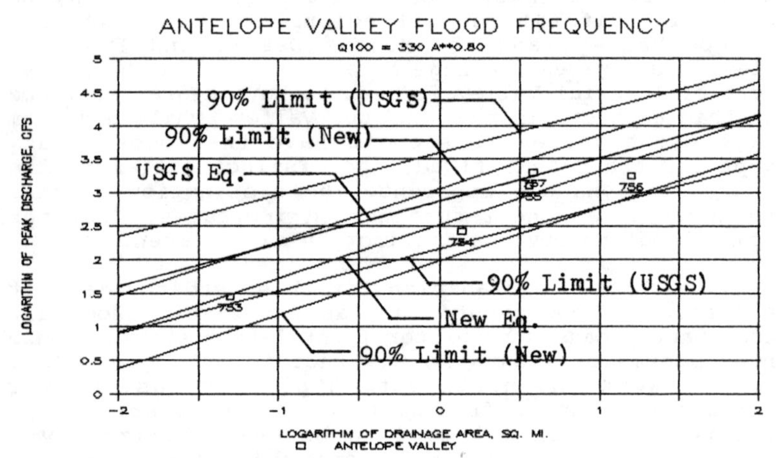

Figure 3. 90% Confidence Limits for the USGS and the New Equations for the 100-Year Flood

higher flood peaks. As shown in Fig. 2 for the 100-year flood, only one out of the five data points in the Antelope Valley was slightly above the USGS regional equation. The regional equations, except for the 2-year flood, apparently over estimate flood peaks for the streams in the Antelope Valley. The regional trend in the flood distributions as noted above should be considered in the development of the flood peak estimation. The plot of the data shows that peaks in the desert south are greater than those of the same return period in regions farther north.

Regional Equations for the Antelope Valley

To establish appropriate regional equations for use in the Antelope Valley, a new set of equations must be developed using the flood data collected for the local streams. The gaging station data used were:

Map #	USGS #	Gaging Station	Period of record
753	10264520	Amargosa Cr. Trib. nr. Palmdale	1959-73
754	10264530	Pine Cr.nr. Palmdale	1959-88
755	10264560	Spencer Canyon Cr. nr. Fairmont	1959-86
756	10264600	Oak Cr.nr. Mojave	1958-86
757	10264605	Joshua Cr. nr. Mojave	1959-73

Using the flood peaks for selected return periods for the above five gaging stations, regional flood frequency equations for the Antelope Valley were developed for the 5-, 10-, 25-, 50-, and 100-year floods (Water Resources Council, 1982). The constants in the regional equations were calculated using least squares analysis. For the 2-year flood, the USGS equation was adopted. Like the USGS equations, flood peak was expressed as a function of drainage area. The equations are as follows:

Q2 = 7.3 A**0.30 (Adopted from USGS) (6)
Q5 = 36 A**0.46 (7)
Q10 = 75 A**0.56 (8)
Q25 = 150 A**0.66 (9)
Q50 = 230 A**0.74 (10)
Q100 = 330 A**0.80 (11)

where Q is flood peak in cubic feet per second for the specified return period and A is drainage area in square miles. The superiority of the new equations is shown by comparing the standard errors of estimate for the USGS and the new equations. Table 1 shows that the standard errors for the new equations are smaller than those of the USGS equations. This is so even though the sample size is much reduced, which decreases the degrees of freedom, and tends to make the standard error larger.

Table 1: Standard Errors of Estimate for the Regional Flood Frequency Equations

Return Period		New Equation	USGS Equation
2-Year	0.60	None	
5-Year		0.31	0.35
10-Year		0.29	0.31
25-Year		0.28	0.32
50-Year		0.27	0.33
100-Year		0.27	0.36

The regional equation developed for the Antelope Valley for the 100-year flood is compared with that by the USGS in Fig. 3. The 90 percent confidence limits (2 standard errors of estimate from the equation) for the new equation and for the USGS equation are also shown.

CONCLUSIONS

The regional flood frequency equations for the South Lahontan - Colorado Desert region developed by the U.S. Geological Survey were evaluated. It was found that there exists a regional trend in flood distributions and that the USGS equations generally over estimate flood peaks in the Antelope Valley. A new set of regional equations for the Antelope Valley was developed using local flood data. The new equations exhibited smaller standard errors of estimate than those of the USGS equations.

References

Kern County Water Agency, "Proposed Antelope Valley Flood Control Plan," January 1971.

Kern County Water Agency, "Summary of Antelope Valley Streams at Los Angeles Aqueduct, Revised Hydrology," 1981.

Kern County Water Agency, "Summary of Antelope Valley Streams at Los Angeles Aqueduct, Revised Hydrology," July 1984.

U.S. Department of the Interior, Geological Survey, "Magnitude and Frequency of Floods in California," Water Resources Investigations 77-21, June 1977

U.S. Department of the Interior, Water Resources Council, "Flood Flow Frequency, Bulletin #17B of the Hydrology Subcommittee," March 1982.

A Multi-Level Contour Method for Tracking and Forecasting Rain Fields of Severe Storms by Weather Radars

Z.Q. Chen[1] and M.L. Kavvas[2], Member, ASCE

ABSTRACT

A new automated method is developed in order to track and predict in short-term (15 min -1 hr lead times) the evolution of rain fields in time and space, as observed by weather radars. First the rain field is decomposed into simple, tractable elements. Then a statistical adaptive forecasting scheme is developed in order to predict the changes in each of these elements in time. The composition of these elements forms the complete rain field with respect to its spatial configuration, location and rain intensity texture at each prediction lead time. The method has been applied to the conventional and doppler digital weather radar data of rain fields. Some application results are given.

INTRODUCTION

A short-term remote-sensor based comprehensive prediction scheme which can extrapolate in time and space the evolving spatial configuration, motion and fine intensity texture of a severe rainstorm is yet to be developed in U.S.A. This paper discusses the development of such a predictive scheme. The "short-term prediction" of the rain fields will mean tracking the time-space behavior of these fields at time increments of 5 to 15 minutes and forecasting the future time-space behavior of these fields for the lead times in the range from 5 minutes to at most one hour at each forecasting stage.

In this study, radar reflectivity fields are considered to be equivalent to the rainfall intensity field of two dimensions (simply called as the rain field) since a 1-1 function relation is provided to transform the radar reflectivity into rain intensity (Battan, 1973). The main focus of this study is to represent, track and predict rain fields as they are observed on a radar scope. A study comparing the displacements of consecutive rain fields versus their corresponding radial velocity fields, as observed on a Doppler weather radar, is also performed in order to gain insight into how to incorporate the Doppler radar observed radial velocity fields into the rain field prediction methodology. The pixel format data are converted to binary plans of different intensity levels. Furthermore, each binary plan is fragmented into contours. Thus the rain field is represented by a group of contours of different intensity levels. We choose to work with the rain fields in the contour format.

---

1. Graduate Student, Dept. of Civil Engineering, U.C. Davis, CA 95616
2. Assoc. Prof., Dept. of Civil Engineering, U.C. Davis, CA 95616

CONTOUR REPRESENTATION

Contour modeling is a planar shape analysis problem which has been studied extensively in computer vision and image analysis. In short term rainfall prediction, the measurement of changes of rain field with respect to time is the most important factor to be inspected. Usually such measurement is in terms of a set of parameters from which the rain field can be reconstructed. These parameters should be mathematically and computationally easy to handle.

Any contour can be written in complex parametric form $P(u) = x(u) + j\,y(u)$ where u is a unique index for a specific point (x,y) on the contour. The contour can also be written in polar coordinate form $P(u) = r(u)\exp(j\theta(u))$. The two most popular ways of representing a contour are the spatial domain representation and the frequency domain representation. Polygon contour model is one of spatial domain contour representations. The parameters of the polygon contour model are called the polygon vertices. The contour is modeled by a group of periodic piecewise linear basis functions and the polygon vertices. The contour is written as

$$P(u) = \sum_{i=0}^{n-1} C(i)Q_i(u) \tag{1}$$

where C(i) = { Cx(i), Cy(i)} are the coordinates of the polygon contour vertices

$$Q_i(u) = \begin{cases} (u-t_i)/(t_{i+1} - t_i) & t_i \le u < t_{i+1} \\ (t_{i+2} - u)/(t_{i+2} - t_{i+1}) & t_{i+1} \le u < t_{i+2} \\ 0 & \text{otherwise} \end{cases} \tag{2}$$

The polygon contour model is a good contour representation model with a set of parameters which is suitable for our tracking and prediction algorithms. Geometrically, polygon vertices can be viewed as points whose positions control the shape of the contour that they model locally. The geometrical meanings of the polygon vertices of a contour are clear. Mathematically, they are coefficients of the basis function expansions defining those curves. Each vertex influences curve shape locally and only at the position defined by the two line segments it touches. The property of the local influence makes it possible to track local shape change by tracking the change of the parameters. In this study a new polygon contour model for the rain intensity contour shapes was developed in order to track and predict the evolution of these shapes efficiently with realistic representations which replicate the observed shapes closely.

CONTOUR TRACKING

An edge-based method for tracking the rain intensity contours in radar images is developed. Tracking the rain intensity contours in radar images amounts to the identification of the relationships of contours between two consecutive frames of the rain intensity maps at different times. The relationship among the contours in one frame is called the infrastructure of the rain field of that frame. The infrastructure of the rain field is obtained by a contour extracting algorithm. A similarity matrix which measures the relationship between the contours at any two consecutive time steps is used for contour matching. Information about the infrastructure of the two rain fields is used to construct the similarity matrix. Merging, splitting, birth and dissipation of contours can be detected by the developed matching procedure. If a contour in the second frame is due to the evolution of a single contour in the first frame and the original contour keeps its identity in the second frame, the two contours in the two frames are called to have an offspring-parent relationship. The evolution of contours can be measured in terms of the change in the contour edges and in the centroid between the two contours that have an offspring-parent relationship in the consecutive frames. For new born contours, merged contours and split contours, different sets of initial state parameters are assigned to each contour according to its relation with other alive contours in the frame. These relations have been identified during the contour extraction stage.

STATE-VECTORS MODEL

The model we choose to describe the process which we are observing on the radar screen is a deterministic polynomial function of the time t. This polynomial has been observed in the presence of additive random errors. The polynomials can be used for smoothing over sufficiently short intervals with very little actual knowledge of the true process. It is reasonable to assume that the processes of interest can be adequately described, at least locally, by a polynomial of appropriate degree. A second degree polynomial quadratic is chosen to describe the process of interest. If the evolution of the centroid of a contour is the process of interest, the three parameters of a quadratic model can be viewed as initial position, velocity and acceleration of the centroid.

Let Z(t) denote the process of interest, where t is the time variable. This model can be defined in the form of a differential equation as: $(d/dt)^3 \cdot Z(t) = 0$, and let us define

$$A(s) = \begin{bmatrix} Ao \\ A1 \\ A2 \end{bmatrix}_s = \begin{bmatrix} Z(t) \\ (d/dt) \cdot Z(t) \\ (d/dt)^2 \cdot Z(t) \end{bmatrix}_s \tag{3}$$

where A(s) is called the state-vector of the process Z(•) at time s. As time s progresses through its successive values, A(s) gives a sequence of vectors, namely a multidimensional time series. In fact, A(s) is the Taylor expansion coefficient vector of the polynomial. Then the solution of the process can be written in the vector form

$$A(s+t) = F(t) \cdot A(s) \tag{4}$$

where F(t) is called the transition matrix of the process and is defined as

$$F(t) = \begin{bmatrix} 1 & t & t^2/2 \\ 0 & 1 & t \\ 0 & 0 & 1 \end{bmatrix} . \tag{5}$$

Equation (4) is called the transition equation of the process.

As soon as the estimation of the state-vector A(s) at time s is available, the prediction of the state-vector A(s+t) can be obtained by equation (4) by using the chosen lead time t.

PARAMETER ESTIMATION BY ADAPTIVE EXPONENTIAL SMOOTHING

Suppose that we select m independent parameters to describe the evolution of a rain intensity contour as it is being observed on a weather radar screen. We can use m independent filters to estimate the m state vectors of the m parameters which describe the evolution of the shape of this contour. These filters are completely unconnected, and operated entirely without any reference to each other. All the filters are chosen to be identical for reason of simplicity. Thus, only one filter is to be discussed in the following paragraphs.

The prediction procedure is simply to estimate the state-vector of the given model, i.e. the Taylor expansion coefficients of polynomials. Usually, we expect the estimates of the coefficients to be stable in the presence of noise. At the same time, the algorithm should detect any rapid change of the process and respond to such a change quickly. These two requirements are often conflicting. Therefore, a compromise between the two requirements has to be made. The adaptive exponential smoothing methods can easily deal with both stability and flexibility. The adaptive exponential smoothing puts more weight on the most recent observation. The influence of observation errors on the estimates decreases rapidly with the increasing age of the observations.

From the theory of the adaptive exponential smoothing (Brown, 1962), the state-vector A(s) can be estimated by $A_{s-t,s-t} + H \cdot E_s$, where E_s is the error between the observed value Z(s) at s and the t-step ahead prediction $Z_{s,s-t}$ made at time s-t, i.e. $E_s = Z(s) - Z_{s,s-t}$, and H is the smoothing coefficient vector, defined for the second degree polynomial (quadratic) model, being used to describe the evolution of parameters, by

$$H = \begin{bmatrix} 1-(1-\beta)^3 \\ 3/2 \cdot \beta^2 \cdot (2-\beta) \\ \beta^3 \end{bmatrix} \quad (6)$$

where β is the exponential smoothing coefficient.

It is an advantage to operate the adaptive exponential smoothing algorithm without knowledge of the distribution of the noise present in the prediction procedure. This algorithm is extremely compact and computationally very efficient. For the second degree polynomial (quadratic) model, only four storage spaces are required for the operation. The values of F(t) matrix and H vector can be set at the beginning of the program, then they are all fixed throughout the forecasting operation. The choice of the β value is decided on the basis of stability of the rain field. The initial values of the state-vector can easily be set up for the adaptive exponential smoothing algorithm. At the first time step, Ao is set to be the observation value Z(1) at time 1, A1 and A2 are set to be zero. At the second time step, Ao = Z(2), A1 = Z(2)-Z(1) and A2 = 0. From the third time step on, estimates of the state-vector are obtained directly from equation (4).

After obtaining the state variables of each parameter by the prediction algorithms, the final predicted rain field image is reconstructed by incorporating the predicted parameters and the infrastructure of the observed rain field.

APPLICATION OF THE DEVELOPED METHODOLOGY TO RADAR DATA

The Doppler weather radar data were the reflectivities and doppler radial velocities of the observed rain fields from a severe convective storm in pixel format with size of 256 x 256. Each pixel has a length of 1.25 km. The given data set includes 21 frames of reflectivity pixel maps and 21 frames of velocity pixel maps which were taken from approximately 6:57 to 9:16 EST, 12 July 1988 with a time lag of about 7 minutes.

The case presented here in Figure 1 is the comparisons among the rain fields observed at every 14 minutes, the predicted rain fields with 28 minutes lead time and 14 minutes updating, and the contours modified from the observed rain field at the 'P-M time' by radial doppler velocities on the boundaries of the contours. 'P-M Time' indicates the time at which the prediction was made. 'Obsv. Time' is the observation time associated with the frame of the radar image data indicated by 'Frame No.'. In order to develop a manageable model representation, tracking and prediction of the rain fields, only 6 most representative intensity (reflectivity) levels were selected and, thereby, an approximate model representation of each original image was obtained. The predicted rain field and the observed rain field are shown respectively on the left and the right sides of the upper part of the picture. The radar observes only part of the whole rain field in the given data set. The given radar data covers only a circular region with a diameter of about 120 KM. In order to compare a predicted rain field with the corresponding observed rain field, the part of the predicted contour region that is outside of the radar covering region is skipped. Only the parts of the predicted regions which are located in the radar covering region are displayed. The grey level index table which indicates the correspondence between the reflectivity value in dBz and the grey level is placed in the middle of each picture. Both the observed and the predicted rain fields, as mentioned above, are sliced into six different intensity level plans, which are shown as individual slices with their corresponding reflectivity colors in six different locations on the lower part of the picture. The scale of each rain field slice is one half of the rain fields shown on the upper part of the picture.

As shown in frames 7 and 15 in Figure 1, not only the movement of the rain field, but also the shapes of the intensity contours have been predicted by the developed methodology to be very close to the corresponding features in the observed rain field, especially in the high intensity level such as the white regions. It is also noted that the whole rain field is moving eastward (right side of the radar screen) and is rotating in the clockwise direction. This case is also chosen to show that the radial doppler velocities can be incorporated into the contour model to improve the rain field prediction. The white contours are the observed contours, the contours with radial displacement are the doppler contours. The hatched line area is the contours predicted by the contour method. In the doppler velocity map on the right lower corners of each quadrant in Fig. 1, the positive radial velocities (towards the screen boundary of the radar scope) are indicated by the darker grey levels, and the negative radial velocities (towards the screen center of the radar scope) are indicated by lighter grey level.

It is interesting to note from the doppler velocity map that a clear belt with zero radial velocity separates the negative velocity region on the lower left part of the screen and the positive velocity region on the right part of the screen. This complies with the movement of the whole rain field observed from the reflectivity field. From the observation of two distinct doppler velocity regions, it follows that the observed rain field is rotating clockwise. It is clear that the doppler velocity can be integrated into our model to improve the rain field prediction process. On the other hand, the observed doppler velocity field alone can not predict the movement of the rain field since the doppler velocities are the measurements of the average micro movement of the rain drops in a small volume while our interest is focused on the macro movement of the whole rain field. From the lowest intensity contours in Fig. 1, it can be seen that the doppler velocities near the boundaries of the rain field are close to zero, but the boundaries are actually moving towards right of the screen. Thus, we believe that the model, reported here, when integrated with the doppler velocities will produce predictions better than those by the model alone and those by the observed doppler velocity field alone.

ACKNOWLEDGEMENT. This research work was supported by the U.S. Geological Survey, Department of the Interior, under USGS award number USDI GS 14-08-0001-G1291. The views and conclusions contained in this document are those of the authors and should not be interpreted as necessarily representing the official policies, either expressed or implied of the U.S. Government.

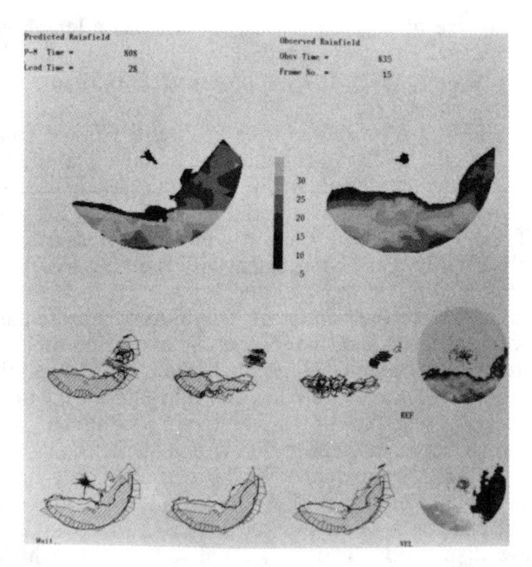

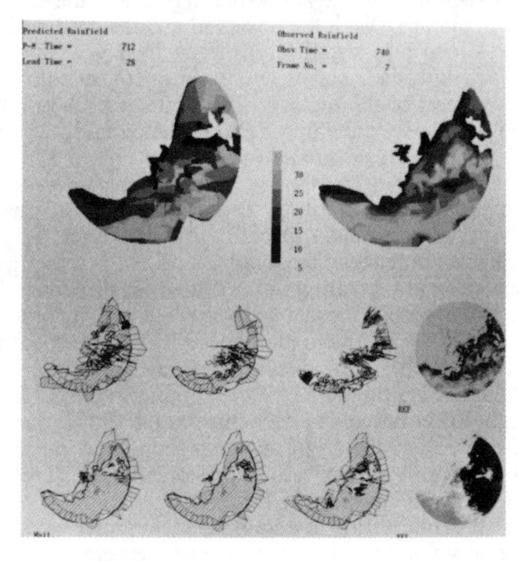

Fig 1. The comparison among the rain fields observed at every 14 minutes, the predicted rain fields with 28 minutes lead time and 14 minutes up-dating, and the contours modified from the observed rain fields at the 'P-M Time" by radial doppler velocities on the boundaries of the contours.

REFERENCES:

Battan, L.B., Radar Observation of the Atmosphere, 2nd ed., University of Chicago Press, Chicago, Ill., 1973.

Brown, R.G., Smoothing Fore-, casting and Prediction of Discrete Time Series,Prentice-Hall , Inc., Englewood Cliffs, 1963.

SIMULATING SPATIALLY VARIED THUNDERSTORM RAINFALL

Martin M. Fogel[1], M. ASCE and Kim Kye Hyun[2]

Abstract

Using data from two independent, relatively dense networks of raingages located in the vicinity of Tucson, Arizona, the spatial distribution of thunderstorm rainfall was determined as was the effect of elevation on the occurrence of rainfall events. Defining an event as the occurrence of a storm center (point of maximum rainfall) somewhere over the watershed, probability distributions were obtained for the time between events, the number of storm cells per event, the location of a storm center as influenced by elevation and the amount of rainfall at the storm center. Simulation techniques were then used to generate a time series of thunderstorm rainfall events.

Introduction

In arid and semiarid regions, hydrologists recognize that the location of a thunderstorm over a watershed has a major influence on the shape of the runoff hydrograph. While this is also true for other regions, the influence is more pronounced in the drier areas as high transmission losses often result in both the overland and channel flow phases from water moving over relatively dry lands. This study attempts to answer three questions, when a storm will occur, where will it happen and how much rain will fall. Answers provide the basis for producing a synthetic record of summer-type rainfall in southern Arizona.

---

[1]Professor of Watershed Management, School of Renewable Natural Resources, University of Arizona, Tucson, AZ 85721.

[2]Graduate Research Assistant, Dept. of Hydrology and Water Resources, University of Arizona, Tucson, AZ 85721.

Spatial Distribution of Storm Rainfall

Analyzing data from a 20 mi^2 (52 km^2) watershed which contained 32 raingages mostly on a one-mile grid, Fogel and Duckstein (1969) adopted the univariate, bell-shaped normal distribution or Gaussian curve to model the depth-area relationship of thunderstorm rainfall. This relationship, first suggested by Court (1961), is given by

$$R = R_o \exp(-\pi x^2 b) \quad (1)$$

where R is storm rainfall depth at a distance from the storm center defined as the point of maximum rainfall, R_o, and b is a shape factor. Analyzing data in addition to that used in the initial study by Fogel and Duckstein (1969), it was found that the relationship between b and R_o is

$$b = 0.20 R_o - 1.30 \quad (2)$$

Based on what is the most intensive set of hydrologic data in the southwestern United States, Osborn and Lane (1972), also, assuming symmetrical isohyets, developed the following depth-area relationship:

$$R = R_o (0.9 - 0.2 \ln A) \quad (3)$$

where R and R_o are expressed in inches of depth and A is the area contained within the R isohyet in square miles for $1 \le A \le 90$.

In a much more theoretical approach, Eagleson et al. (1987) analyzed eight years of summer storm rainfall records from the Walnut Gulch Experimental Watershed located near Tombstone, Arizona, the same watershed network used in the Osborn and Lane (1972) study. They looked at three point process models that "appear capable of reproducing important features of the spatial distribution of total storm precipitation, at least for storm types that are essentially stationary in space." They preferred a quadratic exponential over a simple exponential or a linear model, which is essentially similar to Equation (1).

For this study, Equations (1) and (2) will be used to describe the spatial distribution of thunderstorm rainfall.

Frequency of Occurrence

To generate a time series of thunderstorm rainfall events, two time-related factors need evaluation. One is how to start the season and the other is what is the time between such events, or the interarrival times. In southern Arizona, thunderstorms generally occur during a two- to three-month period which starts in early July following an extended dry period.

In earlier papers Duckstein et al. (1972), Fogel et al. (1974) and Duckstein et al. (1979), the interarrival time between events referred to the time between measurable rain at a point. For this study, the interarrival time refers to the time between storm center occurrences anywhere within the watershed. Historical data from the Atterbury Experimental Watershed indicated that the mean number of storm centers per season per square mile of watershed was 0.27. Assuming a seasonal length of from 60 to 90 days starting July 1, the average interarrival time between storm center occurrences would be calculated by dividing the average number of storm centers per season for the entire watershed into the length of season.

Location of Storm Centers

Osborn and Reynolds (1963) assumed that the locations of thunderstorms in a fairly level country can be considered as being randomly distributed. Analysis of available data confirmed the hypothesis that the occurrence of storm centers in level topography is essentially a chance operation.

In a subsequent study using a different set of data, Duckstein, Fogel and Thames (1972) showed that elevation has a marked influence on the mean number of rainfall events that occur in a season. The study concluded that for an increase in elevation of 1000 feet (305 m), there was a corresponding increase of 3.12 in the mean number of events per season. A study by Osborn and Davis (1977), determined that for stations in Arizona and New Mexico, the increase in the mean number of events per additional 1,000 feet (305 m) in elevation was nearly four. They went on to state that the data set used by Duckstein et al. (1972) was for a period that was in general about 20 percent drier than the data set they analyzed, and by making this adjustment, the results from the two studies were comparable.

Distribution of Storm Center Depths

In the initial study by Fogel and Duckstein (1969), a type 1 extremal distribution function was fit to the data for storm center depths. The use of subsequent data indicated that a gamma distribution function performed equally as well and was more readily applicable to the use of simulation techniques.

Simulation

A previous attempt for simulating rainfall occurrence and amounts for ungaged watersheds up to 150 km^2 was accomplished by Osborn et al. (1980). Theirs is more of a regional model than the approach taken herein. A later study by Hanson et al. (1989) uses a Markov chain-mixed exponential model structure to simulate daily precipitation amounts at a point for mountainous watersheds in Idaho.

Generating a time series of thunderstorm events that may occur over a watershed requires answering three questions, e.g., when, where, and how much.

When Will A Storm Center Occur

The simulated occurrence of the first storm center that falls on a watershed in a season can be estimated by one of two ways. The first method uses statistical data (mean and standard deviation) and then draws from a standardized normal distribution. For the Tucson region, the initial storm has usually occurred sometime during the first three weeks of July with the mean and standard deviation being July 10 and four days, respectively. The other method assumes a fixed starting date for the summer-type thunderstorm season to occur, such as July 1. Then, by assuming a storm occurred on that date, the first storm will occur on a date based on the first random interarrival time drawn from an exponential or equivalent geometric distribution.

Where Will A Storm Center Occur

The first step in answering this question is obviously to divide the watershed under study into sub-areas, each approximately 10 mi^2 (26 km^2). A rationale for choosing this value is that the area of a thunderstorm cell within which most of the rain falls is on order of this magnitude.

To determine the location of a storm center, given that a storm will occur, it can be shown that the probability of a storm occurring in a given sub-area is a function of elevation. The assumption previously made was that this probability increases linearly with elevation, in which there are an increase in the mean number of occurrences of 3.8 per 1000 feet (305 m) increase in elevation.

Monte Carlo simulation is then used to determine on which sub-area the storm center will occur.

Up to now, it has been assumed that there is only

one storm center per event. Data from the Atterbury Experimental Watershed indicates that there were 1.23 centers per event with one center occurring on the 20 mi^2 (52 km^2) watershed 81 percent of the time, two occurring 14 percent and three centers the other 5 percent of the time. A geometric distribution function was fit to these values to be used in the simulation program. It would be difficult to extrapolate much beyond this area and therefore, it is suggested that if the occurrence of multi-cells per event are to be simulated that it be done for moderate-sized watersheds, say less than 100 square miles (360 km^2).

How Much Rain Will Fall

Given that a storm center will occur somewhere in the watershed this last question is answered also by the use of Monte Carlo simulation techniques. In a previous section, it was determined that a gamma distribution function can be used to represent the distribution of rainfall depths at the storm center. Then, given the location and amount of rainfall at the storm center, the spatial variation of rainfall is determined by Equations (1) and (2).

Conclusions

The basis for a simulation program has been developed that can be used to generate a time series of thunderstorm rainfall events that considers spatial variability and the effect of elevation. Data from two raingage networks in the vicinity of Tucson, Arizona were used to obtain parameter estimates for probability distributions to describe the location of a rainfall event, the number of cells per event, the amount of rainfall per event and the time between events.

While the general approach may be applied to other regions, it would be inappropriate to extrapolate the quantitative aspects beyond the southern Arizona area.

References

1. Court, A., "Area-Depth Rainfall Formulas," Journal of Geophysical Research, Vol. 66, 1961, pp. 1823-1831.

2. Duckstein, L., Fogel, M. and Kisiel, C., "A Stochastic Model of Runoff-Producing Rainfall for Summer Type Storms," Water Resources Research, Vol. 8, No. 2, 1972, pp. 410-421.

3. Duckstein, L. Fogel, M. and Thames, J., "Elevation

Effects on Rainfall: A Stochastic Model," Journal of Hydrology, Vol. 19, No. 1, 1972, pp. 21-35.

4. Eagleson, P., Fennessey, N., Qinliang, W. and Rodriquez-Iturbe, I., "Application of Spatial Poisson Models to Air Mass Thunderstorm Rainfall," Journal of Geophysical Research, Vol. 92, No. D8, 1987, pp. 9661-9678.

5. Fogel, M. and Duckstein, L., "Point Rainfall Frequencies in Convective Storms," Water Resources Research, Vol. 5, No. 6, 1969, pp. 1229-1237.

6. Fogel, M., Duckstein, L. and Sanders, J., "An Event-Based Stochastic Model of Areal Rainfall and Runoff," Proceedings, Symposium of Statistical Hydrology, USDA/ARS Misc. Pub. No. 1275, 1974, pp. 247-261.

7. Hanson, C., Osborn, H. and Woolhiser, D., "Daily Precipitation Simulation Model for Mountainous Areas," Transactions of the ASAE, Vol. 32, No. 3, 1989, pp. 865-873.

8. Osborn, H. and Davis, D., Simulation of Summer Rain Occurrence in Arizona and New Mexico. Hydrology and Water Resources in Arizona and the Southwest, Vol. 7, 1977, pp. 153-162.

9. Osborn, H. and Lane, L., "Depth Area Relationships for Thunderstorm Rainfall in Southwestern Arizona," Transactions of the ASAE, Vol. 15, No. 4, 1972, pp. 670-673.

10. Osborn, H. and Reynolds, W., "Convective Storm Patterns in the Southwestern United States," Bulletin, International Association of Scientific Hydrology, Vol. 8, 1963, pp. 73-83.

11. Osborn, H., Shirley, E., Davis, D. and Koehler, R., "Model of Time and Space Distribution of Rainfall in Arizona and New Mexico," USDA/SEA, ARM-W-14, 1980, 27 p.

Rainfall-Sampling Impacts on Runoff

by D.C. Goodrich[1], D.A. Woolhiser[2], and C.L. Unkrich

Abstract

The sensitivity of computed peak-runoff rates to rainfall-sampling frequency over three semiarid watersheds in southeastern Arizona of increasing size is investigated using the methodology of Woolhiser (1986). The impacts of rainfall sampling on runoff computations are assessed using the distributed rainfall-runoff model KINEROS. Results provide guidelines for the sampling frequency required to maintain accurate peak computations as a function of basin scale.

Introduction

In arid and semiarid regions flash floods are caused by high intensity, short duration storms with a high degree of spatial variability (Renard, 1977). The runoff hydrographs from these storms typically exhibit very short rise times, even for large catchments. From a study in progress on distributed, physically-based rainfall-runoff modeling, we have found that the outflow hydrographs are more sensitive to the rainfall input, including it's spatial variability, than to model parameters. Because distributed rainfall data are usually not available, stochastic generation techniques will be required to provide input for design purposes. The question then arises as to what rainfall sampling frequency is appropriate. This problem has been studied by Eagleson and Shack (1966), Harley et al. (1970), Bras (1979) and Woolhiser (1986).

Approach

To address the central issue of defining an appropriate temporal rainfall sampling frequency over a range of watershed scales, three subwatersheds from the USDA-ARS Walnut Gulch Experimental Watershed were selected for the study. Lucky Hills Watershed 106 (LH-106) has an area of 0.89 acres (0.36 ha). LH-104 has an area of 10.87 (4.40 ha) and contains the smaller LH-106 watershed. Walnut Gulch 11 (WG-11) has an area of 1560 acres (631 ha). All watersheds are highly

1: Assoc. Member, 2: Member; all authors from USDA-Agr. Res. Service, Aridland Watershed Management Research Unit, 2000 E. Allen Rd., Tucson, AZ, 85719 dissected. The vegetation on the Lucky Hills watersheds is predominantly brush, while WG-11 has a mixture of brush and grass cover. Rainfall is measured with two weighing-type raingages on the LH watersheds and with 10 similar gages on WG-11.

KINEROS was used to model watershed runoff. A detailed description of this model is provided elsewhere (Woolhiser, et al., 1990). Watershed geometry is represented as a cascade of planes and channels and the Smith-Parlange infiltration model (Smith and Parlange, 1978) is used in an interactive manner at each computational node to determine surface runoff rates. Interactive infiltration differs from an infiltration excess computation by allowing upslope inflow to infiltrate even if rainfall ceases. Channel infiltration is handled in a similar manner. Parameters required for each plane element include: interception depth, saturated hydraulic conductivity, coefficient of variation of hydraulic conductivity, soil porosity, rock content, Manning's n, and the length, width and slope of the plane. Similar data are required for channel elements. The daily water balance component of the chemical transport model CREAMS (Knisel, 1980) was run for the period of record at each gage to provide an estimate of the initial soil water content for each storm.

The geometric characteristics of the watershed were estimated from topographic maps. Soil texture and rock content were sampled in the field and initial estimates of saturated conductivity were obtained from regression relationships and tables presented by Rawls et al. (1982). Manning's n values were estimated from tables in Woolhiser et al. (1990). Channel cross section geometry was measured in the field. Ten runoff events were selected as an optimization set. They were chosen to cover a range from small to large storms, from dry to wet initial conditions and from simple to complex rainfall intensity patterns. It was assumed that the relative values of saturated hydraulic conductivity and Manning's n were correct so the parameters to be optimized were multipliers of these values and a global coefficient of variation of saturated hydraulic conductivity. The efficiency criterion of Nash and Sutcliffe (1970) was used as the objective function and high efficiencies (> 90%) were achieved for runoff peaks and volumes (Goodrich 1990). This high degree of fit establishes confidence in using the model to assess the effects of rainfall sampling interval on peak runoff rates and volumes.

The most detailed temporal representation of the rainfall intensity at a gage is provided by the "breakpoint" accumulated rainfall depth data. This data is obtained from analog charts by digitizing at irregular time intervals corresponding to changes in the rainfall intensity. The sampled storm data were obtained by placing sampling grids at increments of 5, 10, 15, and 20 minutes randomly over the breakpoint data for each storm. The starting time of the breakpoint data was distributed as a uniform random variable over the first sampling interval. The same random time shift was used for all gages for a single storm. This simulates the effects of a random start time within a given uniform sampling time interval. An interpolation procedure has been developed to describe the spatial and temporal variability of rainfall input to individual plane elements.

Twenty runoff-producing storms were used in this study. KINEROS was first run for each storm for each watershed using the breakpoint rainfall as input. Runs for each storm were then repeated using rainfall sampled at 5, 10, 15 and 20 minute intervals. In many modeling applications we are interested in reproducing the distributions of peak rates or volumes so the sampling effects were examined by creating quantile-quantile (Q/Q) plots of sampled versus breakpoint results. If the points on the Q/Q plot lie along the 1:1 line, then the two populations are identical. If they depart from the line in a systematic manner, additional insight can be gained.

Results

The effect of watershed size on appropriate rainfall sampling rate is demonstrated by the ranked peak rate Q/Q (dashed lines are +/- 10% slope of 1:1 line) plots shown in Fig. 1 for impervious watershed representations. As the sampling intervals increase, peak runoff rates are reduced. For the LH-106 watershed, which has a kinematic time to equilibrium on the order of 3 minutes, even the 5 minute sampling time results in a serious bias in calculated peak rates. The 5 minute sampling time is marginal for LH-104 but appears to be perfectly adequate for WG-11.

In Fig. 2 the same plots are presented for infiltrating cases. The same trends are apparent, but infiltration tends to make the runoff rates slightly less sensitive to rainfall sampling rate for the small watersheds but more sensitive for WG-11 for comparable ranges of peak rates (0-20 mm/hr on WG-11). The 5 minute sampling rate is now acceptable for LH-104. Note also that one event with rainfall sampled at 10 minutes has a higher peak rate than that calculated with breakpoint data. This phenomenon appears to be related to the way the infiltration algorithm handles a rainfall hiatus. In contrast, the 15 minute sampling interval, which was adequate for an impervious WG-11 is no longer adequate. An examination of the runoff volumes (not shown) reveals that they decrease with increasing sampling interval.

Harley et al. (1970) recommended a sampling interval equal to the basin kinematic time to equilibrium divided by 3.2. Using this criterion the calculated sampling intervals for LH-106, LH-104 and WG-11 based on a rainfall excess rate of 2 inches/hr (50.8 mm/hr) are 1.0, 2.7, and 13.3 minutes respectively. This rainfall excess rate is reasonable for the two smaller watersheds but is very conservative for WG-11 because rainfall excess rates per unit area decrease with basin size. The criterion of Harley et al. (1970) appears to be appropriate for the LH watersheds but is too large for WG-11, apparently due to the infiltration effect.

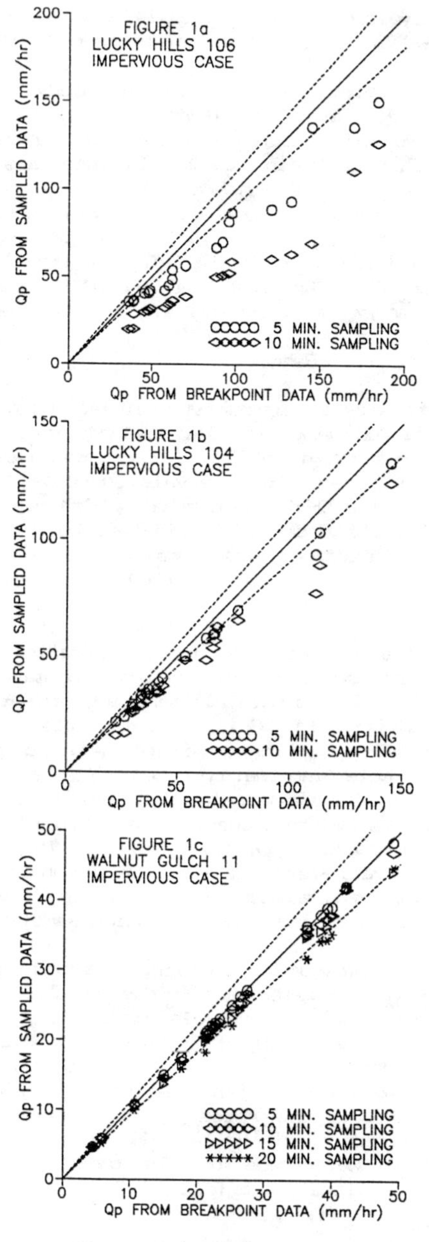

Figure 1 - Impervious Cases

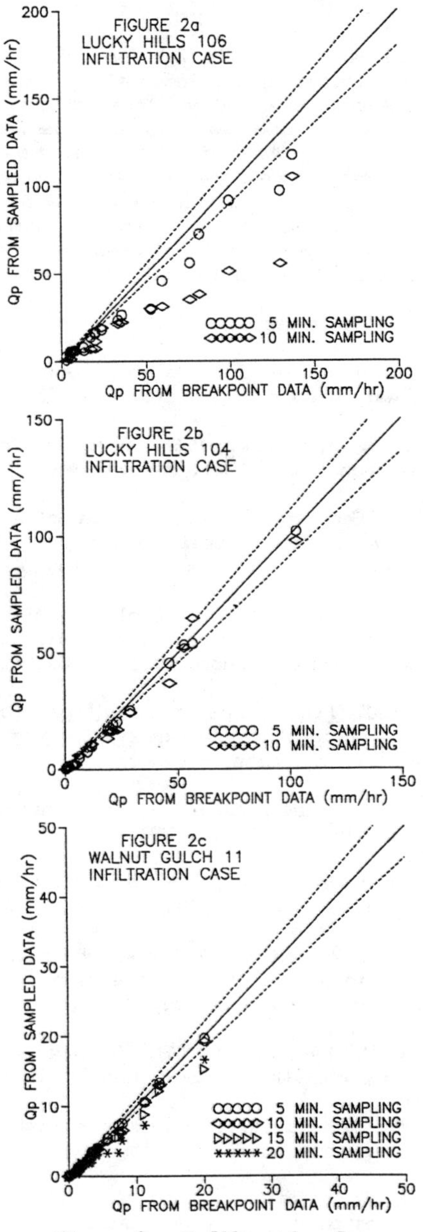

Figure 2 - Infiltrating Cases

Discussion and Conclusions

The appropriate rainfall-sampling interval for aridland watersheds depends on many factors including the temporal pattern of rainfall intensity, watershed response time and infiltration characteristics. The results of this study in which real watershed geometry was used agree with the analysis made by Wooliser (1986) based on runoff from a single plane. It is recommended that either breakpoint rainfall data or data sampled at uniform time increments according to the criterion of Harley et al. (1970) be used for watersheds with equilibrium times smaller than about 15 minutes and that a maximum interval of 5 minutes be used for more slowly responding basins.

References

Bras, R. L., 1979. Sampling of interrelated random fields: The rainfall-runoff case. Wat. Resour. Res. 15(6):1767-1780.

Eagleson, P. S., and W. J. Shack, 1966. Some criteria for the measurement of rainfall and runoff. Wat. Resour. Res. 2(3):427-436.

Goodrich, D. C., 1990. Geometric simplification of a distributed rainfall-runoff model over a range of basin scales. Ph.D. Diss., Dept. of Hydrol. and Water Resour., Univ. of Ariz., Tucson, AZ.

Harley, B. M., F. E. Perkins, and P. S. Eagleson, 1970. A modular distributed model of catchment dynamics. Report No. 133, Ralph M. Parsons Laboratory, Dept. of Civil Engr., M.I.T., Cambridge, Mass.

Knisel, W. G. (ed.), 1980. CREAMS: A Field Scale Model for Chemicals, Runoff, and Erosion from Agricultural Management Systems, U. S. Dept. of Agric., Sci. and Ed. Admin., Cons. Research Report 26, 643 pp.

Nash, J. E., and J. V. Sutcliffe, 1970. River flow forecasting through conceptual models, J. of Hydrol., 10:282-290.

Rawls, W. J., D. L. Brakensiek and K. E. Saxton, 1982. Estimation of soil water properties. Trans. ASAE, 25(5):1316-1320, 1328.

Renard, K. G., 1977. Past, present and future water resources research in Arid and Semiarid areas of the Southwestern United States. Australian Instit. of Engr., 1977 Hydrology Symp., Brisbane, p 1-29.

Smith, R. E. and J.-Y. Parlange, 1978. A parameter efficient hydrologic infiltration model. Wat. Resour. Res. 14(3):533-538.

Woolhiser, D. A., 1986. Sensitivity of calculated peak-runoff rates to rainfall-sampling frequency. IAHS Publ. No. 158, pp. 161-171.

Woolhiser, D. A., R. E. Smith and D. C. Goodrich, (1990). KINEROS, a Kinematic Runoff and Erosion Model: Documentation and Users Manual. U.S. Dept. of Agriculture, Agricultural Research Service, ARS-77.

Precipitation on Arid or Semi-Arid Regions of the Southwestern United States: Research Needs from a Consultant's Perspective

Michael E. Zeller, Member, ASCE[1]

Abstract

In terms of both instantaneous flow peaks and total runoff volumes, a key element in the quantification of stormwater runoff from arid and semi-arid regions of the southwestern United States is the proper characterization of the amount and spatial/temporal distribution of precipitation that occurs during individual rainfall events--especially the high-intensity, convective thunderstorms which are typical of these locales.

In recent years, the advent of the personal computer has enabled both engineers and hydrologists to mathematically model the physical processes of rainfall/runoff in ever increasing detail; yet the basic input parameter of "precipitation", and its influence upon the output of such mathematical models, still lacks the necessary standardization to produce useable results with a high degree of confidence when calibration data is either limited or altogether lacking, as is often the case within arid and semi-arid regions of the southwestern United States.

This paper is intended to present one consultant's perspective on the research needs for characterizing precipitation on arid and semi-arid regions of the southwestern United States. Its focus will be upon calling attention to the need for development of improved rainfall/runoff algorithms through research of the interaction between high-intensity, convective thunderstorm precipitation (amount and spatial/ temporal distribution) and the land surface, including vegetation, soil types, and land use--particularly urbanization by man.

----

[1]Principal-In-Charge, Arizona Region, Simons, Li & Associates, Inc., P.O. Box 2712, Tucson, Arizona 85702-2712.

Introduction

Many areas lying within southwestern regions of the United States are described as having either arid or semi-arid environments. What this generally means is that annual precipitation is too low to support significant amounts of vegetation, thus exposing the native soil complexes to severe erosion from wind and rain. Adding to this problem is the fact that when precipitation does occur within such climates, it is often in the form of high-intensity, convective thunderstorms. These storms can account for up to 50%, or more, of the average-annual precipitation for a specific locale during a single, one- to three-hour time period. For example, Yuma, Arizona, experienced two extreme precipitation events on July 27 and August 9, 1989. During the July event, 77 mm of precipitation fell in approximately one hour; and during the August event, 130 mm of precipitation fell in approximately two hours (City of Yuma 1989). Yet, the average-annual precipitation for Yuma, Arizona, is only 67 mm (Yuma Daily Sun 1989)! Other accounts of similar extreme precipitation events have been documented in the literature for locations in Arizona, California, Colorado, New Mexico and Utah (National Weather Service 1977).

Because precipitation on arid and semi-arid regions of the southwestern United States may be of such an extremely violent nature when it does occur, "conventional" methods used for the prediction of both instantaneous flow peaks and total runoff volumes in the eastern or midwestern areas of the United States are not generally appropriate for use by engineers and hydrologists who wish to quantify such parameters for design purposes within these former regions. However, for lack of better methodologies, these procedures are often "misused" to do just this very thing.

Present Status

In terms of both instantaneous flow peaks and total runoff volumes, a key element in the quantification of stormwater runoff from arid and semi-arid regions of the southwestern United States is the proper characterization of the amount and spatial/temporal distribution of precipitation that occurs during individual rainfall events--especially the high-intensity, convective thunderstorms which are typical of these locales. Yet, even with the advent in recent years of the personal computer, enabling both engineers and hydrologists to mathematically model the physical processes of rainfall/runoff in ever increasing detail, the basic input parameter of "precipitation" (and its influence upon the output of such mathematical models) still lacks the necessary standardization to produce useable results with a high degree of confidence when calibration data is either limited or altogether lacking, as is often the case within the arid and semi-arid regions of the southwestern United States. A recent paper (Reich 1988) discusses, in great length, the shortcomings of our present knowledge concerning precipitation in the western United States. It describes our present reliance upon precipitation data which has not been updated for over 20 years. With the population explosion that has occurred within the southwestern United States during the past 20 years exposing more and more people to devastating floods, the present status of reliable precipitation data to aid in the prediction of 100-year floods, or greater, with any meaningful precision must be questioned.

Research Needs

Engineers and hydrologists need simple, yet reliable, rainfall/runoff algorithms for use in their prediction models for arid and semi-arid regions of the southwestern United States, whether they be empirical procedures or mathematical models. These algorithms should incorporate parameters which reflect the interaction between high-intensity, convective thunderstorm precipitation (amount and spatial/temporal distribution) and the land surface, including vegetation, soil types, and land use--particularly urbanization by man.

To accomplish the above, research should focus upon establishing a clear relationship between rainfall intensity and topographic relief during thunderstorm precipitation. Past research has indicated that the maximum point rainfalls and rainfall intensities associated with convective thunderstorms within arid or semi-arid regions of the southwestern United States may not increase directly as a simple function of elevation (i.e., topographic relief), but rather may actually "peak" at the mid-level elevations associated with the bajadas and foothill areas typical of these locales. More definitive criteria with regard to this influence upon precipitation within the southwestern United States should be developed.

In addition, areal reduction of thunderstorm precipitation as a function of area versus maximum point rainfall should be better quantified, and a relationship between maximum peak discharges from small drainage areas versus storm duration (e.g., 1-hour, 2-hour, 6-hour, 24-hour) should be properly characterized. Many engineers and hydrologists are very familiar with the depth-area curves that were developed by the U.S. Weather Bureau (now called the National Weather Service) for use on a nationwide basis; and which were presented in their landmark publication titled "Technical Paper No. 40, Rainfall Frequency Atlas of the United States" (Hershfield 1961).

Unfortunately, these depth-area curves do not properly characterize the spatial distribution of thunderstorm rainfall. However, a more recent publication titled "Depth-Area Ratios In The Semi-Arid Southwest United States" (Zehr and Myers 1984) makes an advancement in this regard, but still is only a start in the right direction. Because the dominant flood-producing events on small watersheds within arid or semi-arid regions of the southwestern United States are convective thunderstorms, the need to better quantify the temporal and spatial distribution of thunderstorm rainfall is vital, and research in this area needs to be accelerated. It is especially critical that design rainfall patterns for thunderstorm durations of three hours, or less, be developed as a substitute for the present practice of embedding thunderstorm rainfall distributions within longer-duration, 24-hour storms for application on small watersheds.

There is also a real need to define the temporal distribution of thunderstorm rainfall, and to establish amounts of such precipitation versus time in terms of actual thunderstorm events, rather than from data sets used to develop 6-hour and 24-hour precipitation amounts (Miller et al. 1973). Rainfall amounts contained within one-hour to three-hour storms, but which occur during time increments of much less than one hour, can produce extremely high rainfall intensities during convective thunderstorms; and the proper placement of these "bursts" of rainfall within design storms is a key element for properly characterizing stormwater runoff from arid or semi-arid areas of the southwestern United States.

Finally, and perhaps most importantly, there needs to be further research into the reaction of land surfaces to extremely high-intensity thunderstorm rainfall--specifically runoff response to such storm events. Consequently, of top priority is the need to better quantify the effect of rainfall intensity and urbanization upon both the peak discharge and total storm volume of convective thunderstorms. For example, recent investigations have revealed an increased runoff to rainfall ratio in response to high-intensity thunderstorm versus frontal-type precipitation on the arid and semi-arid regions of the southwestern United States. The combined effects of this phenomenon and the impacts of urbanization are likely to result in the production of much larger flood peaks and flood volumes than present "Eastern" or "Midwestern" methodologies would otherwise predict for thunderstorm-generated runoff from small, southwestern watersheds. Therefore, the continuance of research in this area of precipitation must especially be encouraged, and the results of such research must be integrated into new and improved rainfall/runoff algorithms that can be easily adapted for use by practicing engineers and hydrologists.

Conclusions

There is a very real need to improve upon our knowledge of the character of thunderstorm precipitation on arid and semi-arid regions of the southwestern United States. Engineers and hydrologists are being asked almost daily to predict the magnitude of a 100-year flood, or beyond. The lives of many individuals may depend upon the reliability of such predictions. Without adequate precipitation data to properly characterize the amount and spatial/temporal distribution of thunderstorm rainfall, such predictions will, at best, remain questionable. Therefore, it behooves all levels of the public sector to join forces with the private sector and work toward development of improved rainfall/runoff algorithms for characterizing high-intensity, convective thunderstorm precipitation on the arid and semi-arid regions of the southwestern United States.

Appendix I. References

City of Yuma (1989). Written communication to Simons, Li & Associates, Inc., dated November 3, 1989.

Hershfield, David M. (1961). Technical Paper No. 40, "Rainfall Frequency Atlas of the United States", U.S. Government Printing Office.

Miller, J. F., et al. (1973). "Precipitation Frequency Atlas of the Western United States", NOAA Atlas 2, National Weather Service, Silver Spring, MD.

National Weather Service (1977). Hydrometeorological Report No. 49, "Probable Maximum Precipitation Estimates, Colorado River and Great Basin Drainages", U.S. Government Printing Office.

Reich, Brian M. (1988). "Need for New Rainfall Intensity Atlas Analyses in the West", Transportation Research Record, 1201, Transportation Research Board, Washington, D.C., 22-29.

Yuma Daily Sun (1989). "Wind and Water! A Diary of Destruction".

Zehr, Raymond M., and Myers, Vance A. (1984). NOAA Technical Memorandum NWS HYDRO-40, "Depth-Area Ratios in the Semi-Arid Southwest United States", available from National Technical Information Service.

Longitudinal Surface Profiles of Granular Debris Flows

Chyan-Deng Jan[1] and Hsieh Wen Shen[2]

ABSTRACT

Longitudinal surface profiles of granular material flowing down an inclined plane were studied by means of theoretical analysis and laboratory experiments. There are two driving forces acting on the granular material. One is a gravity force component along the inclined plane and the second is a force induced by lateral pressure gradient. The lateral earth pressure theory used in soil mechanics was expanded for granular debris flow. The total resistant stress acting on granular flows was treated as the linear sum of a rate-independent, contact friction part and a rate-dependent, viscous part. The former was expressed as a Mohr-Coulomb relationship. The latter was expressed as a friction slope relating to flow velocity and flow depth. According to this forgoing assumptions and the balance of longitudinal momentum, a method to calculate the longitudinal surface profiles was developed, by using a moving coordinate.

Experiments with 5 mm glass beads were conducted in a conveyor-belt flume. Preliminary results suggest that the slope of the surface profile of granular flow decreases as the flume slope increases, and the slope of the surface profile increases as the belt speed increases. Experimental results were also compared with theoretical results.

1. INTRODUCTION

Granular debris flows exhibit non-Newtonian behavior, and thus a rheological model relating stresses, strain and other variables is needed. Theoretically there are two approaches capable of formulating a rheological model for granular flows. One is the microscopic approach, based on the dynamics of individual particle collision using statistical mechanics to develop a rheological model. The other is a macroscopic approach, based on continuum mechanics to describe the flow behaviors of granular debris flows without using any detailed reference on the microstructure of debris material. Despite a conceptual difference in the two approaches, the final expressions of the rheological model obtained from either approach can be readily shown to relate to the other. Rapid advances have been made in the microscopic approach (Ling and Chen, 1989), but their results are mainly suitable to the rapid granular flows where the contact force between particle plays little significance. The contact force is more

---

[1]Graduate Student, U.C. Berkeley, 310 O'Brien Hall, Berkeley, CA 94720

[2]Professor, Dept. of Civil Engineering, U.C. Berkeley, Berkeley, CA 94720

important in granular debris flows since most particles are in contact with each other during debris flowing.

Chen (1986) proposed that the one-dimensional unsteady flow equations for debris flow are identical to those for clear-water flow, except for different value in flow parameters, such as the momentum correction factor and the resistance coefficient. If one assume no erosion and deposition of sediment in the transport process, the dynamic equations for debris flow routing are continuity equation,

$$\frac{\partial h}{\partial t} + \frac{\partial (hU)}{\partial x} = 0 \tag{1}$$

and depth-averaged momentum equation,

$$\frac{\partial U}{\partial t} + (2\beta - 1)\, U \frac{\partial U}{\partial x} + (\beta - 1)\, \frac{U^2}{h} \frac{\partial h}{\partial x} = g\sin\theta - \frac{\partial P}{\partial x} - \frac{\tau_0}{\rho} \tag{2}$$

in which h = flow depth; t = time; U = depth-averaged velocity; x = space distance in the longitudinal direction of flow; β = momentum correction factor; g = gravity acceleration; θ = angle of inclination of channel bed; τ_0 = bed shear and ρ is bulk density of the debris mixture. For solving these equations; one need a rheological model expressing the bed shear stress τ_0.

The term of shear stress has entirely different expressions in solid physics and in hydraulics. In solid physics, the shear stress τ is defined as normal stress σ times a friction factor tan ϕ ($\tau = \sigma \tan \phi$), in which ϕ is a friction angle of the solid grains. In hydraulics, on the other hand, the shear stress over a flow boundary does not relate to any fluid stress normal to the boundary, but to the square of some representative flow velocity U. Thus $\tau = \rho C_d U^2$. The flow velocity of a fluid is independent of any normal stress it may exert on the flow boundary.

Granular debris is a mixture of solid grains, soil and fluid, an applicable model describing the shear stress caused by debris flows should possess two types of stress: One is dependent on normal stress relating to interparticle contact friction and friction on flow boundary. The other is dependent on flow velocity, relating to interparticle collisions.

2. LONGITUDINAL SURFACE PROFILES OF GRANULAR DEBRIS FLOW

Consider debris flows down an inclined plane with angle θ and let ξ be the moving coordinate parallel to the inclined plane. The longitudinal surface profile is designed as $Z = h(\xi)$, in which Z is perpendicular to the longitudinal moving coordinate ξ-axis, pointing toward the surface of debris flow. $\xi = x - Ut$, where x is the fixed coordinate along the inclined plane; U is the same as the traveling velocity of the front and t is the time. A definition sketch is shown in Figure 1.

2.1 Driving Forces Acting on Granular Debris Flows

There are two driving forces acting on debris flow. One is the gravity force component along the inclined plane and the other is induced by lateral pressure gradient. Granular debris flow is a discrete system that can be treated as a continuum with suitable statistical averaging. However, the pressure acting on debris is not isotropic, the lateral pressure is not exactly equal to normal pressure. The lateral earth pressure theory used in soil mechanics is expanded for granular debris flow. The lateral pressure P is expresses as

$$P = K\rho g(h-z)\cos\theta \tag{3}$$

in which K is active/passive pressure coefficient and it could be larger or less than unity. According to the Rankin's theory, the pressure coefficients are expressed as

$$K = tan^2(45 \pm \phi/2) \tag{4}$$

in which ϕ is the static friction angle of soil material; positive/negative signs express passive/active pressure coefficients, respectively.

By analogy with Rankin's relationship the pressure coefficients for debris or granular debris flow can be expressed as

$$K = tan^2(45° \pm \phi_D/2) \tag{5}$$

in which ϕ_D is the dynamic friction angle of moving debris.

2.2 Resistance Stress

The resistant shear stress in granular debris flow is caused by both the interparticle contact friction and the interparticle collisions. In this study, the stress induced by interparticle contact friction is assumed to be independent of shear rate while the stress induced by interparticle collisions is dependent on shear rate. The total resistant stress is treated as the linear sum of a rate-independent contact part and a rate-dependent viscous part. Based on Mohr-Coulomb relationship, the rate-independent resistance stress τ_I can be expressed as

$$\tau_I = \sigma \tan\phi_D \tag{6}$$

in which σ is normal stress and can be expressed as

$$\sigma = \rho g(h-z)cos\theta \tag{7}$$

The rate-independent resistance shear acting on the inclined plane can be written as

$$\bar{\tau}_I = \rho gh cos\theta\ tan\phi \tag{8}$$

The rate-dependent shear acting on a inclined plane can be expressed with a friction slope S_f as follows

$$\bar{\tau}_D = gS_f \tag{9}$$

The friction slope S_f can be expressed in terms of U, h and a resistance coefficient using the uniform flow formula (Chen, 1986). Use of different flow formulas result in the various expressions of S_f, such as the Darcy-Weisbach equation, Manning's equation, Chezy's formula, etc. For Bingham plastic fluid, Chen (1986) showed that the friction slope S_f can be expressed as

$$S_f = \frac{CU}{h^2} \tag{10}$$

in which C is a resistance coefficient. According to Bagnold's experiments (1954), one can see that the expression of S_f in Eq. (10) seems reasonable for granular flow in the range of laminar flow. In this study, experiments with 5 mm glass beads showed that granular debris flow established in a conveyer-belt flume were laminar flows, so Eq. (10) is used in following theoretical analysis.

2.3 Longitudinal Surface Profiles

The established longitudinal surface profiles of granular debris flow can be theoretically derived by solving the one-dimensional equations of continuity and motion shown in Eqs. (1) and (2), which are formulated based on the assumption of no erosion and deposition of debris in the transport process. Substituting Eqs. (3), (8) and (9) into Eq. (2) yields

$$\frac{\partial U}{\partial t} + (2\beta-1)U\frac{\partial U}{\partial x} + (\beta-1)\frac{U^2}{h}\frac{\partial h}{\partial x} = g\sin\theta - Kg\cos\theta\frac{\partial h}{\partial x} - g\cos\theta\,\tan\phi_D - gS_f \tag{11}$$

Under the consideration of the motion in a system of coordinate moving at a velocity U which is the same as the traveling velocity of the front and after combination with continuity equation, the equation of motion yields

$$\left(1 + \frac{h_*}{h}F^2\right)\frac{dh}{d\xi} = \frac{\tan\theta - \tan\phi_D}{K} - \frac{CU}{Kgh^2\cos\theta} \tag{12}$$

in which $F = U\sqrt{gh_*\cos\theta/(\beta-1)}$, being a modified Froude number for uniformly progressive flow; h_* is a characteristic flow depth.

From Eq. (12), one can see that longitudinal surface profiles are mainly depends on the angle of inclination of plane, the dynamic friction angle of debris, flow depth, and flow velocity. Theoretically, the surface profiles can be classified to five cases as described in the following cases:

CASE 1: As $\theta = \phi_D$, $\frac{dh}{d\xi} < 0$ always

$$\frac{\xi}{h_*} = -\frac{1}{3}\left(\frac{h}{h_*}\right)^3 - \frac{1}{2}\left(\frac{h}{h_*}\right)^2 F^2 \tag{13}$$

in which $h_* = [CU/Kg\cos\theta]^{1/2}$ (14)

CASE 2: As $\theta < \phi_D$, $\frac{dh}{d\xi} < 0$ always

$$\frac{\tan\theta - \tan\phi_D}{K}\frac{\xi}{h_*} = \frac{h}{h_*} - \tan^{-1}\left(\frac{h}{h_*}\right) + \frac{F^2}{2}\ln\left[1 + \left(\frac{h}{h_*}\right)^2\right] \quad (15)$$

in which $h_* = [CU/gc\cos\theta(\tan\phi_o - \tan\theta)]^{1/2}$. (16)

CASE 3: As $\theta > \phi_D$, but $(\tan\theta - \tan\phi_D)/K < CU/Kgh^2\cos\theta$,

then $\frac{dh}{d\xi} < 0$ and $h_* > h$ always

$$\frac{\tan\theta - \tan\phi_D}{K}\frac{\xi}{h_*} = \frac{h}{h_*} - \frac{1}{2}\ln\left(\frac{1 + h/h_*}{1 - h/h_*}\right) + \frac{F^2}{2}\ln\left[1 - \left(\frac{h}{h_*}\right)^2\right] \quad (17)$$

in which $h_* = [CU/g\cos\theta(\tan\theta - \tan\phi_D)]^{1/2}$ (18)

CASE 4: As $\theta > \phi_D$ and $(\tan\theta - \tan\phi_D)/K = CU/Kgh^2\cos\theta$,

then $\frac{dh}{d\xi} = 0$ everywhere,

$$h = [CU/g\cos\theta(\tan\theta - \tan\phi_D)]^{1/2} \quad (19)$$

in which flow depth h is independent of space distance ξ so the flow is in uniform.

CASE 5: As $\theta > \phi_D$ and $(\tan\theta - \tan\phi_D)/K > CU/Kgh^2\cos\theta$, then

$\frac{dh}{d\xi} > 0$ *and* $h > h_*$ with a boundary condition $h = h_o$ *as* $\xi = \xi_o$

$$\frac{\tan\theta - \tan\phi_D}{K}\frac{\xi - \xi_o}{h_*} = \frac{h - h_o}{h_*} + \frac{1}{2}\ln\left[\frac{h/h_* - 1}{h_o/h_* - 1}\right] - \frac{1}{2}\ln\left[\frac{h/h_* + 1}{h_o/h_* + 1}\right] + \frac{F^2}{2}\ln\left[\frac{(h/h_*)^2 - 1}{(h_o/h_*)^2 - 1}\right] \quad (20)$$

in which h_* is the same as Eq. (18) but could be with different values of resistance coefficient C and friction angle ϕ_D.

If $K = 1$, $\phi_D = 0$ and $F = 0$ ($\beta = 1$), Eq. (17) reduces exactly to that obtained by Yomaoka (1981). If $K = 1$ and $\phi_D = 0$, Eq. (17) reduces to that obtained by Chen (1986).

3. COMPARISON WITH EXPERIMENTS

A conveyor–belt flume with effective inner dimensions of 388 cm long, 50 cm high and 30 cm wide was used in this study. By adjusting the angle of inclination of the flume and the speed of the conveyor belt, a simulated debris flow in the flume can be maintained at the steady state, thereby allowing the measurements of longitudinal surface profiles through the transparent side walls.

Experiments with 5 mm dry glass beads were conducted in the conveyor–belt flume. The slope of the flume was tilted from 0 to 20 degrees and the speed of the conveyor belt was adjusted from 0.5 m/sec to 3.5 m/sec in the experiments. The dynamic friction angle ϕ_D is about 13° for 5 mm glass beads based on experiment results conducted in a ring–shear apparatus (Shen and Jan, 1989). Various surface profiles of debris flow established in the conveyor–belt flume were observed and measured. These surface profiles may be classified into three major categories based on the difference between the flume slope θ and the dynamic friction angle ϕ_D and based on the conveyor–belt speed.

(a) $\frac{dh}{d\xi} < 0$: The flow depth decreases from upstream to downstream. This type of surface profile usually occurred as $\theta<\phi_D$ no matter what the belt speed and as $\theta \geq \phi_D$ at higher belt speed.

(b) $\frac{dh}{d\xi} \approx 0$: The flow depth is almost uniform throughout the flume with the exception of the upstream and downstream ends. This case usually occurred as the angle of the flume was close to the dynamic friction angle at a suitable belt.

(c) $\frac{dh}{d\xi} > 0$: The flow depth increases from upstream to downstream. As the angle of the flume was greater than the friction angle, this type of surface profiles occurred usually as the conveyor belt was at a lower speed.

Experimental results showed that the longitudinal surface profile slopes decreased as the tilted angle of the flume increased and profile slopes increased as the belt speed increased. Surface profiles derived theoretically in the previous section were in reasonable agreement with these experimental surface profiles as shown in Fig. 2.

4. CONCLUSIONS

Longitudinal surface profiles of granular debris flows have been theoretically developed under the assumption that the lateral earth pressure theory which used soil mechanics can be expanded for granular debris flows and the total resistant stress acting on granular debris flows is the linear sum of a normal stress–dependent, contact friction part and a rate–dependent viscous part.

Theoretically, longitudinal surface profiles of granular debris flows are mainly dependent on the difference between the angle of the inclined plane and the dynamic friction angle of debris as well as on rate–dependent viscous stress.

The surface profiles of debris flows flattened as the flume slope increased and the profiles grew steeper as the belt speed increased. Theoretically, surface profiles were in reasonable agreement with experimental surface profiles.

5. REFERENCES

Bagnold, R.A., "Experiments on a gravity-free dispersion of large solid spheres in a Newtonian fluid under shear," *Proc. of the Royal Society of London, Ser. A.*, 225, 49-63, 1954.

Chen, C.L., "Viscoplastic fluid model for debris flow routing," *Proc. of ASCE Conference for World Water Issues in Evolution, Long Beach, CA*, August 4-6, 1986. pp 10-18.

Ling, C.H. and Chen, C.L., "Idealized debris flow in flume with bed driven by a conveyor belt," *National Conference on hydraulic engineering*, ASCE, New Orleans, LA, August 1989, pp. 1144-1149.

Shen, H.W., and Jan, C.D., "Fundamental study of debris flow processes - phase II," Draft of Final Report, Submitted to *U.S. Geological Survey under U.S.G.S. 14-08-001-A-0367*, February, 1990.

Yamaoka, I., "An experimental research on mean velocities of Usu volcanic ash-pumice mudflows in open channels," *Experimental Research Report 485135*, College of Engineering, Hokkaido University, Sapporo, Japan, 1981.

6. ACKNOWLEDGEMENT

The authors are grateful to the U.S. Geological Survey for providing equipment for experiments in this study.

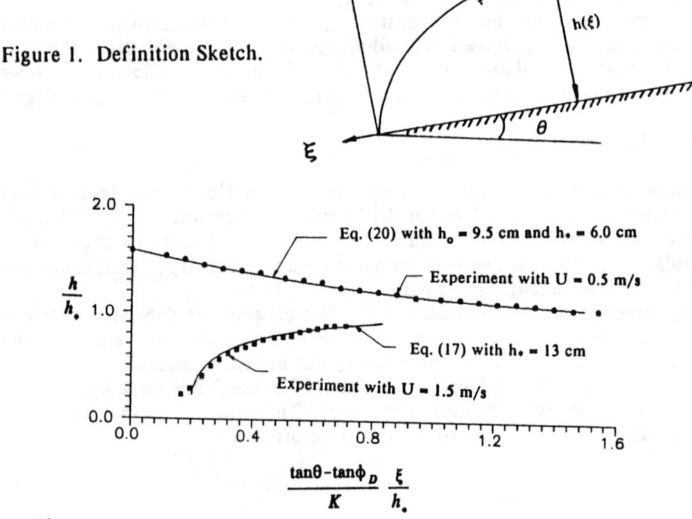

Figure 1. Definition Sketch.

Figure 2. Comparison of Experimental and Theoretical Results with $\theta = 14°$ and $\phi_D = 13°$.

Model of Deposition of Grains from Debris Flow

Haruyuki Hashimoto[1] and Muneo Hirano[2]

Abstract

Rapid deposition of grains from debris flow occurs in a channel whose slope angle decreases abruptly. A model of the deposition is presented on the basis of experiments and theoretical analysis of the movement of grains during deposition. From the model an equation of bed changes due to the deposition is derived. The calculated bed profile is compared with experimental data.

Introduction

There occurs rapid deposition of sediment from debris flows at the mouth of mountain canyon because of rapid flattening of the slope. Tsubaki and Hashimoto (1984) dealt with a simple case of such deposition, which is caused by the rapid decrease in channel slope without the expansion of channel width. They experimentally and theoretically examined trajectories of grains during deposition. Using this result they proposed the method of calculating changes in bed profile due to the deposition. However, there remained a difficulty in this method because of the movement of the boundary in the upstream direction due to the extension of deposition area.

The present work is based on that of Tsubaki and Hashimoto (1984). We first review their paper and then propose an improved calculation method of development of longitudinal bed profile due to the deposition.

Review of the Paper on Grain Movement

Tsubaki and Hashimoto (1984) performed a series of experiments on deposition of grains from debris flow.

---

[1]Associate Professor, Department of Civil Engineering (Suiko), Kyushu University, Fukuoka 812, Japan.
[2]Professor, Department of Civil Engineering(Suiko), Kyushu University, Fukuoka 812, Japan.

The flume used in the experiments was 13.5 m long and 10 cm wide, as shown schematically in Fig.1. The slope of the flume changes abruptly from a steep slope to a gentle slope at station O. Gravel with a diameter of 7 mm and a specific gravity of 2.63 was used as a bed material for the movable bed. The fixed bed was roughened with this gravel. The steep flume was placed at the slope angle θ_u of 18°, while the gentle flume was set at any one of the four different slope angles θ_d of 2°, 4°, 6° and 8°. Water was supplied at a rate of $q_{w0} = 2200\ cm^3/s$ at upstream end of the flume. Debris flow occured on the movable bed, and then moved rapidly on the steep fixed bed. From a close-up view of the moving gravels taken with two 16 mm high-speed cameras they measured the trajectories of gravels during deposition.

Relationship between incoming velocity u of gravels before passing the y_d-axis and moving distance x_{sp} from the y_d-axis to their rest location was examined. The examination showed that there was some scatter in the experimental data. Hence the range of incoming velocity measured was divided into five subranges; within each subrange the mean value $\bar{x}_{sp}$, standard deviation s and variation coefficient α of moving distance were determined. The result is shown in Fig. 2. It was found that mean moving distance becomes large with an increase in incoming velocity and variation coefficient is approximately constant; its value is $\alpha = 0.42$.

Tsubaki and Hashimoto(1984) derived simple analysis to explain the above experimental results.

Consider a grain in steady and uniform debris flow in a channel with steep slope and the grain at the moment of its arrival in the downstream channel with gentle slope. Forces acting on the grain in the flow direction are gravity component and the fluid forces F^f that cause motion and the intergranular forces F^g that resist motion respectively. Hence the equation of motion of a grain of

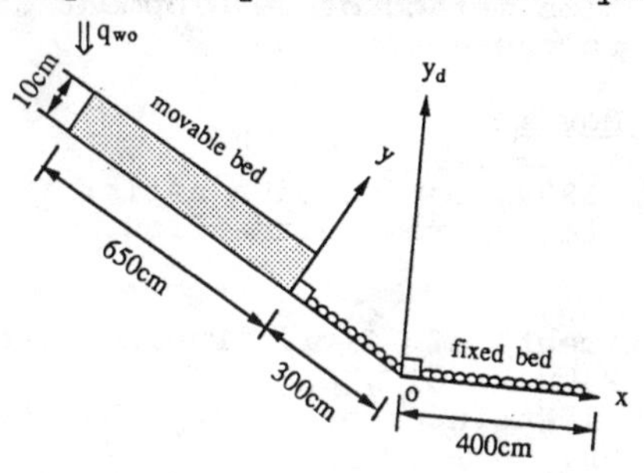

Fig.1. Elevational View of the Flume

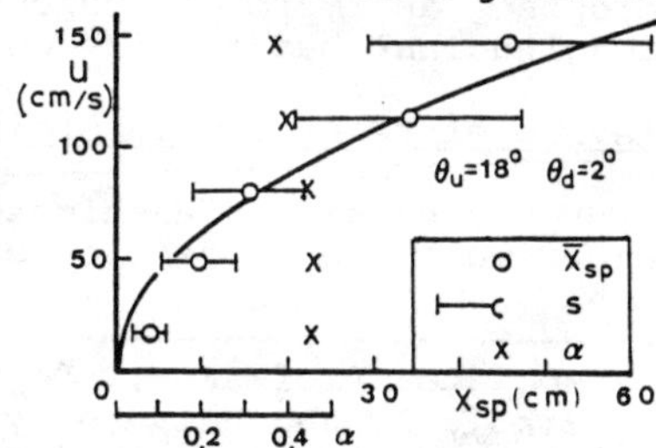

Fig.2. Mean Value, Standard Deviation and Variation Coefficient of Moving Distance of Gravels

mass m in the steep channel can be written as

$$0 = -F_u^g + mg\sin\theta_u + F_u^f \quad \text{.................................(1)}$$

For the grain just arriving in the downstream gentle channel, on the other hand, we have

$$m\left(1+\frac{\rho}{2\sigma}\right)\frac{d^2x}{dt^2} = -F_d^f + mg\sin\theta_d + F_d^f \quad \text{..........................(2)}$$

where subscript u and d denote the upstream steep and downstream gentle channels, respectively.

An arbitrary grain changes its velocity from u(y) at an elevation y in the steep channel to $u_d = u(y)\cos(\theta_u - \theta_d)$ at the elevation $y_d = y/\cos(\theta_u - \theta_d)$ in the gentle channel. Grain concentration in the neighbour of the grain, however, is assumed to be unchanged. From the expression for intergranular forces (Hashimoto and Tsubaki 1983) and Eq.1 they found that

$$F_d^g = F_u^g \cos^5(\theta_u - \theta_d) = (mg\sin\theta_u + F_u^f)\cos^5(\theta_u - \theta_d) \quad \text{..................(3)}$$

Substituting this equation into Eq. 2 yields approximately

$$\frac{d^2x}{dt^2} = -\frac{g}{1+\frac{\rho}{2\sigma}}\left\{\sin\theta_u\cos^5(\theta_u - \theta_d) - \sin\theta_d\right\} \equiv -G \quad \text{..................(4)}$$

Since the initial time is defined as the one when a grain pass the y_d-axis, the initial condition is

$$x = 0 \quad \text{and} \quad \frac{dx}{dt} = u(y)\cos(\theta_u - \theta_d) \quad \text{.............................(5)}$$

Solving Eq. 4 under the condition, we have equations for grain velocity and for grain trajectories. Since grains stop when the velocity becomes zero, from these equations we obtain the moving period t_{sp} and distance x_{sp} of grains from their arrival at the y_d-axis to their rest:

$$t_{sp} = \frac{u(y)\cos(\theta_u - \theta_d)}{G} \quad \text{and} \quad x_{sp} = \frac{u^2(y)\cos^2(\theta_u - \theta_d)}{2G} = \frac{G}{2}t_{sp}^2 \quad \text{......(6)}$$

Calculated values from Eq. 6 gave the curve shown by the solid line in Fig.2, which indicates a very good agreement between the calculated and the measured mean moving distance. Therefore, Tsubaki and Hashimoto (1984) concluded that grains with same incoming velocity u(y) are deposited at random about the mean distance given by Eq.6.

Using the above results they obtained an equation for development of the longitudinal bed profile. However, since rapid deposition from debris flow makes the boundary of deposition move in the upstream direction, there remained a computational difficulty in their equation. In order to avoid this difficulty we need to introduce a coordinate system moving with the boundary of deposition.

Grain Movement in a Moving Coordinate System

We consider now grain movement in a coordinate system moving with the upstream end of deposits. Since the moving velocity of the upstream end was found approximately constant from the experiments (Tsubaki and Hashimoto, 1984), a transformation of the system of x coordinate can be expressed as

$$x_1 = x + v_B \cos(\theta_u - \theta_d)\, t \quad \cdots\cdots (7)$$

The equation of motion of a grain in debris flow is same as Eq.4. Initial time is defined as the one when grains pass the y_{1d} -axis (Fig.3). Therefore, the initial condition is

$$x_1 = 0 \quad \text{and} \quad \frac{dx_1}{dt} = [\, u(y) + v_B \,] \cos(\theta_u - \theta_d) \quad \cdots\cdots (8)$$

Grains stop at the velocity of $dx_1/dt = v_B \cos(\theta_u - \theta_d)$. Solving the equation of motion under the initial condition yields moving distance x_{1sp} of grains from their passage past the y_{1d} -axis to thier rest :

$$x_{1sp} = [\, u(y) + 2v_B] \frac{u(y)}{2G} \cos^2(\theta_u - \theta_d) \equiv \bar{x}_1 \quad \cdots\cdots (9)$$

Derivation of an Equation for Bed Profile

In a similar way to that presented by Tsubaki and Hashimoto (1984), the deposition of grains is modeled as shown in Fig. 3. Grains with velocity of $u(y)+v_B$ and concentration of $C(y)$ coming through the interval between y and y+dy from the upstream steep channel settle somewhere about the mean distance in the downstream gentle channel. The moving distance of grains with the same incoming velocity is regarded as a random variable and its mean $\bar{x}_1$ is given by Eq. 9.

Denoting the probability density function of the

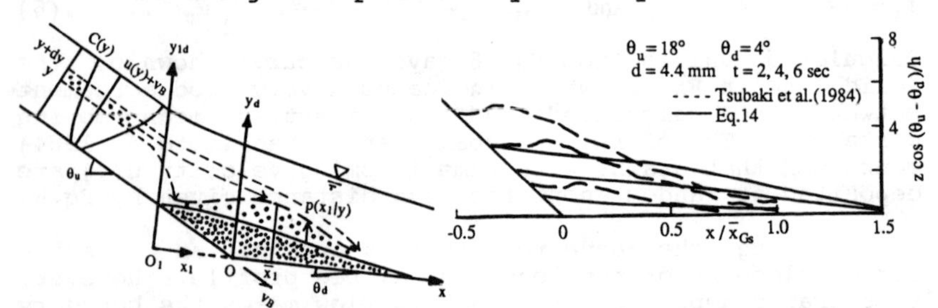

Fig.3. A Model for the Deposition of Grains from Debris Flow

Fig.4. Longitudinal Bed Profiles at Different Times

random variable, bed elevation of deposited grains originating in the elevation between y and y+dy, and volumetric concentration of grains in bed by $p(x_1|y)$, z_y and C_* respectively, we can express the mass conservation of deposited grains as

$$\frac{\partial}{\partial t} z_y + v_B\cos(\theta_u-\theta_d)\frac{\partial}{\partial x_1} z_y = \left[(u+v_B)\frac{C}{C_*}dy\right] p(x_1|y)\, s(x_1-v_B\cos(\theta_u-\theta_d)\, t_{sp}) \quad (10)$$

where $s(x_1-v_B\cos(\theta_u-\theta_d)\, t_{sp})$ is the unit step function. Rearranging Eq.10 by using Eq.7 gives

$$\frac{\partial}{\partial t} z_y = \left[(u+v_B)\frac{C}{C_*}dy\right] p(x_1|y)\, s(x_1-v_B\cos(\theta_u-\theta_d)\, t_{sp}) \quad (11)$$

Integrating Eq.11 with respect to t and rewriting the resulting equation by using Eq. 7, we have

$$z_y = \int_{x_1-v_B\cos(\theta_u-\theta_d)(t-t_0)}^{x_1-v_B\cos(\theta_u-\theta_d)t_{sp}} \left[(u+v_B)\frac{C}{C_*}dy\right] p(x_1|y) s(x_1-v_B\cos(\theta_u-\theta_d)\, t_{sp}) \frac{dx_1}{v_B\cos(\theta_u-\theta_d)} \quad (12)$$

where

$$t_0 = -\frac{x}{v_B\cos(\theta_u-\theta_d)} s(-x) \quad (13)$$

Further integrating Eq. 12 with respect to y from the bottom to the surface, we obtain

$$z = \int_{\bar{x}_{1G_0}}^{\bar{x}_{1G}} \frac{C}{C_*} \frac{G}{v_B \frac{du}{dy}\cos^3(\theta_u-\theta_d)} \left[\int_{x_1-v_B\cos(\theta_u-\theta_d)(t-t_0)}^{x_1-v_B\cos(\theta_u-\theta_d)t_{sp}} p(x_1|y)\, s(x_1-v_B\cos(\theta_u-\theta_d)\, t_{sp})\, dx_1\right] d\bar{x}_1 - x\tan(\theta_u-\theta_d)\, s(-x) \quad (14)$$

Here the equation

$$dy = \frac{G}{(u+v_B)\frac{du}{dy}\cos^2(\theta_u-\theta_d)} d\bar{x}_1 \quad (15)$$

derived from Eq.9 is used. The rate of increase in bed elevation is also obtained by integrating Eq. 11.

$$\frac{\partial z}{\partial t} = \int_{\bar{x}_{1G0}}^{\bar{x}_{1G}} \frac{C}{C_*} \frac{G}{\frac{du}{dy}\cos^2(\theta_u-\theta_d)} p(x_1|y)\, s(x_1-v_B\cos(\theta_u-\theta_d)\, t_{sp}) d\bar{x}_1 \quad (16)$$

The calculation of Eqs. 14 and 16 requires the evaluation of the moving velocity v_B of the upstream end of deposits. The extension of deposition area in the upstream direction is due to the deposition at the upstream end $x_1 = 0$. Hence the following relation is deduced :

$$v_B \cong \frac{1}{\sin(\theta_u - \theta_d)} \left.\frac{\partial z}{\partial t}\right|_{x_1=0} \quad \ldots\ldots\ldots\ldots\ldots\ldots\ldots\ldots\ldots\ldots\ldots(17)$$

Since the value of $\partial z/\partial t$ becomes maximum in the immediate vicinity of $x_1 = 0$, it is appropriate that $\partial z/\partial t|_{x_1=0}$ in the above equation is replaced by $\partial z/\partial t|_{max}$.

The random variable x_1 is assumed to be normally distributed. The variation coefficient $\alpha = 0.42$ obtained from the experiments of grain trajectories is used.

Calculation of Longitudinal Bed Profile

First, substituting the given values of flow depth h and surface velocity u_s in the steep channel into the equation for velocity profile (Tsubaki, Hashimoto and Suetsugi 1982), we determine the values of du/dy and u(y) for every elevations y. Second, substituting this result into Eqs.6 and 9, we calculate t_{sp} and $\bar{x}_1$ for every y. Third, using Eqs. 16 and 17, we can obtain the value of v_B and finally, using Eq. 14, we can calculate the longitudinal bed profile for a given value of t.

Comparison of the calculated and measured bed changes is shown in Fig.4. The deviation of the calculated curve from experimental data is ascribed to the escape of water from debris flow in deposits in the experiments.

Conclusions

The coordinate system moving with the boundary of deposition in the upstream direction was introduced. In the moving coordinate system the moving period and distance of a grain was estimated. From this result and the conservation law of grain mass, equations for the rate of deposition and for the longitudinal bed profile were derived. The velocity of moving boundary was estimated from the rate of deposition at the boundary and the calculated bed profile gives reasonable result.

Appendix References

Hashimoto, H., and Tsubaki, T. (1983). " Reverse grading in debris flow ". Proc.JSCE, 336, 75-84 (in Japanese).

Tsubaki, T., Hashimoto, H., and Suetsugi, T. (1982). " Grain stresses and flow properties of debris flow ". Proc.JSCE, 317, 79-91 (in Japanese).

Tsubaki, T., and Hashimoto, H. (1984). " Deposition of debris flow due to abrupt change of bed slope ". Proc. 28th Japanese Conf. Hydraulics, JSCE, 711-716 (in Japanese).

GRAPHICAL VERIFICATION OF FLOOD FREQUENCY ANALYSIS

Darde G. de Roulhac, P.E., A.M. ASCE *

Abstract: In the analysis of flood frequency data, a proper distribution must be chosen to fit the data being analyzed. A properly applied distribution can minimize errors in the extrapolation of extreme events. A major drawback to mathematically fitting curves to flood data is that preselection of a statistical model is necessary. A model which is inappropriate for the specific watershed being analyzed may easily and unknowingly be chosen. Too often, a model is improperly applied or is selected solely because of its widespread use. To assist in the graphical verification, a computer program has been developed to generate graphical displays of observed data plotted on different probability "papers," and curves fitted to the data.

INTRODUCTION: In flood frequency analysis, computer programs can easily be developed to produce tables of mathematically fit data for the various probability distributions which are in use. With such programs, a statistical model must be preselected. Many probability distributions are available, and the importance of proper model selection is often overlooked. Thus, a model which is inappropriate for a particular set of data may easily be chosen. A major indicator of the validity of a particular distribution is the linearity of the plot of observed data points. Linearity facilitates the accurate projection of extreme events. The data should initially be plotted on several different probability papers in order to evaluate the linearity of the plot or the continuity of curvature. Also, statistical parameters for the entire record are often inappropriate to prediction at either end of the probability scale. This may be readily observed in a data plot.

The cost and time involved in manually generating these plots is usually prohibitive. Nevertheless, advance graphical verification and visual comprehension of statistical techniques is needed. This paper explains algorithms used by a computer program to rapidly produce these plots. Many probability distributions may be compared us-

---

* Civil Engineer, Clark County Regional Flood Control District, 301 East Clark Ave., # 300, Las Vegas, NV 89101

ing rapid computer plots. The criteria for judging the appropriateness of various models are also explained.

The program supplements the statistical fitting of extreme value, log extreme value, normal, and log normal with rapid screen plots of the data and fitted curves on these four probability "papers". The program shows several small plots on the screen at one time, or a full-screen higher-resolution plot on one paper. This program assists in the graphical verification that an appropriate statistical method was chosen. Examples of these procedures follow, in the "King's Table" section.

LITERATURE REVIEW: This review is a historical perspective of the analysis of hydrologic data and the projection of extreme events. Traditionally, this type of analysis has been a graphical technique. In more recent years, with the widespread availability of computers, there has been an increased emphasis on the analytical aspects of the analysis. The graphical techniques are still valid, and should still be used in conjunction with numerical techniques to verify that the chosen statistical distributions are the most appropriate for the data being analyzed.

In 1914, Hazen suggested the plotting of hydrologic data on log normal paper. Another paper, the extreme value paper, also plays an important role in hydrologic frequency analysis. The extreme value distribution is currently used by the National Weather Service for all rainfall analysis. This distribution was first developed by Fisher and Tippett (1928) in England, and first applied to hydrologic analysis in the United States by Gumbel (1941). Gumbel (1954) stated the following: "In the applications we are usually dealing with extremes of large samples... The founders of the calculus of probabilities were too occupied with the general behavior of statistical masses to be interested in the extremes... the normal distribution is very important. But it is erroneous to assume that all distributions are normal or tend to normality. Within the theory of extreme values, most distributions are skewed, and do not cease to be skewed if the sample sizes are increased." Gumbel developed a "cook book" approach to hydrologic frequency analysis. His biggest contribution probably was to be able to fit a single straight line through a plot to project extreme events while computing the mean and standard deviation of the series. The procedure used in standard statistical regression analysis is to regress y on x, as well as x on y. Gumbel developed a line which is a mathematical compromise between the two regressions. He also transformed the probability of exceedance into a value known as the "reduced variate", which is a linear scale on the abscissa. The observed data is plotted on the ordinate. The equation of the reduced variate is as follows:

y = reduced variate = $-\ln[-\ln(1 - P_e)]$, where
P_e = Probability of exceedance.

Chow (1954) further simplified the mathematics of the extreme value distribution, developing "K" values which relate to a specific plotting position formula. He applied the slope-intercept equation for a straight line to this distribution.

$X = b + my$, where
X = observed annual maxima plotted on ordinate;
y = reduced variate plotted on abscissa;
m = slope of line, function of standard deviation;
b = constant.

The X corresponding to a probability of exceedance of 0.4292 is equal to the mean of the X values for an extreme value distribution.

Gumbel (1941) and Chow (1954) applied this theory to flood frequency analysis. Hershfield (1961), using Gumbel's theory, developed a rainfall atlas for the United States, published as TP-40.

Gringorten (1963) proposed another plotting position to replace the Weibull plotting position used by Gumbel. Using Gringorten's plotting position, the "Gumbel" analysis is called "extreme value", since it is a slight deviation from the procedure that Gumbel proposed.

In 1967, the Water Resources Council, in an attempt to promote a consistent approach to frequency analysis, published Bulletin 15 (1967), and later versions, Bulletin 17 (1976), Bulletin 17A (1977), and Bulletin 17B (1981). It discourages the use of Gumbel analysis in favor of the Log Pearson III distribution, and a special case where the skew is zero, the log normal distribution.

Cunnane (1978) established a compromise plotting position, unbiased towards any particular distribution, that could be used for plotting on either the extreme value or normal papers.

King (1971) presented a graphical method of choosing the proper paper on which to plot. Figure 1 is an extract from King's table. Reich (1981) also gives a discussion of the use of King's table.

DESCRIPTION OF "KINGS TABLE": This table of probability distributions on different types of paper (Figure 1) gives clues as to which plot could be the most appropriate. The basic principle is that each data set approximates one frequency distribution. On the paper corresponding to the most appropriate frequency distribution, the plot should be approximately linear. Confusion may arise if more than one plot exhibit an approximately equal degree of linearity. In that case, the most appropriate frequency distribution may be determined based on the shapes of the plots on the other papers. The shape of these other plots will vary according to the distribution which should be used. The most appropriate probability distribution scale is on the left side of the table. The assumed probability distribution scale is along the top of the table. As an example, if the plot of the annual maxima on extreme value paper has an "S" shape, the plot would be approximately linear if plotted on log normal paper. An illustration of this shape is shown in the plot on the second row from the top, and the third column from the left. As a second example, if another plot of observed annual maxima on log normal paper decreases in slope towards the right side of the plot, the plot would be approximately linear if plotted on normal paper. The observed annual maxima would therefore be normally distributed. This second example is shown in the plot on the first row from the top, and on the second column from the left.

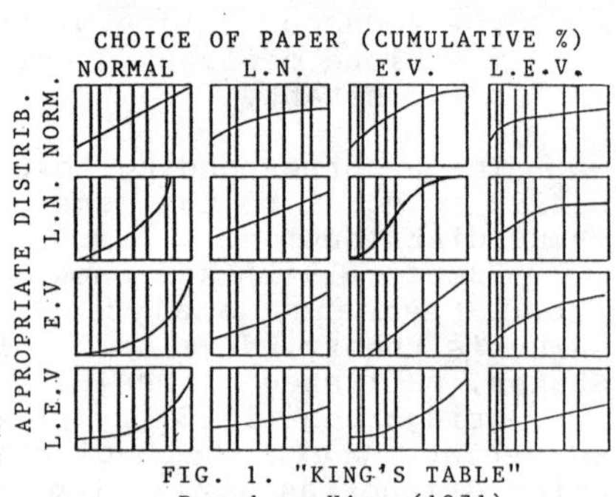

FIG. 1. "KING'S TABLE"
Based on King (1971)

Ideally, the plot of the points should be linear on one, and only one of the papers. If none of the plots are linear, then the plot should at least show oscillations or random scatter about a straight line. The actual deviation of the plotted points from the mathematically-fitted line should also be examined. It should be noted that the scale compression on the logarithmic plots may create an erroneous illusion of a small deviation of the observed data points from the mathematically-fitted line.

It may be possible that a plot produces two linear segments where the slopes of the segments may be different, or where three segments form a zigzag. Physical changes in a watershed during the time period being analyzed might produce such plots. If this is the case, the earlier (pre-change) points should be discarded from the analysis in order to more accurately project extreme events in the current watershed.

Since plots of actual data do not always approximate ideal conditions, a degree of engineering judgment is sometimes required in choosing an appropriate distribution.

COMPUTER IMPLEMENTATION: A computer program has been developed to rapidly generate the required plots. The plotting program uses an input file containing the results from a frequency analysis program. The plotting program then generates data points and mathematically-fit lines for the extreme value, log extreme value, normal, and log normal papers. This allows the user to check for the best distribution to use, based on the criteria previously described. The data may be plotted on four papers simultaneously. Any single paper may be selected for a plot, and the math fit may be plotted with the data. Thus the user has the availability of rapid screen plots (and hardcopy, if desired) so that all possibilities may be explored. This graphical verification technique assists in the proper selection of a distribution.

Normal Plotting Algorithm: The following section describes the algorithm to plot on normal or log normal "paper." To plot an observed data point, the points are first ranked according to their magnitude, in descending order. Then a plotting position is assigned per the Cunnane compromise plotting formula, which is as follows:

$$PP = (m - 0.4) / (N + 0.2)$$, where

PP = the Cunnane plotting position;
m = the rank of the point within the data set; and
N = the total number of points in the data set.

The location on the normal scale of the plot's abscissa, corresponding to the Cunnane plotting position, must next be found. In the normal probability function, the plotting position is the cumulative area under the normal curve. The probability is therefore an integral, and the abscissa on the plot is the derivative of that probability, which is a linear scale upon which to plot. The normal probability function is as follows:

$$F(y) = \int_{-\infty}^{y} \frac{1}{\sqrt{2\pi}} e^{-0.5t^2} dt$$, where

F(y) = the plotting position, per Cunnane; and
y = the position on the linear scale of the abscissa.

The y-value thus obtained is then plotted on the linear scale of the abscissa. The value of the ordinate is the magnitude of the data for normal paper, or the logarithm (base 10) of the magnitude for log normal paper.

Extreme Value Plotting Algorithm: The following section describes the method of plotting on extreme value "paper." The annual precipitation maxima are first ranked in order of decreasing magnitude. The Cunnane compromise plotting position is used, as with the normal plotting algorithm.

The reduced variate, which is the location on a linearly scaled abscissa, is determined for each data point by the following equation:

$$y = -\ln [-\ln (1 - P_e)]$$, where
y = the reduced variate; and

P_e = the plotting position.

The location on the ordinate is the magnitude of the observed data for the extreme value plot, or the logarithm (base 10) of the magnitude for the log extreme value plot.

SUMMARY AND CONCLUSIONS: This paper presents algorithms for practical computer-generated graphical tools which allow the analyst to quickly choose the proper statistical distribution. The choice of distribution is based on a procedure which has been outlined for analyzing the plots. The emphasis is on graphical techniques based on statistics, rather than on the statistics themselves. This allows the engineer to retain a "feel" for what the data really means. This "feel" can easily be lost when dealing with the numerous calculations involved in flood frequency analyses. Without the control over the data which the graphical tools afford, an improper distribution could easily be chosen.

REFERENCES

Chow, V. T., The Log-Probability Law and Its Engineering Applications, ASCE Proc., 80, 536, 1954.

Cunnane, C. "Unbiased Plotting Positions - A Review." Journal of Hydrology, 37(3), 1978.

Gringorten, I. I., "A Plotting Rule for Extreme Probability Paper." J. Geoph. Res. 68(3), 813, 1963.

Gumbel, E. J. "The Return Period of Flood Flows", Ann. Math. Stat., 12, 163, 1941.

Gumbel, E. J. "Statistical Theory of Extreme Values and Some Practical Applications, U.S. National Bureau of Standards, Appl. Math. Ser. No. 33, 1954.

King, J. R. Probability Charts for Decision Making, Industrial Press, N.Y., 290, 1971.

Reich, B. M. and Renard, K.G., "Application of Advances in Flood Frequency Analysis", Water Resources Bulletin, 17(1), 67, 1981.

U. S. Water Resources Council, A Uniform Technique for Determining Flood Flow Frequency. Bulletin No. 15 of the Hydrology Committee, Washington, D.C., 1967.

U. S. Water Resources Council, Guidelines for Determining Flood Flow Frequency. Bulletin No. 17 of the Hydrology Committee, Washington, D.C., 1976. [Also Bulletin No. 17A, 1977 and Bulletin No. 17B, 1981.]

Multiple Phenomena of Debris-Flow Processes: A Challenge for Hazard Assessments

Christopher C. Mathewson[1], M.ASCE
Jeffrey R. Keaton[2], M.ASCE,

Abstract

Hazardous hydraulic processes associated with arid environments commonly represent significant risks to the safety of the public. However, the public generally considers these risks to be minimal until an event occurs. In the case of debris flows, the determination of recurrence intervals is controlled by numerous interacting factors that generate debris flow events having different recurrence intervals. The engineer or hazard response planner must, therefore, understand the phenomena that controls each individual event to determine the risk to the public. In Davis County, Utah for example, debris flows have estimated average recurrence intervals of about 100, 3,000 and 10,000 years depending upon the debris-flow phenomena selected for evaluation.

Introduction

The general public perception of arid lands is that they are **DRY**. As a result, hazards associated with water in these regions are often ignored or at least downplayed. The engineer or other professional responsible for the protection of the public from these hazardous processes must be able to determine realistic recurrence intervals for damaging events if they are to successfully fulfill this responsibility.

In the case of flooding, the triggering event is a meteorological process that can be evaluated stocastically using the existing record. Debris flows, however, are hazardous hydraulic processes that are controlled by interrelated factors such as rainfall/snow melt rates, slope aspect, slope geomorphology, thickness of the colluvium, vegetative cover, season, ground water regime, bedrock geology and seismicity, among others. As a result, debris flows at any specific location can be initiated by more than one triggering event, which makes them non-stocastic processes unless each event can be individually evaluated and assessed and expressed in terms which may be combined.

---

[1]Director, Center for Engineering Geosciences and Professor, Department of Geology, Texas A&M University, College Station, TX 77843-3115
[2]Vice President and Senior Engineering Geologist, Sergent, Hauskins and Beckwith, 4030 S. 500 West, Suite 90, Salt Lake City, UT 84123

Sources of water to initiate debris flows have included excess precipitation (Campbell, 1975), cloudburst thunderstorms (Woolley, 1946; Butler and Marsell, 1972; Costa, 1984; among others) and rapid snowmelt (Paul and Baker, 1923; Marsell, 1972; Anderson, et al., 1984). Elevated pore water pressures needed to initiate slope failures which mobilize into debris flows have been developed when the precipitation rate exceeds the rate of internal drainage (Campbell, 1975) or from excess drainage from an underlying bedrock aquifer (Eisenlohr, 1952; Hack and Goodlett, 1960; Wilson and Dietrich, 1987; Mathewson, et al., 1990). Debris flows have also been caused by the erosion of abused mountain lands during thunderstorms (Crawford and Thackwell, 1931). In addition to various initiation mechanisms, geomorphic thresholds also influence debris flows; for example, a minimum thickness of the colluvium is required for any given slope condition before failure can occur.

Wasatch Front, Utah - Case History

Numerous damaging debris flows occurred in the western United States during the springs of 1983 and 1984 (Anderson, et al., 1984). The debris flows that occurred along the Wasatch Front in Davis County, Utah, were not considered unique because debris flows also occurred there during the 1920s and 1930s. A review of the 140-year historical record of debris flows in Davis County would suggest that the 1983/1984 events should have a recurrence interval of approximately 100 years (Figure 1). This suggestion, however, is based on the assumption that the initiation mechanisms of the debris flows is irrelevant.

Stratigraphic analyses of post-Lake Bonneville alluvial fan deposits show that only 3 to 5 prehistoric debris-flow events occurred in the past 12,000 years (Keaton, 1988). These prehistoric debris-flow deposits include no soil profile development between deposits of successive events, suggesting that they occurred over a geologically short time span. These flows probably occurred in the early Holocene following melting of the mountain glaciers when unvegetated frost-shattered slopes were subjected to cloudburst rainfall events which eroded the materials resulting in debris flows (Keaton, 1988). Following the 3 to 5 events in early Holocene time, the mountain block would have been stripped of most of the available post-glacial debris, thereby reducing the potential for future events until a sufficient thickness of colluvium recevelopes.

Weathering of bedrock and revegetation of the developing colluvium permitted the mountain slopes to become tree covered. The complex geologic history of the Wasatch Front produced a structural fabric that supports a bedrock ground-water system. Shallow ground water in developing colluvial material supports vegetative cover which inhibited significant erosion and debris-flow events until the arrival of settlers and changed land use patterns.

The development of cattle and sheep grazing in the Wasatch Range brought about abusive land use practices in the early 1920s. Land clearing and burning set the stage for subsequent debris-flow events. Historical accounts of the 1920s and 1930s debris flows clearly indicate that they were triggered by cloudburst thunderstorms (Paul and Baker, 1293; Crawford and Thackwell, 1931; Wooley, 1946; Butler and Marsell, 1972; and Marsell, 1972). The slopes had been cleared of dense vegetation to improve grazing and burned to encourage new growth of grasses. The debris flows of the 1920s and 1930s were triggered by intense rainfall on abused slopes which led to rapid runoff and sheetwash erosion resulting in debris flows and hyperconcentrated floods at the canyon mouths.

Following the early historical debris flows, erosion control contour trenches were constructed on the upper slopes of the Wasatch Range. These trenches trapped precipitation causing it to infiltrate into the underlying fractured bedrock system (Coleman, 1989). A gradual increase in the moisture balance in the Salt Lake basin since the mid-1960s recharged the bedrock aquifer and supported a shallow ground-water system in the colluvium (Figure 2). Phreatophyte vegetation and springs developed on the mountain slopes in response to the availability of shallow water.

The debris-flow events of 1983 and 1984 were totally different from those of the 1920s and 1930s. These debris flows resulted from a rapid melting of the above average snow pack in the Wasatch Range (Wieczorek, et al., 1989). Snowmelt floods also occurred in Davis County in 1952, when a thick winter snow pack rapidly melted; however, no significant debris flows were generated (Marsell, 1972). Mathewson and others (1990) suggested that excess pore water pressures developed within the bedrock below the colluvium due to rapid melting of the record snowpacks in the springs of 1983 and 1984. These elevated porewater pressures, when transmitted into the saturated colluvium, resulted in small slope failures which mobilized into debris flows, many of which extended beyond the canyon mouths and into urbanized development on the alluvial fans.

Field observations indicate that a significant bedrock aquifer exists that supports springs and seeps in the Wasatch Front. Following the debris-flow events, numerous new springs and seeps were exposed in the slope failure and debris-flow scars. These are now draining the ground-water reservoir in the fractured bedrock. This drainage has also caused a regional change in the hydrology of the mountain block.

Analyses of the initiation and kinematics of the 1983/1984 debris flows suggest that these flows were unique to the Holocene (last 10,000 years). This conclusion is supported by the stratigraphic record of the post-Bonneville debris flows and the historical record of the recent events. The geologic and hydrologic conditions required for the initiation of numerous small slides, some in canyons that had not experienced a debris flow in historic times, were not developed until 1983/1984. These conditions include:

1. Soil conservation measures to rehabilitate the Wasatch Front and allow infiltration of precipitation into the bedrock aquifer.
2. Gradual recharging of the bedrock aquifer in response a climatic change that started in the mid-1960s.
3. Heavy fall precipitation followed by an above normal snow pack.
4. Rapid snow melt in the late spring.

These conditions resulted in a fully charged bedrock aquifer that became over pressurized. Excess porewater pressures in the already saturated colluvium allowed small landslides to form and mobilize into debris flows (Mathewson, et al,. 1990).

At least two distinctly different phenomena have produced damaging debris flows at the canyon mouths along the Wasatch Front in Davis County, Utah. Debris flows may be triggered by

1. intense rainfall on susceptible slopes through a sheetwash mechanism or
2. elevated pore-water pressures within the colluvium from ground water supplied by the underlying fractured bedrock aquifer.

To complicate matters further, the 1983/84 debris flows produced oversteepened colluvial slopes and channel banks that are likely to become unstable when subjected to other meteorological or tectonic processes, thereby producing debris flows by a possible third mechanism (Figure 3).

The average recurrence intervals for the Davis County debris flows can be set at about 100 years based on the historical record, at about 3,000 years based on the stratigraphic record, at about 10,000 years based on the phenomena causing the 1983/84 flow events, or at some even longer interval based on the possibility of future events triggered by another sequence of processes.

Conclusions

The hydraulic engineer and emergency response planner must recognize that the geologic system is complex. Damage reduction and disaster response must be based on a sound understanding of the geologic phenomena which control the hazardous processes. Most hazardous processes can be triggered by more than one phenomena, as is the case with Wasatch Front debris flows. Uncertainty in estimating future debris-flow magnitude and frequency based on the historical record must be supplemented by detailed knowledge of the geologic processes which are best viewed from

1. interpretation of the stratigraphic record preserved in alluvial fan deposits and
2. examination of conditions in source areas for potential future debris flows.

The phenomena that control debris-flow processes must be understood if we are to accurately assess the hazard.

References

Anderson, L. R., Keaton, J. R., Saarinen, T. F. and Wells, W. G. II, 1984, *The Utah Landslides, Debris Flows and Floods of May and June 1983*: National Academy Press, Washington, DC, 96 p.

Butler, E. and Marsell, R. E., 1972, *Cloudburst Floods in Utah, 1939-69*: Utah Department of Natural Resources, Division of Water Resources, Cooperative Investigations Report No. 11, Salt Lake City, UT, 103 p.

Campbell, R. H., 1975, *Soil Slips, Debris Flows, and Rainstorms in the Santa Monica Mountains and Vicinity, Southern California*: US Geological Survey Professional Paper 851, 51 p.

Coleman, W. K., 1989, Role of Contour Trenching in the Alteration of Hydrogeologic Conditions of the Wasatch Front: Abstract, *Abstracts and Program*, Annual Meeting, Association of Engineering Geologists,Vail CO, p.57.

Costa, J. E., 1984, Physical Geomorphology of Debris Flows: in Costa, J. E. and Fleisher, P. J., (editors), *Developments and Applications of Geomorphology*, Springer-Verlag, New York, pp. 268-317.

Crawford, A. C. and Thackwell, F. E., 1931, *Some Aspects of the Mud Flows North of Salt Lake City*: Utah Academy of Sciences Proceedings, Vol. 8, pp. 97-105.

Eisenlohr, W. S. Jr., 1952, Floods of July 18, 1942, Pennsylvania, in *Notable Floods of 1942-42*: USGS Water Supply Paper 1134B, US Geological Survey, pp. 75-78.

Hack, J. T. and Goodlett, J. C., 1960, *Geomorphology and Forest Ecology of a Mountain Region in the Central Appalachians*: US Geological Survey Professional Paper 347, pp. 45-47

Keaton, J. R., 1988, *A Probabilistic Model for Hazards Related to Sedimentation Processes of Alluvial Fans in Davis County, Utah*: Unpublished Doctoral dissertation, Texas A&M University, 441 p.

Marsell, R. E., 1972, *Cloudburst and Snowmelt Floods*: in Environmental Geology of the Wasatch Front, Utah Geological Association Publication No. 1, pp. N1-N18.

Mathewson, C. C., Keaton, J. R. and Santi, P. M., 1990, Role of Bedrock Ground Water in the Initiation of Debris Flows and Sustained Post-Flow Stream Discharge: *Bulletin of the Association of Engineering Geologists,* Vol XXVII, No. 1, February, pp. 73-83.

Paul, J. H. and Baker, F. S., 1293, *The Floods of 1923 in Northern Utah*: University of Utah Bulletin, Vol. 15, Salt Lake City, UT, 20 p.

Wieczorek, G. F., Lips, E. W. and Ellen, S. D., 1989, Debris Flows and Hyperconcentrated Floods Along the Wasatch Front, Utah, 1983 and 1984, *Bulletin of the Association of Engineering Geologists*, Vol. XXVI, No. 2, pp 191-208.

Wooley, R. R., 1946, *Cloudburst Floods in Utah, 1850-1938*: USGS Water Supply Paper 994, US Geological Survey, 128 p.

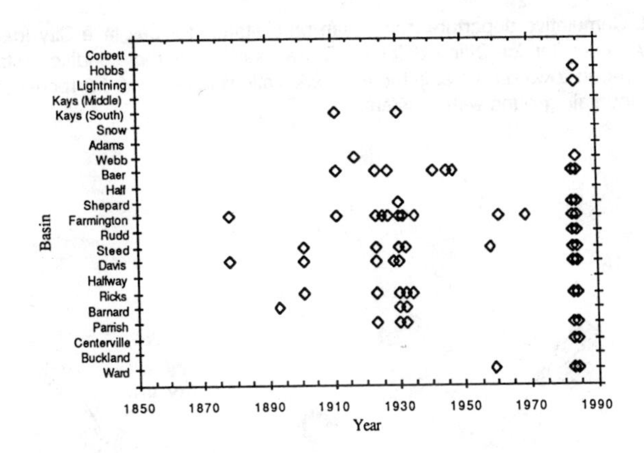

Figure 1. Temporal distribution of historic flood and debris flow events in Davis County, Utah (from Keaton, 1988). Note that this distribution indicates that the 1983-84 events have a recurrence interval of about 100 years.

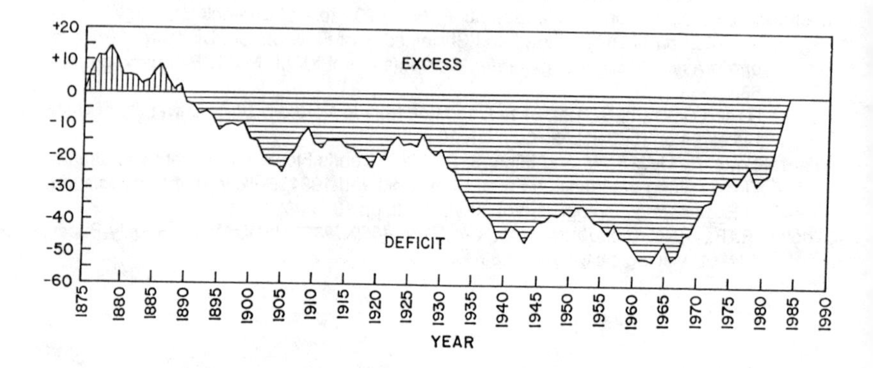

Figure 2. Cumulative departure from mean precipitation for Salt lake City (data from Weather Service Office, Salt Lake CIty Airport). Note that conditions started to a return toward normal in the mid-1960s allowing 20 years for recharge of the mountain ground water system.

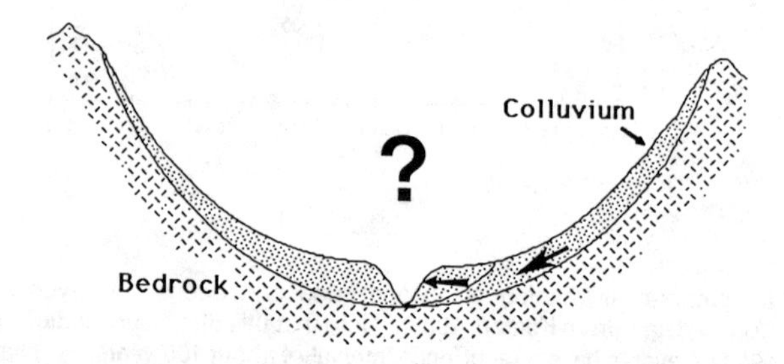

Figure 3. Schematic cross section across a Wasatch Front canyon following the 1983-84 debris-flow events. Note the incised channel and the potential for debris flows to initiate from failures of the canyon walls.

Recurrence of Debris Flows on an Alluvial Fan in Central Utah

Elliott W. Lips[1] and Gerald F. Wieczorek[2], M.ASCE

Abstract

In the spring of 1983 a large debris flow occurred in the drainage of Birch Creek, near Fountain Green in central Utah. During the debris-flow activity, a new channel was incised into part of the alluvial fan at the mouth of the canyon. This channel was up to 20 meters deep, approximately one kilometer long, and exposed layers of unconsolidated materials representing distinct depositional events.

Based on morphology and analysis of sedimentological parameters, at least 17 deposits were identified as paleo-debris-flows in the upper part of the alluvial fan. By dating deposits, the average recurrence interval for debris flows to reach this location was determined to be 206 years during the period 4210 to 710 years B.P. At a point lower on the alluvial fan, where eight paleo-debris flows were identified, the average recurrence interval was established to be 296 years during the period 4470 to 2105 years B.P. Exceedance probabilities were calculated for debris flows to reach these two positions on the fan during different intervals of late Holocene time. The techniques used at this site could be applied to other alluvial fans to estimate the hazard from debris flows in probabilistic terms.

Introduction

In late May and early June of 1983, a rapid and sustained snowmelt triggered hundreds of landslides in the mountainous regions of Utah (Wieczorek and others, 1989; Anderson and others, 1984; and Lips, 1985). These were not isolated events; in fact, debris flows have occurred repeatedly throughout historic and pre-historic time in Utah (Bailey and others, 1934; Croft, 1967; and Keaton and others, 1988). Often these debris flows have inundated residential communities causing severe damage and threatening life safety. The recent

---

[1] Engineering Geologist, JBR Consultants Group, 1952 E. Fort Union Blvd. Suite 209, Salt Lake City, Utah 84121

[2] Deputy for Engineering, U.S. Geological Survey, National Center, MS 905, Reston, Virginia 22092

activity, and recurrent nature of these events, demonstrate the potential risk to residential communities juxtaposed to mountainous regions. Two important aspects of assessing potential risks on alluvial fans are the ability to identify and distinguish debris flows from floods in the subsurface, and to establish the probability of occurrence. This paper presents the results of an investigation conducted on an alluvial fan in central Utah which utilized a combination of sedimentological parameters and dating of paleo-debris-flow deposits to establish recurrence intervals and exceedance probabilities.

Site Description

One of the 1983 debris flows in Utah occurred in the Birch Creek drainage, approximately 5 km southwest of the town of Fountain Green. This debris flow began on the eastern flanks of the San Pitch Mountains from the mobilization of a landslide involving approximately 167,000 cubic meters of colluvium and weathered bedrock. The flow traveled down the hillside where it intersected and followed Birch Creek before it ultimately flowed beyond the canyon mouth onto an alluvial fan in Sanpete Valley for a total travel distance of 5700 m.

Evidence at this site suggests that several pulses of debris flows, as well as high runoff of snowmelt water occurred. One of the debris-flow pulses blocked the established channel, resulting in the formation of a new fluvial channel, which then incised the alluvial fan. This newly incised channel was up to 20 m deep and 30 m wide. The channel cut approximately one kilometer of the fan and exposed several distinct layers of unconsolidated materials representing separate depositional events.

Three types of deposits were identified within the subsurface of the exposed alluvial fan. One type consisted of angular clasts, up to boulder size, with woody debris randomly scattered in a matrix of fine-grained material. These deposits were similar to recent (1983) debris-flow deposits in morphology as well as field-observed grain size distribution and sorting, and were, therefore, preliminarily identified as paleo-debris-flow deposits. The second type of material consisted of well-sorted, rounded sand and gravel in stratified deposits devoid of large clasts and woody debris. These were identified as paleo-flood (possibly hyperconcentrated) deposits. The third type of material consisted of organic-rich, dark gray to black, peaty soils which probably represent soil horizons rich in organic material that developed on the paleo-debris-flow or flood deposits.

Physical Characteristics of the Deposits

In order to quantitatively confirm that the paleo-debris flows were correctly identified in the field, two samples of the deposits were collected for comparison with 19 samples taken from the 1983 debris-flow and hyperconcentrated-flood deposits at this site. The physical characteristics of the deposits were analyzed by considering the grain-size distribution of the matrix, which was determined in the laboratory by sieve and hydrometer techniques. For practical purposes in the sampling and analyses, particles larger in size than 2 mm were omitted from the debris-flow deposits.

Figure 1 shows the envelopes of grain-size distribution curves for the 1983 debris-flow and hyperconcentrated-flood deposits. Two distinct groups of curves can be observed on these graphs. The first are the poorly sorted curves of the debris-flow deposits; the second are those from the deposits of the hyperconcentrated floods, which are visually better sorted. The curves would have been further separated if the large particles from the debris flows were considered. Also shown are the curves for the two deposits from the presumed paleo-debris flows which more closely resemble the 1983 debris flows than the hyperconcentrated floods.

The deposits from the presumed paleo-debris flows are further identified and distinguished from flood deposits by examining scatter diagrams of selected sedimentological parameters of the deposits. Parameters considered are the Graphic Mean, Inclusive Graphic Standard Deviation, and Graphic Kurtosis according to Folk (1980). In addition, percent clay was considered in the analyses. Similar plots and analyses have been used by Pierson (1985), Wells and Harvey (1987), and Lips (1990) to distinguish debris flows from hyperconcentrated floods.

Figure 2 shows a scatter plot of standard deviation (sorting) versus mean grain size for the 1983 deposits. Two distinct fields are evident on this scatter plot; one corresponding to the debris-flow deposits, and one to the hyperconcentrated-flood deposits. The 1983 deposits can also be visually separated into fields on scatter plots of standard deviation (sorting) versus clay content (Figure 3) and on scatter plots of kurtosis versus standard deviation (sorting) as shown in Figure 4. The positions of the presumed paleo-debris-flow deposits on these scatter plots corroborates their initial identification.

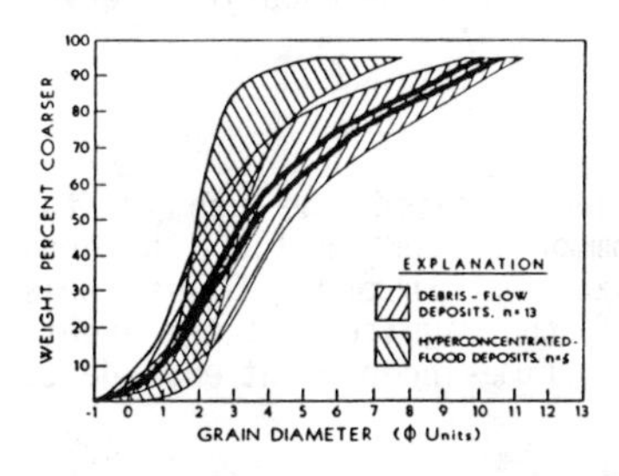

Figure 1. Envelopes of grain-size distribution curves of the 1983 debris-flow and hyperconcentrated-flood deposits. Solid lines are the distribution curves for the paleo-debris-flow deposits.

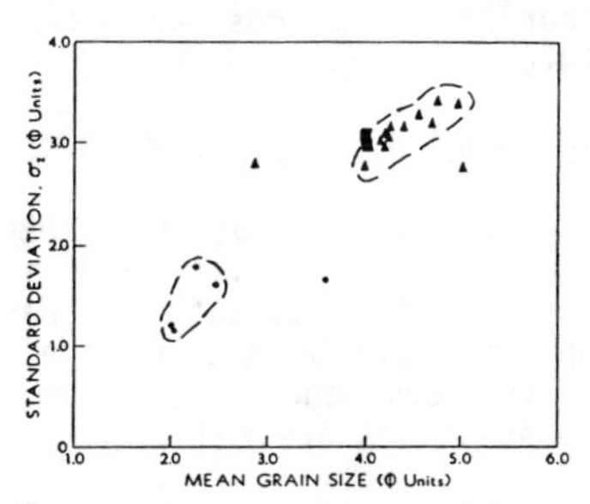

Figure 2. Scatter plot of standard deviation versus mean grain size for the debris-flow (triangles) and hyperconcentrated-flood (circles) deposits. Larger squares are the paleo-debris-flow deposits.

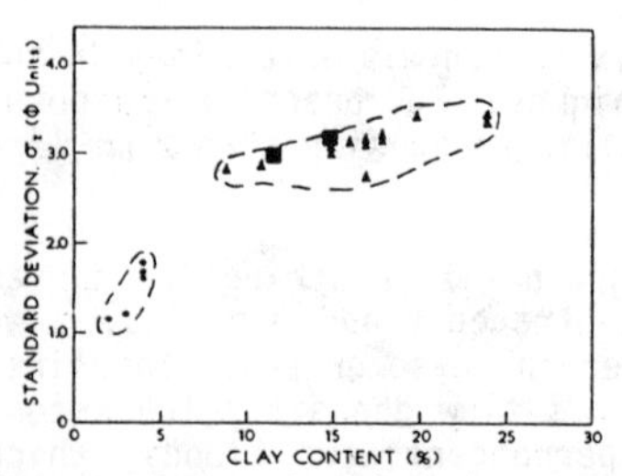

Figure 3. Scatter plot of standard deviation versus clay content for the debris-flow (triangles) and hyperconcentrated-flood (circles) deposits. Larger squares are the paleo-debris-flow deposits.

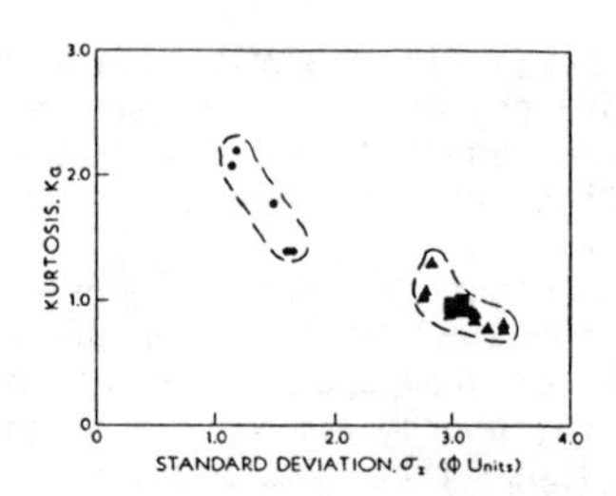

Figure 4. Scatter plot of kurtosis versus standard deviation for the debris-flow (triangles) and hyperconcentrated-flood (circles) deposits. Larger squares are the paleo-debris-flow deposits.

Radiocarbon Dating

Dating of organic material immediately below the lowest identifiable paleo-debris-flow deposit establishes the maximum age for this flow at 4210 ± 60 years B.P (W-1822). Organic material immediately above the uppermost paleo-debris flow was dated at 710 ± 50 years B.P. (W-1823), which established its minimum age. Exposed in this channel were at least 17 paleo-debris flows. Using these maximum and minimum ages, the average recurrence for paleo-debris flows to reach this location on the upper part of the alluvial fan is approximately 206 years for the 3500-yr interval (4210 - 710 yrs B.P.).

Using a similar sampling and dating technique at a point approximately 400 m lower on the fan, the uppermost paleo-debris-flow deposit was dated at two locations at 2110 ± 40 years B.P. (W-1819A) and 2100 ± 60 years B.P. (W-1819B). The lowermost paleo-debris-flow deposit was dated at 4470 ± 70 years B.P. (W-1821). At this location only 8 paleo-debris flows were identified; therefore, assuming an average age for the upper deposit of 2105 years B.P., the average recurrence is at most approximately 296 years for the 2365-yr interval (4470 - 2105 yrs B.P.).

Exceedance Probabilities

Keaton and others (1988) have demonstrated the use of the exceedance probability approach, as commonly used in flood hazards analyses, for assessing debris flow hazards. In order to use this approach, two factors must be considered; the magnitude of the event, and the recurrence interval of the event. Time-independent exceedance probabilities are calculated as

$$P(e \geq M,t) = 1-[1-(1/RI_M)]^t \qquad (1)$$

where $P(e \geq M,t)$ is the probability of occurrence of an event greater than or equal to magnitude M within time t years, and RI_M is the average recurrence interval in years for events of magnitude M.

It is generally accepted that large debris flows occur less often and cover larger areas than small ones. This is evident at Birch Creek by fewer debris flows, and therefore longer recurrence intervals at the distal fan locations. Thus, a debris flow of sufficient volume to leave a distinguishable deposit can be substituted for magnitude when calculating exceedance probabilities.

Equation 1 was used to calculate the exceedance probabilities for different time intervals at two locations on the fan. The results of these calculations are summarized in table 1. At the site on the lower fan, one paleo-debris flow was identified between the 2105 B.P. horizon and the 1983 debris flow. Thus, there were two debris flows at this location in the interval 2105 B.P. to 1983, whereas for the location on the upper fan only the 1983 debris flow was observed above the 710 B.P. horizon.

Table 1. Summary of representative intervals of Holocene time, recurrence intervals, and exceedance probabilities for debris flows at Birch Creek fan.

Location	Interval of Holocene Time	Number of Debris Flows	Recurrence Interval (yr)	Exceedance Probability for Time Period of Interest		
				10 yr	50 yr	100 yr
Upper Fan	4210 - 710 B.P.	17	206	0.05	0.22	0.39
	4210 B.P. - 1983	18	236	0.04	0.19	0.35
	710 B.P. - 1983	1	743	0.01	0.07	0.13
Lower Fan	4470 - 2105 B.P.	8	296	0.03	0.16	0.29
	4470 B.P. - 1983	10	450	0.02	0.11	0.20
	2105 B.P. - 1983	2	1069	0.01	0.05	0.09

Discussion

Debris flows may have occurred more frequently at either of the fan locations; for example, it is possible that some paleo-debris flows exist, but were not exposed in the channel, or that some flows existed but were subsequently eroded. In fact, 1952 aerial photographs show a recent debris-flow deposit on the upper fan; however it was not evident in the incised channel, and was not included in the calculation of the recurrence intervals. Therefore, the recurrence intervals presented in table 1 are considered maximum values for these time intervals.

There appears to be a period of time, beginning at about 4.5 ka, of frequent debris flows at the Birch Creek fan. This is in close proximity to the early Holocene period (> 6 ka) of heavy sediment accumulation along the Wasatch Front identified by Keaton and others (1988). They attribute this heavy sedimentation to a combination of the drainage basins being dominated by glacially weathered rock debris and solifluction mantles, and frequent, intensive cloudburst storms associated with the Milankovitch-type summer insolation abnormality. The period of frequent debris flows and heavy sedimentation at Birch Creek may be attributed to these same conditions. However, small topographic and/or climatic differences between the San Pitch Mountains and the Wasatch Front (150 km north) may have caused the slightly later period (< 4.5 ka) of debris flow activity at Birch Creek.

The recurrence intervals and exceedance probabilities at the Birch Creek fan depend on the interval of time considered. During more recent time, there appears to be longer recurrence intervals. This could be associated with either climatic changes, or due to the fact that much of the accumulated sediment in the basin was previously removed, and that sufficient time has not elapsed for the basin to recharge. These varying recurrence intervals and exceedance probabilities illustrate the possible error introduced when assuming a time-independent model, and from not fully understanding the conditional probability that control the processes generating debris flows.

Summary

Debris flows can be identified in exposed channels (or trenches) on alluvial fans based on field observed characteristics. These observations can be verified, and the deposits distinguished from those resulting from floods, by analyzing sedimentological parameters.

Recurrence intervals can be established by dating soil horizons immediately above and below a series of debris-flow deposits. This can be done at several locations to determine the exceedance probabilities at each location. The exceedance probabilities provide a quantitative measure of the hazard from debris flows, but is dependent on the interval of time considered. Therefore, care must be taken when applying a time-independent model to calculating the probability of having a flow. This type of analysis can aid in quantifying the amount of risk to which a facility located on an alluvial fan may be exposed.

References

Anderson, L. R., Keaton, J. R., Saarinen, T. F., and Wells, W. G. III, 1984, The Utah landslides, debris flows, and floods of May and June 1983: Washington, D. C., National Academy Press, 96 p.

Bailey, R. W., Forsling, C. L., and Becraft, R. J., 1934, Floods and accelerated erosion in Northern Utah: U.S. Department of Agriculture Miscellaneous Publication No. 196, 21 p.

Croft, A.R., 1967, Rainstorm debris flows, a problem in public welfare: University of Arizona, Agricultural Experiment Station Report 248, 35 p.

Folk, R. L., 1980, Petrology of sedimentary rocks: Austin, Texas, Hemphill's, 182 p.

Keaton, J. R., Anderson, L. R., and Mathewson, C. C., 1988, Assessing debris flow hazards on alluvial fans in Davis County, Utah, in, Fragaszy, R. J., ed., Twenty-fourth Annual Symposium on Engineering Geology and Soils Engineering: Pullman, Washington, Publications and Printing, Washington State University, p. 89-108.

Lips, E. W., 1985, Landslides and debris flows east of Mount Pleasant, Utah, 1983 and 1984: U.S. Geological Survey Open-File Report 85-382, 12 p.

Lips, E. W., 1990, Characteristics of debris flows in central Utah, 1983: Fort Collins, Colorado, Colorado State University M.S. thesis, 66 p.

Pierson, T. C., 1985, Initiation and flow behavior of the 1980 Pine Creek and Muddy River lahars, Mt. St. Helens, Washington: Geological Society of America, v. 96, p. 1056-1069.

Wells, S. G., and Harvey, A. M., 1987, Sedimentological and geomorphic variations in storm-generated alluvial fans, Howgill Fells, northwest England, Geological Society of America, Bulletin, v. 98, no. 2, pp. 182-198.

Wieczorek, G. F., Lips, E. W., and Ellen, S. D., 1989, Debris flows and hyperconcentrated floods along the Wasatch Front, Utah, 1983 and 1984: Bulletin of the Association of Engineering Geologists, v. 26, no. 2, p. 191-208.

Design Cloudburst and Flash Flood Methodology for the Western Mojave Desert, California

By Wesley H. Blood[1] and John H. Humphrey[1], Member, ASCE

Abstract

A methodology was required for estimating flow in ungaged drainages. An analysis of floods in the western Mojave Desert showed that the largest events on basins below 1,500 meters MSL were caused by mesoscale convective complexes (cloudbursts). Regional hydrologic methods were not applicable since peak flow measurements were inaccurate due to sediment bulking and alluvial fan and channel losses. A methodology was developed which determined representative spatial and temporal distributions of precipitation from analysis of historic cloudbursts in California, Nevada and Arizona. Templates of spatial distribution of point gage statistics were used in the HEC-1 hydrologic model to simulate peak flows. A comparison was made between simulated peak flows and observed peak flows at gages.

Introduction

Failure of regional hydrologic methods in the Mojave Desert required development of models for cloudburst flash floods. Regional methods failed primarily because (1) streamgage records were limited in number, short in duration, and at non-representative locations, (2) approximately 60% of the annual peak flows were zero due to channel losses, (3) slope-area estimates of the magnitude of larger events (debris floods) were too high due to sediment bulking, and (4) incorrect assumptions were made that basin elevation or mean annual precipitation were required independent variables in the regressions.

A spatial and temporal model of cloudburst precipitation was developed for various recurrence

---

[1]Hydmet, Inc., 9855 Meadowlark Way, Palo Cedro, CA 96073

intervals. This precipitation was used in the HEC-1 hydrologic model to simulate peak flows and hydrographs.

Cloudburst Methodology

Local convective storms dominate precipitation events in most of the southwestern United States interior valley and foothill areas. The largest events are called cloudbursts (or more scientifically, mesoscale convective complexes), which are severe thunderstorms having locally intense precipitation, generally causing flash floods. Cloudbursts are most likely to occur in the Mojave Desert in July and August, but have been observed in all months except November, December and January.

A review of reports and data concerning flood events in the desert Southwest found that cloudbursts were responsible for high recurrence interval peak flows on all lower elevation stream basins. The cloudburst methodology was developed for application to the western Mojave Desert valley and foothill watersheds east of the Tehachapi Mountains below 1500 meters elevation MSL.

Table 1 describes historic cloudbursts in the Southwest. This data was obtained from Osborn and Renard (1969), U. S. Army (1948), U. S. Army (1976), U. S. Department of Commerce (1972) and unpublished analyses by one of the authors (Humphrey).

Table 1. Historic Cloudburst Storms

Date	Location	Precipitation (mm)	Duration (min.)	Area (sq.km.)
18 Jul 1922	Cajun Pass	127	90	260
30 Sep 1932	Tehachapi	111	420	650
10 Aug 1942	Valyermo	51	40	NA
4 Mar 1943	Sierra Madre	58	60	650
9 Sep 1946	Cucamonga	89	60	520
9 Jul 1958	Barstow	62	75	390
30 Jul 1958	Morongo Valley	57	45	NA
4 Aug 1961	Vicentes	76	90	NA
22 Jul 1964	Walnut Gulch	64	60	390
10 Sep 1967	Walnut Gulch	88	50	260
3 Jul 1975	Las Vegas	76	240	550
17 Sep 1977	Redding	38	45	390
1 Mar 1983	Santa Ana	46	60	390
18 Feb 1986	Roseville	36	45	520

Based on the storms in Table 1 and comparisons to the Probable Maximum Precipitation (U. S. Department of Commerce, 1972) and Standard Project Storms (U. S. Army,

1976), the size of the 100-year design cloudburst storm is approximately 500 square kilometers with a duration of 60 minutes and the size of the 10-year cloudburst storm is 130 square kilometers with a duration of 45 minutes.

Observed short-duration precipitation depth-duration frequency data from the California Department of Water Resources (1986) were used to determine areal averaged station point statistics for the western Mojave Desert. Short-duration rainfall at a fixed location within a watershed has a high probability of being less than the greatest precipitation for the same duration and frequency occurring somewhere within that watershed. With increasing area, the probability that a point station measured the maximum center intensity precipitation will decrease. Corrections to point precipitation statistics for hydrologic basin area, based on studies in Arizona (Osborn and Laursen, 1973 and Osborn and Lane, 1981), are shown in Table 2.

Table 2. Cloudburst Point Precipitation Gage and Area Relationships

Area	Edge Ratio					Center
(sq km)	100-Yr	50-Yr	25-Yr	10-Yr	5-Yr	Ratio
3	1.00	.99	.99	.98	.97	1.00
5	.96	.93	.90	.84	.81	1.07
13	.88	.85	.82	.76	.70	1.16
26	.78	.75	.72	.66	.60	1.23
39	.71	.68	.65	.59	.53	1.27
52	.67	.64	.61	.55	.49	1.30
65	.63	.60	.57	.51	.45	1.32
78	.59	.57	.54	.48	.42	1.34
91	.57	.54	.51	.45	.39	1.35
104	.55	.52	.49	.43	.37	1.37
117	.53	.50	.47	.41	.35	1.38
130	.51	.48	.45	.39	.33	1.39
155	.49	.46	.43	.37	.31	1.41
181	.46	.43	.40	.34	.28	1.42
207	.44	.41	.38	.32	.26	1.43
233	.42	.39	.36	.30	.24	1.44
259	.40	.39	.36	.30	.24	1.45
389	.33	.30	.27	.21	.15	1.50
518	.29	.26	.23	.17	.11	1.52

Center Ratio, last column in Table 2, is used for correcting depth-duration-frequency point values from gage statistics to maximum center intensity values for the drainage area. Ratios of central rainfall intensity to edge area intensity are shown in Table 2 and may be used to construct cloudburst spatial templates.

A review of the literature found that there were no significant elvation effects on cloudburst central precipitation intensity or area for basins with elevations below 1,500 meters MSL. Analysis of cloudburst events showed that they characteristically had an elliptical shape with major:minor axis ratio of 2:1. Orientations of cloudbursts were aligned with the flow pattern of the upper air or aligned with associated weather fronts. In the western Mojave Desert cloudburst orientations were similar to upper air wind directions, southwest through south to southeast.

Time distribution of thunderstorm precipitation in other studies have shown that maximum point intensities have the greatest likelihood of occurrence in the first or second quartile of the 60 minute most intense precipitation period. Duration of precipitation time steps should approximate minimum subbasin lag time divided by 5.5 with values of decreasing intensity arranged alternatively around the largest value (U. S. Bureau of Reclamation, 1989).

Hydrologic Modeling of Cloudbursts

A case study of the cloudburst methodology for Rosamond Wash, Kern County, California was performed. Figure 1 shows Rosamond Wash basin with the maximum peak producing 100-year recurrence interval cloudburst centering. The area of Rosamond Wash is 530 square kilometers. Surface drainages originate in the Tehachapi Mountains in the northwest and proceed southeasterly across alluvial fans and in poorly defined channels.

The HEC-1 Hydrologic Computer Program (U. S. Army Crops of Engineers, 1982), was used for this case study. Subbasin and stream channel delineation was based on U. S. Geological Survey 1:24,000 mapping. Synthetic unit hydrographs were developed from procedures described in U. S. Department of Interior (1989). Basin infiltration and channel losses were based on field reconnaissance and U. S. Soil Conservation Service soil mapping.

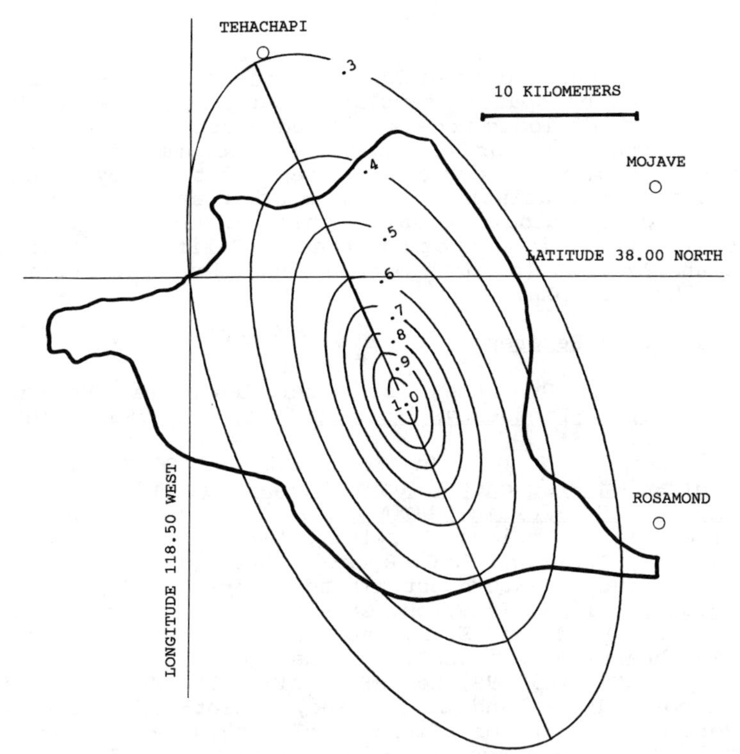

Figure 1. Rosamond Wash Basin and 100-Year Cloudburst Centering Producing Maximum Flow at the Mouth

Table 3 shows results of modeling the 100-year cloudburst in three gaged basins which are part of Rosamond Wash basin. As expected, the HEC-1 peak flows are usually lower than those estimated from a annual peak flow frequency analysis of gage records due to the influence of sediment bulking.

Table 3. Comparison of 100-Year Recurrence Interval HEC-1 Model Peak Flows With Frequency Analysis of Gage Records

Location	Gage (cu m/s)	HEC-1 (cu m/s)
Cottonwood Creek	230	200
Oak Creek	290	280
Joshua Canyon	130	90

Conclusions

Cloudburst precipitation in the Mojave Desert can be described by spatial templates and produce reasonable results in hydrologic models. Cloudbursts are similar in most of the interior Southwest with storm differences between regions adequately represented by point precipitation statistics. With appropriately set up or calibrated hydrologic models, this methodology may be suitable for use in most of non-coastal California, Nevada, Utah and Arizona for determining peak flow statistics or design hydrographs.

Appendix I. - References

1. California Department of Water Resources, Rainfall Depth-Duration-Frequency for California, Sacramento, California, 1986.
2. Miller, J. F., R. H. Frederick and R. J. Tracey, Precipitation-Frequency Atlas of the Western United States, Volume XI, California, NOAA Atlas 2, National Weather Service, Silver Springs, Maryland, 1973.
3. Osborn, H. B. and K. G. Renard, "Analysis of Two Major Runoff-Producing Southwest Thunderstorms", Journal of Hydrology, Vol. 8, 1969, pp. 282-302.
4. Osborn, H. B. and E. M. Laursen, "Thunderstorm Runoff in Southeastern Arizona", Journal of the Hydraulics Divison, ASCE, Vol. 99, No. HY7, July 1973, PP.1129-1145.
5. Osborn, H. B. and L. J. Lane, "Point-Area-Frequency Conversions for Summer Rainfall in Southeastern Arizona", Hydrology and Water Resources in Arizona and the Southwest, Vol. II, Proceedings of the American Water Resources Association, Arizona Section, May 1981.
6. U. S. Army Corps of Engineers, Prado Dam Design Hydrology, Los Angeles, California, 1948.
7. U. S. Army Corps of Engineers, Antelope Valley Streams Survey Report, Hydrology, Part I, Los Angeles, California, 1976.
8. U. S. Army Corps of Engineers, HEC-1 Flood Hydrograph Package Users Manual, The Hydrologic Engineering Center, Davis, California, 1982.
9. U. S. Department of Commerce, National Weather Service, Probable Maximum Thunderstorm Precipitation Estimates, Southwest States, 1972.
10. U. S. Department of the Interior, Bureau of Reclamation, Flood Hydrology Manual, Denver, Colorado, 1989.

Western Surface Climate and Streamflow and the El Nino/Southern Oscillation

Kelly T. Redmond[1] and Roy W. Koch[2], M. ASCE

INTRODUCTION

Climate variability has direct social and economic impacts the most direct of which occur through the hydrologic cycle. Of primary concern to society are the hydrologic extremes of floods and droughts. An improved knowledge of climate variability and its relationship to hydrologic variability is a first step in better planning, design and operation of the systems which are both directly and indirectly climate dependent. In considering climatic and hydrologic variability, there are many aspects which can be addressed. In this study we focus our analysis on the spatial patterns of variability of surface climate (precipitation and temperature) and streamflow across the western United States and the relationship of these patterns to a large scale circulation in the atmosphere, the Southern Oscillation.

THE EL NINO/SOUTHERN OSCILLATION

The Southern Oscillation (SO) is an equatorial circulation pattern which has wide reaching effects on global climate. It is characterized by a gradient in surface pressure between the eastern and western subtropical Pacific Ocean just south of the equator. Typically, a center of high pressure is located in the south Pacific near Tahiti while a low pressure area is located over Indonesia and northern Australia. Monthly sea level pressure departures from long-term means at these two locations exhibit a strong negative correlation with each other. Periods when the east-west pressure gradient is weak are often associated with El Nino, a condition of higher than average sea surface temperature in equatorial waters off of South America. Taken together, these

[1]Kelly T. Redmond, Western Regional Climate Center, Desert Research Institute, P.O. Box 60220, Reno, Nevada, 89506.
[2]Roy W. Koch, Department of Civil Engineering and Systems Science Program, Portland State University, P.O Box 751, Portland, Oregon.

two events are called El Nino - Southern Oscillation (ENSO). There have been many comprehensive descriptions of the phenomenon, including the one by Philander (1989).

DATA

The analyses presented here are based on average temperature and total precipitation of the climate divisions in the western U.S. and streamflow at selected locations which roughly correspond to selected climate divisions. Streamflow data have been taken from watersheds of moderate size where the flows have not been appreciably affected by either diversion or storage. The surface climate data set consists of the mean monthly divisional temperature and monthly divisional precipitation from the 84 climatic divisions lying within the study area. The period of coverage is 1931-1984. Monthly streamflow data for 14 river basins was also assembled for a similar period. A number of simple indices of the Southern Oscillation have been formulated. We use the version of the Southern Oscillation Index (SOI) presently reported monthly by the NOAA Climate Analysis Center, defined as the standardized monthly sea level pressure departure from average at Tahiti minus the standardized monthly departure at Darwin, Australia. A low (negative) value of the index corresponds to a weaker east-west pressure gradient, and thus relatively lower pressure in the eastern South Pacific and/or relatively higher pressure over Indonesia. Low SOI values are also closely linked to El Nino episodes. The coincident records of surface climate and SOI span a 50 or 51 year period (depending on the season).

Monthly climate data were aggregated into six and twelve month intervals. The principal interval selected was the 12-month period October-September, known as the "water year." Streamflow also is traditionally summarized in terms of the water year. The water year can conveniently be split into two six-month sub-intervals, October-March and April-September. Neither the use of the entire water year nor the use of these two six-month sub-intervals result in an awkward splitting of the natural precipitation cycle into two adjacent years. Only annual streamflow volumes are analyzed.

We focus on results for the winter half-year, where the strongest associations were seen. In addition, this is the part of the year most important to surface water supplies in the western U.S. Analyses for the summer half year data are not presented since results were not statistically significant.

ASSOCIATION OF SURFACE CLIMATE WITH SOI

To explore the relationship between the Southern Oscillation and western U.S. climate, correlation coefficients of the SOI with annual and seasonal precipitation and temperature were computed. Annual and six-month divisional averages of precipitation and temperature were correlated with the SOI at lags from -6 (SOI leads surface climate) to +6 months (SOI lags surface climate). The magnitude and sign of the correlation coefficients are used to determine whether a relationship exists and if so, to aid in understanding the nature and strength of the association. In addition, the samples of climate data were split into three categories based on the SOI averaged over the period which produced the strongest correlation, overall, for the entire region. One category was associated with SOI values in excess of 0.5 while the other was based on SOI values less than -0.5. This division isolates the more extreme phase of the SOI while still preserving a reasonable number of values in each category.

An evaluation of all of the correlation results showed that, in general, the highest correlation coefficients between SOI and precipitation occurred when the SOI averaging period led the climate data by 4 months. Figure 1 shows the pattern of the correlation between the June-November averaged SOI and October-March precipitation for the 51 years of coincident data. Negative correlations dominate in southern portions of the domain and positive correlations exist in the Pacific Northwest. Many of the correlation coefficients are significantly different from zero (the null hypothesis) at the 0.1% level. As much as 33 percent of the variance ($r = -0.57$) of individual climate divisions is associated with the SOI in the southwest, and up to 26 percent ($r = 0.51$) in the Pacific Northwest. Since a negative anomaly in the SOI is strongly related to the occurrence of an ENSO event, it can be inferred that below average precipitation might be expected during ENSO years. The opposite is true for the desert Southwest with above average winter precipitation expected in ENSO years.

To further clarify the precipitation response in relation to the SOI, only the most positive and most negative events are considered. Differences were found between October-March precipitation in those years when the June-November SOI averaged +0.50 or higher (11 cases) and the precipitation in those years when the SOI averaged -0.50 or lower (15 cases). A t-test of the differences in means was performed in each division. The results are shown in Figure 2. Significant differences are found in both the lower Colorado River Basin and the Columbia River Basin, many of the latter significant at the 0.1% level. Low SOI (El Nino) is associated with dryer than average winters in the Columbia Basin and wetter that average winters in the Colorado Basin. There is some tendency for the mountainous climate divisions in the Pacific Northwest to exceed significance criteria, in contrast to nearby more arid intermontane divisions dominated by valley stations.

The behavior just described can be further illustrated using the results from two individual divisions. Washington division 2 in northern Puget Sound lies in the center of the area with positive precipitation-SOI correlations ($r=0.46$). Average October-March monthly precipitation in the 11 years with SOI values of at least +0.50 is 81 mm, compared with an average of 59 mm in the 15 years with SOI values of -0.50 or less. For Arizona division 6 (South Central), located near the center of the strongest negative correlations with the

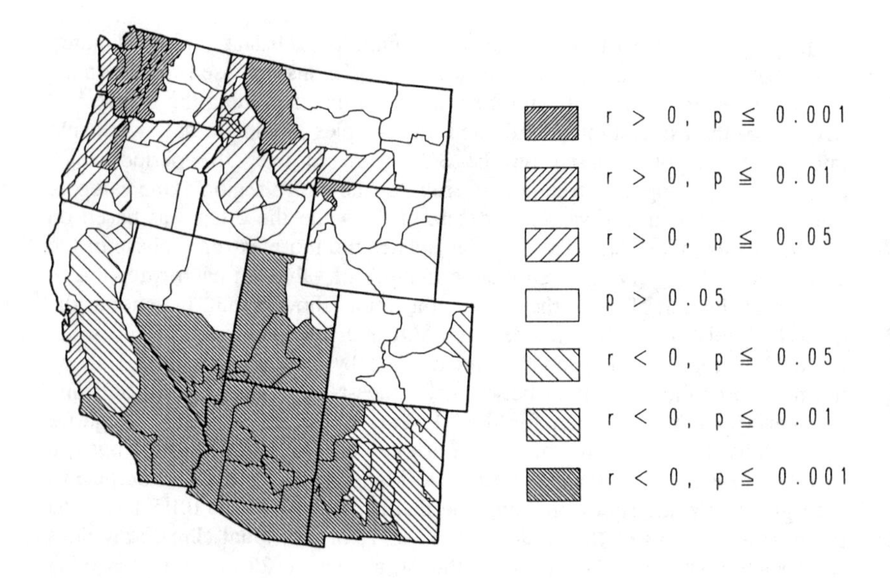

Figure 1. Correlation between June-November SOI and October-March precipitation.

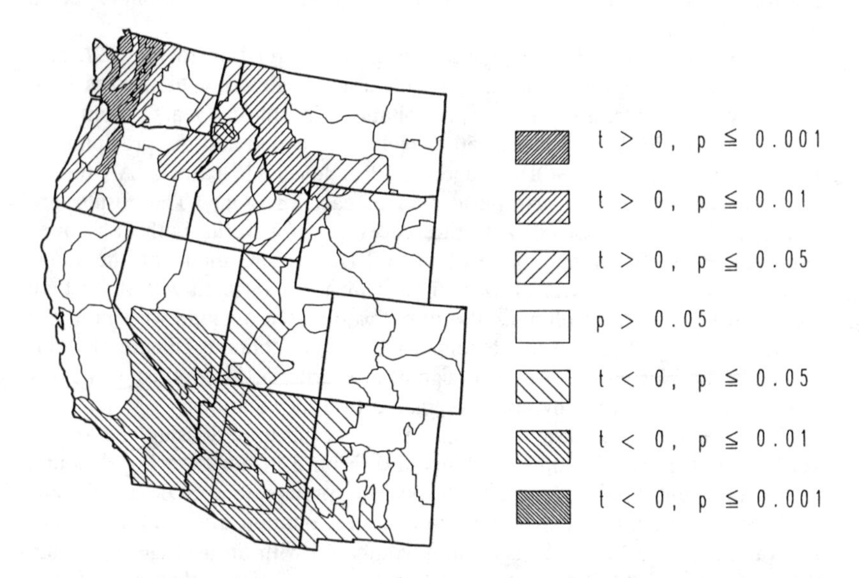

Figure 2. Results of split sample tests for October-March precipitation.

SOI (r=-0.54). For years with SOI values of 0.50 or greater, average precipitation is 29 mm, in contrast to those years with SOI values of -0.50 or less, when average precipitation is just 15 mm.

Most of the statistically significant correlation coefficients between October-March averaged temperature and June-November averaged SOI occurred in the Pacific Northwest where the sign was negative. In the desert Southwest, most divisions were positively correlated with SOI but not at a statistically significant level. Based on the split sample analysis for those years the northwestern portion of the domain is warmer when the SOI is low, while the southwestern portion is cooler.

ASSOCIATION OF STREAMFLOW WITH SOI

Of primary interest to the hydrologist is how these associations of surface climate and SOI result in variability of streamflow. region. A set of the 14 river basins were used in this analysis. The watersheds are primarily headwater streams and are not uniformly distributed throughout the region. However, basins such as these are the primary source areas for major river systems throughout the western U.S and are of particular interest from the point of view of water supply forecasting.

Table 1. Correlation and split sample results for streamflow.

River Basin	Correlation Coefficient	Ratio
Animas River, CO	-0.319	0.87
Bear Creek, CO	-0.289	0.84
Boise River, ID	0.282	1.33 **
Clearwater River, ID	0.498	1.45 ***
Gila River, NM	-0.392	0.90 *
John Day River, OR	0.182	1.43 **
Lochsa River, ID	0.522	1.42 ***
Merced River, CA	-0.249	0.90
Selway River, ID	0.498	1.36 ***
Skykomish River, WA	0.533	1.35 ***
Smith River, CA	0.148	1.19
Umpqua River, OR	0.351	1.40 ***
Weber River, UT	-0.003	1.16
Wilson River, OR	0.401	1.27 ***

Statistically significant at: * - 5%, ** - .1%, *** - .1%

Annual streamflow for these watersheds were subjected to correlation and split sample analyses. Similar but even more striking results were obtained in many cases. Correlation coefficients of annual streamflow with SOI averaged over the previous June-November period are presented in Table 1. Fairly large, positive correlations are the rule in the Pacific Northwest (Idaho, Oregon and Washington), some insignificant correlations occur in the Great Basin and central California area, and correlation coefficients become larger and negative in the lower Colorado River Basin. The strongest correlations occur in the rivers with headwaters in the mountains of central Idaho (the Clearwater, Lochsa and Selway Rivers) and the Coastal and Cascade Range of Washington and Oregon (Skykomish, Umpqua and Wilson Rivers). These results also suggest that during an ENSO event, below average flows can be expected throughout the Pacific Northwest. The two rivers in the Southwest (Animas and Gila Rivers) both exhibit statistically significant negative correlations with June-November averaged SOI.

The correlation results are reinforced by the split sample analyses (Table 1). All of the rivers in the Pacific Northwest show statistically significant differences in the mean values of streamflow between years when the SOI is in the high phase in comparison to the years when the SOI was low. Streamflows in the Southwest did not show such a strong relationship.

DISCUSSION

The major conclusion from this investigation is that there are statistically significant associations between the surface climate in the western United States and atmospheric circulation in the equatorial Pacific Ocean. The associations are not uniform but are strongest in two well defined areas located in the Pacific Northwest and desert Southwest. Even within these areas, the strongest associations, particularly for precipitation, occur in the mountainous divisions. Analyses of streamflow data show even stronger associations in the Pacific Northwest suggesting that a combination of precipitation and temperature effects may combine to impact snow accumulation. Furthermore, there is a suggestion of cause-effect relationship, particularly as related to the SOI and ENSO events. Since the highest correlation between both surface climate and streamflow with SOI occurred when SOI led the climate and hydrologic response by several months, improved predictive ability is suggested.

REFERENCES

Philander, G. (1989), El Nino and La Nina, American Scientist, Volume 77, No. 5, pp. 451-459.

Redmond, K.T. and R.W. Koch (1990). Surface Climate and Streamflow Variability in the Western United States and their Relationship to Large Scale Circulation Indices, submitted for review to Water Resourc. Res.

Spectral Analysis of Annual Time Series of Mountain Precipitation

Christopher J. Duffy[1], Ying Fan[2], and Upmanu Lall [2]

Abstract

In this paper we examine the statistical moments of long record, annual, precipitation time series for 19 west facing stations along the Wasatch Front, northern Utah. The mean, variance, autocorrelation, and spectrum for each time series, along with the time scale of fluctuation, are estimated and compared. Although it is well known that mean precipitation in the Wasatch is positively correlated with altitude (Peck and Brown, 1962), the objective of this study is to determine whether an orographic relationship also exists for the variance, covariance, and spectral distributions as well. This study is an initial effort to catalog space-time variability of hydrologic processes in mountainous terrain.

Introduction

There is currently a great deal of interest within the research community for making hydrologic predictions at large spatial and temporal scales. In the western United States it is well known that mountain precipitation and high elevation climate conditions, are the most important factors supplying river flows within the interior basins. In addition to long term average conditions, year to year fluctuations in basin precipitation and the time scale of these fluctuations, also have a large impact on man and environment. The spirit of this brief paper is to present an analysis of the mean and fluctuation of annual precipitation time series along the Wasatch front, northern Utah, with particular emphasis on determining patterns of precipitation with altitude. The data consists of 19 stations located along the western slope of theWasatch Range. The stations are from National Weather Service and Soil

---

[1]Member ASCE, Dept. of Civil Engineering, Pennsylvania State University, 212 Sackett Bldg, University Park, PA 16802. [2]Utah Water Research Laboratory, Utah State University, Logan, Utah 84322.

Conservation Service snow-course stations. The length of record for the annual data-base is variable, and basic statistics were determined from the original records. For time series analysis, the standardized record length was 1875-1988, and missing data were reconstructed by multiple correlation with adjacent stations.

Precipitation Time Series and Moments

Our stated goal in this analysis is to examine the characteristic fluctuations of annual precipitation, and to search for patterns in the first two moments of the time series, especially as these patterns relate to altitude. Eight of the 19 time series, covering a range of altitudes are illustrated in figure 1. Also included in the figure are the mean and variance of the series. Several stations indicate a slight downward trend over the period of record, and this trend was removed for time series analysis. The mean and standard deviation of the time series are plotted versus altitude in figures 2a and b. Although there is substantial scatter, an exponential-type fit for the mean illustrates a slightly nonlinear rate of increase in precipitation with elevation. The standard deviation also shows a slightly nonlinear rate of increase with elevation. It is interesting to note that the rate of change in the mean and standard deviation with elevation is quite similar. This is illustrated in figure 2c, where the coefficient of variation of precipitation is observed to have a nearly constant value (~0.2) for all elevations.

Estimating the Spectrum and Covariance

The covariance and spectrum of precipitation provide information on the degree of correlation and frequency content of fluctuations in the record. Since low elevation precipitation along the Wasatch is more likely to fall as rain, and high elevation precipitation is dominated by snow, one might expect that the pattern of correlation would change with altitude. Likewise for the spectrum, the distribution of variance with frequency (cycles/year) in the record, should also exhibit altitude differences. The spectrum is a property of the autocovariance of the time series x(t), and we define the autocovariance in terms of lag τ and mean μ

$$R_{xx}(\tau) = E[(x(t) - \mu)(x(t+\tau) - \mu)] \qquad (1)$$

where $E[\cdot]$ indicates expectation and $i=\sqrt{-1}$. The spectrum-covariance relation is determined by the Fourier transform pair

$$\phi_{xx}(f) = \int_{-\infty}^{\infty} e^{-i2\pi f\tau} R_{xx}(\tau)\, d\tau\, ; \quad R_{xx}(\tau) = \int_{-\infty}^{\infty} e^{i2\pi f\tau} \phi_{xx}(f)\, df \qquad (2)$$

The autocorrelation is defined

$$\rho_{xx}(\tau) = \frac{R_{xx}(\tau)}{\sigma^2} \tag{3}$$

where σ^2 is the variance. The autocorrelation and spectra were estimated by constructing (1) and applying a discrete Fourier transform (Jenkins and Watts, 1968). The 'time scale' or integral scale of precipitation fluctuations is a measure of the persistence, or average number of years over which precipitation is correlated.

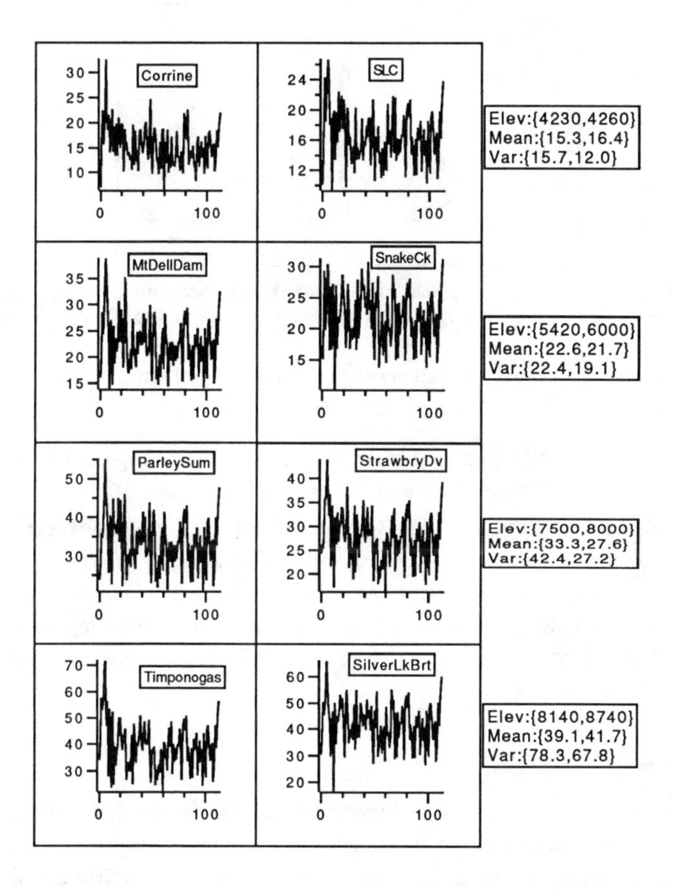

Figure 1. Annual precipitation time series (1875-1988) for 8 stations at various elevations (feet above sea level) along the west-facing Wasatch Front, northern Utah.

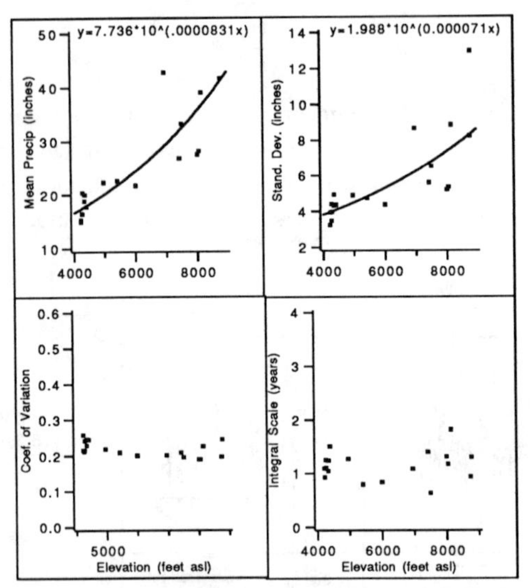

Figure 2. a) The mean, b) standard deviation, c) coefficient of variation, and d) integral scale of annual precipitation versus elevation along the Wasatch Front.

We assume an exponential correlation structure for the annual series

$$\rho_{xx}(\tau) = e^{-\tau/\lambda} \tag{4}$$

where the 'time scale' or integral scale of fluctuation is λ. Figure 3 illustrates estimates of the autocorrelation and spectra for the 8 sites of figure 1. The autocorrelation falls off very quickly, with the estimated λ to be on the order of one year for all sites (figure 2d). In keeping with the rapid decay of correlation, the spectral estimates of the annual series are relatively flat, which indicates a uniform contribution from precipitation fluctuations at all frequencies. The effect of altitude on the spectra is to displace the curve upward, reflecting the higher variance at higher elevation stations.

Summary and Conclusions

The moment analysis of annual precipitation for the period 1875-1988 on the Wasatch Front, indicate a substantial orographic effect for the mean and variance or standard deviation. However, the coefficient of variation is nearly a constant, indicating that the orographic effect is about the same for the mean and the fluctuation. From time series analysis it was found that all records lack significant autocorrelation, with the integral scale not significantly different from the sampling interval. The spectra are relatively flat and increase in proportion to the variance change with elevation. The correlation-spectral structure is not unlike a 'white-noise' process, which would indicate that year-to-year variation in the record is statistically independent. The analysis is part of a long-term effort to study the effects of topography on hydrologic process.

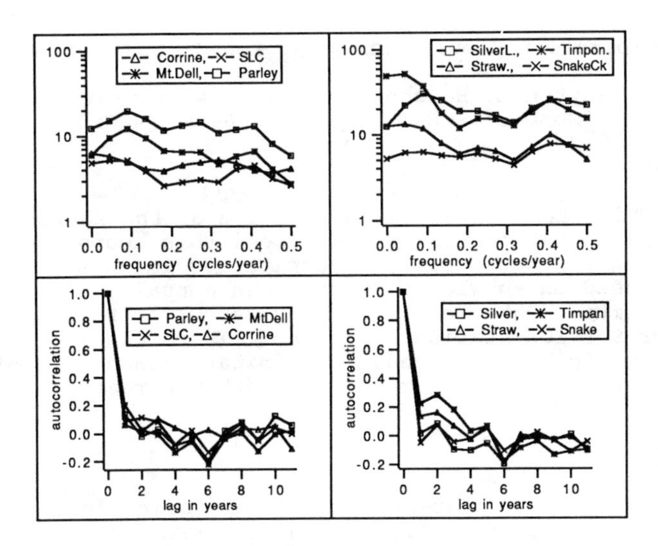

Figure 3. Spectrum and autocorrelation of annual time series illustrating the general increase in 'power' with altitude, and the similarity of short-range correlation structure at all elevations.

Acknowledgements

Partial support for this work came from the U. S. Department of the Interior, Geological Survey grant no. 14-08-0001-G1630, and U. S. Department of the Army, Army Research Office, grant no. DAAL03-90-G-0075. The content of this paper does not necessarily reflect the policy or position of these federal agencies and no endorsement is implied.

References

Peck, L. P and M. J. Brown, 1962, An approach to the development isohyetal maps for mountainous areas. Journal of Geophysical Research 67(2), 681-694.

Jenkins, G. M., and D. G. Watts, 1968, Spectral Analysis and its Applications, Holden-Day 525.

Precipitation Simulation Model for Mountainous Areas

Clayton L. Hanson[1] Member ASCE

David A. Woolhiser[1] Member ASCE

ABSTRACT

This study investigated the use of the Markov chain-mixed exponential (MCME) model as a means of obtaining synthetic daily precipitation amounts at a site, based on knowledge of the mean annual precipitation of the site. Data for this study were from 25 National Weather Service stations in Idaho. There was a relationship between annual precipitation and some of the parameters in the MCME model, and other parameters could be assumed constant.

INTRODUCTION

Daily precipitation is a primary climatic variable in several hydrologic and natural resource models. The Markov chain-mixed exponential (MCME) model was developed to simulate daily precipitation series in areas such as the western United States where there are few measuring stations. In the mountainous areas of the western United States, most precipitation stations are located at valley sites which do not necessarily represent the climatic conditions at higher elevations. The effects of elevation on precipitation occurrence and amount have been investigated by several scientists (Hanson et al., 1989). In this paper, the relationships between mean annual precipitation from 22 stations in Idaho and MCME model parameters discussed by Hanson and Woolhiser (1990) were used to evaluate how well the MCME model represented daily precipitation from three Idaho stations that were not in the original analysis.

----

[1]Agricultural Engineer, USDA-Agricultural Research Service (USDA-ARS), 270 South Orchard, Boise, ID 83705, and Hydraulic Engineer, USDA-ARS, 2000 East Allen Road, Tucson, AZ 85719.

MODELING DAILY PRECIPITATION

The daily precipitation process was described by the MCME model (Woolhiser and Roldan, 1986). Precipitation occurrence was described by a first-order Markov chain specified by parameters $P_{00}(n)$, the probability of a dry day on day "n" given that day n-1 was dry, and $P_{10}(n)$, the probability of a dry day on day "n" given that day n-1 was wet.

The amount of precipitation on a "wet" day was simulated with the mixed exponential distribution:

$$f(x) = \frac{\alpha(n)}{\beta(n)} \exp\left[\frac{-x}{\beta(n)}\right] + \frac{1-\alpha(n)}{\delta(n)} \exp\left[\frac{-x}{\delta(n)}\right] \qquad (1)$$

where the mean precipitation per "wet" day, $\mu(n)$, equals $\alpha(n)\beta(n) + [1-\alpha(n)]\delta(n)$. The parameter α is usually assumed constant throughout the year, but β and μ may vary seasonally. Note that if $\alpha(n) = \alpha$ = a constant and $\beta(n)$ and $\mu(n)$ are specified by Fourier series, $\delta(n)$ is determined by the above relationship.

The seasonal variations in the parameters P_{00}, P_{10}, β, and μ were described by the polar form of a finite Fourier series which was limited to six harmonics (Woolhiser and Pegram, 1979).

Data for this study were from 25 National Weather Service stations in Idaho (Table 1 in Hanson and Woolhiser, 1990). Based on data availability, these daily precipitation records were as close to 40-year periods, beginning March 1, 1940 as possible. Data from 22 stations were used for parameter estimation and data from three stations were used for testing the accuracy of the model. The three stations used for testing were selected a priori to represent different climatic regions of the state.

Based on each record, constant terms in the Fourier series, $\bar{P}_{00}$, $\bar{P}_{10}$, significant amplitudes and phase angles, as determined by the Akaike information criterion (Akaike, 1974), and the number of wet days were determined for the occurrence (Markov chain) portion of the MCME model. The constant terms $\bar{\alpha}$, $\bar{\beta}$, and the mean precipitation per wet day, $\bar{\mu}$, and significant amplitudes and phase angles were determined for the mixed exponential portion of MCME. MCME parameters were estimated by maximum likelihood techniques described by

Woolhiser and Roldan (1986) who also discussed data requirements for obtaining reliable parameter estimates. All statistical tests were at the 0.05 level of probability.

ANALYSIS

Analysis of the records by Hanson and Woolhiser (1990) showed that there were significant linear relationships between most of the parameters in MCME and mean annual precipitation. In this paper we used the individual relationships between the parameters P_{00}, P_{10}, α, β, and μ and mean annual precipitation to generate daily precipitation for the three Idaho sites not used in developing the relationships. This approach is used because mean annual precipitation values are available for many locations or estimates can be calculated from other sources (Hanson, 1984).

Occurrence of Wet Days

Because values of $\bar{P}_{00}$, $\bar{P}_{10}$, and $\bar{\alpha}$ vary between 0 and 1, the logit transformation (Hanson et al., 1989) of their values was used to develop relationships between each of them and mean annual precipitation. This transformation was used to prevent computing unrealistic values of the parameters at sites where the mean annual precipitation is considerably more or less than at the sites used to develop the relationships.

The following linear relationships between $\bar{P}_{00}$ and $\bar{P}_{10}$, and mean annual precipitation were used to calculate the values of $\bar{P}_{00}$ and $\bar{P}_{10}$:

$$\Theta_1 = 2.046 - 0.0011X \qquad r = 0.815 \qquad (2)$$

$$\Theta_2 = 0.648 - 0.0013X \qquad r = 0.927 \qquad (3)$$

where Θ_1 and Θ_2 are the logit transformations of $\bar{P}_{00}$ and $\bar{P}_{10}$, respectively, and X is mean annual precipitation in mm.

P_{00} varied seasonally with the first five harmonics being significant. The amplitudes of the first harmonics (C_{P001}) and second (C_{P002}) increased linearly with increasing mean annual precipitation. The data from the other three amplitudes and all of the phase angles associated with P_{00} had a considerable amount of scatter so they were assumed constant. The relationships between C_{P001} and C_{P002}, and mean annual precipitation were:

$$C_{P001} = 0.0344 + 0.0001X \quad r = 0.631 \tag{4}$$

$$C_{P002} = 0.0117 + .00003X \quad r = 0.625 \tag{5}$$

P_{10} varied seasonally with the first three harmonics being significant. None of the amplitudes or phase angles were related to mean annual precipitation and were assumed constant.

Amount of Precipitation on Wet Days

There was a significant linear decrease in $\bar{\alpha}$ and it could not be shown that α varied seasonally. Values of $\bar{\alpha}$ were computed for the test sites by the following equation:

$$\Theta_3 = -0.379 - 0.0012X \quad r = 0.411 \tag{6}$$

After values of $\bar{\alpha}$ were computed for each site, new values of $\bar{\beta}$, $\bar{\mu}$, and the Fourier coefficients for β and μ were obtained.

There was a significant linear increase in $\bar{\beta}$ with increasing mean annual precipitation which resulted in the following equation.

$$\bar{\beta} = 0.736 + 0.0021X \quad r = 0.401 \tag{7}$$

Only the first harmonic of β was significant and neither the amplitude or phase angle were related to mean annual precipitation, so constant values were used for generating daily precipitation.

There was a significant positive linear increase in $\bar{\mu}$ with mean annual precipitation which is:

$$\bar{\mu} = 2.231 + 0.0046X \quad r = .891 \tag{8}$$

The first three harmonics of μ were significant. However, only the first phase angle ($\phi_{\mu 1}$) was related to mean annual precipitation as shown by equation (9), and the other phase angles and amplitudes were set to constants for genenerating precipitation.

$$\phi_{\mu 1} = -0.411 - 0.0033X \quad r = -0.586 \tag{9}$$

TEST OF ESTIMATION PROCEDURE AND DISCUSSION

Two sets of 50-year, daily precipitation records were generated for the three test stations, Coeur d'Alene, Idaho Falls, and Riggins (Table 1). The first record was generated using maximum likelihood (ML) parameters computed from each station's record and the second record using parameter sets estimated from regional mean values or equations (2) through (9).

Table 1. Sample monthly and annual summary of 50-year daily precipitation (mm) simulations.

	January		July		Annual	
Coeur d'Alene, ID	Ave.	Std.	Ave.	Std.	Ave.	Std.
Historical	91	45	19	19	660	122
Simulated (ML)*	85	29	27	21	671	85
Simulated (est.)**	76	29	22	23	638	98
Idaho Falls, ID	Ave.	Std.	Ave.	Std.	Ave.	Std.
Historical	20	14	11	9	240	63
Simulated (ML)	24	14	16	17	250	47
Simulated (est.)	29	14	17	16	263	53
Riggins, ID	Ave.	Std.	Ave.	Std.	Ave.	Std.
Historical	34	23	18	17	429	77
Simulated (ML)	36	19	23	19	428	69
Simulated (est.)	48	16	16	15	409	65

*Simulations using ML parameters.

**Simulations using regional mean parameters.

Historical and simulated precipitation for January, July, and mean annual and the standard deviations are shown in Table 1. Simulations of mean annual precipitation using ML parameter sets varied from the same precipitation at Riggins to 4% greater than the historical record at Idaho Falls. Simulated annual precipitation, using the estimated parameter set, varied from 5% less than the historical record at Riggins to 9% greater at Idaho Falls. There was no pattern of over- or under-estimating monthly precipitation from either parameter set and simulated monthly values followed the seasonal trends at the three test sites.

Standard deviations of the simulated annual precipitation means were considerably less for both parameter sets than that from the historical record. The annual standard deviations were greater for two of the three test stations for simulations which used the estimated parameter set than simulations using the ML parameters.

The range of simulated monthly high and low values was similar to that of the historical record for the three sites. The range of simulated annual precipitation was about the same for two of the three sites for both parameter sets, but both simulations underestimated the range at one site by about 30%. The average number of wet days was simulated within three days by the ML parameter set. The simulation based on the ML parameter set overestimated the number of wet days by 3 out of 86 days at the site with the least number of wet days which was the greatest difference between historical and simulated of the three sites. The simulation based on the estimated parameter set underestimated the average number of wet days by four days at two sites and nine days at the site with the least number of wet days.

The ML model preserved the important statistics within a year using both sets of parameters, but as shown in a previous study (Hanson et al., 1989) caution should be taken when using the model to study annual phenomena.

REFERENCES

Akaike, H. 1974. A new look at the statistical model indentification. IEEE, Transactions on Automatic Control AC-19(6):716-723.

Hanson, C. L. 1984. Annual and monthly precipitation generation in Idaho. Transactions of the American Society of Agricultural Engineers 27(6):1792-1797, 1804.

Hanson, C. L. and D. A. Woolhiser. 1990. Annual precipitation and regional effects on daily precipitation model parameters. IAHS (in press).

Hanson, C. L., H. B. Osborn, and D. A. Woolhiser. 1989. Daily precipitation simulation model for mountainous areas. Transactions of the American Society of Agricultural Engineers 32(3):865-873.

Woolhiser, D. A. and G. G. S. Pegram. 1979. Maximum likelihood estimation of Fourier coefficients to describe seasonal variations of parameters in stochastic daily precipitation models. Journal of Applied Meteorology 18(1):34-42.

Woolhiser, D.A. and J. Roldan. 1986. Seasonal and regional variability of parameters for stochastic daily precipitation models: South Dakota, U.S.A. Water Resources Research 22(6):965-978.

Temporal Characteristics of Aridland Rainfall Events

Virginia A. Ferreira[1]

Abstract

Rainfall intensity data are required to drive infiltration-based runoff models. Because data are sparse and cumbersome, a model to generate reasonable sequences of data is needed. Studying available data and characterizing rainfall events are the first steps in developing such a model. This study analyzes aridland rainfall traits: depth, duration, interlude, maximum intensity, and time to maximum intensity. The effect of chart speed on event statistics is shown. Sensitivity of an infiltration model to the time interval of synthetic input data is demonstrated.

Introduction

Several comprehensive simulation models have been developed by the USDA Agricultural Research Service (ARS) to assess the hydrologic and chemical response of agricultural fields to rainfall input under various management practices. For many studies of nonpoint-source pollution from small agricultural areas, a daily time step is too coarse to provide insight into management problems. For example, a 100-mm storm will yield different chemical fate if it occurs over a 24-hr period than if it occurs in 1 hr. CREAMS (USDA, 1980), Opus (Smith, in preparation), RZWQM (DeCoursey et al., 1989), and KINEROS (Woolhiser and Smith, 1989) all can employ breakpoint rainfall data as input. Opus, an agricultural ecosystem model with an infiltration-based hydrology option, is utilized to simulate field response to rainfall input of various time intervals.

The objective of this study is to characterize aridland rainfall events. This is the first step in developing a model to generate rainfall data. The approach is to determine statistics describing several rainfall characteristics, using data from four USDA-ARS research watersheds in arid climates.

[1]Mathematician, USDA-Agricultural Research Service, P.O. Box E, Ft. Collins, CO 80522.

Procedure

The definition of a rainfall event affects many of the event statistics. In this study, an event is defined as a continuous period of rain. Thus, a 1-min hiatus ends the event. The variables used to describe event characteristics are illustrated in Fig. 1, a plot of rainfall intensity versus time. Total event rainfall accumulation, P (area under the curve), is an important factor in data-generating schemes. It is expected to be related to other variables, including maximum rate and duration. P is the most precisely-measured variable, because it does not depend on chart time precision. The

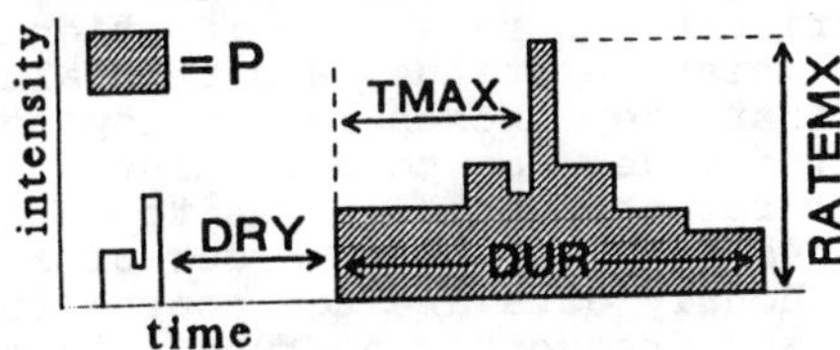

FIG. 1. Event Variable Definitions

hiatus between events (time since previous rain), DRY, is important in infiltration models; of particular interest are short interludes, as will be demonstrated. Event duration, DUR, is expected to be important in determining the magnitude and distribution of rates within events. The maximum rain rate, RATEMX, is a necessary component, strongly affecting predicted runoff hydrographs. Another variable used to characterize events is the time to the maximum rate, TMAX.

Available Data

The USDA-ARS has a unique data resource from watershed studies that began as early as the 1920's. Periods of record in excess of 40 yrs are available from weighing, recording gages with time and depth resolutions of 1 min and .25 mm (.01 in). The data, of high quality and distributed within various climates, are in the form of analog pen traces on charts. These have been digitized by reading time-depth pairs at points on the traces where the rainfall rate changes (breakpoints). In this study, breakpoint data from four ARS watersheds are analyzed: Albuquerque, NM, Reynolds Creek, ID, Safford, AZ, and Walnut Gulch, AZ. These locations are considered "arid" for the purpose of this conference. The average annual rainfall of each location for the period of record is listed in Table 1.

Model Sensitivity to Rainfall Time Scale

Runoff prediction with infiltration-based models can be extremely complex. Opus, for example, uses time scales which adapt to current conditions (i.e., the time

TABLE 1. Mean Annual Rainfall by Location

Location	Record Period	No. Events	Mean Annual Rainfall (mm)
Albuquerque, NM	1939-74	1340	183
Reynolds Creek, ID	1962-81	5344	349
Safford, AZ	1939-75	1461	185
Walnut Gulch, AZ	1968-77	727	273

step is dynamic, changing within simulations). Runoff is a function of both rainfall rate and infiltration rate, and infiltration rate is a function of soil type, current soil moisture status, and rainfall rate.

The first step chosen in developing a rainfall-data generating model was to determine an appropriate time interval for the size of area to be simulated. For Opus, watershed area is on the order of a few hectares; in this case hourly data are too coarse for the small space scale, and frequently no runoff is predicted, due to extreme damping of rainfall rates.

Given available data recorded at 1-min intervals, a computer program synthesizes data that would have been recorded by a gage every 10 or 15 min. Opus runoff volume predictions from simulations using 1-min rainfall data are compared to those from the other time intervals. Opus simulations of 3-yr duration were run on a 1.3-ha watershed; space limitations preclude detailed discussion of the conditions simulated.

Table 2 shows that predicted annual runoff totals are usually smaller for longer time intervals than for shorter intervals. However, several events in the 3-yr simulation produced more runoff (and sometimes higher peaks) from the longer-intervals. Fig. 2 illustrates event runoff volumes predicted for the 3-yr period for all events where runoff exceeded 2.5 mm. Each event is represented as three bars, one for each time interval used to represent the event. The differences among input scales are strikingly varied and, as shown in Table 2, make significant differences in annual totals. The rationale is that just as longer time steps generally lower rain rates, sometimes they artificially raise

TABLE 2. Runoff Predicted Using Opus with Different Precipitation Time Intervals

Yr	Rain	Annual Total Runoff (mm) ----From Time Interval----		
	(mm)	1-min	10-min	15-min
1	709	97	85	89
2	1020	164	155	140
3	1226	132	131	128
Total	2955	393	372	354

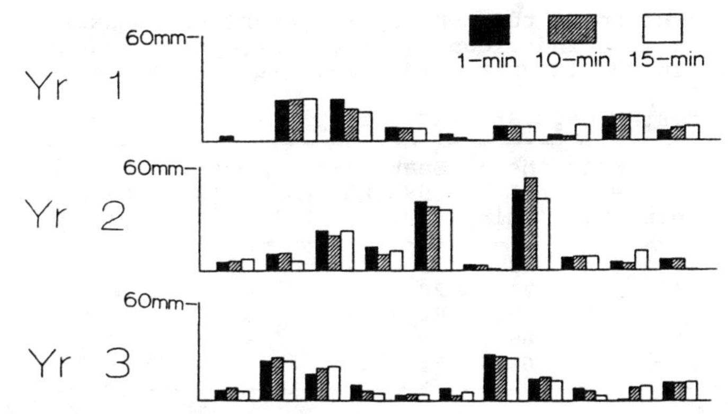

FIG. 2. Predicted Runoff from Rainfall Input of Various Time Scales

others. Thus the distribution of intensities within events strongly affects both runoff prediction and sensitivity to input scale. It is generally obvious why larger time intervals (lower intensities) produce less runoff, but it is not so obvious why larger scales can in some cases produce more runoff. In some events, high intensity rain falls on nearly dry soil and can be absorbed. This situation is accurately simulated by Opus using a 1-min description of the rain. If the high-intensity burst was split by the partitioning into longer time intervals, then it is possible that the first part of the burst would wet the soil. The simulation would then show the second part of the burst producing more runoff than it should have. These results indicate that event characterizations should be performed at the smallest possible time interval consistent with watershed response.

Event Statistic Sensitivity to Chart Scale

The above discussion focuses on the time interval of data to be generated by the rainfall model under development. The following addresses the variation in rain event statistics due to the chart speed (time scale) of the recorder (i.e., number of hours per revolution). Thus we change focus from the scale of proposed product to the scale of available data. The effect of chart speed on event statistics is investigated using data from five gages on the Walnut Gulch AZ watershed. The gages are chosen to minimize differences caused by elevation and distance. The greatest between-gage distance is 1.3 km; the elevation range is 34 m. Table 3 presents several event statistics for these gages. They clearly describe two separate populations: gages with 6-hr charts and those with 24-hr charts. The finer resolution of the 6-hr charts enables the user to detect many more individual events, with lower average rainfall, in the 10-yr record. The means of all event

TABLE 3. Nearby Gage Statistic Comparison

Gage	Events	Mean P (mm)	Mean ELAP (min)	Mean DRY (hr)	Mean RATEMX (mm/hr)	Mean TMAX (min)	Chart Scale (hr/rev)
384	1549	1.5	23	31	11	11	6
386	1716	1.5	21	28	11	11	6
23	696	3.8	68	89	13	24	24
27	676	3.8	69	90	15	25	24
83	727	3.8	69	84	13	26	24

characteristics indicate that chart speed plays an important role in event characterization.

Arid Climate Rainfall Events

Aridland precipitation in this study is characterized by statistical analysis of data from locations listed in Table 1. The data were recorded on 24-hr gages. Events are defined as periods of continuous precipitation (a 1-min hiatus indicates a new event).

Fig. 3 presents the monthly mean number of events per year, and the mean values of P, DUR, DRY, RATEMX, and TMAX. Strong seasonal trends are apparent in all variables. Climate similarities among the three Southwestern watersheds (Albq, Saff, and WG) are apparent. The number of events, P, and RATEMX increase dramatically in the summer. TMAX decreases in the summer, showing that if the event intensity distribution is modeled as a triangle, the peak is sharply skewed toward the start time in the summer.

The Idaho (Reynolds Creek) climate is shown to be different, with fewer events during summer and fall, and fairly constant event P throughout the year. RATEMX increases only slightly in the summer. Event hiatus (DRY) is relatively low and seasonally constant.

Conclusions

Opus simulations using synthetic rainfall data indicate strong sensitivity of runoff predictions to time interval of input rainfall data. Contrary to common belief, there are cases where runoff volume <u>increases</u> with increased rain event time interval.

Chart rotation speed strongly affects event statistics. Users of rainfall data must be aware of chart speeds when using data to develop rainfall and runoff models or to drive runoff models.

Strong seasonal trends were discovered in all variables. Event statistics for an Idaho watershed

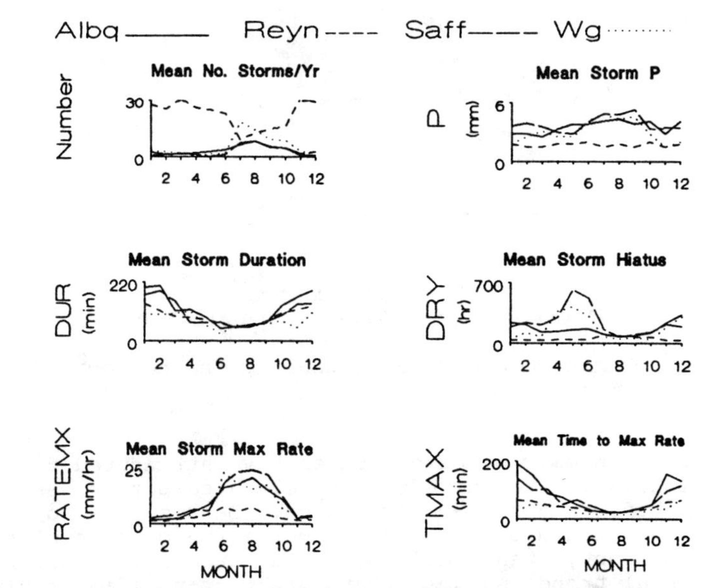

FIG. 3. Statistics of Aridland Event Variables for Albuquerque, NM (Albq), Reynolds Creek, ID (Reyn), Safford, AZ (Saff), and Walnut Gulch, AZ (WG).

indicate that although arid, the climate is quantifiably different from that of three Southwestern watersheds.

Acknowledgments

USDA-ARS data for this study were supplied by: J. Thurman, Beltsville, MD; C. Hanson, Boise, ID; and F. Lopez, Tucson, AZ. Without the continuing support of USDA-ARS, this immense resource would not exist. The assistance of Jeff Marshall, Colorado State University, in data manipulation and reduction is appreciated.

References

DeCoursey, D. G., Rojas, K. W., and Ahuja, L. R. (1989). "Potentials for nonpoint-source groundwater contamination analyzed using RZWQM." ASAE Paper No. 892563.

Smith, R. E. (in press). "Opus: an advanced simulation model for nonpoint-source pollution transport at the field scale, Vol. I. Model documentation." USDA-ARS.

USDA. (1980). "CREAMS: chemicals, runoff and erosion from agricultural management systems." W. G. Knisel, ed. USDA-SEA Conservation Research Report No. 26.

Woolhiser, D. A., Smith, R. E. , and Goodrich, D. C. (1990). "KINEROS, a kinematic runoff and erosion model: documentation and user manual." USDA-ARS-77.

Structural Control of Ground-Water Induced Debris Flows

Souren Ala[1] and Christopher C. Mathewson[2]

Abstract

A number of case studies have attributed debris flow initiation to the effect of elevated pore pressure from a bedrock ground-water source. Debris flows on slopes underlain by Precambrian metamorphic rocks of the Wasatch Front, Utah, fall into this category. Bedrock structural trends suggest that northwest-trending faults are avenues for ground-water flow. This may explain the higher number of debris flows on slopes intersecting fault traces.

Introduction: Failure Mechanisms

Antecedent rainfall of at least 25cm followed by storms with a rainfall intensity of 0.6cm/hr or greater initiated a series of damaging debris flows in the Santa Monica mountains of southern California (Campbell, 1975). These events took place in colluvial soils underlain by sedimentary, volcanic and low-grade metamorphic rocks ranging in age from Quaternary to Triassic (State of California Department of Natural Resources, 1954). The observed failure mechanism was a critical reduction of effective stress, due to a rise in pore pressure brought about by continued infiltration of surface water into saturated colluvium, at a rate which exceeded the hydraulic conductivity of the underlying bedrock.

An alternative mechanism for the initiation of debris flows involves the contribution of *upwelling*

---

[1]Graduate Student, Center for Engineering Geosciences, Department of Geology, Texas A&M University.
[2]Professor, Department of Geology and Director, Center for Engineering Geosciences, Department of Geology, Texas A&M University, College Station, TX 77843-3115.

ground water from permeable zones in the bedrock, rather than downward infiltration of water through the soil. If regional ground-water flow lines are projected onto a slope, the lower section of the slope is in a zone of discharge. Where low-permeability rock units or clays prevent discharge, pore pressure rises and the potential for slope failure is increased (Figures 1A and 1B). Hicks (1988) has observed a somewhat different mechanism in the Rogue River National Forest, Oregon, shown schematically in Figure 1C.

In cohesionless soils without a low-permeability basal layer, upwelling ground water can cause piping (Deere and Patton, 1971). This process has been recognized as a contributor to slurry flows (Howard and McLane, 1988).

Slope failures initiated by ground water from bedrock have generally been associated with heavy rainfall. Eisenlohr (1952) correlated these "blowouts" with layers of shattered rock recharged by rainfall on higher ground. Hack and Goodlett (1960) found "water blowouts" along the lower contact of an impermeable diabase sill within a hillside composed mainly of permeable clastic sedimentary rocks.

Everett (1979) observed that landslide sources on forested slopes in Mingo County, West Virginia were associated with the *upper* surfaces of relatively less permeable sandstones, interbedded with highly fractured coal beds. These events were, therefore, associated with perched rather than artesian water table conditions.

Evidence exists that artesian ground-water conditions helped initiate debris flows on slopes underlain by Precambrian metamorphic rocks of the Wasatch Front, Utah (Mathewson et al, 1990). In May and June of 1983 and 1984, the Wasatch Front was the site of numerous debris flows and floods. Many of these originated as small blowouts which gathered material during their progress down the channel. Failures were not correlated with heavy rainfall, but with rapid spring snowmelt. Several debris flow scars exposed new springs, which flowed for up to six months after failure. Mathewson et al (1990) proposed that hydrostatic head in the fractured bedrock, combined with local topographic drive, led to elevated pore pressures in the axes of upper mountain swales (Figure 1B). This hypothesis has been confirmed in at least one case: a study by Monteith (1989) showed that a landslide in this area was initiated under artesian water table conditions.

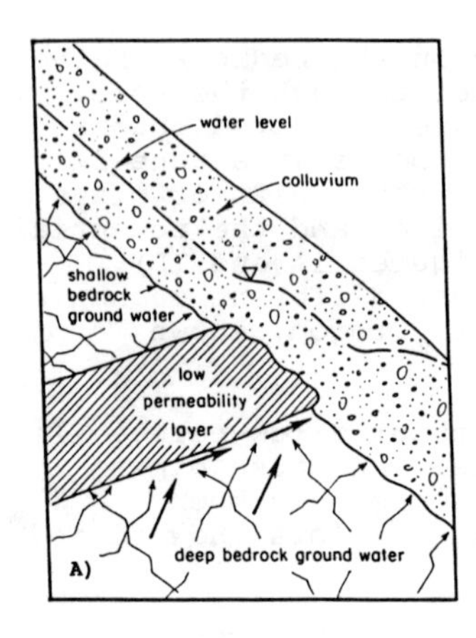

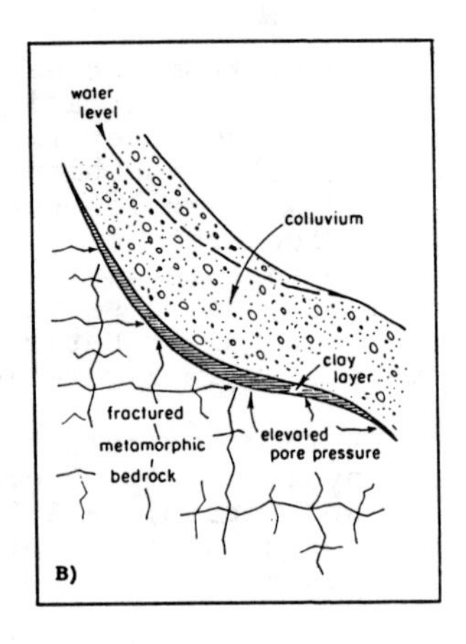

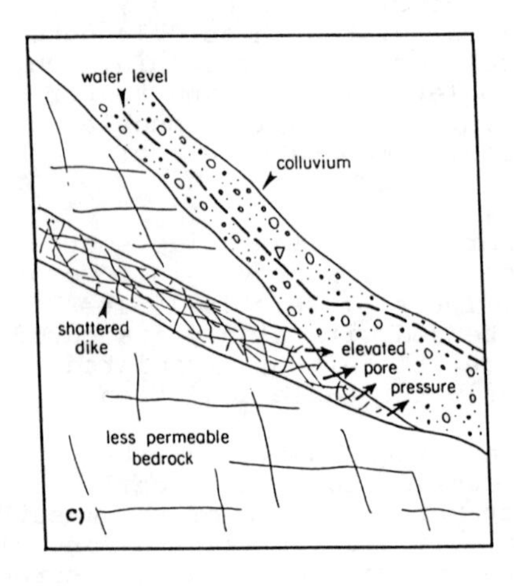

Figure 1. Elevated pore pressure due to ground-water source: **A)** Changes in bedrock permeability, and **B)** Topographic drive; adapted from Mathewson et al, 1990. **C)** A highly permeable layer oriented downslope acts as a conduit for ground water.

<u>The Study Area</u>

The Wasatch Front comprises the western flank of the Wasatch Range, located in north-central Utah. A series of steep westward-draining canyons have eroded down into this mountain block; those included in this study are shown in Figure 2.

The bedrock of this part of the Wasatch Front is comprised of the Precambrian Farmington Canyon Complex (FCC), a heterogeneous suite of competent to highly sheared and weathered metamorphic rocks (Bryant, 1989).

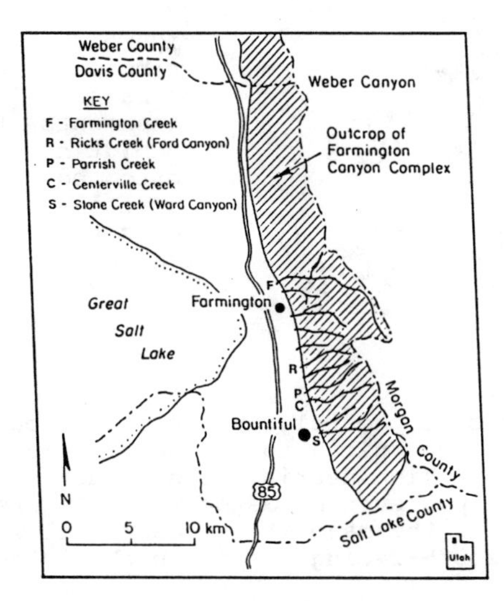

Figure 2. The study area is located between Farmington and Ward Canyons, on the Wasatch Front, Utah.

These rocks have been metamorphosed to a crystalline texture; porosity is defined by fractures and partings along foliation planes.

A geologic history of repeated deformation of the FCC has created through-going faults and/or fracture systems that may conduct large volumes of water. The mountain block supplies abundant recharge to alluvial and lacustrine aquifers at its base. Gateway tunnel, dug through bedrock of the FCC at the base of the Wasatch Front in 1953, had a continuous ground-water yield of between 19 and 38 l/s (Feth, 1954).

Bedrock Structural Trends

An investigation was conducted by Ala (1990) to characterize bedrock structure (fractures, faults and foliation) and its influences on the hydrologeology of the FCC. The goal of this study was to contribute to a predictive model of debris flow susceptibility in the area shown in Figure 2.

Photogeologic analysis indicated that major lineaments (interpreted to be faults) in the study area generally trended toward azimuths 175°, 185° and 295° (Figure 3A). In contrast, the principal trend of fractures was toward azimuth 223° (Figure 3B).

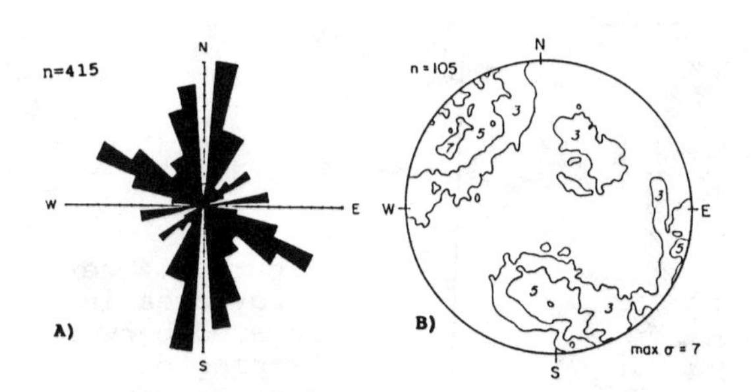

Figure 3. A) Principal trend of major lineaments. **B)** Schmidt net of Contoured densities of poles to fractures.

Discussion

In order to assess the relative contribution of fractures and faults to the initiation of debris flows, the aspects of slopes on which debris flows occurred have been measured for 83 mapped events, as shown in Figure 4. The results indicate that the majority of debris flows occurred on generally south-facing slopes, probably due

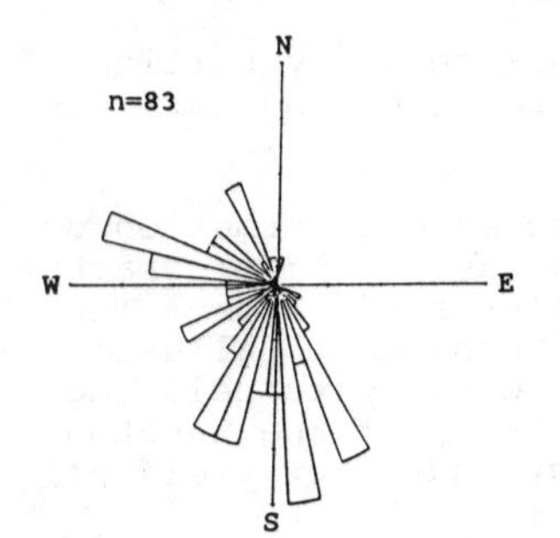

Figure 4. Rose diagram showing the aspect of slopes on which a debris flow (or shallow soil slip leading to debris flow) took place.

to climatic conditions. However, within this general trend, the number of debris flows is greater on slopes facing azimuths 175° and 290°. These correspond to the orientation of major lineaments in the study area. There is no increase in debris flow occurrences for slopes facing azimuth 223°, the main trend of fractures.

Summary

Bedrock ground water is a contributor to the debris flow process in mountainous terrain. The distribution of shallow soil slips and debris flows on slopes underlain by the FCC apppears to be partially controlled by anisotropic ground-water flow parallel to major structural lineaments. The orientations of fractures at the surface do not seem to be related to cross-slope movement of ground water.

Appendix A - References

Ala, S., 1990, Analysis of Bedrock Structural Fabric on the Wasatch Front, Utah, in Robinson, L. (editor), Engineering Geology and Geotechnical Engineering, Proceedings No.26, Idaho State University, Pocatello, Idaho, p.22-(1-17).

Bryant, Bruce, 1988, Geology of the Farmington Canyon Complex, Wasatch Mountains, Utah, U.S. Geological Survey Professional Paper 1476, Washington, D.C.

Campbell, R.H., 1975, Soil Slips, Debris Flows, and Rainstorms in the Santa Monica Mountains and Vicinity, Southern California, U.S. Geological Survey Special Paper 851, U.S. Geological Survey, Denver, CO, 51 p.

Deere, D.H. and Patton, F.D., 1971, Slope Stability in Residual Soils. In Proceedings of the Fourth Panamerican Conference on Soil Mechanics and Foundation Engineering, Vol. 1, American Society of Civil Engineers, New York, p.87-170.

Eisenlohr, W.S. Jr., 1952, Floods of July 18, 1942, Pennsylvania. In Notable Floods of 1942-3, U.S.Geological Survey Water Supply paper 1134-B, Washington, DC, p.75-79.

Everett, A.G., 1979, Secondary Permeability as a Possible Factor in the Origin of Debris Avalanches Associated with Heavy Rainfall, Journal of Hydrology, No. 43, p.347-354.

Feth, J.H., 1964, Hidden Recharge, U.S. Geological Survey Water Resources Division, p.14-17.

Hack, J.T. and Goodlett, J.C., 1960, Geomorphology and Forest Ecology of a Mountain Region in the Central Appalachians, U.S. Geological Survey Professional Paper #347, p.41-56.

Hicks, B.G., 1988, Geologist, Rogue River National Forest, Medford, Oregon, Personal Communication.

Howard, A.D., and McLane, C.F. III, 1988, Erosion of Cohesionless Sediment by Groundwater Seepage, Water Resources Research, Vol. 24, No. 10, pp.1659-74.

Mathewson, C.C.; Keaton, J.R.; and Santi, P.M., 1990, Role of Bedrock Ground Water in Debris Flows and Sustained Post-Flow Stream Discharge, The Bulletin of the Association of Engineering Geologists, Vol. XXVII, No.1.

Monteith, S., 1988, Stability Analysis of the Steed Canyon Landslide, Unpublished Master of Science Thesis, Department of Civil and Environmental Engineering, Utah State University, Logan, Utah.

State of California Department of Natural Resources, 1954, Geology of Southern California, Bulletin 170, Division of Mines, San Francisco.

INITIATION AND LAWS OF MOTION OF DEBRIS FLOW[1)]

Wang Zhaoyin[2)] and Zhang Xinyu[3)]

Abstract

The mechanism of initiation and laws of motion of debris flow are studied experimentally, by allowing water or clay muds, with different clay concentrations and at different flow rates, to flow over a flume bed piled with gravels (d=4-25 mm), and observing and measuring movement of gravels, flow of muds and interaction of the two phases. As the flow rate is high enough gravels on the bed are scoured and then move in saltation or roll on the bed. A large amount of gravels concentrates in the front of the flow wave and forms a high head of debris flow like a bulldozer. Initiation of debris flow depends mainly on the energy of liquid phase supplying to gravels. The concentration of gravels in the head is much higher than that in trunk zone while speed of the head is much smaller than velocity of liquid or gravels in trunk zone. Large gravels are acted by a large tractive force from the liquid phase and a small resistance force from collisions with other particles, therefore, they move faster than small ones and concentrate in the head consequently.

Experiment procedures

The experiment was conducted in an 8.7m-long, 10cm-wide and 20cm-deep plexiglass flume. The bed slope could be adjusted in the range of J=0-0.2. The liquid phase was water or muds with different concentrations of clay. The rheological properties of the clay muds can be approximately characterized by the equation

$$\tau = \tau_B + \eta \dot{\varepsilon} \qquad (1)$$

where τ and τ_B are shear stress and yield shear stress, respectively; η is rigidity coefficient; $\dot{\varepsilon}$ is shear rate which equals velocity gradient. τ_B and η increase with increasing clay concentration and roughly observe the following empirical formulas

$$\tau_B = 5.4 \times 10^5 C_{V0}^{4.1} \ (\mathrm{dyn/cm^2}) \qquad (2)$$

$$\eta = \eta_0 (1 + 10^{15 C_{V0}}) \qquad (3)$$

---

1) The project supported by National Natural Science Foundation
2) Dr. Eng., Senior Engineer, Institute of Water Conservancy and Hydroelectric Power Research (IWHR), P.O.Box 366, Beijing, China
3) B. Sc., Engineer, IWHR, P.O.Box 366, Beijing, China

where η_0 is viscosity of water and C_{vo} is volume concentration of clay.

Gravels with diameters from 4 mm to 25 mm and median diameter d_{50}= 10 mm, were put on the flume bed and formed a 10 cm-thick gravels bed before the experiment. Water or clay muds flowed over the gravels bed and took some gravels into the flow in the experiment. The gravels carried by the flow were separated from the liquid in a tail tank and the liquid was recirculated by a pump. A magnetic flowmeter was used to record the flow rate. Velocity of the liquid phase was measured by an electro-magnetic velocity meter and gravels velocity was analysed by means of photography.

Initiation of debris flow

As the flow rate is high enough the turbulent liquid flow can trigger a mass of gravels to enter into motion. Many gravels concentrate in front of the flow and form a steep, high bore head of debris flow. The speed of the head is much lower than velocities of liquid phase and gravels in trunk zone of the debris flow. As the flow rate is low and the slope of the channel is gentle or the clay concentration is very high, the flow becomes laminar and very few gravels join the flow. The front of the flow is low, flat and free of gravels in this case. The speed of the front is higher than liquid in trunk zone. The former described above is refered to debris flow and the latter is atributed to ordinary sediment-laden flow. Initiation of debris flow depends mainly on flow rate of liquid in unit width,q, and slope of the channel bed, J. Fig.1 shows critical conditions for initiation of debris flow, in which some points from Takahashi (1978) are also plotted for comparison. It can be seen that the curves to divide debris flow zone and sediment-laden flow zone in q-J coordinates plane for different liquid phases are hyperbolas, which suggests a criterion of initiation of debris flow, K=γqJ. The critical condition of debris flow could be given by

$$\gamma q J \geqslant K_c \qquad (4)$$

where γ is specific weight of the liquid phase.

γqJ can be interpreted in view of

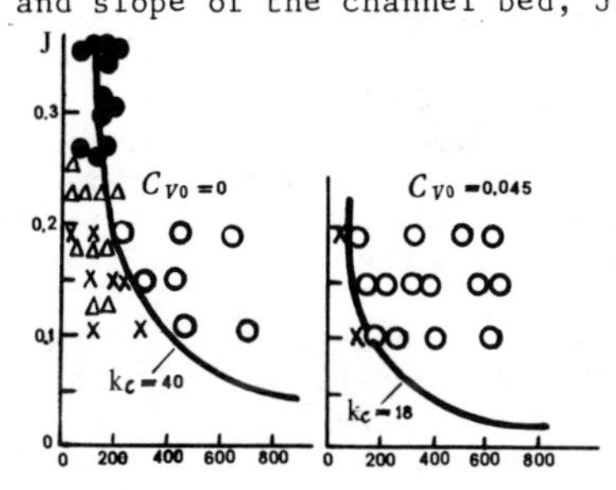

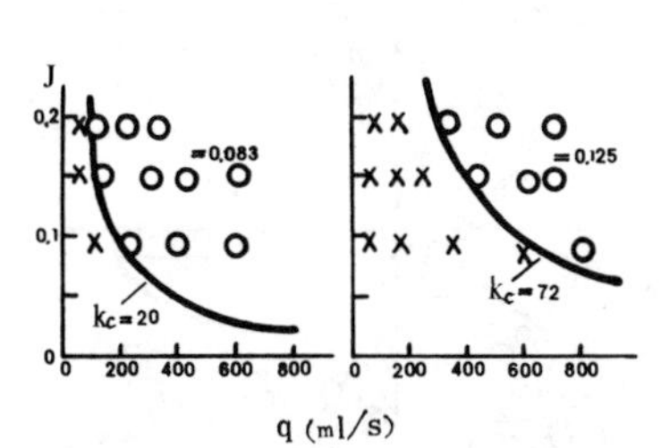

Debris flow (o Wang et al.
● Takahashi)
Sediment-laden flow (x Wang et al.
▲ Takahashi)

Fig.1 Critical conditions of initiation of debris flow

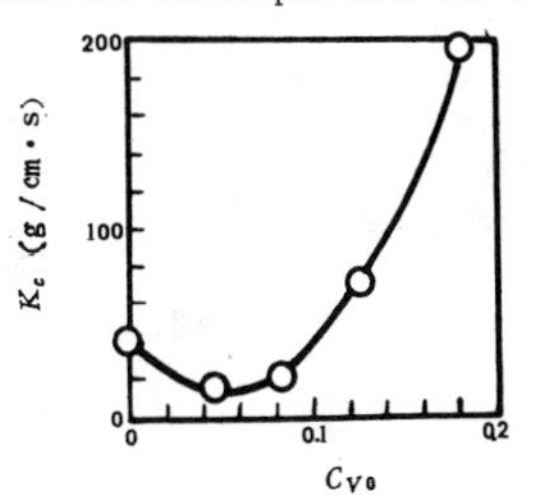

Fig.2 Criterion K_c versus clay concentration C_{vo}

physical meaning as energy of the liquid supplying to the flow in unit time and K_c is the minimum energy for triggering debris flow. From those shown in fig.1, K_c varies with clay concentration C_{vo}. Fig.2 shows K_c versus C_{vo}. Gravels in the flow move as bed load, they collide with each other consecutively during their movement. The collisions result in a dispersive force to support their effective weight, and at the same time result in a large resistance force (Wang and Qian, 1985). The liquid phase acts a tractive force on the gravels to balance the resistance force, which is given by

$$F = 3\pi d\eta_e u_r \qquad (5)$$

where η_e $(=\eta + \tau_B/\dot{\varepsilon})$ is effective viscosity of the liquid, d is diameter of gravels and u_r is relative velocity between the particle and ambient liquid. Increase in C_{vo} enhances the tractive force on gravels and consequently reduces K_c, so that K_c reduces with increasing C_{vo} as C_{vo} is less than 0.06.

As C_{vo} is larger than 0.06, however, K_c increases with increasing C_{vo} very fast, which means that the thicker the mud is, the more difficult the initiation of debris flow will be. This could be interpreted from the fact that the thicker the mud is, the more liable the flow to maintain in laminar state. Energy diffusion in laminar flow is much weaker than in turbulent flow. Only the liquid near the bed and contacting with gravels delivers kinetic energy to gravels, so that the upper part of the flow consumes little energy on bed load movement and flows at a high velocity. Consequently a higher total energy of the flow is needed for triggering debris flow as C_{vo} is over 0.06.

The smallest K_c falls in the range of C_{vo}=0.04-0.08. The Jiangjia Gully in South China is a notorious debris flow gully. Several tens debris flows occur annually in the gully. It was recorded that most of these debris flows have liquid phases of clay concentration in the range of C_{vo}=0.04-0.09 which strongly supports the conclusion from fig.2 (Kang,1985).

Fig.3 presents velocity distributions of gravels (u_b) and liquid (u) in turbulent debris flow and laminar sediment-laden flow. The flow rates in the two runs are nearly the same. The flow with thin mud (C_{vo}=0.045) is turbulent, it scours the gravels bed into a considerable depth and as a result the whole flow depth is nearly doubled. The liquid flow consumes a lot of energy on carrying gravels and its velocity distribution is very ununiform. The flow with thick mud (C_{vo} =0.181) is laminar, its upper part is a flow core with zero velocity gradient and a high, uniform velocity, u_c. Only gravels on the bed surface are acted by shear stress of the flow and very few particles are carried by the flow.

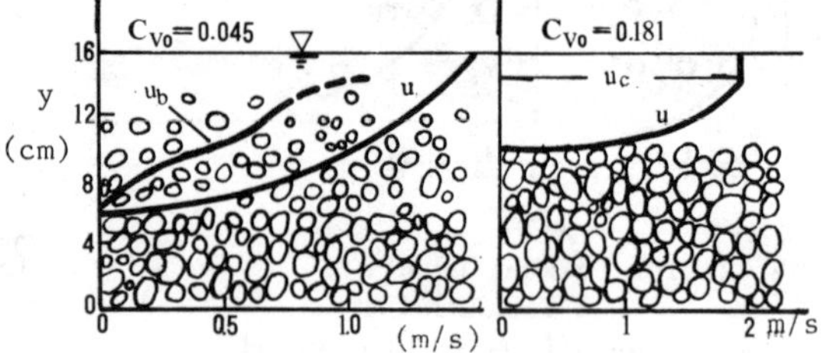

(a) Turbulent flow (b) Laminar flow
Fig.3 Comparison between velocity profiles in turbulent debris flow and laminar mud flow

These results indicate that the initiation of debris flow depends mainly on impact of turbulent flow on bed materials and internal energy diffusion of the liquid flow, but not only on tractive shear stress.

Mechanism of debris flow

(1) **Basic physical picture** A debris flow could be divided into three parts, namely bore head, trunk zone and tail zone (Wang and Qian, 1989). Liquid and gravels in the trunk zone move faster than the bore head in debris flow. The gravels catch up with and roll over the head and then stop over on the bed. Liquid in the trunk zone flows at a velocity about double of the speed of the head (see Table 1), it flows through the bore head and delivers its kinetic energy to gravels and then falls on the bed and scours bed gravels. The bore head rolls forward like a bulldozer. Collisions of gravels in the bore head make noise and consume a lot of energy. The liquid flow functions like a energy conveyer belt and transports energy to the head continuously. We checked the theory by cutting supply of liquid just after a debris flow had been initiated and found that the bore head reduced gradually and gravels in the head stopped.

(2) **Development of bore head of debris flow** Following liquid flow scouring the gravels bed the front of the flow develops into a bore head. Fig.4 shows the growth of debris flow head in pace with propagation of the flows in several runs. The height of debris flow head increases fast in the first 2-4 meters and from thereon keeps constant. It seems that the larger the C_{vo}, the longer the travel distance for the debris flow head to reach into equilibrium. The flow with C_{vo}=0.181 is laminar and there is no gravel in the front. It propagates fast and the height of the front reduces along its course.

(3) **Gravels concentration in bore head** Gravels concentration in bore head of debris flow, S_b,(in kg/m^3), is 30-60% higher than that in the trunk zone. S_b increases with increasing J and q in general, as shown in fig.5. S_b decreases with increasing C_{vo}, especially as C_{vo} is larger than 0.083, which implies that the scouring capability of bore head reduces with increasing clay concentration of the liquid phase.

(4) **Propagating speed of debris flow** Bore head of debris flow rolls forward along the channel. Its speed, U_b, is much less than the velocity of li-

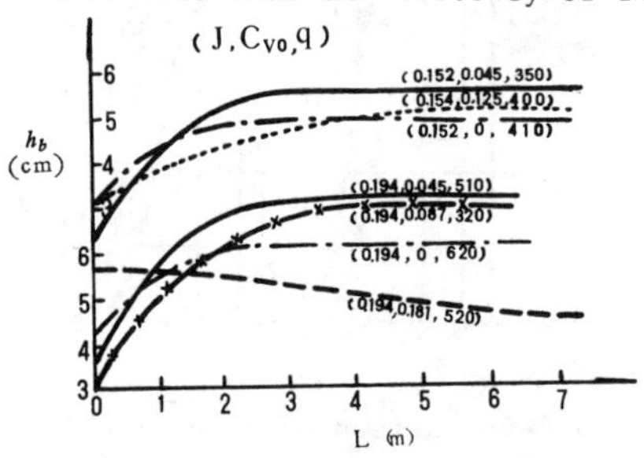

Fig.4 Development of height of head (h_b) along the flow course

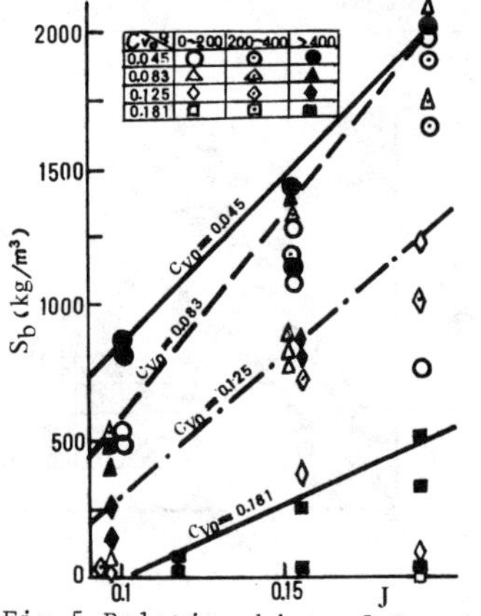

Fig.5 Relationships of S_b with J, q and C_{vo}

Table 1 Velocity of debris flow and relative parameters

J	C_{V0}	q (ml/s·cm)	u_m (m/s)	U_b (m/s)	S_b (kg/m³)	S (kg/m³)	h_b (cm)	h (cm)	d_{b50} (mm)	d_{50} (mm)	q/h_b (ml/s·cm²)
0.103	0.045	140	0.85	0.21	540	317	4.0				35
		250	1.21	0.49	533	432	5.0	4			50
		420	1.77	0.66	807	477	6.3	5.0			67
		600	1.83	0.68	814	573	7.1	5.8			85
0.096	0.079	190	1.46	0.56	528	271	3.8	2.5			50
		410	1.83	0.85	402	321	4.8				85
		620	2.02	0.99	494	331	5.5				113
		730	2.21	1.21			6.0				122
0.152	0.045	200	1.26	0.41	1119	785	4.8	3.5	12.7	11.4	42
		350	1.32	0.66	1299	927	5.5	3.8	13.0	10.4	64
		630	2.21	0.85	1154	827	6.5	4.8	11.2	9.2	97
		570	2.27	0.79	1241	944	6.0		11.0	9.7	95
0.152	0.079	85	0.79	0.32	805	565	4.0				22
		180	0.97	0.33	847	548	5.0	3.0	14.1	10.6	36
		285	1.31	0.62	1338	866	5.2		12.0	10.5	55
		440	1.77	0.72	872	615	5.6	4.0	13.4	9.7	79
		590	2.50	1.20	1406		6.5	5.0	13.0	9.3	91
0.154	0.125	150	1.72	0.86	873	545	4.0	2.5			37
		400	1.82	1.09	769	635	5.1	3.5			80
		600	2.51	1.28	857	653	6.6	5.0			91
		650	2.54	1.64	765	514	6.6	5.0	11.8	10.0	98
0.194	0	430		0.69	1890		5.5	3.0			78
		620		1.15			6.2	5.0			100
0.191	0.045	510	3.21	1.30	2000		7.2	4.0			71
		300		0.81	2000		5.8	2.5			52
0.194	0.086	180		0.90	1700		5.0	2.5			36
		320		1.00	2010		7.0	3.0			45
0.191	0.132	320	1.74	0.99	1090	851	5.5	3.0			58
		500	2.26	1.29	1218	792	6.8	3.5			74
0.191	0.125	715	2.35	1.32	1216	625	7.6	5.0	12.2	10.3	94

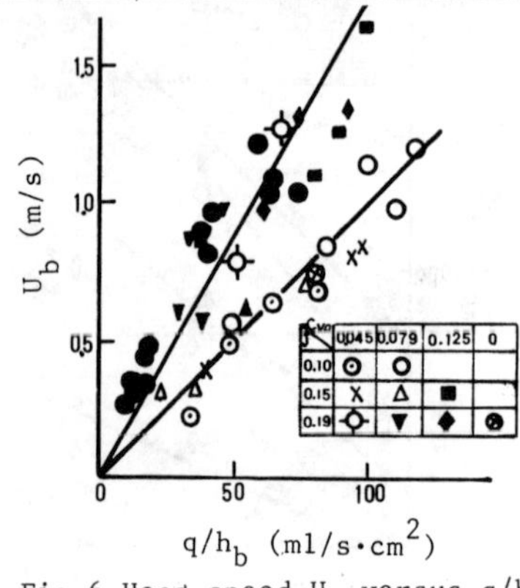

Fig.6 Head speed U_b versus q/h_b

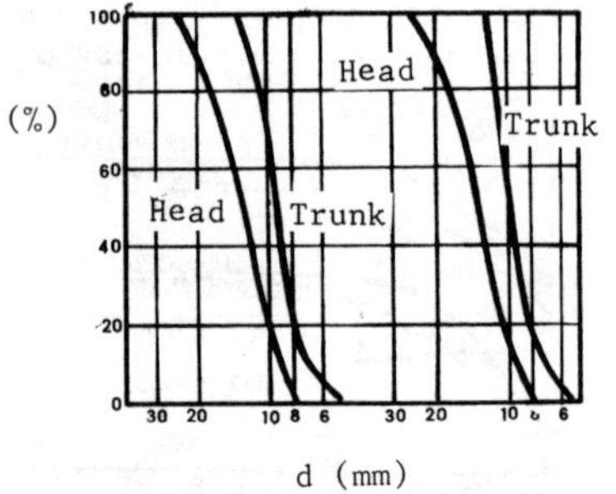

Fig.7 Comparison of size distributions of gravels in bore head and trunk zone of debris flow

quid or individual gravel. Table 1 presents measured U_b and relative parameters, in which u_m is surface liquid velocity in the trunk zone; h and h_b are average flow depth and height of bore head, respectively; S_b and d_{b50} are concentration and median diameter of gravels in bore head; S and d_{50} are concentration and median diameter of gravels in trunk zone. It can be seen that the propagating speed of debris flow is only one quarter to one half of the surface liquid velocity, and it increases with increasing q and decreases with increasing h_b. Fig. 6 shows U_b versus q/h_b. It is obvious that U_b varies linearly with q/h_b, and the larger the sum of J and C_{vo}, the larger the proportion of U_b to q/h_b.

(5) **Mechanism of large gravels concentrating in bore head** Many witnesses of debris flow reported that gravels in bore head of debris flow are much coarser than gravels in trunk zone. Although the gravels used in this experiment are quite uniform, the phenomenon of large gravels concentrating in bore head is apparent. Fig.7 gives a comparison of size distributions of gravels in bore head and in trunk zone, which suggests that most particles larger than 12 mm concentrate in bore head. We designed a special experiment to investigate the mechanism of large gravels concentrating in bore head by using tracer particles. The main results are briefed as follows: Large gravels roll on the bed at a stable velocity and their movement is little affected by collisions with small particles, while small particles move in saltation, sometimes jump into upper zone and move at a high velocity and sometimes fall in lower zone and move at a low speed. Once a small particle collides with a large gravel, it may slows down sharply or possibly stops. Gravels can move in debris flow because they are acted by a tractive force from the liquid phase. Eq. (5) indicates that the tractive force is in proportion to gravel's diameter. On the other hand, resistance against gravel's motion mainly comes from collisions with bed gravels. Collision of large gravel with bed gravels does not change the gravel's momentum in flow direction a lot, but a small particle may totally lose its momentum component in flow direction at one collision with bed gravel, as shown in fig.8. Large gravels are acted by larger tractive force and smaller resistance force, therefore, they move faster than small ones and concentrate in bore head.

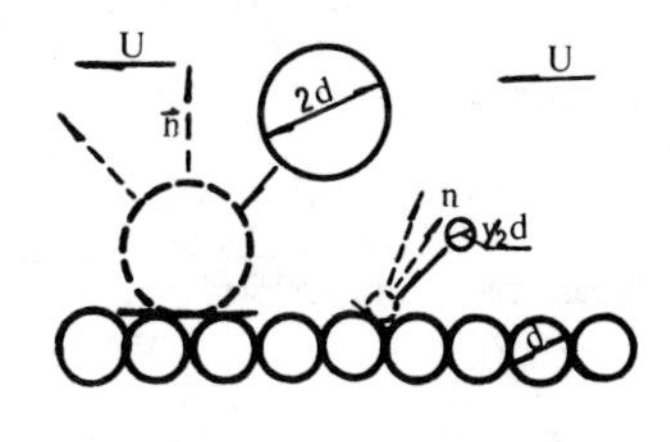

Fig. 8

References

(1) Kang Zhicheng, Characteristics of flow patterns of debris flow at Jiangjia Gully of Dongchuan in Yunnan, Science Press, 1985, pp.97-107 (in Chinese).

(2) Takahashi, T., Mechanical characteristics of debris flow, J.of Hydraulic Division, ASCE, 1978, HY8, pp. 1153-1169.

(3) Wang Zhaoyin and Qian Ning, A preliminary investigation on the mechanism of hyperconcentrated flow, Proc. of Intern. Workshop on Flow at Hyperconcentration of Sediment, IRTCES, 1985, pp.II4-1-16.

(4) Wang,Z. and Qian N., Characteristics and mechanism of debris flow, Proc. 4th Symp. on River Sedimentation, IRTCES, 1989, 722-729.

Geomorphology and Sedimentology of a Valley Fan, Southern Utah

K.J. Fischer[1], M.D. Harvey[2]

Abstract

Sink Valley, located in semi-arid west-central Kane County, Utah, lies on an erosional surface of Cretaceous age Tropic Shale. The valley contains an example of a previously undocumented fan type, a valley fan. Morphologic, morphometric, and sedimentologic characteristics of the valley fill confirm the valley fan interpretation. Cross sections are convex in shape and the longitudinal profile is segmented. The relationship between fan area and source area for the Sink Valley fan and two tributary fans is intermediate between that of fans of the humid eastern U.S. and the more arid western U.S. Sediment transport is dominated by sheet flooding in the distal region of the fan, whereas mudflows and debris flows are significant depositional processes in the proximal and medial regions of the fan.

Introduction

Alluvial fans are generally considered to be cone-shaped depositional features that are located where streams emanate from high relief sediment source areas onto a piedmont surface. Fans form at the margins of steep basins due to a decreased confinement of streamflow (1). Sink Valley is atypical of a valley formed in a uniform lithologic setting either by fluvial processes, or tectonic deformation. The shape of the valley is predominantly controlled by the resistance to erosion of the rocks that comprise the valley margins. Mass wasting of the bounding Cretaceous-age Tropic shale has formed an anomalously wide valley in a region which is dominated by narrow canyons bounded by indurated sandstones. A topographic gradient of 600 ft/mile extends from Sink Valley north to the Paunsaugunt Plateau. The rim of the plateau consists of a nearly vertical 400 ft high escarpment (Sunset Cliffs) of Eocene-age Wasatch

---

[1,2]Geomorphologists, Water Engineering & Technology, Inc., P.O. Box 1946, Fort Collins, CO 80522, USA.

Formation. Massive debris flow and landslide deposits derived from erosion of the cliffs have formed a prograding valley fan in Sink Valley which is laterally constrained by the Cretaceous outcrop. Surveyed profiles and 37 measured stratigraphic sections were obtained for Sink Valley to evaluate the processes responsible for the deposition of the valley fill (Fig. 1).

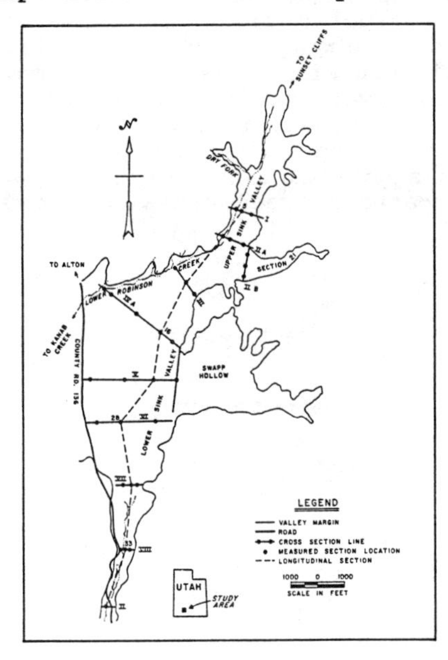

Figure 1. Location map showing surveyed cross sections and measured stratigraphic sections.

Fan Morphology

The upper reach of Sink Valley at the confluence of Dry fork and Water Canyon (Fig. 1) is narrow (900 ft) and the valley margins are composed of resistant sandstones of the Cretaceous-age Straight Cliffs Formation. Tropic Shale forms the basin boundary for almost the remainder of Sink Valley in the down-valley (southern) direction. Valley width within the confines of the Tropic Shale, increases from 900 feet at cross section I to 3200 feet at cross section IV and then decreases to about 700 feet at cross section VIII (Fig. 1). The Dakota Formation crops out down-valley from cross section VIII; valley width at cross section IX is about 500 feet. The first morphologic indication that the floor of Sink Valley might be composed of a valley fan is seen on the Alton 7.5 min. Quadrangle (USGS, 1966). Broadly convex 40-foot contours suggest that a fan-shaped body has prograded down valley from the confluence of Dry Fork and Water Canyon.

To substantiate the valley fan hypothesis, longitudinal and cross-section profiles of Sink Valley were constructed from surveyed data. A longitudinal profile extending from the base of the Sunset Cliffs down through the valley was constructed from the surveyed cross sections and the Alton Quadrangle map (Figure 2). The profile is highly irregular in Water Canyon, reflecting the presence of over-lapping debris flow lobes. The profile of the valley floor shows that the toe of the fan is located in the vicinity of cross section V, where there is a significant change in slope. The crest of the fan is located between cross sections III and IVA.

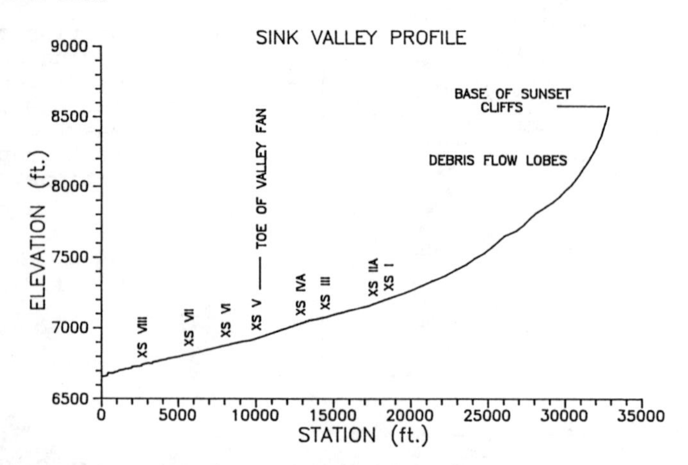

Figure 2. Longitudinal profile of Sink Valley.

The Sink Valley cross section profiles are consistent with those of previously described fans (Bull, 1964). The surface profiles of Cross Sections I and IIA are highly irregular due to the presence of debris flow and mudflow deposits in the upper fan region. A distinct convexity is present in Cross Sections III and IVA.

Fan Morphometry

Several investigators have demonstrated that the size of drainage basins and the surface area of associated alluvial fans are related (1, 2). The fan area-drainage basin area relationships for three fans in the location of Sink Valley were determined from the USGS topographic map: (1) Sink Valley fan, (2) Section 21 fan, and (3) Swapp Hollow fan. The best-fit line for the Sink Valley data is subparallel to the line of Bull (Fresno, CA), and lies above the best-fit line for data from Nelson County, VA (Fig. 3).

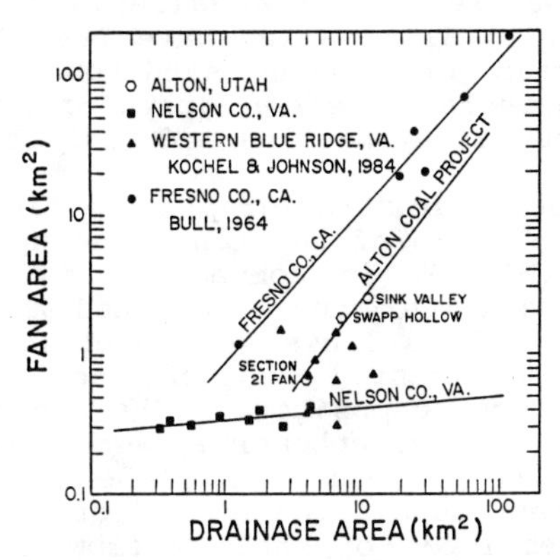

Figure 3. Fan area vs. drainage basin area for data sets from Fresno, CA, Nelson County, VA and Sink Valley, Utah (3).

The data in Figure 3 show that there is a systematic relationship between fan and drainage basin areas in Sink Valley. The down-valley fan margins were determined from topographic and sedimentologic evidence. Although lateral confinement caused by the gradually widening valley has caused the three fans to prograde down valley, a fan area-drainage area relationship has been maintained. In addition, the difference in the fan area-drainage basin area relationship between Sink Valley (UT), Fresno (CA), and Nelson County (VA), suggests that climatic differences may in part control the relationship, as annual precipitation in Alton, Utah is 16.7 inches, which is intermediate between the other two sites.

Fan Sedimentology

In order to determine the sedimentologic characteristics of the valley fill deposits of Sink Valley, 37 stratigraphic sections were measured and described along 9 cross sections which were surveyed across the valley in orientations perpendicular to depositional strike (Figure 1). Two of the cross sections were surveyed across the mouths of tributary fans. Because Tropic shale weathers to dark brown silt and clay, weathered Cretaceous sandstones are yellow-grey, and sediment derived from the Eocene Wasatch Formation (Sunset Cliffs) are orange-red-brown in color, the three sediment source areas were readily discernable in the valley fill stratigraphy. This source area delineation is critical to the determination of sedimentation processes in Sink Valley. The Wasatch-derived sediments in Sink Valley have been transported down-valley whereas Cretaceous sandstone and Tropic Shale-derived sediments reflect a lateral contribution of sediments from the valley margins and tributary fans.

Figure 4 shows four stratigraphic sections that lie on the depositional axis of the fan. The upper part of the valley fan (MS #5) is composed of predominantly Wasatch-derived, relatively coarse-grained sediments that were deposited by debris flows and sheetfloods. Minor channel fills are located within the horizontally laminated sheetflood deposits. The channels were probably small-scale and ephemeral, and they probably represent a braided pattern during sheetflood events. Coarse-fine couplets consisting of mud drapes over sand units record sheetfloods of shallow flow depths. The medial portion of the valley fan is represented by MS #16. The sediments were deposited by mudflows, debris flows and sheetfloods. However, there is a dilution of the Wasatch-derived sediments by Tropic-derived and Cretaceous sandstone-derived sediments. These sediments probably originated from the tributary fans and the valley margins. Further downvalley, MS #28 marks the approximate location of the distal portion of the fan.

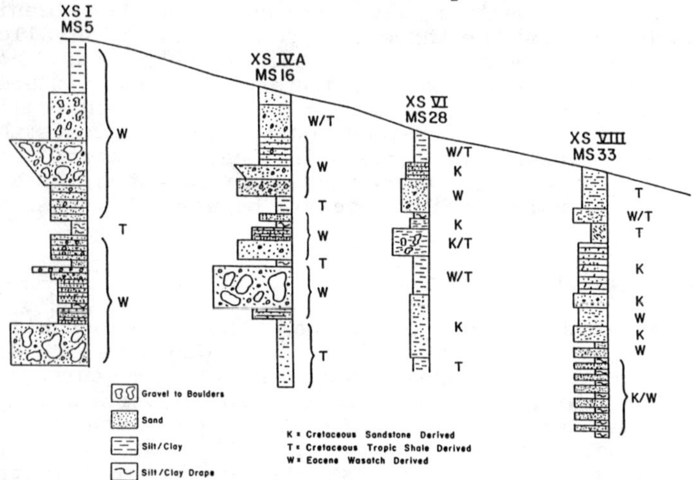

Figure 4. Stratigraphic sections on depositional axis of Sink Valley fan.

Both the volume and caliber of the Wasatch-derived sediments have declined significantly. The Wasatch-derived sediments are fine-grained and document discrete down-valley flood events. Beyond the toe of the fan, MS #33 contains sediment which is derived predominantly from the Cretaceous age sandstones. The increasing dominance of these sands and the concomitant reduction in the amount of Wasatch-derived sediments may reflect the fact that the valley fan did not prograde far enough down-valley to deliver significant volumes of sediment to the lower valley. As the Wasatch source area has since been captured by Lower Robinson Creek, fan progradation has ceased.

Summary and Conclusions

The combined morphologic, morphometric, and sedimentologic data indicate that the floor of Sink Valley is composed of a laterally confined alluvial fan. The shape of the deposit is convex upwards in cross section and segmented in longitudinal profile. The main valley fan and two tributary fans have maintained a consistent fan area-drainage basin area relationship through progradation. Stratigraphic data show that whereas the upper fan is characterized by debris flow and sheetflood deposits which are derived almost entirely from the Sunset Cliffs, the lower fan contains relatively fine grained, thinly bedded sediments derived from primarily Tropic Shale and Cretaceous age sandstone. These sediments were deposited during large flood events or represent lateral input from the valley margin. Failure to recognize valley fans in the past may have led to an underestimation of the importance of alluvial fans and processes in the semi-arid west.

References

1) Bull, W.B., 1977, The alluvial fan environment. Progress in Physical Geography, v. 1, p. 222.

2) Bull, W.B., 1964, Alluvial fans and near-surface subsidence in western Fresno County, California: U.S. Geol. Surv. Prof. Paper 437-A, pp. 79-129.

3) Kochel, R.C. and Johnson, R.A., 1984, Geomorphology and sedimentology of humid-temperate alluvial fans, central Virginia, _in_ Kosta, E.H. and Steel, R.J. (eds.), Sedimentology of Gravel and Conglomerates, Can. Soc. Petrol. Geol. Mem. 10, p. 109-122.

PREDICTING ALLUVIAL-FAN SEDIMENT-WATER SLURRY CHARACTERISTICS AND BEHAVIOR FROM SEDIMENTOLOGY AND STRATIGRAPHY OF PAST DEPOSITS

Jeffrey R. Keaton[1], M. ASCE

Abstract

Sediment-water "slurries" theoretically can range from clear water (0% solid) to pure solid (0% water), and their behavior varies with sediment concentration. At low sediment concentration, slurry behavior is Newtonian and its ability to transport sediment is a function of velocity or strain rate. As a Newtonian fluid slows, coarser grains settle before finer grains, resulting in a clast-supported, normally graded (fining upward) fluvial deposit. As sediment load increases to hyperconcentration, a threshold is exceeded and the slurry no longer behaves strictly as a Newtonian fluid. So much sediment is included in the slurry that coarser grains cannot settle as it slows, resulting in a clast-supported, but ungraded, deposit. Flows of higher sediment concentration are transitional between hyperconcentrated and debris flow and produce deposits that are partly clast-supported and partly matrix-supported. At higher sediment concentrations, but in which the water content still exceeds the liquid limit, flows behave as debris flows and deposits are matrix-supported and commonly contain megaclasts if they are available in the source areas of the flows.

Slurry characteristics can be predicted by a combination of field and laboratory analyses. Field observations are needed to determine deposit stratigraphy and to estimate the abundance of clasts too large to evaluate in the laboratory (> 50 mm diameter). Laboratory analyses are needed to estimate (1) clast density, (2) grain size distribution, and (3) water content at which the slurry matrix becomes fluid. Laboratory grain size distributions can be corrected for large (> 50 mm diameter) clasts to reflect the characteristics of the actual slurry rather than the matrix only. Slurry density can be calculated from matrix water content at the laboratory threshold of fluid behavior and the abundance of coarse clasts. Flow velocity can be estimated from the geomorphology of deposits in super-elevated curves; velocity gradients can be estimated from the geomorphology of alluvial fans beyond canyon mouths. Flow inertia and impact forces can be estimated from flow velocity, slurry density, and clast diameters.

The principal law of engineering geology is "the recent past is the key to the near future." Deposits comprising alluvial fans document the characteristics of recent flows; therefore, future flows should be expected to create similar deposits. Analyses of alluvial-fan sedimentology and stratigraphy provide a geologic basis for calibrating engineering models.

---

[1] Sergent, Hauskins & Beckwith, 4030 S. 500 W., Suite 90, Salt Lake City, Utah 84123.

Introduction

Quantitative assessment of hazards associated with sedimentation events on alluvial fans should be based on the magnitude and frequency of the events. The magnitude of a sedimentation event can be expressed in terms of the volume of sediment discharge, the area buried by sediment, the velocity of the discharge, the depth of the discharge, the inertia of the flow, and impact forces. The frequency of events can be developed from historical data only in those locations where significant sedimentation events occur at least every few years; otherwise, geologic data must be used to develop event recurrences (Keaton and Mathewson, 1988). Currently, this is problematic because the Federal Emergency Management Agency (FEMA) does not allow consideration of geologic data in evaluation of flood hazards.

Sedimentation events on alluvial fans consist of slurries of sediment mixed with water. Geomorphic and sedimentologic features of sedimentation processes should be sufficiently unique to provide unequivocal means of classifying the deposits on the the basis of dominant process (Keaton, 1988). Clear water floods have generally low shear strength which increases slowly with increasing sediment load and can be described by conventional hydraulic formulae based on Newtonian fluid behavior. Up to fluid densities of about 1.5 to 1.8 g/cm^3 (14.7 to 17.6 kN/m^3) flood flows transport sediment by turbulence, shear, lift, and drag forces (Costa, 1984). At some critical sediment concentration, hence fluid density, shear strength increases rapidly and is accompanied by an apparently irreversible entrainment of sediment (Pierson and Scott, 1985).

Alluvial-fan deposits traditionally have been subdivided into two classifications -- streamflow (or water-laid) and debris flow (or mudflow); a third classification, transitional between water-laid and debris flow, has been included by some workers (cf., Costa, 1984). Alternative evaluations of alluvial-fan sediments have incorporated rheological properties of moving sediment-water mixtures (Costa, 1984; Johnson and Rodine, 1984; Nemec and Steel, 1984; Wiezcorek, 1986; Pierson and Costa, 1987). The expected character of sedimentary deposits due to a range of rheological properties has been considered by Lowe (1982) and Nemec and Steel (1984).

Theoretical Relationships

Streamflows exhibit Newtonian behavior and have strength characteristics dependent on the velocity (strain rate) of the flow. Thus, swift water has greater competence than slow water, and as water velocity slows, the largest particles are deposited first, creating clast-supported, stratified, graded (fining upward) bedding. Debris flows exhibit non-Newtonian behavior and have strength characteristics which are independent of the velocity of the flow. Plastic deformation is dominant in slower-moving flows while visco-plastic deformation is dominant in faster-moving flows. Thus, debris-flow deposits commonly are ungraded, unsorted, unstratified sediment accumulations (diamictons) which are matrix-supported (Keaton, 1988).

Processes transitional between streamflow and debris flow should be expected because the ratio of sediment to water is a continuum from 0 to 1. At some threshold ratio, deformation processes change from debris flow to landslide (Obrien and Julien, 1985). Sediment concentrations as high as 0.88 by weight in debris flows have been measured by Morton and Campbell (1974) in California and by Pierson (1985) in

Utah. The transition from debris flow processes to extreme concentrations of sediment in flowing water has been assigned to sediment concentrations ranging from 0.8 to 0.4 by weight by Beverage and Culbertson (1964) and has been called hyperconcentrated flow.

Relationships among Cv, sediment concentration by volume, Cw, sediment concentration by weight, w, water content, and e, void ratio, for a specified unit total volume, V, are summarized below and shown graphically on Figure 1:

$$Cv = \frac{Vs}{V} = \frac{\frac{Ws}{G}}{\frac{Ws}{G} + wWs + Va} = \frac{1}{1 + Gw} \text{ [if saturated]} \quad (1)$$

$$Cw = \frac{Ws}{W} = \frac{1}{1 + w} \quad (2)$$

$$Cw = \frac{CvG}{1 + Cv\,(G-1)} \text{ [if saturated]} \quad (3)$$

$$w = \frac{Ww}{Ws} = \frac{1 - Cw}{Cw} = \frac{1 - Cv}{CvG} \quad (4)$$

$$e = \frac{Vv}{Vs} = Gw \text{ [if saturated]} = \frac{1 - Cv}{Cv} \quad (5)$$

where Vs is the volume of solids, Ws is the weight of solids, G is the specific gravity, Va is the volume of air, W is the total weight, Ww is the weight of water, and Vv is the volume of voids. Sediment concentrations and the values of selected parameters for the range of flow behaviors of sedimentation events on alluvial fans are summarized in Table 1.

Slurry Characteristics

Slurry characteristics can be estimated from field and laboratory measurements on alluvial-fan deposits. Representative samples of deposit material < 50 mm in diameter are collected for laboratory testing; the number and size of all clasts ≥ 50 mm within a representative area are recorded. Grain size distributions of laboratory samples may be combined with field observations using a two-dimensional representation of a procedure developed by Williams and Guy (1973) in which the area-percentage of surface clasts is proportional to the weight-percentage of clasts in a randomly distributed three-dimensional sample if all grains have the same specific gravity. The specific gravity of clasts can be determined from their weights and volumetric displacements using Archimedes' principle.

Field samples reconstituted in the laboratory permit determination of specific slurry parameters. Johnson and Rodine (1984) determined the density of flowing debris including coarse clasts with the following relationship:

$$\rho_d = \rho_s + \left[\frac{V_c}{V_c + V_f}(\rho_c - \rho_s)\right] \quad (6)$$

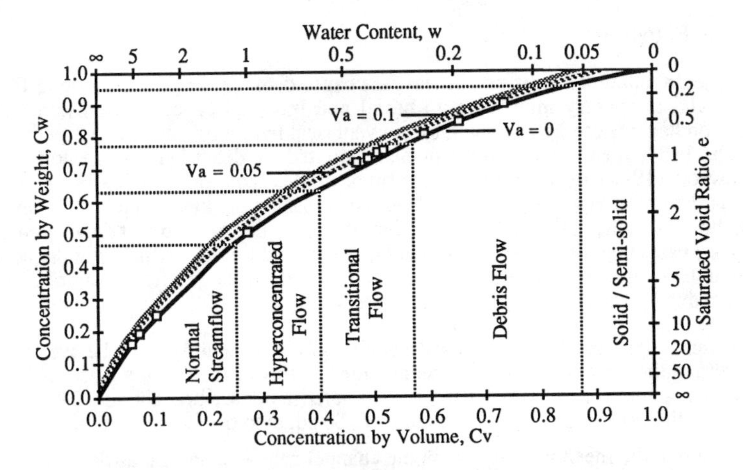

Figure 1. Theoretical relationship between sediment concentration and slurry behavior for G = 2.65. Data points (squares) from Pierson (1985).

Table 1. Sediment concentrations and selected parameters for the range of flow behaviors of sedimentation events on alluvial fans. Unit volume, saturated conditions, and a specific gravity of 2.65 were assumed for calculation of parameters.

Flow Behavior	Solid & Semi-solid	Debris Flow	Transitional Flow	Hyperconcentrated Flow	Normal Streamflow	
Sediment Concentration						
by Volume (Cv)	1.00	0.87	0.57	0.40	0.25	0
by Weight (Cw)	1.00	0.95	0.78	0.64	0.47	0
Water Content (w)	0	0.056	0.285	0.566	1.132	∞
Saturated Void Ratio (e)	0	0.148	0.755	1.500	3.000	∞
Weight of Solids (Ws) (g)	2.650	2.308	1.510	1.060	0.663	0
Weight of Water (Ww) (g)	0	0.129	0.430	0.600	0.750	1.000
Total Unit Weight (W)						
(g/cm^3)	2.650	2.437	1.940	1.660	1.413	1.000
(kN/m^3)	25.98	23.89	19.02	16.28	13.85	9.805

where ρ_d is the density of flowing debris, ρ_s is the density of the slurry reconstituted in the laboratory, V_c and V_f are the volumes of coarse (> 19 mm) and fine fractions, respectively, and ρ_c is the density of the coarse fraction. The value of $\frac{V_c}{V_c+V_f}$ can be taken as the weight percentage of the deposit coarser than 19 mm (Johnson and Rodine, 1984, use 20 mm, but the opening size of a standard 3/4-inch sieve is 19 mm) if the average specific gravity of both size fractions are equal. Thus, for a laboratory slurry density of 2.074 g/cm^3, a clast density of 2.65 g/cm^3, and 57 percent coarser than 19 mm, the debris flow density would be 2.402 g/cm^3 or 23.55 kN/m^3.

Damage Potential

Sedimentation events on alluvial fans cause damage ranging from flood-like erosion and temporary inundation to burial and impact. Flood-like erosion and inundation hazards can be evaluated by conventional hydraulic engineering methods (French, 1987). Burial hazards can be evaluated as described by Keaton and Mathewson (1988) by geomorphic analyses of the fans and application of the principal law of engineering geology, "the recent past is the key to the near future." Thus, those portions of the fans with stratigraphic evidence of sediment deposition in the recent past should be considered likely to receive additional sediment in the near future, unless some significant change (e.g., a fanhead trench or a flood control dam) has altered the possible flood paths on fan surface.

Impact hazards caused by non-Newtonian flowing sediment can be evaluated by assessing the flow density or unit weight, the velocity of the flowing mass, and the likely flow paths. The flow density is estimated as described above. The flow velocity at the apex of a fan can be assumed to be equal to the velocity in the channel above the fan; the mean velocity, $\overline{v}$, in the channel can be estimated as described by Johnson and Rodine (1984) from the geomorphology of deposits in super-elevated curves using the following relationship:

$$\overline{v} = \sqrt{g\, r_c \cos\delta \tan\beta} \qquad (7)$$

where g is gravitational acceleration, r_c is the radius of curvature of the center-line of the channel controlling the flow going into the curve, δ is the channel slope, and β is the angle of super-elevation of the deposits marking the surface of the flow going around the curve.

Velocities across fan surfaces should range from a maximum at the apex to zero where the flows stop. These velocity gradients can be estimated from the geomorphology of the alluvial fans below canyon mouths. Impact pressure, P_i, can be estimated as described by Costa (1984) from the following relationship:

$$P_i = \alpha \left(\frac{\rho_d\, \overline{v}^2}{g} \right) \qquad (8)$$

where α is the velocity-head coefficient (2.0 for laminar flow), ρ_d is the debris flow density, $\overline{v}$ is the mean velocity, and g is gravitational acceleration. The impact pressure can be converted to a force by multiplying by the area of the object (e.g., a wall) on the fan. The impact force of particular megaclasts can be estimated similarly.

Conclusions

Deposits comprising alluvial fans document the characteristics of recent sedimentation events. Since "the recent past is the key to the near future," future sedimentation events should be expected to be similar to those creating past deposits. Analyses of alluvial-fan sedimentology and stratigraphy provide a geologic basis for calibrating engineering models. Theoretical relationships governing sediment-water "slurries" can be used to predict slurry characteristics and behavior for hazard analysis and engineering design.

Acknowledgments

This paper is based on the author's Ph.D. research which was supported by the U.S. Geological Survey Landslide Hazard Reduction Program and the Utah Geological and Mineral Survey. The author benefitted from discussions with Christopher Mathewson, Loren Anderson, Gerald Wieczorek, Elliott Lips, and Paul Santi regarding sedimentation processes and hazards.

Appendix A - References Cited

Beverage, J. P., and Culbertson, J. K., 1964, Hyperconcentrations of suspended sediment: American Society of Civil Engineers Journal of the Hydraulics Division, v. 90, no. HY6, p. 117-126.

Costa, J. E., 1984, Physical geomorphology of debris flows, *in* Costa, J. E., and Fleisher, P. J., eds., Developments and applications of geomorphology: Berlin, Springer-Verlag, p. 268-317.

French, R. H., 1987, Hydraulic processes on alluvial fans: Amsterdam, Elsevier, 244 p.

Johnson, A. M., and Rodine J. R., 1984, Debris flow, *in* Brunsden, D., and Prior, D. B., eds., Slope stability: Chichester, England, John Wiley and Sons, p. 257-361.

Keaton, J. R., 1988, A probabilistic model for hazards related to sedimentation processes on alluvial fans in Davis County, Utah [Ph.D. thesis]: College Station, Texas A&M University, 441 p.

Keaton, J. R., and Mathewson, C. C., 1988, Stratigraphy of alluvial fan flood deposits, *in* Abt, S. R., and Gessler, J., eds., Proceedings of the 1988 National Conference on Hydraulic Engineering: New York, American Society of Civil Engineers, p. 149-154.

Lowe, D. R., 1982, Sediment gravity flows II -- Depositional models with special reference to the deposits of high-density turbidity currents: Journal of Sedimentary Petrology, v. 52, p. 279-297.

Morton, D. M., and Campbell, R. H., 1974, Spring mudflows at Wrightwood, Southern California: Quarterly Journal of Engineering Geology, v. 7, p. 377-384.

Nemec, W., and Steel, R. J., 1984, Alluvial and coastal conglomerates -- Their significant features and some comments on gravelly mass-flow deposits, *in* Koster, E. H., and Steel, R. J., eds., Sedimentology of gravels and conglomerates: Calgary, Alberta, Canadian Society of Petroleum Geologists, Memoir 10, p. 1-31.

Obrien, J. S., and Julian, P. Y., 1985, Physical properties and mechanics of hyperconcentrated sediment flows, *in* Bowles, D. S., ed., Delineation of landslide, flash flood, and debris flow hazards in Utah: Logan, Utah, Utah State University, Utah Water Research Laboratory publication G-85/03, p. 260-279.

Pierson, T. C., 1985, Effects of slurry composition on debris flow dynamics, Rudd Canyon, Utah, *in* Bowles, D. S., ed., Delineation of landslide, flash flood, and debris flow hazards in Utah: Logan, Utah, Utah State University, Utah Water Research Laboratory publication G-85/03, p. 132-152.

Pierson, T. C., and Costa, J. E., 1987, A rheologic classification of subaerial sediment-water flows, *in* Costa, J. E., and Wieczorek, G. F., eds., Debris flows/avalanches -- Process, recognition, and mitigation: Geological Society of America Reviews in Engineering Geology, v. 7, p. 1-12.

Pierson, T. C., and Scott, K. M., 1985, Downstream dilution of a lahar -- Transition from debris flow to hyperconcentrated streamflow: Water Resources Research, v. 21, p. 1511-1523.

Wiezcorek, G. F., 1986, Debris flows and hyperconcentrated streamflows: American Society of Civil Engineers, Proceedings, Water Forum '86: World Water Issues in Evolution, p. 219-226.

Williams, G. P., and Guy, H. P., 1973, Erosional and depositional aspects of Hurricane Camille in Virginia, 1969: Washington, D.C., U.S. Geological Survey Professional Paper 804, 84 p.

RUNOFF AND EROSIONAL PROCESSES IN AN ARID WATERSHED

Peter F. Lagasse, M. ASCE[1], James D. Schall, M. ASCE[1], and Stephen G. Wells[2]

Abstract

This paper reports the results of geomorphic, hydrologic, and hydraulic analyses that integrated a physical process computer model of rainfall/runoff processes with a geomorphic process/response model of contemporary arroyo behavior for an arid watershed in New Mexico.

Sensitivity modeling demonstrates the importance of headwater areas in producing runoff in comparison to valley areas nearer the principal arroyos of the drainage system. Based on these results and observations, a process/response model of contemporary arroyo behavior is presented. This model unifies the different processes operating in different geographic locations of the watershed.

Introduction

Arroyo processes in an arid watershed reflect a natural balance between the driving forces of climate and the resisting geologic environment. These processes operate at different rates and magnitudes according to their hierarchy in the fluvial system. In this paper a physical process computer model of rainfall/runoff processes is combined with a geomorphic process/response model of contemporary arroyo behavior for the Zuni River Basin, an arid watershed in New Mexico. Based on field data, physical process simulation modeling, and geomorphic analyses a unifying model of processes

---

[1] President and Senior Water Resource Engineer, respectively, Resource Consultants, Inc., Fort Collins, Colorado.

[2] Head, Department of Geology, University of New Mexico, Albuquerque, New Mexico.

operating in different geographic locations of the arid watershed is developed.

<u>The Geomorphic Setting</u>

The Zuni River and its major tributaries in west-central New Mexico flow through broad, gently sloping valleys. In many places the Zuni River and its tributaries are entrenched several meters below the level of the valley floor. Based on the accepted definition of an arroyo as a channel of an ephemeral or intermittent stream, usually with vertical banks of unconsolidated material 2 ft. (0.6 m) or more high (3), most entrenched reaches of the Zuni River system can be classified as arroyos.

The Zuni River drainage basin is characterized by two major landscapes: bedrock uplands and topographically lower valley floors. The vast majority of arroyos in the Zuni River watershed, including the main trunks of the major drainages, are developed in alluvium filling valley and canyon floors or thin alluvial mantles on bedrock uplands.

To analyze the relationships between the driving forces of climate and the resisting geologic environment it is helpful to classify arroyos and the drainage network of the Zuni River watershed. These arroyos can be classified into three groups based on their hydrogeologic conditions, associated drainage basin size, and geomorphic position in the drainage network.

The largest arroyos are the trunk channels (A-scale arroyos - Figure 1) that receive water and sediment from tributary arroyos. Major tributaries to these A-scale arroyos are classified as B-scale arroyos. The B-scale arroyos receive water and sediment from smaller tributaries which typically head in upland areas (C-scale arroyos - Figure 1) and drain bedrock areas.

For the A-scale arroyos runoff is generated by snowmelt in the Zuni Mountains as well as convective thunderstorm activity. The A-scale channel generally receives recharge from major springs and flow can be locally perennial. In the B- and C-scale arroyos runoff is generated primarily from convective thunderstorms and flows are ephemeral. The drainage area for a B-scale arroyo is between 8 and 190 sq. mi. (20.7 and 492 km^2), while a C-scale arroyo has less than 8 sq. mi. (20.7 km^2) of drainage area.

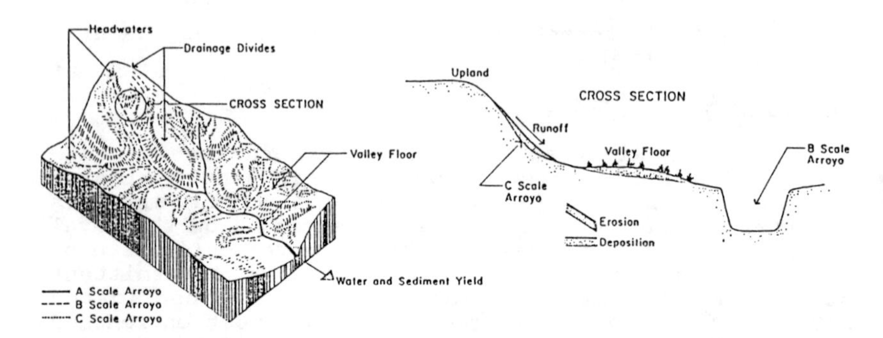

Figure 1. Schematic of Geomorphic Features in an Arid Watershed.

A characteristic of all sizes of arroyos in the Zuni River drainage is that the channels can become discontinuous; that is, incised channel reaches are separated by unincised reaches, the result of erosion in one reach and deposition in another. Over time incised channel reaches can migrate up valley through a process known as headcutting, and deposition below a headcutting reach can lead to wide, shallow unincised reaches where the channel bed coincides with the valley floor. These processes are illustrated in the cross section of Figure 1 which shows a C-scale arroyo eroding toward an upland area. Deposition of sediments from the eroding C-scale arroyo results in aggradation on the valley floor, the C-scale arroyo becomes discontinuous, and sediments from the upland area do not reach the larger B-scale arroyo.

Analysis of Arroyo Processes

Arroyo processes have been measured in the field. In 1966, Leopold and others (4) analyzed channel and hillslope processes in a semiarid region of New Mexico. They found that the largest sediment source related to erosional processes operating in this terrain was derived from "a small percentage of the basin area near the basin divides." The dominant erosional processes supplying the sediment was sheet erosion, that is, erosion of the land surface outside of incised channels. However, they noted that not all the sheetwash derived sediment reaches the channel, "but is temporarily stored in thin deposits widely dispersed over the colluvial area" (4). A similar response was documented in erosion studies on a semiarid watershed northwest of Gallup, New Mexico (5).

To quantify the relationship between energy input and erosional response, a physical process mathematical model of the rainfall/runoff/sediment process was applied under conditions representative of a portion of the Zuni watershed. The analysis was supported by rainfall simulation studies for calibration of selected parameters.

MULTSED, a physical process simulation model, was adopted for this purpose (2). The model was applied to a representative 20 sq. mi. (51.8 km^2) drainage subunit of the 810 sq. mi. (2098 km^2) Zuni River drainage. This drainage is located in the eastern portion of the Zuni River drainage and has slope, geometric, and cover characteristics typical of surrounding areas.

MULTSED is a single event simulation model which routes water and sediment from relatively large, complex watersheds to produce runoff hydrographs and determine sediment yield. Physical processes considered in the water and sediment routing scheme include rainfall, interception, infiltration, raindrop splash detachment, overland surface water routing, soil detachment by runoff, overland sediment yield by size fractions, and channel water and sediment routing (by size fractions).

The model was applied to provide a sensitivity analysis of the relative importance of headwater subareas and valley floor subareas in generating runoff. This analysis clearly demonstrated the importance of headwater areas in producing runoff on the eastern portion of the Zuni watershed. Figure 2 compares direct runoff per unit area for a headwater subarea adjacent to the drainage divide (with no upstream inflow) with a valley subarea adjacent to the main arroyo near the modeled watershed outlet. Headwater areas are significantly more important to generation of runoff than are the valley areas. For the 2-year event, runoff from the valley area is nearly zero, resulting in a factor of 68 increase for the headwater area over the valley area. Essentially, during a relatively small thunderstorm any runoff occurring is generated in the headwater areas. Even for a relatively large storm (e.g., a 10-year event) when a measurable amount of runoff occurs from the valley areas, the headwater area still contributes five times as much water per unit area. Unit discharge hydrographs show that the headwater areas not only produce greater volumes of runoff per unit area, but also a much more "flashy" or quickly occurring runoff. The valley runoff occurs more slowly and uniformly over time.

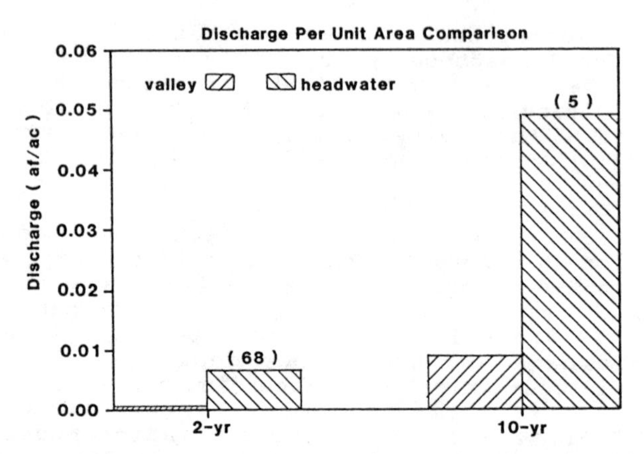

Figure 2. Discharge per unit area for valley and headwater areas (Note: Values in parentheses are factor increases of upland over valley areas).

Total runoff at the outlet of the modeled watershed is a function of how much area in the watershed is classified as headwater compared to valley. Based on subdivision of the representative watershed for modeling, 6,160 acres or about 48 percent of the drainage was delineated as subwatershed or "headwater" units. For the 2-year storm, most all of the runoff from the modeled watershed is generated in the headwater areas. During larger storms (e.g., 10-year), headwater areas will contribute about 80 percent of the total runoff even though these areas comprise only about 48 percent of the total area. In the modeled watershed the headwater areas are far more important in generation of total runoff than are the valley areas.

Unifying Model of Arroyo Processes

In some studies of arroyo initiation and development, researchers have made basic assumptions concerning changes in the erosional regime of the valley floors. First, they postulate that arroyos are initiated in response to increased erodibility of the valley-fill alluvium, and secondly, arroyos are initiated due to increased erosiveness of flows over the valley floors (1). Thus, valley floor conditions have been considered primary by some researchers in controlling runoff and incision.

In contrast, the MULTSED modeling results have demonstrated that bedrock uplands are the primary source of runoff provided to valley floors and that arroyo development is primarily by runoff from the uplands. Between 60 and 90 percent of all arroyos in the Zuni watersheds studied head on bedrock uplands. Based on field observation and aerial photographic analyses, those heading on bedrock uplands are longer than any heading on valley floors and many are discontinuous and do not intersect mainstem arroyos.

Based upon these results and observations, a process/response model of contemporary arroyo behavior can be developed. This model unifies the different processes operating in different geographic locations within an arid watershed. The processes can be described with reference to Figure 1. They are driven by precipitation falling upon the uplands and sediment production from the uplands. Valley floor aggradation occurs in response to excess sediment derived from sheet erosion and discontinuous C-scale arroyos, responding to the more permeable and rough surfaces of the valley floor. Arroyo cutting and filling responds to flow and sediment supplied from continuous arroyos heading in the uplands.

This process/response model is consistent with the contemporary rates of processes observed in the Zuni River watershed. Faster rates of change of C- and B-scale arroyos occurs in response to more frequent and erosive discharges, (flashy flows) from uplands. The longer term stability of the larger A-scale arroyos, occurs in response to the long-term sediment storage in the valleys and alluvial fans of the B-scale arroyos. Reduced sediment supply allows the A-scale arroyo to remain relatively stable in its configuration over time periods involving centuries.

Conclusions

Arroyo processes within the Zuni River watershed reflect a natural balance between the driving forces of climate and the resisting geologic environment. These processes respond predominantly to conditions within the headwaters in the bedrock upland areas. Arroyo processes operate at different rates and magnitudes according to their hierarchy in the fluvial system. Smaller C-scale arroyos of the headwater areas cut and fill at a faster rate than B- or A-scale arroyos due to their proximity to the runoff and sediment generating areas, the bedrock uplands. Large-scale storage of sediment and the hydrologic discontinuities of the B-scale arroyos and their valleys enhance the long-term stability of the A-scale arroyos.

References

1. Cooke, R. U., and Reeves, R. W., "Arroyos and Environmental Change in the American Southwest," Oxford, Clarendon Press, 1976, 213 pp.

2. Fullerton, W. T., "Water and Sediment Routing from Complex Watersheds and Example Application to Surface Mining," Masters Thesis, Colorado State University, Fort Collins, Colorado, 1983.

3. Howell, J. V. (coordinating chairman), "Glossary of Geology and Related Sciences, with Supplement," The American Geological Institute, second edition, Washington, D.C., 1962, 325 pp.

4. Leopold, L. B., Emmett, W. W., and Myrick, R. M., "Channel and Hillslope Processes in a Semiarid Area, New Mexico," U. S. Geological Survey Professional Paper 352-G, 1966, 253 pp.

5. Wells, S. G., and Gardner, T. W., "Geomorphic Criteria for Selecting Stable Uranium Tailings Disposal Sites in New Mexico," New Mexico Energy Research and Development Institute Report 2-69-1112, v. 1, 1985, 353 pp.

Application of Geological Information to Arizona Flood Hazard Assessment

V.R. Baker[1], K.A. Demsey[2], L.L. Ely[3], J.E. Fuller[4], P.K. House[3], J.E. O'Connor[3], J.A. Onken[3], P.A. Pearthree[5], and K.R. Vincent[3]

Abstract

Hydrological modeling procedures applied to regulatory flood-hazard zonation can be misapplied when assumptions concerning flood-hazardous processes are violated. Geomorphological mapping of the Tortolita Mountain piedmont in southern Arizona reveals extensive high-standing nonhazardous inactive, relict Pleistocene fan surfaces within zones mapped by FEMA as subject to active alluvial fan processes and 100-year flooding. Paleoflood analysis of upstream mountain canyons documents that maximum flood discharges for the past century have been about 50% lower than the regulatory (100-year) flood discharges. Geological studies are essential complements to engineering models in order to generate public confidence that regulatory requirements derive from knowledge of real rather than idealized arid-region flood-hazard processes.

Introduction

Arid region flooding profoundly challenges flood-hazard management because of locally unstable channel systems, high spatial and temporal variability of flood

---

[1]Regents Professor, Dept. of Geosciences, Univ. of Arizona, Tucson, AZ, 85721

[2]Geologist, Arizona Geological Survey, 845 N. Park, Tucson, AZ, 85719

[3]Research Assoc., Dept. of Geosciences, Univ. of Arizona, Tucson, AZ, 85721

[4]Principal Hydrologist, Pima County Dept. of Transportation and Flood Control Dist., 32 N. Stone, Suite 300, Tucson, AZ, 85701

[5]Research Geologist, Arizona Geological Survey, 845 N. Park, Tucson, AZ, 85719

processes, and inadequate conventional hydrological data bases. Standard flood-hazard engineering practice involves the application of models that predict parameters used as the basis for design or regulatory control. Model misapplication occurs when real-world processes for the modeled setting deviate from those arbitrarily assumed in the modeled idealization of the hazard.

Geological flood studies infer causative flood processes from the effects of those processes on landscapes, sediments, and vegetation (Baker et al., 1988). Clearly such studies should precede the application of models used for regulatory zonation. Inadequate attention to geological evidence prior to regulatory zonation can result in wide disparities between the hazards assumed for regulation and those actually occurring. The resulting public perception that regulation is arbitrary and violates common sense may defeat the purpose of regulation by encouraging noncompliance, litigation, or political agitation. This problem is illustrated by an analysis of Flood Insurance Rate Maps (FIRMs) mandated by the Federal Emergency Management Agency (FEMA) for an area in southern Arizona.

Piedmont Flood Hazards: FEMA Approach

A piedmont area south of the Tortolita Mountains near Tucson, Arizona, was classified by FIRMs according to FEMA specifications based on the procedure developed by Dawdy (1979) for flood hazard assessment on active alluvial fans. The Dawdy method for flood-frequency estimation on alluvial fans is generally considered to be the best available for delineating flood hazard zones in these dynamic environments (French, 1987). In order to quantify the highly variable flow characteristics, the Dawdy method translates probabilities of given discharges (derived from empirical formulae) at fan apices into probabilities of associated flow depths and velocities on fan areas below the apices.

A fundamental assumption for the above procedure is that flow be largely restricted to self-formed channels which are randomly located below the fan apex and free to migrate laterally. The assumption requires that certain precautions be taken in site selection to achieve realistic results. The most pertinent requirements are the accurate delineation of active fan surfaces and location of their associated apices. These two procedures are essential because the magnitude of the flood hazard at a point on the active fan surface is directly related to its location below the apex. Thus, it is essential to understand the spatial relationships of active fluvial processes on the piedmont in order to appropriately designate the associated hazardous zones.

The above procedures can fail in their purpose because of either (1) errors in following the established procedure, or (2) misapplication of the procedure to a situation for which its basic assumptions do not apply. Both failures can either subject life and property to unrecognized hazards or subject landowners to undue financial costs when sites are incorrectly designated as flood prone. Unfortunately, type (2) errors receive far less attention than type (1).

Geomorphological Mapping

General misapplication of the FEMA active alluvial fan procedure to the Tortolita piedmont is demonstrated by geomorphological mapping of active versus inactive fan surfaces. The surfaces were distinguished on the basis of topography, vegetation characteristics, and degree of soil development. Mapping results (Fig. 1) indicate that

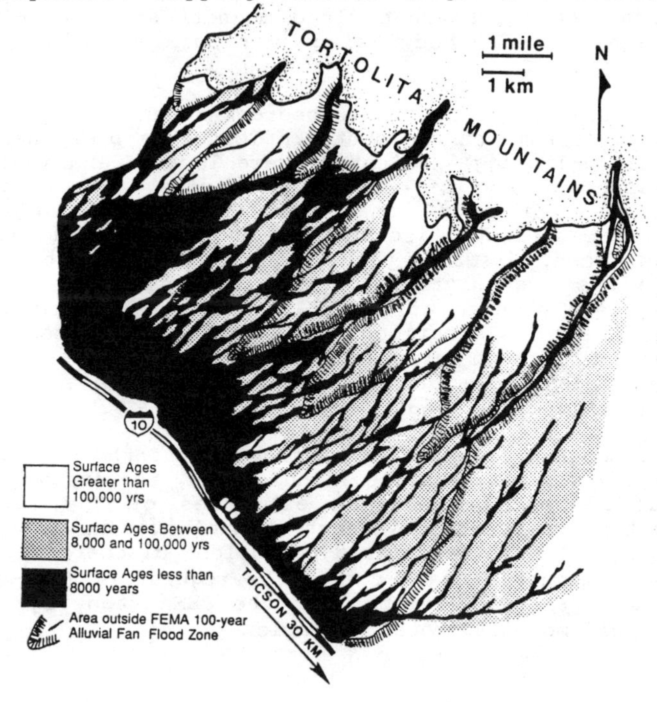

Figure 1: Generalized geomorphic surface-age map and FEMA-mapped 100-year flood zone for Tortolita fan area.

within FIRM zones designated as subject to active alluvial fan processes are large areas of relatively high-standing relict Pleistocene fan surfaces that have not been subject to flooding from the main channel for at least 10,000 years. The disparity between the FIRM zonation and the geomorphological reality results from inappropriate delineation of active fan surfaces and associated apices coupled with the implementation of rainfall-runoff models that significantly overestimate magnitudes of rare floods. The latter problem was addressed by a program of paleoflood hydrology.

Paleoflood Hydrology

Mountain canyons supplying flood flows to the Tortolita piedmont contain geological evidence of floods occurring over the past century. The geological evidence was analyzed by slackwater deposit-paleostage indicator (SWD-PSI) procedures described by Baker (1987). The SWD-PSI technique relates paleoflood stages to the heights of specific indicators and employs hydraulic modeling techniques to reconstruct flood magnitudes. Paleoflood ages are established using various techniques of geochronology.

In well-confined, stable bedrock reaches of the five principal streams on the southern Tortolita piedmont, all potential paleostage indicators were located and included in detailed surveys of channel geometries. The types of paleostage indicators that were most often utilized included scour lines, isolated slackwater flood deposits, scarred trees, distinct plant distributions, silt lines, and various accumulations of flood transported debris. Reaches were modeled using HEC-2 (Hydrologic Engineering Center, 1982). Paleoflood discharges were subsequently determined through graphical comparison of paleostage indicator elevations and computed water surface elevations (O'Connor and Webb, 1988).

Although a detailed reconstruction of the paleoflood chronology was not feasible in this area, the analysis did provide reasonably accurate estimates of the largest events to have occurred in each channel over at least the past century. All calculated paleoflood discharges (Q_{max}) were significantly smaller than the regulatory (100-year) magnitudes (Q_{100}) developed for the FEMA study by rainfall-runoff modeling methods (Table 1).

Table 1. Comparison of FEMA 100-year discharges and maximum discharges estimated with paleoflood techniques. All discharges are reported in cubic meters per second.

Canyon	FEMA Q_{100}	SWD-PSI Q_{max}
Wild Burro	270	90-100
Ruelas	170	90-100
Cochie	190	120-150
Prospect	185	50-70
Canada Agua	125	30-50

Flood Process Studies

The geomorphological effects of a 1988 high-magnitude flood are exceptionally well preserved along one of the mountain/piedmont streams emanating from Wild Burro Canyon. The event magnitude of 80 to 100 m^3s^{-1} was reconstructed by HEC-2 analysis of stable bedrock reaches containing abundant well-preserved flood deposits. Thus, the flood is of the scale of the real hazard posed by streams in this study area (Table 1). The widespread effects of this event on the piedmont were mapped, distinguishing channelized flow and sheetflow. Field observations indicate that this is the largest flow event to have occurred for the past several hundred years. The flow map provides an excellent graphical depiction of the dynamics of an individual event of known magnitude offering considerable insight into the real flood processes that occur in southern Arizona piedmont environments.

Discussion

The rainfall-runoff models used to predict regulatory flood values (Table 1) encounter severe problems of credibility in arid regions because of the high spatial and temporal variability of rainfall and runoff, the diversity of runoff-producing processes, and the inadequacy of conventional data for model verification (Pilgrim et al., 1988). We advocate a scientific approach, which requires the testing of idealized models against measured reality. The modeled FEMA design discharges for the Tortolita piedmont are clearly disparate from geological evidence of real floods experienced in the area (Table 1). Moreover, the design flows are routed by an idealized model (Dawdy, 1979) that assumes active alluvial fan processes in large portions of a mountain piedmont comprised mostly of inactive, relict fans (Fig. 1).

Because many piedmont environments in the southwestern U.S. are characterized by pediments and relict, inactive alluvial fans, there are extensive possibilities for misapplication of FEMA active alluvial fan hazard procedures. Public reaction to unreasonable imposition of the idealized models imposed by FEMA in erroneous contexts will defeat the purpose of the FEMA regulations. If geological flood evidence in a region is ignored prior to the imposition of regulations, that evidence will remain for contestants to use in legal, political, and scientific efforts to defeat those regulations that violate a reasoned treatment of the phenomenon that the regulators purport to wisely control. The irony of the present regulatory inattention to geological information is that in the public debate over responses to potential hazards, the documented occurrence of an ancient (but real) cataclysmic process is likely to have more positive impact than is the arbitrary assertion that the hazard is defined by various hypothetical frequency distributions, random diversions of flow, and idealized channel behavior.

Acknowledgments: Research supported by (1) Engineering Directorate, Natural and Manmade Hazards Mitigation Program, National Science Foundation, Grant BCS-8901470, (2) Pima County Department of Transportation and Flood Control District, and (3) Arizona Geological Survey.

References

1. Baker, V.R., "Paleoflood Hydrology and Extraordinary Flood Events," Jour. Hydrology, Vol. 96, 1987, pp. 79-99.
2. Baker, V.R., Kochel, R.C., and Patton, P.C., Eds., Flood Geomorphology, Wiley, N.Y., 1988.
3. Dawdy, D.R., "Flood-Frequency Estimates on Alluvial Fans," Journal of the Hydraulics Division, ASCE, Vol. 105, No. HY11, Dec., 1979, pp. 1407-1413.
4. French, R.H., "Flood Hazard Assessment on Alluvial Fans: An Examination of the Methodology," Application of Frequency and Risk in Water Resources, V.P. Singh, Ed., D. Reidel, Boston, 1987, pp. 361-375.
5. Hydrologic Engineering Center, "HEC-2 Water Surface Profiles Program Users' Manual," U.S. Army Corps of Engineers, Davis, California, 1982.
6. O'Connor, J.E., and Webb, R.H., "Hydraulic Modeling for Paleoflood Analysis," Flood Geomorphology, V.R. Baker, R.C. Kochel, and P.C. Patton, Eds., Wiley, N.Y., 1988, pp. 393-402.
7. Pilgrim, D.H., Chapman, T.G., and Doran, D.G., "Problems of Rainfall-Runoff Modelling in Arid and Semiarid Regions," Hydrological Sciences Journal, Vol. 33, 1988, pp. 379-400.

LAND USE CHANGE AND SOIL EROSION HAZARD IN THE ZAMBEZI BASIN

H. Leenaers[1] & K. Salewicz[2]

Abstract

An experimental model for estimating soil erosion hazard in southern Africa (SLEMSA) was linked to a raster based Geographical Information System. Soil erosion hazard maps were constructed by running a series of five land use scenarios. White's vegetation map of Africa was used as the basis for predicting soil erosion hazard under 'natural' vegetation. Four land use scenarios representing subsequent stages of increasing land pressure (due to cropping, grazing and deforestation) were used to delineate areas currently or potentially susceptible to soil erosion.

Introduction

Soil erosion and associated land degradation are a major problem in many areas of the African continent and this problem is growing year by year in response to increasing population pressure on agricultural land. In the Zambezi drainage basin two main issues must be faced from the water management point of view: (1) the soil erosion problem due to intensive deforestation and (2) reservoir operations in order to optimize the hydroelectric power generation (Pinay, 1988). These two main issues are interconnected in the sense that erosion processes may entail an increase in silt deposition in reservoirs diminishing their real storage capacity.

A model for the estimation of soil erosion hazard in southern Africa (SLEMSA) has been developed and validated by Stocking et al. (1988). Applying this empirical model soil erosion hazard maps were constructed for the eight countries in the Zambezi River basin and these maps will soon be published. However, physical planners may want to estimate the impact of future land use changes on erosion hazards. Therefore, they need a means of data storage and manipulation that allows them to repeat the mapping excercise on a scenario basis. This paper describes how SLEMSA was linked to a PC-based GIS and how five land use scenarios were constructed and erosion hazard maps were produced for each. The vegetation map of Africa (White, 1983) provided the 'natural vegetation' scenario. Four land use scenarios representing subsequent stages of increasing land pressure were used to map areas that are potentially susceptible to soil erosion.

---

1) Department of Geography, Utrecht University, P.O. Box 80115, NL-3508 TC Utrecht, The Netherlands

2) IIASA, A-2361 Laxenburg, Austria

The Zambezi drainage basin

The Zambezi basin lies between 20-38° E and 12-20° S (Figure 1), and is the largest of the African river systems flowing into the Indian Ocean. The 2,660 km long river rises on the Central African Plateau and flows southward. The catchment covers 1,570,000 km^2 (Balek, 1977). The climate in this region is continental in character with appreciable seasonal variations in temperature. Mean annual temperature varies from 18 to 24 °C. Rainfall is between 250 and 2,000 mm per year and in general decreases from north to south. There are three main seasons.

During the past 20 years large scale deforestation has occured in Zambia, Angola and Mozambique and the area covered by arable land has grown in all countries in the catchment. In the current situation, 17 % of the communal lands in Zimbabwe is very severely eroded, 10 % severely eroded, 13 % moderately eroded and 60 % in relatively good condition (Whitlow, 1988). The damage to the water resources is large: 12-13 % of the dams in Masvingo Province are now totally useless due to siltation and 50 % of all dam structures have less than half of their full capacity left (Elwell, 1985). At the present rate of sediment input, the dead storage of Lake Kariba will be filled in 1,600-16,000 years and therefore the effect of sediment on the operation of the project may be ignored (Bolton, 1984). However, the sediment input rate to the Cabora Bassa reservoir appears to be a factor three larger (Pinay, 1988). Consequently, the dead storage would be filled in 60-600 years and the reservoir operation may be affected much sooner than expected.

SLEMSA: Soil Loss Estimation Model for Southern Africa

SLEMSA is thoroughly described by Stocking et al. (1988) and has been validated for conditions in the Zimbabwe Highveld. Four broad factors are used to summarize erosion hazard: (1) rainfall, (2) soil, (3) vegetation and (4) relief. These factors are described by five control variables, which can be expressed numerically: seasonal rainfall energy, E (in $J/m^2/y$); soil erodibility, F (as an index); seasonal energy intercepted by the crop, i (in %); slope steepness, S (in %); and slope length, L (in m). These control variables have been arranged into three submodels: a principal submodel, K, yielding estimates of soil losses from bare fallow land at a specified slope steepness and slope length; a crop canopy cover model, C, giving a ratio to adjust from bare fallow to a specific crop type; and a topographic model, X, giving another ratio which enables soil losses to be estimated from slopes other than those specified in the K submodel. The first two submodels were developed from a limited amount of field plot data supplemented by expert opinion. The third submodel was derived from the slope factor relationship of the Universal Soil Loss Equation. The submodels for the Zimbabwe Highveld are formulated as folows:

$$K = \exp((0.4681 + 0.7663*F)*\ln(E) + 2.884 - 8.2109*F) \quad [1]$$

$$C = \exp(-0.06*i) \text{ for } i < 50 \quad [2.1]$$
$$C = (2.3 - 0.01*i)/30 \text{ for } i >= 50 \quad [2.2]$$

$$X = \sqrt{L}*(0.76 + 0.53*S + 0.076*S^2)/25.65 \quad [3]$$

The main model expresses the relationship between the submodels and takes the form of their product:

$$Z = KCX, \quad [4]$$

where Z is predicted mean annual soil loss (t/ha/y), K is the mean annual soil loss (t/ha/y) from a standard field plot 30 m x 10 m at a 4.5 % slope for a soil of known erodibility F under a weed-free bare fallow surface, C is the ratio of soil lost from a cropped plot to that lost from bare fallow land, X is the ratio of soil lost from a plot of length L under slope percent S, to that lost from the standard plot. The results for Z were segmented and expressed in dimensionless Erosion Hazard Units (EHU) on a scale of 1 (low erosion hazard) to 5 (high erosion hazard). These EHU's provide a relative idea of the degree of soil loss that might be expected on a field under the mean conditions used for the calculations.

The input data

In the Global Environmental Monitoring System (GEMS) of UNEP several datasets were available for the Zambezi region. The datapoints are located on the intersections of a grid with a cell size of c. 10 km and are held in the Miller Oblated Stereographic Projection. The following GEMS data layers were used to run SLEMSA: mean annual rainfall (mm), topsoil texture (according to the FAO texture diagram), the vegetation map of Africa (White, 1983) and the FAO land use map. Slope steepness was derived from topographical data provided by NOAA/NGDC. All data layers were fed into the dBase3 Plus data base management system linked to the grid based Geographical Analysis System IDRISI (Eastman, 1988). Equation [1] to [4] were formulated in the programming language provided by dBase3 Plus. After each run of SLEMSA maps displaying 5 erosion hazard classes were produced (Figure 2).

Land use scenarios

The vegetation factor in SLEMSA is measured by the mean seasonal interception of rainfall by vegetation. This gives the proportion of the erosive rainfall, i (in %), that is intercepted by a growing crop or vegetation in a growing season. The value of i takes into account the influence of crop type, planting date, plant density and management. A vegetal cover data bank is available for the region (Elwell and Wendelaar, 1978), which gives values of i for a whole range of crops. The land use scenarios employed were constructed by changing the value of i in each gridcell according to Table 1.

Scenario I represents 'natural' vegetation, i.e. the vegetation cover as described by White (1983). White's map units were grouped into (1) forest (i=90%), (2) woodland and thicket (i=70%) and (3) grassland (i=70%). For detailed characterization of the map units, refer to White (1983). Decreasing i, four other

Table 1: Percentage rainfall energy interception, i, for the five land use scenarios (*) depends on mean annual rainfall).

landuse/ vegetation	energy interception, i (%), for scenario: I	II	III	IV	V
forest	90	90	70	50	0
woodland	70	70	50	30	0
grassland	70	60/40*)	30/20*)	30/20*)	0
cropland	-	36	20	20	0
cropland/ natural mosaic	-	43	23	23	0

scenarios were designed, simulating increasing intensity of cultivation, increasing grazing pressure and increasing deforestation (Table 1). Scenario II represents a situation where well-yielding crops are grown in the cultivated areas (e.g. 4000 kg/ha maize), the grazing pressure is moderate in grassland areas (i.e. 100-300 cattledays/ha) and no significant deforestation occurs. Scenarios III and IV represent situations where crops produce moderate yields (e.g. 1000 kg/ha maize), the grazing pressure is heavy (> 300 cattledays/ha), and deforestation progresses. In scenario V all vegetation is cleared.

Results and discussion

Figure 2 provides soil erosion hazard maps for scenario 1, 3 and 5. When the catchment is covered by natural vegetation the erosion hazard is generally low, with exceptions for areas with poor soils in Angola and few small areas characterized by steep slopes in the eastern part of the catchment. In scenario III the land pressure is increased and deforestation takes place in the evergreen forests and the woodland areas. The results of these activities are clearly shown in Figure 2c: large areas in Angola, Zambia, Malawi, Mozambique and Zimbabwe fall in erosion hazard classes 3, 4 and 5. Moreover, contrary to the situation in scenario I, the areas susceptible to soil erosion are no longer restricted to areas with steep slopes or highly erodible soils. Namibia and Botswana remain unaffected by the land use changes in this scenario. Scenario V represents a situation where all vegetation is cleared and suggests that almost the entire basin has a potentially high erosion hazard. Malawi and Mozambique fall almost entirely in erosion hazard class 5. The lack of steeper slopes explain why Namibia and Botswana have a relatively low erosion potential.

The routing of soil derived from erosion processes in the catchment may lead to reservoir siltation. Notable differences can be observed between the erosion hazards in the upstream catchment areas of the two main reservoirs in the Zambezi basin. Even in scenario I the drainage area of Lake Cabora Bassa exhibits locally medium to high erosion hazards. These areas grow in size as the vegetation cover is decreased (Figures 2c-2d). These estimates confirm the reported excessive occurrence of soil erosion in the entire Luanawa basin due to high natural rates of

Figure 1: The Zambezi river basin

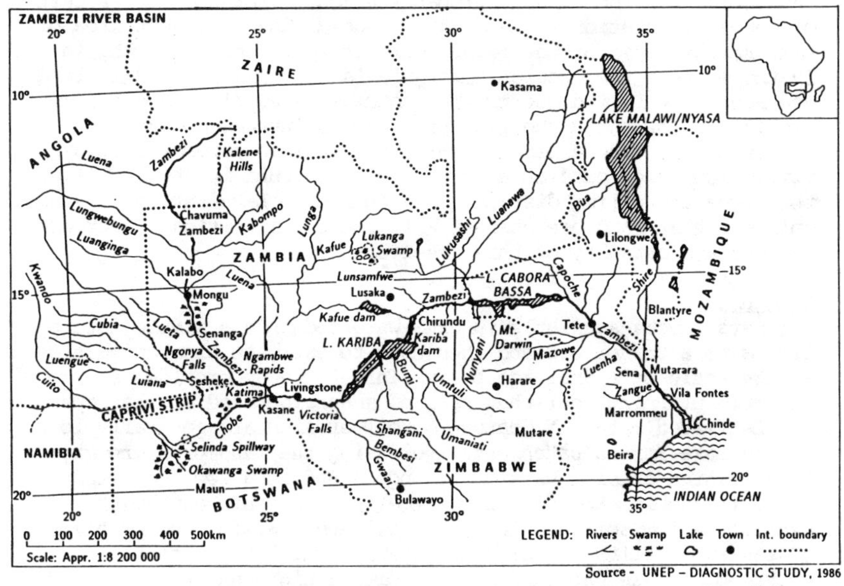

Figure 2: (a) State boundaries; Erosion hazard maps for (b) scenario 1 ('natural' vegetation), (c) scenario 3 and (d) scenario 5 (bare soil).

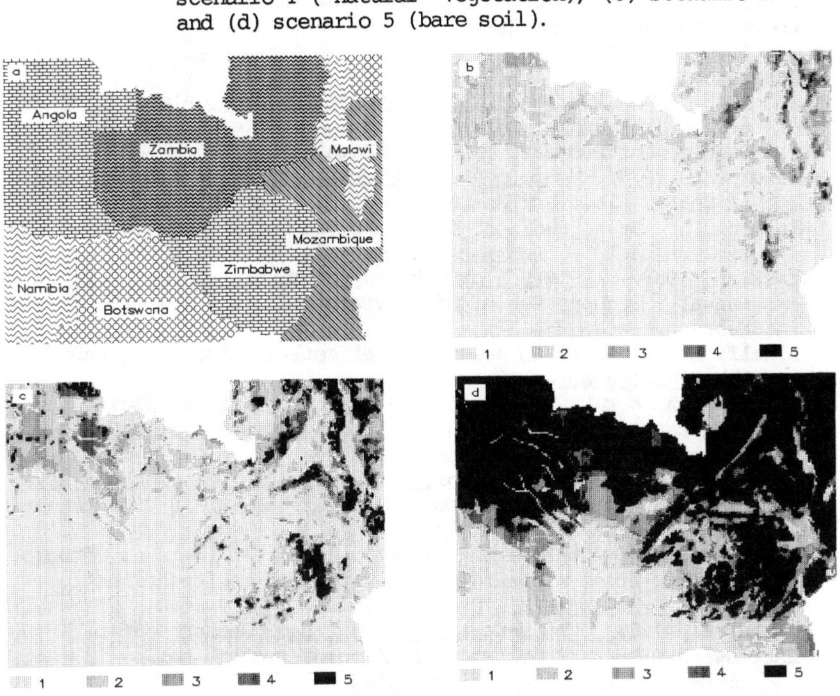

erosion and human activities (Albrecht, 1973). Bolton (1984) also suggested that the Luanawa probably accounts for the larger part of the sedimentation in Lake Cabora Bassa. The catchment area of Lake Kariba lies for the larger part in west Zambia and Angola. In these areas the erosion hazard falls in class 1 and 2 in land use scenarios I, indicating that natural rates of soil erosion are fairly low. When the land pressure is increased during scenarios III, the erosion hazard in west Zambia remains relatively low. High erosion hazards are only found in Angola. Given the large travelling distance from Angola to Lake Kariba it seems unlikely that land use changes will cause a significant decrease the storage capacity of this reservoir.

Conclusions

Data layers available in GEMS were fed into a simple database linked to a raster GIS and then used to estimate erosion hazards at the scale of a drainage basin. This PC-setup provided a flexible planning tool that allowed scenario-based simulations.

Estimated rates of 'natural' soil erosion are generally low. In the Lake Kariba catchment, increasing the land pressure may lead to high erosion hazards in Angola and in a few scattered areas in Zambia. However, the larger part of this catchment exhibits a low soil erosion potential and therefore no problems of reservoir siltation are anticipated. The Luanawa basin upstream of Lake Cabora Bassa exhibits a high 'natural' erosion hazard. Land use changes in this area may contribute to a very large extent to the sediment load of the Zambezi River as it flows into Lake Cabora Bassa.

References

Albrecht, R.W., 1973, 'Land use development proposals for the Mid and Upper Luangwe watershed', Luangwe Valley Conservation and Development Project, UNDP/FAO, Rome.

Balek, J., 1977, 'Hydrology and Water Resources in Tropical Africa', Elsevier, Amsterdam, 208 p.

Bolton, P., 1984, 'Sediment deposition in major reservoirs in the Zambezi basin', IAHS Publication, 144, 559-567.

Eastman, 1988, 'IDRISI: a grid-based geographic analysis system', Manual, Graduate School of Geography, Clark University, Massachusetts, USA.

Elwell, H.A., 1985, 'An assessment of soil erosion in Zimbabwe', The Zimbabwe Science News.

Elwell, H.A. & F.E. Wendelaar, 1978, 'To initiate a vegetal cover data bank for soil loss estimation', Research Bulletin, Dept. Conservation and Extension, Zimbabwe.

Pinay, G., 1988, 'Hydrobiological Assessment of the Zambezi River Basin: A Review', IIASA Working Paper, no. WP-88-089, Laxenburg, Austria, 116 p.

Stocking, M., Q. Chakela & H. Elwell, 1988, 'An improved methodology for erosion hazard mapping, part 1: the technique', Geografiska Annaler, 70A, 3, 169-180.

White, F., 1983, 'The Vegetation of Africa', Unesco, 356 p.

Whitlow, R., 1988, 'Soil erosion and conservation policy in Zimbabwe', Land Use Policy, 419-433.

Experimental Investigation of Density-Driven Hyperconcentrated Flows

P. D. Scarlatos[1] and B. J. Wilder[2]

Abstract

Lock-exchange type of laboratory experiments have been conducted in order to evaluate the dynamical characteristics of density-driven hyperconcentrated fluid mud. Fluid mud, confined in one end of the channel was released through a gate opening into the receiving fresh water, and its density-driven motion was monitored. Experiments included various concentrations of water-kaolinite and natural mud mixtures. The results documented a very thin hyperconcentrated wedge moving along the bottom, and a much thicker turbidity current flowing over the wedge. The turbidity current eventually covered the entire water mass of the channel through diffusion processes. Shear instabilities of the Kelvin-Helmholtz type were recorded at the interface of the propagating turbid cloud and the overlaying water.

Introduction

Near-bottom, high concentrations of liquified cohesive materials (i.e. natural clays, silts, or organic matter) constitute a potential environmental hazard. Due to their small particle size, cohesive materials adsorb contaminants, and those contaminants are eventually introduced into the water column when the mud is resuspended. Fluid mud is usually separated from the overlaying water column by a steep density gradient, the lutocline. For resuspension to occur, some turbulent or shear energy should be provided along the lytocline. Resuspension is characterized with breaking of internal waves generated along the interface. The wave breaking can be either in the form of rolling, stretching, and folding (K-H type of instability) or in the form of small

---

[1]Associate Professor, [2]Graduate Research Assistant, Department of Ocean Engineering, Florida Atlantic University, Boca Raton, Florida 33431.

bursts from the wave tips (Holmboe type of instability) (Broward and Wang, 1974). Shear flow experiments in a recirculating flume indicated that for a pure kaolinite mud the rate of entrainment is inversely proportional to some power of Richardson's number (Shrinivas, 1989). Therefore, mud entrainment follows the same behavior as other stratified systems (salt or temperature stratification), i.e. is proportional to shear production and inversely proportional to the stabilizing buoyancy effects.

The purpose of this study is to investigate with lock-exchange type of experiments the propagational, resuspension, and entrainment characteristics of hyperconcentrated fluid mud. Shear is generated at the surface of the density-driven mud wedge as it moves under gravity through the resting, receiving water. As a result, the interface erodes and the fluid mud entrains into the water column. Both the hyperconcentrated wedge and the overlaying turbidity current are monitored and their motion is properly analyzed.

Experimental Set-up

The experimental channel was made of plexiglass of 1.27-cm (1/2-in) thickness. The dimensions of the channel are 2.43-m (8-ft) long, 0.30-m (1-ft) wide, and 0.30-m (1-ft) deep. At the one end of the channel fluid mud was confined at a compartment separated from the rest of the channel by a vertically removable partition. The compartment had the same width and depth as the rest of the channel and a length of 0.10-m (1/3-ft). The water depth for all of the experiments was 0.20-m (2/3-ft). A schematic representation of the experimental channel is given in figure 1. A rectangular grid of 1 x 1 cm was placed along the one side of the channel in order to facilitate quantification of the observations. Since the experiments lasted no longer than 25 seconds, they were videotaped and then analyzed in slow motion.

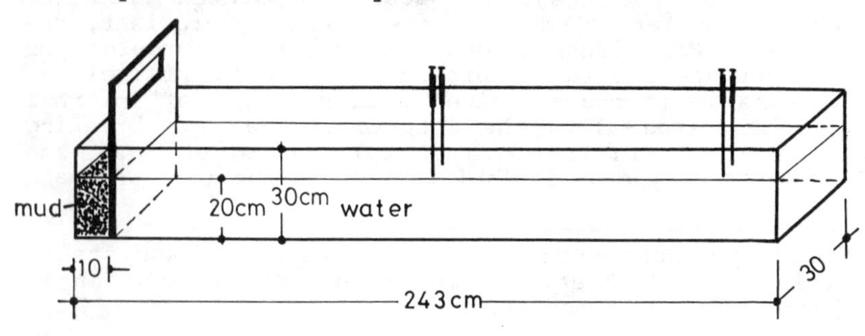

Figure 1. Experimental Channel.

Two types of fluid mud were used for the experiments. One was artificially made by mixing kaolinite with fresh water, and the other was natural mud composed of organic matter (43%), silt (42%) and fine sand (15%). The mud concentration in the experiments exceeded 200 gr/l. In order to maintain uniformity of the mixture, the fluid mud was thoroughly stirred before it was placed into the end compartment within the channel. To avoid depositional effects, the partition was lifted immediately after the mud was placed into the compartment. The purpose of using both kaolinite and artificial mud was to investigate the ability of artificial mud to reproduce the behavior of natural mud. Experiments were conducted for three different mud concentrations of approximately 200, 400 and 600 gr/l, for the kaolinite and the natural mud mixture. More specifically, data for five experiments are given in table 1.

Table 1. Experimental Data

Experiment	Type of Mud	Initial Concentration
Ia	Kaolinite	228 gr/l
Ib	Kaolinite	413 gr/l
Ic	Kaolinite	542 gr/l
IIa	Natural mud	200 gr/l
IIb	Natural mud	420 gr/l

Other than the videotaping, the only direct data that have been collected were water samples at two different locations (x = 100-cm; x = 200-cm) and two different depths. One sample was taken directly from the bottom and the other 5-cm above the bottom. Also, in some experiments dye was used to visualize any return current motion. The water samples were collected right after the passage of the toe of the propagating mud wedge.

Experimental Results

Once the partition was lifted, the mud started to propagate along the bottom of the channel as a density current. The basic qualitative observation from all the experiments was the formation of two moving layers; one thin layer at the bottom with concentration of the same order with the original fluid mud, and one thick layer of very low concentration (cloud) moving on top of the bottom layer. The bottom layer in some cases arrested before it reached the end of the channel (experiments Ib-Ic). On the other hand, the cloud always propagated all the way to the end. The cloud was the result of solid particle entrainment due to shear production at the interface between the moving fluid mud wedge and the overlaying motionless water. Internal waves were observed at the upper surface of the cloud. The cloud grew and eventually covered the entire water mass. For concentrations of the order of 400 mg/l and less, visualization of the two layers was not possible for the reason that the bottom layer was very small (< 1-cm) and its interface was totally diffused with the cloud. However, from bottom sediment samples and observations taken after part of the cloud had deposited, the existence of the thin hyperconcentrated bottom layer was evident. For all the experiments, except experiment Ic, the bottom layer was evenly distributed along its length. In experiment Ic, the bottom layer profile was slightly tapered. Profiles of the movement of the turbid cloud and the bottom layer, for the five experiments listed in table 1., are given in figure 2. Whenever possible, the bottom layer is given in figure 2 by dotted lines.

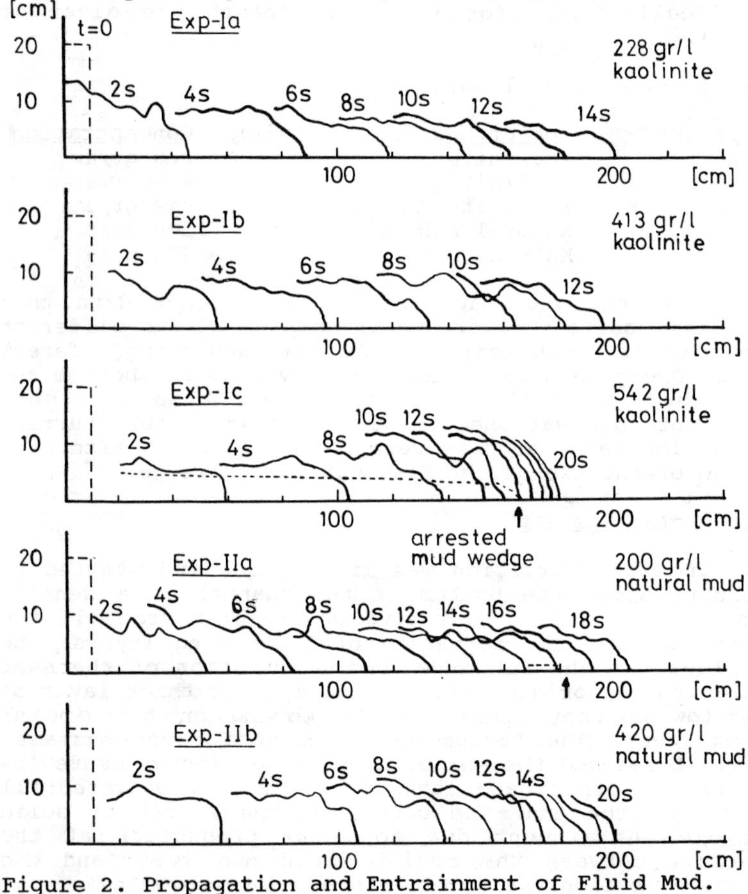

Figure 2. Propagation and Entrainment of Fluid Mud.

Analysis of the Experimental Data

Qualitatively, the phenomenon of fluid mud dynamics in lock-exchange type of experiments can be explained as follows: The densimetric potential energy of the fluid mud is converted, after the gate opening, into kinetic energy which eventually dissipates under the action of boundary and internal friction. As a result of the interfacial shear the fluid mud entrains into the water column. The behavior of the bottom hyperconcentrated layer, for the kaolinite mixtures, resembles Bingham fluid with yield strength τ_y = 1.5 N/m^2 and plastic viscosity μ_p = 1.5 N/m^2 (Williams, 1986). Assuming for the kaolinite experiments a bottom layer depth between 1-cm ≤ h ≤ 2.5-cm, and a shear stress between 0.25 N/m^2 ≤ τ ≤ 2 N/m^2, then the Reynold's and Hedstrom's numbers can be estimated correspondingly as: $\mathbf{R_e}$ = 1.7 x 10^3 - 4.2 x 10^3, $\mathbf{H}$ = 1.1 x 10^4 - 5.8 x 10^5. Those values indicate that the flow of the thin bottom layer is laminar (Jeyapalan, et. al., 1983). To investigate the propagational characteristics of the fluid mud, the movement of the mud wedge tip was plotted in terms of distance vs. time and velocity vs. time (figures 3). From these figures it is evident that the wedge decelerates at an exponential rate, and this in qualitative agreement with theoretical results (Jeyapalan, et. al., 1983).

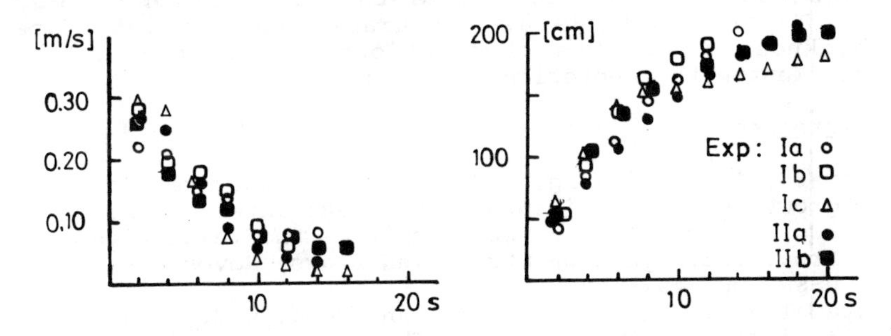

Figure 3. Propagation of the Mud Wedge Tip.

Assuming that the velocity of the mud wedge tip is representative of the shear production at the interface between the lower dense layer and the turbid cloud, and utilizing the fact that during the initial time period of propagation the mud velocity is approximately constant, the entrainment velocity V_e can be estimated as h/t ≈ 0.04 m/s, where h is the thickness of the turbid cloud and t is the time required for its growth. The corresponding Richardson's number $\mathbf{R_{iu}}$ = g $(\Delta\rho/\rho)h/U^2$ is of the order of 1. Internal waves were observed at the interface between the turbid cloud and the overlaying water. Their wave length varied between 5 and 20 cm for the various experiments. A relation between the wave length λ of those internal waves and the interfacial thickness δ is given as $2\pi\delta/\lambda \approx 0.8$. Therefore, $\delta \approx 0.64 - 2.54$ cm which is comparable with the amplitude of the internal waves. Based on the interfacial thickness δ, a Richardson's number defined as $\mathbf{R_i} = (g/\rho)(\Delta\rho'\delta)/U^2$ can be estimated, where $\Delta\rho'$ is the density difference between the cloud and the clear water and U is the horizontal velocity of the cloud. For all of the experiments, the value of the $\mathbf{R_i}$ were much less than 0.3, which is the critical value for instability to occur (Wyatt, 1978).

Summary and Conclusions

Lock-exchange type of experiments were conducted utilizing high concentrations of fluid mud, composed either from pure kaolinite or natural mud mixtures. The conclusions derived from this study are:
a) When the fluid mud is released two propagating layers are formed: one thin, high concentrated, near the bottom layer, and one cloud of very low concentration moving over the bottom layer; b) The flow of the lower layer is laminar and decelerates exponentially; c) The entrainment of the bottom layer is of the order $V_e/U \approx 0.16$ and the Richardson's $\mathbf{R_{iu}} \approx 1$; d) Unstable internal waves were formed on the cloud surface which had a length of 5-20 cm. Their amplitude is comparable to the interface thickness δ; e) The observations are in qualitative agreement with theoretical results.

References

Broward, F.K. and Wang, Y.H., 1972. An experiment on the growth of small disturbances at the interface between two streams of different densities and velocities. Proc. Int. Symp. on Stratified Flows, Novosibirsk, USSR, pp. 491-498.

Jeyapalan, J.K., Duncan, J.M., and Seed, H.B., 1983. Analyses of flow failures of mine tailings dams. Journal of Geotechnical Engineering, ASCE, Vol. 109, No. 2, pp. 150-171.

Srinivas, R., 1989. Response of Fine Sediment-Water Interface to Shear Flow. M.Sc. Thesis, Dept. of Coastal and Oceanographic Engineering, University of of Florida, Gainesville, Florida, pp. 114.

Williams, D.J.A., 1986. Rheology of cohesive suspensions. In: Estuarine Cohesive Sediment Dynamics, A.J. Mehta, Ed., Lecture Notes on Coastal and Estuarine Studies, Vol. 14, Springer-Verlag, pp. 110-125.

Wyatt, L.R., 1978. The entrainment interface in a stratified fluid. Journal of Fluid Mechanics, Vol. 86, Part 2, pp. 293-311.

ACTIVATION AND DEGENERATION OF TURBIDITY CURRENTS

By Muneo Hirano1, Kesayoshi Hadano2,
Juichiro Akiyama3 and Takashi Saitou4

ABSTRACT

Flow properties of the front of a turbidity current are investigated using the two layers model which takes into account of pick up and deposition of sediment. The amount of pick up rate was estimated through laboratory experiments. Governing equations of the flow thickness, density and velocity of the head of turbidity currents are derived by means of the characteristic curve method. From inspection of the resultant equations, division of progress and decay of turbidity currents is obtained.

INTRODUCTION

A turbidity current, which derives its driving force from sediment in suspension, is a family of a gravity current. Transport, erosion, and deposition process of sediment due to turbidity currents are of great practical significance. Since the presense of suspended sediment has a major effect on the dynamics of the flow, the hydrodynamics of turbidity currents is very complex. If suspended sediment is lost by sedimentation or added by resuspension from the bed, the driving force of the turbidity current changes.

In this paper, an integral model is proposed on the motion of the head of a turbidity current with erosion and deposition of uniform sediments. Since the densimetric Froude number governs the flow velocity itself, and bottom shear stress governs pick up rate of sediment, the analysis is directed towards the progress and decay of the current owing to the changes of these parameters.

---

1 Prof., Civil Eng.(Suiko),Kyushu Univ., Fukuoka, Japan.
2 Assoc. Prof.,Civil Eng.,Yamaguchi Univ., Ube, Japan.
3 Assoc. Prof.,Civil Eng., Kyushu Inst. of Technology, Kitakyushu, Japan.
4 Prof., Civil Eng., Yamagichi Univ., Ube, Japan.

THEORY

GOVERNING EQUATIONS

The flow treated here is a two-dimensional unsteady gravity current with pick up and deposition. As in Fig.1, the two layers model is employed. Governing equations are those of conservations of volume, mass, and the x-component of equation of motion, such that

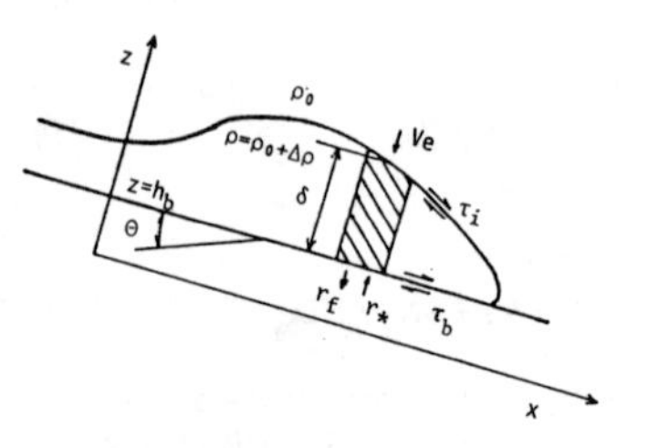

Fig.1 Definition

$$\frac{\partial \delta}{\partial t} + \frac{\partial q}{\partial x} = v_e + r_* - r_f \tag{1}$$

$$\frac{\partial(\rho\delta)}{\partial t} + \frac{\partial(\rho q)}{\partial x} = \rho_o v_e + \{\sigma(1-\lambda) + \rho_o \lambda) \tag{2}$$

$$\int_{h_b}^{h_b+\delta} \rho(\frac{\partial u}{\partial t} + u\frac{\partial u}{\partial x} + w\frac{\partial u}{\partial Z})\, dZ = -\int_{b_b}^{h_b+\delta} (\frac{\partial P}{\partial x} + \rho g \, \text{Sin}\theta)\, d Z + \tau_i - \tau_b \tag{3}$$

where, $q = q(x,t) = \int_{h_b}^{h_b+\delta} u(x,z,t)dz$, Ve = the entrainment velocity, r* and rf = the degrading and aggrading rates of the bed due to erosion and deposition of sediment, respectively. σ = the density of solid particle, λ = the percentage of void, τ_i and τ_b = the interfacial and bottom shear stresses, respectively.

After changed by means of continuity equation, Boussinesq approximation and assumption of hydrostatic pressure, these equations are expressed as below along the characteristic curve $dx/dt=\bar{u}$.

$$\frac{d\delta}{dx} = E - \frac{\delta}{\bar{u}}\frac{\partial \bar{u}}{\partial x} + \frac{r_* - r_f}{\bar{u}} \tag{4}$$

$$\frac{d\Delta\rho}{dx} = -\frac{\Delta\rho}{\delta}\{E + \frac{r_* - r_f}{\bar{u}}\} + \frac{(1-\lambda)(\sigma-\rho_o)}{\delta}\cdot\frac{r_* - r_f}{\bar{u}} \tag{5}$$

$$\frac{d\bar{u}}{dx} = -\frac{\bar{u}}{\delta}\{\frac{\rho_o}{\rho}(1-k)E + \frac{\sigma(1-\lambda) + \rho_o\lambda}{\rho}\frac{r_* - r_f}{\bar{u}} + f_i + f_b\} + \frac{\Delta\rho}{\rho}\frac{g}{\bar{u}}\sin\theta - \frac{g}{\rho\delta\bar{u}}\cos\theta\frac{\partial}{\partial x}(\frac{1}{2}\Delta\rho\delta^2) - 2(\beta-1)\frac{\partial\bar{u}}{\partial x} - (\beta-1)\frac{\bar{u}}{\delta}\frac{\partial\delta}{\partial x} - \bar{u}\frac{\partial\beta}{\partial x} \tag{6}$$

where, $\Delta\rho = \rho - \rho_o$, E = the entrainment coefficient($=Ve/\bar{u}$), $k=u|_{z=hb+\delta}/\bar{u}$, b = the momentum correction factor, and fi and fb = the coefficients of the interfacial and bottom shear stresses defined respectively by $-\tau_i=\rho f_i\bar{u}^2$ and $\tau_b=\rho f_b\bar{u}^2$..

Partial differentiated terms on the right hand side of Eq.(6) are difficult to be evaluated properly. Formerly Hadano(1981) verified experimentally that the shape of the head of a gravity current without pick up and deposition remains similar figures and that its maximum thickness increases almost in proportional to moved distance, and made an analysis by putting m, defined by the following form;

$$m=E-\delta/u\ \partial u/\partial x \tag{7}$$

to be constant and focusing on the section of the maximum thickness of the head. The same procedure will be employed now. In the section of the maximum thickness of the head $\partial\delta/\partial x=0$. Furthermore, we assume that $\partial\Delta\rho/\partial x=0$ in this section. Finally Eqs.(4), (5) and (6) are written into the following dimensionless form;

$$\frac{dD}{dx} = m + \frac{r_* - r_f}{\bar{u}} \tag{8}$$

$$\frac{dR}{dx} = -\frac{R}{D}\{E + \frac{r_* - r_f}{\bar{u}} - (1-\lambda)\ F_o^2 \frac{\delta_o}{d} \frac{f_b}{\tau_{*o}} \frac{r_* - r_f}{\bar{u}} \frac{1}{R}\} \tag{9}$$

$$\frac{dU}{dx} = -\frac{U}{D}\{\frac{\rho_o}{\rho}(1-k)E + \frac{\sigma(1-\lambda) + \rho_o\lambda}{\rho}\frac{r_* - r_f}{\bar{u}} + f_i + f_b\} + \frac{R}{U}F_o^{-2}\sin\theta + 2(\beta-1)(m-E)\frac{U}{D} + U\frac{\partial\beta}{\partial x} \tag{10}$$

where, $X = x/\delta_o$, $D = \delta/\delta_o$, $R = \Delta\rho/\Delta\rho$, $U = \bar{u}/\bar{u}_o$, $F_o = \bar{u}_o\sqrt{\Delta\rho_o g\delta_o/\rho_o}$ and $\tau_{*o} = f_b\ \bar{u}_o^2/\{(\sigma/\rho_o - 1)gd\}$.

. Analytical solution can be obtained if r*=rf and E, Fi, fb and k are constant.

ESTIMATION OF r* AND rf

The rate of degrading of the bed resulting from erosion, r*, will be evaluated with the concept of sediment pick up rate. Pick up rate, Ps, is defined as a probability that one particle located on the bed would be entrained by the flow per unit time. In the case of simple cubic piling the volume of sediment picked up during unit time from unit area of bed surface, rs, is expressed using Ps in the following form;

$$r_s = P_s \pi\ d/6 \tag{11}$$

where, $\pi/6$ = the solid volume percentage, $(1-\lambda)$, in this piling condition, and d = the diameter of particle. Thus r*, rs and Ps are related by the following relation.

$$r_* = P_s d = r_s/(1-\lambda) \tag{12}$$

Laboratory experiments were conducted to quantify the pick up rate of fine sediment entrained by the heads of bottom density currents made of brine solution flowing over the bed composed of sillica sand with specific gravity $\sigma/\rho_0 = 2.65$ and medium diameter d50 = 47µm (Akashi(1980). The result is given in Fig.2. For comparison, the Tsujimoto's (1978) formula on the pick up rate is also added to Fig.2. The formula is given as Eq.(13) using the dimensionless tractive force (τ_*) and the critical value for incipient motion of sediment particle ($\tau*c$).

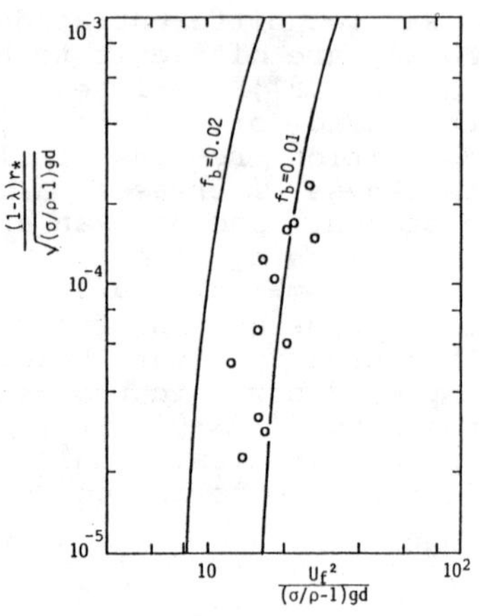

Fig.2 Pick up rate

$$p_s\sqrt{\frac{d}{(\sigma/\rho-1)g}} = 0.03\ \tau_*(1-\tau_{*c}/\tau_*)^3\ ;\ \tau_* = \frac{u_*^2}{(\sigma/\rho-1)gd} \tag{13}$$

The value of τ_{*c} is estimsted from Iwagaki's (1956) formula. Bottom shear stress is estimated from experimental results on plane wall jet by Sigalla(1958), and fb is evaluated referring to measurements of velocity profile of gravity currents (e.g., Fukuoka et al.(1978), Akashi(1980)). Fig.2 implies that Eq.(13) is valid for the pick up rate of fine sediment by a bottom density current. From Eq.(13) r*/u is written into the following form;

$$r^*/\bar{u} = 0.03\ \sqrt{f_b}\ \tau_*\ (1-\tau_{*c}/\tau_*)^3 \tag{14}$$

The aggrading rate of the bed, rf, may be estimated by

$$r_f = \frac{1}{1-\lambda}\,cw_0 = \frac{1}{1-\lambda}\,\frac{\Delta\rho}{\sigma-\rho_0}\,w_0 \tag{15}$$

where, wo = the settling velocity of a particle, C = the volumetric concentration. Accordingly, rf/u may be represented by the following form using the previously defined nondimensional quantities.

$$\frac{r_f}{\bar{u}} = \frac{F_0^{-2}}{1-\lambda}\,\frac{\tau_{*0}}{f_b}\,\frac{d}{\delta_0}\,\frac{w_0}{\bar{u}_0}\,\frac{R}{U} \tag{16}$$

RESULTS AND DISCUSSION

Flow properties of turbidity currents will be investigated. When neither pick up nor deposition of sediment occurs, flow properties of such conservative bottom density currents primarily depend on the initial value of the densimetric Froude number Fo(Hadano(1981)). In the case of turbidity currents with erosion or deposition, entrainment and deposition of sediment are an additional important factors. As obvious from Eq.(13), pick up of sediment depends on whether the tractive force is larger than the critical value of the tractive force τ_{*c}.

When Fo is small so that $dU/dx|_{x=0} > 0$ and $\tau_{*0} > \tau_{*c}$, U and therefore τ_* will increase and pick up will always occur. Once pick up of sediment becomes active, both density and velocity increase and hence pick up of sediment is activated. By these multiplier effects density and velocity increase rapidly. The condition that $\tau_* > \tau_{*c}$ and $dU/dX > 0$ at X=0 becomes as Eq.(17), when we put D = U = R = 1 and $\sigma/\rho = \sigma/\rho_0$ in Eq.(10).

$$\tau_{*0}/\tau_{*c} > 1$$

and

$$F_0^2 < \frac{\sin\theta + \frac{1}{1-\lambda}\frac{\tau_{*o}}{f_b}\frac{d}{\delta_o}\frac{w_o}{\bar{u}_o}\frac{\sigma(1-\lambda)+\rho_o\lambda}{\rho_o}}{(1-k)E + 0.03\{\frac{\sigma(1-\lambda)+\rho_o\lambda}{\rho_o}\}\sqrt{f_b\,\tau_{*o}}\,(1-\frac{\tau_{*c}}{\tau_*})^3 + f_i + f_b} \tag{17}$$

The range of Fo and τ_{*0}/τ_{*c} to satisfy Eq.(17) is indicated as the region A in Fig.3. Turbidity currents which start from this region will become active and pass into mud flows.

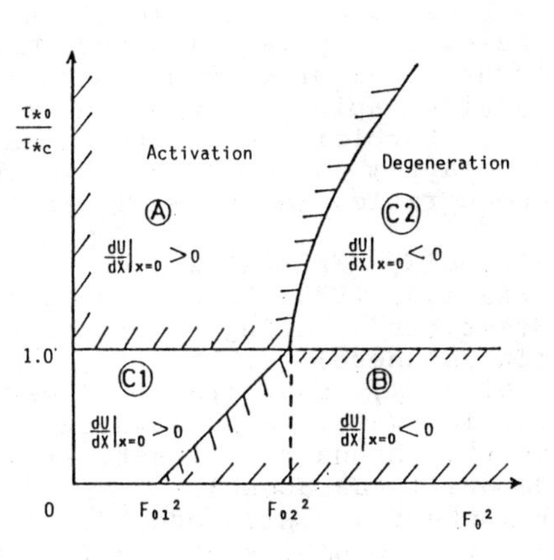

Fig.3 Regional division

When Fo is large so that $dU/dX|_{x=0} < 0$ and $\tau_{*0} < \tau_{*c}$, velocity and density monotonously decrease. In this case, exchange of sediment at the bed surface is limited to deposition only. It suggests that density current will ultimately vanish. Condition that $\tau_{*0} < \tau_{*c}$ and $dU/dX < 0$ at X=0 becomes as follows:

$$\tau_{*0}/\tau_{*c} < 1$$

and

$$F_0^2 > \frac{\sin\theta + \frac{\sigma(1-\lambda)+\rho_0\lambda}{\rho_0}\frac{1}{1-\lambda}\frac{\tau_{*o}}{f_b}\frac{d}{\delta_o}\frac{w_o}{\bar{u}_o}}{(1-k)E + f_i + f_b} \tag{18}$$

The range of Fo and τ_{*0}/τ_{*c} to satisfy Eq.(18) is indicated as region B in Fig.3.

The flows which start from the region neither A nor B in Fig.3 will pass into one of these regions after acceleration or deceleration, and will grow or decay, respectively. Which course the turbidity current would take may depend on the initial condition Fo, τ_{*0}/τ_{*c} ; the value of E, m, $\sin\theta$ etc.; and parameters concerning the sediment such as wo/uo and d/δ_0 . Noticed that the values of F_{o1}^2 and F_{o2}^2 in Fig.3 are as follows:

$$F_{01}^2 = \frac{\sin\theta}{(1-k)\ E + f_i + f_b}$$

$$F_{02}^2 = \frac{\sin\theta + \dfrac{1}{1-\lambda}\dfrac{\tau_{*c}}{f_b}\dfrac{d}{\delta_o}\dfrac{w_o}{\bar{u}_o}\dfrac{\sigma(1-\lambda)+\rho_o\lambda}{\rho_o}}{(1-k)E + f_i + f_b} \qquad (19)$$

CONCLUSIONS

Flow property of the unsteady state fronts of turbidity currents with erosion and deposition was investigated focusing on the dependence on Fo and τ_{*0}/τ_{*c}. The main results are as follows; When Fo is smaller than F_{o2} indicated by Eq.(19) and $\tau_{*0}/\tau_{*c} > 1$, pick up always occurs and gravity currents increases its velocity and density rapidly. When Fo is larger than Fo2 and $\tau_{*0}/\tau_{*c} < 1$, turbidity currents loses its velocity and density. A diagram describing the regions of activation and degeneration was shown in Fo - τ_{*0}/τ_{*c} plane.

APPENDIX, REFERENCES

Akashi,J.(1980)."Study on Turbidity Currents flowing into Reservoir", M.Eng. Theisis, Kyushu Univ., Fukuoka, Japan (In Japanese).

Ashida,K., Egashira,S. & Nakagawa,H.(1981)."Behavior of Turbid Water Mass Released Into Two Dimensional Reservoir", Annuals. Disast. Prev. Inst. Kyoto Univ., Kyoto, Japan.(in Japanese).

Fukuoka,S., Mizumura,M. & Kano,T.(1978)."Fundamental Study on Dynamics of the Head of Gravity Currents, Proc. of JSCE, 274, 41-55.(in Japanese).

Hadano, K.(1981)."Study on the Front of a Turbidity Current", Memo. of the Faculty of Eng., Kyushu Univ., Fukuoka, Japan, Vol.41, No.4, 307-317.

Iwagaki,Y.(1956)."Hydrodynamical Study on Critical Tractive Force, Trans. of JSCE., Vol.41, 1-21.(in Japanese).

Sigalla,A.(1958)."Measurement of Skin Friction in a Plane Turbulent Wall Jet", J.Aeroneut.Sci.,62, 873-877.

Tsujimoto, T.(1978)."A probablistic Model on Bed Load Process and Its Application to The Problems of a Movable Bed, D.Eng.Theisis, Kyoto Univ., Kyoto, Japan. (in Japanese).

Resistance Consisting of Both Viscous Force and Dispersive Force*

Wan Zhaohui[1] Hua Jingsheng[2]

Abstract

A set of specially designed experiments have been carried for studying the resistance of a flow consisting of water and both fine and coarse particles. The viscous force caused by fine clay particles and the reduced dispersive force are linearly addible. The dispersive force caused by collision of coarse particles reduces to a great degree due to the cushioning effect of the flocculent structures formed by fine particles. The preliminary result is presented in this paper and further study is needed.

Introduction

Debris flow or hyperconcentrated flow usually carries sediment with wide range of sizes, including a certain amount of clay, silt, fine and coarse sand, gravel and even large stones. Study on resistance of flow with both fine and coarse particles and the interaction between these two kinds of particles is meaningful. For this purpose a set of specially designed experiments have been carried out and a preliminary analytical result is presented in this paper.

Experiments

Experiments were carried out in a capillary viscometer. The length and the inner diameter of the capillary are 100.3 cm and 5 mm, respectively. Rheological properties of clay slurry, consisting of water and pure clay particles, with a specific gravity 1.41, which was

---

1. Professor, Senior Engineer, Institute of Water Conservancy and Hydroelectric Power Research (IWHR), Box 366, Beijing, China
2. Engineer, IWHR
* The Project Supported by National Natural Science Foundation of China

equal to the specific gravity of coal grains, was measured at first. Then uniform coal grains with a diameter of 0.9 mm were added gradually and rheological properties of clay slurries containing different amount of coal grains were measured. The content of coal grains was up to 0.352, in volume. The settling of coal grains in clay slurries could be avoided due to the same specific gravity. The rheological property of pure clay slurry could be well described by Bingham fluid. The existence of coarse coal grains induced the increase of shear stress at the same shear rate. But yield stress did not change obviously. Part of the results of rheological measurement is shown in Fig. 1.

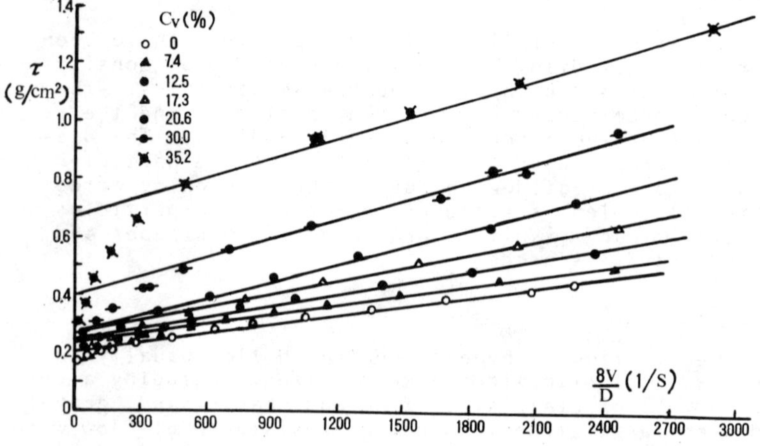

Fig. 1 Results of Rheological Measurement

Mechanism and Hypothesis

Based on the results of rheological measuement, the following hypothesis are made. 1. In a flow with both fine and coarse grains viscous force and dispersive force coexist simutaneously. 2. Viscous force mainly caused by fine clay particles can be described by Bingham model. 3. Dispersive force mainly caused by mutual collision of coarse grains and collision between the boundary and coarse grains can be described by Bagnold's model. 4. The viscous force and the dispersive force are linearly addible in principle. 5. The flocculent structures formed by fine clay particles absorb shocks between coarse particles and produce a cushioning effect, thus dispersive force caused by mutual collision reduces to a certain degree, which can be reflected by as coefficient K. The foregoing idea can be written as:

$$\tau = \tau_B + \eta \frac{du}{dy} + KT \tag{1}$$

in which τ is the shear stress at the point y from the boundary where the flow velocity is u, τ_B and η are yield stress and rigidity of the clay slurry. T is the dispersive force caused by mutual collision of coarse particles.

Expressions of Dispersive Force

Dispersive force can be described by Bagnold's formula, which is based on a set of specially designed experiments. (Bagnold, R., 1956).

In viscous region, that is, $N = \dfrac{\lambda^{1/2} \rho_s D^2 \frac{du}{dy}}{\eta} < 70$

$$T = 2.2 \lambda^{3/2} \eta \frac{du}{dy} \tag{2}$$

In inertial region, N > 400

$$T = 0.013 (\lambda D)^2 \left(\frac{du}{dy}\right)^2 \tag{3}$$

ρ_s and D are density and diameter of coarse grains, λ is the linear concentration defined by Bagnold as:

$$\lambda = \left[\left(\frac{C*}{Cv}\right)^{1/3} - 1\right]^{-1} \tag{4}$$

Cv is the volumetric concentration of coarse particles and C* is its maximum value, or packing concentration, taken as 0.65.

Bagnold did his experiments with Newtonian fluid, in that case viscosity was the only rheological parameter. In our case experiments were carried out with clay slurry, so effective

viscosity $\eta_e = \eta \left(1 + \dfrac{\tau_B}{\eta \frac{8V}{d}}\right)$ was used to replace η

in Eq. (2). Here V and d are the average velocity and the inner pipe diameter.

Basic Equations

The rheological equation of a fluid can be written as:

$$\frac{du}{dy} = f(\tau) \tag{5}$$

The discharge Q of a laminar flow in a pipe can be obtained by the following integration.

$$\frac{Q}{\pi R^3} = \frac{1}{4}\left(\frac{8V}{d}\right) = \frac{1}{\tau_w^3}\int_0^{\tau_w} \tau^2 f(\tau)\, d\tau \qquad (6)$$

in which τ_w is the shear stress at the boundary and R is the inner pipe radius. Only if the expression of $f(\tau)$ is given, this integration can be obtained.

As discussed above, in the case of a laminar flow in pipe, containing both fine and coarse particles, shear stress can be summarized as:

$$\tau = \begin{cases} \tau_B - \eta \frac{du}{dr} + K_1 0.013 \rho_s (\lambda D)^2 \left(\frac{du}{dr}\right)^2 & \text{(for N>400)} \\ \tau_B - \eta \frac{du}{dr} - K_2 2.2 \lambda^{3/2} \eta \frac{du}{dr} & \text{(for N<70)} \end{cases} \qquad (7)$$

Here K_1 and K_2 are used in different regions. (inertial, viscous) r is the distance from the pipe center.

In viscous region (N<70) the following equation is deduced from Eq. (7).

$$f(\tau) = -\frac{du}{dr} = \frac{\tau - \tau_B}{\eta (1 + K_2 2.2 \lambda^{3/2})} \qquad (8)$$

Substituting Eq. (8) into Eq. (6) and assuming to be constant as a first approximation, we can easily get:

$$\frac{8V}{d} = \frac{\tau_w}{\eta (1 + K_2 2.2 \lambda^{3/2})} \left\{1 - \frac{4}{3}\frac{\tau_B}{\tau_w} + \frac{1}{3}\left(\frac{\tau_B}{\tau_w}\right)\right\} \qquad (9)$$

In inertial region (N>400) the situation is a little more complicant.

$$\tau = \tau_B - \eta \frac{du}{dr} + K_1 0.013 \rho_s (\lambda D)^2 \left(\frac{du}{dr}\right)^2 \qquad (10)$$

It is a two order equation of (- du/dr). Let b = η and a = K_1 0.013 ρ_s $(\lambda D)^2$, as (-du/dr) is always positive, we can write down:

$$f(\tau) = -\frac{du}{dr} = \frac{-b + \sqrt{b^2 - 4a(\tau_B - \tau)}}{2a} \qquad (11)$$

Substituting Eq. (11) into Eq. (6) and integrating, we obtained:

$$\frac{Q}{\pi R^3} = \frac{1}{4}\frac{8V}{d} = \frac{b}{2a}\cdot\frac{\tau_B^3 - \tau_w^3}{3\tau_w^3}$$

$$+ \frac{1}{6720 a^4 \tau_w^3}\left\{240a^2\left[\tau_w^3\sqrt{(4a\tau_w + c)^3} - \tau_B^2\sqrt{(4a\tau_B + c)^3}\right] - 48ac\left[\tau_w\sqrt{(4a\tau_w + c)^3}\right.\right.$$

$$\left.\left. - \tau_B\sqrt{(4a\tau_B + c)^3}\right] + 8c^2\left[\sqrt{(4a\tau_w + c)^3} - \sqrt{(4a\tau_B + c)^3}\right]\right\} \qquad (12)$$

Eq. (10) and Eq. (12) are basic equations of flow containing both fine and coarse particles in viscous region and in inertial region, respectively. Both of them are functions of ρ_s, d, λ, τ_B, η, τ_w and coefficient K_1, or K_2.

Verification

In our experiments d, D and ρ_s were given. Rheological parameters of pure clay slurry τ_B and η were obtained by rheological measurement at first. Then in each run of rheological measurement of fluid containing both fine and coarse particles Q(V), $\lambda(C_v)$ and τ_w at each point were measured.

Based on listed above parameters, a measured 8V/d, that is, $(8V/d)_{mea}$ could be obtained. Provided coefficient K_1 (or K_2) is given, we can also calculate a 8V/d, that is, $(8V/d)_{cal}$ according to either Eq. (10) or Eq.(12). In our treatment we got K_1 or K_2 by trial. It is found that K_1 or K_2 varies with concentration C_v, as shown in Fig. 2.

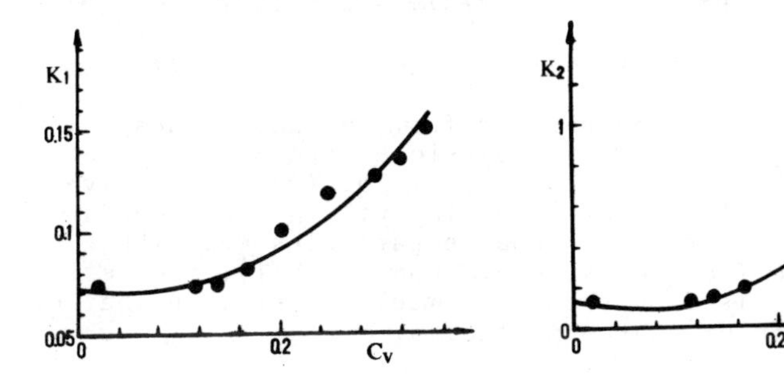

Fig. 2 K 1 ~ (K 2) - C v

Once K_1 or K_2 was determined, $(8V/d)_{cal}$ could be calculated and they were compared with $(8V/d)_{mea}$ obtained by direct measurement, see Fig. 3. The agreement between them is quite satisfied.

Discussion and Conclusions

1. In a flow containing fine and coarse particles both fine and coarse particles contribute to the flow resistance. The total flow resistance can be assumed to be the sum of viscous force caused by clay particles and reduced dispersive force caused by collision between coarse particiles.

2. Viscous force can be described by Bingham model. Dispersive force can be described by Bagnold's formula. The flocculent structures formed by fine partiles absorb shocks between coarse particles and produce a cushing effect. Thus the dispersive force reduces to a certain degree correspondingly.

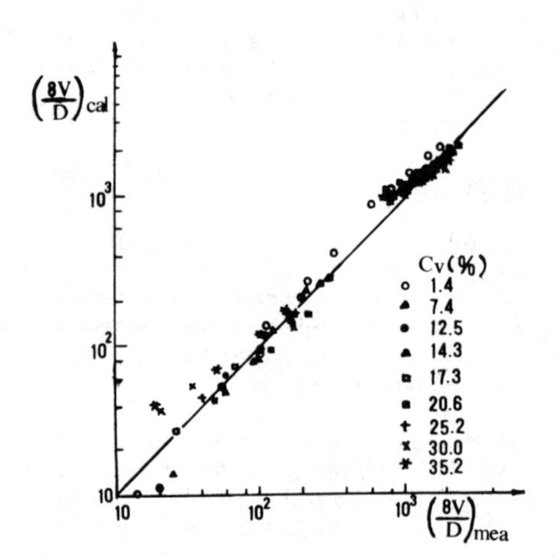

Fig. 3 Comparison between $(8V/d)_{cal}$ and $(8V/d)_{mea}$

3. The reduction of dispersive force due to the flocculent structures is quite obvious. K_1 varies in the range of 0.075 - 0.15, K_2 varies in the range of 0.15 - 0.95.

4. Some concepts can be deduced from the above idea. In the case of debris flow both particle size D and concentration λ are large. As seen from Eq. (3), dispersive force should be large correspondingly. In other words, in the case of debris flow coarse particles contribute greatly to the flow resistance. Consequently, the resistance of a debris flow should be much larger than that of a mud flow without coarse particles.

5. Compared to that of a water-stone debris flow, the resistance of a mud-stone debris flow reduces greatly due to the cushioning effect of the flocculent structures formed by clay particles.

It explains the following fact in nature: water-stone debris flow usually occurs in gullies with longitudinal slope larger than 0.05, and mud-stone debris flow may occur in gullies with slope of 10^{-3} - 10^{-4}.

Reference

Bagnold, R.A., 1956, The Flow of Cohesionless Grains in Fluids, Royal Society of London.

LIMIT CONCENTRATION OF SUSPENDED SEDIMENT[1)]

Wang Zhaoyin[2)]

Abstract

It is found from experiments that there is a limit on suspended sediment concentration. The limit concentration of uniform sediment varies with sediment diameter, and it is about 1000 kg/m^3 for cohesionless sediment finer than 0.15 mm and it reduces from 1000 to 200kg/m^3 as sediment diameter increases from 0.15 to 5 mm. The limit concentration of sediment of diameter d is not affected by existence of suspended sediment finer than d/10, so that the limit concentration of nonuniform sediment is larger than that of uniform sediment. The concept of limit concentration is verified in a flume experiment. The capacity of flow to carry suspended sediment increases with flow rate in general cases. As the concentration reaches the limit concentration, however, further increasing flow rate and adding sediment raise nothing in suspended sediment concentration. On the other hand, the total concentration can be easily increased to 1200 kg/m^3 at the same flow rate if much coarser sediment is added into the flow saturated with finer sediment. The wider the range of sediment size distribution, the easier the flow to carry more suspended sediment. This interpretes the phenomenon of increase in median diameter of suspended load with increasing concentration in hyperconcentrated flow in the Yellow River and its tributories.

Concept of limit concentration of suspended sediment

Hyperconcentrated flood often occurs in the middle reach of the Yellow River and its tributories. The highest concentration of suspended sediment recorded in the Yellow River is 933 kg/m^3. The Kuye River watched a hyperconcentrated flood of suspended sediment 1500 kg/m^3 in 1964, and the Jingbian Gauging Station on the Luhe River recorded 1540 kg/m^3 of suspended sediment concentration in Aug., 1969, which is still higher than concentration of sand compacted in the river bed. The extreamly high concentration is usually accompanied by very wide range of sediment size distribution.

On the other hand, laboratory experiments suggest that the concentration of suspended uniform sediment is limited to 1000 kg/m^3. An

---

1) The project supported by National Natural Science Foundation

2) Dr. Eng., Senior Engineer, Institute of Water Conservancy and Hydroelectric Power Research, P.O.Box 366, Beijing, China.

experiment was conducted in a close-circuit pipe of cross-section $0.18 \times 0.10\ m^2$ (Wang and Qian, 1984). An uniform sand with diameter d= 0.15 mm and specific weight 2.64 g/cm^3 was used as suspended load in the experiment. Average discharge and average concentration of suspended sediment were measured by using a gauging tank. Several manometers were employed for measuring frictional loss of head. The average velocity varied between 0.5 and 5 m/s. It was found that as the average velocity was over 2 m/s, all sediment particles suspended in the flow. Following adding more sediment into the recirculating system the average concentration of suspended sediment increased while the head loss also increased but at a very low rate. As the concentration was over 900 kg/m^3, however, to further increase concentration was very difficult and was accompanied by high-rate increase in head loss. The maximum concentration measured in the experiment was 980 kg/m^3. To attempt to get higher concentration we added more sediment into the system. It turned out that the head loss increased sharply and pipe-clogging occured.

The similar phenomenon was observed in a flume experiment. The suspended load was a sand with d_{50}=0.31 mm, d_{90}/d_{10}=5.3 (fig.1) and the liquid phase was dilute clay mud. The concentration of suspended load increased fast as the concentration was not very high, and it kept constant as it had reached 1000 kg/m^3 no matter how the flow rate increased further and much more sediment was put into the flume.

The facts mentioned above suggest a limit concentration of suspended load. It seems that the limit concentration is about or less than 1000 kg/m^3. Nevertheless the highest concentration recorded in river is much higher than 1000 kg/m^3, which comes into conflict with the results from the experiments. To answer the problem a special experiment is designed and conducted.

Limit concentration of different sediments

The experiment is carried out in a plexiglass agitator tank as shown in fig.2. 11 uniform sediments, with diameter d=0.015, 0.051, 0.15, 0.23, 0.31, 0.56, 0.82, 0.9, 1.0, 2.2, 5.0 mm, are used as suspended load, in which the sediments with d=0.015 and 0.051 mm are made by milling quartz sand into powder and the others are got by screening from natural sand. Water in the tank is made fully-turbulent by turning blades of the agitator. The experiment is conducted by

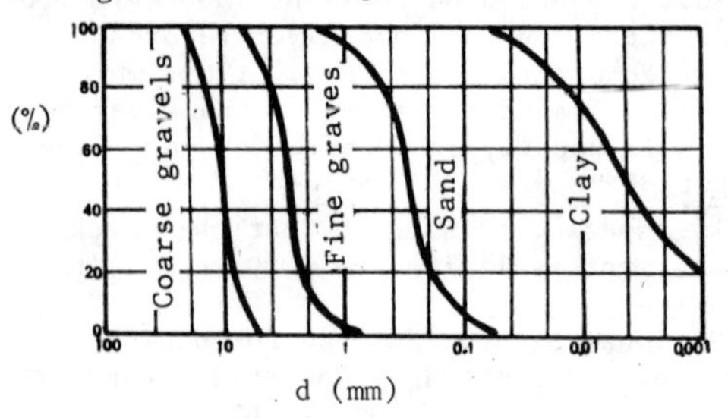

Fig.1 Size distributions of sediments

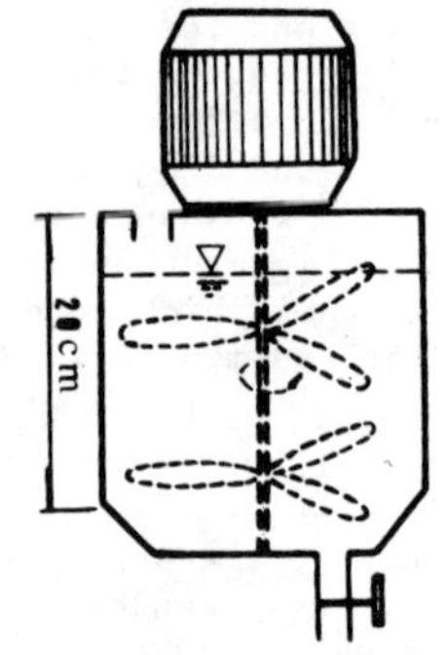

Fig.2 Agitator tank

putting each sediment into the tank, measuring concentration and observing suspension or deposition of sediment. As the concentration is lower than the limit concentration, the sediment can suspend completely and uniformly in the tank. As the concentration reaches the limit concentration, adding more sediment results in deposition of sediment on the bottom of the tank and no further increase in concentration. Fig.3 presents the limit concentration, S_m, versus diameter d. It can be seen that S_m maintains about 1000 kg/m^3 as d is smaller than 0.15 mm and reduces very fast with increasing d as d is larger than 0.2 mm.

The limit concentration of sediment with diameter d is not affected by existence of much finer suspended sediment. Table 1 gives the limit concentration of d=0.15mm sediment in water and clay muds (Size distribution of the clay is shown in fig.1). Table 2 presents limit concentrations of coarse sediments in fine sediments suspensions. These results prove that the limit concentration of sediment of diameter d is not affected by existence of sediment finer than d/10.

Table 1 Limit concentration of 0.15mm sediment in water or muds

Volume concentration of clay mud	0.0	0.025	0.038	0.053	0.083	0.113
Limit concentration S_m (kg/m^3)	987	989	1000	991	1007	995
Relative error $\lvert S_m-995 \rvert/995$	0.008	0.007	0.005	0.012	0.007	0.00

Table 2 Limit concentration of coarse sand in fine sediment slurry

Diameter of coarse sand (mm)	0.56	0.56	0.56	2.25	2.25	2.25
Diameter of fine sediment (mm)		0.05	0.05		0.05	0.05
S of fine sediment suspension (kg/m^3)	0	700	900	0	300	700
S_m of coarse sediment (kg/m^3)	670	690	680	270	280	285

It is deduced from the property of limit concentration that the limit concentration of non-uniform sediment is higher than that of uniform sediment. The limit concentration of 0.3 mm uniform sediment is 880 kg/m^3 (fig.3), but the highest concentration of sediment with d_{50}=0.3 and d_{90}/d_{10}=5.3 is 1010 kg/m^3. It is safe to say that the concentration of suspended sediment with continuous and wide range size distribution is possibly much higher than S_m of all uniform sediment. It is proved by an experiment with mixed sediment. Table 3 gives proportions of different sediments in the mixed sediment, in which clay concentration is taken as 250 kg/m^3 and other sediment's dosages are calculated according to S_m of uniform sediment and taking the space occupied by coarser particles into consideration.

Table 3 Proportions of sediments in the mixed sediment

Diameter d (mm)	2.5–5.0	0.4–0.6	0.04–0.06	$<$0.01 (Clay)
S_m (kg/m^3)	160	740	1015	250
Dosage (kg/m^3)	160	695	688	104
Percentage (%)	9.71	42.19	41.77	6.33

By putting the mixed sediment into the agitator tank while maintaining sediment suspension fully-turbulent, a maximum suspended sediment concentration of 1550 kg/m^3 is obtained, which coincides with the maximum concentration recorded in hyperconcentrated floods in rivers. It has to be noted that the size distribution of the mixed sediment is not continuous and its component sediments may not suit each

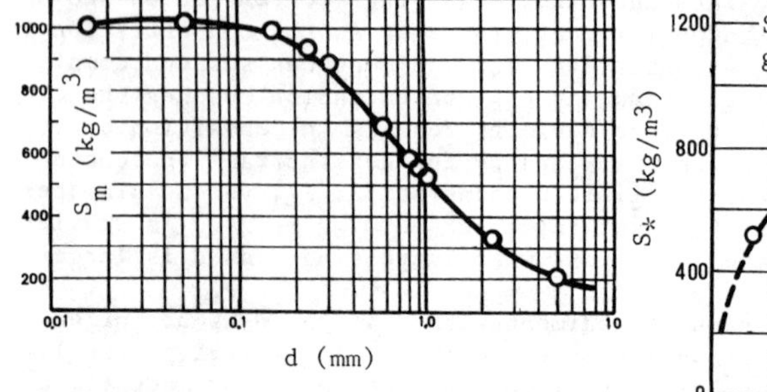

Fig.3 Limit concentration of uniform sediment

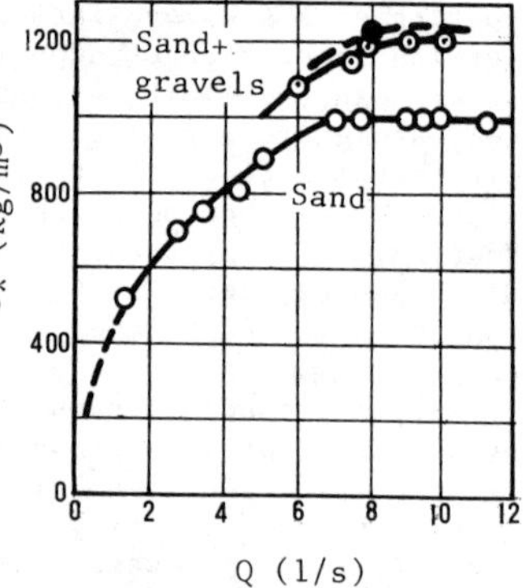

Fig.4 Sediment-carrying capacity versus discharge

other best. It is very possible that a fully-turbulent flow, supplied with aboundant sediments of different sizes can reach a concentration a little bid higher than 1550 kg/m^3.

The effect of the limit concentration on sediment-carrying capacity is studied in a flume experiment. The flume is 8 m-long and 10 cm wide. The suspended load are sand (d_{50}=0.31 mm), fine gravels (d_{50}= 3 mm) and coarse gravels (d_{50}=10 mm) as shown in fig.1. The liquid phase is clay mud with volume concentration of clay C_{vo}=0.067. The group fall velocity of sand in clay mud observes the empirical formula

$$\omega = \omega_o (1 - 9C_{vo})^{2.5}\, e^{0.0019S} \qquad (1)$$

where S is the concentration of sand in mud and ω_o=3.1 cm/s is the fall velocity of single particle in clear water.

In the experiment the liquid and suspended sediment is recirculated by a slurry pump and sediment is put into the system at the entrance of the flume. Sediment concentration is measured by sampling and weighing at the outlet of the flume. The sand is added into the flume until there appears deposited sediment on the bed of the flume. The concentration of suspended sediment in this case is so called sediment-carrying capacity of the flow and is denoted by S_*. Fig.4 shows the measured S_* versus flow discharge Q. As the discharge increases from 1.35 to 7 l/s, S_* increases from 510 to 1000 kg/m^3. Further increase in flow discharge, however, could not enhance S_* again because the suspended sediment has reached its limit concentration. Adding more sand results in more deposited sediment on the bed. Nevertheless about 300 kg of fine gravels are put into the system, they are easily dispersed and suspended in the flow already saturated with sand. The total concentration this time is 1200 kg/m^3, in which fine gravels concentration is 250-300 kg/m^3 and the concentration of sand remains at 1000 kg/m^3. The total concentration could be further increased to 1230 kg/m^3 as coarse gravels are also put into the flow.

Mechanism of limit concentration

Suspended load is carried by turbulent eddies in the flow. As an eddy carries particle P_a from point A to B, the water lump at B has to move away because of the continity of the fluid. Particle P_b at B is carried away by the water lump at the same time. If sediment concentration is over S_m, the water lump enclosing particle P_b is too small to carry P_b, P_a and P_b will collide with each other. Collisions between particles result in a dispersive force to support the effective weight of particles instead of turbulent diffusion and at the same time they also result in a resistance force which is much larger than turbulent shear stress and viscous frictional force. The suspended load motion transforms into laminated load motion in this case (Wang and Qian,1987). Only in debris flow and high-pressure hydro-transport pipeline can maintain such kind of extreamly high concentrated two phase flow.

Consider a particle P moving together with an eddy at velocity u. The eddy speeds up to $u+\Delta u$ at time $t=0$, which results in a relative velocity Δu between the eddy and P. The eddy subjects a drag force F on P

$$F = C_D\,(\pi d^2/4)\,(\rho u_r^2/2) \qquad (2)$$

where u_r is the relative velocity between the eddy and P and $u_r=\Delta u$ at $t=0$; C_D is drag coefficient and ρ is density of the liquid. Acted by the force P is accelerated to reduce the relative velocity u_r. The equation of motion of the particle P in the eddy is

$$\frac{\pi}{6}\rho_s d^3\frac{du_r}{dt} + C_D\frac{\pi d^2}{4}\,\frac{\rho u_r^2}{2} = 0 \qquad (3)$$

where ρ_s is the density of the particle. By employing the initial condition $u_r=\Delta u$ at $t=0$, the solution of the equation could be found as

$$\frac{u_r}{\Delta u} = \left(1 + \frac{3}{4}C_D\frac{\rho}{\rho_s}\,\frac{\Delta u t}{d}\right)^{-1} \qquad (4)$$

Eq.(4) interpretes that $u_r\rightarrow 0$ as $t\rightarrow\infty$. Nevertheless the eddy may change its velocity again after it has traveled an average distance ℓ, which is in fact the mixed length presented by Prandtl. A characteristic time is taken as $\tau=\ell/\Delta u$. If particle P remains in the eddy in the time $0-\tau$, it can be carried by the eddy and avoid direct collision with other particles. Therefore the water lump enclosing each particle should have the size larger than L in suspended load motion,

$$L=d+\int_0^{\tau}u_r dt = d\,\left(1+\frac{4}{3}\,\frac{1}{C_D}\,\frac{\rho_s}{\rho}\,\ln\left(1+\frac{3}{4}\,\frac{\rho}{\rho_s}\,\frac{\ell}{d}C_D\right)\right) \qquad (5)$$

Eq.(5) can be simplified into

$$L = d(1+kd) \qquad (6)$$

by taking $C_D\sim 1/d$ into consideration. There could be L^{-3} water lumps in unit volume, or so many suspended particles can exist in unit volume of liquid. Therefore the limit concentration of suspended sediment can be roughly given by

$$S_m \doteq \pi d^3 g\rho_s/6\ d^3(1+kd)^3 = \pi g\rho_s/6(1+kd)^3 \qquad (6)$$

It may be seen that S_m veries very little as d is very small ($kd<1$), and S_m reduces very fast with incresing d as d is large enough. These are coincident with that observed in the experiments (see fig.3).

There are many eddies of different sizes in turbulent flow. Small eddies could carry small particles while large eddies carry large particles, so that the limit concentration of coarse sediment is not affected by existence of fine suspended sediment.

Implications of limit concentration in hyperconcentrated flow

Quite a few researchers reported that the median diameter of suspended sediment increases with concentration in hyperconcentrated flow. In fact the increase in concentration is mainly attributed to widening of range of size distribution of sediment. Fig.5 shows the percentage of sediment coarser than 0.05 mm verus the total concentration in hyperconcentrated flows in the Yellow River and its tributories (Qian and Wan, 1983). The percentage of coarse sediment increases linearly with the total concentration, which is because only with wide range size distribution can the sediment be easily carried by the flow in extreamly high concentration. As the total concentration reaches 1550 kg/m^3, sediment coarser than 0.05 mm is about 60% of the total which is the same as that in Table 3.

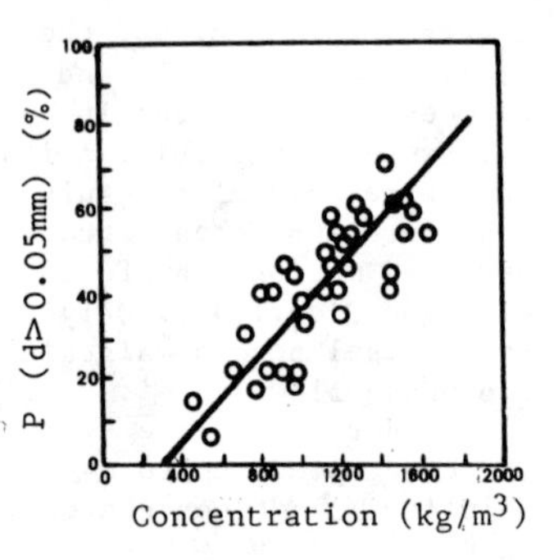

Fig.5 Percentage of coarse sediment

Hyperconcentrated flood often brings about serious erosion to the river bed. Table 4 presents measured data of serious erosion at Longmen Section of the Yellow River. It seems that the higher the concentration the more serious the erosion. But no erosion took place in the Uranmlun River as a hyperconcentrated flood with concentration as high as 1510 kg/m^3 and average velocity 7.37 m/s occured in 1966, as shown in fig.6. This is because the limit concentration confines sediment-carrying capacity of the flow.

Table 4 Serious erosion at Longmen section of the Yellow River

Date	S (kg/m^3)	U (m/s)	d_{50} (mm)	Erosion (m)	Date	S (kg/m^3)	U (m/s)	d_{50} (mm)	Erosion (m)
6,7,1964	711	7.2	0.050	3.5	18,7,1966	812	6.8	0.120	7.5
27,7,1969	601	8.0	0.038	3.0	2,8,1970	772	6.6	0.062	9.0
6,7,1977	635	8.5	0.031	4.0	6,8,1977	821	6.5	0.060	2.0

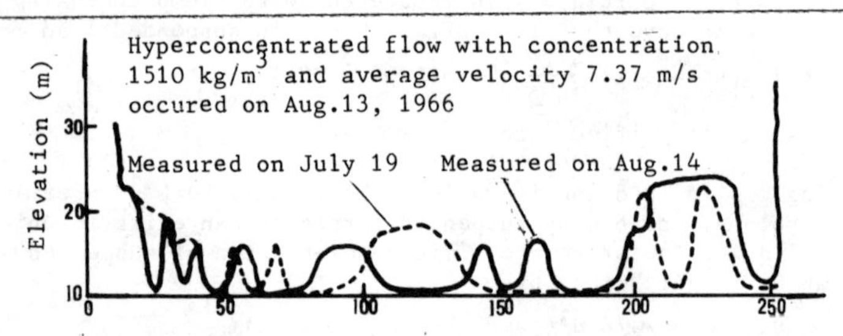

Distance from the left bank (m)

Fig.6 Cross-sections of the river bed before and after a hyperconcentrated flood

References

(1) Qian Ning and Wan Zhaohui, Mechanics of Sediment Movement, Science Press, 1983 (in Chinese).

(2) Wang Zhaoyin and Qian Ning, Experimental study of two-phase turbulent flow with hyperconcentration of coarse sediment, Scientia Sinica, Ser.A, Vol.27,1984, No.12, pp.1317-1327.

(3) Wang,Z.and Qian N., Intern. J. of Sedi. Research, No.1,1987.

Sediment Transport Characteristics of Hyperconcentrated Flow with Suspended Load

Cao Ruxuan[1] Qian Shanqi[2]

Abstract: In this paper, based on the analysis of the ovserved data, a set of theories is proposed of the transport of heperconcentrated sediment, the basic points of which are: (1) The hyperconcentrated flow can be divided into homogeneous and nonhomogeneous flow. The former is in the sphere of resistance work, while the latter is in the sphere of suspension work.

With the hyperconcentrated flow in natural rivers, including volcanic lava and debris flows, the theory in sediment transport follow the non-equilibrium hyperconcentrated sediment transport theory.

Introduction

The trunk streams and tributaries on the middle reach of the Yellow River and some of the sediment-laden rivers in the northern and northeastern parts of China originate and flow through mostly in the areas of soil erosion. As a result, the sediment concentration of the river is great, and the sediment concentration of a flood may reach more than 1000 kg/m^3.

It is shown from the observed data that there is a distinctive difference between the basic nature of the hyperconcentrated flow and the ordinary sediment-laden flow. In the late 1970s, effective investigations were made by some Chinese scholars into the basic characteristics of the hyperconcentrated flow, with some important results achieved (4)-(5). In this paper, based on the analysis of the prototypes and experimental data, sediment transport characteristics of the hyperconcentrated flow are studied.

Basic Characteristics of Hyperconcentrated Flow

Physical characteristics.— Studies indicate that the motion viscosity of the hyperconcentrated fluid increases

[1]Senior Engineer, Shaanxi Mechanical Engrg. Inst., Xi'an, China
[2]Associate Professor, Shaanxi Mechanical Engrg. Inst. Xi'an, China

with the increase of the concentration and varies in inverse proportion to the size of the particle. The clay content plays an important role in the flow pattern of the fluid particularly. The flow pattern of the hyperconcentrated fluid can be described by the Bingham model.The pattern of the homogeneous fluid with a great clay content is of the time-dependent Non-Newtonian fluid.

Hydrometric characteristics.— The particle size carried in the natural hyperconcentrated flow becomes larger with the increase of the sediment concentration. The two events change synchronously.

Setting Characteristics. — Studies indicate that the settling of composite sediment can be divided into three types: (a) in the process of low concentration settling, there is a gradient of concentration and particle size along the course, indicating that this is the ordinary sediment laden flow; (b) when the concentration increases to a certain value, the coarse particles of sediment settle first selectively, while the sediment remaining in the pipe forms stable suspension particles, which has a distinctive interface and settles uniformly, denoting that this is a hyperconcentrated non-homogeneous flow ; (c) when the concentration gets higher, the sediment does not settle, but solidifies, signifying that the flow is a hyperconcentrated homogeneous flow. The foundary of the above three fluids depends on the coarseness gradation and the clay content.

Clearly, for type (a), the settling velocity of sediment can be determined with the influence of concentration ignored. For type (b), there is a limited particle diameter d_o and the sediment with the particle diameter greater than d_o is called "load". The settling velocity should be determined with the concentration considered. The sediment with the particle diameter less than d_o is called "carrier"(2), type (c) presents no notion of sediment-carrying.

Because of the great viscosity, the flow pattern is sometimes of the type of laminar flow, the velocity distribution of which has a flow nucleus. Under the turbulent condition, however, the distribution of flow velocity still conforms to the logarithmic formula and the distribution of the sediment concentration conforms to the diffusive equation.

In the field investigations and laboratory experiments, two forms of sedimentation have been observed. The sedimentation form for the homogeneous flow is of the type of complete standstill during which the discharges decreases along the course and the concentration and particle diameter presents no change along the course. In the non-homogeneous flow, it is manifested that the concentration and particle diameter decrease along the course, while the discharge remains unchanged.

Sediment-Carrying Capacity of Hyperconcentrated Flow

Investigation has been done by the first author of this paper into this kind of force (1). The conclusion arrived at is as follows: The hyperconcentrated flow, which does not have the concept of sediment carrying, is in the sphere of resistance work. From the equation

$$\gamma_m ASV = \frac{A}{h} \tau_B \, v \qquad (1)$$

We have the flow condition

$$\gamma_m hS = \tau_B \qquad (2)$$

The non-homogeneous flow is still in the sphere of suspension work, in which all the sediment is divided into two parts, the "load" and the "carrier" with d_o as the boundary, when the settling velocity has been corrected in concentration and size distribution, and effect of unit weight has been introduced into calculation, the sediment carrying capacity of a hyperconcentrated flow still obeys the formula of an ordinary sediment-laden flow. Fig.1 (a) and Fig.1(b) show the comparison of sediment carrying capacity between calculated values from Chih Ted Yang Equation (3) and $C_* \sim V^3/gR\omega$ to measured ones respectively, both values are in good agreement with the hyperconcentration characteristics considered.

Theory of Hyperconcentrated Non-Equilibrium Sediment Transport

The hyperconcentrated flow in natural rivers, which is characteristic of its sudden rise and fall, is difficult to reach the saturation state in a short period through the erosion and sedimentation adjustment, thus making it difficult to effect the equilibrium sediment transport.

With the hyperconcentration characteristics considered, formula of the along-course change of the concentration for the load-sediment

$$C_o = C_* + (C_i - C_*) \sum_{k=1}^{n} \Delta p_k \, e^{-\frac{\alpha \omega_k \Delta x}{q}} \qquad (3)$$

can still be used to describe the hyperconcentrated sediment transport. Fig. 2 indicates the comparison between the measured and calculated data with hyperconcentrated sediment transport characteristics considered and Fig.3 with the hyperconcentrated sediment transport characteristics ignored.

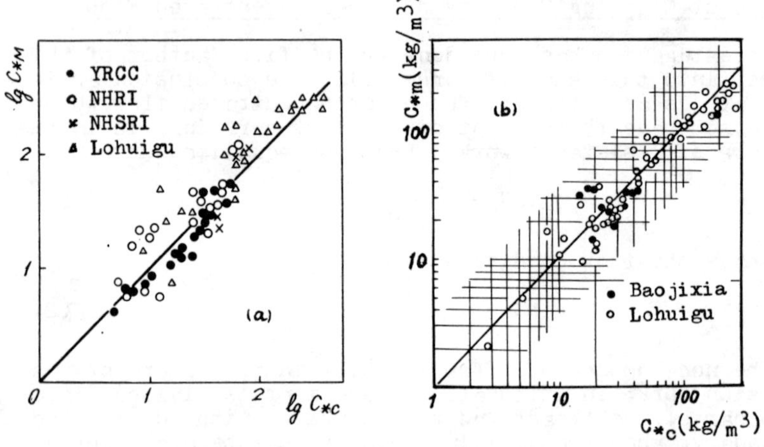

Fig. 1 Comparison of sediment carrying capacity between calculated values to measured ones, (a) from Yan's Equation, (b) from $C_* \sim V^3/gR\omega$

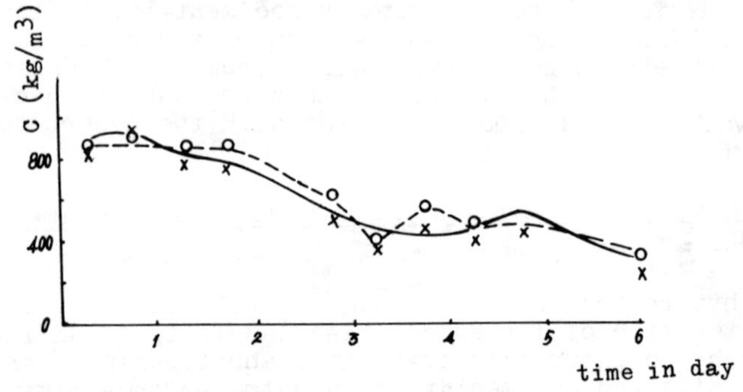

Fig. 2 The hydrograph of concentration

Explanation

—— Measured at inlet

o Measured at outlet

× Calculated at outlet

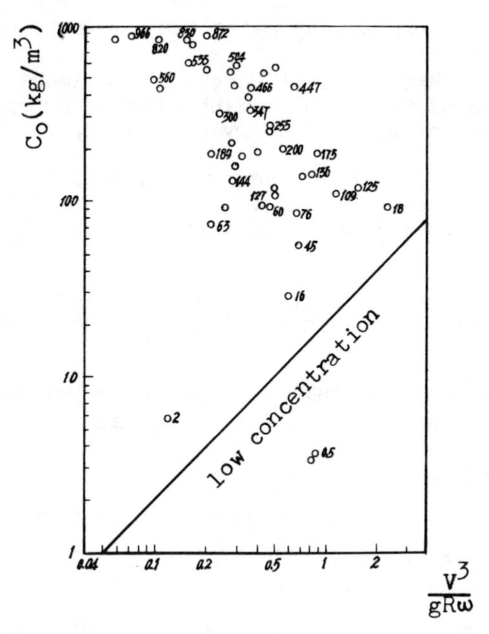

Fig. 3 Relationship of

$$C_o \sim \frac{V^3}{gR\omega}$$

Explanation
Number by the point are concentration at inlet.

Conclusions

1. The computational formulas, in which ω and d_o are determined with the hyperconcentration charateristics considered, are of a general nature, thereby unifying the sediment transport laws for both the low and high concentration of sediment.

2. The hyperconcentration sediment transport laws in rivers and reservoirs conform to the non-equilibrium sediment transport theory for hyperconcentration instead of the equilibrium sediment transport theory.

Appendix I-References

1. Cao Ruxuan, Preliminary Study of Sediment-Carrying capacity of Hyperconcentrated Flow, Water Conservancy and

Hydro-Electrical Technology, No.5, 1979.

2. Cao Ruxuan, Mathematical Model for Sediment Transport Competancy of Canals Diverting Flow at Hyperconcentration, Journal of Water Conservancy, No.9, 1987.

3. Chih Ted Yang, Rate of Energy Dissipation and River Sedimentation, Proceedings of the Second International Symposium on River Sedimentation, Oct. 1983, Nanjing, China.

4. Proceedings of International Workshop on Flow at Hyperconcentration of Sediment, IRTCES, Sept. 1985. Beijing, China.

5. Qian Ning and Wan Zhanohui, Sediment Motion Mechanics, Science Press, 1986, Beijing, China.

Appendix II-Notation

The following symbols are used in this paper:

A = cross-sectional area

c = sediment concentration

C_i, C_o = sediment concentration at inlet and outlet

C = sediment-carrying capacity

d_o = limited particle diameter

h = water depth

p_k = percentage by weight taken by the k-group sediment

q = unit width discharge of flow

R = hydraulic radius

S = bed slope

v = average flow velocity

α = coefficient

γ_m = unit weight of muddy water

τ_B = yield stress

ω = settling velocity

A MATHEMATICAL MODEL OF EROSION AND SEDIMENTATION OF HYPERCONCENTRATED FLOW IN RESERVOIRS

Qian Shanqi[1] Cao Ruxuan[2] Wang Xinhong[3]

Abstract

The motion of the flow with hyperconcentration has its unique law. In this paper, based on the analysis of the motion of the flow with hyperconcentration of sediment and the sediment transport law, a mathematical model of erosion and sedimentation of the hyperconcentrated flow has been established. The system of Saint-Venant equations has been used as the flow equation and the unbalance sediment transport theory for hyperconcentration has been used as the equation for sediment motion so as to make it possible to reflect the law of sediment transport. The model was verified by the data obtained in the Sanmenxia Reservoir of the Yellow River, indicating that there is a relatively high accuracy. The model can be used for determining the erosion and sedimentation of debris flow and volcanic ashes which get into rivers and reservoirs.

Introduction

In the arid regions in China, sediment-laden rivers are characteristic of the large amount of sediment load and hyperconcentration. The hyperconcentrated flow in reservoir are in different patterns in sediment transport, they are the open-channel flow in backwater, the density current, the flow in muddy storage, the progressive erosion and the retrogressive erosion. The formation condition and process of sediment transport for each pattern are different (1).

With the sediment-laden rivers, the formation of new river beds is relatively rapid. The perennial average amount of sediment entering into the Sanmenxia Reservoir of the Yellow River is 1.6 billion tons. The reservoir,

[1]Associate Professor. Shaanxi Mechanical Engrg. Inst., Xi'an, China

[2]Senior Engineer, Shaanxi Mechanical Engrg. Inst. Xi'an, China

[3]Teaching Assistant, Shaanxi Mechanical Engrg. Inst., Xi'an, China

which was built up and put into use in 1960, formed a completely sedimented delta in 1961.

Model Establishment

The hyperconcentrated flow may be divided into homogeneous flow and non-homogeneous flow. The motion condition for the former is $\gamma_m hS = \tau_B$. Since the transversal distribution of the sediment concentration is very uniform, the lagging $h_s = \tau_B / \gamma_m S$ in the reservoir is basically the same everywhere. In the latter, there is a limited particle diameter d_o, which can be determined by the decision number of suspension $\omega / kV_* = 0.06$. The sediment with $d < d_o$ is the wash load, which forms slurry with muddy water. The sediment with $d > d_o$ is the bed material load, which is carried by the muddy-water slurry, the transport following the unbalance sediment transport law for the case of hyperconcentration (3).

Flow Control Equation. —— We have the continuous equation of flow

$$B \frac{\partial z}{\partial t} + \frac{\partial Q}{\partial x} = 0 \qquad (1)$$

and the equation of flow motion

$$\frac{\partial Q}{\partial t} + A \frac{\partial}{\partial x} \left(\frac{Q^2}{2A^2}\right) + gA \frac{\partial z}{\partial x} + g \frac{|Q| \quad QN^2}{AR^{4/3}} = 0 \qquad (2)$$

when considering long distances and periods, the flow can be simplified as a constant one and as far as a flood in short period is concerned, it can be treated as an unsteady flow.

Sediment Control Equation. —— We have the continuous equation of sediment

$$(Q_s - Q_{so}) \Delta T = \Delta A \, \Delta x \, \gamma'_s \qquad (3)$$

and the equation of sediment motion

$$C = C_* + (C_o - C_*) \sum_{k=1}^{E} \Delta p_k e^{-\frac{\alpha \omega_k \Delta x}{q}} \qquad (4)$$

We also have the equation of settling velocity of sediment in the muddy-water slurry

$$\omega = \omega_o \left(1 - \frac{C}{C_m}\right)^n \qquad (5)$$

in which ω_o is determined by the Stokes equation.

The equation for determining the limited particle diameter d_o is as follows:

$$d_o = \frac{0.41(hS)^{0.25}\nu^{0.5}}{\left(\frac{\gamma_s-\gamma_m}{\gamma_m}\right)^{0.5}\left(1-\frac{C}{C_m}\right)^{0.5n}} \tag{6}$$

As for the equation of sediment carrying capacity, we adopt the equation with the characteristics of hyperconcentration of sediment considered (3).

$$C_* = K\left(\frac{\gamma_m}{\gamma_s-\gamma_m}\ \frac{V^3}{gR\omega}\right)^m \tag{7}$$

Determination of Density Current in Reservoirs.—— The plunging condition (2) is:

$$h_o = \left[0.365+2\left(\frac{g\tau_s}{\gamma_m V_o^2}\right)^{0.82}\right]\frac{q^{2/3}}{\left(\frac{\Delta\gamma}{\gamma_m}gs\right)^{1/3}} \tag{8}$$

The depth of the density flow is

$$h = \left(\frac{\lambda}{8s}\ \frac{Q}{\frac{\Delta\gamma}{\gamma_m}gB}\right)^{1/3} \tag{9}$$

The sediment transport equation of the density flow is

$$C=Co\sum_{k=1}^{E}\Delta p_k e^{\frac{-\alpha\omega_k\Delta x}{q}} \tag{10}$$

Relevant problems can be treated as follows:

1. Simplified Profiles. —— The simplified profile should be made based on the mode of sedimentation in the reservoir region. For example, since the mode of sedimentation in the Sanmenxia Reservoir is of the type of high flood plain and deep channel, the profile can be simplified as a step type.

2. Roughness. —— The resistance of the river bed is determined by the roughness with the discharge as its parameter.

$$N = \frac{R^{2/3}}{V\Delta x}\left[\Delta\left(\frac{v^2}{2g}+z\right)\right] \tag{11}$$

According to which the relation curve between the roughness and the discharge can be drawn through dotting by use of the observed data and hence the roughness N can be determined by the curve.

3. The distribution of the discharge and sediment on the flood plain and in the main channel. —— The distribution of the discharge can be dealt with by the method for module distribution of the discharge on the flood plain and in the main channel. The concentration C_n on the flood plain and the concentration C_p in the main channel can be determined by $C_n = K_s C_p$. The value K_s can be determined by use of the data of the overbank flood. From the continuous equation, we have

$$C_p = \frac{Q}{Q_p + K_s Q_n} C \tag{12}$$

The model has been verified with the measured data obtained from Sanmenxia Reservoir over a ten-year period (from 1974 to 1984), the verified items including the accumulated hydrograph of the total erosion and sedimentation amount in reservoir (Fig.1), the longitudinal profiles of the river bed in different years (Fig.2) etc..

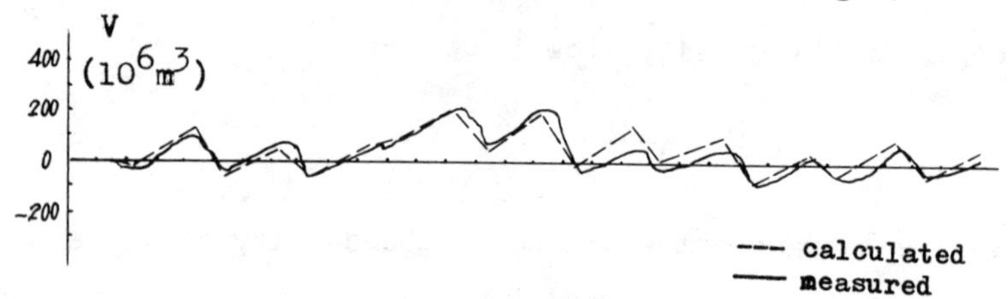

Fig.1 The accumulated hydrograph of the total erosion and sedimentation amount in Sanmenxia Reservoir

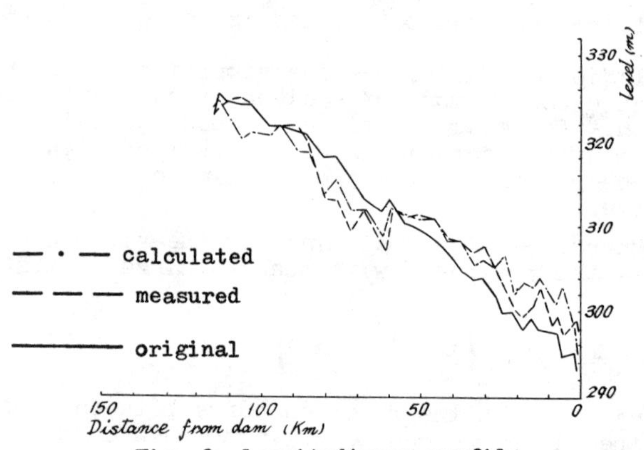

Fig. 2 longitudinary profile

The results of verification show that the computational results is in good agreement with the measured data.

Conclusions

1. Since the sediment factor in the model is treated with the influence of the sediment concentration, the model can be used for both of the flows with high or low concentration of sediment.

2. The model, which has considered different flow patterns in sediment transport, can be used for determining the erosion and sedimentation in reservoirs and rivers.

Appendix I-References

1. Cao Ruxuan and Chen Jiangliang, Erosion and Sedimentation of Flow at Hyperconcentration in Reservoirs, Proceedings of International Symposium on River Sedimentation, Guanghua Press, 1980, Beijing, China.

2. Gao Ruxuan and Ren Xiaofeng, Conditions of Formation and Continuous Motion of Density Current with Hyperconcentration, Proceeding of International Workshop on Flow at Hyperconcentration of Sediment, IRTCES, 1985, Beijing , China.

3. Cao Ruxuan, et at., A Mathematical Model for Sediment Transport Capacity of Hyperconcentration Flow in Diversion Canals, Journal of Hydraulic Engineering No. 9 , 1987, Beijing, China.

Appendix II - Notation

The following symbols are used in this paper:

A = cross-sectional area

B = width of river

C_o, C = sediment concentration at inlet and outlet section

C_m = maximum sediment concentration

C = sediment carrying capacity

d_o = critical particle diameter

K = constant

g = acceleration of gravity

m = empirical constant

n = empirical constant

Q, q = discharge, unit discharge

Q_{so}, Q_s = sediment trasport rate at inlet and outlet

R = hydraulic radius

S = bed slop

v_o = velocity of plunging section

V = shear velocity

z = water level

ω_o, ω = settling velocity of clear and muddy water

γ_m, γ_s = unit weight of muddy water and sediment

ν = kinematic viscosity coefficient of clear water

Water Quality Study in an Arid Region Lake: Lake Bosten, Northwest China

Zhu Dongwei,[1] Steve C. McCutcheon[1] M. ASCE, and Zhang Guo-an[2]

Abstract

This paper gives a brief description of Lake Bosten, its importance to Xinjiang Province of the PRC, objectives of an on-going study, and some preliminary results. This on-going study is intended to determine the feasibility of integrating modeling studies into resource management in this area of the world. This site seems well suited for such an evaluation. Some data have been collected since the 1950s and extensive data collection and study have been going on since 1987. The specific goals of this study are: a) evaluate important factors in the water and salt balances in the Lake; b) calibrate and validate hydrodynamic and water quality models for the Lake; and c) simulate management scenarios for the Lake.

Physical Setting and Problems

Lake Bosten is one of the largest fresh water lakes in the People's Republic of China. It is composed of a large lake area, a small lake area, and a wetland area. The surface area of the large lake is about 1,000 km^2 with about 55 km long in east-west orientation and 20 km wide in north-south direction. The volume is 8,100 million m^3 with an average depth of 8.15 m and maximum depth of 16.5 m. The water surface elevation is about 1046 m above the mean sea level. Around it, the small lake and wetland areas make up about 400 km^2, of which, 53 km^2 is the small

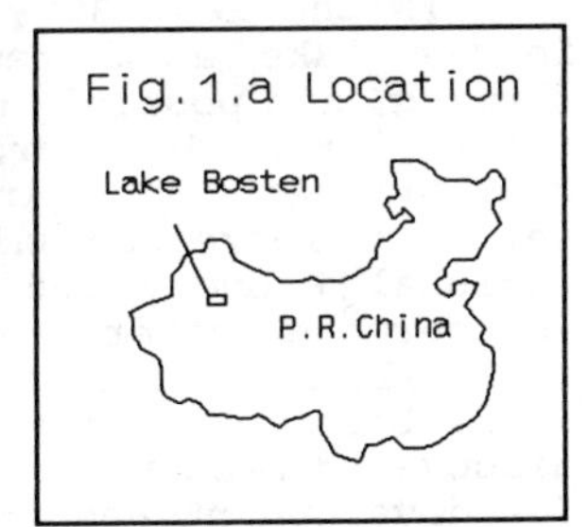

Fig.1.a Location

---

[1]Center for Exposure Assessment Modeling, U.S. EPA, Environmental Research Laboratory, Athens, GA 30613.

[2]Xinjiang Institute for Environmental Protection, Urumqi, P.R. China.

lake area that contains a large amount of aquatic plants, especially reeds.

Lake Bosten is at the lowest elevations of the Yanji Basin in northwest China with the longitude of $86^{\circ}40'$-$87^{\circ}26'$, and latitude of $41^{\circ}50'$-$42^{\circ}14'$(Fig.1.a and 1.b). Yanji Basin is south of the Tianshan Mountains from which the melting snow provides almost all of the water to Lake Bosten and the surrounding region mostly through the Kaidu River. These waters are used extensively for irrigation before reaching the lake. Losses of water in the lake include evaporation and a large amount that is pumped out into the Peacock River to provide valuable water supply to other parts of the Yanji Basin and to the arid Kurle Plain downstream. In the Yanji Basin, the annual average temperature is about 8°C, annual rainfall is 50 to 80 mm, and evaporation is 1,800 to 2,500 mm, which is 30 times that of rainfall. There is an apparent imbalance between precipitation and evaporation. The limited rainfall means that Lake Bosten must be a vital water resource to the region. It provides water for domestic use, industry, irrigation, fishery activity, and reed production. It also receives industrial waste waters.

Fig.1.b Lake Bosten

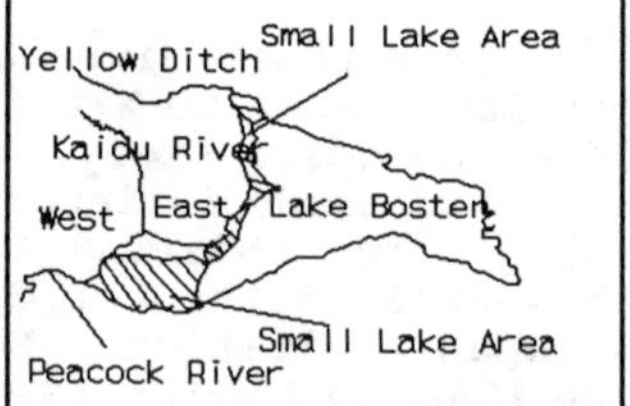

In the last 30 years, serious changes have occurred in the lake as a result of human activities, such as farming, and possibly climate changes around the lake. The water level has dropped 3 m, and dissolved solid concentrations have increased 3 times. The dropped water level and increased salinity threaten water quality and the regional economy and ecosystem that the lake supports. Furthermore, water demand has been increasing steadily and is likely to be intensified due to the population increase, recreation activities, industrial development, and the oil discovered recently in the region, which will require water for development and transportation.

Many events and processes have possibly contributed to the changes in the lake. Farming had been increasing in the region until 1980. The total area of farm land increased to 40 k-hectares in 1958, 60 k-hectares in 1960, and to the maximum of 80 k-hectares in 1980, and has not increased since then. As a result of the intensified farming, water tables in the region were elevated by a maximum of 2 to 3 m in 1980 and salinity in the soil had also increased greatly. To ease these problems, drainage canals were built beginning in 1962 to drain the return flow to the lake. As the drainage system became more complete, the amount of salt brought into the lake through these canals increased rapidly until the 1980s. Since 1980, the salts going into the lake have decreased because of the blockage of drainage canals by sedimentation.

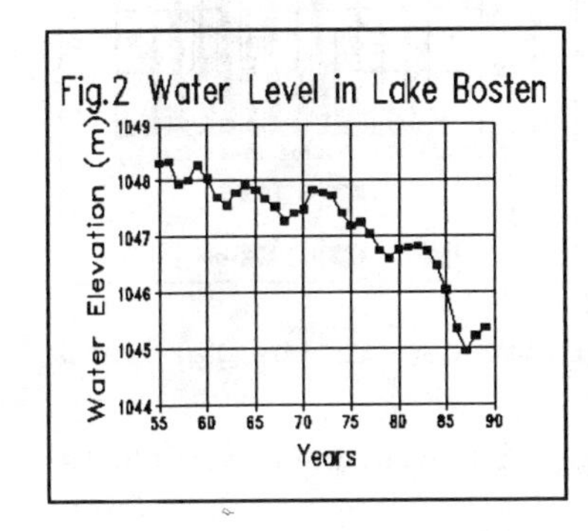

A control gate was built in 1958 on Kaidu River, which makes up 85% of the total surface water runoff in Yanji Basin, to control the relative amount of water divided into west and east branches of the river. The west branch goes into the small lake area, while the east goes into the large lake area directly. This gate brings the average percentage of water going to the east branch from 78.5% down to 51.5%, and therefore greatly reduces the amount of water supply to the large lake.

In 1981, a pumping station began operation to supply the Peacock River because natural outflows were threatened by dropping lake levels. This station improves the water supply to the remainder of the Yanji Basin and the Kurle Plain, and might have improved the salinity cycle in the lake.

The general trends of water quantity and salinity in Lake Bosten are illustrated in the following figures. Fig.2 shows the water level changes from 1955 to 1989. It can be seen that the water level has been dropping steadily for most years since the 1950s, and decreased more rapidly from 1983 to 1987. The total dissolved solid concentrations in the lake also show a dramatic increase in the late 1970s and a slow decrease since 1987. This is shown in Fig.3 through Fig.5

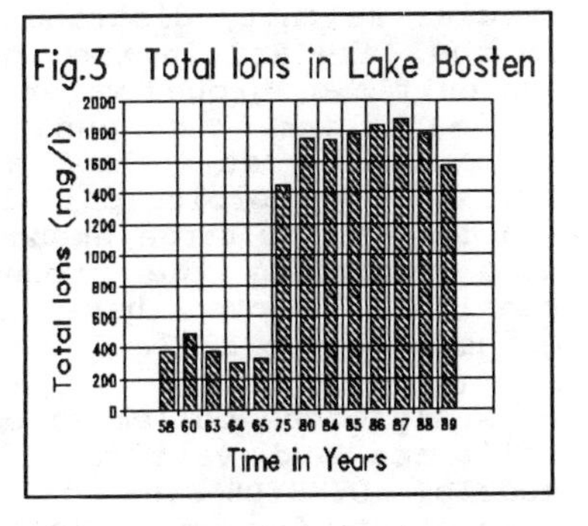

Ground and subsurface waters might also play an important role in the water quantity and salinity balances of Lake Bosten.

Goals and Approaches of the Study

The goals of the study can be described as:

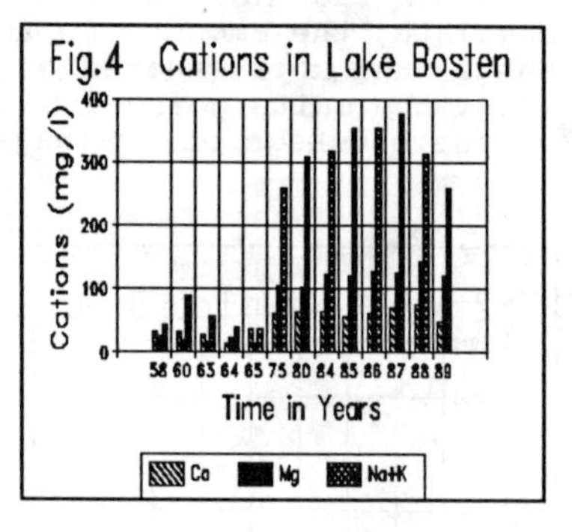

a) Identify and evaluate the important factors controlling the water and salinity balances of the lake.
b) Set up, calibrate and validate hydrodynamic and water quality models for Lake Bosten.
c) With the help of the models, analyze water and salinity processes affecting the lake, simulate and assess management scenarios, and provide advice for better management of water resources in the Lake and the region.

The approach of this study is designed in three phases:

a) Collect the accumulated historical data, consider the lake as a completely mixing unit, make a crude water and salt balance and evaluate the processes that can influence the lake. Information obtained through this stage was used to design further data collection and revise further study approach.

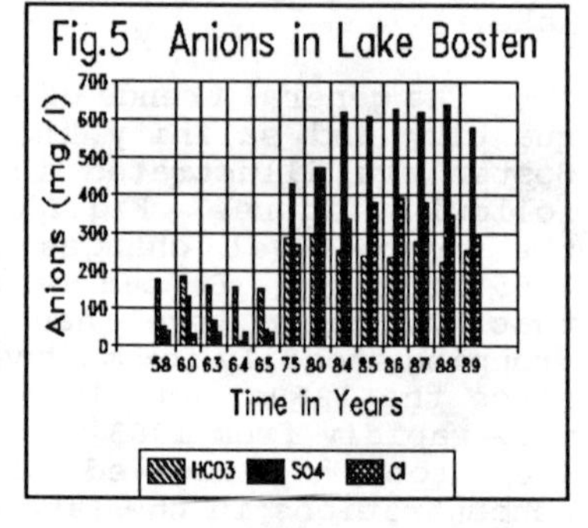

b) Conduct surveys for inflow and outflow rates, temporal and spacial salinity distributions in the lake and in the inflow and outflows, ground water flow regime and composition, evaporation at the water surface, meteorological data, and so on. A large amount of data regarding the processes mentioned above have been gathered since 1987.

c) Analyze the collected data, calibrate and validate hydrodynamic and water quality models(DYNHYD5 and WASP4 are currently used), make simulations for future scenarios and analysis for important factors or processes with the calibrated models. In this step, the mechanisms of water quantity and salinity balances should be studied and understood. Evaluations for water resource management practices in the region will be made. This step is designed to be on two different time scales: long term and short term. For the former scale, long term trends of water quantity and salinity changes that have taken place or might take place in the future will be analyzed. Important parameters that can affect those changes, including climate change if appropriate, will also be evaluated. For the short term on the other hand, the much more detailed data collected recently will be used. From shorter time scales and more detailed spatial distributions of salinity in the lake, the dynamic processes should be investigated more accurately.

Preliminary Results on Water and Salinity Balances

Water balance in Lake Bosten is governed by a rather complex system. As noted above, water from the Kaidu River is divided into both large and small lake areas. Agricultural runoff discharges considerable amounts of water into the lake. Possible groundwater interactions between large and small lake areas, evapotranspiration in the marshes, and the interaction with ground and subsurface waters may all be important factors to the water balance. Effort is being made to set up a more accurate water balance for Lake Bosten with the new data base. However, a preliminary balance was attempted using historical records and some approximations. Table 1 gives the results of this preliminary balance.

Table 1 Water Balance for Large Lake Area (10^8 m^3)

	Inflows and Outflows	1983	1984	1985
IN	Kaidu River	8.755	10.222	11.526
	Agricultural Runoff	2.237	1.400	1.268
	Rain Fall	0.830	0.432	0.497
	Ground Water	0.581	0.461	0.581
OUT	Evaporation	8.775	9.006	9.103
	Peacock River	5.831	7.336	8.530
	Subsurface Water	0.357	0.357	0.357
	Balance	-2.560	-4.185	-4.119

It can be seen that the Kaidu River and agricultural runoff seem to be the most important recharges, while evaporation from the lake surface and Peacock River outflow are the most important discharges from the lake. Apparently, more accurate data concerning surface evaporation and other processes are needed before more reliable and detailed conclusions can be drawn.

The salinity balance in Lake Bosten is more complex due to extra factors that can affect salinity content in the lake. These include the high concentrations in agricultural runoff and the uncertainty in its measurement or estimation, and the wind-carried salts from the surrounding desert that may deposit into the lake. A preliminary estimation was also made which is shown in Table 2. The budget in Table 2 shows that agricultural runoff may be the most important factor to the salt balance because Kaidu River is actually bringing in clean water to the Lake. An imbalance of ions suggests that precipitation of insoluble minerals may occur in the lake water and this will be investigated further with geochemical calculations if needed.

For the quality assurance of the salinity data, the balance between the gram-equivalents of cations and anions of the annually averaged data was examined. The average difference was found to be less than 5%. This indicates, from one aspect, that the salinity data are reliable.

Table 2 Salt Balance in Large Lake Area (10^4 ton/year)

	Inflows and Outflows	1983	1984	1985
IN	Kaidu River	28.842	35.223	34.135
	Agricultural Runoff	37.130	32.380	28.183
	Ground Water	1.471	1.167	1.471
	Yellow Ditch Dredge	50.962	0.000	0.000
	Rain Deposition	0.520	0.291	0.236
	Wind Delivered	6.248	6.186	6.075
OUT	Peacock River	77.710	100.660	100.560
	Subsurface Water	6.430	6.570	6.250
	Balance	+41.033	-31.983	-36.710

The study is on-going and additional results will be presented at the Symposium. Final results will be published later.

HYDROLOGIC ANALYSIS MODEL FOR SCREENING WETLAND RESTORATION SITES IN THE RAINWATER BASIN, NEBRASKA

Douglas J. Clemetson[1], A.M. ASCE

ABSTRACT

Historically, the Rainwater Basin in south central Nebraska contained approximately 4,000 wetlands. Today, over 90 percent of those wetlands have been lost, primarily due to drainage for agricultural purposes. With an increasing interest in the restoration of wetland habitats, a need has arisen for prioritizing candidate wetland restoration sites, which would ensure expenditure of funds on the best available sites. A hydrologic model has been developed as a tool to assist in screening candidate wetland sites for restoration in the Rainwater Basin. The Wetland Hydrologic Analysis Model (WHAM) was developed to perform daily analysis of the hydrologic budget of individual wetland basins. Inflows to the wetland, which are input data for the WHAM, were derived with the Streamflow Synthesis and Reservoir Regulation (SSARR) model, which was developed by the North Pacific Division of the U.S. Army Corps of Engineers. Statistical analyses were performed on the simulated hydrologic parameters including surface area, water surface elevation, storage volume, and average depth to evaluate the hydrologic effectiveness of the proposed restoration for each site. Priorities for acquisition and restoration can be assigned to the individual wetlands based on the results of these analyses.

INTRODUCTION

Located in south central Nebraska, the Rainwater Basin covers portions of 17 counties, an area of approximately 4,200 square miles, which is classified, topographically, as the Loess Plains region. Two separate regions divided at about 98° 30′ West longitude, as shown on Figure 1, comprise the area designated as the Rainwater Basin (Gilbert,1989). Soil types in the eastern region are primarily silty clays to silty clay loams, while those in the western region are generally silty clay loams to silt loams. Normal annual precipitation ranges from about 21 inches per year on the western boundary to about 30 inches per year along the eastern boundary. Historically, the Rainwater Basin contained nearly 4,000 wetlands, which ranged in size from less than one acre to about 1,000 acres, for a total area exceeding 94,000 acres. Today, it has been estimated that over 90 percent of those wetlands have been lost, primarily due to drainage for agricultural purposes (Gersib et al.,1990). A hydrologic modeling system was developed to assist in screening candidate wetland sites for restoration in the Rainwater Basin. Components of this system include a runoff model, a wetland routing model, and a statistical processor. For the runoff model, the Streamflow Synthesis and Reservoir Regulation (SSARR) model was selected to derive synthetic inflows to the wetland. The SSARR model was developed by the North Pacific Division of the U.S. Army Corps of Engineers (USACE,1987). A new model, the Wetland Hydrologic Analysis Model (WHAM), was developed to perform the routing through the wetland and the daily accounting of the hydrologic budget. Statistical processing included duration

[1]Hydraulic Engineer, Hydrology & Meteorology Section, Omaha District, U.S. Army Corps of Engineers, Omaha, Nebraska

analysis of the hydrologic parameters computed by the WHAM for various seasons of the year.

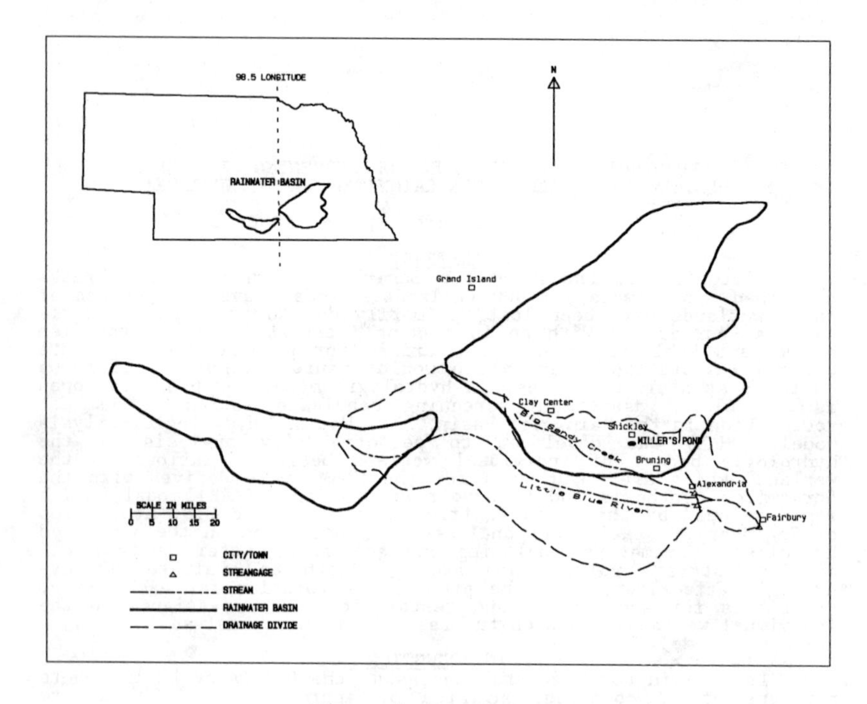

Figure 1. Location Map

SSARR MODEL CALIBRATION

Database Development. Thirteen stream gaging stations, all with drainage areas exceeding 300 square miles, exist in the vicinity of the Rainwater Basin. Since drainage areas for individual wetland sites at which the model is to be applied are generally less than 10 square miles, it was decided to calibrate the model to the incremental runoff on the Little Blue River between the gages near Alexandria and Fairbury, a total incremental area of 186 square miles. Streamflows during the period of 1980 through 1985, which contained annual runoff volumes ranging from 47 percent of average to 238 percent of average, were used to calibrate the SSARR model. Once the model was calibrated to observed streamflows during the six year calibration period, synthetic streamflows were derived with the calibrated model based on meteorological records for the total study period of 1949 through 1986. Daily precipitation records at Bruning and Fairbury were selected for use in the SSARR model calibration. Precipitation weightings of 50 percent were used for both stations, since Thiessen polygons indicated each gage had primary influence over about half of the incremental drainage area. Temperature records, which include daily maximum and minimum, at the Fairbury weather station were used in the SSARR model to determine whether the precipitation on a given day is rain or snow and to determine the amount of degree-days for the snowmelt computations. Evaporation records at the Clay Center, Grand Island, and Lincoln weather stations were used with precipitation data to compute the daily change in the soil moisture index. Average monthly values (converted to inches per day), which were computed by the National

Weather Service (USDOC,1982) using the Penman equation and meteorological measurements at the Omaha weather station, were substituted for missing records. A weighting of 73 percent was used to convert the pan evaporation data to free water surface evaporation.

Model Parameters. In computing the total volume of runoff, the SSARR model uses a relationship between soil moisture index and runoff in percent of moisture input. This relationship is one of the many calibration parameters in the SSARR model. A single curve can be specified to define this relationship or a family of curves can be used to define relationships which vary with precipitation intensity. For the Rainwater Basin, a family of curves were calibrated, since the area is subject to greatly varying precipitation intensities. In the simulation, the soil moisture index is adjusted daily based on the total moisture input and evapotranspiration. Evapotranspiration is reduced as the soil moisture index decreases to account for lesser evapotranspiration losses from dryer soils by specifying a relationship between soil moisture index and effective evapotranspiration. Once the total volume of runoff is computed, the SSARR model distributes the runoff into surface flow, subsurface flow and baseflow. The amount of runoff contributing to baseflow is determined first using a relationship between baseflow infiltration index and percent of runoff to baseflow. Another calibration parameter is the baseflow input limit, which was specified as a function of the baseflow infiltration index. After the amount of runoff contributing to baseflow is determined, the remaining runoff is divided into surface runoff and subsurface runoff. Routing parameters, including the number of routing phases and time of storage for each phase, were specified for surface runoff, subsurface runoff, and baseflow. These parameters may be calibrated or estimated from observed streamflow records. Since the primary goal of this study was to compute runoff volumes, routing parameters were not an important calibration parameter. Also, it was assumed that baseflow would be lost to deep groundwater storage, since the depth to the water table throughout most of the study area exceeds 50 feet. Therefore, the routing parameters for baseflow were specified as the maximum possible, which results in the baseflow not returning to the stream during the simulation. A constant melt rate of 0.15 inches per degree-day was calibrated for the snowmelt computations. A base temperature of 35°F was specified to compute the number of degree-days, while 32°F was used for the rain-freeze temperature.

Results of SSARR Model Calibration. Approximately 50 different simulations were required to calibrate the SSARR model. Following each simulation, the computed streamflows were plotted with the observed streamflows and inspected for "goodness of fit" and the observed and computed runoff volumes were computed for each year and compared. Model parameters were adjusted by trial-and-error following each iteration until satisfactory results were obtained. The primary goal of the calibration was to provide a reasonable reconstitution of the observed cumulative runoff mass curve, while reproducing the observed annual runoff volumes in a reasonable fashion. Shorter duration refinements were not considered in this study since the cumulative runoff is the most important factor affecting the long term condition of the wetland. Figure 2 displays the observed and computed accumulated runoffs for the calibration period.

MILLER'S POND ANALYSIS

Basin Description. Miller's Pond, as it was referred to prior to drainage for agricultural purposes, is located approximately 2 miles south of Shickley. The depression area which formed the historical wetland area covers approximately 500 acres, with an average depth of about 2.5 feet and a maximum depth (prior to ditch construction) of about 5 feet. Historically, Miller's Pond was known for abundant waterfowl production as it lured wealthy hunters from the East to shoot waterfowl (Farrar,1982). Although the exact date is not known, in the late 1800's or early 1900's a 24-inch diameter clay tile pipeline was constructed near the southern edge of Miller's Pond to drain the wetland for agricultural purposes. The drain is about 2400 feet in length and conveys waters, which naturally would be retained in Miller's Pond, south to a tributary of Big Sandy Creek. Presently, the watershed area which drains into Miller's Pond encompasses about 1725 acres of primarily agricultural land.

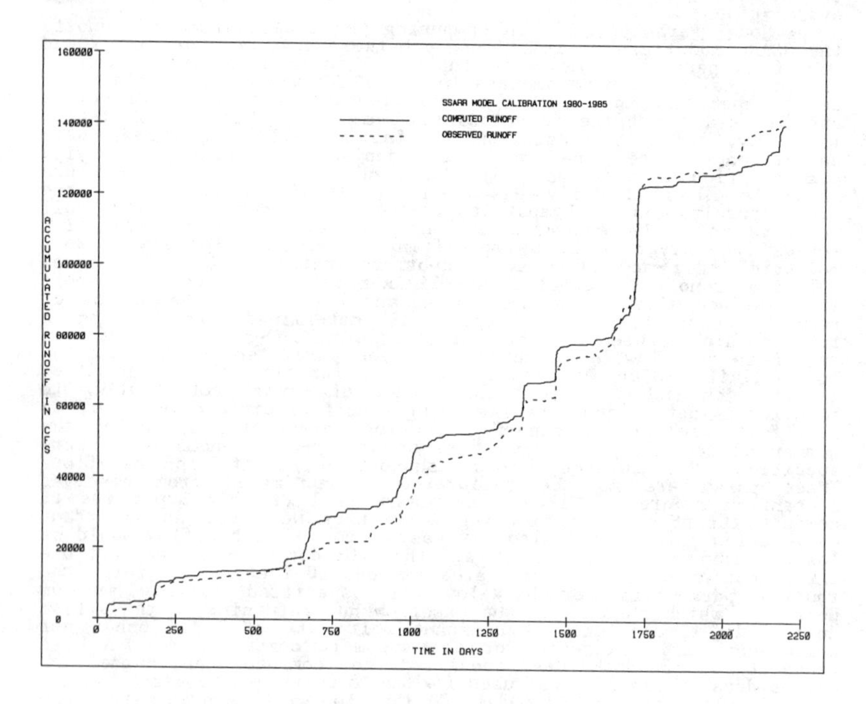

Figure 2. Results of SSARR Model Calibration

<u>Hydrologic Budget Analysis.</u> Simulation of the daily hydrologic budget of Miller's Pond over a historical period of record was performed with the Wetland Hydrologic Analysis Model (WHAM). The WHAM was developed by the Omaha District, U.S. Army Corps of Engineers to evaluate inflow, outflow, change in storage, evapotranspiration, and seepage of individual wetland sites. Input data requirements for the WHAM include elevation-surface area-capacity-outflow relationships for the wetland site, monthly seepage rates, daily precipitation, daily evaporation, monthly evaporation factors, monthly outflow factors, and daily inflows. Daily inflow values for Miller's Pond were synthesized with the calibrated SSARR model, as discussed previously, during the historical period of water years 1949 through 1986. Elevation-area-capacity relationships were derived for Miller's Pond based on the Conic method. Surface areas were digitized from 1-foot contour mapping. Average depth throughout the wetland was computed as capacity divided by surface area. Evapotranspiration rates were based on recorded pan evaporation data at National Weather Service stations in the region as used in the SSARR model, discussed previously. Daily precipitation values recorded at the Fairbury and Bruning weather stations were averaged to obtain the precipitation on the water surface of the wetland. Seepage losses of 0.059 inches per day were based on seepage losses at existing reservoirs in the region. Outflow capacity for the existing conditions analysis was based on a hydraulic analysis of the 24-inch clay tile outlet pipe.

For the restored conditions analysis, it was assumed that the outlet pipe did not exist.

Results of Hydrologic Budget Analysis. Results of the historical daily simulation of the hydrologic budget are summarized in Table 1 for the existing and restored conditions. Average values are provided for water surface area, water surface elevation, storage contents and average depth. Comparison of restored and existing conditions monthly mean water surface elevation are shown on Figure 3.

Table 1. Miller's Pond Historical Simulation

Condition	Average Area (acres)	Average WS Elev (ft-msl)	Average Storage (ac-ft)	Average Depth (feet)
Existing	2	1633.0	1	0.01
Restored	180	1636.7	195	0.69

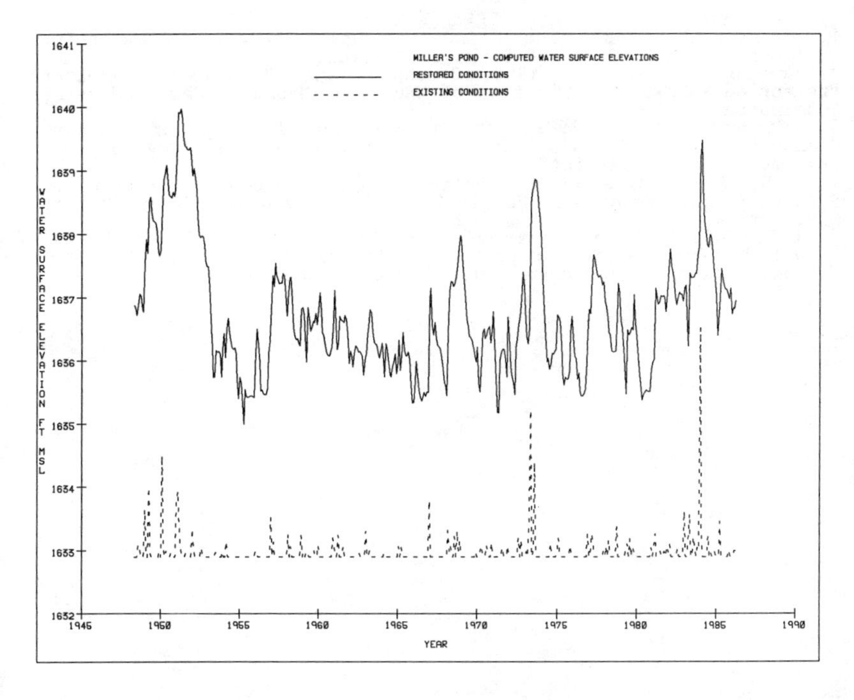

Figure 3. Miller's Pond Computed Water Surface Elevations

Statistical analyses were performed on the computed hydrologic parameters to determine the percent of time that a given value would be equalled or exceeded. These duration analyses were completed for water surface elevation, surface area, and average depth for separate seasons including annual (Oct 1 to Sep 30), spring migration (Mar 1 to Apr 30), summer nesting (May 1 to Jul 31), and fall migration (Oct 1 to Oct 31). Results of this analysis are shown in Table 2 for the surface area duration analysis.

Table 2. Restored Conditions Duration Analysis

Season	Water Surface Area in Acres for Given Percent of Time Exceeded		
	10	50	90
Annual	411	199	32
Spring	390	189	39
Summer	401	232	57
Fall	435	219	19

SUMMARY

A modelling system has been developed to analyze the hydrologic conditions of wetlands in the Rainwater Basin in Nebraska. A test application of the model has been completed at the Miller's Pond site to determine the hydrologic budgets for existing and restored conditions. It is planned to analyze additional sites with the model in order to assist in the prioritizing of candidate wetland sites for acquisition and restoration under a Joint Venture Program as part of the North American Waterfowl Management Plan.

REFERENCES

1. Farrar, J., 1982. The Rainwater Basin: Nebraska's vanishing wetlands. Nebraska Game and Parks Commission.
2. Gersib, D., et al., 1990. Concept Plan for Waterfowl Habitat Protection Rainwater Basin Area of Nebraska. Nebraska Game and Parks Commission.
3. Gilbert, M.C., 1989. Ordination and Mapping of Wetland Communities in Nebraska's Rainwater Basin Region. U.S. Army Corps of Engineers, Omaha District.
4. U.S. Army Corps of Engineers, 1987. SSARR Model Streamflow Synthesis and Reservoir Regulation, User Manual. North Pacific Division.
5. U.S. Department of Commerce, 1982. Mean Monthly, Seasonal, and Annual Pan Evaporation for the United States. NOAA Technical Report NWS 34. National Oceanic and Atmospheric Administration.

ARID ZONE HYDROLOGIC IMPLICATIONS FOR THE PROPOSED YUCCA MOUNTAIN REPOSITORY

John W. Fordham, Member ASCE[1]

Abstract

The U.S. Department of Energy (DOE) has chosen Yucca Mountain, Nevada for characterization to determine if it is suitable to become the Nation's first High-Level Radioactive Waste Repository. This paper examines the historic background for the choice, the regulatory requirements, and the studies made necessary by its arid setting to determine its viability as a permanent repository

Introduction

The need for permanent disposal of high-level radioactive wastes has been evident for decades. As early as the mid-1950's the Atomic Energy Commission (AEC) began to consider alternatives to the improvised waste management methods used during the development of atomic bombs to handle large volumes of wastes from the expected commercial nuclear energy industry. Deep geologic disposal was among these alternatives considered, but since the problem was then perceived to be far enough into the future, 20 to 30 years, no real policy decisions were made. Efforts were undertaken during the late 50's and early 60's to examine a variety of waste-processing techniques mostly related to solidifying high-level wastes. Oak Ridge National Laboratory began investigating storage in bedded salt deposits in the mid-60's under project Salt Vault, which led to the announcement by the AEC in 1970 that the salt mines at Lyons, Kansas was a tentative choice for a permanent repository. During that time period, the AEC developed a waste policy that would have the government assume full responsibility for ultimate disposal of wastes from reprocessors (Nuclear Regulatory Commission, 1970). However, the tentative choice of the salt mine at Lyons proved to be premature and subsequent studies of the site integrity in 1971 caused the site to be judged unacceptable. Subsequently, a change in leadership at the AEC resulted in a change in emphasis to plans for a retriev-

[1] Water Resources Center, Desert Research Institute, University of Nevada System, P.O. Box 60220, Reno, Nevada 89506

able surface storage facility although some work continued to investigate the geologic disposal option, primarily in bedded salt near Carlsbad, New Mexico. This retrievable surface storage concept died and the AEC itself was abolished in 1974 to be succeeded by the Energy Research and Development Administration (ERDA) which proposed as many as six geologic repositories for disposal of high–level and long–lived wastes associated with the nuclear fuel cycle. These were to be in salt and other rock types and in operation by the 1990's. However, with the 1977 decision by President Carter to defer reprocessing, disposal concepts had to be rethought. By 1977, DOE, successor to ERDA, adopted the concept of permanent geologic disposal focusing on several salt sites as well as the Hanford and Nevada Test Site reservations. The eventual outcome of the political process guided by changes in administration and the numerous interest groups was the Nuclear Waste Policy Act of 1982 (PL 97–425) which specified the process for selecting a repository site. The Act in Section 111(b)(1) directs DOE "to establish a schedule for the siting, construction, and operation of repositories that will provide a reasonable assurance that the public and the environment will be adequately protected from the hazards posed by high–level radioactive waste and such spent nuclear fuel as may be disposed of in a repository."

Subsequently, DOE identified nine sites as potentially acceptable, recommended three sites for full characterization. Congress then singled out the Yucca Mountain site as the only candidate for characterization (Nuclear Waste Policy Amendments Act of 1987). That process has began with the preparation and release of the "Site Characterization Plan, Yucca Mountain Site" (SCP) (DOE, 1988).

Regulatory Arena

The SCP, prepared in accordance with requirements of the Nuclear Waste Policy Act Section 113(b)(1)(A), summarized the data collected to date and presented plans for obtaining the geologic/hydrologic information to demonstrate the suitability of the site for the repository. The standards which determine the suitability of the site are concerned with protecting the health and safety of the public from hazards associated with the wastes to be emplaced. These are promulgated by the Environmental Protection Agency (EPA) in 40 CFR 191 and specify: a) the amount of radioactivity that may enter the environment for 10,000 years after disposal; b) limits on radiation dose to the public for 1,000 years after disposal; and c) protection of groundwater. The EPA standards are implemented and enforced by Nuclear Regulatory Commission regulations in 10 CFR 60 which contains the procedures for licensing the repository, as well as the technical criteria to be used in evaluation of the license application. In particular, 10 CFR 60 requires a minimum 1,000–year groundwater travel time from the disturbed zone to the accessible environment.

In addition to these regulations, the DOE developed general siting guidelines, 10 CFR 960, which were used in the screening and selection of the site and post-closure guidelines related to long-term performance of the repository.

The Site

Yucca Mountain is in the southern part of the Great Basin about 160 km northwest of Las Vegas on land controlled by DOE, the U.S. Air Force, and the U.S. Bureau of Land Management. The area is characterized by north to northwest trending mountain ranges separated by intermontane sediment filled structural basins. The mountain itself is an irregularly shaped volcanic upland with approximately 600 m relief. It is composed of eastward dipping volcanic and volcaniclastic strata.

The hydrologic system in the southern Great Basin is characterized by low precipitation, thick unsaturated zones, closed topographic basins restricting surface water flow and large regional groundwater basins underlying several topographically closed basins. Groundwater is recharged through porous surface materials and fractured rock overlying the aquifer system. Most recharge is thought to take place at upper elevations although significant recharge is postulated to occur along ephemeral stream channels as a result of infrequent flow events (Czarnecki and Waddell, 1984). The average precipitation at the site is 150 mm/yr (DOE, 1988) with only a small fraction estimated to be recharge (Montazer and Wilson, 1984).

The hydrologic conditions at the site are critical to long-term performance at Yucca Mountain since the movement of groundwater is the most likely mechanism to transport radionuclides to the accessible environment. The proposed location of the repository deep within the thick unsaturated zone has some definite advantages, but there are also disadvantages, primarily those associated with our lack of understanding of flow processes into and within the unsaturated zone. DOE has specifically pointed out a number of performance issues and needs for information which require thorough understanding of the site.

The strategy for filling these information needs is an extensive program of investigation to develop a complete and accurate description of the components of the hydrologic system and their interrelationships. The program is intended to develop data and associated evaluations which will result in hydrologic models for surface water and both the unsaturated and the saturated zones. Each of these models must be capable of not only describing the existing conditions, but reasonable postulated future conditions over the life of the repository, 10,000 years.

Specific Hydrologic Concerns

At first glance, it would appear that the concept of deep geologic disposal of high-level radioactive waste in an arid environment has numerous advantages since groundwater flow is the most likely mechanism for transport of hazardous material away from a repository. However, these advantages quickly become uncertain when we consider that natural variations in both arid climatic factors and the geologic medium combine to produce a hydrologic system which is extremely difficult to adequately characterize. Our knowledge is limited with respect to quantifying recharge in arid environments, and the movement of water within unsaturated fractured rock is poorly understood. Yet, these are two key hydrologic processes at Yucca Mountain which must be quantified with a high degree of certainty if the site is to meet the currently promulgated licensing criteria.

Several studies are proposed for Yucca Mountain which focus on aspects of arid–zone hydrology. These are: 1) to define recharge mechanisms under current climatic conditions together with identifying spatial and temporal variability of recharge; 2) to quantify water movement in the unsaturated zone and to understand the physical relationships governing that flow regime; and 3) to quantify influences of climate change on site hydrology. Proper evaluation of the site will require a considerable increase in our understanding of all of these basic hydrologic processes.

Preliminary studies at the site have shown a wide variation in hydrologic properties among the various hydrologic units of the unsaturated zone as well as considerable variation in surficial characteristics. Studies have, therefore, been designed to identify which hydrologic characteristics and environments at the site contribute to, or take away, from waste isolation. The current level of understanding of water movement in the unsaturated zone indicates that for the site to meet one key criterion, pre–waste emplacement groundwater travel time, flow in the unsaturated zone must essentially be limited to the rock matrix with little or no fracture flow. For this to happen, the unsaturated zone flux must be extremely small, i.e., less than 1 mm per year (Peters et al., 1986). To be assured that such a flux rate currently exists and will remain small, it will be necessary to demonstrate that net recharge at the site under present and expected climatic conditions is less than 1 mm per year. The site's physical characteristics and the large variation seen in the geologic materials will make actual quantification of recharge extremely difficult.

The DOE approach is to define the upper boundary flux under present–day and wetter climatic conditions (DOE, 1988; Flint, 1989). To achieve this, a three–part program has been proposed to characterize infiltration to the unsaturated zone (Flint, 1989). The first program element is to characterize surficial materials based on: a) measured physical and hydrologic properties; b) surface and borehole geophysics; and c) mapping the site to define hydrologic response units. To achieve this, it will be necessary to undertake extensive sampling to define units based on geomorphic processes and characteristics, including pedologic and hydrologic factors such as soil thickness, slope, aspect, texture, density, water content, and water potential. Areas of similar hydrologic properties will be defined to extrapolate data from the natural and artificial infiltration studies.

The second program element is to characterize natural infiltration by examining the basic water balance components using water balance approach but not solving specifically for net infiltration where net infiltration is defined as infiltrated water which has percolated below the zone of evapotranspiration (i.e., the zero flux plane). This value is the potential recharge. The net infiltration at the site is highly spatially and temporally variable and will involve measurement of precipitation, runoff and evapotranspiration in a number of different environments. The activity will also involve extensive neutron hole logging to estimate changes in volumetric water content as well as isotope analyses to attempt to trace the origin and age of water at various depths within the unsaturated zone. Techniques such as tritium profiling will be used to estimate water flow velocities.

The third program element is to characterize infiltration under wetter than present–day conditions. These studies are intended to characterize the range and spatial variability of infiltration rates, flow velocities, and flow pathways in the upper 5 m of both consolidated and unconsolidated surficial materials and to develop relationships between precipitation runoff, infiltration, and evaporation for at least one site in each hydrogeologic surficial unit.

Discussion

Use of the water balance approach,

$$\text{infiltration} = \text{precipitation} - \text{evapotranspiration} - \text{runoff} +/- \text{change in storage},$$

has inherently large sources of error when considering the measurement of each parameter. These combined errors, even from carefully controlled experiments using the best available instrumentation, will likely exceed the net infiltration values expected in this environment. Preliminary data suggest that the potential evapotranspiration is an order of magnitude greater than the 150 mm yearly precipitation, which is approximately two orders of magnitude greater than the flux values which would ensure matrix flow and the long travel times necessary to meet licensing criteria (Leffler, 1989; DOE, 1988). A compounding factor is the extreme spatial variation due to differences in properties of the surficial units, topography and vegetative cover which yield an unevenly distributed infiltration and resultant recharge over the site.

A related concern, given the long–term waste–isolation requirements for the repository, is that relatively small changes in either precipitation or temperature and their temporal distributions could result in a significant increase or decrease in recharge. Estimates made during preliminary investigations suggested that a doubling of precipitation at the site could increase the recharge manyfold (Czarnecki, 1985); therefore, reasonable estimates of future precipitation increases would be expected to increase recharge significantly. Evidence of relatively rapid short–term climatic perturbations, on the order of hundreds of years, does exist for the southern Great Basin (Wigand, 1990). Such changes could result in significant increases in recharge at the site; therefore, the existing preferred conceptual model of matrix flow in the unsaturated zone would no longer hold and the thickness of the unsaturated zone could be significantly reduced resulting in groundwater travel time being greatly reduced. Such a change could result in the site not meeting the existing criteria for licensing.

Conclusion

To address the question of present–day, as well as future recharge and unsaturated flow conditions at the site, a number of studies have been recommended in the SCP (DOE, 1988). Although these studies are numerous, complex and expensive, it will be extremely difficult to fully understand existing and predict future conditions in a definitive manner. It may well be that in an arid environment such as Yucca Mountain, the temporal and spatial variability of the natural properties and processes, together with limitation on measurement techniques and time available to do the studies, make true characterization an impossible task.

References

Czarnecki, J.B. and R.K. Waddell, 1984. "Finite Element Simulation of Ground–Water Flow in the Vicinity of Yucca Mountain, Nevada–California", USGS–WRI–84–4349, U.S. Geological Survey.

Czarnecki, J.B., 1985. "Simulated Effects of Increased Recharge on the Ground–Water Flow System of Yucca Mountain and Vicinity, Nevada–California", USGS–WRI–84–4344, U.S. Geological Survey.

Flint, A., 1989. "Characterization of Infiltration," Presentation to the Nuclear Waste Technical Review Board.

Montazer, P. and W.E. Wilson, 1984. "Conceptual Hydrologic Model of Flow in the Unsaturated Zone, Yucca Mountain, Nevada", USGS–WRI–84–4345, U.S. Geological Survey.

Peters, R.R., J.H. Gauthier and A.L. Dudley, 1986. "The Effect of Percolation Rate on Water–Travel Time in Deep, Partially Saturated Zones", SAND 85–0854, Sandia National Laboratory, Albuquerque, New Mexico.

Wigand, P., 1990. Personal Communication regarding pollen cores from Southern Great Basin.

U.S. Nuclear Regulatory Commission. "Disposal of High–Level Radioactive Wastes in Geologic Repositories", 10 CFR 60.

U.S. Nuclear Regulatory Commission. "Policy Relating to the Siting of Fuel Reprocessing Plants and Related Waste Management Facilities", 10 CFR 50.

U.S. Department of Energy. "General Guidelines for the Recommendation of Sites for Nuclear Waste Repositories", 10 CFR 960.

U.S. Environmental Protection Agency. "Environmental Radiation Protection Standards for Management and Disposal of Spent Nuclear Fuel, High–Level and Transuranic Radioactive Wastes", 40 CFR 191.

U.S. Department of Energy, Office of Civilian Radioactive Waste Management, 1988. "Site Characterization Plan, Yucca Mountain Site, Nevada Research and Development Area, Nevada", DOE/RW–0199.

Leffler, P., 1989. "Characterization of Present and Possible Future (Wetter) Climatic Regimes Utilizing the Historical Record and an Analogue Technique in the Southern Great Basin", unpublished Masters Thesis, University of Nevada, Reno.

Nuclear Waste Policy Act of 1982 (NWPA) Public Law 97–425, 96 Stat, 2201, 42 U.S.C., 10101, January 7, 1983.

Nuclear Waste Policy Amendments Act of 1987 (NWPAA) Public Law 100–203, December 22, 1987.

A STUDY ON SEDIMENT TRANSPORTATION IN DEBRIS FLOW

Kang Zhicheng[1]

Abstract

By analysing the observation data of debris flows at the Jiangjia Ravine, Dongchuan, Yunnan Province, China in 1982-1986, this paper comprehensively and systematically describes the sediment transportation properties, rates and quantities of viscous debris flow, fluid debris flow and high concentration flow to show that debris flow is a high-intensity sediment carrier, which is more than 40 times that of the Yellow River under the same discharge conditions. All the above distinct sediment transport properties will promote both theoretical and experimental studies on debris flow.

INTRODUCTION

Debris flow is different from ordinary sediment-laden rivers in sediment origin. According to data analysis at the Jiangjia Ravine, more than 90% of sediments in debris flows are fed by collapses and landslides on both sides of ravines, while less than 10% of fine particles are produced by slope surface erosion (soil and water losses)[1]. Sediments in sediment-laden rivers originate largely from slope erosion caused by rainstorms; they go with the ground surface runoff into tributaries and then into rivers. The former is debris flows fed by collapses mainly under the gravitational effects of loose and saturated earth and rock materials; the latter is that ground surface runoff caries sediments into rivers. Therefore, the former has a maximum concentration of 2180kg/m³; the latter 1600kg/m³. The grain size of coarse particles carried by debris flow can be hundreds times larger than that by the Yellow River. The movement of extremely thick

----

Director, Dongchuan Debris Flow Observation & Research Station, Chinese Academy of Sciences, Chengdu P.O.Box417, Sichuan 610015, China sediments and rocks and the sediments transportation in high gradient ravines in mountain areas are worthy of note, because they are one mode of sediment origin for rivers.

THE EVOLUTION PROCESS OF SEDIMENT CONCENTRATION IN DEBRIS FLOW

In order to explain this question on sound grounds, 2 detailed observations and sample analyses were conducted on debris flows; figures drawn for the changing processes of sediment concentration in 3 debris flows that occurred on Aug.13, 19 and 28, 1983 respectively (Fig.1, 2 and 3). On the basis of the variations of debris flow processes, debris flow is divided into the sediment concentration process of early fluid debris flow, the sediment concentration process of viscous debris flow, the sediment concentration process of late fluid debris flow, and the sediment concentration process of sediment-laden water flow, which were drawn on a coordinate paper for comparison (Fig.4). The process lines show that there is a certain law in the variation of sediment concentration in the evolution process of debris flow, which is closely related with the occurrence of debris flow. The sediment concentration process of early fluid debris flow, i.e., from the normal sediment concentration (159kg/m^3) to the sediment concentration of viscous debris flow, is nery short. Usually, it ranges from minutes to half an hour; sometimes, there is almost no early process. Therefore, this process line rises up sharply in the form of a straight line; soon the sediment concentration can reach 1900kg/m^3 almost without a gradually changing process of sediment concentration. Fig.2 shows that the duration of the early fluid debris flow process only accounts for 5% of the whole process. the reason is that the supplement of solid materials in the occurrence zone is associated with the occurence of collapses and landslides. Fig. 2 also indicates that the line of the sediment concentration changing process of viscous debris flow is tooth-like with the sediment concentration ranging between 1588kg/m^3 and 2180kg/m^3. It is the stage of the maximum sediment concentration during the process of sediment transport in debris flow, also the major sediment transport process of debris flow. This process is rather stable. Since the sediment concentration is relatively stable, its sediment transport rate is only the function of discharge, which can be called the stable sediment transport process. The duration of this process depends on the precipitation and intensity of the debris flow. The sediment transport process of the late fluid debris flow is the stage that means that the debris flow is about to end. Its process line is a sharply descending one (Fig.1),i.e.,

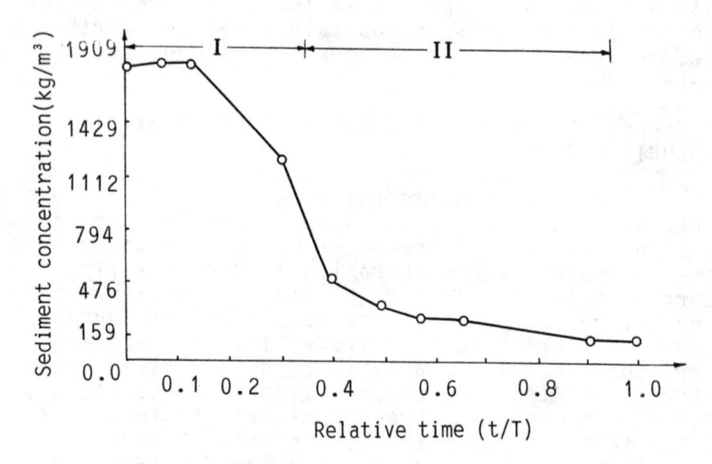

Fig.1 Varied sediment transport process of fluid debris flow(I) and sediment-laden water flow(II) on Aug.13, 1983(t-duration at gauges; T-total duration)

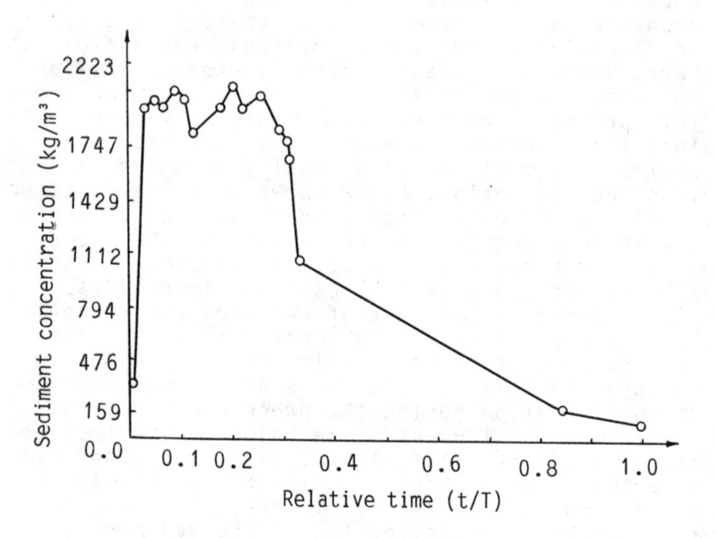

Fig.2 Varied sediment concentration processes of a debris flow on Aug.19, 1983 (t-duration at gauges; T-total duration)

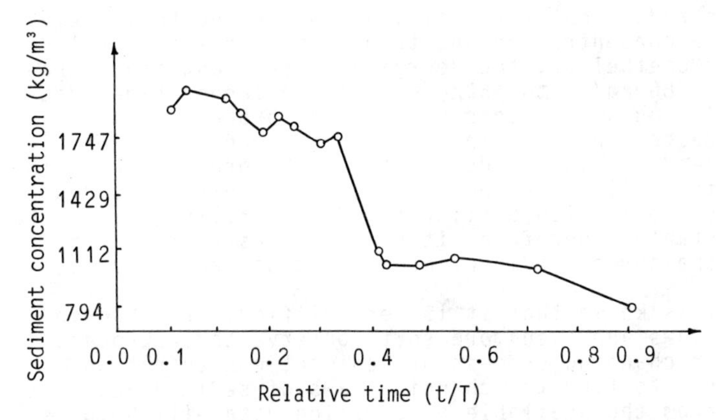

Fig.3 Varied sediment concentration processes of a debris flow on Aug. 28, 1983 (t-duration at gauges; T-total duration)

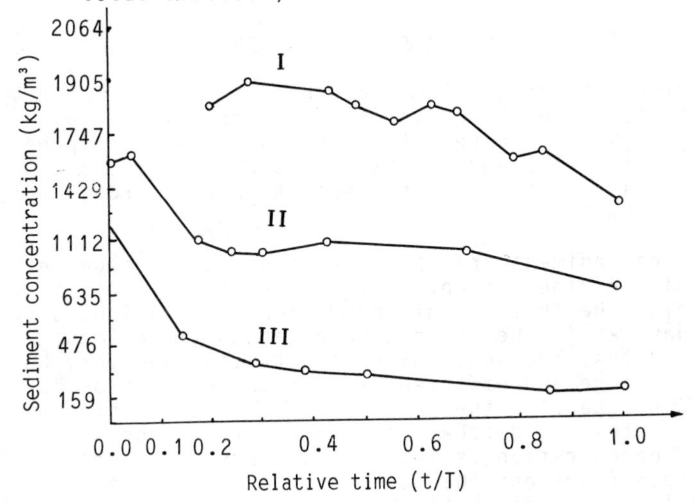

Fig.4 A comparison between the sediment transport process lines of diffrent debris flows (t-duration at gauges; T-total duration; I-viscous debris flow; II-fluid debris flow; III-sediment-laden water flow)

the sediment concentration decreases from 1600kg/m³ to 800kg/m³ at a high speed. However, the sediment concentration exists in a gradually changing process, which is different from the variation of the early sediment concentration. The main reason for the occurrence of this process is that when a debris flow enter the ending stage, large quantities of caorse particles settle down due to a sudden reduction of velocity and discharge to quickly lower the concentration and transform into sediment-laden floods. Nonetheless, the decrease of sediment concentration from 800kg/m^3 to 150kg/m^3 needs a longer time, usually 8-10 hours. For this reason, the process line is a slowly descending one. In fact, this process is that after the end of viscous debris flow, the ordinary flow in the ravine will scour and carry the fine particles deposited by debris flow until the bed load fully coarsens to become stable. Therefore, it needs to last a long time before the normal sediment concentration can recover.

It is known that it is very difficult to observe debris flows and even more so to observe their process and their changing process of sediment concentration. Therefore, to find the changing laws of sediment concentration from the available observation data will help us evaluate the neglected information in the process of debris flow calculation and push the research work a step further.

SEDIMENT TRANSPORTATION PROPERTIES OF DEBRIS FLOW

Sediment transportation in debris flow is affected by many factors. On the basis of the observation data at the Jiangjia Ravine in recent years, the relationship between sediment rates and discharges is drawn (Fig.5). Fig.5 reveals the following sediment transport properties of debris flow:

1. The sediment transport rates of debris flows at the Jiangjia Ravine are not only the function of discharge, but also the function of sediment concentration. Both factors have an influence on the variation of sediment transport rates. The sediment concentration, however, has a regular change and different stages in debris flow movement. Consequently, the relationship between sediment transport rates and discharges is 3 kinds of process with sediment concentration as their parameters. The three lines in Fig.5 represent the sediment transport rates of sediment-laden flood(I), fluid debris flow(II) and viscous debris flow(III) respectively.

2. The three stages of sediment transport process are valid only under certain discharge conditions, otherwise they will enter or decrease to other process. As for the sediment transport process of sediment-laden floods at the Jiangjia Ravine, their discharge, as a rule, does not exceed 50m^3/s.; fluid debris flows do not have a discharge higher than 300m^3/s.; viscous debris flows usually have a discharge above 50m^3/s. with a maximum observed discharge up to 2820m^3/s. and sediment transport rate

close to 5000×10^3 kg/s. The above three processes indicate that viscous debris flow is the main sediment transport process, and the other two processes can be regarded as the sediment transport process of the transitional stage.

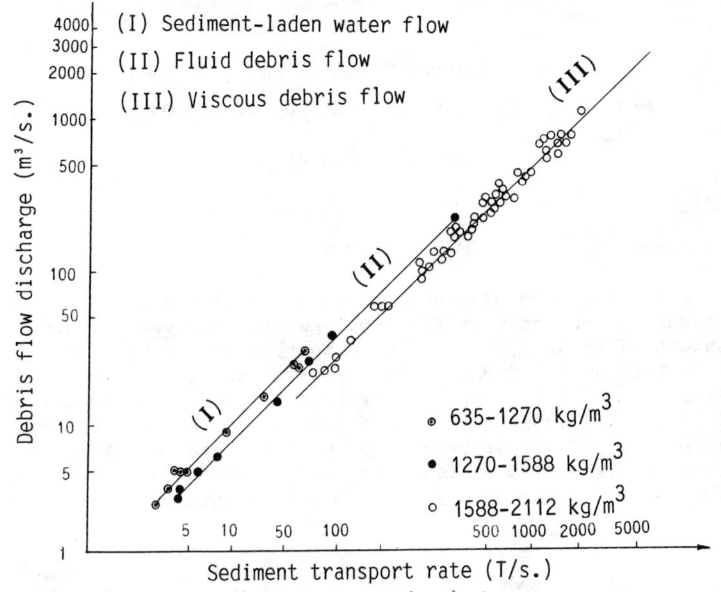

Fig.5 Relationship between sediment transport rates and discharges of debris flows

3. Debris flow has a high-intensity sediment transport capability. In comparison with the maximum sediment transport rate on the lower reaches of the Yellow River, if their discharges are $2000m^3/s.$, the sediment transport rate of the Yellow River would be 90×10^3kg/s., while that of the Jiangjia Ravine would be 3800×10^3kg/s., more than 40 times the Yellow River [2]. Obviously, debris flow is a high intensity sediment transport process. It will be inevitable task for researchers in theory to study the laws.

References

[1] Kang Zhichengand Zhang Shucheng, 1984, An Analysis of Sediment Transport Catchment Experiments in Fluvial Geomorphology, University Press, Cambridge, A 77-488.

[2] Mai Qiaowei and others, 1980, Sediment Problems in the Channels on the Lower Reaches of the Yellow River, Proceedings of International Symposium on River Sedimentation, Guanghua Press, pp. 397-404.

SEDIMENT TRANSPORT MODELS FOR SMALL GULLIES IN LOESS HILL AND GULLY REGIONS

Cao Ruxuan   Fan Erlan   Qin Yi

Abstract

Three items are studied: the equation of the relation between discharge and sediment for small gullies derived by using the theory of river dynamics; the extension of the Nash theory of transient unit hydrograph to the obtaining of the transient unit hydrograph for the sediment transport; the establishment of models of the discharge-sediment response function. The two models obtained can be used to predict the sediment transport process at the outlet of small gullies. The computational results are in good agreement with the obtained data used for verification.

Introduction

On the upper and middle reaches of the Yellow River is located the largest loess plateau in the world with an area amounting to 0.58 million square kilometers. On the plateau, there is a loess hill and gully region with an area of 0.236 million square kilometers, where soil erosion is extremely serious and the annual erosion module works out to be above 10,000 T/km^2, constituting the major source of the sediment of the Yellow River .Since the 1950s. the experimental run-off stations set up by the Yellow River Water Conservancy Commision have observed a great deal of data and made a lot of the research achievements (3)-(4). Based on the data, the model of sediment transport is established.

Discharge-Sediment Relation

The comprehensive actions of various natural factors

---

[1]Senior Engineer, Shaanxi Mechanical Engrg.Inst., Xi'an, China
[2]Lecturer, Shaanxi Mechanical Engrg.Inst., Xi'an, China
[3]Lecturer, Shaanxi Mechanical Engrg.Inst., Xi'an, China

and human activities create the feature that flood peaks at the outlet of small gullies in the loess hill and gully region have sudden rises and falls, but the sediment peaks have sudden rises and slow falls. The relation between the rate of sediment transport and the discharge has a feature as shown in Fig.1. It shows that, for valleys with different area, there is a critical discharge Q_s, when $Q > Q_c$, there is a unified discharge-sediment relation in all valleys, which can be denoted as follows:

$$Q_s = 550\, Q^{1.06} \qquad (1)$$

When $Q < Q_c$, each valley has its own relation

$$Q_s = KQ^n \qquad n > 1.06 \qquad (2)$$

and coefficients K,n varies with the specific features of the valley.

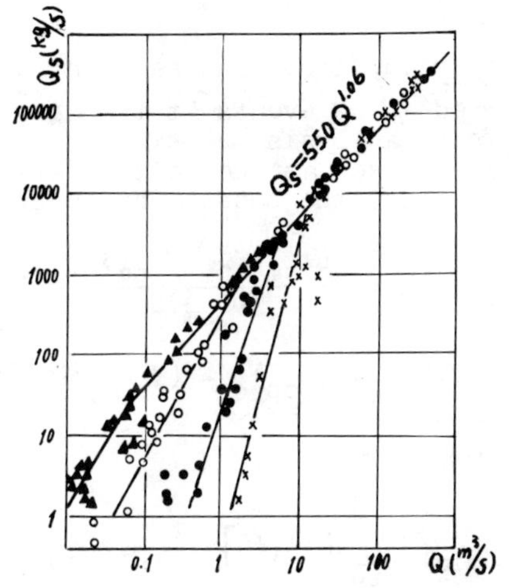

Fig. 1 Relationship of Q and Q_s

It has been shown through studies that the above feature is based on the fact that when $Q > Q_c$, the flow has a high concentration, creating a hyperconcentrated homogeneous flow and the sediment transport characteristics are in the sphere of resistance work (1). From the flow condition $\gamma_m hS = \tau_B$, emprical relation of yield stress $\tau_B = 0.615 \times 10^{-4} c^5/d_{50}^3$, and the gully profile analytical equation $h = K_1 Q^m$, when taking the average values of gullies S = 0.01 - 0.03, d_{50}=0.04 - 0.05mm, we may get the unified relation of Equation (1).

When $Q < Q_c$, n and K varies with the features of valleys. This is due to the fact that in various valleys, there are rains in different places and at different times and there is time difference in causing the flow to get together.

Analysis, Determination and Vertification of Transient Sediment Transport Unit Hydrograph

1. Derivation of parameters for transient sediment Transport Unit Hydrograph

When extending the mode of the Nash transient unit hydrograph to the derivation of the transient sediment transport unit hydrograph, the basic formula is as follows (2):

$$U(o,t) = \frac{1}{K_s \Gamma(n_s)} \left(\frac{t}{K_s}\right)^{n_s - 1} e^{\left(-\frac{t}{K_s}\right)} \quad (3)$$

Let's take the rain, flood and sediment data of an area of 21 Km^2 observed in 12 events at the Liujiagou Station as the materials of analysis.

Two minutes are chosen as the unit duration. When optimizing the parameters, the constraint functions adopted are

$$S = \frac{(Q_{sm})_c - (Q_{sm})_p}{(Q_{sm})_p} \leqslant \pm 10\% \quad (4)$$

$$\Delta T_{sp} = (T_{sp})_c - (T_{sp})_p \leqslant \pm 8 \text{ min.} \quad (5)$$

and the target function is

$$S_{min} = \sum_{t=1}^{T} \left\{ \left[Q_s(t)\right]_c - \left[Q_s(t)\right]_p \right\}^2 \quad (6)$$

The parameters n_s, k_s are optimized according to the observed data. Suppose that the relation between unit hydrograph lag time $m_{sI} = n_s \cdot k_s$ and the yield rate of sediment I_s is as follows

$$m_{sI} = a_s I_s^{-b_s} \quad (7)$$

Let's take the data of 8 out of 12 events for computation. When $b_s = 0.2$, $a_s = 9.0$ and $a_s = 9.75$, the rates of acceptance are all 75%.

2. Verification of Transient Sediment Transport Unit Hydrograph

Verification was made by using the data of another four events from the 12 events according to $S \leqslant \pm 20\%$, $T_{sp} \leqslant \pm 8$ min. and $m_{sI} = 9\ I_s^{-0.2}$. The results are shown in Table 1, the rate of acceptance being 100%.

Table 1 Verified Results of Transient Sediment Transport Unit Hypograph in Liujiagou Valley

Date	R_s (mm)	I_{sp} mm/min	m_{sI}	Q_{max} comp.	Q_{max} obser.	$\Delta Q_{sm}/Q_{sm}$ %	Peak Occ. Time comp.	Peak Occ. Time obser.	ΔT_{sp} min
July 17, 1967	1.537	0.085	14.7	21.0	23.5	-10.9	18	16	2
July 15, 1968	1.065	0.107	14.1	18.5	16.8	-10.5	14	14	0
May 11, 1969	2.630	0.146	13.2	39.5	40.71	-3.2	18	20	-2
Aug. 20, 1969	0.5	0.044	16.8	7.9	7.9	-6.6	16	14	2

Establishment of Discharge-Sediment Model

1. Model Establishment

Since there is a fixed relation between discharge and sediment in the loess hill and gully region, the transition process of the rain, flood and sediment transport rate can be considred to be a system which is supposed to be linear and functions as the discharge-sediment function. If we take Q and Q_s as the object of investigation, the relation between the input Q and the output Q_s and the action of the system will satisfy the convolution integration

$$Q_s(t) = \int_0^t Q(t-I)H_s(I)d\tau \qquad (8)$$

$$Q_{sr} = \sum_{i=1}^{r} Q_i H_{s(r+1-i)} \qquad (9)$$

The discharge-sediment response function $H_s(t)$ can be determined by the observed data Q and Q_s.

2. Model Application

Verification was carried out by using the data of 4 events

observed at the Caoping Station regarding a valley area of 187 km^2. The difference of occurred time between calculated and observed sediment peak e_t, the relative error of the sediment peak e_p and the determinate coefficient R^2 were taken as the accuracy indexes.

$$e_t = |(T_{sp})_p - (T_{sp})_c| \leq 3\Delta t \qquad (10)$$

$$e_p = \frac{|(Q_{sm})_p - (Q_{sm})_c|}{(Q_{sm})_p} 100\% \leq 20\% \qquad (11)$$

$$R^2 = 1 - \frac{\sum_{i=1}^{N}\left[(Q_{si})_p - (Q_{si})_c\right]^2}{\sum_{i=1}^{N}(Q_{si} - \overline{Q}_s)^2} 100\% \qquad (12)$$

The verified results are shown in Table 2

Table 2 Verified Results of Models

No.	e	e	R
1	0	2.44	90.8
2	0	3.77	92.46
3	0	3.73	78.67
4	0	2.14	99.31

Appendix - I - References

1. Cao Ruxuan and Cian Shanqi, Sediment Transport characteristics of Gullies in Loess Hill and Gully Regions, ACTA conservations soil ET aquae sinica,No. 4, 1988
2. Fan Erlan, A study of Transient Unit Hydrograph of Suspended Particles, Sediment Research, No.2, 1988
3. Kung Shiyang and Chiang Techi, Soil Erosion and its Control in Small Gulley Watersheds in the Rolling Loess Area on the Middle Reaches of the Yellow River, Sciences of China, No.6, 1978
4. Qian Ning and Zhang Ren, The Hyperconcentrations Flow in the Main Stem and Tributaries of the Yellow River,Proceeding of International Workshop on Flow at Hyperconcentrations of Sediment,Sept.10-14, 1985,

IRTCES

Appendix II- Notation

The following symbols are used in this paper

C = sediment concentration

$Q_s, (Q_s)_p, (Q_s)_c$ = rate of sediment transport and its observed and calculated values

$Q_{sm}, (Q_{sm})_p, (Q_{sm})_c$ = maximum rate of sediment transport and its observed and calculated values

S = bed slope

$T_{sp}, (T_{sp})_p, (T_{sp})_c$ = peak occurrence time, its observed and determined values

$U(0,t)$ = transient sediment unit hydrograph

ESTIMATING SEDIMENT DELIVERY AND YIELD ON ALLUVIAL FANS

Robert C. MacArthur[1], M.ASCE, Michael D. Harvey[2], and
Edward F. Sing[3], M.ASCE

Abstract

This paper summarizes the procedures used for computing the basinwide annual yields and single event sediment production for ephemoral channels located on an incised alluvial fan in Central California. Unique geomorphic characteristics of the basin and alluvial fan are discussed in light of data and analytical methods necessary to compute sediment delivery and yield at a proposed damsite.

Introduction

A Sediment Engineering Investigation (SEI) of the Caliente Creek watershed (470 sq. mi.) in Kern County, California was conducted to determine the watershed sediment yield upstream from a proposed flood detention reservoir located on the Caliente Fan. Previous studies estimated annual sediment yields at the proposed reservoir site based on traditional soil loss methods and sediment accumulation rates observed in impoundments along the Sierra Nevada, Tehachapi and Transverse Mountain Ranges. Initial project feasibility was considered based on preliminary cost/benefit analyses using the rough sediment yield estimates. Further review of the potential annual maintenance requirements led to the conclusion that the economic viability of the project depended heavily on annual O & M costs potentially required to remove the yearly accumulation of sediment within the proposed reservoir. Accurate estimates for the average annual sediment yield and single event sediment delivery were essential.

Further studies were undertaken to (1) identify specific geomorphic characteristics of the stream channels and watersheds upstream from the proposed flood control reservoir that could effect the sediment yield at the damsite, and (2) to relate channel and basin processes to sediment production and yields for various frequency precipitation and flood flow events in the watershed. This paper summarizes the procedures used for computing the basinwide annual yields and single event sediment production, along with conclusions and recommendations for other project design modifications.

---

[1] Hydraulic Engineer, U.S. Army Corps of Engineers, Hydrologic Engineering Center, 609 Second Street, Davis, CA 95616 U.S.A.

[2] Principal Geomorphologist, Water Engineering & Technology, Inc., 419 Canyon, Suite 225, Fort Collins, CO 80521, U.S.A.

[3] Hydraulic Engineer, U.S. Army Corps of Engineers, Sacramento District, 650 Capitol Mall, Sacramento, CA 95814, U.S.A.

Approach

A two element SEI was conducted to address the sediment yield question: (1) geomorphic analyses (Harvey et al., 1990) were conducted to determine those unique characteristics of the basin and channels important to estimating sediment yield, and (2) sedimentation analyses (HEC, 1990) were conducted to determine the sediment yield in light of the findings from the geomorphic analyses.

To determine the amount of sediment that can possibly enter the proposed reservoir during its design life (100 years), both the average annual sediment yield and single event sediment yields are estimated using a variety of sediment engineering procedures as reported in EM 1110-2-4000, "Sediment Investigations of Rivers and Reservoirs," (COE, 1989) and recommended by others. Available scientific and engineering literature was reviewed, a three-day field reconnaissance and sediment data collection investigation was conducted, persons familiar with the Caliente Creek Project and watershed were interviewed, and a series of sediment engineering analyses to determine the possible sedimentation characteristics of the drainage basin at the damsite were carried out. Morphometric data for the alluvial fan in the vicinity of the proposed reservoir site were obtained from 2-foot contour mapping. Sixteen bed and bank material samples and two Wolman Counts were collected at representative locations throughout the drainage basin.

Average Annual Sediment Yield - The possible range of average annual sediment yield at the proposed reservoir site is estimated from the results from eight different sources of data and/or methods for estimating sediment yield. The following sources of data and procedures were used: (1) Previous reports and publications were thoroughly reviewed, (2) U.S.D.A. (1977) reservoir sedimentation rates were examined, (3) recent COE reservoir sedimentation survey data were analyzed, (4) sediment yield maps for the Western United States (U.S.D.A., SCS, 1975) were examined, (5) the average annual sediment yield was estimated from computations of the total event sediment volumes for single events ranging from the 2-year event up to the PMF based on channel transport capacity rather than watershed sediment production and delivery, (6) a similar flow duration and sediment load curve integration method (see EM 1110-2-4000, COE, 1989) was used to estimate the average annual sediment production and yield to the reservoir site, (7) the Pacific Southwest Inter-Agency Committee (PSIAC) method was used to estimate basin-wide sediment yield from the entire watershed, and (8) the Dendy and Bolton (1976) Regional Analysis Method for sediment yield was applied. Results from these analyses are discussed next. Detailed procedures for conducting such investigations are presented in the references cited and in Engineering Manual 1110-2-4000 (COE, 1989).

Table 1 presents the estimated sediment yields computed using the various computational procedures listed above and from measured reservoir surveys conducted by the Corps of Engineers and SCS. Based on measured sediment accumulation rates recorded in the six Tulare, Kings, and Kern County reservoirs, the approximate range of observed sediment yields is from 0.2 AF/sq mi/yr to 2.2 AF/sq mi/yr with an average of approximately 1.0 AF/sq mi/yr. Sediment yield rates determined for the Western United States are reported by the U.S.D.A., SCS (1975). From the mapping of yield rates, it appears that the upper Caliente watershed area has sediment yield rates from 0.2 to 0.5 AF/sq mi/yr, with pockets as high as 0.5 to 1.0 AF/sq mi/yr. In the lower portions of the basin, on the valley floor and on portions of the broad alluvial fan, the estimated yields are reported to be in the 0.1 to 0.2 AF/sq mi/yr range. Using area weighting methods to sum the yields from contributing subbasins, the approximate annual yield appears to range from 0.2 to about 0.75 AF/sq mi/yr, with an average of about 0.47 AF/sq mi/yr for the entire watershed.

Harvey et al., (1990) determined that the sediment delivery and yield at the damsite depends on the channel transport capacity in the fan area upstream from the reservoir rather than the watershed production of sediment. The broad (3,000 to 6,600 feet wide) alluvial fan contains an unlimited supply of easily mobilized sediment materials. This result lead to the following approach based on the transport capacity of the channels in the supply reach. The supply reach is a 4-mile section of the channel considered to be representative of the channel hydraulic conditions and sediment transport characteristics

TABLE 1

Sediment Surveys for Reservoirs in the Vicinity of Caliente Creek, Kern County, California, and Estimated Sediment Yields Based on Various Computational Methods

Data Source	See References	Drainage Basin, Reservoir or Computational Method Used	Drainage Area (sq mi)	Yield (AF/sq mi/yr)
SCS	10	Blackburn	7.1	2.20
SCS	10	Antelope Canyon	4.4	1.50
CESPK	5	Isabella	2,074	0.37
CESPK	9	Pine Flat	1,542	0.20
CESPK	9	Success	393	0.76
CESPK	9	Terminus	560	0.75
SCS	8	SCS Yield Map of Western US (HEC)	470	0.47
Computed	7	Integration of the Event Volume vs. Frequency Curve (HEC)	470	0.55
Computed	7	Flow Duration Method (HEC)	470	0.90
Computed	7	Dendy & Bolton Method (HEC)	470	0.71
Computed	4	PSIAC Method (HEC)	470	0.75
Computed	6	Kern County Water Agency Study (SLA)	470	0.97

upstream from the dam site. Single event total sediment volumes were computed for each of the 2, 5, 10, 20, 50, 100, SPF, and PMF events. The total sediment production for each event was based on the sediment transport capacity of the alluvial channel (supply reach) upstream from the reservoir and the flow hydrographs used for each of the flood events evaluated.

A total sediment load versus percent exceedance curve was developed from these data and the area under the total load frequency curve was computed to give an estimate for the expected average annual sediment delivery to the reservoir based on channel transport capacity upstream from the reservoir. Two different transport relationships were used to develop the total load curves. The resulting average annual sediment delivery ranged from 0.1 AF/sq mi/yr to 1.0 AF/sq mi/yr due to the difference in transport capacity computed with the transport functions. Using these results as a representative range in expected yields based on channel capacity, an average of the two yields seems reasonable. Therefore, based on the channel transport capacity above the reservoir site and the estimated total sediment production from a range of single events, an approximate sediment yield at the reservoir is 0.55 AF/sq mi/yr. This method does not account for the additional contribution of sediment from dry ravel erosion, wind-blown sand transport into the channel or reservoir, channel bank caving, local scour, or toe failure that may occur along the Sand Hills. Therefore, the sediment yield to the reservoir may be as high as the higher of the two transport functions predicts, especially during periods of exceptionally wet years.

The "flow duration sediment discharge rating curve method," (COE, 1989) is a simple method where the flow duration curve is integrated with the sediment discharge rating curve developed for the damsite. It is very similar to the method just described, however, the average annual sediment yield is based on the transport capacity and flow duration relationship at the damsite rather than the total event volume frequency. The resulting annual sediment yield is approximately 438 AF/year, or 0.9 AF/sq mi/yr.

Further examination of the U.S.D.A., SCS (1975) "Sediment Yield Rates for the Western United States" shows areas in the vicinity of the proposed damsite with estimated yields from 0.5 to 1.0 AF/sq mi/yr. These areas may correspond to the broad floodplain channels (4000 to 6500 feet wide) immediately upstream from the proposed reservoir site. If that is the case, then the higher yield values estimated with the channel transport capacity method (1.0 AF/sq mi/yr) and the flow duration method (0.9 AF/sq mi/yr) are supported by SCS yield mapping estimates.

The Dendy and Bolton (1976) method produces an average annual sediment yield of approximately 0.71 AF/sq mi/yr for the Caliente Basin at the Sivert damsite, while the application of PSIAC procedures to the Caliente Creek watershed produces an estimated average annual sediment yield of 0.75 AF/sq mi/yr at the dam site. These values are right in line with the range of values predicted from the channel capacity approach and the measured reservoir accumulation results from Tulare County.

Others (Simons, Li & Associates, 1989) conducted an independent assessment of the proposed Caliente Creek Project. The authors report the arithmetic average of their yield estimates (**0.97 AF/sq mi/yr**) in **Table 1. Figure 1** shows all thirteen yield values and the drainage basin area associated with each yield. A best fit line through these data points gives an average annual sediment yield of **0.75 AF/sq mi/yr**. This is more than twice the original annual estimate.

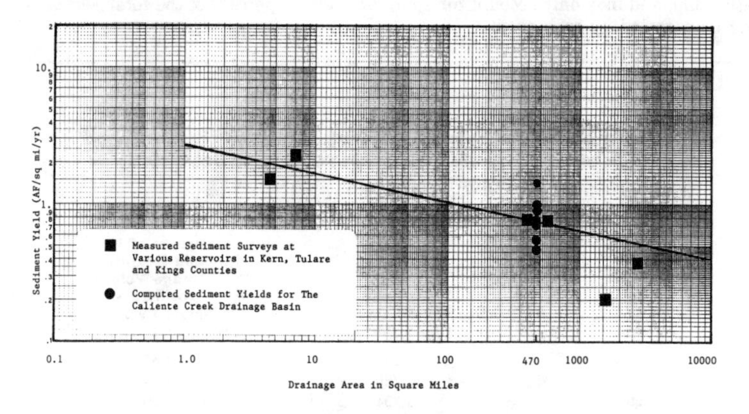

Figure 1

Measured and Computed Values of Average Annual Sediment Yield Versus Drainage Basin Area

It is important to note that arid and semi-arid basins, such as Caliente Creek, are very episodic in nature. During dry years (perhaps even normal years) the sediment production and delivery (and, therefore, annual yield) is small. During large runoff events the sediment production and delivery can produce tremendous loads of sediment in the channels. The annual yield during an excessively wet year can be quite high. Therefore, the presentation of a single average annual yield value may be misleading. For planning purposes, the consideration of the range of possible annual yields is more meaningful.

Single Event Analyses

In addition to the average annual sediment yield, it is important to estimate the sediment production and delivery from possible single events ranging from small 5-year flows to the design event (100 year flood) and, perhaps the SPF and PMF. It is possible that one or more single events during the design life of the project can significantly affect the operation and maintenance of the reservoir.

The study reach upstream from the proposed damsite was partitioned into four different zones or subreaches based on distinct hydraulic and geomorphic characteristics. The transport capacity is computed for each reach and is compared to the others with different hydraulic and geomorphic characteristics. The channel averaged sediment grain size and averaged channel hydraulic conditions for a range of discharges are used with several different total bed material load transport functions to develop representative water discharge versus total bed material load relationships for each of the subreaches and flow conditions.

Table 2 presents the computed sediment inflow to the proposed damsite for the various flood events. The 100 year flood event can possibly produce enough sediment during the single design event to remove 43.7 percent of the gross pool storage capacity (6992 AF). It also suggests that events greater than about the 15 year event can possibly remove 10 percent or more of the gross pool storage in one 5 day period. This indicates that the present design capacity of the reservoir may be undersized. The computed total sediment loads account for the total bed material load with an additional 15 percent estimated for the wash load. Typical wash loads can account for as much as 90 to 95 percent of the total load in most sand bed rivers (Vanoni, 1975). However, in the Caliente River Basin the availability of fines (silts and clays) may be limited due to the nature of the granitic parent materials throughout the basin (see Harvey et al., 1990). The authors postulate that the wash load near the damsite will have an inverted bed load/wash load relationship, and may only account for approximately 15 percent of the total sediment load being transported by each event.

TABLE 2

Computed Single Event Sediment Inflow to the Proposed Reservoir and Comparison to Planned Detention Storage Volume of 16,000 Feet

Event	Total Load Per Event (acre-feet) [dry volume]	Percent of the Planned Detention Storage Volume Associated with Single Event Sediment Delivery
5	245	1.5%
10	760	4.8%
20	1,794	11.2%
50	4,709	29.4%
100	6,992	43.7%
SPF	11,615	72.3%
PMF	29,440	184.0%

Harvey et al., (1990) estimate that there may have been approximately 9 inches of sediment deposited in the reach upstream from the Highway 58 crossing during the 1983 flood event. That event is estimated to be approximately a 50 year event according to the Kern County Water Agency. Comparing the total sediment loads entering and leaving the reach it is seen that there is approximately 575 acre feet more sediment transported into the reach from the upstream supply reach than leaves the reach. The approximate surface area of the reach is one square mile (640 acres). Assuming that the 575 acre feet of sediment deposits uniformly over the reach, this gives an approximate sediment deposition thickness of 10.8 inches. This matches the observed deposition depth for a 50 year event reasonably well.

Large events such as a 50 year flood or greater may produce large amounts of sediment material that enter the water course due to mass wasting, channel bank failure and erosion of prograded alluvial fans that often extend into the channel in the upper basin. It may be that single event sediment production can contribute significant quantities of sediment materials to the reservoir in a short period of time (a few days) and affect the operation and storage characteristics of the project.

Conclusions

The following conclusions are drawn from the results of this investigation:

1) The morphology of the Caliente Creek drainage basin and the nature of the sediments delivered to the channels and the potential for sediment storage within the drainage basin are controlled by the basin geology (Harvey et al., 1990).
2) Sediment transport in the basin is episodic and is governed by the occurrence of large runoff events. Sediment is stored in the broad valley washes (3000 to 6600 feet wide) in the lower portions of the Caliente Basin. There is sufficient material located in these expansive washes to provide sediment supply to the lower fan areas somewhat independently of the production and delivery of sediments from the upper watershed areas. Therefore, sediment yield at the proposed damsite may be more dependent upon the transport capacity of the channels and washes upstream from the damsite, than the watershed production of sediment materials during a flood event.
3) Examination of eight different sources of yield data and methods for estimating yield at the damsite concludes that the approximate average annual sediment yield at the Sivert Reservoir is 0.75 AF/sq mi/yr. This is more than twice the initial yield estimate developed during the planning studies. Annual sediment yields can range from 0.47 AF/sq mi/yr to approximately 1.5 AF/sq mi/yr.
4) Single event floods may produce significantly more sediment per event than the annual sediment yield would indicate. As much as 43 percent of the total gross pool storage volume (16,000 AF) may be lost due to sediment deposition during a 100 year event. This would necessitate the removal of approximately 7,000 AF of sediment material (dry volume) from the reservoir prior to the next flood season. It also indicates that the design capacity of the reservoir may be undersized.

Acknowledgements

The study reported herein was conducted by the Hydrologic Engineering Center, Davis, California at the request of the Sacramento District, Corps of Engineers. The views expressed herein are those of the authors and not necessarily those of the U.S. Army Corps of Engineers.

References

1. Dendy, F.E. and G.C. Bolton (1976). "Sediment Yield - Drainage Area Relationships in the United States," Journal of Soil and Water Conservation.
2. Harvey, Michael D., E.F. Sing and Robert C. MacArthur (1990). "Sediment Sources, Transport and Delivery to An Alluvial Fan, Caliente Creek, CA," Proceedings of the International Symposium on Hydraulics and Hydrology of Arid Lands," ASCE, July 30 - Aug 2, San Diego, CA.
3. Hydrologic Engineering Center (1990). "Phase I Sediment Engineering Investigation of the Caliente Creek Drainage Basin," Final Project Report No. 90-03, prepared for CESPK, Sacramento, CA.
4. Pacific Southwest Inter-Agency Committee (1968). Report of the Water Management Subcommittee on "Factors Affecting Sediment Yield in the Pacific Southwest Areas and Selection and Evaluation of Measures for Reduction of Erosion and Sediment Yield," Sedimentation Task Force, PSIAC.
5. Personal Communication with CESPK-ED-H/Herb Hereth (11/21/89).
6. Simons, Li & Associates, Inc. (1989). "Sedimentation Study of Caliente Creek Watershed, Kern County, CA," a report prepared for Kern County Water Agency, Bakersfield, CA.
7. U.S. Army Corps of Engineers (1989). "Sedimentation Investigations of Rivers and Reservoirs; Engineering Manual EM 1110-2-4000, Office of the Chief of Engineers, Washington, D.C.
8. U.S. Department of Agriculture (1975). "Erosion, Sediment and Related Salt Problems and Treatment Opportunities," Special Projects Division, Soil Conservation Service, Golden, CO.
9. U.S. Department of Agriculture (1977). "Supplement to Sediment Deposition in U.S. Reservoirs: Summary of Data Reported Through 1975," Miscellaneous Publication No. 1362 compiled by the Sedimentation Committee, Water Resources Council at the USDA Sedimentation Laboratory, Oxford, MS.
10. U.S. Department of Agriculture (1980). "Watershed Plan and Environmental Impact Statement - Tehachapi Watershed," Soil Conservation Service, Tehachapi, CA.
11. Vanoni, Vito A., ed. (1975). Sedimentation Engineering, American Society of Civil Engineers, New York, NY.

Sediment Sources, Transport and Delivery to an Alluvial Fan, Caliente Creek, California

M.D. Harvey[1], E.F. Sing[2] MASCE and R.C. MacArthur[3] MASCE

Abstract

Sediment delivery to a proposed flood detention reservoir site in the medial region of the Caliente Creek alluvial fan is dependant on the magnitude and duration of flood flows. Because the flows in the basin are ephemeral estimates of annual sediment delivery rates are somewhat meaningless. However, stratigraphic evidence showed that a 50- year recurrence interval flood in 1983 deposited an average of 9 inches of sediment on the fan surface upstream of the dam site. An incised channel exposure on the fan indicated that previous floods had deposited similar depths of sediment. Sediment deposited on the valley floors upstream of the fan during lower magnitude and higher frequency events are the primary source of sediment transported during flood flows.

Introduction

The Caliente Creek drainage basin (470 mi^2) is located at the southern end of the San Joaquin Valley, southeast of the city of Bakersfield in Kern County, California (Fig. 1). A flood detention reservoir is planned for Caliente Creek on the Caliente alluvial fan (Fig. 1). The structure is to be located about 2 miles downstream of Highway 58 crossing in the medial region of the alluvial fan. The present fan surface is inset about 40 feet below the surface of a Pleistocene age fan, the slope of which was about twice that of the modern fan (0.01).

---

[1]Geomorphologist, Water Engineering and Technology, Inc., P.O. Box 1946, Fort Collins, CO 80522, USA.

[2]Hydraulic Engineer, U.S. Army Engineer District, Sacramento, 650 Capitol Mall, Sacramento, CA 95814, USA.

[3]Research Hydraulic Engineer, U.S. Army Corps of Engineers, Hydrologic Eng. Center, Davis, CA 95616, USA.

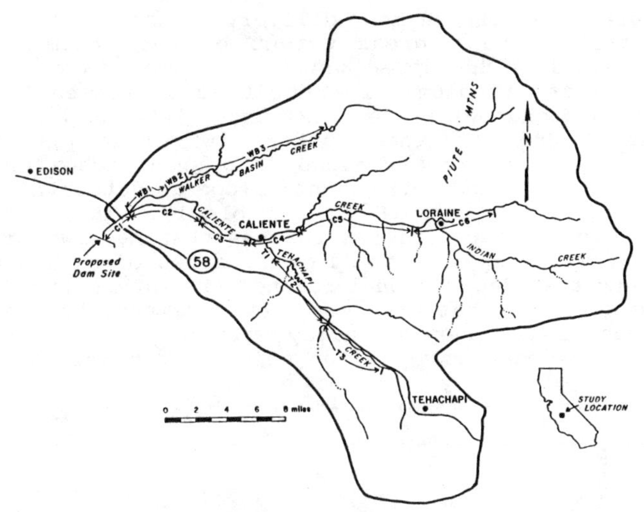

Figure 1. Location map for Caliente Creek Basin.

A feasibility phase investigation of the project indicated that the economics of the project were very sensitive to the watershed sediment yield at the dam site. The objectives of this investigation (4), that was conducted concurrently with a sedimentation analysis (2), were twofold: 1) to understand the dynamics of the watershed and 2) to relate the watershed dynamics to sediment delivery from the basin to the proposed dam site on the alluvial fan.

Watershed and Channel Morphology

The morphology of the drainage basin is controlled by geology and structural setting. The lower- elevation portions of the basin are composed of Pleistocene and Recent-age alluvial fans. Tertiary -age non-marine rocks separate the fans from the majority of the basin that is underlain by quartz diorites. Slopes as steep as 35 degrees are characteristic of the diorites, whereas the slopes on the Tertiary- age rocks are generally in the range of 10 to 15 degrees. The lithological control of the basin morphology is overidden by structural control in the upper reaches of both the principal tributaries, Walker Basin and Tehachapi Creeks. Both valleys are depositional centers that impede sediment delivery to the lower basin. Active faults (White Wolf, Breckenridge, Edison) traverse the basin, and earthquakes as recent as 1952 have caused mass wasting of slopes and surface deformation.

The potential for sediment delivery to the alluvial fan is controlled to a great extent by the sediment storage potential in the upper basin. Storage potential is a function of lithological control as expressed by valley width and channel slope. Caliente, Tehachapi and Walker Basin Creeks, the latter two below the depositional centers, were divided into sub-reaches on the basis of valley width and channel slope as determined from USGS topographic maps (Fig. 1; Table 1). The data show that there are depositional reaches that have wider valleys and lower slopes in the lower reaches of each of the creeks and that depositional reaches are interspersed with canyon sections that have very low sediment storage potential because of narrow valley widths and steeper slopes. Reach boundaries are lithologically controlled.

Table 1. Morphometric data for reaches of Caliente, Tehachapi and Walker Basin Creeks.

Reach	Boundaries (RM - RM)*	Mean Valley Width (ft)	Average Bed Slope (ft/ft)	Reach Description
Caliente Creek				
C1	0 - 4	5,334	0.0097	Caliente alluvial fan
C2	4 - 8.4	1,775	0.0084	Depositional reach
C3	8.4 - 11.2	333	0.0135	Canyon reach
C4	11.2 - 14	1,050	0.0135	Depositional reach
C5	14 - 24	248	0.021	Canyon reach
C6	24 - 27.5	725	0.009	Depositional reach
Tehachapi Creek				
T1	0 - 2.3	850	0.023	Depositional reach
T2	2.3 - 7.8	157	0.037	Canyon reach
T3	7.8 14	338	0.031	Depositional reach
Walker Basin				
WB1	0 - 2.5	760	0.0091	Depositional reach
WB2	2.5 - 3.5	225	0.027	Depositional reach (minor)
WB3	3.5 - 13.5	144	0.038	Canyon reach

*RM0 is dam site

Channel morphology within the basin is also indirectly controlled by the basin lithology. In the canyon sections bedrock control causes the flows to be perennial, but the morphology of the channel is a function of the recent flood history. However, within the depositional reaches flows are ephemeral, and channel morphology reflects the last flow that was experienced. Significant infiltration losses occur in the depositional reaches, and this reinforces sediment storage during lower magnitude events.

Watershed Hydrology

Normal annual precipitation varies from about 6 inches at the lower elevations to about 45 inches at the higher elevations. About 90 percent of the precipitation occurs during winter frontal rainstorms from November to April. During the remainder of the year, spatially limited but very intense thunderstorms can occur. Floods are generated by both types of precipitation. The Standard Project Flood (SPF) for thunderstorm events and frontal storms have been calculated to be 56,500 cfs and 17,500 cfs, respectively (3). Significant frontal rainstorm type floods occurred in 1937, 1943, 1944, 1966, 1969, 1978 and 1983 (15,800 cfs 50-yr. RI) Significant thunderstorm type floods occurred in 1932 (flood of record, 21,000 cfs at dam site) and 1945. The presence in many of the drainages of very coarse grained water-flood deposits and the fact that the basin characteristics fit the profile for extreme flood discharges (1) may indicate that the peak discharges have been underestimated. This question can only be resolved by undertaking paleohydraulic studies.

Watershed Sediment Sources

Sediment sources within the basin can be grouped broadly into hillside and channel sources. In the upper basin where the slopes are steep and poorly vegetated dry ravel of the grus derived from weathering of the diorite delivers significant amounts of sediment to the valley floors. Hillslope derived sediments accumulate on the valley floor and are episodically transported by debris flows. Rilling and gullying do occur on the slopes, but they are a minor sediment source generally related to flow concentration by roads, railroad tracks and rock outcrop. Undercutting of slopes during flood flows introduces considerable volumes of colluvial material to the valley floors. Because of the ephemeral nature of the flows sediment delivery to the lower basin and the fan is dependent on the occurrence of flood flows and individual floods may transport several decades worth of accumulated materials on the valley floor.

Channel sources of sediment during flood flows include the sediment stored on the valley floor and terraces and tributary alluvial fans. Flood flows erode the terraces and the margins of the alluvial fans. Some of the tributary fans have incised and have already supplied large volumes of sediment to the higher order channels. Others are not incised and are actively storing sediment. Sediment delivery to the dam site during an individual event depends on the simultaneous occurrence of flood flows within the individual basins and in Caliente Creek.

Sediment samples were obtained from the channels in the basin in 1989, and these are representative of the sediments transported in the 1983 flood because there have been no significant flow events in the intervening period. Boulders as large as 700 mm (intermediate axis) were transported in Indian Creek, a tributary to Caliente Creek. In reach C4 (Table 1), which is a depositional zone downstream of a canyon reach (C5), both coarse and finer sediments were deposited. The median grain size (D_{50}) of the coarser deposits was 70.5 mm and that of the finer deposits was 9.0 mm. Similarly, bimodal sediments were deposited at the downstream end of reach C3 (D_{50}'s of 33 mm and 1.2 mm). In reach C2, which is a depositional reach, the D_{50} was 0.9 mm which is the same as that at the dam site in reach C1. Silt-clay content of all the samples was less than 2 percent, which reflects the lack of chemical weathering of the diorite in a semi-arid climate.

Sediment Delivery

Sediment delivery from the Caliente Creek drainage basin to Reach C1 on the Caliente alluvial fan on which the dam site is proposed is dependent on the episodic occurrence of flood flows and to some extent is independent, or at least lagged, from sediment production in the watershed. Reach C2 on Caliente Creek, which is a depositional reach during low magnitude events, is in fact the primary sediment supply reach for reach C1 during flood flows. Stratigraphic evidence indicates that the amount of sediment deposited during the 1983 flood was dependent on the valley width. In reach C2 18 inches of sediment was deposited where the valley was 600 feet wide. In reach C1 9 inches of sediment were deposited where the valley was 4,000 feet wide. Seven flood deposits were identified in a cutbank exposure in an incised channel segment downstream of Highway 58. The individual flood deposits fined upwards in terms of both grain size and thickness. The average thickness of the 3 lowest flood deposits was about 8 inches which provides an average value for the thickness of individual flood deposits on the fan. However, the presence of Highway 58 embankment (since 1970) has altered the patterns of sediment deposition on the fan. Approximately 10 feet of sediment has been deposited upstream of the embankment(7 inches/yr). Sediment transport modeling (2) indicated that about 10 inches of sediment would be deposited on the fan surface upstream of the embankment during a 50-yr. event and this correlates reasonably well with the stratigraphic evidence.

Conclusions

Floods that eventuate from both winter frontal and summer thunderstorm rainfall are the primary sediment delivery mechanisms to the proposed dam site on the Caliente alluvial fan. Stratigraphic evidence on the fan indicates that a 50-yr. event in 1983 deposited an average of 9 inches of sediment upstream of the dam site. Sediment stored on the valley floor upstream of the fan during lower magnitude events is the primary sediment source during flood events. Traditional methods for estimating annual sediment yields from the watershed will not provide a reasonable basis for design because sediment delivery is episodic and lagged from watershed sediment production.

Acknowledgements

This study was performed under Contract No. DACW05 - 88-D-044 for the U.S. Army Corps of Engineers. The views expressed herein are those of the authors, and not necessarily those of the U.S. Army Corps of Engineers.

References

1. Costa, J.E., 1987b. A comparison of the largest rainfall-runoff floods in the United States with those of the Peoples Republic of China and the World, J. of Hydrology, v. 96, p. 101-115.

2. MacArthur, R.C., Harvey, M.D. and Sing, E.F., 1990. Estimating sediment delivery and yield on alluvial fans. This volume.

3. U.S. Army Corps of Engineers, 1980. Caliente Creek stream group, California, Hydrology Office Report. Department of the Army, Sacramento District, Sacramento, California.

4. Water Engineering and Technology Inc., 1989. Caliente Creek, California Project, Geomorphic Analysis. Report to Sacramento District, Corps of Engineers, Contract No. DACW05-88-D-0044, November, 1989, 76 p.

PREDICTION OF DEBRIS FLOW PRONE AREAS AND DAMAGE

Takahisa Mizuyama[1] and Yoshiharu Ishikawa[1]

ABSTRACT

It is very important when drawing hazard maps to estimate damage to houses from debris flow of certain fixed quantities. A debris flow disaster, which occurred in Kake Town, Hiroshima Prefecture in July 1988 was studied in a field survey and aerial photograph ciphering to clarify the relation between the fluid force of debris flow and the degree of damage to houses. The distribution of the fluid force was obtained by numerical simulation. The method can be used to estimate the degree of damage to houses in debris flow flood areas more objectively and precisely.

1. INTRODUCTION

The district surrounding Kake Town, Hiroshima Prefecture suffered from torrential rains in a Bai-u front from July 20th to 21st, 1988. Ten torrents caused damage to houses and 11 precious lives were lost.
After the disaster, field surveys were carried out to determine flow-down traces of debris flows, thickness of sediment deposition, gradient of ground, degree of damage to houses by debris flow, etc. with respect to the ten torrents in which houses were damaged. The degree of damage to houses were classified into five grades i.e. totally collapsed, half-collapsed, damaged, submerged and no damage.

2. FIELD INVESTIGATION

2·1 Degree of Damage to Houses

The degree of damage to houses located within

---

[1] Erosion Control Division, Public Works Research Institute, Ministry of Construction, Tsukuba City, Ibaraki, 305, Japan

the debris flow flood area in each torrent surveyed, is indicated in Table 1.

As 94% of the houses surveyed were of wooden construction, it was not possible to obtain a clear conclusion regarding difference in the degree of damage due to type of constrution.

Table 1 Degree of damage to houses in torrents
(Unit: House)

No., Torrent name	Totally collapsed	Half-collapsed	Damaged	Submerged	No damage	Total
(1) Egouchidani River	19	4	12	1	16	52
(2) Tao River	1	0	1	0	2	4
(3) Udosezawa River	1	0	0	1	2	5
(4) Kisakanishi River	1	1	2	0	0	4
(5) Nakanishihiratani River	9	2	1	0	0	12
(6) Yamashiro River	0	1	0	0	0	1
(7) Ueharadani River	1	1	1	0	6	9
(8) Nakaodani River	1	1	1	0	0	3
(9) Jyanotani River	2	1	2	0	5	10
(10) Senbon River	4	3	0	0	0	7
Total	39	15	20	2	31	107

2·2 Horizontal Yield Strength Possessed by Houses

For the purpose of collecting basic data mainly to examine earthquake resistance of wooden buildings, several tests applying static horizontal forces have been conducted using full-scale buildings. The maximum horizontal yield strength possessed by wooden houses are shown in Table 2. As a criterion for judging the degree of damage to wooden houses by the fluid force of debris flow, 5kN per 1 m of width and 10 kN per 1 m of width are indicative of damage or half-collapse respectively.

Table 2 Horizontal yield strength of wooden houses
(from the results of previous full- scale experiments)

Name of house	Loading direction	Max. yield strength Pmax(N)	Width (m)	Yield strength per unit width (N/m)	No. of stories	Years elapsed
Mitakadai Metropolis managed houses	Ridge derection	29,400	4.55	6,462	One-story	22 years
	Span	33,810	7.28	4,644		
Shizuoka Prefecture managed houses	Span	25,480	7.735	3,294	One-story	25 years
Minamisuna Town building A	Span	137,200	10.01	13,706	Two-story	Immediately after construction
Minamisuna Town building B	Span	284,200	10.01	28,392	Two-story	
Fully 2-story test house	Span	79,380	7.28	10,904	Two-story	
Semi 2-story test house	Span	78,400	7.28	10,769	Two-story	
One-story, average				4,800		
Two-story, average				15,943		
Average				11,167		

2·3 Fluid Force by Debirs Flow Acting on Houses

When a debris flow collides with a house, the fluid force per 1 m of width acting on the house can be expressed by the following formula.

$$F = \frac{w}{g} hv^2 \quad \text{------------------------(1)}$$

where, F= fluid force per unit width (N/m), w= bulk density (specific weight) of debris flow (N/m^3), g= acceleration of gravity (m/s^2), h= flow depth of debris flow(m), v= velocity of debris flow(m/s).

The velocity of debris flow is approximately expressed by using Manning's formula, and the above formula is transformed into the following equation.

$$F = \frac{w}{n^2 g} h^{7/3} I \quad \text{-------------------(2)}$$

where, n= Manning's coefficient of roughness, I= gradient of riverbed (ground).

The distribution of fluid force per unit width

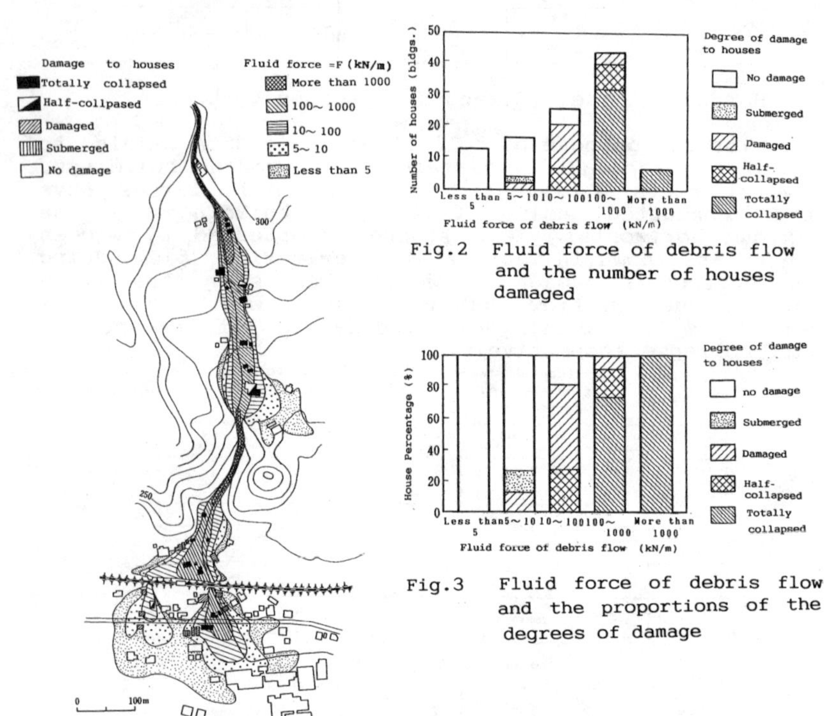

Fig.2 Fluid force of debris flow and the number of houses damaged

Fig.3 Fluid force of debris flow and the proportions of the degrees of damage

Fig.1. Fluid force distribution and degree of damage to houses from results of debris flow and deposition survey in the flood area

F in the flood area in the Egouchidani River are shown in Fig.1. In this case, the flow depth of debris was determined based on the flood marks at the site and eyewitness reports. In the same manner, the distribution of the fluid force per unit width F was determined for 9 other torrents where the houses were damaged. The number of damaged houses for each magnitude class of fluid force per unit width F in the ten torrents and a classification of the degree of damage are shown in Fig.2. The numbers of the houses expressed as percentages are indicated in Fig.3.

From these Figures, the fluid force of debris flows is proportional to the degree of damage to the houses. Also, within the range where the fluid force of debris flows is less than 5kN/m, no houses were worse than "damaged"; within the range less than 10kN/m, no houses were worse than "half-collapsed". Thus, the values can be utilized as an index for judging the degree of damage to wooden houses after debris flows.

3 DISTRIBUTION OF FLUID FORCE BY NUMERICAL SIMULATION

3·1 Simulation Model of Debris Flow

The debris flow which occurred in the Egouchidani River was simulated by the simulation model of debris flow derived from the numerical metod for two-dimentional overland flow by Iwasa et al.,1979. The calculation consisted of a numerical solution of the following equations.

(1)Momentum conservation equation in the x direction.

$$\frac{\partial M}{\partial t}+\frac{\partial}{\partial x}(\beta\cdot u\cdot M)+\frac{\partial}{\partial y}(\beta\cdot v\cdot M)=-g\cdot h\cdot\frac{\partial H}{\partial x}-\frac{F_x}{\rho_o} \quad (3)$$

(2)Momentum conservation equation in the y direction.

$$\frac{\partial N}{\partial t}+\frac{\partial}{\partial x}(\beta\cdot u\cdot N)+\frac{\partial}{\partial y}(\beta\cdot v\cdot N)=-g\cdot h\cdot\frac{\partial H}{\partial y}-\frac{F_y}{\rho_o} \quad (4)$$

(3)Volume continuous equation for a continuous fluid

$$\frac{\partial h}{\partial t}+\frac{\partial M}{\partial x}+\frac{\partial N}{\partial y}=0 \quad (5)$$

where: $M= u\cdot h =$ flux of flow in the x direction
$N= v\cdot h =$ flux of flow in the y direction
$\beta =$ momentum correction factor
$u=$ mean velocity in the x direction
$v=$ mean velocity in the y direction
$H= Zb + h =$ elevation of water surface
$h=$ depth of flow
$Zb =$ elevation of riverbed
$\rho_o =$ density of debris flow including sediment

Fx and Fy are the shear resistance in the x and y directions.

(4)Calculation of the amount of sediment transportation

Amount of sediment transportation is calculated as follows. Setting the critical gradient ic, when the gradient Sij between meshes is

$Sij \geqq ic$, sediment concentration of debris flow is calculated by Takahashi's equation (1977)

$Sij < ic$, sediment transport rate was calculated as bed-load transport by Ashida, Takahashi, Mizuyama's formula in 1978.

(5)Conservation equation of sediment

Riverbed fluctuation was calculated by the following two dimensional conservation equation of sediment.

$$\frac{\partial Zb}{\partial t} + \frac{1}{1-\lambda} \cdot \left(\frac{\partial qBx}{\partial x} + \frac{\partial qBy}{\partial y}\right) = 0 \quad \text{-----(6)}$$

where: Zb= elevation of riverbed

λ = void ratio

qBx= sediment discharge in the x direction

qBy= sediment discharge in the y direction

The calculation of debris flow flooding simulation was carried out through the difference calculus for the equations (3)-(6) by the Leapfrog method.

3·2 Fluid Force by Numerical Simulation and Degree of Damage to Houses

Through the calculation of debris flow flooding simulation, the velocity, depth and specific weight of debris flow were obtained for each mesh per minute. Applying these values to equation (1) fluid force of debris flow per unit width in each mesh was obtained. The distribution of maximum fluid force and the degree of damage to houses are shown in Figure 4. The number of damaged houses for each magnitude class of fluid force of debris flow is indicated in Figure 5 and the number of houses expressed by percentage is shown in Figure 6.

It was concluded that the distribution of fluid force by means of debris flow flooding simulation calculation is effective for the estimation of the degree of damage to houses by debris flow.

4 CONCLUSION

The degree of damage to houses in the debris flow hazard zones was estimated objectively and precisely by acquiring the distribution of fluid force by means of the calculation of debris flow flooding simulation. This method was utilized to draw a precise hazard map for debris flow, through which a warning and evacuation system for a debris flow hazard zone, land-use planning and transfer of houses can be examined.

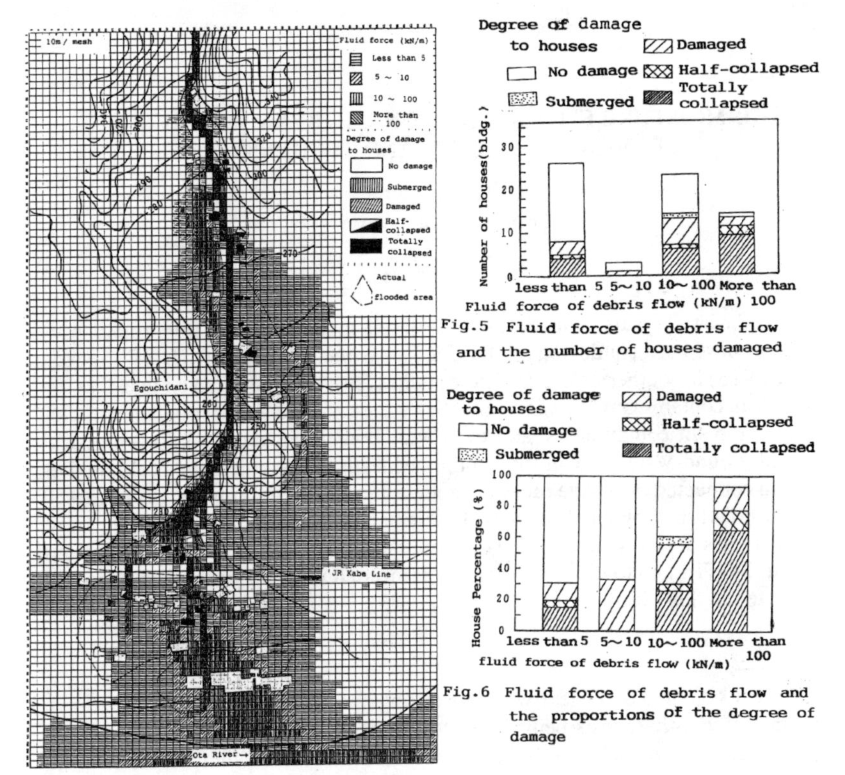

Fig.5 Fluid force of debris flow and the number of houses damaged

Fig.6 Fluid force of debris flow and the proportions of the degree of damage

Figure 4 Fluid force distribution and degree of damage to houses from results of debris flow flooding simulation

REFERENCES

1. Architectural Institute of Japan (1981). Retained yield strength and deformation performance in Earthquake-resistant design of building, p490.
2. Ashida,K., Takahashi,T., and Mizuyama,T.(1978). "Study on bed load equation for mountain streams". Journal of the Japan Society of Erosion Control Engineering, Vol.30, No.4, pp.9~ 17.
3. Iwasa,Y., Inoue,K., Mizushima,M. (1980). "Hydraulic analysis of overland flood flows by means of numerical method". Kyoto Univ. Disaster Prevention Research Institute annuals, Vol.23, B-2, pp.305~ 317
4. Takahashi,T. (1977). "A mechanism of occurrence of mud-debris flows and their characteristics in motion". Kyoto Univ. Disaster Prevention Research Institute annuals, Vol.20, B-2, pp.405 ~ 435.

SOIL-MOISTURE FLUX STUDIES ON THE NEVADA TEST SITE A REVIEW OF RESULTS AND TECHNIQUES

Scott W. Tyler[1]
Roger L. Jacobson[1]

Abstract

Over the last 20 years, numerous studies of soil-moisture flux and/or ground-water recharge have been conducted on the U.S. Department of Energy's (DOE) Nevada Test Site (NTS) in southern Nevada. These studies have been undertaken to assess the magnitude of contaminant transport potential in the soil zone, as water balance studies in arid zone plant communities, and to provide boundary conditions for numerical models of ground-water flow in the aquifers beneath the testing area. These studies have been conducted on a wide range of geomorphic and topographic environs as well as in areas disturbed by man's activity. In this paper, a review of several of the most relevant studies are presented to demonstrate the wide range of recharge conditions found on the NTS and to discuss the errors associated with the techniques employed to estimate the flux of soil water.

Introduction

Since 1963, all U.S. nuclear weapons testing has been conducted underground, with a majority of these tests being held at the NTS. By conducting each detonation in the subsurface, the likelihood of immediate release of radioactivity to the environment is greatly reduced. Subsurface detonation, however, does not preclude radionuclide release. One of the most likely pathways for release to the accessible environment from testing is through the ground water. Since many of the tests are conducted above the zone of ground-water saturation, partially saturated flow, or soil-water flow, must be understood if these pathways are to be fully evaluated.

At first glance, the NTS appears to offer little, if any, opportunity for deep soil-water movement. The area is one of the driest in the continental United States (Winograd and Thordarson, 1975). Mean daily maximum temperatures may range from 13 to 40°C. Combining the low precipitation and the high potential evapotranspiration into a water budget model, it has been generally thought that little or no ground-water recharge can occur. Recent studies (Gee and Kirkham, 1984; Tyler et al., 1986) have shown, however, that simple water balance approaches to soil-water move-

---

[1] Water Resources Center, Desert Research Institute, University of Nevada System, Reno and Las Vegas, Nevada

ment and ground-water recharge may be insensitive in arid regions. These insensitivities arise from the conflict between the long time scales used in water budget models and the short time frames over which precipitation, infiltration, and evaporation take place in arid regions. These time scales may be on the order of minutes to a few hours, while most modeling approaches use daily, weekly, or even monthly estimates. In addition, studies by Clebsch (1961), Thordarson (1965), and Tyler et al. (1986) have suggested that deep soil-water movement (recharge) is occurring in selected areas of the NTS. These studies indicate that areas such as fractured bedrock surfaces, wash bottoms, and man-made depressions may allow deep soil-water migration. These geomorphic features satisfy two criteria, each locale receives or concentrates precipitation in small areas (ponds, channels, or fractures) and each locale allows for the rapid migration of potential recharge below the depth of active evaporation and transpiration because the surficial materials are highly permeable. These two criteria, source water and rapid transport, are crucial in estimating the potential recharge in arid areas.

The goal of this paper is to review some of the data base regarding soil-water movement on or near the NTS. Due to constraints of length, three techniques to estimate recharge will be reviewed: hydraulic properties; natural tracers; and *in situ* flux measurements. Due to the fairly limited data base, very simplistic models will be used to predict the magnitude of soil moisture flux. As more detailed data are collected, more sophisticated approaches, including a wide range of processes, should be developed. The results of this study, however, should provide a clearer understanding of the geomorphic nature of the recharge areas, point out data gaps, and define the errors associated with these techniques.

Hydrogeologic Setting

From a vadose zone standpoint, the NTS also contains some of the thickest unsaturated zones in the United States. Depth to water in certain locales may exceed 600 m (Winograd, 1980). The great depth to the water is a result of the following: areas of moderate to high relief; relatively permeable strata within the saturated zone; regional aquifers with topographically low discharge points; and little recharge (Winograd and Thordarson, 1975).

These climatic and hydrologic considerations support the conclusion that water movement through the unsaturated zone should be slight. The lack of perennial streams and the small number of springs found in the area tend to support this conclusion. In many arid locales, there would be little incentive to study site-specific soil-moisture movement and recharge. At the NTS, a detailed knowledge of soil-water flux is necessary to assess the potential for transport of radionuclides from weapons testing to the biosphere.

Flux Analysis Using Soil Hydraulic Properties

Soil hydraulic data may be used in an indirect method to calculate soil-water migration. Soil water, under most deep-soil conditions, is driven by two forces, capillary and gravity. In the near-surface soils, thermal forces may affect the movement both by inducing vapor transport and through thermal convection. For deep migration, we shall assume that soil water is unaffected by thermal gradients, gradients are vertically downward, and the soil water obeys Darcy's equation for flux. Such an assumption may not be applicable in regions of very low soil–moisture flux.

To calculate the flux, the hydraulic conductivity (a function of the soil texture and water content) and the driving forces, capillary and gravity, must be known. Field and laboratory measurements of these properties have been reported by several authors (Mehuys et al., 1975; Kearl, 1982; Romney et al., 1973; Tyler et al., 1986). In each of these studies, soils used in the analyses were taken from near the surface (< 50 m) and were collected in the alluvial basins.

The most complete set of soil hydraulic data was published by Mehuys et al. (1975) on soil samples from Rock Valley (Area 25). In their study to determine the effects of stone content on hydraulic properties, conductivity measurements on gravelly sandy loam were collected over the range of –0.05 to –50 bars of capillary pressure. In addition, water retention data were also collected.

To estimate a range of water fluxes using the Rock Valley soil, the conductivity data must be combined with hydraulic gradient data via Darcy's law. Unfortunately, no *in situ* matric potential or gradient data were presented by Mehuys et al. (1975). Other studies (Case et al., 1984; Kearl, 1982; Tyler et al., 1986) on similar textured soils report *in situ* matric potentials of between –8 and –40 bars. Using the hydraulic conductivity data of Mehuys et al. (1975), this corresponds to a conductivity range of 10^{-6} to 10^{-7} cm/hr. Hydraulic gradients in similar settings of the NTS reported by Case et al. (1984) and Tyler et al. (1986) ranged from 77 m/m upward to roughly 40 m/m downward. For this simple analysis, the range of gradients was chosen between 0.5 and 40 m/m downward. The four possible data combinations may be combined to produce a range of fluxes and velocities as shown in Table 1.

TABLE 1. ESTIMATED FLUX AND VELOCITY FROM ROCK VALLEY SOIL.

Vol. Water* Content	Potential** (bars)	K* (cm/hr)	dh/dz	Flux (cm/yr)	Velocity (cm/yr)
0.09	-8	10^{-6}	0.5	4.4×10^{-3}	4.9×10^{-2}
0.09	-8	10^{-6}	40	3.5×10^{-1}	3.9×10^{0}
0.06	–40	10^{-7}	0.5	4.4×10^{-4}	7.3×10^{-3}
0.06	–40	10^{-7}	40	3.5×10^{-2}	5.8×10^{-1}

* estimated from Mehuys et al., 1975
** Tyler et al., (1986); and Kearl (1982)

Although the results presented in Table 1 may not truly represent the actual field conditions, the highest estimate is a flux of 0.35 cm/year, with an average pore–water velocity of 3.9 cm/year. This indicates that flux through these sediments, assuming steady–state conditions, may be quite small. This flux is roughly two percent of the average annual precipitation in the valleys of the NTS.

Flux Estimates Using Tritium Tracers

Since 1952, atmospheric testing of nuclear weapons has produced concentrations of tritium in precipitation that have been well above levels that existed prior to 1952. The result of this dramatic increase in tritium levels in precipitation has been to used to detect the depth of migration of post–1952 infiltration (Isaacson et al., 1974; Anderson and Sevel, 1974). This approach has been used to determine the average soil–water velocity by noting the depth of penetration of the tritium in the soil profile. In the simple analysis that follows, the tritium transport is treated as a piston–like displacement beginning in 1952. Vapor transport, which may be dominant in the near surface, is ignored.

At least two published reports describe post–bomb tritium migration in soils on the NTS. Hansen (1978) describes the results of soil coring and soil–water tritium analysis from several areas of the NTS. Samples were collected in alluvial and playa sediments of both Yucca and Frenchman Flats. Unfortunately, the tritium analysis techniques used in Hansen's work have high errors associated with any samples containing less than 100 TU (3.2×10^{-1} pCi/ml). As a result, samples containing less than 100 TU may be interpreted as either old (pre–1952) or younger water. In each of his shallow borings (<1 m), tritium in excess of pre–1952 levels and well above the 100 TU experimental error level was found throughout the soil profiles. Maximum tritium levels were generally found at or near the soil surface. These results indicate that recent (less than 26 years old at the time of sampling) precipitation had migrated to at least 1 m below the soil surface. Since no pre–1952 levels were detected in these shallow sites, no estimates of soil–moisture velocities may be made except that the velocity was at least 1 m in 26 years or a yearly average of 3.8 cm/year.

Hansen (1982) also presents tritium data from several deep boreholes. Borehole YF–10, completed in playa sediments in Yucca Flat, showed tritium to a depth of about 6 m. Borehole FFA–A1, drilled to a depth of 3.3 m in alluvial fan sediments of Frenchman Flat, shows a rapid decline in soil–water tritium from the surface to 1 m. The tritium levels in FFA–A1 are slightly higher than those of YF–10, indicating that recent moisture may have penetrated to at least 3.3 m. The uncertainties in analytical techniques may, however, mask the actual tritium distribution at the bottom of the borehole.

The data from these deeper boreholes are quite enlightening. Using the data presented from YF–10, post–1952 water had conservatively migrated to a depth of 6 m. Simplistically assuming that recharge occurred uniformly from 1952, the average soil–moisture velocity (V) may be estimated from the following simple mass–balance equation:

$$V = \text{Depth of Penetration}/(\text{Date of Sample}-1952) \quad (1)$$

Using equation (1), the soil–water velocity at YF–10 may be calculated at 23.4 cm/year. From the same approach, if one assumes a uniform water content with depth, the recharge flux may also be calculated by multiplying the velocity by the average volumetric water content. Using an estimated bulk density of 1.6 g/cm^3 and the gravimetric water content reported by Hansen of 10.4 percent, the flux is calculated as approximately 3.9 cm/year. This represents approximately 25 percent of the average yearly precipitation at Yucca Flat. Although this may seem rather high, the site is in playa sediments and is periodically inundated by ponded water. The velocity estimated at FFA–A1 using the same approach is 13 cm/yr, while the flux is at least 1.3 cm/yr.

Buddemeier and Isherwood (1985) present tritium profiles in soils from Frenchman Flat. Borehole #1 was drilled 15 m from an unlined ditch containing high levels of tritiated water, while borehole #2 was drilled 30 m from the same ditch; both boreholes were drilled to a depth of 6.2 m. This site is underlain by alluvial soils classified as sand to sandy loams. Coarse, cobble layers were also encountered during drilling.

Post–1952 precipitation had migrated no deeper than 2.5 m in the sandy alluvial soils, particularly in the borehole completed furthest from the ditch. Using the same flux approach as applied to Hansen's data, the soil–moisture velocity at this site can be estimated to be 8 cm/year. Assuming a soil volumetric water content of 10 percent, the average downward flux is estimated to be 0.8 cm/year. An alternative hypothesis for this distribution indicates that 2.5 m is the maximum depth of storage of precipitation. Precipitation, held in storage in the upper 2.5 m of soil, is evaporated and transpired at a rate equal to yearly flux into the soil surface. Hence, the 2.5 m depth may represent a soil–moisture flow divide and the deep flux may be much less. Further study, and the inclusion of nonisothermal liquid and vapor transport, must be conducted before either of the hypotheses are chosen.

The simple models and conservative estimates developed to use the tritium data indicate that the soil–water movement in alluvial soils is slow in the few areas tested. The piston–flow model suggests soil–moisture fluxes ranged from 0.8 to 3.9 cm/year. As expected, the depth of recent soil–water penetration was greater in the one area where periodic ponding had occurred.

In Situ Measurements/Monitoring

Long–term direct monitoring of soil–moisture movement is extremely difficult in arid environments, due to very small rates of fluid movement. Errors associated with input signals (precipitation), evaporation, transpiration, and the small changes in soil–moisture storage, are each potentially larger than the magnitude of the flux term. Nevertheless, several authors (Kearl, 1982; Case et al., 1984; Morgan and Fischer, 1984; Tyler et al., 1986) have reported data on *in situ* moisture and flux measurements at or near the NTS.

Tyler et al. (1986), in an investigation of potential downward soil–moisture movement from ponded water in nuclear subsidence craters in Yucca Flat, present data from two deep (> 30 m) boreholes. The first borehole (N–1) was completed within a subsidence crater from nuclear testing while the second hole (N–2) was completed in nearby undisturbed alluvial sediments.

Core analysis from borehole N–1 indicated high moisture contents and very low capillary pressures (> –0.1 bar). A unit hydraulic gradient was observed throughout the borehole. Using core data from beneath the crater and a simple one–dimensional flow model, calculated fluxes ranged from 1.7 to 29,000 cm/yr. In actuality, the upper flux is limited to the total yearly precipitation falling on the crater catchment, which yields an approximate upper bound of flux of 1,700 cm/yr. Such a wide range of estimated fluxes shows the difficulty of using Darcy–type analysis in unsaturated, heterogeneous soils. In the undisturbed area, core data showed much lower moisture contents and matric potentials of between -8 and -36 bars. Gradients calculated in the undisturbed zone were erratic with depth, indicating both upward and downward movement.

Calibrated neutron logs were run on both boreholes and indicated a significantly higher stored moisture content in the soils below the crater. In the undisturbed borehole, 50 percent lower moisture storage was observed as well as much lower variability in moisture content.

Although simplified methods could not clearly define the infiltration below the crater environment, it is clear that concentration of precipitation in the man-made depression significantly increased the recharge potential. The intermittent ponding, combined with a highly permeable soil (dessicated fine soils overlying sandy soils), allows the occasional precipitation to deeply infiltrate the soil where it is able to move below the zone of active evapotranspiration.

Conclusions

The study of recharge in relation to ground-water flow systems is often quantified as an areally distributed process. In traditional studies, these approximations are deemed valid in light of the discretization and inherent averaging of hydraulic properties. In general, these estimates have proved useful for ground-water management. In recent years, however, the need to quantify contaminant transport processes in ground water has led to the realization that variability in parameters such as recharge is critical to understanding these processes. Studies at the NTS have shown that recharge (soil-moisture flux) is a highly variable phenomenon.

Using simple models combined with the limited data suggests that significant recharge occurs in areas combining intermittent ponding and highly permeable soils. In vegetated non-inundated areas, soil-moisture flux is low and extremely difficult to quantify due to the errors associated with parameter estimation. It is likely that fluxes are in the range of 1 cm/yr or significantly less. To further refine these estimates, more sophisticated process-based models and field experiments will need to be developed.

Due to space constraints, only a limited number of studies have been presented in this paper. The reader is referred to Tyler (1986) for a more complete review. Additional studies, in support of the high-level nuclear waste repository and the weapons testing programs, are currently on-going and are likely to provide a more complete data base regarding recharge on the Nevada Test Site.

References

Anderson, L.J. and T. Sevel, 1974. Profiles in the unsaturated and saturated zones. Gronhog, Denmark, in: Isotope Techniques in Groundwater Hydrology, 1. IAEA, Vienna, pp. 3–20.

Buddemeier, R.W. and D. Isherwood, 1985. Radionuclide Migration Project 1984 Progress Report. Lawrence Livermore National Laboratory. UCRL-53628, p. 71.

Case, C., J. Davis and R.F. French, 1984. Site Characterization in Connection with the Low-Level Defense Waste Management Site in Area 5 of the Nevada Test Site, Nye County, Nevada. Desert Research Institute, Pub. #45034. University of Nevada System, Reno, p. 157.

Clebsch, A., 1961. Tritium Age of Ground Water at the Nevada Test Site, Nye County, Nevada, in: Geological Survey Research, pp. C–122 - C–125.

Gee, G.W. and R.R. Kirkham, 1984. Transport Assessment–Arid: Measurement and Prediction of Water Movement Below the Root Zone. Presented at Sixth Annual DOE Low–Level Waste Management Program, Participants' Information Meeting, Denver, Colorado. September 11–13, 1984.

Hansen, D.S., 1978. Tritium Movement in the Unsaturated Zone, Nevada Test Site. Master's Thesis, University of Nevada, Reno.

Isaacson, R.E., L.E. Brownell, R.W. Nelson and E.L. Roetman, 1974. Soil Moisture Transport in Arid Site Vadose Zones, in: Isotope Techniques in Groundwater Hydrology, 1. IAEA. Vienna, pp. 97–114.

Kearl, P.M., 1982. Water Transport in Desert Soils. Desert Research Institute, Pub. #45024. University of Nevada System, Reno, p. 126.

Mehuys, G.R., L.H. Stolzy, J. Letey and L.V. Weeks, 1975. Effect of Stones on the Hydraulic Conductivity of Relatively Dry Desert Soils. Soil Sci. Soc. Amer. Proc., Vol. 39, pp. 37–42.

Morgan, D.S. and J.M. Fischer, 1984. Unsaturated–Zone Instrumentation in Coarse Alluvial Deposits of the Amargosa Desert near Beatty, Nevada, in: Proceedings of the Sixth Annual Participants' Meeting, DOE Low–Level Waste Management Program. September 11–13, 1984. Denver, Colorado, pp. 617–630.

Romney, E.M., V.Q. Hale, A. Wallace, O.R. Lunt, J.P. Childress, H. Kaaz, G.V. Alexander, J.E. Kinnear and T.L. Ackerman, 1973. Some Characteristics of Soil and Perennial Vegetation in Northern Mojave Desert Areas of the Nevada Test Site. UCLA/12–916, TID–4500.

Thordarson, W., 1965. Perched Ground Water in Zeolitized Bedded Tuff, Rainier Mesa and Vicinity, Nevada Test Site. U.S. Geological Survey Report TEI–862, p. 90.

Tyler, S.W., W.A. McKay, J.W. Hess, R.L. Jacobson and K. Taylor, 1986. Effects of Surface Collapse Structures in Infiltration and Moisture Redistribution. Desert Research Institute, Pub. #45045, p. 48.

Winograd, I.J., 1980. Radioactive Waste Disposal in Thick Unsaturated Zones. Science, 212–(4502), pp. 1457–1464.

Winograd, I.J. and W. Thordarson, 1975. Hydrogeologic and Hydrochemical Framework, South–Central Great Basin, Nevada – California, with special reference to the Nevada Test Site. U.S. Geological Survey Prof. Paper 712–C, p. 126.

Vertical Transport in the Vadose Zone by Fingering

Steven J. Wright[1]

Abstract

Water containing contaminants may infiltrate through the vadose zone as a series of fingers which result from a hydrodynamic instability initiated at an interface in which a fine textured soil overlies a coarser one. Although a number of theoretical analyses have been developed to predict the onset of the instabilities, experimental investigations have been severely limited by the difficulty in obtaining the necessary measurements. The present study makes use of neutron radiography to visualize both the onset of the instabilities and their subsequent propagation. Results obtained to date are in general agreement with some of the proposed theories.

Introduction

The transport of contaminants through the vadose zone typically originates in discrete infiltration events, either directly from a spill on the ground surface or due to solution in downward percolating rainwater. Infiltration models are generally one-dimensional in formulation and could be coupled with appropriate mass transport models to estimate transport times through the vadose zone, etc. However, if the transport process itself is not one-dimensional, these estimates may bear no relationship to the actual transport itself. A phenomenon of much interest in the soil science and petroleum literatures is *fingering* in which an initially plane front degenerates into discrete large amplitude disturbances hereafter referred to as fingers. Finger formation is not necessarily related to lateral heterogeneities in the porous medium, but is initiated by hydrodynamic instabilities that may be due to a variety of mechanisms that are not entirely understood at present. Possible scenarios that may result in finger formation include the immiscible displacement of a more viscous fluid by a less viscous one (important in water flooding of petroleum reservoirs to extract residual oil), the downward migration of a more dense fluid through a less dense one, and gravitational drainage of water through a vadose zone that has horizontal layers of varying properties. Soil layering is a consequence of various depositional environments and this latter fingering phenomenon may be of importance in many applications and is considered in this paper.

Assoc. Prof. of Civil Engrg, 113 Eng. 1A, Univ of Michigan, Ann Arbor MI, 48109-2125.

The exact mechanisms for finger formation during infiltration into a layered soil have not clearly been defined. Peck (1965) observed fingers while investigating infiltration into an unsaturated sand bounded at the base by an impervious seal and the mechanism for fingering was assumed to be the air pressure build-up below the wetting front. In most other investigations (e.g. Hill and Parlange, 1972), fingers were observed to be initiated at the interface between a fine (low permeability) sand overlying a more permeable one. Theoretical analyses have been presented by a variety of investigators to define the necessary conditions to initiate a hydrodynamic instability at such an interface. Most analyses, such as the one by Parlange and Hill (1976) assume a steep (non-diffuse) wetting front in order to simplify the analysis, although Diment and Watson (1983) have explicitly included the soil moisture diffusion at the wetting front in their formulation. Although the details of the different analyses vary, a common feature is that the models are only intended to predict the onset of fingering and not the characteristics of the fingers as they continue to propagate.

Some experimental investigations have been performed to examine the issues associated with finger formation. Most investigations have involved observations at the sidewalls of a transparent container, raising the issue of wall effects on both the finger formation and propagation. For example, Baker and Hillel (1990) conducted their studies in a 1.34 cm wide medium leading to the possibility that not only were the fingers two-dimensional in nature due to the wall influence, but the soil packing is also strongly influence by the walls. Observations by Hill and Parlange (1972) suggested that the fingers propagate at the saturated hydraulic conductivity of the coarse layer (gravity drainage condition) and that the size of the finger was more or less constant with depth and independent upon the infiltration rate applied at the upper surface. Starr, et al (1986) in a study of water with a chloride tracer indicated that the chloride moved downwards approximately three to four times more rapidly than would occur in a one-dimensional transport. They also indicated that a finger would maintain a constant lateral dimension until it arrived at the capillary zone above the water table after which it would spread rapidly laterally. It appears that most such studies have been performed in air dry soils and thus may not be representative of actual field conditions where at least the residual saturation (field capacity) is present. According to Diment and Watson (1983), this would affect the onset of fingering and from the results of Starr, et al, the propagation characteristics may be altered as well.

The present study relied upon a novel experimental technique to provide a visual record of the finger formation and propagation in the absence of wall influence. The neutron radiography facility at the Phoenix Memorial Laboratory at The University of Michigan was used to generate real time two-dimensional (horizontal-vertical) images of infiltration into layered soil systems. After a brief description of the experimental methodology, some typical experimental results are presented to describe some of the most significant findings in the present investigation.

Neutron Radiography

Neutron radiography has many parallels to the more common and established procedure of X-ray radiography (and the associated technique of computed tomography). In either case, a sample is exposed to a beam of radiation and the geometric pattern of transmitted radiation intensity is recorded on a detector. The intensity is related to the attenuation characteristics of the sample. The advantage to using a neutron source is that the attenuation coefficients of the elements vary abruptly from one another for neutrons, while the attenuation to X-rays increases monotonically with atomic number. Therefore, neutrons can be used to differentiate between two neighboring elements while X-rays may provide little or no contrast. Some materials will be virtually opaque to neutrons whereas others, such as silicon, are nearly transparent to neutrons.

One of the most important reasons for selecting a neutron source for investigation of porous media flows is that the hydrogen atom is relatively opaque to neutrons. Therefore, hydrogen containing substances, such as water, can be easily tracked with neutron radiography. Silicon containing substances, such as sand, are much more transparent to neutrons. This consideration has led to the widespread application of conventional neutron probes to measure in-situ water contents. Real time neutron radiography applications utilize a nuclear reactor as the source. These systems use neutron scintillators that yield light when irradiated with thermal neutrons. The resultant light can be amplified by image intensifiers and the dynamic event is recorded by capturing the real-time images on videotape.

Experiments

Most experiments were performed in an aluminum box approximately 10 cm wide by 20 cm long by 25 cm high; it was oriented so that the neutron beam passed through the 10 cm width. This dimension was selected to be greater than previously reported estimates of finger widths on the order of 5 cm. The sands used were various sizes of commercially available quartz or aluminum oxide (*alundum* grinding powder); both were about equally transparent to the neutron beam. Most experiments were performed with air dry sands and the box was filled by a "raining" technique in which the sand was dropped through several layers of screens. In all cases, the coarse sand layer was approximately 20 cm deep with a 2-3 cm layer of fine sand covering it. Initial experiments indicated difficulties with water flowing preferentially along the side walls so a funnel device that applied the water uniformly over the central portion of the box to within about 2 cm of the side walls was installed in the fine sand. This did not extend through the fine sand but simply ensured that the box perimeter would receive water last and resulted in the formation of fingers in the interior of the box. If the gap between the edges of the funnel and the box sides was allowed to remain open to air flow, any fingers that did form always moved more or less horizontally to the box walls. On the other hand, the fingers propagated primarily vertically if the only path for air escape was through the box bottom. All experiments reported herein are for this configuration.

Water was applied at a fixed rate to a region above the top of the fine sand. This rate was sufficient to maintain a small (1-2 cm) ponding depth on the surface of the fine sand. The box was placed so that the fine sand layer was near the top of the viewing port which had a dimension of approximately 15 cm diameter so that most of the length and depth of the coarse sand layer were visible. In experiments where fingers did form, only one or occasionally two were initiated so that the radiography image was clear with respect to finger detail. In a few instances, the sand column was disturbed immediately after the completion of the experiment to ensure that the fingers were propagating internally within the box and not along the walls; this was found to be the case.

Although most experiments were performed with air dry soil, a few were performed with initial moisture present in the coarse sand. Water was mixed with the sand prior to placement in order to ensure a homogeneous initial moisture. The test cells were prepared just prior to the experiment to prevent gravity drainage. For the cases reported, the initial moisture contents were about 11 and 33 percent; the latter is close to saturation for the sand packing employed in the experiment (a porosity of 0.45 was obtained for the air dry sand). A set of three experiments were conducted in which equivalent systems were set up except for the initial moisture content. One problem is that the partially saturated sands could not be prepared with the same packing arrangement and so the same porosities were not likely attained in the three experiments.

Hydraulic conductivities of the various sands utilized in the study were measured with an falling head permeameter. The method of packing the permeameter could vary the measured conductivity by nearly a factor of two, so the same sifting procedure as was used in the filling of the test cells was employed. The unsaturated soil moisture characteristics for these sands were not measured.

Experimental Results

A variety of different experiments were performed and only representative results are presented herein. Fig. 1 presents photographs of the video monitor at two different times for the finger that was formed for water infiltrating through a coarse aluminum oxide layer ($K \approx 0.016$ cm/s) from a fine aluminum oxide layer ($K \approx 0.0015$ cm/s). The maximum width of the finger is about 4-5 cm and this was a characteristic width for all experiments run with the same media. When a similar experiment was performed with an initial moisture content of about 11 percent, two fingers formed with roughly the same diameter. However, when the higher moisture content of about 33 percent was investigated, no tendency for finger formation was observed.

Fig. 2 presents a time history of the front presented in Fig. 1 and also for the two higher moisture contents. For the intermediate moisture content, two fingers were observed and time histories for both are presented. One of these propagated faster than the other in the latter stages of the experiment. The time and space origins of the individual traces are arbitrary and they are shifted so that they correspond at an early time. It can be seen that in both cases, the fingers propagate faster

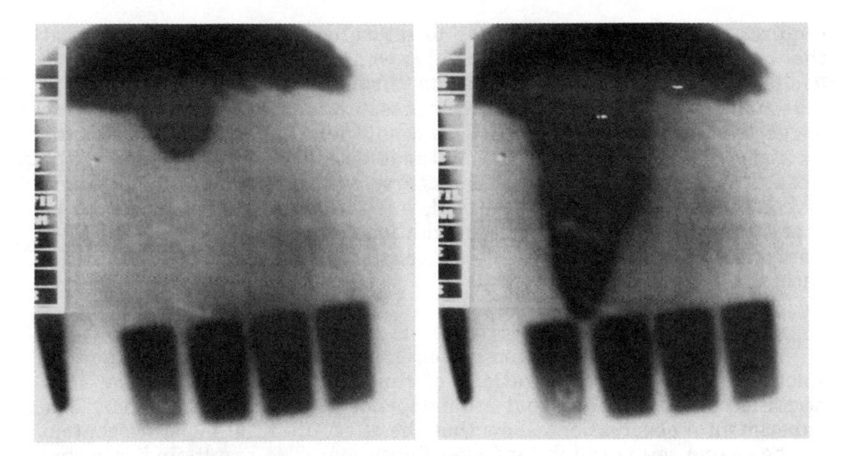

Figure 1. Finger approximately a.) 45 s and b.) 360 s after formation

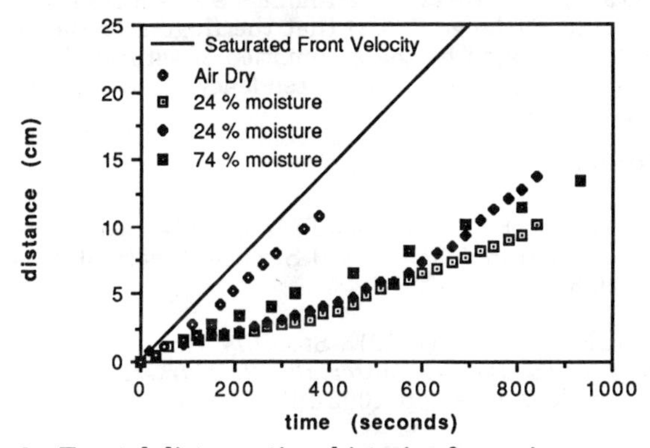

Fig. 2. Frontal distance-time histories for various experiments.

by about a factor of two than the plane front. Hill and Parlange (1972) indicate that the fingers propagate at approximately the saturated hydraulic conductivity divided by the effective porosity. The solid line indicated in Fig. 2 indicates that this relation is reasonably well satisfied, probably within the uncertainty in the determination of the porosity for the specific experiment. There is some tendency for the fingers to slow down with time that was observed in many experiments, apparently a result of the soil moisture diffusion laterally from the finger.

The theories of Parlange and Hill (1976) and Hillel and Baker (1988) regarding the required conditions for the onset of fingering yield quite different criteria. The former compares the vertical infiltration rate to the saturated hydraulic conductivity of the coarser soil while the latter requires the comparison to the the unsaturated hydraulic conductivity at the water entry head. This latter value would be somewhat lower. Using a sharp front model for infiltration through the upper layer with a reasonable estimate for the air entry pressure indicates that the criterion suggested by Parlange and Hill would not yield finger formation for the conditions indicated in Fig. 1. Without detailed measurements of the unsaturated characteristics of the coarse soil, the theory of Hillel and Baker cannot be rigorously checked, but the observations appear to be consistent with their criterion. An additional pertinent factor is that their analysis indicates a qualitatively consistent influence of the initial saturation in the coarse layer.

Conclusions

The technique of neutron radiography was applied to make the first known quantitative observations of fingering during infiltration in a stratified porous medium that are independent of the wall influence. The experimental observations show that the effect of initial moisture content in the coarse soil is to inhibit finger formation. The results indicate that the analysis developed by Hillel and Baker (1988) tends to better describe the required conditions for finger formation than the theory presented by Parlange and Hill (1976) and others. The findings also substantiate the observations by Hill and Parlange (1972) that the fingers propagate at velocities that may be estimated by the assumption of saturated gravity drainage at unit hydraulic gradient. Some tendency for the fingers to slow down with time was noted.

APPENDIX - REFERENCES

Baker, R.S. and D. Hillel, (1990) *Laboratory Tests of a Theory of Fingering During Infiltration into Layered Soils,* Soil Science Society of America Journal, Vol. 54, pp. 20-30.

Diment, G.A. and K.K. Watson, (1983) *Stability Analysis of Water Movement in Unsaturated Porous Materials, 2. Numerical Studies,* Water Resources Research, Vol. 19, pp. 1002-1010.

Hill, D.E. and J.-Y. Parlange, (1972) *Wetting Front Instability in Layered Soils,* Soil Science Society of America Journal, Vol. 36, pp. 697-702.

Hillel, D. and R.S. Baker, (1988) *A Descriptive Theory of Fingering During Infiltration into Layered Soils,* Soil Science, Vol. 146, pp. 51-56.

Parlange, J.-Y. and D.E. Hill, (1976) *Theoretical Analysis of Wetting Front Instability in Soils,* Soil Science, Vol. 122, pp. 236-239.

Peck, A.J., (1965) *Moisture Profile Development and Air Compression During Water Uptake by Bounded Porous Bodies, 3: Vertical Columns,* Soil Science, Vol. 100, pp. 44-51.

Starr, J.L., J.-Y. Parlange, and C.R. Frind, (1986) *Water and Chloride Movement through a Layered Field Soil,* Soil Science Society of America Journal, Vol. 50, pp. 1384-1390.

NITRATE AND BROMIDE TRANSPORT THROUGH A VADOSE ZONE IN NEW JERSEY

Deva K. Borah[1], M. ASCE and Sunil M. Gupte[2]

Abstract

On going field experiments on monitoring transport of Sodium Nitrate and Potassium Bromide in the vadose zone over a shallow groundwater table in the coastal plains of New Jersey are discussed. The chemicals were mixed with lake water, and introduced into the surface of a field strip whose soil profile was continuously monitored for moisture content variations and movement of the chemicals. Monitoring was done from a trench dug along the strip. A new soil solution sampler was developed, and used in this experiment in collecting soil solution samples from the soil profile. Results from a preliminary run are presented.

Introduction

Contamination of groundwater with shallow water tables due to agricultural pollutants has become a major problem. Nitrate is one of the major pollutants (Ritter et. al., 1989). Understanding the movement of chemical solutions through the unsaturated zone is critical in evaluating groundwater contamination. Due to unavailability of experimental data under field conditions, many of the theoretical models for unsaturated zone transport remain untested (Dagan and Bresler, 1979).

Many lysimeter studies have been reported which show varying degree of nitrate leaching (Haman et al., 1987). Although utmost care is taken in laying the soil layers on lysimeters to preserve the field conditions, it naturally destroys the structural characteristics such as cracks, root canals, worm holes, etc., which may be responsible for preferential flow of contaminants through the soil profile.

In order to develop a data base for nitrate-nitrogen transport through the vadose zone under undisturbed soil conditions, this study was developed to conduct field experiments on leaching of

---

[1]Assistant Research Professor and [2]Graduate Assistant, Department of Biological and Agricultural Engineering, Cook College, Rutgers University, New Brunswick, New Jersey

nitrate-nitrogen and bromide through the unsaturated zone by applying water solution of these chemicals at the soil surface, and monitoring moisture content variations, and nitrate and bromide concentrations along the soil profile. Bromide is a conservative tracer, and as it does not occur naturally in its free form, it is easier to monitor. In addition to these, fluctuations of the groundwater table is also monitored to study the effect of preferential flow.

In this paper, preparation of the field experiments, and results from a preliminary run are presented.

Soil Solution Sampler

Soil solution samplers are the most important component in the study of contaminant transport through the vadose zone. Six commercial soil solution samplers were tested in the field. None of the samplers were found suitable to collect soil solution samples in the experimental site. Some of the samplers disturbed more soil than the others and still yielded small amount of sample. The others had very high porosity with low bubble pressure, and could not hold any solution.

In this study, a new type of soil solution sampler was developed using ceramic cups based on their proven performance in the piezometers used in earlier study (German and Gupte, 1988). Figure 1 shows cross sectional view of the sampler. Construction of the sampler is very simple, and it uses a similar concept as in some of the existing samplers. Vacuum is created within the sampler and maintained for a period of time during which soil solution moves into the sampler. Then a pressure is applied to force the solution out through a tube extending to the bottom of the sampler. The samplers are completely air tight, and are always in close contact with the surrounding soil surface.

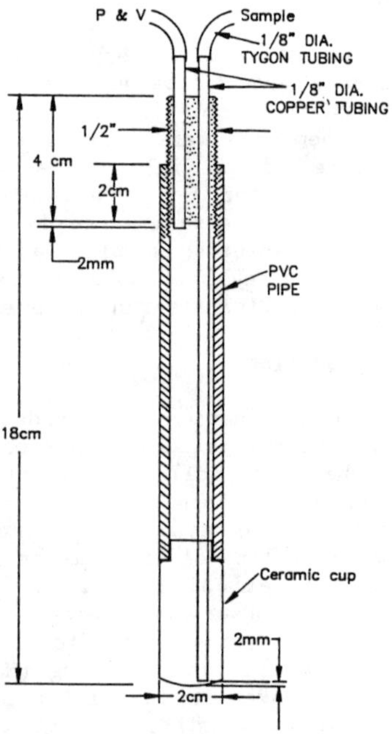

Fig. 1. Soil Solution Sampler

Experimental Set-up

The experiments have been conducted at Adelphia, New Jersey on the fields of Soils & Crops Research Experimentation Center of Rutgers The State University of New Jersey, where the groundwater table is approximately 2 m below the ground surface. A 0.1-m deep,

0.5-m wide and 21-m long strip of soil with no slope was isolated at a distance of approximately 150 ft (46 m) from a lake using sheet metal foil. Holes of 2-in (51-mm) diameter were drilled up to the groundwater at 1-m spacing. Piezometers whose tips were mounted close to the groundwater and hydraulically connected to piezoresistive pressure transducers were installed in 2-in (51-mm) PVC pipes placed in these holes. The piezoresistive transducers were calibrated in the laboratory for their pressure response. All these pressure transducers were connected to a Campbell data logger to monitor fluctuations of the groundwater table. The entire field strip was used in an earlier experiment (Germann and Gupte, 1988) to study rapid rise of shallow groundwater table due to preferential flow.

In this study, the above strip was divided into three smaller experimental substrips of 5-m each with two buffer substrips of 3-m each. The buffer substrips provide protection from contamination of the adjacent experimental substrips. At each experimental substrip, double ring infiltrometer and tension infiltrometer experiments (Watson and Luxmoore, 1986) were conducted at every meter distance to determine infiltration characteristics of the soil.

The soil on three-fourth of the strip length is Freehold loamy sand underlain by uniform greensand. The last quarter length of the strip is a medium textured surface soil with poorly drained fine clay at a depth of 30 cm. The current experimental substrip is located in the loamy sand area.

Water is lifted from the lake into a tank of 460 gallons (1741 litters) capacity using a centrifugal pump, and known quantities of the chemicals are mixed. A 2-in (51-mm) PVC main is used to carry the chemical solution to the experimental substrip. The main has a flow meter installed near the tank bottom to monitor the quantity of flow passing through it. Two 1.5-in (38-mm) submains cover the entire length of the strip. Water is siphoned into the strip using rubber tubes. Two gate valves on the submain control the depth of water ponded on the experimental substrip.

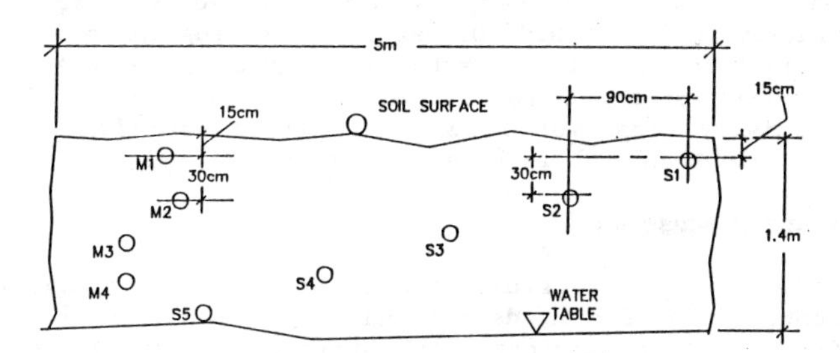

Figure 2. The Soil Profile Showing Positions of Soil Solution Samplers and Moisture Blocks

A trench of 5x5x2 m^3 was dug at 1 m away from the experimental substrip. Five soil solution samplers, S1-S5, were installed under the substrip through the vertical soil surface of the trench diagonally at various locations as shown in Figure 2. The samplers were kept air tight using MTV-2 one-way valves. A 1/8-in (3-mm) Tygon tubing (Fig. 1) was used to connect the samplers to the vacuum pump. A valve manifold was designed to allow control of individual samplers as well as all the samplers together. A positive displacement vacuum pump was used to create the required vacuum and pressure.

In order to monitor the continuous changes in moisture content of the soil at different depths, four Model 227 Delmhorst Cylindrical Moisture Blocks, M1-M4, were placed in the soil profile through the vertical soil surface as shown in Figure 2. The moisture blocks were connected to a Campbell data logger to monitor the fluctuations at every 2-minute time interval.

Experimental Method

The moisture measurement and groundwater table monitoring was started 30 minutes prior to the start of an experimental run in order to get the initial conditions. Water containing 120 ppm of nitrate ($NaNO_3$) and 60 ppm of bromide (KBr) was applied at the substrip surface and continued for 2 hours. At any given time, the depth of water ponded on the substrip was 1 inch, which was controlled using the submain valve. Soil solutions were extracted from the sampler S5 (Fig. 2) after 15 and 30 minutes of solution application. Soil solution extraction was resumed at 35 minutes starting at sampler S1 and continued towards S5.

At each sampler, a vacuum pressure of 0.76 bar (23-in of Hg) was created for a maximum of 5 minutes and the collected sample was flushed out of the ceramic suction cups using air pressure of 0.9 to 1 bar (13 to 15 psi) for 3 minutes. The sampler was flushed clean of any solution for next 2 minutes by blowing air through it. Sampling was continued till 5 hrs from the start of irrigation. Soil solution samples were stored in an air-tight, thermally insulated storage box and were immediately transferred to a cold storage.

The soil solution samples were tested for NO_3 and Br concentrations using a 2010i Dionex Ion Chromatograph equipped with an electrical conductivity detector and a standard HPIC-AS4A column generally used for soil anions. The test flow rate was 2 mL/min and the sensitivity range was 1 mg/L for strong acid ions. A sample volume of less than 1 mL is required for the analysis.

Results and Discussion

Variations of moisture condition (m.c.) throughout the experiment, at different depths are shown in Figure 3. The reduction in tension indicates the arrival of water front at that depth. As expected, m.c. increased first at depth 15 cm. Arrivals of moisture at 15 cm and 45 cm depths are simultaneous which may be an indication of fast flow through the soil zone between these two depths. It took approximately 90 minutes for the water front to reach depth 75 cm. The fast increase in tension after the arrival of water front indicates a characteristic of sandy soil.

Figure 4 represents the change in concentration of the

contaminants with respect to time at different depths. The concentration graphs at various depths do not show a consistent pattern which may be attributed to conversion of nitrates to nitrite and other forms. The fluctuations of concentrations may also indicate channeling of flow.

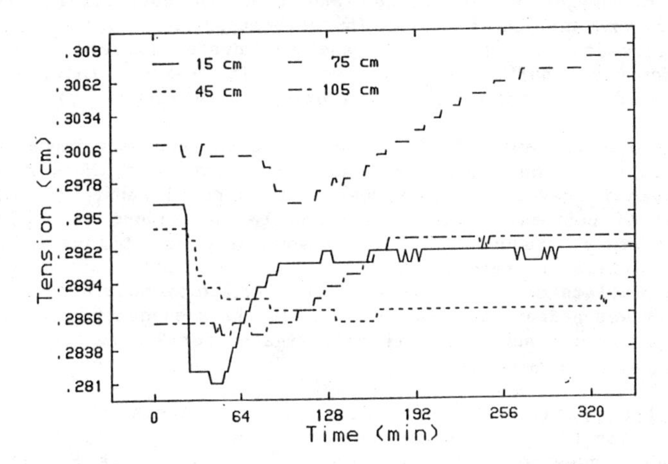

Figure 3. Moisture Content Variations at Different Depths

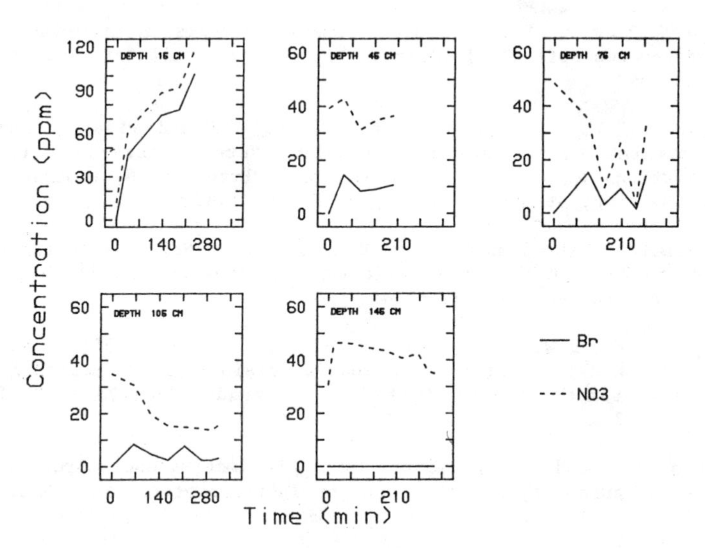

Figure 4. Temporal Changes of Nitrate and Bromide Concentrations at Different Depths

Conclusions

Procedure and preliminary results on an experimental field study of nitrate and bromide transport through the unsaturated zone of a coastal plain field in New Jersey are presented. A soil solution sampler was developed and used in collecting samples. Samples were analyzed using an Ion chromatograph. Moisture content variations in the soil profile and groundwater table fluctuations were monitored during the experiment. The results indicate wide variations of contaminant concentrations with respect to time and space.

Sampler placement in the vadose zone is extremely important in order to track the changes in contaminant transport. More research is needed in developing instruments which could sense the dynamic changes of contaminant concentrations in the vadose zone. In the present study, frequent and simultaneous sampling at many locations in the profile are needed.

A complete mass recovery was not possible in this experiment as no water was added after chemigation. In the subsequent experiments, the experimental substrip will be irrigated for 2 more hours to get a complete mass recovery.

Acknowledgments: The authors wish to express their sincere appreciation to P. F. Germann who began the study in 1988, for his valuable suggestions and guidance. The guidance of H. Motto in analyzing the solute samples is also highly appreciated. NJAES Paper No. J-03149-06-90 supported by U.S. Hatch Act Fund.

References

Dagan G. and E. Bresler. 1979. Solute dispersion in unsaturated heterogeneous soil at field scale. 1. Theory. Soil Sci. Soc. Am. J. 43, 461-467.

Germann P. F., and S. M. Gupte. 1988. Rapid rise of shallow groundwater tables on infiltration - Preliminary results and implications on aquifer contamination. Proc. of Nat. Water Well Asso. Meeting, Stamford, CT, August 27-29, 1988.

Haman D. Z., K.L. Campbell, and D.A. Graetz. 1987. Lysimeter study of micro irrigation and fertigation of tomatoes. Soil Crop Sci. Soc. Fla. Proceedings. 47, 45-49.

Ritter W. F., R.W. Scarborough, and A.E.M. Chirnside. 1989. Nitrate leaching under irrigation on a coastal plain soil. Presented at the ASAE Int. Summer Meeting, Quebec City, Canada. June 25 - 28, Paper No. 89-2027.

Watson, K.W. and R.J. Luxmoore. 1986. Estimating macroporosity in a forest watershed by use of a tension infiltrometer. Soil Sci. Soc. Am. J. 50, 578-582.

Graphics Advances Aid Flood Engineers

Brian M. Reich and Kenneth G. Renard, F.ASCE[1]

Abstract: In the last decade options[2] have been suggested to the Log-Pearson III (LP3) computation for predicting design flood behavior in semi-arid conditions. Observations from an ongoing study of seven watersheds are presented. Areas ranged from 25 to 3,200 sq. mi., with record lengths from 40 through 73 years. Computer graphics on log-normal (LN), extreme value (EV), and log-extreme value (LEV), along with mathematical estimates for 10- and 100-year floods, are presented. Differences between the behavior of the largest 1/5 of annual floods, from more frequent annual maxima, are illustrated.

Background: During the 1940's and 1950's many agencies concerned with 100-year flood predictions (Q100) were using Gumbel's National Bureau of Standards, and Columbia University work. Gumbel's application on the EV distribution appeared in hydrology texts and was used by the USGS, Bureau of Reclamation, Bureau of Public Roads, and others. Even today, the National Weather Service uses EV as its probability model for estimating storm rainfall which can produce Q100. In 1967 the Water Resources Council (WRC), under pressure from the Bureau of the Budget, recommended in Bulletin 15 that the 1924 Log-Pearson III (LP3) statistical analysis be used. This technique normally forces convex or concave statistical curves through data which plot into very different patterns. We used the method of moments described in Bulletin 15 for our calculations.

It may be that LP3 is particularly unsuitable in semi-arid regions because annual floods contain many low maximum flows, which can cause strongly negative coefficients of skewness of the logs (CSL). CSL also varies greatly between closely spaced stream gages. This makes the WRC Bulletin CSL map of questionable value. Some flood experts can fine-tune regional and weighted CSL's to produce reasonable Q100's from LP3. In all cases, the fitted curve and annual maxima should be plotted by Cunnane's compromise formula.

---

[1] Consulting Engineer, 2635 E. Cerrada Adelita, Tucson, AZ 85718, and Research Hydraulic Engineer, USDA-ARS, 2000 E. Allen Road, Tucson, AZ 85719.

[2] Space does not permit a bibliography. Some references are located in "Linearizing Large Flood-peaks on Small Arid Watershed," by these authors and F.A. Lopez in the ASCE 1990 National Conference on Hydraulic Engineering.

Unfortunately, novices may rely on simple numerical computer output. Busy engineers may be called upon to speedily determine the return period of an oversized flood. Immediately following a flood disaster, local flood agencies may need to estimate the event return period. If a local agency has a personal computer database, appropriate software, and trained personnel, estimates can be completed within a few days. This paper shows how much variation can be expected after executing different numerical computations for semi-arid watersheds. New computer graphics programs offer an opportunity to appreciate Q100 variation in semi-arid regions. The authors wish to thank Gert Aron and Dard de Roulhac for providing the computational and graphics programs, respectively, and F.A. Lopez for modifying them.

Relative Consistency Among Four Statistical Models: Table 1 was generated by dividing each 10-year flood estimate (Q10) by Q10 obtained from LP3, using station skew. In this way LP3 was always assigned an arbitrary 1. LN, EV, and LEV can readily be compared to LP3 and each other as ratios. The ratios do not vary much across the seven watersheds, but do increase systematically from LP3, to LN, to LEV, to EV as shown at the bottom of Table 1. Big floods are seldom measured within this accuracy.

Table 1. Relative size of Q10's from three methods, expressed as ratios to LP3 estimate.

Watershed	Square Miles	Years	LN	EV	LEV	Aver.
Sabino Creek, Tucson, AZ	25.5	57	1.06	1.05	1.14	1.08
Santa Cruz River, Lochiel, AZ	82.2	40	1.52	1.35	1.67	1.54
Eagle Creek, Morenci, AZ	613	45	1.01	1.16	1.09	1.09
Rillito Creek, Tucson, AZ	918	68	1.00	1.11	1.05	1.05
San Pedro River, Charleston, AZ	1,219	78	0.97	1.40	1.01	1.13
Santa Cruz River, Continental, AZ	1,662	45	1.01	1.23	1.11	1.12
Gila River, Virden, NM	3,203	62	0.98	1.30	1.01	1.10
Average			1.0	1.23	1.15	

The last column shows the averages of LN, EV, and LEV Q10's on each river are 5, 8, 9, 10, 12, 13, and 54% greater than the LP3's. This is another indication that LP3 under predicts even 10-year floods. There is insufficient evidence to determine whether LN, EV, and LEV, are sensitive to watershed size.

"Increases" in Flood Estimates with Watershed Size: Engineers are interested in how Q10 and Q100 estimates change with watershed size. Our sample is too small to aid them with such relationships. Some appreciation for the unusual properties of semi-arid desert flood series can be gained by discussing the specimen computations.

Table 2 summarizes results from computational fitting of LN, EV, LEV, and LP3. Q10's for the 1,662 and 3,203 sq. mi. watersheds are smaller than Q10 from 918 or 1,219 sq. mi. This phenomena is related to channel losses and/or areal distribution of storm rainfall in semi-arid areas. Whereas Q10's from all four models increased from 35 through 600 sq. mi., a general decrease occurred as drainage areas increased towards 3,000 sq. mi., particularly for Q100.

Table 2. Computational estimates, in cfs, from four statistical models.

Watershed	Drainage Area Sq. Mi.	Computer Q10 cfs				Computer Q100 cfs			
		LN	EV	LEV	LP3	LN	EV	LEV	LP3
Sabino Creek, Tucson, AZ	25.5	4,262	4,170	4,582	4,029	13,439	7,580	36,381	9,693
Santa Cruz River, Lochiel, AZ	82.2	6,540	5,807	7,218	4,315	23,718	10,734	75,278	5,141
Eagle Creek, Morenci, AZ	613	12,703	14,537	13,895	12,522	44,441	27,581	134,757	40,715
Rillito Creek, Tucson, AZ	918	14,927	16,640	15,694	14,927	35,648	29,848	75,067	35,648
San Pedro River, Charleston, AZ	1,219	17,169	25,590	17,880	17,792	35,590	48,118	66,188	50,388
Santa Cruz River, Continental, AZ	1,662	14,444	17,504	15,544	14,275	40,282	32,742	99,924	37,494
Gila River, Virden, NM	3,203	16,143	21,299	17,057	16,413	40,169	39,915	88,102	45,631

Instability Grows for Q100 Estimates: Examination of table 3 is one way of appreciating the variability between LN, EV, LEV, and LP3 in the case of Q100. For each river the first three estimates were divided by the corresponding LP3. These ratios, like 1.39, 0.78, and 3.75 in table 3, give an indication of how three well known statistical determinations compare with each other and LP3. More sophisticated examinations could be used if a full-scale investigation was undertaken. The case of the Gila River, on the last line of table 3, shows simply that LEV's Q100 was virtually twice that of LP3. LN and EV were slightly smaller, which may persuade some analysts to use them. Others may question using LEV, though theoretical reasons exist why LEV may give oversized Q100 estimates.

Table 3. Ratios of Q100 to Station-LP3 for three models.

Watershed	LN	EV	LEV	Average
Sabino Creek, Tucson, AZ	1.39	0.78	3.75	2.0
Santa Cruz River, Lochiel, AZ	4.61	2.09	14.64	7.1
Eagle Creek, Morenci, AZ	1.09	0.68	3.31	1.7
Rillito Creek, Tucson, AZ	1.00	0.84	2.11	1.3
San Pedro River, Charleston, AZ	0.70	0.95	1.31	1.0
Santa Cruz River, Continental, AZ	1.07	0.87	2.67	1.5
Gila River, Virden, NM	0.88	0.87	1.93	1.2
Average, excluding Santa Cruz River, Lochiel, AZ	1.02	0.83	2.51	

The Santa Cruz at Lochiel gave anomalous scatter between the four models, which require further study. Values listed in these tables should not be used in estimating the best Q100. The authors would appreciate correspondence from practitioners with established rules-of-thumb relating Q100 to Q10 in similar regions. Our tentative average for six rivers suggest that LN and LP3 produce similar Q100's. EV estimates may be 15% smaller; LEV averages about 2.5 times EV's value. Another approach taken was to express Q100 as a multiple of Q10. Such Q100/Q10 for our seven watersheds were 1.78, 1.64, 1.90, 1.99, 1.88, 1.87, and 1.87, respectively, which averaged 1.85 for EV.

<u>Searching Alignment of Large Observed Flood Series</u>: Recently, the difficult task of estimating Q500 was mandated for flood plain administrators. We estimated Q500 for our watersheds; table 4 summarizes the results in the form of two multipliers. The first Q100/Q10 could bring a user from the relatively stable Q10 to a less stable Q100. Q500 averages 1.6 times Q100. The last column shows how variable the overall ratio to estimate Q500/Q10 really is. Over or underestimating Q500 can have serious societal consequences.

Table 4. Q100/Q10 and Q500/Q100 for some Arizona watershed LP3s.

Watershed	Q100/Q10	Q500/Q100	Q500/Q10
Sabino Creek, Tucson, AZ	2.41	1.50	3.6
Santa Cruz River, Lochiel, AZ	1.19	1.02	1.2
Eagle Creek, Morenci, AZ	3.25	1.83	6.0
Rillito Creek, Tucson, AZ	2.39	1.58	3.8
San Pedro River, Charleston, AZ	2.83	1.88	5.3
Santa Cruz River, Continental, AZ	2.63	1.64	4.3
Gila River, Virden, NM	2.78	1.76	4.9

<u>Attention to the Plotted Largest Observations</u>: Improved understanding of how larger floods can behave will come by integrating patterns in observed flood plots with statistical analyses. In arid areas, a 70-year record may only contain 5 or 10 items involving the same runoff processes that operate during design floods. Figure 1 shows two rivers with 40- and 45-year records. In both cases, the mathematically fitted line (shown as the solid line) did not represent the trend of the larger quarter data points. In one case, this computed line passes significantly below the three largest points which, together with the next three, are closely fitted by an eye-fitted straight line (shown as a dashed line). Q100 changes from 27,600 cfs computed to 75,000 cfs eye-fitted. In another watershed the computed Q100 changed from LN 23,720 cfs to an eye-fitted LN of 13,000 cfs.

Dashed lines were drawn through the linear configurations of big floods. We consider eye-fitted lines give far better estimates of Q10, Q100, and Q500 than can be done by statistical fitting for these examples.

Figure 2 shows that all seven watersheds display separation and misalignments of large floods. In the Santa Cruz at Lochiel case, the influence of different plotting paper was also shown.

Summary: Computations from four recognized statistical models can produce somewhat similar Q10's for semi-arid regions. LP3 answers are smaller than LN, EV, or LEV. In the case of Q100, three or four statistical distributions should be computed and, more importantly, graphical examination should be added to show variation, along with consideration of individual flood incidents, physical runoff properties, and partial-duration series.

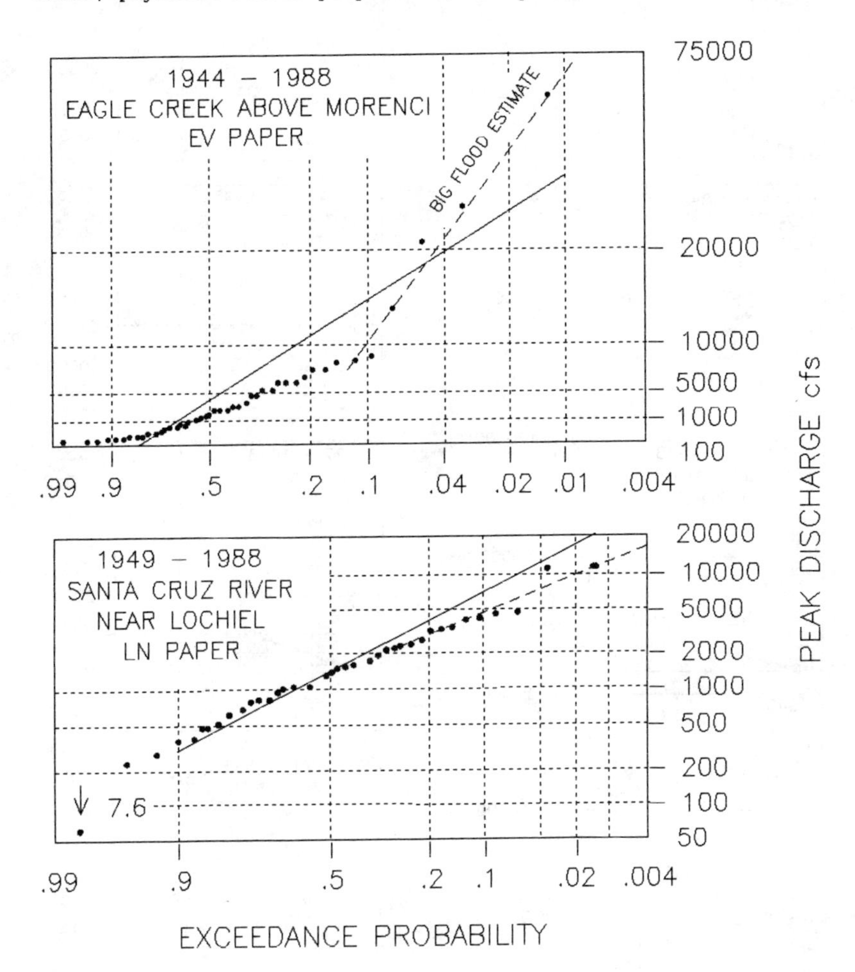

Figure 1. Two Arizona examples where the alignment of the larger events were overlooked by computations of total annual series.

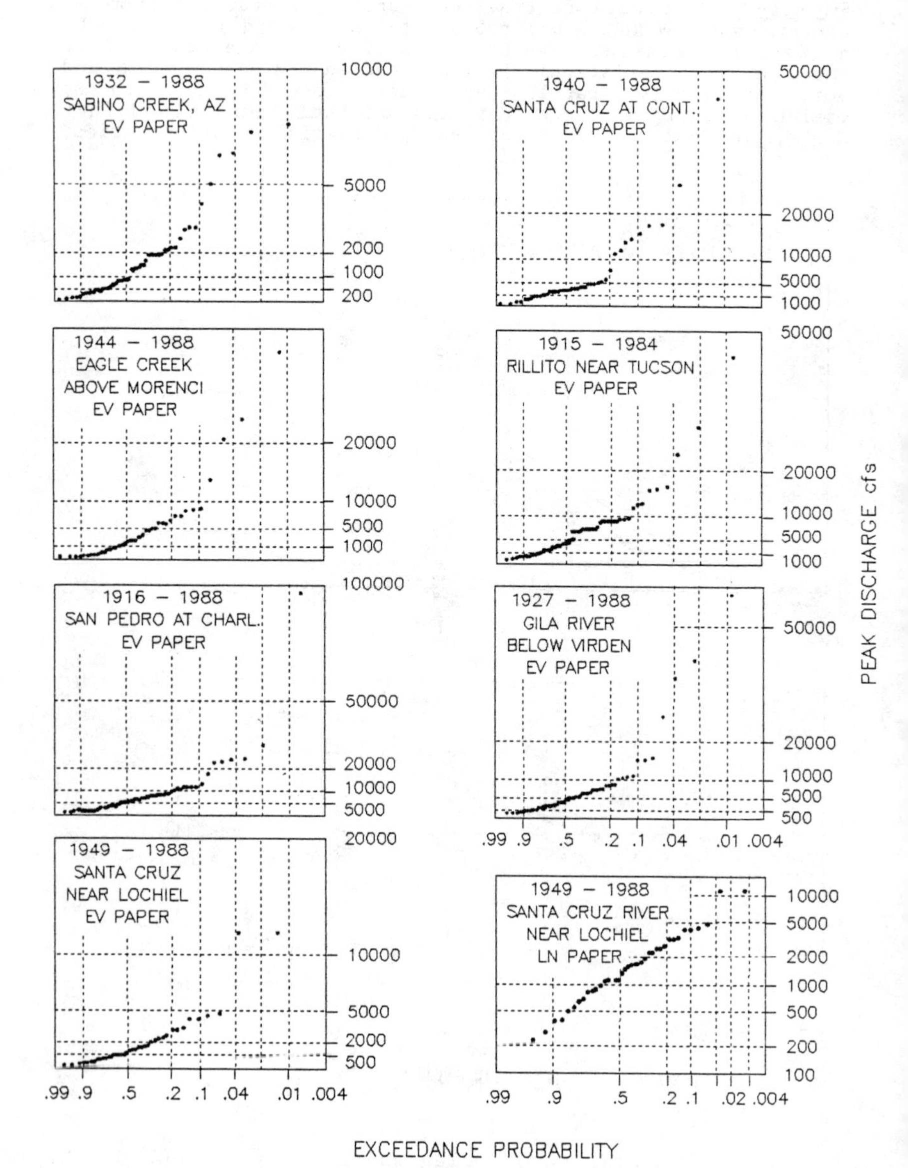

Figure 2. Long series on seven Arizona Watersheds displaying different orientations of large floods than do the more numerous non-flood annual maxima.

THE SORRY STATE OF FLOOD HYDROLOGY IN THE ARID SOUTHWEST

by
David R. Dawdy[1], Aff. M., ASCE

Abstract

Flood hydrology for urban areas of the arid and semiarid Southwest is in a sorry state. It is based in large part on precomputer manual methods. The state-of-the-art should move into the 20th century before the 21st century is upon us.

Introduction

There appears to be a continuing interest in the development of hydrologic techniques appropriate for use by local governmental jurisdictions in the arid Southwest for urban land use planning and management. That is an encouraging sign, and I am in complete accord with the goal of finding appropriate tools for use for land use planning and management in such a difficult environment. A short review of my recent experiences may help to place in perspective my suggestions for the next necessary steps for improving the hydrologic tools used for urban planning in our Southwest.

Organizations involved in urban water planning should be concerned with problems of master planning. The cumulative impacts and the assessment of downstream impacts in a consistent manner should be the objective. There are serious problems in the use of most presently accepted techniques, particularly in their use for master planning. That problem exists in Southern California and similar areas in general. In my opinion, there is a need for a rational approach to hydrology of the semi-arid and desert areas of the American Southwest. It is appropriate for the affected local jurisdictions to undertake a review of desert hydrologic methods and their utility for master planning in the urban environment.

What are the problems involved in design storm analysis in general, and in the arid Southwest in particular? The first question is how to determine the design storms and resulting peaks. It has been shown

----

[1]Consulting Hydrologist, 3055 23d Ave., San Francisco

that the U. S. Geological Survey regional flood frequency estimates are the best estimators for natural conditions. However, they have two problems. First, they are for natural conditions. In urban hydrology we often want to study man-affected systems. We want to study how proposed changes will change the response of the basin. Also, we often want to model the total storm hydrograph in order to study the potential effect of proposed ameliorative measures, such as detention basins. This naturally leads to the need to develop rainfall-runoff models for use in urban master planning. There are four major problems to be solved in developing appropriate hydrologic models for urban master planning. They are the choice of a design storm, determination of the infiltration rate, the effect of impervious area as it increases in a basin, and the choice of a routing model to develop the resulting flood hydrograph.

The Problem

The development of the design storm rainfall is perhaps the most straightforward problem of all. Often, state agencies have analyzed all rainfall records on a regional basis and have developed a set of statistics for duration-depth-frequency relations. The choice of return period --- 100 years for major problems --- is almost automatic as a result of FEMA and the Flood Insurance Program. However, the economic and esthetic and environmental consequences of that automatic choice should be considered. If protection to a 70 year storm level will preserve a particular historic site or environmental amenity, there is nothing in the law that says flood insurance considerations should override all others. The choice of storm duration may have a major impact on detention basin size. Many agencies require storms of several durations to be simulated to determine which storm requires the largest amount of detention. With "proper" choice of "nesting" pattern the choice of duration becomes less important. However, the later in the storm the higher intensities occur, the higher the resulting peaks. Also, the flood peak should not vary as the duration of the storm is varied. Thus, arrangement of the design rainfalls within a storm should be consistent from duration to duration.

A more important consideration because usually less well understood and less well handled is the determination of the infiltration rates. Infiltration should be related to soil types and should be calibrated to field data --- rainfall and runoff measured in the area. Many hydrology manuals in the arid Southwest are not. Often they are based on SCS curve number methods. That is a crime. This method was developed for stock pond design in the Midwest. The basic curves on which it is based were published in the Soil Conservation Service

Engineering Handbook at least by 1957 and have not been revised since. Early on one was cautioned that the curves were for estimation of storm volumes and were not to be used for incremental storm amounts. They were so used, and use engendered justification. However, the only justification for their use in the urban environment and, in particular, in the arid Southwest, is that the SCS saw that people were moving to the cities and it wanted a piece of the action. Like the giant in Jack and the Beanstalk, they stretched or cut off the problem to fit the tool.

There are better ways to estimate infiltration. There are several models of infiltration available, varying from constant infiltration rates to Green-Ampt models where infiltration is inversely related to soil moisture content. Which to use should depend upon the use for the model and the availability of data for calibration of the model. The determination of the effect of impervious area is particularly important. Impervious area increases volume and increases velocity of the water, both of which tend to increase peak flows. However, the effect of increased impervious area depends upon its location in the basin. Also, The effect depends upon the "connectedness" of the impervious area to the channel --- runoff from impervious areas not directly connected to the channel system must flow over pervious areas, and thus contribute less runoff.

Certainly, the definition of the runoff response function of the basin is the best defined portion of the problem, if a physically based model is used. There is no reason not to use a model directly related to the physically measured or estimated characteristics of the basin --- slope, length, resistance, etc. A distributed parameter model, such as HEC-1 with the kinematic wave option, rather than a generalized unit hydrograph, should be used for the surface water response of the basin. Generalized regional unit hydrographs no longer are needed, even for relatively small jobs. Yet, most hydrology manuals still use 1940's hydrolgic tools to define a unit hydrograph type of basin response as we approach the 21st century.

Any model should be calibrated to field data. The reason for the requirement of so little data for the determination of the response function for the physically based models is that all the description of the model is based on field data. There is a transfer of information about the physical process which is substituted for the need for rainfall and runoff data.

The needs of the local jurisdictions with their oversight duties must be kept in mind. However, the approach to calibration to field data should be fairly straightforward, but is seldom used in the design of hydrology manuals. A new start and a new look are needed. Calibration at the basin scale can be used to model at smaller scale to develop a tool, such as the Rational Method, which is calibrated by proxy and is consistent with the results from the calibrated basins.

To illustrate the need for data-based design storms, a comparison was made of the effect of different SCS storm distributions for a small basin in San Diego County. The SCS models require input of a 24-hour design storm, then the time distribution of the rainfall during that 24 hours is internally derived by the models. For various sub-basins within the Kit Carson Creek drainage, the following table compares peak runoff rates resulting from the use of two different rainfall distributions.

Flow Point	Drainage Area	Rainfall Distribution Type B cfs	in/hr	Type 1 cfs	in/hr
Upper KCC	0.6	612	1.58	1140	2.94
Lower UKCC	1.2	1180	1.53	1950	2.52
Combined flow	1.8	1780	1.53	2930	2.52
Middle KCC	1.2	1130	1.46	1630	2.11
Combined	3.0	2880	1.49	4010	2.07
Lower KCC	.4*	368	1.42	707	2.74
Combined	3.4	3130	1.43	4290	1.96

* 24-hour 100-yr design rainfall 8", others 9".

Thus, for the area receiving 9" of rainfall, the runoff rate varies from 1.46 to 1.58 inches per hour, a ratio of 8 percent when a Type B distribution is used. For Type 1, the variation is from 2.07 to 2.94, or 42 percent. A basin must be much larger than (the lag time must be much longer than for a basin of) 3.4 square miles before the results of Type B and Type 1 are similar for a similar total design storm. Runoff rates per unit area should not be constant or nearly so for smaller basins. The nested rainfall pattern for shorter time intervals should be chosen so that it does not produce anomalous results when applied to smaller drainage basins, such as are of interest for urban hydrology, to develop peak discharges from regional rainfall relations. More important, a method which "builds in" a rainfall distribution is inadequate, and cannot be calibrated to local conditions. Even though the 24-hour rainfall is based on local data, the way in which it is used eliminates the intrinsic meaning of the data. A similar situation was found in Placer County, where the choice between a Type 1 and a Type 1A storm was necessary. The resulting peak varied by fully as much as the San Diego County case, and the line dividing the region of applicability of the two storm types went through the center of the county.

As stated earlier, the time increment used for precipitation input and, thus, for determination of the flood peak should be much shorter than the time response of the basin. The peakedness of the output hydrograph should be determined by the excess precipitation during the period equivalemt to the time response of the basin. The time increment chosen for modeling of design storms always should be determined by a consideration of the time response of the system modeled. If a nested group of basins are being modeled in order to do basin master planning, the time increment for modeling should be based on the time response of the smallest segment of the basin which is of interest in the basin master plan. The depth-duration-frequency data must be determined down to the duration for that smallest subbasin. There is no excuse for a hydrologist to choose a one-hour time step to model urban basins. Or even a 15-minute time step. There always should be a computation which justifies the choice of the appropriate time step. Better too short (which makes the computer run longer) than too long (which helps the hydrologist to think less).

The most important lack in urban hydrology in general is the use of unit hydrograph analysis based on precomputer methods. For example, Orange County and several other counties in Southern California use a dimensionless unit hydrograph based on the Snyder synthetic unit hydrograph approach of the 1930's. The basic approach combines a "lag time" with a dimensionless S-curve to derive the unit hydrograph. An S-curve is a precomputer manual method for deriving a T-hour unit hydrograph. The equation for lag time is:

$$Lag = C\ ((LL_{ca}/S^{0.5})^{0.38}$$

where L is stream length, L_{ca} is length along the stream to the centroid of the basin, S is watershed slope measured somehow, and C is a "basin roughness coefficient". The effect of urbanization or development is reflected in C, and there is no means to differentiate between development in different parts of the basin. A basin can be subdivided in order to study the effect of partial development, but there is no guarantee that the results of subdivision will agree with the results of simulation without subdivision. In fact, the method is almost guaranteed to have inconsistent results with different degrees of subdivision. On the other hand, a physically based model, such as the HEC-1 kinematic wave overland flow combined with a channel routing, will be consistent. Results should not change with the scale of the modeling effort, as long as the modeling is consistent. The coefficient C is a butch factor which can be, and at times is, varied to obtain the peaks desired --- whether by the developer or the regulatory agency. After all, the results have to be "reasonable", that is, they should agree with our preconceived notions or our preordained values for a 100-year flood.

Conclusions

There is no scientific or engineering justification for the use of precomputer methods during the age of the desktop computer in every office. Any method will work if it is properly calibrated to field data, and if it is used within the range of data over which it is calibrated. However, there is little or no evidence that the methods used in the Southwest are calibrated, and there are no data that show how C is calibrated to estimate the varying effects of urbanization and development. Even more important, the S-curve approach assumes that all basins have the same response, regardless of the shape of the basin. All variation is built into the lag time, which in itself is not all too hot as a tool of modern, scientific hydrology.

So what is the answer? The community of hydrologists should get together with the users --- the public agencies in particular --- and develop a decent set of tools for use in the determination of urban runoff in the arid and semiarid regions of our Southwest. We have the tools. If we don't have the data, we should define what is needed and encourage the agencies to obtain the data for their own good. As concerned professionals, we should demand that our profession move into the modern age. Professional ethics demands no less.

Stochastic Modeling of Monthly Flows in Streams of Arid Regions

Jose D. Salas[1], Member ASCE and Mohamed Chebaane[2]

Abstract

A new stochastic model is presented herein which is applicable to simulating monthly flows of streams in arid regions. The model enables one to reproduce the percentage of zero flows in each month, the monthly mean, variance, and lag-one autocorrelation of the intermittent flows as well as the average skewness coefficient throughout the year. The model considers the intermittent monthly flow process as a product of a periodic binary discrete process times a periodic continuous process. Both discrete and continuous processes are periodic first order autoregressive. The model developed was fitted to the monthly intermittent streamflow data from Arroyo Trabuco Creek in California. It has been shown that the flows generated based on the proposed model resemble closely the key monthly historical statistics.

Introduction

Monthly streamflows are generally periodic-stochastic processes in which the mean, variance, skewness and the month-to-month correlation vary along the year. However, in arid and semi-arid regions monthly flows may be in addition, intermittent, i.e., flow pattern is an alternating sequence of non-zero and zero values. A noticeable characteristic of this type of process is that they are usually highly variable with high coefficient of variation and high skewness coefficient as compared to monthly flows in humid or temperate regions which are perennial and relatively less variable and less skewed. Thus, models developed for humid or temperate regions cannot simply be adapted to arid zones. For example, the so-called Thomas-Fiering model has been proven to be successful for modeling monthly flows in temperate zones. However, the same model is in general, unsatisfactory when used for monthly ephemeral flows.

---

[1]Professor and Program Leader, Hydrology and Water Resources Program, Department of Civil Engineering, Colorado State University, Fort Collins, CO 80523.

[2] Senior Hydrologist, Public Authority for Water Resources, P.O. Box 5575, Ruwi, Sultanate of Oman

The major problems encountered in modeling monthly ephemeral flows are the handling of the zero flows and the high variability of flow magnitudes. Clarke (1973), Beard (1973), and Srikanthan et al. (1980) attempted to model monthly ephemeral flows by using Thomas Fiering model with some modifications. However, the model generated a significant number of negative flows. Srikanthan et al. (1980) tried to remedy this problem. Although, they succeeded to preserve the overall number of zero flows, the monthly variation of the number of zero flows was not preserved.

Jacobs and Lewis (1978 a,b,c, 1983) introduced the discrete autoregressive moving average, DARMA, models to describe stationary dependent discrete random variables having a specified marginal distribution and serial correlation structure. A particular version of this model is the discrete autoregressive (DAR) model. Buishand (1978) and Chang et al. (1982, 1984) used this approach to model wet/dry rainfall sequences. Chang et al. (1984) analyzed daily rainfall process in terms of locally stationary seasons (90 days per season) and assumed that seasons are independent from each other. However, this assumption is not always valid. For instance, monthly intermittent flows are, as stated earlier, periodic dependent processes. The approach to be taken here considers the monthly ephemeral flow process as a product of a periodic binary (1,0) process and a periodic continuous process. The monthly binary (1,0) process is a periodic DAR(1) process with Bernoulli marginal distribution. The continuous process is a periodic AR(1) process with lognormal or Gamma marginal distribution.

Model Description

The product, (binary) discrete-continuous periodic model can be expressed as

$$Y_{\nu,\tau} = X_{\nu,\tau} \, Z_{\nu,\tau} \qquad (1)$$

where $Y_{\nu,\tau}$ = non negative intermittent flow variable during month τ of year ν; $\tau=1,\ldots,\omega$ and $\nu=1,\ldots,N$ where ω is the number of months in the year and N is the number of years of data; $Z_{\nu,\tau}$ = positive continuous variable and $X_{\nu,\tau}$ = binary (1,0) variable having a Bernoulli marginal distribution. All three variables X, Y and Z are variables with periodic properties. Equation (1) can also be written as $Y_{\nu,\tau} = Z_{\nu,\tau}$ if $X_{\nu,\tau} = 1$ and $Y_{\nu,\tau} = 0$ if $X_{\nu,\tau} = 0$. Furthermore, the processes $X_{\nu,\tau}$ and $Z_{\nu,\tau}$ are assumed to be mutually independent.

The mathematical model of the binary (1,0) process can be written as

$$X_{\nu,\tau} = V_{\nu,\tau} \, X_{\nu,\tau-1} + (1-V_{\nu,\tau}) \, U_{\nu,\tau} \qquad (2)$$

where $X_{\nu,\tau}$ is a periodic dependent binary (1,0) Bernoulli process; τ and ω as defined previously and $X_{\nu,\tau-1}$ is replaced by $X_{\nu-1,\omega}$ for τ=1. Furthermore, $V_{\nu,\tau}$ and $U_{\nu,\tau}$ are each independent Bernoulli processes with

$$P\{V_{\nu,\tau}=1\} = \gamma_\tau \quad \text{and} \quad P\{U_{\nu,\tau} = 1\} = \delta_\tau \tag{3}$$

where $0 \le \gamma_\tau \le 1$ and $0 \le \delta_\tau \le 1$. The initial variable $X_{0,\omega}$ is Bernoulli distributed with parameter $\mu_\omega(x)$, the mean of the $X_{\nu,\omega}$ sequence. In addition, $V_{\nu,\tau}$ and $U_{\nu,\tau}$ are mutually independent processes. Likewise, the variables $V_{1,1}$ and $X_{0,\omega}$, $V_{\nu,1}$ and $X_{\nu-1,\omega}$ and $V_{\nu,\tau}$ and $X_{\nu,\tau-1}$ are independent between them. The model given by Eq. (2), denoted here as the PDAR model, accounts for periodic dependence and periodic variation of zero flows.

In regard to the remaining term $Z_{\nu,\tau}$ of Eq. (1), it may be assumed that it is described by a periodic autoregressive (PAR) process lognormally distributed with two parameters. Thus, $Z_{\nu,\tau}$ may be expressed as

$$Z_{\nu,\tau} = \exp\{Z^*_{\nu,\tau}\} \tag{4}$$

where

$$Z^*_{\nu,\tau} = \mu_\tau(z^*) + \phi_\tau(z^*)\,[Z^*_{\nu,\tau-1} - \mu_{\tau-1}(z^*)] + \epsilon_{\nu,\tau} \tag{5}$$

in which $\mu_\tau(z^*)$ and $\phi_\tau(z^*)$ are respectively the periodic mean and autoregressive coefficient of $Z^*_{\nu,\tau}$ and $\epsilon_{\nu,\tau}$ is an independent normal noise with mean zero and periodic variance $\sigma^2_\tau(\epsilon)$.

Alternatively, the process $Z_{\nu,\tau}$ may be assumed to be a periodic gamma autoregressive process (PGAR) as developed by Fernandez and Salas (1986). The PGAR process is expressed as

$$Z_{\nu,\tau} = \phi_\tau\, Z_{\nu,\tau-1} + Z^{\theta_\tau}_{\nu,\tau-1}\, \epsilon_{\nu,\tau} \tag{8}$$

where $Z_{\nu,\tau}$ is a continuous positive variable defined for season τ of year ν; $Z_{\nu,0} = Z_{\nu-1,\omega}$; ϕ_τ and θ_τ are seasonal (periodic) autoregressive coefficients; and $\epsilon_{\nu,\tau}$ is a random noise term. In addition, Z is assumed to be gamma distributed with scale parameter α_τ and shape parameter β_τ.

Properties and parameter estimation by the method of moments for the foregoing models have been derived by Chebaane et al (1990).

Application

Both PDAR-PAR and PDAR-PGAR models are fitted to monthly intermittent streamflows of Arroyo Trabuco Creek at San Juan, California. The period of record is 42 years (1939 - 1970). The flow data is intermittent with (1,0) occurrences which are seasonal and dependent as shown in Fig. 1. The overall percentage of zero flows is around 35% which is a sign of relatively high intermittency. The observed monthly mean, standard deviation and lag-one autocorrelation exhibit a strong seasonality as shown in Figs. 2, 3 and 4, respectively. The observed skewness also exhibits some seasonality. The monthly lag-1 autocorrelation are positive, the monthly skewness varies between 2.76 to 6.21 and the monthly coefficient of variation ranges from 2.17 to 4.42. The adequacy of the PDAR-PAR and PDAR-PGAR models as applied to the Arroyo Trabuco Creek monthly flows are checked by simulation experiments. For this purpose, the parameters of the models are first estimated, then a number of synthetic flow traces are generated and subsequently analyzed to be compared with historical statistics.

The adequacy of the models is evaluated by comparing a number of statistical properties derived from 500 traces of synthetic flows and from their corresponding historical values. A set of basic characteristics which were explicitly parameterized in the model are compared with the historical characteristics. This comparison is often referred to as model verification (Stedinger and Taylor, 1982). Such basic characteristics are: the monthly mean and lag-one autocovariance of the binary (1,0) process; and the monthly mean, standard deviation and lag-one autocorrelation of the intermittent flow process.

Graphical display of historical (observed) and generated statistics are shown in which each generated statistic is the average of 500 sample values. In addition, plus and minus one standard deviation relative to such generated statistics are shown. For instance, the historical and generated monthly mean and lag-one autocovariance of the binary (1,0) process corresponding to the monthly flows of Arroyo Trabuco Creek are displayed in Fig. 1. This figure clearly shows that the PDAR model closely reproduces the seasonal pattern of the referred statistical properties.

In relation to the mean monthly flows, Fig. 2 indicates that both PDAR-PAR and PDAR-PGAR models reproduce very closely the monthly pattern of the mean flows with small or negligible bias. In relation to the monthly standard deviation, Fig. 3 indicates that the seasonal pattern is well reproduced by both models. In addition, it may be observed from Fig. 4 that both of the aforementioned models adequately reproduce the seasonal lag-one autocorrelation of the observed flows with a slightly better performance, of the PDAR-PGAR model specially in terms of bias. However, the monthly skewness, which is not explicitly parameterized in the models, is underestimated. Obviously, this feature is a shortcoming of the models.

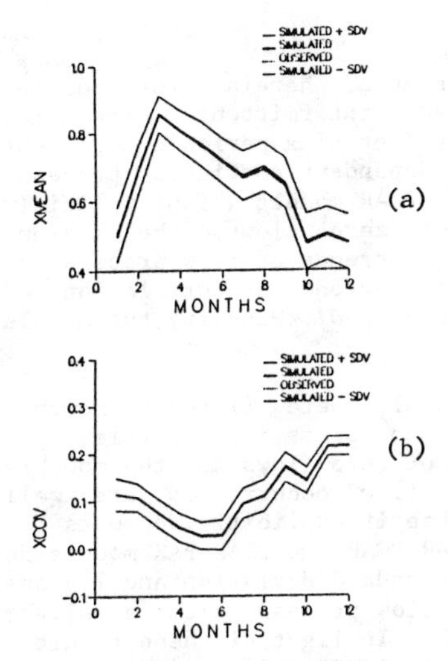

Fig. 1. Monthly parameters of the binary (1,0) process at Arroyo Trabuco: (a) mean and (b) lag-one covariance.

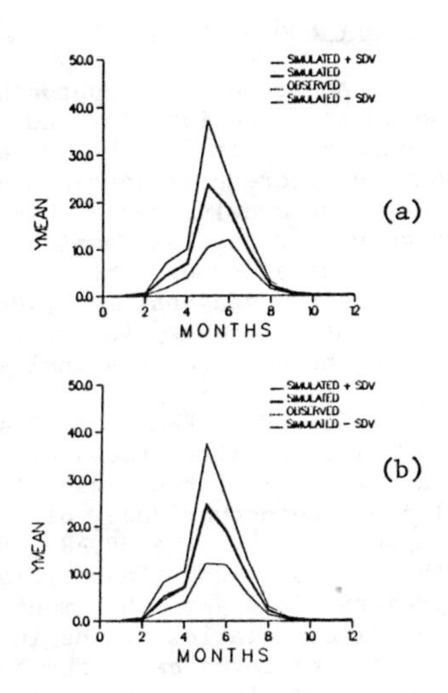

Fig. 2. Monthly mean of the intermittent streamflows at Arroyo Trabuco: (a) PDAR-PAR model and (b) PDAR-PGAR model.

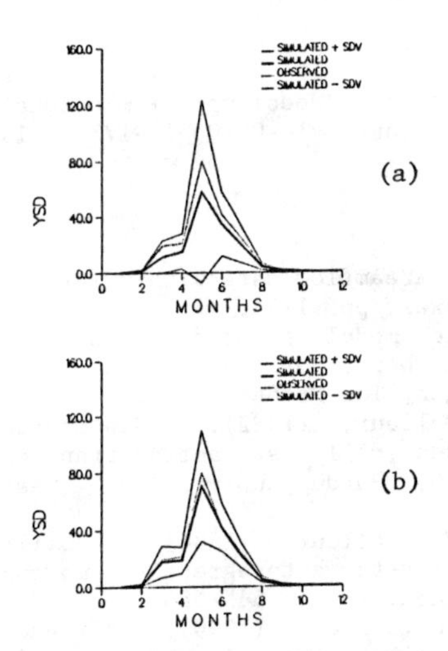

Fig. 3. Monthly standard deviation of the intermittent streamflows at Arroyo Trabuco: (a) PDAR-PAR model and (b) PDAR-PGAR model.

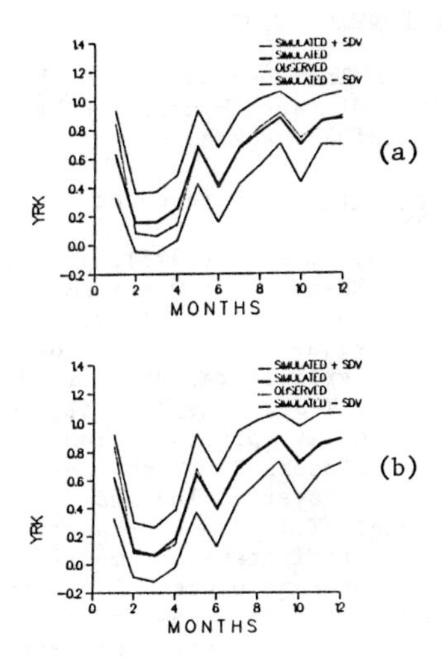

Fig. 4. Monthly lag-one autocorrelation of the intermittent streamflows at Arroyo Trabuco: (a) PDAR-PAR model and (b) PDAR-PGAR model.

Summary and Conclusions

A stochastic approach was presented herein, for modeling seasonal (periodic) and correlated intermittent streamflow processes. It is based on the product of a periodic dependent binary discrete process and a periodic dependent continuous process. Thus, two models, the PDAR-PAR and PDAR-PGAR models which explicitly account for the seasonal variation of zero flows, the lag-one autocovariance of the occurrence/nonoccurrence of flow process, as well as the seasonal mean, variance, and lag-one autocorrelation of the intermittent flow process were developed. However, the models do not account for seasonal skewness.

The PDAR-PAR and PDAR-PGAR models were fitted to monthly intermittent streamflows of Arroyo Trabuco Creek, California. It was found that the seasonal variation of zero flows and the monthly lag-one autocovariance of the (1,0) flow occurrences are well reproduced by the PDAR process since it explicitly accounts for these statistics. In addition, the PDAR-PGAR and PDAR-PAR models do preserve closely the monthly mean, standard deviation and lag-one serial correlation of the intermittent flow process with a slight better performance of the former model. In light of these results, one may conclude that the proposed periodic discrete/continuous processes appear to be useful for modeling seasonal dependent intermittent hydrologic processes such as streamflow which are typical of arid regions. Further testing and application of these models to other intermittent streamflow processes are underway and will be reported elsewhere.

Acknowledgments

The support of the USGS project "Modeling of Seasonal Intermittent Hydrologic Processes," Grant 14-08-0001-G1737 is gratefully acknowledged.

APPENDIX I: References

Beard, L.R., (1973). "Transfer of streamflow data within Texas." Texas Water Development Board, Texas, pp. 1-24.

Buishand, T.A., (1978). "Stochastic modeling of daily rainfall sequences." Dept. of Math and Dept. of Land Water Use, Agricultural University, Wageningen, The Netherlands.

Chang, T.J., M.L. Kavvas, and J.W. Delleur, (1982). "Stochastic daily precipitation modeling and daily streamflow transfer process." Technical Report 146, Purdue University, West Lafayette, Indiana.

Chang, T.J., M.L. Kavvas, and J.W. Delleur, (1984). "Daily precipitation modeling by discrete autoregressive moving average process." Water Resour. Res. (6) 5, 565-580.

Chebaane, M., Salas, J.D. and Boes, D.C., (1990). "Product autoregressive process for modeling intermittent monthly streamflows." Submitted to Water Resour. Res., 35 p.

Clarke, R.T., (1973). "Mathematical model in hydrology." FAO Irrigation and Drainage Paper, No. 19, FAO, Rome.

Fernandez, B., and J.D. Salas, (1986). "Periodic gamma autoregressive processes for operational hydrology." Water Resour. Res., 22(10), 1385-1396.
Jacobs, P.A., and P.A.W. Lewis, (1978a). "Discrete time series generated by mixtures, 1, correlation and run properties." J.K. Stat. B. Vol. 40, part 2, pp. 91-105.
Jacobs, P.A., and P.A.W. Lewis, (1978b). "Discrete time series generated by mixtures, 2, asymptotic properties." J.R. Stat. Soc. B. Vol. 40, 2, pp. 222-228.
Jacobs, P.A., and P.A.W. Lewis, (1978c). "Discrete time series generated by mixtures, 3, autoregressive process [DAR(p)]." Tech. Rep. WPS 55-77, Nav. Postgrad. School, Monterey, California.
Jacobs, P.A., and P.A.W. Lewis, (1983). "Stationary discrete autoregressive moving average time series generated by mixtures."
Srikanthan, R., and T.A. McMahon, (1980). "Stochastic generation of monthly flows for ephemeral streams." J. Hydrol., 47, pp. 19-40.
Stedinger, J.R., and M.W. Taylor, (1982). "Synthetic streamflow generation, 1, model verification and validation." Water Resour. Res., 18(4), pp. 909-918.

Way Against Debris Flows at China's Mountain Towns

Zhou Bifan*

Abatract

For importance of protected objectives the measures against debris flow at mountain towns should ensure a all-round reliability and adopt a comprehensive system of engineering projects combined with planting. perfect predict and alarm equipments are also necessary to carry out a emergent refuge plan for dispersing the population and properties in time.In oder to ensure the safety of the towns as well as people's life, a series of administrative laws must be formulated, and the reasonable use of land and environment conservation should also be done.

Introduction

Debris flow is a moving phenomenon of a mixture of loose soil with water along a slope under gravity . This phenomenon often occurs in mountainous area in rainy season and causes a disaster . In China's mountainous district are there more than 70 towns threatened by debris flow,accounted for 3 percent of the totality of the country's towns. Occurrence of a debris flow usually leads a financial loss of million chinese yuan (about $ 200,000) and even makes a death toll of hundred people at times,making people upset and want to leave from there. How to prevent debris flow has become a urgent task for developing these districts.

Mountain towns generally are political, economic and cultural centers of the districts, and have clouded population and great properties. Once a debris flow occurs there, the damage is often serious and the effect on all aspects is very big.To ensure the safety of towns

---

*Zhou Bifan, Assoc. Prof., Chendu Institute of Mountain Disasters and Environment, Chinese Academy of Sciences, Chengdu, Sichuan, P.R. China

a comprehensive measure, including engineering combined with planting, predicting and warning system as well as the effective administrative laws, is generally adopted. The reliability of the measures must be ensured. To protect a important town, model testing should be done for better designing measures or revising and renovating existed ones.

Measures Against Debris Flows

Based on gained experiences from preventing debris flow, the following measures should be given priority to consideration:

(1)Choosing a favorable site for a town. Due to limited landform, it is quite difficult to choose a town site in mountainous districts. The site is generally located on a fan made from debris flows of one or more gullies, so a debris flow disaster at the town becomes possible and a way aginst it is usually concerned with utilizing the fan. During choosing the site, the following should be paid to attention:

1) To consider a fan type. Generally are there four kinds of type:

Rising type. There is no clear or unchanged flow course on a fan. Due to great deposition of debris flows on it,the fan tends to rising and expanding increasingly.

Scouring type. On a fan is there a clear flow route tending to deepening. General defris flows are drained through the route,only a very big one will flood over the beach beneath the fan.

Balancing type. There is a stable flow way on a fan,neither appearing sensible rising and expanding nor scouring.

Scouring and silting type. A fan may be clearly divided into two parts. The upper,with steeper slope and deeper flow course, approaching the gully mouth often exists slides and collapses providing material for debris flows. The lower with flat and less gradient is equal with rising type.

As a town site, the scouring and balancing types are favorable. They indicate that upper and middle environment of a fan is quite well and debris flows at it do not occur for long time. As long as the environment is not worsened, the debris flow disaster generally will not occur there. The rising and silting types show that the debris flow is vigorous, so they are unfavorable for town sites. But unfortunately, a lot of towms had bulit on them, and therefore it is necessary to carry out work preventing debris flow disasters.

2)To distinguish deteriorated districts by debris flows. According to data attained from field survey, references as well as model testing, the harmed districts can be distinguished. A non-harmed zone is a excellent town site. A general zone,which suffers debris flows but can become a non-harmed or slightly harmed one through remedying, may be chosen as a town site. A heavily deteriorated area is unfavorable for a town site.

(2)The framework against debris flows. The following frameworks may be provided for choosing:

1) Comprehensive cure emphasized in stopping and checking measures. This framework is suitable for a

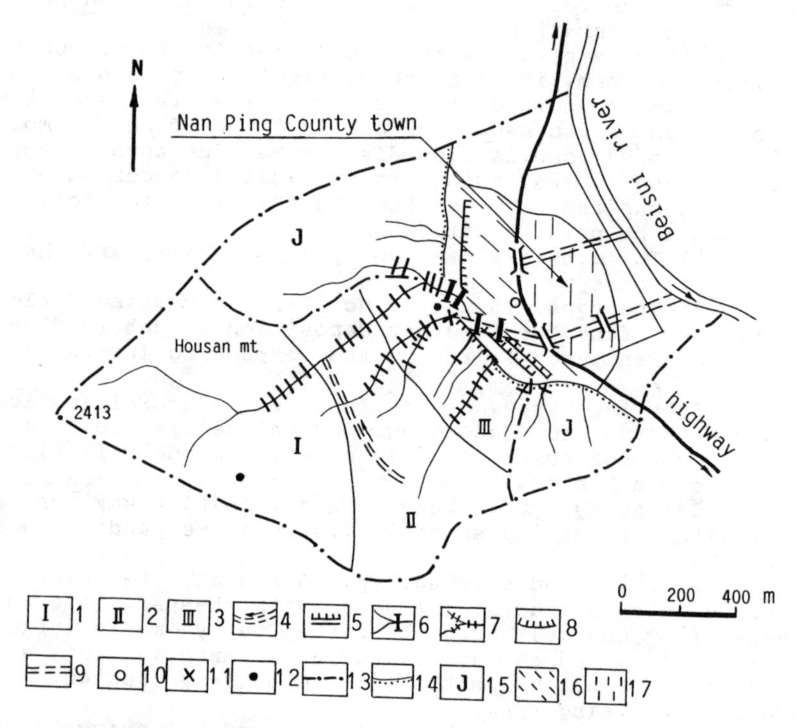

1 Forest area for containing water resource,2 Economic plant area,3 Forest area for conservation of water and soil, 4 Channel for checking water flow, 5 Wall for stopping soil, 6 checking dams, 7 Steel barrier dams, 8 Silting depots for debris flows, 9 Channel for draining floods,10 Warning center,11 Warning equipment of debris flow level, 12 Rainfall observation,13 Limit of basin, 14 Line along mountain toe,15 Bared bedrock, 16 Depositing zone, 17 Deteriorated zone.

Figure 1.Scheme of Comprehensive cure for Debris Flows at Nanping County Town, Sichuan Province of China.

rising or silting type fan. That is, in the upper and middle of basin,to build checking dams and works stabilizing landslides for stopping or reducing debris flows and becoming debris flows into general floods, and on the fan or in the lower, to build a manned channel for draining floods away. As an example, This frame has been adopted for remedying debris flows at Nanping county town, Sichuan province of China (Fig.1) .

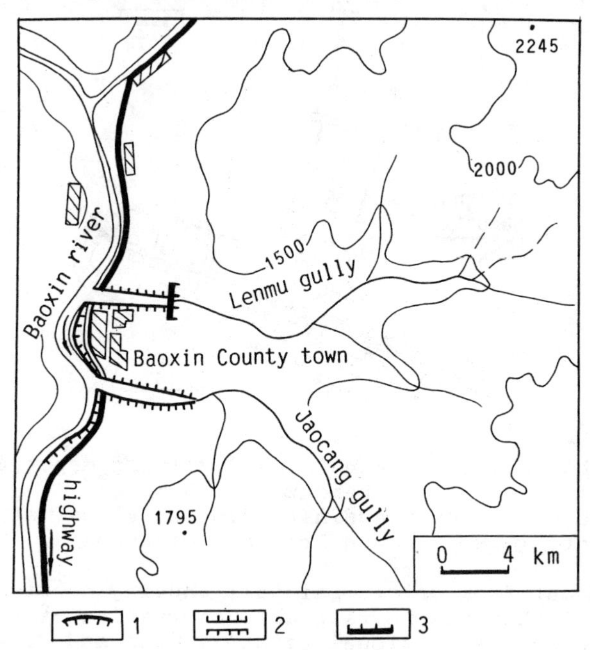

1 Embank for preventing floods of Baoxing river, 2 Channels for draining debris flows, 3 Concrete barrier dam
Figure 2. Scheme Draining Debris Flows and Floods at Baoxing County Town, Sichuan Province of China

2) Draining measures. This frame is favorable for a cutting or balancing fan.That is to build a manned channel on a fan for smooth draining debris flows or floods. An example is shown as Fig.2.

3)Stopping with draining measures. The frame is used on a scouring and silting fan. That is,in the upper and middle of a basin, to build works for stabilizing slope and riverbed, on the fan to build a manned channel for draining floods or debris flows. Fig.3 shows an example of this frame.

(3) To set forecast and alert equipments and work out refuge plan. Generally, a debris flows forecast is

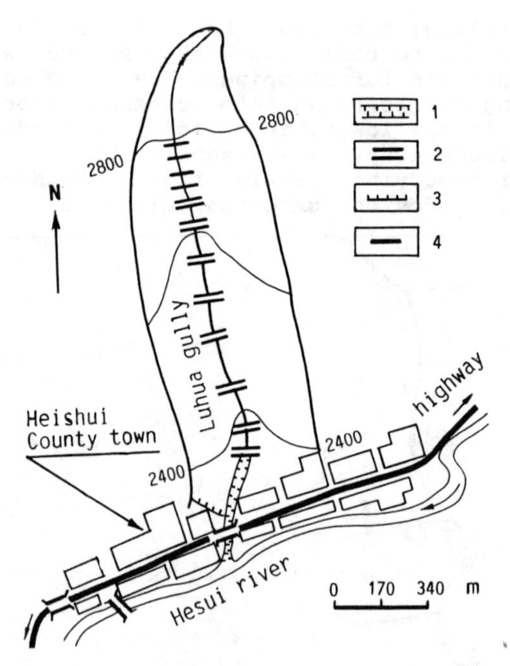

1 Channel, 2 Checking dams, 3 Silting depot, 4 Low dams with smaller height than 5 meter
Figure 3. Scheme against Debris Flows at Hesui County Town, Sichuan Province of China

carried out from critic rainfall causing debris flow at a bain.If there is no rainfall observation at a gully, a relation between regional rainfall data and debris flows at the gully may be first utilized as preliminary forcast, then carrying out the rainfall observatioon at the basin and revising the relatioon.

Warning is conducted from moving characteristics of debris flows, such as a debris flow level, velocity, quake, ect.. At a due place, touching or untouching sensors are set for searching these charateristics and transfering them to warning center, once reaching a dangerous value, the warning is sounded.The value is determined from field surrvey, historic material and modle testing.

Based on divided dangerous zones, a urgent refuge plan is worked out. When sounding the warning, people evacuate from the zone along given routes.

(4) Ministerial and legal measures. In order to fulfil above mentioned technical measures well and prevent manned debris flows by unreasonable human

activities, the following should be done:

1) To raise people's knowledge on debris dlows for effectively preventing them.

2) To establish a unit against debris flow. That is usually consisted of workers from concerned departments and its director is charged by one of the town's governors.Its task is to manage the remedied zone for full winning achievements of the engineering against debris flows, and carry out observation for collecting scientific data.

3)To issue concerned regulations.Such as managing the protected zone, controlling pasture and mining, mobilizing people for sympathizing and taking part in the work against debris flows,etc..

Notable Problems Concerned with a Town Plan

A town plan at a debris flow area should mainly pay attention to the following:

(1) It is necessary to leave some spaces along both sides of a debris flow channel at the town for possibility of taking urgency measures during the urgent term.

(2) A bridge acrossing a debris flow channel should be set with a enough span and no pillar within the channel for avoiding being damaged and impinged by debris flows.

(3) Important projects, such as supplying water and electricity ceter, etc., should not be set in the potentially attacked area by debris flows.

(4) Units with close connection should be settled on the same side of the channel for avoiding losing their communications during debris flows.

(5) Measures against debris flows should combine with town plan and construction. For example, the plant on mountain side for conservation of water and soil may be done as log and economy forest for increasing people's income, and some greening places may be also used as pleased gardens.

(6) To settle concerned people's living and producing problems well.Measures against debris flows at a town are usually concerned with many people's benefits , such as housing, land use and properties, etc., only these problems are well settled by the town government, can the measures be smooth realized.

conclusion

Debris flows have been giving people at mountain towns of china and other concerned countries a lot of damage, but people are able to prevent them with favorable measures adopted.

SUBJECT INDEX

Page number refers to first page of paper.

AUTHOR INDEX

Page number refers to first page of paper.